Pharmacodynamic
Basis of
HERBAL
MEDICINE

Pharmacodynamic Basis of
HERBAL MEDICINE

Manuchair Ebadi, Ph.D., F.A.C.C.P.
School of Medicine and Health Sciences
University of North Dakota
Grand Forks

CRC PRESS

Boca Raton   London   New York   Washington, D.C.

Library of Congress Cataloging-in-Publication Data

Ebadi, Manuchair S.
   Pharmacodynamic basis of herbal medicine / by Manuchair Ebadi.
     p. cm.
   Includes bibliographical references and index.
   ISBN 0-8493-0743-0 (alk. paper)
    1. Herbs—Therapeutic use. I. Title.

   RM666.H33 E23 2001
   615′.321—dc21

2001043781

Visit the CRC Press Web site at www.crcpress.com

© 2002 by CRC Press LLC

No claim to original U.S. Government works
International Standard Book Number 0-8493-0743-0
Library of Congress Card Number 2001043781
Printed in the United States of America   2 3 4 5 6 7 8 9 0
Printed on acid-free paper

In books lie the soul of the whole past time,
the articulate audible voice of the past,
when the body and material substances
of it have altogether vanished like a dream.

Thomas Carlyle

I dedicate this book to my beloved grandchildren,

Jesse Manu Ebadi,
Caylin Jane McCormick,
Christine Pari McCormick,
and Jolie Manizeh Ebadi.

PREFACE

Hippocrates (460–377 B.C.), who used many of the herbal medicines described in this book, lamented, "Life is short, and the art long; the occasion fleeting; experience fallacious; and judgment difficult."

Herbal remedies have become a major component of American health care. Botanicals like **ginseng**, **ma huang**, **St. John's wort**, and **valerian** are now household words throughout the world, and the sales of herbal medicine are increasing exponentially.

Alternative therapies include acupuncture, energy healing, folk medicines, herbal medicines, homeopathy, massage, and megavitamins, to name only a few. Millions of people in the third world have used and will always use herbal medicines because they believe in them and regard them as "their" medicine, in contrast to the "allopathic" (conventional Western) system of medicine brought in from "outside." These medicinal herbs are available locally and are prescribed by traditional practitioners of medicine who are part of the community and in whose presence the patient feels comfortable. In Western countries, there is now an increased use of herbal medicines, largely because of a belief that powerful synthetic agents used in Western medicine can exert more unwanted side effects and are too often used indiscriminately and irrationally. Many of our present medicines are derived directly or indirectly from higher plants. Although several classic plant drugs have lost much ground to synthetic competitors, others have gained a new investigational or therapeutic status in recent years. In addition, a number of novel plant-derived substances have entered into Western drug markets.

The word **pharmacodynamic** may be defined as the study of the actions and effects of drugs on organ, tissue, cellular, and subcellular levels. Therefore, pharmacodynamics provides us with information about how drugs bring about their beneficial effects and how they cause their side effects. By understanding and applying the knowledge gained in studying pharmacodynamics, physicians and other members of the health-care delivery team are able to provide effective and safe therapeutic care to their patients. Western physicians prescribing synthetic drugs may want to believe in herbal medicines and use them in their patients, but wish to see scientific documentation as they have learned in their pharmacology and therapeutic courses in medical school. Eastern physicians yearn to find out how their drugs, proven efficacious for thousands of years in ancient civilizations such as Chinese, Indian, Persian, and Egyptian, work and bring about their beneficial effects. This book bridges the past to the present and shows at the molecular level how the herbal medications most often used work. For example:

- This book shows how Eastern physicians used **meadow saffron**, which is good for gout, but also provides evidence that meadow saffron possesses **colchicine**, an **antigoutic medication**, and provides the mechanisms of action for it.
- This book shows how Eastern physicians used **belladonna** alkaloid for **Parkinson's disease** but also provides evidence that belladonna alkaloid contains **anticholinergic drugs**, which counterbalance the dopamine deficiency syndrome seen in Parkinson's disease.
- This book shows how Indians used ***Rauwolfia serpentina*** to reduce blood pressure and as an antipsychotic, but also shows that it contains **reserpine**, which depletes norepinephrine in the periphery bringing about its antihypertensive effects and depletes dopamine in the mesocortical system causing tranquility.
- This book compares the actions of **fluoxetin**, the most frequently used antidepressant in the world, with those of **St. John's wort**, in altering the uptake of serotonin.
- This book compares the hypnotic and sedative actions of **valerian** used for thousand years to treat insomnia with those of **benzodiazepine derivatives** in altering GABAergic transmission.
- This book describes the existence of drugs isolated from food substances such as **horseradish**, **garlic**, and **rhubarb**, and provides their mechanisms of action.
- This book describes the efficacy of **Shing Jing** in male infertility and **erectile dysfunction** and compares its efficacy with those brought about by levodopa, amylnitrite, vitamin E, and **sildenafil** (Viagra).

The majority of believers in alternative medicine are more educated, but they report poorer health status. They take herbal medicine not so much because they are dissatisfied with conventional medicines, but largely because they find these health-care alternatives to be more congruent with their own values, beliefs, and philosophical orientations toward health care and life in general.

M. Ebadi
Grand Forks, ND

ACKNOWLEDGMENTS

The author expresses his appreciation to Liz Covello, the former publisher of life sciences for CRC Press LLC, and Barbara Ellen Norwitz, the current publisher of life sciences for CRC Press LLC for a gracious invitation to prepare a book on herbal medicine. The author acknowledges the support of Tiffany Lane, editorial assistant, and the magnificent contribution of Gail Renard, production editor, for polishing and refining the book.

The author extends his expression of admiration to Debra Jean Kroese, Dawn Halvorson, Lacy Kay Boushee, and Haley Ann Kroese for gathering reference materials and for typing certain sections of the book.

The author remains indebted to Victoria Swift, the director of the art department at the University of North Dakota School of Medicine and Health Sciences for her marvelous artistic talent in completing many of the diagrams; and to Betty Ann Karolski, the associate director, Biomedical Communications Information Technology Services at the University of Nebraska Medical Center for her support in completing this book.

The author extends his grateful appreciation and eternal sense of indebtedness to KayLynn Marie Bergland for her magnificent dedication to her job, marvelous work ethics, and incredible skills in typing, reading, editing, proofing, and revising the entire book.

The author pays an affectionate tribute and extends his heartfelt gratitude to H. David Wilson, M.D., the eminent Dean and Vice President for Health Affairs, University of North Dakota School of Medicine and Health Sciences, for his unyielding support, Solomonic wisdom, and genuine friendship in facilitating the completion of this book.

THE AUTHOR

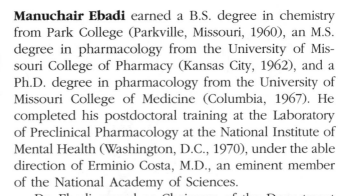 **Manuchair Ebadi** earned a B.S. degree in chemistry from Park College (Parkville, Missouri, 1960), an M.S. degree in pharmacology from the University of Missouri College of Pharmacy (Kansas City, 1962), and a Ph.D. degree in pharmacology from the University of Missouri College of Medicine (Columbia, 1967). He completed his postdoctoral training at the Laboratory of Preclinical Pharmacology at the National Institute of Mental Health (Washington, D.C., 1970), under the able direction of Erminio Costa, M.D., an eminent member of the National Academy of Sciences.

Dr. Ebadi served as Chairman of the Department of Pharmacology at the University of Nebraska College of Medicine from 1970 until 1988, and subsequently as Professor of Pharmacology, Neurology, and Psychiatry from 1988 through 1999.

During his academic career, Professor Ebadi has received 32 awards including the Burlington Northern Faculty Achievement Award (1987), the University of Nebraska's systemwide Outstanding Teaching and Creative Activity Award (1995); and was inducted into the Golden Apple Hall of Fame (1995) for having received 11 Golden Apple awards. He is a member of 16 research and scholarly societies including Alpha Omega Alpha Honor Medical Society.

In 1976, Dr. Ebadi became the Mid-America State Universities Association (MASUA) honor lecturer; in 1987, he received an award for Meritorious Contributions to Pharmaceutical Sciences from the University of Missouri Alumni Association; in 1995, he was honored by a Resolution and Commendation of the Board of Regents of the University of Nebraska for having developed a sustained record of excellence in teaching, including creative instructional methodology; and in 1996, he received the Distinguished Alumni Award from Park College, his alma mater.

In July 1999, he was appointed Professor and Chairman of the Department of Pharmacology and Toxicology at the University of North Dakota School of Medicine and Health Sciences. In September 1999, Dr. Ebadi became Professor and Chairman of the Department of Pharmacology, Physiology, and Therapeutics; in November 1999, he became Professor of Neuroscience; and in December 1999, he was appointed Associate Dean for Research and Program Development. In September 2000, Dr. Ebadi was appointed Director of the Center of Excellence in Neurosciences at the University of North Dakota School of Medicine and Health Sciences.

CONTENTS

INTRODUCTION

"The Lord has created medicines out of the earth and he that is wise will not abhor them."

— Ecclesiastes, Chapter 38, Verse 4

In actuality, 120 medicinal plants are found in the Bible including **aloe**, **rue**, **madder**, **frankincense**, **myrrh**, and **marigold** (***Calendula officinalis***). With its therapeutic efficacy on cough and bronchitis, dyspepsia, liver and gallbladder maladies, and as an appetite stimulant, **marigold**, known in the Persian language as **Hamisheh Bahar** (meaning "always spring") is a sacred flower, and is respected and honored in such Eastern religions as:

- Buddhism
- Hinduism
- Taoism

Marigold, a sacred flower, always follows the sun, needs little care, but gives so much beauty. Hence, it is used as a garland to adorn dignitaries and religious leaders. The saffron-colored robes of the Eastern monks have the color of marigold.

Traditional Chinese medicines have developed over a period of at least 5000 years. The earliest known work on Chinese herbs is:

Shing Nung Bon Cas Chien, or the herbal classic of the **Divine Plowman**. According to this work, Sheng Nung, the divine plowman, tested and recommended a total of 365 herbs — one for each day of the year.

The Chinese are proud of their heritage, especially in the uses of natural substances, plants, chemicals, and animal products.

For example, **acupuncture** performs certain functions in traditional Chinese medicine. Acupuncture:

- Regulates the flow of **chi** through the channels and organs
- Removes blockage

- Strengthens the body's protective chi
- Lessens the virulence of excesses

The most sophisticated and state-of-the-art science at the molecular level has shown that acupuncture:

- Releases pain-killing **endorphin** and **enkephalin**
- Releases **immunoenhancing cytokines**
- Releases **neurotrophins** capable of repairing damaged tissue
- Suppresses the elaboration and release of **inflammatory and necrotizing peptides**

Physicians and scientists are beginning to appreciate the therapeutic efficacy of acupuncture. Dr. Ted J. Kaptchuk, who holds a doctorate of Eastern medicine and is the author of a book entitled ***The Web That Has No Weaver***, tells a story about a Chinese peasant who had worked as a maintenance man in a newly established Western missionary hospital. The peasant took some antibiotics from the hospital and treated people in his village who had fever. Many of them were cured. The peasant knew nothing about the pharmacology of the antibiotics, but that was irrelevant; the sick got better. The physicians of yesteryear could not explain the efficacy of acupuncture, but acupuncture worked. The mighty oak tree stands tall and firm in the ground because of its roots. One does not see the roots but they are there.

In the last 30 years, significant events occurred that have already produced fundamental changes in the attitude of both the public and scientists toward herbal medicines.

- In the first place, laypersons discovered the utility of whole plant drugs or herbs.
- In the second place, dissatisfaction with the effectiveness and especially the cost of modern medicines caused an appreciation for all things "**natural**" and "**organic**," which inspired millions of people throughout the world to gain a deep appreciation for the use of classical plant drugs for the treatment of many ailments.

The **"green" revolution** in terms of herbal medicine has now achieved astonishing popularity in the United States, although it is not yet encouraged by an ultraconservative **Food and Drug Administration** that classifies most plant drugs as dietary supplements or food additives, which places severe limitation on labeling. However, in fairness to the FDA, because a plant drug is "natural" does not necessarily mean that it is safe.

Primeval humans in search of food undoubtedly experienced much poisoning. We learned by trial-and-error that eating certain mushrooms, berries, and roots could produce various degrees of gastrointestinal discomfort or death, whereas others could be ingested safely. Certain lessons were learned quickly, and primitive food-gathering humans soon became toxicologists of no mean ability. Eventually, this knowledge was made to work for humans; we prepared **arrow poisons** from plant extracts to bring down game or foe. We threw crushed leaves of particular plants into water and with little effort quickly obtained a bountiful supply of stupefied fish, and we learned also to wash poison (e.g., cyanide) from a number of common, staple foods to make them edible. It probably took *Homo sapiens* but a short time to learn how to gather food with relative ease and to participate in the good life.

Humans also learned that plants could be used to poison one another. In classical Rome, for example, mushrooms were the poison of choice and were expertly used by **Agrippina, wife of Emperor Claudius and mother of Nero**. Agrippina had Lollia Paulina put to death because Claudius, in a careless moment, remarked on Lollia's beauty.

Higher plants were among those studied by **Cleopatra** in her search for a suicidal poison. Using her prisoners and slaves as guinea pigs, she was quite systematic. It was reported that she was not satisfied with the effects of either **henbane** (*Hyoscyamus niger*) or **belladonna** (*Atropa belladonna*), for they produced too much pain in spite of their rapid actions. She was further disappointed with ***Strychnos nux-vomica*** from which **strychnine** was eventually extracted; although its action was instantaneous, it produced convulsions that left distorted facial features at death. Finally, she selected the bite of the asp (Egyptian cobra), which produced a serene and prompt death!

The FDA could do better, and there is little doubt that consumer demand will promote an ever-increasing interest in herbal medicine. Although some plants are poisonous, the majority are not and should be used as therapeutic agents.

Alfred North-Whitehead said: "The rejection of any source of evidence is always treason to that ultimate rationalism which urges forward science and philosophy alike."

Practitioners throughout the world have used **marijuana**. The most sophisticated pharmacological studies have shown for more than two decades that the active ingredients of marijuana are efficacious in:

- Cancer chemotherapy
- Glaucoma
- Epilepsy
- Multiple sclerosis
- Paraplegia and quadriplegia
- AIDS
- Chronic pain
- Migraine
- Rheumatic diseases (osteoarthritis and ankylosing spondylitis)
- Pruritus
- Premenstrual syndrome, menstrual cramps, and labor pains
- Depression and other mood disorders

Only recently has marijuana become available as a drug for patients with cancer.

Would the FDA approve chocolate, which is related to cocaine? *Theobroma cacao* or the chocolate tree, named by the great Swedish naturalist Linnaeus (1707–1778) gives us chocolate, also known by the Aztec as "**the food of the gods**," which has acquired immense importance socially, religiously, medically, economically, and, of course, gastronomically.

Natural products are extremely important for four reasons:

1. In the first place, they provide a number of extremely useful drugs that are difficult, if not impossible, to produce commercially by synthetic means. These include such diverse groups of compounds as the **alkaloids** of the opium poppy, of **ergot**, and of **solanaceous plants**; the cardiotonic glycosides of **digitalis**; most of the antibiotics; and all of the serums, vaccines, and related products.
2. Natural sources also supply basic compounds that may be modified slightly to render them more effective or less toxic. The numerous variations of the morphine molecule serve as examples here.
3. A third role of natural products is their utility as prototypes or models for synthetic drugs possessing physiological activities similar to the original drug. **Procaine** and similar local anesthetics are commonly cited representatives of this category.
4. There is a fourth role for natural products that is quite different from the above, but which is nonetheless important. Some natural products contain compounds that demonstrate little or no activity themselves but which can be modified by chemical or biological methods to produce potent drugs not easily obtained by other methods. For example, **taxol** may be synthesized from baccatin III, which occurs more or less abundantly in the leaves of various **yew species**, whereas taxol itself is found only in the bark of the scarce **Pacific yew**. Proper chemical and biological treatment of **stigmasterol**, which occurs abundantly in **soybean oil**, permits the large-scale production of **hydrocortisone** or related corticosteroids, compounds that occur in nature in only small amounts. The importance of natural products as precursors of significant drugs cannot be overemphasized.

One should consider that plant-derived steroids alone account for about 15% ($22 billion) of the $150 billion world pharmaceuticals market and that the annual market for taxol exceeded $1 billion in the year 2000. The antineoplastic agents **vinblastine** and **vincristine** have sales amounting to $100 million per year; the market for **psyllium seed products** amounts to some $300 million annually; and **nicotine** and **scopolamine** patches now have combined sales of more than $1 billion per year. It is obvious that natural products continue to play important economic as well as therapeutic roles in modern medicine.

Why then, do Western physicians and scientists resist the **chi** vitality and **yin** and **yang** theory?

Although yin and yang can be distinguished, they cannot be separated. They depend on each other for definition. And the things in which yin and yang are distinguished could not be defined without the existence of yin and yang qualities. Certainly, in Western sciences, there are millions of examples of yin and yang, such as:

■ Adrenergic–cholinergic systems controlling heart beat
■ Dopaminergic–cholinergic systems controlling movements
■ Glutamatergic–GABAergic systems controlling motion

Nevertheless, Western culture does not fully understand chi. In this cause, **Lao Tzu** (Iaozi), the eminent and reputable founder of **Taoism**, came to illuminate and to show the way.

In **Tao-Te-Ching**, the classic of the **Tao** and its virtue, he says, "**Yin and Yang** control each other by stating":

> Being and non-being produce each other;
> Difficult and easy complete each other;
> Long and short contrast each other;
> High and low distinguish each other;
> Sound and voice harmonize each other;
> Front and back follow each other.

Lao Tzu describes the transformation process of yin and yang poetically by stating:

> He who stands on tiptoe is not steady.
> He who strides forward does not go.
> He who shows himself is not luminous.
> He who justifies himself is not prominent.
> He who boasts of himself is not given credit.
> He who brags does not endure for long.

Therefore, chi in Chinese medicine, known in India as **prana** (life force) and in Japan as **ki**, gives us a measure of the vitality of a person, object, or state.

Appropriate drug therapy can improve the quality of life, whereas injudicious drug therapy may be harmful. Medications are given for a variety of diagnostic, prophylactic, and therapeutic purposes. These prescribed medications bring about the desired effects in most patients, but they may also prove to be inert and ineffective in some. They may even evoke totally unexpected responses and precipitate serious reactions in others.

Pharmacology can be defined as the study of the selective biologic activity of chemical substances in living matter. Often, but not always, these selective biologic activities are triggered by very small amounts of drugs. Because it is selective, the drug should produce responses in only some of the cells.

The use of drugs in the treatment of a disease is termed **pharmacothera-peutics**. However, the use of drugs is not always necessary in managing a disease. A drug may be used substitutively, supportively, prophylactically, symptomatically, diagnostically, or correctively. For example, in **juvenile-onset diabetes mellitus (Type I, insulin-dependent diabetes mellitus)** and **Addison's disease**, **insulin** and **cortisone acetate** are used, respectively, as substitutes or supplements for substances that either were never produced or were once there but are not now produced. In adult-onset diabetes mellitus, oral antidiabetic agents support the physiological function of the body by stimulating the synthesis and release of insulin.

Oral contraceptive tablets are used to **prevent pregnancy**. **Isoniazid** may be used to prevent the development of active **tuberculosis** in those individuals who have been exposed to the disease but show no evidence of infection, in those who test positively for it but have no apparent disease, and in those with once active but now inactive disease.

Drugs may eliminate or reduce the symptoms of a disease without influencing the actual pathology. For example, fever may be associated with respiratory tract infection, bacterial endocarditis, biliary tract disorders, tuberculosis, carcinoma, cirrhosis of the liver, collagen diseases, encephalitis, glomerulonephritis, Hodgkin's disease, hysteria, malaria, leukemia, measles, mumps, and plague, to name a few. **Aspirin** can reduce the fever in these disorders but cannot alter the disease processes themselves.

A drug may also be used to diagnose a disease. **Histamine** has been used to assess the ability of the stomach to secrete acid and to determine parietal cell mass. If an acidity or hyposecretion occurs in response to histamine administration, this may indicate **pernicious anemia**, **atrophic gastritis**, or **gastric carcinoma**; a hypersecretory response may be observed in patients with **duodenal ulcer** or with the **Zollinger–Ellison syndrome**.

In most cases, drugs do not cure diseases, but do ease or eliminate the associated symptoms. For example, no drugs exist that cure essential hypertension, but there are some that lower blood pressure. In reducing symptoms, drugs never create new functions. They can only stimulate, or depress, the functions already inherent in the cells.

In alleviating symptoms, drugs may also induce adverse effects, which may or may not be acceptable to patients. For example, numerous agents with anticholinergic properties cause dry mouth, which is easily correctable and hence is acceptable to patients. Conversely, some **antihypertensive medications** cause **impotence** in male patients, which they may find unacceptable, and this side effect may thus lead to lack of compliance with the prescribed medication.

Drugs can therefore resemble a double-edged sword in that they can both help and harm the patient. By fully appreciating the nature of **pharmacokinetics**, **pharmacodynamic** principles, and **drug–drug interactions**, practitioners can drastically reduce unwanted side effects and at the same time enhance the therapeutic efficacy and usefulness of drugs.

Pharmacodynamics may be defined as the study of the actions and effects of drugs on organ, tissue, cellular, and subcellular levels. Therefore, pharmacodynamics provides us with information about how drugs bring about their beneficial effects and how they cause their side effects. Pharmacodynamics considers the sites, modes, and mechanisms of action of drugs. For example, if a patient with multiple fractures receives a subcutaneous injection of 10 to 15 mg of morphine sulfate, analgesia, sedation, respiratory depression, emesis, miosis, suppression of the gastrointestinal tract, and oliguria may ensue. These diversified effects occur at multiple peripheral and central sites and through the influence of numerous modes and mechanisms of actions.

Site of Action. The receptor sites where a drug acts to initiate a group of functions is that drug's site of action. The central sites of action of morphine include the cerebral cortex, hypothalamus, and medullary center.

Mode of Action. The character of an effect produced by a drug is called the mode of action of that drug. Morphine, by depressing the function of the cerebral cortex, hypothalamus, and medullary center, is responsible for decreasing pain perception (**analgesia**), inducing narcosis (**heavy sedation**), depressing the cough center (**antitussive effect**), initially stimulating then depressing the vomiting center, and depressing respiration.

Mechanism of Action. The identification of molecular and biochemical events leading to an effect is called the mechanism of action of that drug. For example, morphine causes respiratory depression by depressing the responsiveness of the respiratory center to carbon dioxide.

Medicinal plants have been used for healing and preventative health for thousands of years all around the world. For examples, the therapeutic efficacy of the following herbs are well known (see Kovach, 2001):

ALOE VERA LEAF

(*Aloe ferox, A. barbadensis*)

- Heals sunburn and other skin problems
- Eases constipation
- Aids ulcers
- Exhibits antibacterial, antiviral, and antifungal activity

ASHWAGANDA

(*Withania somnifera*)

- Reduces stress
- Has anti-inflammatory effect
- Boosts sexual energy in men (**aphrodisiac**)

ASTRAGALUS LEAF

(*Astragalus membranaceus*)

- Enhances immune response
- Heals burns and abscesses
- Relieves effects of chemotherapy
- Helps lower blood pressure
- Acts as a heart tonic, protecting the heart against viral damage

BILBERRY FRUIT

(European blueberry — *Vaccinium myrtillus*)

- Relieves varicose veins
- Exhibits antidiarrhetic effects
- Enhances circulation
- Improves night blindness, cataracts, and diabetic retinopathy

CAT'S CLAW BARK

(*Uncaria tomentosa*)

- Stimulates immune response
- Exhibits anti-inflammatory effects
- Exhibits antiviral effects

CAYENNE PEPPERS

(*Capsicum* spp.)

- Relieves painful muscles
- Acts as a cardiovascular tonic
- Eases stiff, painful arthritic joints
- Exhibits an antibacterial effect
- Aids weight loss

CHAMOMILE

(*Matricaria recutita*)

- Is calmative
- Acts as a diuretic
- Promotes wound healing
- Exhibits an anti-inflammatory effect

CHASTE BERRY

(*Vitex agnus-castus*)

- Aids menstrual disorders
- Eases premenstrual syndrome and menopause symptoms
- Helps hot flashes

BLACK COHOSH ROOT

(*Cimicifugae racemosae rhizoma*)

- Eases menopause
- Relieves premenstrual syndrome
- Aids dysmenorrhea

CRANBERRY FRUIT

(*Vaccinium macrocarpon*)

- Treats urinary tract infections
- Is a good source of vitamin C

DEVIL'S CLAW ROOT

(*Harpagophytum procumbens*)

- Exhibits an anti-inflammatory effect
- Acts as an analgesic
- Relieves tendonitis, acute lower back pain
- Relieves loss of appetite
- Treats skin lesions

EVENING PRIMROSE OIL

(*Oenothera biennis*)

- Relieves premenstrual syndrome symptoms
- Reduces cholesterol
- Fights heart disease
- Acts as a liver tonic
- Relieves skin conditions

FLAXSEED OIL

(*Linum usitatissimum*)

- Aids endometriosis
- Relieves premenstrual syndrome and menopausal symptoms
- Helps psoriasis
- Lowers cholesterol and blood pressure
- Protects against heart attack and stroke

GINGER

(*Zingiber officinale*)

- Relieves nausea, vomiting, and motion sickness
- Helps stomach upset
- Exhibits an anti-inflammatory effect

GINKGO BILOBA

(*Gingko*)

- Aids age-related memory loss
- Eases dizziness
- Relieves tinnitus (ringing in the ears)
- Improves circulation and oxygen delivery to the brain
- Acts as a potent antioxidant, stronger than vitamin E

GINSENG ROOT, ASIAN

(*Panax*)

- Enhances memory
- Fights fatigue
- Protects against stress damage, both physical and emotional
- Has anti-aging properties

GINSENG ROOT, SIBERIAN

(*Eleutherococcus senticosus*)

- Is an immunostimulant
- Exhibits anti-inflammatory effects
- Prevents colds and flu

GOLDENSEAL ROOT

(*Hydrastis canadensis*)

- Relieves cold and flu symptoms
- Prevents or treats "traveler's diarrhea"
- Acts as an antiseptic

GRAPE SEED EXTRACT

(*Vitis vinifera*)

- Potent antioxidant
- Treatment of varicose veins and other vein disorders
- Cardiovascular protectant
- Anti-inflammatory agent
- Antihistamine
- Wrinkle treatment

GREEN TEA LEAF

(*Camellia sinensis*)

- Acts as an antioxidant
- Exhibits chemopreventative effects
- Stops tooth decay

HAWTHORN BERRIES, LEAVES

(*Crataegus monogyna, C. laevigata,* and related species)

- Dilates coronary arteries
- Improves blood flow to the heart
- Decreases blood cholesterol levels
- Acts as a hypotensive agent

LICORICE ROOT

(*Glycyrrhiza glabra*)

- Soothes sore throat
- Relieves coughs
- Treats peptic ulcers
- Relieves canker sores
- Aids viral infections
- Eases skin conditions
- Helps premenstrual syndrome and menopausal disorders

MILK THISTLE FRUIT

(*Silybum marianum*)

- Acts as a liver tonic and protectant

PEPPERMINT

(*Mentha piperita*)

- Exhibits antispasmodic effects
- Calms the stomach and intestinal tract
- Relieves the common cold
- Aids muscle pains
- Eases headaches
- Helps irritable bowel syndrome
- Soothes gallstones

RED CLOVER BLOSSOMS

(*Trifolium pratense*)

- Relieves menopausal symptoms
- Exhibits antitumor properties
- Reduces swollen glands
- Acts as an expectorant
- Soothes skin problems

SAW PALMETTO BERRIES

(*Serenoa repens*)

■ Relieves benign prostate hyperplasia (enlarged prostate)

UVA URSI

(*Arctostaphylos uva-ursi*)

■ Aids urinary tract infections and inflammation
■ Eases kidney and bladder stones
■ Reduces heavy, painful menstrual periods

VALERIAN

(*Valeriana officinalis*)

■ Relieves insomnia
■ Aids restlessness
■ Eases nervous conditions
■ Acts as a sedative

The Pharmacodynamic Basis of Herbal Medicines has attempted to provide detailed molecular mechanisms on how these and other herbal plants bring about their beneficial therapeutic effects, in relieving the mental and physical sufferings of humans.

1

ALTERNATIVE THERAPIES

- Acupuncture
- Energy Healing
- Folk Medicines
- Herbal Medicines
- Homeopathy
- Massage
- Megavitamins
- Self-Help Group Treatments (see Hammond, 1995; Lawless, 1995; Douggans, 1996; Hoffmann, 1996; Thomas, 1997; Mitchell, 1997; Warrier and Gunawant, 1997; MacDonald, 1998; and Liechti, 1998).

WHY PATIENTS USE ALTERNATIVE MEDICINE

Along with being more educated and more likely to report their health status, the majority of alternative medicine users appear to turn to it not so much because they feel dissatisfied with conventional medicine but largely because they find these health care alternatives to be more congruent with their own values, beliefs, and philosophical orientations toward health and life.

Herbs and related products are commonly used by patients who also seek conventional health care. All physicians, regardless of specialty or interest, care for patients who use products that are neither prescribed nor recommended.

The earliest evidence of human use of plants for healing dates to the Neanderthal period. In the 16th century, botanical gardens were created to grow medicinal plants for medical schools. Herbal medicine practice flourished until the 17th century when more "scientific" pharmacological remedies were favored (see Thomas, 1997; Fontanarosa, 2000).

The following items provide short descriptions of alternative medicine.

ACUPRESSURE

Acupressure involves the application of fingertip or nail pressure to acupuncture points on the body to remove blockages or pain and enhance the **flow of chi**.

In Chinese medicine acupressure may be used in its own right or incorporated into an acupuncture treatment (Thomas, 1997).

ACUPUNCTURE

There are over 350 **acupoints on the meridians of the body**. Selected points are stimulated by inserting fine acupuncture needles to improve the **flow of chi or qi** in the meridians and to restore balance and healthy functioning to the internal organs of the body. Treatment points are selected on the basis of pulse and tongue diagnosis, examination, and questioning and categorized according to their effects on specific body systems and organs. The insertion of the needles is quick and virtually painless, and often a comfortable and relaxed feeling follows (Thomas, 1997).

Aromatherapy

Aromatherapy is the use of **essential oils from plants** to enhance general health and appearance. The term was coined by its originator, the French chemist **René-Maurice Gattefosse.** The following summarizes the ten most useful oils and their *proposed* effects (Thomas, 1997).

Camomile	Calming
Eucalyptus	Antiseptic
Geranium	Mildly astringent
Jasmine	Antidepressant
Lavender	Mildly analgesic
Marjoram	Mildly analgesic
Neroli	Mildly sedative
Rose	Antiseptic
Rosemary	Mild stimulant
Sandalwood	Antiseptic

AURICULAR THERAPY

Ear acupuncture is based on the idea that each part of the ear is a mirror of the body as a whole. There are, for example, **kidney meridian points on the ear.** As a result, almost any ailment can be treated with ear acupuncture. Sometimes it is used together with body acupuncture, but some practitioners use ear acupuncture on its own for diagnosis and treatment. Acupuncture points on the external ear can be used to treat disease, relieve pain, and promote anesthesia. The technique is over 1000 years old but has been further developed in both France and China in recent years. Points are located in the ear by pressing with a fine, blunt instrument to locate sensitive spots or by using an electrical sensor to find points that have a low resistance to electricity. The point is then stimulated either electrically or with small conventional needles. The needles are kept in for about 15 minutes and manipulated every now and then. Alternatively, a small ring-needle or seed may be fixed to the point in the ear and retained for several days, providing continuous treatment lasting for a short period of time. These are held in place with small patches of adhesive bandage. The needles work in the same way as traditional acupuncture, but the seed operates more as a mild form of acupressure (Thomas, 1997).

AYURVEDA

Ayurveda, meaning the "**Science of Life**," is said to be the oldest and most complete medical system in the world and dates back to 5000 B.C. Its roots are in ancient Indian civilization and Hindu philosophy, and it has been an important influence on the development of all the other Eastern medical systems. The original source of Ayurveda is the *Vedas* and the texts known as the *Sambitas,* which give a treatise on health care and describe medical procedures, including surgery and a form of massage of vital energy points (similar to Chinese acupressure). In Ayurvedic medicine, the life force is **prana**, similar to the Chinese **chi**, or **qi**. As in Chinese medicine, the functioning of the body is controlled by immaterial forces, linked to physical substances. These substances are the three basic forces or **doshas** that exist in all things:

Pitta, the force of heat and energy, linked with the sun, that controls digestion and all biochemical processes in the body
Kapha, the force of water and tides, influenced by the moon, the stabilizing influence that controls fluid metabolism in the body
Vata, linked to the wind, the force that controls movement and the functioning of the nervous system in the body

When "not abnormal" these three forces ensure that the body is healthy, but when they are "abnormal" or unbalanced, disease follows (Thomas, 1997).

BIODYNAMIC MASSAGE

Developed by Norwegian therapist **Gerda Boyesen** out of the integrated form of physical and psychological therapy known as **bioenergetics**, **biodynamic massage** combines massage with exercise to release emotional tension. Participants are taught to "connect with" and experience the energy flowing through them in much the same way as they might experience an **orgasm**. Because it may have a liberating effect on the mind, as well as the body, it is known also as **dynamic psychology** (Thomas, 1997).

BIOENERGETIC THERAPY

Bioenergetic therapies are based on the concept of the **essential involvement of the body in personal development.** They stem from the work of **Wilhelm Reich** (1897–1957) during the 1930s and 1940s, in Austria and the United States. He was fascinated by the flow of energy through the body and wanted to find out why it became blocked and how to release it. Reich believed that **tensing our muscles** is a natural way to defend ourselves against **unendurable mental pain**, in particular the fear of **losing the love of our protectors** by displaying unacceptable emotions. Painful experience is split off from the conscious and kept separate by physical "**armoring**," which also affects our breathing rhythms. Therapists are trained to read the messages of the body — which, it is said, never lies. Repressed traumas (in primal work such traumas include those that occur at birth) can be brought to awareness, released, and resolved, using a combination of physical and psychological techniques, once the central focus of the trauma has been determined. **Physical exercises**, **breathing exercises**, and **various forms of massage** may be used, usually in group therapy sessions. This liberating process can also bring back pleasurable memories of past experiences, especially childhood delights (Thomas, 1997).

CHI KUNG MEDICINE

Although chi kung involves physical activity, its aim is to develop and conserve the internal energy and integrate body and mind. Chi kung is one of the most effective of all Eastern methods for combating stress and is also becoming highly popular as a method of enhancing and managing sexual potential. Taoist "secrets" of how to conserve sexual energy are now receiving serious attention as an important part of resistance to the aging process. An ancient Chinese system for the development of internal energy, **chi kung** or **qi gung** ("chee goong") is, like **t'ai chi**, based on the principles of the **Taoist religion** as extended to Eastern medicine. Unlike **yoga**, chi kung puts less emphasis on stretching movements and postures, and more on learning how to feel and move energy inside the body. Some of the basic chi kung drills involve standing still for anything from minutes to hours at a time, sensing the movement of energy in the body.

Other chi kung exercises involve gentle, rhythmic swinging or stretching movements to generate and conserve energy, but not to burn calories. The chi kung practitioner aims to avoid stress on joints and muscles and directs his or her awareness inside the body rather than concentrating on building up what the Chinese refer to as "external" strength in the form of tight, well-developed muscles. As a result, chi kung practice develops enormous inner power in a pliable, flexible, and relaxed body. An experienced practitioner can use the therapy not only for healing but also in a preventive capacity (Thomas, 1997).

CHINESE HERBALISM

The Chinese herbal tradition is believed to date back some 4000 years to the **Emperor Shen Nong** (or **Chi'en Nung**). He is said to have described over 300 medicinal plants and their uses in a book called the **PenTsao**. Although versions of several ancient pharmacopoeia still exist today, the main surviving text on herbalism is the book by the physician **Li Shih-chen** describing almost 2000 herbs and 10,000 herbal remedies, and written in the 16th century (Thomas, 1997).

CHIROPRACTIC MEDICINE

In 1895, **Daniel David Palmer**, the founder of chiropractic medicine, treated his office janitor for deafness by realigning some small bones in his spine. Fascinated by this discovery, Palmer moved from his native Canada to Davenport, Iowa, where he developed the principles upon which chiropractic is based. *Chiropractic* — a term that derives from two ancient Greek words meaning "manually effective" — is technically described as "**the diagnosis, treatment, and rehabilitation of conditions that affect the neuromusculoskeletal system.**" Through a series of special examination and manipulative techniques, chiropractors can diagnose and treat numerous disorders associated with the nerves, muscles, bones, and joints of the body. The emphasis chiropractors put on the spine has led many people to believe that the therapy is useful only for treating back pain. In fact, skilled therapists can treat almost every structural problem from headaches to ankle problems (Thomas, 1997).

CRYSTAL AND GEM THERAPIES

More than 200 gem and mineral essences have recently been created in the belief they will assist the healing of specific mental and emotional states. Designed to be taken orally, they are made by immersing the stones in pure water exposed to sunshine, which is said to instill the water with their "**healing vibrations**." Healers believe that gems and crystals placed on and around their patients can focus and enhance healing energy. Placing crystals around the home is said to improve the atmosphere, absorbing negativity, but they should be washed regularly under cold running water and "**recharged**" in sunlight (Thomas, 1997).

CUPPING THERAPY

Cupping is particularly helpful for conditions such as **rheumatism, lumbago**, and **stiff neck** and **shoulders** as it increases circulation and the mobility of affected areas. Cupping has been used in China since the third century B.C. It involves lighting a match in a small rounded "cup," made of glass, bamboo, metal, or pottery, and then removing it quickly and applying the cup to the skin. The flame creates a vacuum, and the cup sticks tightly to the skin. Several cups may be applied at any one time to a particular part of the body such as the back. The cups are left in position for 10 to 15 minutes while the vacuum inside the cup produces strong suction on the skin and increases the blood flow and circulation. The cup is released by pressing the skin next to the edge of the cup so that the vacuum is broken. Cupping is slightly uncomfortable but not painful (Thomas, 1997).

DIETARY AND NUTRITIONAL THERAPIES

Diet as therapy has been practiced for centuries. The father of medicine, **Hippocrates**, wrote extensively about the therapeutic use of diet, yet until relatively recently modern medicine has largely forgotten the overwhelming role of diet, except as related to such problems as obesity and diabetes.

With the growth of nutritional science in the 20th century, particularly since World War II, specific foods have been recognized as risk factors in disease. For example, **too much fat**, and saturated animal fat in particular, is now widely recognized as a risk factor of heart disease and some cancers. Equally, too much refined food and **too little fiber** causes a range of digestive and bowel disorders, from constipation to irritable bowel syndrome — and may even be a cause of some cancers. **Too much salt** may exacerbate high blood pressure and reactions to food-trigger allergies.

Most alternative therapists believe that everybody can benefit from dietary self-help for both prevention and treatment of disease. Many therapies recommend **fasting**, perhaps drinking only fruit and vegetable juices, or eating salads or single fruits. **Fasting** is probably the oldest therapy known to humankind. Primitive people, and animals, instinctively stop eating when they are ill, probably because digestion takes up energy and puts extra strain on the body. The aim of fasting and dietary therapy is to rid the body of the toxins that are said to accumulate from the wrong diet, so that it can function at its optimum level.

Nutrients are the chemical components of diet and are essential to life and health. Nutrients are classed as either **macronutrients** or **micronutrients:**

- Macronutrients are **carbohydrates** (sugars and starches), **fats** (including essential fatty acids), **proteins** (including essential amino acids), and fiber.
- Micronutrients are **vitamins**, **minerals**, and **trace elements** that cannot be manufactured in the body, and so must be consumed daily.

If micronutrients are absent or too low, illness results. **Scurvy**, for example, is a disease resulting from **lack of vitamin C**. It was once the curse of sailors who had to make long trips at sea with no access to fresh fruit or vegetables. Once the connection between scurvy and fresh fruit had been made, and the sailors issued **lime juice** to drink, scurvy virtually disappeared.

Other factors can also increase the need for vitamins and minerals. Examples are poor digestion and absorption, smoking, and drinking. Rapid growth in childhood and **adolescence**, **pregnancy**, **lactation**, and **old age** all increase the need for nutrients. Requirements are also raised during illness, when we are under stress, taking drugs, or affected by environmental toxins and pollutants. It is in these situations that nutrients such as vitamins, minerals, amino acids, and essential fatty acids may be recommended and prescribed (Thomas, 1997).

EASTERN HERBAL MEDICINE

The use of herbs once belonged partly to the realm of magic, in the healing rites performed by **shamans** (men and women of "natural wisdom"); the use also stems from observations of the way in which animals treat themselves to various plants when sick or wounded. But careful study of herbs and their properties over thousands of years has developed Eastern herbal medicine into a highly refined and complex discipline. In Chinese medicine, diet is a matter of great importance in preventing and treating disease, and the Eastern therapist also gives instructions on how to correct the diet and lifestyle. Acupuncture may also be recommended. Western research now suggests that many of the substances that are used in Eastern medicine may in fact be capable of forming part of the treatment of cancers and other diseases. Eastern medicine herbs are generally used in combinations. There are approximately 300 herbal ingredients regularly used in prescriptions, and about half of these would be considered essential in any Chinese herb store.

- **Ginseng** is a well-known herbal tonic in Eastern countries and its properties are also recognized in the West.
- **Betel nuts**, *Bing Lang*, are used for intestinal worms and irregular bowel movements.
- **Oyster shells** are the remedy source for **Mu Li**, which has a sedative effect.
- The medicinal qualities of *Cinnamomum zeylanicum* as a stomachic and carminative are recognized in the West and East alike (Thomas, 1997).

ENVIRONMENTAL THERAPIES

Air pollution is one of the worst problems caused by industries, agriculture, power stations, aerosols and other chemicals, and coal and other fires. As a result we suffer from headaches, respiratory tract infections and ailments, asthma, bronchitis, emphysema, eye problems, and eventually a profound breakdown in health manifested by various cancers. Other environmental hazards include lead from gasoline — which causes hyperactivity and birth defects — acid rain, carbon monoxide poisoning, water pollution, and radiation. Clinical ecologists treat illnesses and disorders that they believe stem from an individual's reaction to these environmental factors. They practice what is known as **environmental medicine**, and they estimate that between 10 and 30% of the population suffers from some form of ecological illness (Thomas, 1997).

FLOWER AND TREE REMEDIES

Although flowers have played a role in healing for centuries — Australian Aborigines and Native Americans were using remedies made from flowers to ease emotional upsets and achieve peace of mind thousands of years ago — the healing power of flowers was only rediscovered in the West in the 1920s when it was revived by **Dr. Edward Bach**, who provided the following attributes:

Agrimony	Worry hidden by a career mask, apparently jovial but suffering
Aspen	Vague, unknown, haunting apprehension and premonitions
Beech	Intolerant, critical, fussy
Centaury	Kind, quiet, gentle, anxious to serve
Cerato	Distrust of self and intuition, easily led and misguided
Cherry Plum	For the thought of losing control, of doing dreaded things
Chestnut Bud	Failing to learn from life, repeating mistakes, lack of observation
Chicory	Demanding, self-pity, self-love, possessive, hurt and tearful
Clematis	Dreamers, drowsy, absent-minded
Crab Apple	Feeling unclean, self-disgust, small things out of proportion
Elm	Capable people with responsibility who falter, temporarily overwhelmed
Gentian	Discouragement, doubt, despondency
Gorse	No hope, accepting the difficulty, pointless to try
Heather	Longing for company, talkative, overconcern with self

Holly	Jealousy, envy, revenge, anger, suspicious
Honeysuckle	Living in memories
Hornbeam	Feels weary and thinks cannot cope
Impatiens	Irritated by constraints, quick, tense, impatient
Larch	Expect failure, lack of confidence and will to succeed
Mimulus	Fright of specific, known things — animals, height, pain, etc.
Mustard	Gloom suddenly clouds one for no apparent reason
Oak	Persevering, despite difficulties, strong, patient, never giving in
Olive	Exhausted, no more strength, need physical and mental renewal
Pine	Self-critical, self-reproach, assuming blame, apologetic
Red Chestnut	Worry for others, anticipating misfortune, projecting worry
Rock Rose	Feeling alarmed, intensely scared, horror, dread
Rock Water	Self-denial, stricture, rigidity, purist
Scleranthus	Cannot resolve two choices, indecision, alternating
Star of Bethlehem	For consolation and comfort in grief, after a fright or sudden alarm
Sweet Chestnut	Unendurable desolation
Vervain	Insistent, willful, fervent, overstriving
Vine	Dominating, tyrannical, bullying, demands obedience
Walnut	Protection from outside influences, for change and stages of development
Water Violet	Withdrawn, aloof, proud, self-reliant, quiet grief
White Chestnut	Unresolved, circling thoughts, constant worry
Wild Oat	Lack of direction, unfulfilled, drifting
Wild Rose	Lack of interest, resignation, no love or point in life
Willow	Dissatisfied, bitter, resentful, feeling life is unfair and unjust

Remedies from flowers and trees are subtle "**elixirs**" that claim to be able to help rekindle a feeling of mental and emotional, as well as physical, well-being. Some aim to bring relief from unsettling moods and emotions such as anxiety, fear, guilt, and anger, and others encourage people who use them to recognize and let go of deep-seated behavioral patterns that give rise to such feelings. Above all, they claim to be able to help people feel calm and content in times of stress. Flower and tree remedies are so safe and harmless that they are often used as self-treatment. They are not targeted at particular symptoms but at states of mind accompanying illness (Thomas, 1997).

FOOD AND DIET THERAPY INCLUDING FASTING

Naturopathy encourages us to take responsibility for our own health by encouraging sensible diet and lifestyle management. This is a principle with which few conventional medical practitioners now argue. Diet is becoming rapidly and widely accepted as much of the basis of good health. In naturopathic terms, a good diet is whole food, comprising "**live foods**" — that is, foods that have not been processed or refined and are mostly organic. Such foods, believe naturopaths, "**fuel vitality**" and stimulate the "**vital force.**" Diets must also provide the necessary materials or "nutrients" on which the body relies for good health.

Fasting means abstaining from solid food for a specified period of time. It does not mean starving completely or going without fluids. Liquids must always be taken regularly during any fast. Neuropaths believe that fasting:

- Cleanses the system of poisons accumulated from bad eating habits, a poor environment, and suppressed or repressed emotions
- Enhances immune functioning and speeds up healing
- Gives the digestive system a well-earned and often much-needed rest

Fasting is particularly beneficial in the treatment of fevers and acute problems such as skin rashes or digestive upsets. Naturopaths also recommend fasting 1 or 2 days a month regularly as an important part of preventive health care (Thomas, 1997).

GEOPATHIC THERAPY

The **electromagnetic field** generated around **pylons** has caused headaches, nausea, and general debility in the unfortunate people who live near or underneath them. The Earth's magnetic field generates powerful unseen energy forces that have been recognized for thousands of years. These Earth energies were probably easier to detect in the preindustrial age, and it is thought that ancient cultures sited their standing stone circles on sites where the Earth's energy could be felt particularly strongly. Not all natural energies are beneficial; for example, **granite rock strata** emit the noxious gas **radon**. Other negative energies emanate from the Earth and artificial contributions include the local electrical fields generated by power cables. Prolonged exposure to these energies — a condition known as **geopathic stress** — contributes to illness and general debility by weakening the body's defenses against disease and potentially harmful substances in the environment (Thomas, 1997).

Feng shui is the ancient Chinese art of balancing energies by integrating people, buildings, and landscape to create a harmonious whole. Only when balance and harmony are achieved can the **chi**, the energy or life force of the universe, flow freely, resulting in good health, happiness, and prosperity. In Chinese culture this is taken very seriously and alterations to existing buildings or the construction of new ones are only undertaken after consultations with a **feng shui expert**. Correcting bad feng shui is considered as much a part of the healing process as prescribing herbs or applying acupuncture or acupressure (Thomas, 1997).

GESTALT THERAPY

Well-known **Gestalt techniques** include increasing the awareness of **"body language"** and of **"negative internal messages"**; emphasizing the client's self-awareness by making him or her speak continually in the present tense and in the first person; concentrating on a part of a client's personality, perhaps even on just one emotion, and addressing it as if it were sitting by itself in the client's chair; the creation by the therapist of episodes and diversions that vividly demonstrate a point rather than explaining in words.

Psychodrama is a role we all play in life. If that role is painful or frustrating, we may need to express our negative emotions in some way or to create scenarios that make us feel better. Psychodrama, which was developed in the 1920s in Austria and the United States by **Jacob Moreno**, provides an arena for this. The openness to self-discovery that is a feature of psychodrama derives from several factors. Once trust has been created, group dynamics encourage disclosure; acting and direction allow for safe exploration of feelings by trial and error; the rare physical freedom of moving at will across an open space promotes expansiveness. Psychodrama can be used to unlock doors to the past, to explore one person's "dark" side, to empathize with others through role exchange, and to discover and practice fresh ways of relating (Thomas, 1997).

HERBAL MEDICINE

The Chinese, Japanese, Indian, and Native (North and South) American cultures all have traditional systems of **herbal medicine**. In China and Japan, the use of herbal remedies is officially promoted by a government ministry and included in national health systems. In India, herbalism is part of the ancient but still widely used system of **Ayurvedic medicine**. Native Americans use herbs in a spiritual sense, placing emphasis on their purifying and cleansing properties both physically and mentally.

Most **herbal therapists** (sometimes called **herbalists** — but not all herbalists treat clients medically) believe that the body acts in unison with the psyche (spirit) and emotions to maintain the equilibrium necessary for overall health and well-being. With this holistic approach, herbal therapists use their knowledge of plant properties to rebalance a client's life energy levels so that his or her body heals itself.

Generally speaking, conventional medicine takes the view that the mind and the body are separate entities. If there is illness or disease in the body, the symptoms and the affected part are treated in isolation. For example, an allergic reaction might be treated with antihistamines by a conventional doctor, while a therapeutic physician would look for the root of the problem, and address that. Every therapeutic herbalist acknowledges that there is a role for surgical procedures and drugs in some situations. The advances of modern medicine can complement medical herbalism, and vice versa (Thomas, 1997).

The following summarizes some herbs and their uses. Scientific data either verifying or denying their therapeutic efficacies do not exist:

HERBS	Used for:
ANGELICA	The common cold, coughs, respiratory infections; flatulence, indigestion, loss of appetite
ARNICA	Bruises, sprains (not for internal use)
BALM OF GILEAD	Eczema, psoriasis; sore throat, laryngitis
BORAGE	Fever; inflammation; respiratory infections; stress
CHAMOMILE	Anxiety, insomnia; flatulence, indigestion; inflammation; conjunctivitis; healing wounds
COMFREY	Bruises, burns, ulcers, healing wounds (use externally only)

DANDELION	Gall bladder problems, water retention, jaundice; general debility
FENNEL	Colic, flatulence, indigestion; coughs; conjunctivitis
GARLIC	The common cold, coughs, adenoid problems, respiratory infections, other bacterial infections; high blood pressure (hypertension); acne
HYSSOP	The common cold, coughs, sore throat, respiratory infections, loss of appetite; bruises; anxiety
LAVENDER	Nervous exhaustion, insomnia; headache; rheumatic pain
LEMON BALM	Flatulence, indigestion; anxiety, depression
MARIGOLD	Bruises, burns, healing wounds, skin inflammation, conjunctivitis, some fungal infections; adenoid problems
MEADOWSWEET	Fever, nausea, digestive problems, constipation, urinary infections; rheumatic pain
NETTLE	Constipation, water retention; bleeding, eczema; allergies
OREGANO	The common cold, coughs, sore throat, headache; muscular pain; bites and stings, healing wounds
PEPPERMINT	Flatulence, loss of appetite, nausea, colic; headache, menstrual problems; skin irritation
ROSEMARY	Headache, muscular pain, neuralgia, general debility; digestive problems; baldness
SAGE	Laryngitis, inflammation, healing wounds; anxiety, depression; (do not use during pregnancy)
ST. JOHN'S WORT	Burns, healing wounds, shingles; neuralgia, rheumatic pain; anxiety; tension; depression
THYME	Coughs, sore throat, laryngitis, respiratory infections; infections in wounds
VALERIAN	Sedative
WITCH HAZEL	Astringent
YARROW	Menstrual problems
YELLOW DOCK	Skin problems

HOMEOPATHY

Homeopathy, founded by a German doctor and chemist named **Samuel Hahnemann** (1755–1843), is a holistic medicine that aims to help the body heal itself. It works for both acute illnesses and chronic ailments, and the prevention of illness is as crucial to its philosophy as the treatment. The name *homeopathy* comes from the Greek word *homios* meaning "like" and *pathos* meaning "suffering." The word *homeopathy* simply means treating like with like. This means that a substance that causes symptoms of illness in a well person can also be used to cure similar symptoms when they result from illness (Thomas, 1997).

HYDROTHERAPY

Hydrotherapy — water therapy — is the use of water to promote healing. It is one of the oldest, simplest, and most effective of all the natural therapies. Water treatments include taking natural spring water internally for its beneficial mineral content, and external treatments such as bathing, douches, and taking exercise in water (Thomas, 1997).

HYPNOSIS AND HYPNOTHERAPY

Many 20th-century scientists have struggled to explain **hypnotherapy** and how it works. It is widely considered to be a useful method of encouraging healing and altering behavioral states. From the study of hypnosis has come a host of other therapies, including **biofeedback**, **autogenic training**, and **relaxation and meditation**. Many people now use self-hypnosis techniques to manage stress, pain, anxiety, and conditions such as migraine, irritable bowel syndrome, obesity, and addictions.

"**Healing**" a patient who is in a state of trance is one of the oldest therapeutic arts. Ancient cultures all around the world revered individuals deemed to be in contact with supernatural powers and apparently able to use such contacts to cure the sick and distressed while these people were in a state resembling sleep. The supposed connection with the supernatural powers lies behind many of the prejudices and fears about hypnosis — that it is a state that enables inner connections to be made — has at last begun to be universally accepted.

The American **Friedrich Anton Mesmer** (1734–1815), the father of **mesmerism**, tried in the 18th century to harness mental energy — known at the time as "**animal magnetism**" — to effect cures. His results were variable, but he developed a ritual around his treatment that genuinely hypnotized those who came to him for help. His "**mesmerizing**" methods received scientific attention throughout the 1800s. When, in 1841, the Scotsman **Dr. James Braid** saw a demonstration he began to develop his own theories and techniques. He demonstrated that a trance, for which he coined the term "**hypnosis**," could be induced very simply, and that hypnotized subjects could not be made to act against their will. They began to make some use of hypnosis, particularly for anesthesia during surgery (Thomas, 1997).

IRIDOLOGY

Iridology was developed in the 19th century by the Hungarian **Dr. Ignatiz von Peckzely**, who as a boy noticed changes in the eye of an owl with a broken leg as the owl made its recovery. He published his theories in 1981, and soon after a Swedish doctor, **Nils Lilinquist**, added his own observations. But, iridology did not become widely popular until **Dr. Bernard Jensen** pioneered its use in the United States. In 1950, he published a chart that showed the location of every gland and organ reflected in each eye. According to iridologists, the iris of the eye represents a kind of map of the human glands, organs, and systems of the whole human body. Problems show up on the iris as spots, flecks, white or dark streaks, and so on. Texture and color indicate the person's general state of health. Some iridologists claim they can find tendencies toward inherited disease and possible future problems, and some even address emotional and spiritual health problems this way. Common conditions that iridologists claim to be able to diagnose include arthritis, heart disease, skin problems, and allergies (Thomas, 1997).

JAPANESE MEDICINE

Chinese medicine was introduced to the imperial court of Japan in the 5th century A.D. by Korean physicians. Monks and traveling physicians from Korea and China introduced Chinese ideas more generally during the 5th and 6th centuries. Medical works on **acupuncture** and **moxibustion**, with detailed diagrams, were made known in Japan by the Chinese doctor **Zhi Cong** around A.D. 560, and from the early 7th century Chinese medicine began to be adopted systematically under the influence of two Buddhist monks who had spent many years in China. The Japanese adaptation of Chinese medicine is known as **kanpo**, and the main foundations of present practice date back to the 16th and 17th centuries (Thomas, 1997).

LIGHT THERAPY

Insufficient natural light can lead to depression, tiredness, or overeating. In the winter, especially in colder countries, the level of indoor light produces only about a tenth of the illumination of a full day of natural light. A form of "**winter blues**" that can result from this is known as **seasonal affective disorder** (SAD). It affects around 1% of the population (more women than men) but can be successfully treated with special "**full spectrum light units**" fitted into the home or into the office (Thomas, 1997).

MASSAGE THERAPY

Massage is one of the oldest forms of remedial therapy. First practiced in a structured way probably in China and Mesopotamia more than 5000 years ago, the art of massage was already well known to the physicians of ancient Greece when **Hippocrates**, the "father of medicine," wrote in the 5th century B.C. that "**the way to health is a scented bath and an oiled massage every day.**"

Shiatsu

Shiatsu is an ancient form of pressure-point massage, which has been practiced for centuries in Japan. It is based on the principle of applying pressure to key acupuncture points with the purpose of promoting the smooth flow of energy around the body.

Therapeutic Massage

This comforting form of massage consists mostly of soothing strokes and rubbing and is now in wide use in both conventional and unconventional medicine for the relief of pain or physical discomfort. In recent years, it has been shown to encourage recovery after a heart attack and to ease the suffering of patients with some types of cancer.

Reflexology

This is a specialized massage for the hands and feet; it is used for both diagnosing and correcting imbalances in the body.

Sports Massage

This form of therapy features deep tissue massage that aims to ease stiff joints, relax tense muscles, and restore suppleness.

Baby Massage

By no means solely used on babies, baby massage relies on especially gentle strokes for promoting general health and happiness.

Biodynamic Massage

This therapy combines massage with elements of physical exercise and psychological development (Thomas, 1997).

MOXIBUSTION

Moxibustion is the application of heat to specific points on the body to treat diseases and restore the smooth **flow of chi in the meridians**. Generally, the heat is obtained by burning dried **mugwort leaves (*Artemesia vulgaris*,** known as **moxa**) either directly or indirectly on the skin. Sometimes a handful of moxa is lit in a specially designed box that is placed on the back in order to warm a larger area such as the kidneys. Moxa may also be placed on a slice of ginger or garlic, or on salt for more specific effects. **Ginger** helps to promote circulation, while **garlic** has a strong antiseptic effect. Moxa is widely used for conditions such as stiff neck, cold, weak back, frozen shoulder, and fatigue and has an invigorating and warming effect (Thomas, 1997).

MYOTHERAPY

Dr. Janet Travell devised "**Trigger Point Injection Therapy**" (TPIT) during the 1940s, after she published her study of muscular pain caused by trigger points. Throughout life — thanks to the exigencies of birth, active childhood, and the traumas, diseases, and occupational stresses of adulthood — our muscles tend to accumulate tender, irritable spots (**trigger points).** The ancient Chinese were aware of them 3000 years ago, but it has taken Western physicians until the 20th century to devise a successful therapy to relieve trigger points. The elements of **myotherapy** allow a patient to rest on a table. Then, sustained and firm pressure is applied for about 7 seconds at a time to the highly irritable spots in abnormally taut bands of muscle. Once the trigger points have been relieved and the muscles released, the client is encouraged to undertake some passive stretching exercises to retain the flexibility created during the course of the lesson (Thomas, 1997).

NATIVE AMERICAN HERBAL MEDICINE

For modern Native Americans who live on reservations, the use of herbs and other traditional methods of healing remains vitally important and is still preferred to conventional medicine. The Cherokee of the Southeast recognized more than 100 types of medicinal herb. Some were used because they resembled the causative agent of the disease or because they looked like the part of the body affected. Herbs were gathered after ritual prayers and promises not to take more than was needed. **Sage (*Salvia officinalis*) is a sacred herb for many Native Americans** (Thomas, 1997).

NATUROPATHY (*MEDICATRIX NATURAE*)

Naturopathy is an umbrella term used in most Western countries to cover a range of therapies coming under the heading of "**natural medicine**." Originally coined by the German pioneer **Benedict Lust**, naturopathy means, literally, "**natural treatment**," and today its practitioners are generally those trained at specialist colleges in a range of skills that include acupuncture, herbalism, homeopathy, osteopathy, hydrotherapy, massage, nutrition, and diet (Thomas, 1997).

OSTEOPATHY

Osteopathy was devised in 1874 by **Andrew Taylor Still** (1828–1917). His philosophy was that "**structure governs function**," a belief that remains one of the basic principles of modern osteopathy. He claimed that tension in muscles and misaligned bones places unnecessary strain on the body as a whole. The initial strain can be caused by any number of factors, such as physical injury, or habitual poor posture, or by destructive emotions such as anxiety and fear. Adjusting the framework of the body would relieve that strain and enable all the systems to run smoothly so that the body would heal itself. Osteopathy is a manipulative therapy that works the body's structure (the skeleton, muscles, ligaments, and connective tissue) to relieve pain, improve mobility, and restore all-around health (Thomas, 1997).

PERSIAN MEDICINE

Traditional Persian, Arabic, or Islamic medicine became known in India, where it is widely practiced, as **Unani-Tibb**. "**Tibb**" is a Persian word meaning "medicine," while "**Unani**" is thought to be derived from "Ionian" (meaning Greek) — acknowledging the influence of the early Greek healing traditions on this system of medicine. The system dates to the 7th century, when the Arab-Islamic world adopted the traditions of Europe as it expanded into areas that had been part of the Greco-Roman empires. Medical practice and theory were then dominated by the works of the Greek physician **Galen** (A.D. 130–200) who studied anatomy and made use of numerous drugs. The Muslims who invaded India in the 11th century brought their medicine with them, and the system is prominent today, particularly among Muslims, in India and its surrounding countries. It owes most to the work of the 10th-century Persian physician **Ibn Sina**, known in the West as **Avicenna**. A follower of Galen, he considered the physical, emotional, and spiritual aspects of health and developed a system of botanical medicine and dietetics for health.

Unani-Tibb has been influenced by **Ayurvedic medicine**, as well as influencing it. It is a holistic system that treats the imbalances that lead to disease and encourages the patient to adopt a balanced way of life. It incorporates the following concepts:

- **Four elements:** earth and water (heavy) and fire and air (light)
- **Nine temperaments:** one equable (balanced) and eight nonequable and relating to hot, cold, wet, and dry
- **Four humors:** as in ancient Greek medicine — blood, phlegm, yellow bile, and black bile — semigaseous vapors that maintain body fluids and balance digestion (Thomas, 1997).

POLARITY THERAPY

Polarity therapy is based on the concept of the energy flow between the five energy centers or **chakras** in the body. Polarity therapy was developed over the course of 50 years by **Dr. Randolph Stone** (1890–1983), an Austrian-born naturopath, chiropractor, and osteopath, who also took an interest in the theories and practice of Eastern medicine and in spirituality. Polarity therapy is a holistic system of healing, based on Stone's belief that **humans are predominantly spiritual beings** whose health and happiness depend upon the free flow of energy within their bodies (Thomas, 1997).

PSYCHIC HEALING

Spiritual or psychic healing is believed to transfer healing energy by the laying on of hands. Christians say it works through God or "**the Christ energy**" and spiritualists claim that the spirits of physicians heal through them. Some healers say they are helped by **angels**. Many describe themselves as channeling "cosmic" or "**universal energy**," "light," or "**bioenergy**." People may experience this energy as a hot or cool current, relieving pain and promoting well-being and relaxation. Some healers touch their clients; others work only in the energy field. Some combine healing with massage; others use crystals, color, and sound. Instant "**miracle**" cures are rare, but most people receive some benefit, from the relief of stress, depression, and pain, to the alleviation of chronic conditions. Treatment may include examining underlying emotional problems.

Therapeutic Touch, or TT, is a method of hands-on healing that has been taught to thousands of nurses and health-care professionals in the United States and 38 other countries. It was developed in the 1970s by an American nursing professor, **Dolores Krieger**, after seeing the work of Hungarian healer **Oscar Estabany**, and studying with healer **Dora Kuntz**. It became popular in the United States as a way around the strict laws against **psychic healing** imposed in many states. Tests have shown that nurses trained in TT produce considerable improvements in the physical and emotional states of their patients (Thomas, 1997).

PSYCHOLOGICAL THERAPIES

Just like the human body, the human mind is always striving to maintain its health, and the many modern forms of **psychological therapy**, or "**psychotherapies**," aim to stimulate and support this process. The complex relationship between the mind and body has now been acknowledged, and there is strong evidence that many illnesses are caused and exacerbated by psychological factors. All complementary therapies promote the idea of the body as a whole, and promote the idea of treating it "holistically" as a result. Psychological influences, such as stress, worry, depression, and anxiety have an enormous impact on health and are implicated in a number of conditions, including migraine, eczema, asthma, digestive disorders, headaches, and vision problems. Psychological therapies aim to address the parts of the body that some Western medicine has yet to reach successfully — the mind and the sense of well-being necessary to good health (Thomas, 1997).

Analytical Therapy

Instituted at the end of the 19th century and now comprising many variations (including **psychodynamic**), analytical theory regards a person's mind-set as the outcome of conflict between internal forces and seeks answers from the unconscious. It aims to uncover and analyze the effects of **early experience** on present difficulties and to look at ways of working to resolve early blocks (Thomas, 1997).

Behavioral Therapy

Introduced at the beginning of the 20th century and based on learning theories, **behavioral psychology** aims to predict and control behavior in a scientific way. **Cognitive therapy**, which developed in the 1960s, is concerned with perception and belief systems: changing the way we view things can change their outcome (Thomas, 1997).

Humanistic Therapy

This optimistic view emphasizes the **essential goodness of humans** and the belief that we all have choices. To realize our full potential we need to "get in touch with" our inner selves and true feelings, and to be able to express them freely. The attitude of the therapist in achieving and conveying understanding from a position of equality is critical (Thomas, 1997).

Integrative Therapy

Most therapists who work in this way have a humanistic orientation but also use elements of other therapeutic disciplines. Its ruling principle is the integration of "**mind, body, and soul**" to constitute one whole and aware person (Thomas, 1997).

Meditation Therapy

Meditation is a way of contacting the inner energy that powers the natural processes of healing and self-realization. Many cultures preserve some form of ritualistic technique — in one or two cultures dating from thousands of years ago — that promotes change from the normal levels of perception and that may result in feelings of well-being. This is probably why meditation in one form or another is central to the practice of so many of the world's religions. It has been found that meditating causes changes in heart rate and respiration, helping to reduce stress. As the thoughts slow and the level of tension in the body drops accordingly, feelings of calm, detachment, peace, and sometimes joy begin to fill the mind and spread through the body. Meditation promotes a state of deep relaxation, but also mental alertness and openness (Thomas, 1997).

Sound Therapy

Sound therapy is a very ancient method of healing. **Tibetan monks**, for example, have used a method of "**overtone chanting**" for thousands of years for treating illness. The theory is that since everything in the universe is in a constant state of vibration, including the human body, even the smallest change in frequency can affect the internal organs. Modern sound therapists consider there is a natural resonance or "**note**" that is right for each part of the human body, and for each individual, so by directing specific sound waves to specific areas they can affect the frequency at which that part is vibrating and thereby restore it to balance and therefore health (Thomas, 1997).

REFLEXOLOGY (REFLEX ZONE THERAPY)

The therapy has its roots in the practices of the healers of ancient Egypt, Greece, and possibly also of ancient China. Among the pictographs dating from around 2300 B.C. in the tomb of the Egyptian physician **Ankhmahar at Saqqara**, for example, is one that portrays two attendants working on the hands and feet of two "patients." Reflexologists also claim that manipulation of the feet for healing purposes was common among the native peoples of both North and South America. Reflexology may be described as a specialized form of massage of the feet and — less commonly — of the hands. Performed to detect and correct "imbalances" in the body that may be causing ill-health, it is, however, much more than simple massage (Thomas, 1997).

REIKI THERAPY

Treatment by a reiki practitioner is intended to promote physical, emotional, and spiritual well-being. Reiki, the Japanese word for "**universal life energy**," is a form of healing based on tapping into the unseen flow of energy that permeates all living things. It is believed that reiki originally evolved as a branch of **Tibetan Buddhism** and that knowledge of its power and how to use it was transmitted from master to disciple. At some point in the intervening centuries, the secrets of reiki were lost. They were rediscovered in the late 19th century by a Japanese minister, **Dr. Mikao Usui**. He spent 14 years seeking the ability to heal, which he believed could be discovered through studying **Buddhism**, learning Chinese and Sanskrit to help his research. It is claimed that he eventually found the knowledge he sought in an **Indian sutra**, or sacred text.

Then, after a 3-week meditation on a mountain top, he had a vision of four symbols that could be used to enable healing energy to be passed to others. The ability to channel the healing power was achieved by attunement to each of these symbols. Before he died in the 1930s, **Dr. Usui** initiated 16 others into the secret of reiki, teaching them the master attunement (Thomas, 1997).

RELAXATION THERAPIES

Relaxation allows the mind–body complex to proceed with its own healing work, restoring internal harmony and creating afresh the conditions for optimal functioning. It is also a very pleasurable experience. In the therapeutic sense, **relaxation is the release of mental and physical tension**. Simple as that may seem, not many people can release both mental and physical tension at the same time without help or training. So a wide range of techniques have been developed to promote a profound level of relaxation and enhanced psychological integration. Some go back thousands of years; others are continually being created. All utilize the effect the mind can have on the body.

Flotation therapy is used more often in **psychotherapy** to lower stress response levels, thus replacing anxiety with a strong sense of well-being, enhancing creativity and awareness, improving problem solving, and accelerating learning processes (Thomas, 1997).

ROLFING MEDICINE

Rolfing is named after its founder, **Dr. Ida Rolf** (1896–1979), an American biochemist whose therapy was intended to integrate manipulative forms of treatment with **bioenergetics** (the study of energy in living systems). Rolf recognized that when we are well aligned, gravity can flow through us, allowing us to move easily. When the body is balanced, the mind and nervous system, and all the organs and tissues to which they relate, function more efficiently and our innate healing system can work at its optimum. A full course of rolfing involves ten treatments, lasting about an hour each. Each session features a different part of the body, but is meant to fuse the part with the parts that have been treated earlier, ultimately leading to complete integration (Thomas, 1997).

SHAMANISTIC MEDICINE

Shamanism has been found in most tribal cultures in every continent, from Alaska to Borneo. **Witch doctors** or **sangomas** (Africa), **medicine men** (North America), **yogis** or **holy men** (India), and **witches** and **wizards** (Europe) are all shamans who follow more or less the same practices everywhere. The techniques used by a shaman to bring about a state of altered consciousness include drumming, rattling, chanting, dancing, and the taking of natural **hallucinogenic drugs**. Drums are a feature of the northern shamanic traditions, whereas rattles are significant in South America, as are hallucinogenic plants. In Peru, shamans are known as **vegetalistas** because of their skilled use of **dangerous plants** (Thomas, 1997).

T'AI CHI CHUAN MEDICINE

T'ai chi chuan, also known as **t'ai chi**, is an old Chinese system designed to develop **chi** within the body. **Lao Tzu** is the founder of **Taoism**. Both **t'ai chi** and **chi kung** are based on the **Taoist** principles of perfect harmony between the **yin** and **yang** energies of the body and the smooth flow of the life force known as **chi**. It can be used to rejuvenate, to heal and prevent illness and injuries, and also to lead to spiritual enlightenment (Thomas, 1997).

THERAPEUTIC DIETS

Exclusion Diets

Exclusion or elimination diets are used to detect foods suspected of causing food allergies or intolerance, or triggering attacks of illness, such as migraine. Suspected foods are avoided for about 2 weeks and then reintroduced one at a time.

Vegetarian Diet

Vegetarians eat no meat, fish, or poultry, but most eat eggs and dairy products (this is called **lacto-ovo-vegetarianism**). A vegetarian diet followed correctly over a long period can reduce risk of heart disease, cancer, and other major illnesses.

Vegan Diet

Vegans eat no animal products. They need **vitamin B12** from fortified foods or supplements. A **vegan diet** shares most of the benefits of a vegetarian diet when carried out correctly.

Food Combining Diet

Food combining (the "**Hay diet**") advises against combining starch and sugar with protein and acid fruits. At least 4 hours should separate starch and protein meals. Protein, starch, and fats are eaten in small quantities, and all refined and processed foods are prohibited. This diet is said to improve arthritis and digestive problems.

Anti-*Candida* Diet

Anti-*Candida* diets for the **treatment of thrush** avoid yeasts and mold — as in malted cereals, cheeses, fungi — sugar and sugary foods, and peanuts.

Liver Diet

In a liver diet the following foods are avoided because some alternative therapists believe they are difficult for the liver to process: meat, poultry, eggs, sugars and sugary foods, dairy produce, nuts, coffee, tea, alcohol, chocolate, fried food.

Low Blood Sugar Diet

A low blood sugar diet is based on three meals a day, plus small, 2-hourly snacks of nuts or seeds, milk, oatcakes, or whole wheat toast. Sugar and sugary foods must be avoided.

Macrobiotic Diet

A macrobiotic diet classifies all foods as either yin or yang. The aim is to eat a perfect balance, taking into account the individual's different yin/yang needs. If the equilibrium is upset, ill-health results. Food is prepared, cut, and cooked in particular ways to preserve **yin/yang characteristics**. To create balance, people living in a yang environment (hot or dry) need to become more yin (cold and wet), and vice versa. **Yin foods** grow in hot dry climates (such as the Middle East), have stronger smells, are hotter, more aromatic, and contain more water, and are therefore softer and juicier. **Yang foods** grow in cold, wet climates (such as Britain) and are drier, shorter, harder, saltier, and more sour (Thomas, 1997).

TIBETAN MEDICINE

Tibetan medicine is practiced throughout Tibet, India, Ladakh, Nepal, and Bhutan and is now becoming more widely available through Tibetan physicians living in Western countries. Tibetan medicine is based on a unique synthesis of Indian and Chinese traditional medicine and Tibetan Buddhism, with elements of Arabic medicine. As with the Ayurvedic and Chinese systems, it is holistic and takes into account such factors as diet, lifestyle, environment, weather, attitudes, and emotions alongside any symptoms of disease. The theory of meridians or energy channels is particularly highly developed. There is also a strong folk and religious tradition relating to healing, which runs parallel to the more orthodox medical tradition. Tibetan diagnosis is based on pulse-taking, urine analysis (which is exceptionally highly developed and which may stem from medieval European medicine, as introduced by the Persians), tongue diagnosis, and observations. Treatments, which aim to restore the balance of the humors, include herbal medicine, accessory therapies (massage, moxibustion, acupuncture), dietary aids, behavioral advice, religious rituals, and purification techniques (Thomas, 1997).

TRANSCENDENTAL MEDITATION (TM)

In the late 1950s, an Indian monk named **Maharishi Majesh Yogi** began teaching a new form of meditation that could be easily practiced by busy, modern people around the world. He called it transcendental meditation and it was based on the concept that a meditation using **mantras** of short words or phrases, repeated in the mind, could help the user subdue many thought processes and reach a deep level of consciousness. Mantras are selected on the basis of the learner's temperament and occupation. The spirit behind TM is the **Vedanta system of philosophy** that forms the basis for most modern schools of **Hinduism**. The simple technique produces a state of "**restful alertness**" that, it is considered, transcends thinking to reach the source of thought — the mind's own reservoir of energy and creative intelligence. Through this, people can find great relaxation, inner peace, enhanced vitality, and creativity. TM is ordinarily applied for 15 to 20 minutes twice daily.

Much of the research into the psychological and physiological effects of meditation has been carried out in relation to TM. It shows that the level of rest achieved is deeper than sleep. Another major finding is that meditation produces all the body responses opposite to the flight-or-fight response characteristic of stress. During meditation, breathing slows, the heartbeat becomes shallower, the muscles relax, blood pressure normalizes, and there are compositional changes in blood and skin that indicate a reduction in tension.

Siddha meditation is another practice whose origins are in India. It utilizes both mantra and breath control to still the mind, and thus to allow a spontaneous shift in consciousness. The mantra is recited aloud and serves two purposes. Regarded as pure sound, it causes the body and mind to vibrate at a particular frequency that induces the meditative state: the energy resonates in the unconscious, and its words provide a focus to hold the mind and stop it from randomizing its energy through scattered thoughts and images (Thomas, 1997).

TRANSPERSONAL THERAPIES

This is a spiritually oriented form of treatment that combines therapy for the individual with a concern for the whole of creation. The transpersonal dimension emerges in the drive of the individual to connect with **cosmic forces**, and in the impulse to discover an awareness of the soul. In operation, transpersonal therapies involve a wide range of functional techniques. The emphasis is on practical activities rather than verbal interpretations, and so, for example, imagery is used to explore, transform, and expand on ideas. Transpersonal therapies are holistic, aimed at balancing and fulfilling the mind, body, and soul. For this reason, a wide range of conditions may respond to therapy, particularly those that are affected by negative feelings or actions (Thomas, 1997).

VISUALIZATION THERAPY

Visualization is the forming of meaningful images in the mind; it is usually most effective when resting comfortably with the eyes closed. Especially when undertaken in conjunction with hypnosis, this natural ability can achieve therapeutic effects in several ways.

- **Relaxation:** Imaging a pleasant, relaxing place — a sunny beach, perhaps — releases body tensions.
- **Life changes:** To visualize yourself competently dealing with your problems, being successful, challenging other people, or playing a musical instrument well, can overcome blocks to being assertive and achieving what you want.
- **Accessing the past:** Images of childhood may afford access to repressed memories and emotions.
- **Rewriting the past:** An event that at the time was traumatic and damaging can be reviewed — this time with a positive outcome.
- **Spiritual growth:** Once facility in visualizing is achieved, images may occur spontaneously, increasing insight into one's own condition, into the way the world works, and into higher levels of consciousness (Thomas, 1997).

YOGA

Yoga aims to improve overall health and well-being, and through that, a number of conditions may be resolved. The full **lotus position** (**padmasana**), taken by a **yogi** is excellent for meditation. It lengthens the spine, allowing one to control breathing. The mind is better able to focus and concentrate on problems or empty itself of worry. Yoga encourages relaxation. By normalizing high blood pressure, helping stave off anxiety, or dealing with stress, yoga is as effective today as it was thousands of years ago (Thomas, 1997).

2

HERBAL THERAPEUTICS THEN AND NOW

Herbalists use whole plants, and traditional physicians use purified ingredients derived from plants.

The reason that **herbalists** and physicians use different terminology, and the reason that most herbal medicine books are of little use to the physician confronted with a desperately ill patient, is that herbalists and traditional physicians think about diseases (and the medicines used to treat them) differently. Of course, many of the drugs first used in modern medicine are extracted from plants. But herbalists use whole plants, and traditional physicians use purified ingredients derived from plants. Traditional physicians and scientists generally believe that, if a plant has any medicinal value at all, it is because different components of the plant act synergistically (Cupp, 2000).

Some herbal products are very toxic. **Comfrey** may be an effective treatment for bruises and sprains, but it also contains **pyrrazolidine alkaloids** which can cause severe liver damage.

Some ancient herbs have found new uses. A few examples are given here.

AGNUS CASTUS

Agnus castus has been used for menstrual problems resulting from corpus luteum deficiency, including premenstrual symptoms and spasmodic dysmenorrhea, for certain menopausal conditions, and for insufficient lactation.

AGRIMONY

Agrimony is said to possess mild astringent and diuretic properties. It has been used for diarrhea in children, mucous colitis, grumbling appendicitis, urinary incontinence, cystitis, and as a gargle for acute sore throat and chronic nasopharyngeal catarrh.

ALFALFA

Alfalfa is stated to be a source of vitamins A, C, E, and K, and of the minerals calcium, potassium, phosphorous, and iron. It has been used for avitaminosis A, C, E, or K, hypoprothrombinemic purpura, and debility of convalescence.

ALOE

Aloe was first mentioned in Greek literature as a laxative before the first century. In the first century, the Greek physician **Dioscorides** wrote of its use in treating wounds, chapping, hair loss, genital ulcers, hemorrhoids, boils, mouth irritation, and inflammation. In the seventh century, aloe was used in the Orient for eczema and sinusitis. Today, aloe is promoted to heal wounds, burns, skin ulcers, frostbite, and dry skin. It is an ingredient in dieters' teas used for their laxative effect. Aloe is found in other laxative and "**body cleansing**" products as well.

ALOE VERA

Traditionally, aloe vera has been used in ointments and creams to assist the healing of wounds, burns, eczema, and psoriasis.

ANGELICA

Angelica is stated to possess antispasmodic, diaphoretic, expectorant, bitter aromatic, carminative, diuretic, and local anti-inflammatory properties. It has been used for respiratory catarrh, psychogenic asthma, flatulent dyspepsia, anorexia nervosa, rheumatic diseases, peripheral vascular disease, and specifically for pleurisy and bronchitis, has been applied as a compress, especially for bronchitis associated with vascular deficiency.

ANISEED

Aniseed is stated to possess expectorant, antispasmodic, carminative, and parasiticidic properties. Traditionally, it has been used for bronchial catarrh, pertussis, spasmodic cough, flatulent colic, topically for pediculosis and scabies, and specifically for bronchitis, tracheitis with persistent cough, and as an aromatic adjuvant to prevent colic from cathartics. Aniseed has been used as an estrogenic agent. It has been reputed to increase milk secretion, promote menstruation, facilitate birth, alleviate symptoms of the male climacteric, and increase libido.

APRICOT

Traditionally, the oil has been incorporated into cosmetic and perfumery products such as soaps and creams.

ARNICA

Arnica is said to possess topical counterirritant properties. It has been used for unbroken chilblains, alopecia neurotica, and specifically for sprains and bruises. Arnica is mainly used in homeopathic preparations; it is used to a lesser extent in herbal products.

ARTICHOKE

Artichoke is stated to possess diuretic, choleretic, hypocholesterolemic, hypolipidemic, and hepatostimulating properties.

ASAFOETIDA

Asafoetida is said to possess carminative, antispasmodic, and expectorant properties. It has been used for chronic bronchitis, pertussis, laryngismus stridulus, hysteria, and specifically for intestinal flatulent colic.

AVENS

Avens is stated to possess antidiarrheal, antihemorrhagic, and febrifugal properties. It has been used for diarrhea, catarrhal colitis, passive uterine hemorrhage, intermittent fevers, and specifically for ulcerative colitis.

BAYBERRY

Bayberry is stated to possess antipyretic, circulatory stimulant, emetic, and mild diaphoretic properties. It has been used for diarrhea, colds, and specifically for mucous colitis. An infusion has been used as a gargle for sore throat, and as a douche for leucorrhea. Powdered root bark has been applied topically for the management of indolent ulcers.

BLOODROOT

Bloodroot is stated to act as an expectorant, spasmolytic, emetic, cathartic, antiseptic, cardioactive, topical irritant, and escharotic (scab-producing). Traditionally, it is indicated for bronchitis (subacute or chronic), asthma, croup, laryngitis, pharyngitis, deficient capillary circulation, nasal polyps (as a snuff), and specifically for asthma and bronchitis with feeble peripheral circulation.

BLUE FLAG

Blue flag is said to possess cholagogue, laxative, diuretic, dermatological, anti-inflammatory, and antiemetic properties. It has been used for skin diseases, biliousness with constipation and liver dysfunction, and specifically for cutaneous eruptions.

BOGBEAN

Bogbean is stated to possess bitter and diuretic properties. It has been used for rheumatism, rheumatoid arthritis, and specifically for muscular rheumatism associated with general asthenia.

BOLDO

Boldo is stated to possess cholagogue, liver stimulant, sedative, diuretic, mild urinary demulcent, and antiseptic properties. It has been used for gallstones, pain in the liver or gall bladder, cystitis, rheumatism, and specifically for cholethiasis with pain.

BONESET

Boneset is stated to possess diaphoretic and aperient properties. Traditionally, it has been used for influenza, acute bronchitis, nasopharyngeal catarrh, and specifically for influenza with deep aching, and congestion of the respiratory mucosa.

BORAGE

As early as the second century A.D., borage was used to treat sore throat when mixed with honey, and is still recommended by herbalists for this purpose. An infusion of the leaves and stems was once used as a diuretic, diaphoretic, and emollient. Other traditional uses include relief of symptoms of rheumatism, colds, and bronchitis. It is also purported to increase breast milk production. A poultice of fresh leaves has been used to treat inflammation.

Borage is stated to possess diaphoretic, expectorant, tonic, anti-inflammatory, and galactogogue properties. Traditionally, borage has been used to treat many ailments including fever, coughs, and depression. Borage is also reputed to act as a restorative agent on the adrenal cortex. Borage oil (**starflower oil**) is used as an alternative source to evening primrose oil for **gamolenic acid**.

BROOM

Broom is stated to possess cardioactive, diuretic, peripheral vasoconstrictor, and antihemorrhagic properties. It has been used for cardiac dropsy, myocardial weakness, tachycardia, profuse menstruation, and specifically for functional palpitation with lowered blood pressure. Broom is also reported to possess emetic and cathartic properties.

BUCHU

Buchu is stated to possess urinary antiseptic and diuretic properties. It has been used for cystitus urethritis, prostatitis, and specifically for acute catarrhal cystitis.

BURDOCK

Burdock is stated to possess diuretic and orexigenic properties. It has been used for cutaneous eruptions, rheumatism, cystitis, gout, anorexia nervosa, and specifically for eczema and psoriasis.

BURNET

Burnet is stated to possess astringent, antihemorrhagic, styptic, and antihemorrhoidal properties. It has been used for ulcerative colitis, metorrhagia, and specifically for acute diarrhea.

CALAMUS

Calamus has been used for centuries to treat gastrointestinal distress including colic in children, and as a sedative. A rhizome infusion is used to treat fevers and dyspepsia, and chewing the rhizome is recommended to clear the voice, relieve dyspepsia, aid digestion, and remove tobacco odor from the breath. The powdered rhizome is used as a cooking spice, as is **calamus oil**, which is responsible for the plant's odor and taste. In the United States, calamus was once used to flavor tooth powders, beer, bitters, and various tonics. Today, calamus is not used.

Calamus is stated to act as a carminative, spasmolytic, and diaphoretic. Traditionally, it has been indicated for acute and chronic dyspepsia, gastritis and gastric ulcer, intestinal colic, and anorexia.

CALENDULA

Calendula is stated to possess antispasmodic, mild diaphoretic, anti-inflammatory, antihemorrhagic, emmenagogue, vulnerary, styptic, and antiseptic properties. Traditionally, it has been used to treat gastric and duodenal ulcers, amenorrhea, dysmenorrhea, and epistaxis; crural ulcers, varicose veins, hemorrhoids, anal eczema, proctitis, lymphadenoma, and inflamed cutaneous lesions (topically); and conjunctivitis (as an eye lotion).

CAPSICUM

Capsicum is stated to possess stimulant, antispasmodic, carminative, diaphoretic, counterirritant, antiseptic, and rubefacient properties. Traditionally, it has been used for colic, flatulent dyspepsia without inflammation, chronic laryngitis (as a gargle), insufficiency of peripheral circulation, and externally for neuralgia including rheumatic pains and unbroken chilblains.

CASCARA

Cascara is stated to possess mild purgative properties and has been used for constipation.

CASCARA SAGRADA

Cascara sagrada has been used as a laxative in the past and continues to be used as a laxative today.

CASSIA

Cassia is stated to possess carminative, antispasmodic, antiemetic, antidiarrheal, and antimicrobial properties. It has been used for flatulent dyspepsia, flatulent colic, diarrhea, the common cold, and specifically for colic or dyspepsia with flatulent distension and nausea. Cassia bark is also documented to possess astringent properties. Carminative and antiseptic properties are documented for the oil.

CAT'S CLAW

Cat's claw has a long history of use in South America as an anti-inflammatory, antirheumatic, and contraceptive. It is also traditionally used to treat gastrointestinal ulcers, tumors, gonorrhea, dysentery, various skin problems, cancers of the female genitourinary tract, and intestinal disorders. Native South Americans also use cat's claw to "**cleanse the kidneys**" and treat bone pain. Some European reports that it is useful in the treatment of AIDS when used in combination with **zidovudine** (AZT), as well as the purported usefulness of cat's claw tea in the treatment of diverticulitis, hemorrhoids, peptic ulcer disease, colitis, parasites, and "**leaky bowel syndrome**," have fueled demand for the bark in the United States.

CELERY

Celery is stated to possess antirheumatic, sedative, mild diuretic, and urinary antiseptic properties. It has been used for arthritis, rheumatism, gout, urinary tract inflammation, and specifically for rheumatoid arthritis with mental depression.

CENTAURY

Centaury is reputed to act as a bitter, aromatic, and stomachic. Traditionally, it has been used for anorexia and dyspepsia.

CEREUS

Cereus is reputed to act as a cardiac stimulant and as a partial substitute for **digitalis**, although there is no proof of its therapeutic value. Cereus has been used in cases of **dropsy** and various cardiac affections.

CHAMOMILE

Chamomile has been used medicinally since ancient Rome for its purported sedative, antispasmodic, and antirheumatic effects. Today, chamomile is used topically to treat a variety of inflammatory conditions involving the mouth, skin, respiratory tract (via inhalation), and gastrointestinal tract. It is also used internally as a gastrointestinal antispasmodic and anti-inflammatory. Chamomile is purported to have sedative, hypnotic, analgesic, and immunostimulant effects.

CHAMOMILE, GERMAN

German chamomile is listed by the Council of Europe as a natural source of food flavoring (category N2). This category indicates that chamomile can be added to food stuffs in small quantities, with a possible limitation of an active principle (as yet unspecified) in the final product. German chamomile is commonly used in herbal teas.

CHAMOMILE, ROMAN

Roman chamomile is stated to possess carminative, antiemetic, antispasmodic, and sedative properties. It has been used for dyspepsia, nausea and vomiting, anorexia, vomiting of pregnancy, dysmenorrhea, and specifically for flatulent dyspepsia associated with mental stress.

CHAPARRAL

Tea made from boiled leaves has been used to treat sexually transmitted diseases and intestinal cramps, and to stimulate urination. The leaves were soaked in water to produce an extract used as a bath for rheumatism and chickenpox. The dried powdered leaves were used as a dusting powder for sores, and were mixed with **badger oil** to make an ointment used on burns to aid new skin formation. Today, chaparral is promoted as an anticancer agent.

Chaparral has been used for the treatment of arthritis, cancer, venereal disease, tuberculosis, bowel cramps, rheumatism, and colds.

CINNAMON

Cinnamon is stated to possess antispasmodic, carminative, orexigenic, antidiarrheal, antimicrobial, refrigerant, and anthelmintic properties. It has been used for anorexia, intestinal colic, infantile diarrhea, common cold, influenza, and specifically for flatulent colic and for dyspepsia with nausea. Cinnamon bark is also stated to be astringent, and **cinnamon oil** is reported to possess carminative and antiseptic properties.

CLIVERS

Clivers is stated to possess diuretic and mild astringent properties. It has been used for dysuria, lymphadenitis, psoriasis, and specifically for enlarged lymph nodes.

CLOVE

Clove has been traditionally used as a carminative, antiemetic, toothache remedy, and counterirritant. **Clove oil** is stated to be a carminative, occasionally used in the treatment of flatulent colic, and is commonly used topically for symptomatic relief of toothache.

COHOSH, BLACK

Black cohosh is stated to possess antirheumatic, antitussive, sedative, and emmenagogue properties. It has been used for intercostal myalgia, sciatica, whooping cough, chorea, tinnitus, dysmenorrhea, uterine colic, and specifically for muscular rheumatism and rheumatoid arthritis.

COHOSH, BLUE

Blue cohosh is stated to possess antispasmodic, emmenagogue, uterine tonic, and antirheumatic properties. Traditionally, it has been used for amenorrhea, threatened miscarriage, false labor pains, dysmenorrhea, rheumatic pains, and specifically for conditions associated with uterine atony.

COLA

Cola is stated to possess central nervous system (CNS) stimulant, thymoleptic, antidepressant, diuretic, cardioactive, and antidiarrheal properties. It has been used for depressive states, melancholy, atony, exhaustion, dysentery, atonic diarrhea, anorexia, migraine, and specifically for depressive states associated with general muscular weakness.

COLTSFOOT

In addition to treatment of respiratory ailments, coltsfoot has also been used to treat diarrhea, to purify the blood, to stimulate metabolism, to cause diuresis and swelling, and topically as a wound treatment. Today, products containing coltsfoot are promoted as antihistamines, decongestants, and as expectorants.

Coltsfoot is stated to possess expectorant, antitussive, demulcent, and anticatarrhal properties. It has been used for asthma, bronchitis, laryngitis, and pertussis.

COMFREY

Comfrey has been used to treat respiratory problems (bronchitis, catarrh hemoptysis, pleurisy, whooping cough), gastrointestinal diseases (cholecystitis, colitis, dysentery, diarrhea, ulcers, hematemesis), metorrhagia, phlebitis, and tonsillitis. Comfrey is currently promoted for prevention of kidney stones, for treatment of rheumatic and pulmonary disorders, and for treatment of injuries such as burns and bruises.

Comfrey is stated to possess vulnerary, cell-proliferant, astringent, antihemorrhagic, and demulcent properties. It has been used for colitis, gastric and duodenal ulcers, and has been applied topically for ulcers, wounds, and fractures.

CORN SILK

Corn silk is stated to possess diuretic and stone-reducing properties. It has been used for cystitis, urethritis, nocturnal enuresis, prostatis, and specifically for acute or chronic inflammation of the urinary system.

COUCHGRASS

Couchgrass is said to possess diuretic properties. It has been used for cystitis, urethritis, prostatitis, benign prostatic hypertrophy, renal calculus, lithuria, and specifically for cystitis with irritation or inflammation of the urinary tract.

COWSLIP

Cowslip is stated to possess sedative, antispasmodic, hypnotic, mild diuretic, expectorant, and mild aperient properties. It has been used for insomnia, nervous excitability, hysteria, and specifically for anxiety states associated with restlessness and irritability.

CRANBERRY

Cranberry has been used to prevent and treat urinary tract infections since the 19th century. Today, cranberry juice is widely used for the prevention, treatment, and symptomatic relief of urinary tract infections. Also, cranberry juice has been given to patients to help reduce urinary odors in incontinence. Another potential benefit of the use of cranberry is a decrease in the rate of formation of **kidney stones**.

DAMIANA

Damiana is said to possess antidepressant, thymoleptic, mild purgative, stomachic, and reputedly aphrodisiac properties. It has been used for depression, nervous dyspepsia, atonic constipation, coital inadequency, and specifically for anxiety neurosis with a predominant sexual factor.

DANDELION

Dandelion is stated to possess diuretic, laxative, cholagogue, and antirheumatic properties. It has been used for cholecystitis, gallstones, jaundice, atonic dyspepsia with constipation, muscular rheumatism, oliguria, and specifically for cholecystitis and dyspepsia.

DEVIL'S CLAW

Devil's claw is reputed to possess anti-inflammatory, antirheumatic, analgesic, sedative, and diuretic properties. It has been used for arthritis, gout, myalgia, fibrositis, lumbago, pleurodynia, and specifically for rheumatic disease.

DONG QUAI

Dong quai has been used in the treatment of dysmenorrhea, amenorrhea, metorrhagia, menopausal syndromes, anemia, abdominal pain, injuries, migraine headaches, and arthritis. It is also said to ensure healthy pregnancies and easy deliveries. In the United States, dong quai is promoted primarily to alleviate problems associated with menstruation and menopause.

DROSERA

Drosera is stated to possess antispasmodic, demulcent, and expectorant properties. It has been used for bronchitis, asthma, pertussis, tracheitis, gastric ulceration, and specifically for asthma and chronic bronchitis with peptic ulceration or gastritis.

ECHINACEA

Echinacea was originally utilized by Native Americans as a "**blood purifier**" and was used for the treatment of snake bites, infections, and malignancy. Today, echinacea is promoted primarily in oral dosage forms as an immune stimulant that helps increase resistance to colds, influenza, and other infections, although topical products for wounds and inflammatory skin conditions are also available.

Echinacea is stated to possess antiseptic, antiviral, and peripheral vasodilator properties. Traditionally, it has been used for furunculosis, septicemia, nasopharyngeal catarrh, pyorrhea, tonsillitis, and specifically for boils, carbuncles, and abscesses. It is under investigation for its immunostimulant action.

ELDER

Elder is said to possess diaphoretic and anticatarrhal properties. Traditionally, it has been used for influenza, colds, chronic nasal catarrh with deafness, and sinusitis. Elder is also stated to act as a diuretic, laxative, and local anti-inflammatory agent.

ELECAMPANE

Elecampane is stated to possess expectorant, antitussive, diaphoretic, and bactericidal properties. Traditionally, it has been used for bronchial/tracheal catarrh, cough associated with pulmonary tuberculosis, and dry irritating cough in children. **Alantolactone** has been used as an anthelmintic in the treatment of roundworm, threadworm, and whipworm infection.

EUCALYPTUS

Eucalyptus leaves and oil have been used as an antiseptic, febrifuge, and expectorant.

EUPHORBIA

Euphorbia is stated to be used for respiratory disorders, such as asthma, bronchitis, catarrh, and laryngeal spasm. It has also been used for intestinal amoebiasis.

EVENING PRIMROSE

An infusion of the whole plant is reputed to have sedative and astringent properties, and has traditionally been used for asthmatic coughs, gastrointestinal disorders, whooping cough, and as a sedative pain killer. Externally, poultices were reputed to ease bruises and to speed wound healing. Evening primrose oil (EPO) is licensed for the treatment of atopic eczema, and cyclical and noncyclical **mastalgia**. Other conditions in which **evening primrose oil** is used include premenstrual syndrome, psoriasis, multiple sclerosis, hypercholesterolemia, rheumatoid arthritis, **Raynaud's phenomenon**, **Sjögren's syndrome**, **postviral fatigue syndrome**, asthma, and diabetic neuropathy.

EYEBRIGHT

Eyebright is stated to possess anticatarrheal, astringent, and anti-inflammatory properties. Traditionally, it has been used for nasal catarrh, sinusitis, and specifically for conjunctivitis when applied locally as an eye lotion.

FALSE UNICORN

False unicorn is stated to possess an action on the uterus. Traditionally, it has been used for ovarian dysmenorrhea, leucorrhea, and specifically for amenorrhea. It is reported to be useful for vomiting of pregnancy and threatened miscarriage.

FENUGREEK

Fenugreek is stated to possess mucilaginous demulcent, laxative, nutritive, expectorant, and orexigenic properties, and has been used topically as an emollient and vulnerary. Traditionally, it has been used in the treatment of anorexia, dyspepsia, and gastritis, and topically for furunculosis, myalgia, lymphadenitis, gout, wounds, and leg ulcers.

FEVERFEW

The uses of feverfew have included the treatment of menstrual pain, asthma, arthritis, psoriasis, threatened miscarriage, toothache, opium abuse, vertigo, tinnitus, anemia, the common cold, and gastrointestinal disturbances. It has also been used to aid in expulsion of the placenta and stillbirths, and in difficult labor. Feverfew has been planted around houses to act as an insect repellent, as well as for use as a topical remedy for insect bites. Today, it is used in the treatment of arthritis and migraine headache.

Feverfew has traditionally been used in the treatment of migraine, tinnitus, vertigo, arthritis, fever, menstrual disorders, difficulty during labor, stomachache, toothache, and insect bites.

FIGWORT

Figwort is stated to act as a dermatological agent and a mild diuretic, and to increase myocardial contraction. Traditionally, it has been used for chronic skin disease, and specifically for eczema, psoriasis, and pruritus.

FRANGULA

Frangula is stated to possess mild purgative properties and has been used traditionally for constipation.

FUCUS

Fucus is stated to possess antihypothyroid, antiobesic, and antirheumatic properties. Traditionally, it has been used for lymphadenoid goiter, myxedema, obesity, arthritis, and rheumatism.

FUMITORY

Fumitory is stated to possess weak diuretic and laxative properties and to act as a cholagogue. Traditionally, it has been used to treat cutaneous eruptions, conjunctivitis (as an eye lotion), and, specifically, chronic eczema.

GARLIC

Over the centuries, garlic has been used to ward off vampires, demons, witches, and evil beings; as an aphrodisiac to improve performance and desire; and as a cure-all for everything from athlete's foot to hemorrhoids and cancer. Today, garlic is promoted to lower cholesterol and blood pressure, delay atherosclerotic processes, prevent heart attack and stroke, improve circulation, and prevent cancer.

Garlic is stated to possess diaphoretic, expectorant, antispasmodic, antiseptic, bacteriostatic, antiviral, hypotensive, and anthelmintic properties, and to promote leucocytosis. Traditionally, it has been used to treat chronic bronchitis, respiratory catarrh, recurrent colds, whooping cough, bronchitic asthma, influenza, and chronic bronchitis.

GENTIAN

Gentian is stated to possess bitter, gastric stimulant, sialogogue, and cholagogue properties. Traditionally, it has been used for anorexia, atonic dyspepsia, gastrointestinal atony, and specifically for dyspepsia with anorexia.

GINGER

The Chinese utilized ginger for stomachaches, diarrhea, nausea, cholera, bleeding, asthma, heart conditions, respiratory disorders, toothache, and rheumatic complaints. In China, the root and stem are used to combat aphids and fungal spores. Today, ginger is promoted to relieve and prevent nausea caused by motion sickness and morning sickness.

Ginger is stated to possess carminative, diaphoretic, and antispasmodic properties. Traditionally, it has been used for colic, flatulent dyspepsia, and specifically for flatulent intestinal colic. Ginger has also been investigated for the prevention of motion sickness.

GINKGO

The medicinal use of the leaves was recorded by the Chinese in *Chen Houng Pen T'sao* published in 2800 B.C. and a monograph exists in the modern Chinese pharmacopoeia. The leaves are recommended as being beneficial to the heart and lungs; inhalation of a decoction of leaves is used for treatment of asthma; boiled leaves are used against chilblains. Standardized concentrated extracts of *G. biloba* leaves are marketed in several European countries (e.g., as **Tanakan**™ in France, and as **Tebonin**™ and **Rokan**™ in Germany). The seed is used as an antitussive and expectorant in Japan and China.

GINKGO BILOBA

Traditional Chinese physicians used ginkgo leaves to treat **asthma** and **chilblains** (swelling of the hands and feet from exposure to damp cold). Today, *G. biloba* is used to improve blood flow to the brain and to improve peripheral circulation.

GINSENG, ELEUTHEROCOCCUS

Eleutherococcus ginseng does not have a traditional herbal use in the United Kingdom, although it has been used for many years in the former Soviet Union. Like *Panax ginseng*, *E. ginseng* is claimed to be an **adaptogen** in that it increases the body's resistance to stress and builds up general vitality.

GINSENG, PANAX

Ginseng is stated to possess thymoleptic, sedative, demulcent, and stomachic properties, and is reputed to be an aphrodisiac. Traditionally, it has been used for neurathenia, neuralgia, insomnia, hypotonia, and specifically for depressive states associated with sexual inadequency. Ginseng has been used traditionally in Chinese medicine for many thousands of years as a stimulant, tonic, diuretic, and stomachic. Traditionally, ginseng use has been divided into two categories:

- **Short-term** to improve stamina, concentration, healing process, stress resistance, vigilance, and work efficiency in healthy individuals
- **Long-term** to improve well-being in debilitated and degenerative conditions, especially those associated with old age

GOLDEN SEAL

Golden seal is stated to be a stimulant to involuntary muscle, and to possess stomachic, oxytocic, antihemorrhagic, and laxative properties. Traditionally, it has been used for digestive disorders, gastritis, peptic ulceration, colitis, anorexia, upper respiratory catarrh, menorrhagia, postpartum hemorrhage, dysmenorrhea, topically for eczema, pruritus, otorrhea, catarrhal deafness, and tinnitus, conjuctivitis, and specifically for atonic dyspepsia with hepatic symptoms.

GRAVEL ROOT

Gravel root is stated to possess antilithic, diuretic, and antirheumatic properties. Traditionally, it has been used for urinary calculus, cystitis, dysuria, urethritis, prostatitis, rheumatism, gout, and specifically for renal or vesicular calculi.

GROUND IVY

Ground ivy is stated to possess mild expectorant, anticatarrhal, astringent, vulnerary, diuretic, and stomachic properties. Traditionally, it has been used for bronchitis, tinnitus, diarrhea, hemorrhoids, cystitis, gastritis, and specifically for chronic bronchial catarrh.

GUAIACUM

Guaiacum is stated to possess antirheumatic, anti-inflammatory, diuretic, mild laxative, and diaphoretic properties. Traditionally, it has been used for subacute rheumatism, prophylaxis against gout, and specifically for chronic rheumatism and rheumatoid arthritis.

HAWTHORN

Native Americans used hawthorn as a diuretic for kidney and bladder disorders and to treat stomachaches, to stimulate appetite, and to improve circulation. The flowers and berries have astringent properties and have been used to treat sore throats in the form of **haw jelly** or **haw marmalade**. Today, hawthorn is promoted for use in heart failure, hypertension, arteriosclerosis, angina pectoris, **Buerger's disease**, paroxysmal tachycardia, heart valve murmurs, sore throat, skin sores, diarrhea, and abdominal distention.

Hawthorn is stated to possess cardiotonic, coronary vasodilator, and hypotensive properties. Traditionally, it has been used for cardiac failure, myocardial weakness, paroxysmal tachycardia, hypertension, arteriosclerosis, and Buerger's disease.

HOLY THISTLE

Holy thistle is stated to possess bitter stomachic, antidiarrheal, antihemorrhagic, febrifuge, expectorant, antibiotic, bacteriostatic, vulnerary, and antiseptic properties. Traditionally, it has been used for anorexia, flatulent dyspepsia, bronchial catarrh, topically for gangrenous and indolent ulcers, and specifically for atonic dyspepsia, and enteropathy with flatulent colic.

HOPS

Hops are stated to possess sedative, hypnotic, and topical bactericidal properties. Traditionally, they have been used for neuralgia, insomnia, excitability, **priapism**, mucous colitis, topically for crural ulcers, and specifically for restlessness associated with nervous tension headache or indigestion.

HOREHOUND, BLACK

Black horehound is stated to possess antiemetic, sedative, and mild astringent properties. Traditionally, it has been used for nausea, vomiting, nervous dyspepsia, and specifically for vomiting of central origin.

HOREHOUND, WHITE

White horehound is stated to possess expectorant and antispasmodic properties. Traditionally, it has been used for acute or chronic bronchitis, whooping cough, and specifically for bronchitis with nonproductive cough.

HORSE CHESTNUT

Traditionally, horse chestnut has been used for the treatment of varicose veins, hemorrhoids, phlebitis, diarrhea, fever, and enlargement of the prostate gland.

HORSERADISH

Horseradish is stated to possess antiseptic, circulatory and digestive stimulant, diuretic, and vulnerary properties. Traditionally, it has been used for pulmonary and urinary infection, urinary stones, edematous conditions, and externally for application to inflamed joints or tissues.

HYDRANGEA

Hydrangea is stated to possess diuretic and antilithic properties. Traditionally, it has been used for cystitis, urethritis, urinary calculi, prostatitis, enlarged prostate gland, and specifically for urinary calculi with gravel and cystitis.

HYDROCOTYLE

Hydrocotyle is stated to possess mild diuretic, antirheumatic, dermatological, peripheral vasodilator, and vulnerary properties. Traditionally, it has been used for rheumatic conditions, cutaneous affections, and by topical application, for indolent wounds, leprous ulcers, and cicatrization after surgery.

ISPAGHULA

Ispaghula is stated to possess demulcent and laxative properties. Traditionally, ispaghula has been used in the treatment of chronic constipation, dysentery, diarrhea, and cystitis. Topically, a poultice has been used for furunculosis.

JAMAICA DOGWOOD

Jamaica dogwood is stated to possess sedative and anodyne properties. Traditionally, it has been used for neuralgia, migraine, insomnia, dysmenorrhea, and specifically for insomnia due to neuralgia or nervous tension.

JUNIPER

Juniper is stated to possess diuretic, antiseptic, carminative, stomachic, and antirheumatic properties. Traditionally, it has been used for cystitis, flatulence, colic, and has been applied topically for rheumatic pains in joints or muscles.

KAVA

Kava has been during religions ceremonies. Kava is currently promoted for relief of anxiety and stress.

LADY'S SLIPPER

Lady's slipper is stated to possess sedative, mild hypnotic, antispasmodic, and thymoleptic properties. Traditionally, it has been used for insomnia, hysteria, emotional tension, anxiety states, and specifically for anxiety states with insomnia.

LEMON VERBENA

Lemon verbena is reputed to possess antispasmodic, antipyretic, sedative, and stomachic properties. Traditionally, it has been used for the treatment of asthma, cold, fever, flatulence, colic, diarrhea, and indigestion.

LICORICE

Western herbalists recognized licorice as a remedy for "dropsy" as did Pliny, and asserted that the root had emollient, demulcent, expectorant, and diuretic effects. Licorice was probably introduced to Native Americans by the early English settlers, and was subsequently used by medicine men to treat diabetes. In traditional Chinese medicine licorice was considered to benefit all organs of the body. Today, licorice is employed in many capacities around the world. In China, licorice is used to treat a variety of symptoms and diseases, including **Addison's disease**, sore throats, carbuncles, diarrhea due to "spleen deficiency," thirst due to "stomach deficiency," cough due to "dry lungs," and palpitations. Other modern uses include bronchitis and other "catarrhal conditions," gastritis, colic, arthritis, and hepatitis. Licorice contains the natural sweetener **glycyrrhizic acid**, and is used to flavor soy sauce in China. Licorice is widely used in foods as a flavoring agent. Licorice root is listed by the Council of Europe as a natural source of food flavoring (category N2). This category indicates that licorice can be added to food stuffs in small quantities, with a possible limitation of an active principle (as yet unspecified) in the final product. In the United States, licorice is listed as **GRAS** (Generally Regarded As Safe).

LIFEROOT

Liferoot is stated to possess uterine tonic, diuretic, and mild expectorant properties. Traditionally, it has been used in the treatment of functional amenorrhea, menopausal neurosis, and leucorrhea (as a douche).

LIME FLOWER

Lime flower is stated to possess sedative, antispasmodic, diaphoretic, diuretic, and mild astringent properties. Traditionally, it has been used for migraine, hysteria, arteriosclerotic hypertension, feverish colds, and specifically for raised arterial pressure associated with arteriosclerosis and nervous tension.

LOBELIA

Lobelia is stated to possess respiratory stimulant, antiasthmatic, antispasmodic, expectorant, and emetic properties. Traditionally, it has been used for bronchitic asthma, chronic bronchitis, and specifically for spasmodic asthma with secondary bronchitis. It has also been used topically for myositis and rheumatic nodules.

MA HUANG AND THE EPHEDRA ALKALOIDS

The 15th-century Chinese recommended ephedra as an antipyretic and antitussive agent. Modern physicians have used intravenous ephedrine for the prophylaxis and treatment of hypotension caused by spinal anesthesia particularly during cesarean section.

MARSHMALLOW

Marshmallow is stated to possess demulcent, expectorant, emollient, diuretic, antilithic, and vulnerary properties. Traditionally, it has been used internally for the treatment of respiratory catarrh and cough, peptic ulceration, inflammation of the mouth and pharynx, enteritis, cystitis, urethritis and urinary calculus, and topically for abscesses, boils, and varicose and thrombotic ulcers.

MATÉ

Maté is stated to possess CNS stimulant, thymoleptic, diuretic, antirheumatic, and mild analgesic properties. Traditionally, it has been used for psychogenic headache and fatigue, nervous depression, rheumatic pains, and specifically for headache associated with fatigue.

MEADOWSWEET

Meadowsweet is stated to possess stomachic, mild urinary antiseptic, antirheumatic, astringent, and antacid properties. Traditionally, it has been used for atonic dyspepsia with heartburn and hyperacidity, acute catarrhal cystitis, rheumatic muscle and joint pains, diarrhea in children, and specifically for the prophylaxis and treatment of peptic ulcer.

MISTLETOE

Mistletoe is stated to possess hypotensive, cardiac depressant, and sedative properties. Traditionally, it has been used for high blood pressure, arteriosclerosis, nervous tachycardia, hypertensive headache, chorea, and hysteria.

MOTHERWORT

Motherwort is stated to possess sedative and antispasmodic properties. Traditionally, it has been used for cardiac debility, simple tachycardia, effort syndrome, amenorrhea, and specifically for cardiac symptoms associated with neurosis.

MYRRH

Myrrh is stated to possess antimicrobial, astringent, carminative, expectorant, anticatarrhal, antiseptic, and vulnerary properties. Traditionally, it has been used for aphthous ulcers, pharyngitis, respiratory catarrh, common cold, furunculosis, wounds and abrasions, and specifically for mouth ulcers, gingivitis, and pharyngitis.

NETTLE

Nettle is stated to possess antihemorrhagic and hypoglycemic properties. Traditionally, it has been used for uterine hemorrhage, cutaneous eruption, infantile and psychogenic eczema, epistaxis, and specifically for nervous eczema.

PANAX GINSENG

Throughout history, the root has been used as a treatment for asthenia, atherosclerosis, blood and bleeding disorders, colitis, and relief of symptoms associated with aging, cancer, and senility. Ginseng is also widely believed to be an aphrodisiac. Ginseng is promoted as a tonic capable of invigorating the user physically, mentally, and sexually. It is also said to possess antistress activity.

PARSLEY

Parsley is stated to possess carminative, antispasmodic, diuretic, emmenagogue, expectorant, antirheumatic, and antimicrobial properties. Traditionally, it has been used for flatulent dyspepsia, colic, cystitis, dysuria, bronchitic cough in elderly people, dysmenorrhea, functional amenorrhea, myalgia, and specifically for flatulent dyspepsia with intestinal colic.

PARSLEY PIERT

Parsley piert is stated to possess diuretic and demulcent properties, and to dissolve urinary deposits. Traditionally, it has been used for kidney and bladder calculi, dysuria, strangury, edema of renal and hepatic origin, and specifically for renal calculus.

PASSIONFLOWER

Passionflower is stated to possess sedative, hypnotic, antispasmodic, and **anodyne** properties. Traditionally, it has been used for neuralgia, generalized seizures, hysteria, nervous tachycardia, spasmodic asthma, and specifically for insomnia. Passionflower is used extensively in homoeopathy.

PENNYROYAL

Pennyroyal is stated to possess carminative, antispasmodic, diaphoretic, and emmenagogue properties, and has been used topically as a refrigerant, antiseptic, and insect repellent. Traditionally, it has been used for flatulent dyspepsia, intestinal colic, common cold, delayed menstruation, and topically for cutaneous eruptions, formication, and gout.

PILEWORT

Pilewort is stated to possess astringent and demulcent properties. Traditionally, it has been used for hemorrhoids, and specifically for internal or prolapsed piles with or without hemorrhage, by topical application as an ointment or a suppository.

PLANTAIN

Plantain is stated to possess diuretic and antihemorrhagic properties. Traditionally, it has been used for cystitis with hematuria, and specifically for hemorrhoids with bleeding and irritation.

PLEURISY ROOT

Pleurisy root is stated to possess diaphoretic, expectorant, antispasmodic, and carminative properties. It has been used for bronchitis, pneumonitis, influenza, and specifically for pleurisy.

POKEWEED

Pokeweed has various traditional uses from medicinal to industrial. It has been used as cathartic, emetic, narcotic, and gargle. Additional medicinal uses included treatment of various skin diseases, conjunctivitis, syphilis, cancer, parasitic infestations of the scalp, chronic rheumatism, ringworm, dyspepsia, swollen glands, scabies, ulcers, edema, dysmenorrhea, mumps, and tonsillitis. Today, pokeweed is used for its antimicrobial activity and antineoplastic activity.

Pokeweed is stated to possess antirheumatic, anticatarrhal, mild anodyne, emetic, purgative, parasiticidal, and fungicidal properties. Traditionally, it has been used for rheumatism, respiratory catarrh, tonsillitis, laryngitis, adenitis, mastitis, mumps, skin infections (e.g., scabies, tinea, sycosis, acne), mammary abscesses, and mastitis.

POPLAR

Poplar is stated to possess antirheumatic, anti-inflammatory, antiseptic, astringent, **anodyne**, and cholagogue properties. Traditionally, it has been used for muscular and arthrodial rheumatism, cystitis, diarrhea, anorexia with stomach or liver disorders, common cold, and specifically for rheumatoid arthritis. The buds of *Populus tremula* (European white poplar, aspen) and *P. nigra* (black poplar) are used, reputedly as expectorant and circulatory stimulant remedies, for upper respiratory tract infections and rheumatic conditions.

PRICKLY ASH, NORTHERN

Prickly ash is stated to possess circulatory stimulant, diaphoretic, antirheumatic, carminative, and sialogogue properties. Traditionally, it has been used for cramps, intermittent claudication, Raynaud's syndrome, chronic rheumatic conditions, and specifically for peripheral circulatory insufficiency associated with rheumatic symptoms. The berries are stated to be therapeutically more active in circulatory disorders.

PRICKLY ASH, SOUTHERN

Southern prickly ash is stated to possess circulatory stimulant, diaphoretic, antirheumatic, carminative, and sialogogue properties. Traditionally, it has been used for cramps, intermittent claudication, **Raynaud's syndrome**, chronic rheumatic conditions, and specifically for peripheral circulatory insufficiency associated with rheumatic symptoms. The berries are stated to be therapeutically more active in circulatory insufficiency associated with rheumatic symptoms.

PULSATILLA

Pulsatilla is stated to possess sedative, analgesic, antispasmodic, and bactericidal properties. Traditionally, it has been used for dysmenorrhea, orchitis, ovaralgia, epididymitis, tension headache, hyperactive states, insomnia, boils, skin eruptions associated with bacterial infection, asthma and pulmonary disease, earache, and specifically for painful conditions of the male or female reproductive systems. Pulsatilla is widely used in homoeopathic preparations as well as in herbal medicine.

QUASSIA

Quassia is stated to possess bitter, orexigenic, sialogogue, gastric stimulant, and anthelmintic properties. Traditionally, it has been used for anorexia, dyspepsia, nematode infestation (by oral or rectal administration), pediculosis (by topical application), and specifically for atonic dyspepsia with loss of appetite.

QUEEN'S DELIGHT

Queen's delight is stated to possess sialogogue, expectorant, diaphoretic, dermatological, astringent, antispasmodic, and, in large doses, cathartic properties. Traditionally, it has been used for bronchitis, laryngitis, laryngismus stridulus, cutaneous eruptions, hemorrhoids, constipation, and specifically for exudative skin eruption and irritation and lymphatic involvement, and laryngismus stridulus.

RASPBERRY

Raspberry is stated to possess astringent and *partus praeparator* properties. Traditionally, it has been used for diarrhea, pregnancy, stomatitis, tonsillitis (as a mouthwash), conjunctivitis (as an eye lotion), and specifically to facilitate parturition.

RED CLOVER

Red clover is stated to act as a dermatological agent, and to possess mildly antispasmodic and expectorant properties. Tannins are known to possess astringent properties. Traditionally, red clover has been used for chronic skin disease, whooping cough, and specifically for eczema and psoriasis.

RHUBARB

Rhubarb has been used traditionally as both a laxative and an antidiarrheal agent.

ROSEMARY

Rosemary is stated to act as a carminative, spasmolytic, thymoleptic, sedative, diuretic, and antimicrobial agent. Topically, rubefacient, mild analgesic and parsiticide properties are documented. Traditionally, rosemary is indicated for flatulent dyspepsia, headache, and topically for myalgia, sciatica, and intercostal neuralgia.

SAGE

Sage is stated to possess carminative, antispasmodic, antiseptic, astringent, and antihidrotic properties. Traditionally, it has been used to treat flatulent dyspepsia, pharyngitis, uvultis, stomatitis, gingivitis, glossitis (internally or as a gargle/mouthwash), hyperhidrosis, and galactorrhea.

SARSAPARILLA

Sarsaparilla is stated to possess antirheumatic, antiseptic, and antipruritic properties. Traditionally, it has been used for psoriasis and other cutaneous conditions, chronic rheumatism, rheumatoid arthritis, as an adjunct to other treatments for leprosy, and specifically for psoriasis.

SASSAFRAS

Sassafras has been used for stomachache, vomiting, urinary retention, lameness, gout, dropsy, syphilis, scurvy, and jaundice. Sassafras still enjoys a reputation as a spring tonic, stimulant, antispasmodic, blood purifier, and sudorific (sweat producer), and as a cure for rheumatism, skin disease, syphilis, typhus, and dropsy (congestive heart failure).

Sassafras is stated to possess carminative, diaphoretic, diuretic, dermatological, and antirheumatic properties. Traditionally, it has been used for cutaneous eruptions, gout, and rheumatic pains.

SAW PALMETTO

Historically, saw palmetto has also been used to increase sperm production, increase breast size, and increase sexual vigor. Today, saw palmetto is promoted to improve prostate health and urinary flow, and to improve reproductive and sexual functioning.

Saw palmetto is stated to possess diuretic, urinary antiseptic, endocrinological, and anabolic properties. Traditionally, it has been used for chronic or subacute cystitis, catarrh of the genitourinary tract, testicular atrophy, sex hormone disorders, and specifically for prostatic enlargement.

SCULLCAP

Scullcap came to be recognized as a tonic, tranquilizer, and antispasmodic, and was therefore used as an ingredient in many "patent medicines" for "**female weakness**." Scullcap was also combined with other reputedly calming herbs such as hop and valerian and promoted as a sedative or anxiolytic. Other traditional uses include treatment of epilepsy, headache, insomnia, various other neurological and psychiatric disorders, hypertension, fever, rheumatism, and stress.

Scullcap is promoted commercially in the United States as a sedative, anxiolytic, and spasmolytic and is promoted for the treatment of premenstrual syndrome (PMS), menstrual cramps, depression, exhaustion, and muscle pain caused by stress. Other purported uses include headache and epilepsy.

Scullcap is stated to possess anticonvulsant and sedative properties. Traditionally, it has been used for epilepsy, chorea, hysteria, nervous tension states, and specifically for grand mal. In Chinese herbal medicine, the roots of *Scutellaria baicalensis* Georgi have been used traditionally as a remedy for inflammation, suppurative dermatitis, allergic diseases, hyperlipidemia, and atherosclerosis.

SENEGA

Senega is stated to possess expectorant, diaphoretic, sialogogue, and emetic properties. Traditionally, it has been used for bronchitic asthma, chronic bronchitis, as a gargle for pharyngitis, and specifically for chronic bronchitis.

SENNA

Senna has been used as a laxative in the past and continues to be used as a laxative today. Senna is stated to possess cathartic properties (leaf greater than fruit) and has been used traditionally for constipation.

SHEPHERD'S PURSE

Shepherd's purse is stated to possess antihemorrhagic and urinary antiseptic properties. Traditionally, it has been used for menorrhagia, hematemesis, diarrhea, and acute catarrhal cystitis.

SKUNK CABBAGE

Skunk cabbage is stated to possess expectorant, antispasmodic, and mild sedative properties. Traditionally, it has been used for bronchitis, whooping cough, asthma, and specifically for bronchitic asthma.

SLIPPERY ELM

Slippery elm is stated to possess demulcent, emollient, nutrient, and antitussive properties. Traditionally, it has been used for inflammation or ulceration of the stomach or duodenum, convalescence, colitis, diarrhea, and locally for abscesses, boils, and ulcers (as a poultice).

SQUILL

Squill is stated to possess expectorant, cathartic, emetic, cardioactive, and diuretic properties. Traditionally, it has been used for chronic bronchitis, asthma with bronchitis, whooping cough, and specifically for chronic bronchitis with scanty sputum.

ST. JOHN'S WORT

Historically, St. John's wort has been used to treat neurological and psychiatric disturbances (anxiety, insomnia, bed-wetting, irritability, migraine, excitability, exhaustion, fibrosis, hysteria, neuralgia, and sciatica), gastritis, gout, hemorrhage, pulmonary disorders, and rheumatism, and has been used as a diuretic. Some forms of the herb have been used topically as an astringent and to treat blisters, burns, cuts, hemorrhoids, inflammation, insect bites, itching, redness, sunburn, and wounds. Today, St. John's wort is promoted for treatment of depression.

St. John's wort is stated to possess sedative and astringent properties. It has been used for excitability, neuralgia, fibrositis, sciatica, wounds, and specifically for menopausal neurosis. St. John's wort is used extensively in homoeopathic preparations as well as in herbal products.

STONE ROOT

Stone root is stated to possess antilithic, litholytic, mild diaphoretic, and diuretic properties. Traditionally, it has been used for renal calculus, lithuria, and specifically for urinary calculus.

TANSY

Tansy is stated to possess anthelmintic, carminative, and antispasmodic properties and to act as a stimulant to abdominal viscera. Traditionally, it has been used for nematode infestation, topically for scabies (as a decoction) and pruritus ani (as an ointment), and specifically for roundworm or threadworm infestation in children.

THYME

Thyme is stated to possess carminative, antispasmodic, antitussive, expectorant, secretomotor, bactericidal, anthelmintic, and astringent properties. Traditionally, it has been used for dyspepsia, chronic gastritis, asthma, diarrhea in children, enuresis in children, laryngitis, tonsillitis (as a gargle), and specifically for pertussis and bronchitis.

UVA-URSI

Uva-ursi is stated to possess diuretic, urinary antiseptic, and astringent properties. Traditionally, it has been used for cystitis, urethritis, dysuria, pyelitis, lithuria, and specifically for acute catarrhal cystitis with dysuria and highly acidic urine.

VALERIAN

Today, valerian is used for treatment of insomnia. Valerian is stated to possess sedative, mild anodyne, hypnotic, antispasmodic, carminative, and hypotensive properties. Traditionally, it has been used for treatment of fatigue, seizures, hysterical states, excitability, insomnia, hypochondriasis, migraine, cramp, intestinal colic, rheumatic pains, dysmenorrhea, and specifically for conditions presenting nervous excitability.

VERVAIN

Vervain is stated to possess sedative, thymoleptic, antispasmodic, mild diaphoretic, and, reputedly, galactogogue properties. Traditionally, it has been used for depression, melancholia, hysteria, generalized seizures, cholecystalgia, jaundice, early stages of fever, and specifically for depression and debility of convalescence after fevers, especially influenza.

WILD CARROT

Wild carrot is stated to possess diuretic, antilithic, and carminative properties. Traditionally, it has been used for urinary calculus, lithuria, cystitis gout, and specifically for urinary gravel or calculus.

WILD LETTUCE

Wild lettuce is stated to possess mild sedative, anodyne, and hypnotic properties. Traditionally, it has been used for insomnia, restlessness and excitability in children, pertussis, irritable cough, priapism, dysmenorrhea, nymphomania, muscular or articular pains, and specifically for irritable cough and insomnia.

WILLOW

Willow is stated to possess anti-inflammatory, antirheumatic, antipyretic, antihidrotic, analgesic, antiseptic, and astringent properties. Traditionally, it has been used for muscular and arthrodial rheumatism with inflammation and pain, influenza, respiratory catarrh, gouty arthritis, and other systemic connective tissue disorders characterized by inflammatory changes.

WITCH HAZEL

Witch hazel is stated to possess astringent, antihemorrhagic, and anti-inflammatory properties. Traditionally, it has been used for diarrhea, mucous colitis, hemorrhoids, hematemesis, hemoptysis, and externally for external hemorrhoids, bruises, and localized inflamed swellings.

YARROW

Yarrow is stated to possess diaphoretic, antipyretic, hypotensive, astringent, diuretic, and urinary antiseptic properties. Traditionally, it has been used for fevers, common colds, essential hypertension, amenorrhea, dysentery, diarrhea, and specifically for thrombotic conditions with hypertension, including cerebral and coronary thromboses.

YELLOW DOCK

Yellow dock is stated to possess gentle purgative and cholagogue properties. Traditionally, it has been used for chronic skin disease, obstructive jaundice, constipation, and specifically for psoriasis with constipation.

YUCCA

Yucca has been used for the treatment of arthritis, diabetes, and stomach disorders. Concentrated plant juice has been used topically to soothe painful joints (see Cupp, 2000).

3

VITAMINS AND DIET*

- Vitamins
- Vegetables
- Fruits
- Nuts
- Greens

VITAMINS

Vitamins are organic dietary substances necessary for the maintenance of normal metabolic function. Only small amounts of the vitamins are required for normal health. In the body, they act as components of the important enzyme systems that catalyze the reactions by which protein, fat, and carbohydrate are metabolized. Some of the vitamins (e.g., vitamin K) may be formed by bacteria in the gut, while vitamin D is synthesized by exposure of the skin to sunlight. With these exceptions, the vitamins must be ingested in the food, and restricted diets or disorders of the gastrointestinal tract, interfering with absorption, lead to vitamin deficiency. When pronounced, such deficiencies give rise to easily recognizable clinical syndromes (**beriberi, pellagra, rickets, scurvy**), which have long been recognized. Milder forms of avitaminosis are much more common and also give rise to disability and ill-health.

The **fat-soluble vitamins** are A, D, E, and K. The **water-soluble vitamins** are **thiamine** (vitamin B_1), **riboflavin, nicotinic acid** (niacin) and **nicotinamide, pyridoxine** (vitamin B_6), **pantothenic acid, biotin, para-aminobenzoic acid, choline, inositol** and other lipotropic agents, **ascorbic acid** (vitamin C), the **riboflavonoids, folate**, and **vitamin B12** (see Figures 1 and 2, and Table 1).

---

* See also two books published by Reader's Digest Association, Inc., Pleasantville, NY: Weiss, S.E., *Foods That Harm and Foods That Heal,* 1997; Gardner, J.L., *Eat Better and Live Better,* 1982.

VITAMIN D

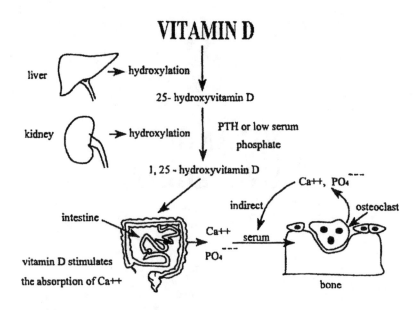

The activation of vitamin D and its action on calcium regulation

---

Figure 1 Vitamins D_3 and D_2 are produced by ultraviolet irradiation of animal skin and plants, respectively. The precursor of vitamin D_3 in skin is 7-dehydrocholesterol, or **provitamin D**. In humans, the storage, transport, metabolism, and potency of vitamins D_2 and D_3 are identical, and the net biologic activity of vitamin D *in vivo* results from the combined effects of the hydroxylated derivatives of vitamins D_2 and D_3.

THE IMPORTANCE OF VITAMIN A IN HUMAN NUTRITION

Night blindness apparently was first described in Egypt around 1500 B.C. Although this disease was not then linked to dietary deficiency, topical treatment with roasted or fried liver was recommended, and **Hippocrates** later recommended eating beef liver as a cure for the affliction. The relationship to nutritional deficiency was definitively recognized in the 1800s. **Ophthalmia Brasiliana**, a disease of eyes that affected primarily poorly nourished slaves, was first described in 1865. In 1887, endemic night blindness was reported to occur among the **Orthodox Russian Catholics** who fasted during the Lenteon period. More pertinent was the observation that the nurslings of mothers who fasted were prone to develop spontaneous sloughing of the cornea (Marcus and Coulson, 1996).

DIETARY SOURCES OF VITAMIN A

The major dietary source for preformed vitamin A is vertebrate animal products that are rich in vitamin A esters (liver, kidney, oil, dairy products, and eggs). Liver and oil, particularly from fish, are the major dietary sources of preformed vitamin A. Levels in milk and eggs depend on dietary **retinoid** and **carotenoid** intake. Freshwater fish are a source of **vitamin A_2 (3,4-dehydroretinol)**, which shows reduced vitamin A activity. Levels of retinal in food are very low, while **retinoic acid** has not been found.

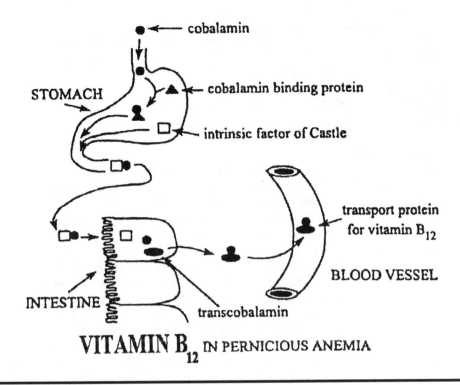

VITAMIN B$_{12}$ IN PERNICIOUS ANEMIA

Figure 2 Both **vitamin B$_{12}$** and **folic acid** are essential for the synthesis of DNA, and this process is impaired in patients with megaloblastic anemia.

Plants are the major source for dietary provitamin A. Since mammals and humans cannot synthesize **carotenoids**, dietary provitamin A is obtained from plant sources that contain carotenoids having **2,6,6-trimethyl-1-cyclohexen-*l*-yl rings, such as β-carotene**. More than 600 carotenoids have been identified in plants and algae, which together biosynthesize about 0.1 billion tons of carotenoids each year. However, only about ten carotenoids, including β-carotene, are nutritionally significant members of the provitamin A class that can be oxidatively metabolized to **retinal** in mammals and humans by such organs as the intestine, liver, and kidney and then reduced to **retinol**.

Provitamin A carotenoids are the major dietary source of retinol for many residents of developed countries, including the United States, and for most residents of poor countries. β-Carotene-rich vegetables and fruits (20 to 85% of carotenoid content) include dark-green leafy, bright yellow, orange, or red vegetables and fruits, whose total carotenoid content can range from less than 0.2 mg/100 g (lettuce leaves) to over 35 mg/100 g (oil from red palm fruit). The major dietary source of vitamin A varies with country. For example, in the United States it is orange carrots (6 to 12 mg/100 g), in Japan yellow-green vegetables, in Singapore dark-green leafy vegetables, and in Brazil and emerging South Asian and African countries unrefined **red palm oil** (37 to 165 mg/100 g). Grains have variable carotenoid content. For example, cereal seed contains only about 10 μg/100 g wet weight of metabolizable carotenoids, in which β-carotene predominates, while the content in corn varies from 1 to 5.5 mg/100 g (field corn) or 200 to 700 μg retinol equivalents (RE)/100 g (yellow sweet corn). (One RE = 1 μg retinol

Table 1   Vitamins: Their Coenzymatic Functions

Fat-Soluble Vitamins

A, retinol	Rhodopsin, visual cycle, night vision
K, menadiol	Blood clotting factors II, VII, IX, and X
D, calciferol	Calcium and phosphorus homeostasis
E, tocopherol	Antioxidant, glutathione oxidase

Water-Soluble Vitamins

C, ascorbate	Antioxidant, regulation of intracellular oxidation-reduction potentials, hydroxylation reactions that require copper or iron
B_1, thiamin	Oxidative decarboxylation of amino acids, transketolase
B_2, riboflavin	Flavin mononucleotide and flavin adenine dinucleotide, essential for oxidative systems and oxygen transport
B_3, pantothenic acid	As CoA precursor, necessary for acyl transfers
B_5, niacin	Endogenous source for tryptophan; component of NAD and its phosphorylate, NADP; assists in hydrogen transfer of glycolysis, fatty acid synthesis, and tissue respiration
B_6, pyridoxine	Nitrogen metabolism: transamination, racemization, decarboxylation, cleavage, synthesis, dehydration, and desulfhydration
B_{12}, cyanocobalamin	Methylation of homocysteine to methionine, conversion of methyl malonyl CoA to succinyl CoA
Biotin	Cofactor for some carboxylases; acetyl CoA carboxylase, pyruvate carboxylase, β-methylcrotonyl carboxylase, and methylmalonyl carboxylase
Folic acid	Transport of single carbon fragments, especially nucleic acid synthesis and metabolism of some amino acids

A vitamin is defined as an organic dietary substance required in trace amounts (µg to mg/day) by humans for health.

or 6 µg β-carotene.) Other sources are sorghum (<31 µg/100 g) and green soybeans, cowpeas, lima beans, and peas (200 to 700 µg/100 g). Root vegetables, such as orange carrots, yams, and sweet potatoes, are a major carotenoid source in which β-carotene predominates (90%). Taro has lower concentrations. Fruits show greater variation in their β-carotene content than vegetables, and their levels are generally lower than those found in leafy vegetables. Other natural sources of carotenoids are lobster, shrimp, and crab, which, incidentally, do not contain retinol or its esters.

Other sources are vitamin A-fortified prepared foods, including sugar, monosodium glutamate, and dried milk, that are distributed to vitamin A-deficient population groups. Creams and ointments containing vitamin A esters are used to treat skin disorders, such as acne and psoriasis. Recently, because of the reported efficacy of topical retinoic acid in the treatment of photoinduced skin damage, retinol esters are added to cosmetics for use against the appearance of age-related wrinkles and lines. Commercially prepared foods (margarine, butter, soft drinks, crustacean products, pastries, cheese, desserts) may contain carotenoids as colorants to increase visual appeal (Dawson, 2000).

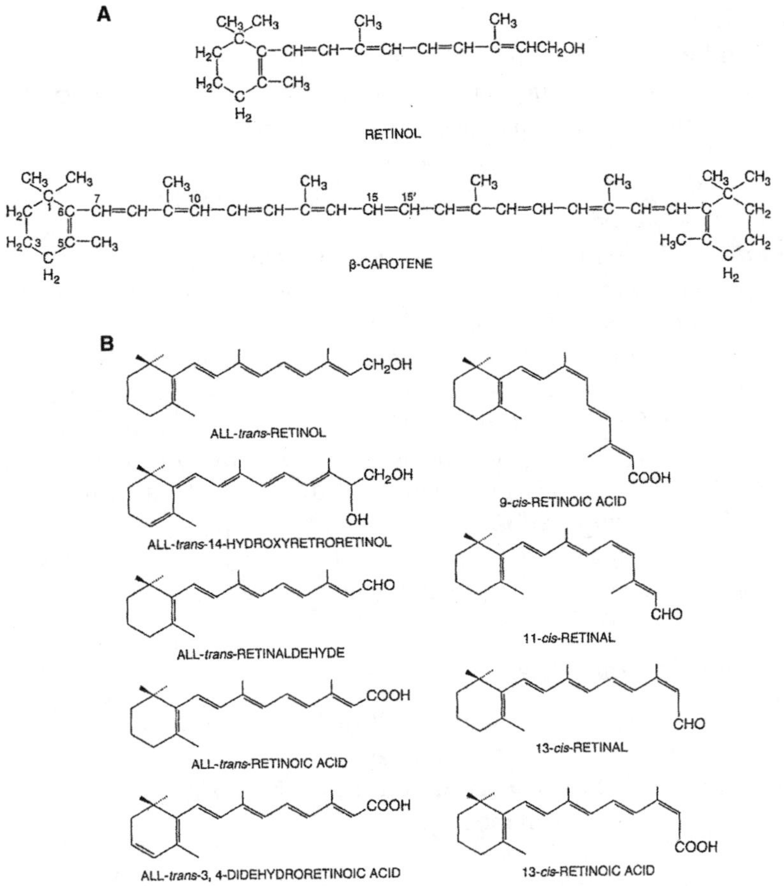

Figure 3   (A) β-Carotene; (B) the vitamin A family.

ACTION OF VITAMIN A

Although the term *vitamin A* has been used to denote specific chemical compounds, such as **retinol** or its esters, this term now is used more as a generic descriptor for compounds that exhibit the biological properties of retinol. **Retinoid** refers to the chemical entity retinol or other closely related naturally occurring derivatives. Retinoids also include structurally related synthetic analogues, which need not have retinol-like (**vitamin A**) activity. The structural formulas for the vitamin A family of retinoids are shown in Figure 3. Retinol (vitamin A_1), a primary alcohol, is present in esterified form in the tissues of animals and saltwater fish, mainly in the liver. A closely related compound, **3-dehydroretinol** (vitamin A_2), is obtained from the tissues of freshwater fish and usually occurs mixed with retinol.

A number of geometric isomers of retinol exist because of the possible *cis-trans* configurations around the double bonds in the side chain. Fish liver oils contain mixtures of the stereoisomers; synthetic retinol is the all-*trans* isomer. Interconversion between isomers readily takes place in the body. In the visual cycle, the reaction between retinal (**vitamin A aldehyde**) and **opsin** to form **rhodopsin** only occurs with the 11-*cis* isomer.

Ethers and esters derived from the alcohol also show activity *in vivo*. The ring structure of **retinol** (β-ionone), or the more unsaturated ring in **3-dehydroretinol** (dehydro-β-ionone), is essential for activity; hydrogenation destroys biological activity. Of all known derivatives, all-*trans*-retinol and its aldehyde, retinal, exhibit the greatest biological potency *in vivo*; 3-dehydroretinol has about 40% of the potency of all-*trans*-retinol.

Retinoic acid (vitamin A acid), in which the alcohol group has been oxidized, shares some but not all of the actions of retinol. Retinoic acid is ineffective in restoring visual or reproductive function in certain species where retinol is effective. However, retinoic acid is very potent in promoting growth and controlling differentiation and maintenance of epithelial tissue in vitamin A-deficient animals. Indeed, all-*trans*-retinoic acid (***tretinoin***) appears to be the active form of vitamin A in all tissues except the retina, and is 10- to100-fold more potent than retinol in various systems *in vitro*. Isomerization of this compound in the body yields 13-*cis*-retinoic acid (***isotretinoin***), which is nearly as potent as tretinoin in many of its actions on epithelial tissues but may be as much as fivefold less potent in producing the toxic symptoms of hypervitaminosis A.

Vitamin A and Carcinogenesis

Because vitamin A regulates epithelial cell differentiation and proliferation, there has been considerable interest in the apparent ability of retinol and related compounds to interfere with **carcinogenesis**. Vitamin A deficiency in humans enhances susceptibility to carcinogenesis.

Vitamin A and Immune Function

It has been known for many years that vitamin A deficiency is associated with increased susceptibility to bacterial, parasitic, and viral infections. Decreased resistance to infections has been demonstrated in numerous animal models of vitamin A deficiency, and even marginal vitamin A status increases the severity and duration of infectious illness.

Symptoms of Deficiency

Tissue reserves of retinoids in the healthy adult are sufficiently large to require long-term dietary deprivation to induce deficiency. Vitamin A deficiency occurs more commonly in chronic diseases affecting fat absorption, such as **biliary tract** or **pancreatic insufficiency**, **sprue**, **Crohn's disease involving the terminal ileum**, and **portal cirrhosis**; following **partial gastrectomy**; or during extreme, chronic dietary inadequacy.

Signs and symptoms of mild vitamin A deficiency are easily overlooked. Skin lesions, such as **follicular hyperkeratosis** and infections, are among the earliest signs of deficiency, but the most recognizable manifestation is **night blindness**, even though its onset occurs only when vitamin A depletion is severe. Children may grow more slowly, although this may be recognized only after correction of the deficiency. In general, rapidly proliferating tissues are more sensitive to vitamin A deficiency than are slowly growing tissues and may revert to an undifferentiated state more readily.

Eye

Keratomalacia, characterized by desiccation, ulceration, and xerosis of the cornea and conjunctiva, is occasionally seen as an acute symptom in the very young who are ingesting severely deficient diets. It is foreshadowed, usually, by night blindness, which appears as the earliest ocular sign of deficiency. Ultimately, severe visual impairment and even blindness result.

Bronchorespiratory Tract

Changes in the bronchorespiratory epithelium from mucus secretion to keratinization lead to increased incidence of respiratory infections in the deficiency state. There also is a decrease in elasticity of the lung and other tissues.

Skin

Keratinization and drying of the epidermis occur, and papular eruptions involving the pilosebaceous follicles may be found, especially on the extremities.

Genitourinary System

Urinary calculi are frequent concomitants of vitamin A deficiency. The epithelium of the urinary tract shares in the general pathological changes of all epithelial structures. Epithelial debris thus may provide the nidus around which a calculus is formed. Abnormalities of reproduction include impairment of spermatogenesis, degeneration of testes, abortion, resorption of fetuses, and production of malformed offspring.

Gastrointestinal Tract

The intestinal mucosa shows a reduction in the number of goblet cells without keratinization. Alterations in intestinal epithelium and metaplasia of pancreatic ductal epithelium are common. They may be responsible for the diarrhea occasionally seen in vitamin A deficiency.

Sweat Glands

These glands may undergo atrophy and keratinizing squamous-cell metaplasia.

Bone

Vitamin A deficiency is associated with faulty modeling of bone, with production of thick, cancellous bone instead of thinner, more compact bone.

Miscellaneous

Often both taste and smell are impaired in vitamin A-deficient individuals, undoubtedly a result of a keratinizing effect. Hearing also may be impaired. Vitamin A deficiency can interfere with erythropoiesis, which may be masked by abnormal losses of fluid. Nerve lesions, increased cerebrospinal fluid pressure, and hydrocephalus have been reported.

Therapeutic Uses

Vitamin A Deficiency Diseases

The normal requirement of vitamin A for adults is supplied by an **adequate diet**. The rational uses of retinol are in the treatment of vitamin A deficiency and as prophylaxis in high-risk subjects during periods of increased requirement, such as infancy, pregnancy, and lactation. Once vitamin A deficiency has been diagnosed, intensive therapy should be instituted. The patient should then be maintained on a proper diet.

There are many types of preparations that contain retinol. Absorption is greatest for aqueous preparations, intermediate for emulsions, and slowest for oil solutions. Whereas oil-soluble preparations may lead to greater hepatic storage of the vitamin, water-miscible preparations usually provide higher concentrations in plasma. Vitamin A is available as capsules. **Tretinoin** (all-*trans*-retinoic acid; Retin A) is available for topical use. **Isotretinoin** (13-*cis*-retinoic acid; Accutane) is available for oral use, as is **etretinate** (Tegison).

During pregnancy and lactation, it is advisable to increase the maternal intake of vitamin A by about 25%. Since the typical North American diet readily provides adequate intake of the vitamin, supplementation is not routinely indicated.

In **kwashiorkor** and other severe vitamin A deficiencies in children, a single intramuscular injection of 30 mg of retinol as the water-miscible palmitate has been advocated, followed by intermittent oral treatment with retinoids.

Dermatological Diseases

Vitamin A may be helpful in certain diseases of the skin, such as acne, psoriasis, **Darier's disease**, and **ichthyosis**.

Cancer and Other Uses

Considerable interest has focused on the possibility that vitamin A and other retinoids may find important roles in **cancer chemoprevention** and therapy (Marcus and Coulson, 1996).

Excessive Vitamin A Intake (Hypervitaminosis A)

High doses of vitamin A can produce the toxic side effects of the acute or chronic hypervitaminosis A syndrome, which in humans is characterized by:

- Hypercalcaemia (osteosclerosis or bone loss, and soft tissue calcification)
- Related bone and joint pain

- Inflammation of the mucous membranes of the lip (**cheilitis**), the mouth (**stomatitis**), the eye (**conjunctivitis**), and skin capillaries (**erythema**)
- Desquamation (scaling of skin)
- Hair loss (alopecia)
- Fatigue
- Irritability
- Headache
- Intestinal disturbances (diarrhea, nausea, and vomiting)
- Low-grade fever
- Liver dysfunction (hypertrigylceridaemia, **Dissë space fibrosis**, and blood flow obstruction)
- Birth defects (teratogenicity and embryotoxicity)

Acute hypervitaminosis A in humans requires doses of at least 600 mg in adults and 100 mg in children; chronic hypervitaminosis A requires 20–50 mg daily for several years (Dawson, 2000).

VITAMIN A

(Retinol) (Aquasol A)

Vitamin A, a fat-soluble vitamin, is indicated for severe vitamin A deficiency with **xerophthalmia**. (See the section on Importance of Vitamin A in Human Nutrition.)

VITAMIN E

(α Tocopherol) (Aquasole E, CEN-E, Eprolin Gelseas, Episilan-M, E-Vital, Pheryl-E 400, Tocopher-Caps, Vita-Plus E, Vitera E)

Vitamin E, a fat-soluble vitamin, is used in vitamin E deficiency in premature infants and in patients with impaired fat absorption (including patients with **cystic fibrosis**) and in **biliary atresia**.

VITAMIN K DERIVATIVES

Menadiol Sodium Diphosphate

(Synkayvite)

NUTRITIONAL ACTIONS OF VITAMINS

Part One: Fat-Soluble Vitamins

Vitamin A: From **retinols** in animal products or **β-carotene** in plant foods.
Retinols: Liver, salmon and other cold-water fish, egg yolks, and fortified milk and dairy products.
Beta carotene: Orange and yellow fruits and vegetables, such as carrots, squash, and cantaloupes; leafy green vegetables.

- Prevents **night blindness**; needed for growth and cell development; maintains healthy skin, hair, and nails, as well as gums, glands, bones, and teeth; may help prevent lung cancer.
- Symptoms of deficiency include night blindness, stunted growth in children, dry skin and eyes, increased susceptibility to infection.
- Symptoms of excess include headaches and blurred vision, fatigue, bone and joint pain, appetite loss and diarrhea, dry, cracked skin, rashes, and itchiness, hair loss. It can cause birth defects if taken in high doses before and during early pregnancy.

Vitamin K: Spinach, cabbage, and other green leafy vegetables; pork, liver, and green tea.

- Essential for proper blood clotting.
- Symptoms of deficiency include excessive bleeding, easy bruising.
- Symptoms of excess: may interfere with anticlotting drugs, possible jaundice.

Vitamin D (calciterol): Fortified milk and butter; egg yolks, fatty fish; fish-liver oils. (Also made by the body when exposed to the sun.)

- Necessary for calcium absorption, helps build and maintain strong bones and teeth.
- Symptoms of deficiency include weak bones, leading to **rickets** in children and **osteomalacia** in adults.
- Symptoms of excess include headaches, loss of appetite, diarrhea, and possible calcium deposits in heart, blood vessels, and kidneys.

Vitamin E (tocopherols): Eggs, vegetable oils, margarine, and mayonnaise; nuts and seeds; fortified cereals; green leafy vegetables.

- Protects fatty acids, maintains muscles and red blood cells, important antioxidant.
- Symptoms of deficiency are unknown in humans.
- Symptoms of excess include excessive bleeding, especially when taken with aspirin and other anticlotting drugs.

Part Two: Water-Soluble Vitamins

Biotin: Egg yolks, soybeans, cereals, and yeast.

- Biotin is necessary for energy metabolism.
- Symptoms of deficiency include scaly skin, hair loss, depression, elevated blood cholesterol levels.
- Symptoms of excess have not been reported.

Folate (folic acid, folacin): Liver; yeast; broccoli and other cruciferous vegetables; avocados; legumes; many raw vegetables.

- It is needed to make DNA, RNA, and red blood cells, and to synthesize certain amino acids.
- Symptoms of deficiency include abnormal red blood cells and impaired cell division, anemia, weight loss and intestinal upsets; deficiency may cause birth defects.
- Symptoms of excess include an inhibition with absorption of **phenytoin**, an antiepileptic agent, causing seizures in epileptics taking this drug; large doses may inhibit zinc absorption.

Niacin (vitamin B₃, nicotinic acid, nicotinamide): Lean meats, poultry, and seafood, milk, eggs, legumes, fortified breads and cereals.

- It is needed to mobilize energy; promotes normal growth. Large doses lower cholesterol.
- Symptoms of deficiency include diarrhea and mouth sores, pellagra (in extreme cases).
- Symptoms of excess include hot flashes, liver damage, elevated blood sugar and uric acid.

Pantothenic acid (vitamin B₅): Almost all foods.

- Aids in energy metabolism, normalizing blood sugar levels, and synthesizing antibodies, cholesterol, hemoglobin, and some hormones.
- Symptoms of deficiency include fatigue, low blood sugar, numbness, digestive problems, and lowered immunity.
- Symptoms of excess include diarrhea and edema.

Riboflavin (vitamin B₂): Fortified cereals and grains, lean meat and poultry, milk and other dairy products, raw mushrooms.

- Essential for energy metabolism; aids adrenal function.
- Symptoms of deficiency include vision problems and light sensitivity, mouth and nose sores, swallowing problems.
- May interfere with cancer chemotherapeutic agents.

Thiamine (vitamin B₁): Pork, legumes, nuts and seeds, fortified cereals, and grains.

- Energy metabolism; helps maintain normal digestion, appetite, and proper nerve function.
- Symptoms of deficiency include depression and mood swings, loss of appetite and nausea, muscle cramps. In extreme cases, muscle wasting and beriberi.
- When given in excess, may cause deficiency of other vitamins.

Vitamin B₆ (pyridoxine, pyridoxamine, pyridoxal): Meat, fish, and poultry, grains and cereals, green leafy vegetables, potatoes, and soybeans.

- Promotes protein metabolism; metabolism of carbohydrates and release of energy, proper nerve function, synthesis of red blood cells.
- Symptoms of deficiency include convulsive seizures, depression and confusion, itchy, scaling skin, red tongue, weight loss.
- Symptoms of excess may include sensory nerve deterioration.

Vitamin B$_{12}$ (cobalamins): All animal products.

- Needed to make red blood cells, DNA, RNA, and myelin (for nerve fibers).
- Symptoms of deficiency may include pernicious anemia, nerve problems and weakness, smooth or sore tongue.

Vitamin C (ascorbic acid): Citrus fruits and juices, melons, berries, and other fruits, peppers, broccoli, potatoes, and many other fruits and vegetables.

- Strengthens blood vessel walls, promotes wound healing, promotes iron absorption, helps control blood cholesterol and prevent atherosclerosis.
- Symptoms of deficiency include loose teeth and bleeding gums, bruises, loss of appetite, dry skin, poor healing. In extreme cases, scurvy and internal hemorrhages.
- Symptoms of excess include diarrhea, kidney stones, urinary tract irritation, iron buildup, bone loss (see Gardner, 1982; Weiss, 1997).

AGING AND DIET

Although aging is inevitable, many of the degenerative changes that prevail among elderly people are not, if preventative steps are taken.

Basic Dietary Recommendations

The National Institute on Aging recommends that an older person's daily diet include the minimum number of servings outlined below. Sugar, salt, and fats should be used sparingly; alcohol should be consumed only in moderation, defined as one drink a day for women and two for men (see Gardner, 1982; Weiss, 1997).

Food Group	Provides
Starches (Complex Carbohydrates)	
At least 5 servings: Whole grains, breads, and cereals; brown rice, kasha, millet, and other grains; dried beans, peas, lentils, and other legumes; potatoes, pasta, and other starchy foods	Thiamine, riboflavin, niacin, folate, B$_6$ and other B vitamins; fiber; and complete protein when grains and legumes are combined

Food Group	Provides
Vegetables and Fruits	
At least 5 servings: Fresh vegetables and fruits, including dark green leafy vegetables, such as broccoli, cabbage, kale, and spinach; yellow vegetables, such as carrots and squash; citrus fruits, berries, tomatoes, bananas, and other fruits	Vitamins C and A, beta carotene, riboflavin, folate, various minerals, and fiber; these are needed to prevent deficiency disorders and may protect against cancer
High-Protein Foods	
At least 2 servings: Lean beef, lamb, chicken and other poultry; fish and other seafood; eggs, tofu, and a combination of grains and legumes	Thiamine, riboflavin, niacin, and vitamins E, B_6, and B_{12}; iron, zinc, and other minerals; canned sardines and salmon with bones provide calcium
Dairy Products	
At least 2 servings: Low-fat milk, cheese, yogurt, and other milk products (Choose lactose-reduced milk and yogurt if one has trouble digesting regular milk)	Protein, calcium, vitamins A and D; needed to prevent the loss of bone minerals that frequently occurs with increasing age

Antioxidants, Protective Plant Chemicals, Preventing Diseases

Because plants are also susceptible to cancer and viruses, they have developed their own protective substances, called **phytochemicals** *(from* phyton, *the Greek term for plant). Mounting research shows that many phytochemicals also protect humans against cancer and other diseases* (Weiss, 1997).

Phytochemicals	Functions	Sources
Allylic sulfides	May stimulate production of protective enzymes	Garlic, onions
Bioflavonoids	Antioxidant; inhibit cancer-promoting hormones	Most fresh fruits and vegetables
Catechins (tannins)	Antioxidant	Berries, green tea
Curcumin	Protects against tobacco-induced carcinogens	Turmeric, cumin
Genistein	Inhibits tumor growth	Broccoli and other cruciferous vegetables
Indoles	Inhibit estrogen, which stimulates some cancers; induce protective enzymes	Broccoli, cabbage, cauliflower, mustard greens
Isoflavones	Inhibit estrogen uptake; destroy cancer enzymes	Beans, peanuts and other legumes, peas

Phytochemicals	Functions	Sources
Isothiocyanates	Induce production of protective enzymes	Horseradish, mustard, radishes
Lignans	Inhibit estrogen and block prostaglandins	Fatty fish, flaxseed, walnuts
Limonoids	Induce protective enzymes	Citrus fruits
Lycopene	Antioxidant, may protect against prostate cancer	Pink grapefruit, tomatoes, watermelon
Monoterpenes	Some antioxidant properties; aid in activity of protective enzymes	Basil, citrus fruits, broccoli, orange and yellow vegetables
ω-3 fatty acids	Inhibit estrogen; reduce inflammation	Canola oil, flaxseed, walnuts, fatty fish
Phenolic acid	Inhibits nitrosamines; enhances enzyme activity	Berries, broccoli, cabbage, carrots, citrus fruits, eggplant, parsley, peppers, teas, tomatoes, whole grains
Protease	Destroys enzyme inhibitors that promote cancer spread	Soybeans
Quercetin	Inhibits cellular mutation, carcinogens, clot formation, and inflammation	Grape skins, red and white wine
Terpenes	Stimulate anticancer enzymes	Citrus fruits

ESSENTIAL TRACE MINERALS

The trace minerals identified as essential to humans and for which deficiency states have been described are **zinc, copper, manganese, selenium, chromium, iodine, molybdenum,** and **iron.** Each of these minerals participates in a variety of biologic functions and is necessary for normal metabolism. Other trace minerals essential to humans but for which deficiency states have not been recognized include **nickel, vanadium, cobalt,** and **silicon** (Table 2).

Zinc deficiency is clinically characterized by the development of a moist **eczematous dermatitis** most apparent in the nasolabial folds and around orifices. Other presenting signs and symptoms may include **hypogeusia** (blunted sense of taste), **alopecia,** diarrhea, rash (which may vary from papular, scaly lesions to weeping, open erosions), apathy, and depression. Clinical zinc deficiency occurs most frequently in the setting of abnormal losses, such as in **Crohn's disease,** malabsorption states, and fistula losses, or from prolonged inadequate intake, such as with zinc-free parenteral nutrition.

Copper deficiency may present as hematological changes (**anemia, leukopenia,** and **neutropenia**) and skeletal demineralization. In severe cases, such as in **Menkes' syndrome,** copper deficiency is further manifested as hypothermia, depigmentation of hair and skin, progressive mental deterioration, and growth retardation. Factors predisposing to copper deficiency include malabsorption

Table 2 Assessment of Trace Mineral Status

Trace Minerals	Signs of Deficiency
Chromium	Glucose intolerance, peripheral neuropathy, increased free fatty acid levels, low respiratory quotient
Copper	Neutropenia, hypochromic anemia, osteoporosis, decreased hair and skin pigmentation; dermatitis, anorexia, diarrhea
Iodine	Hypothyroid goiter, hypothyroidism
Manganese	Nausea, vomiting, dermatitis, color changes in hair, hypocholesterolemia, growth retardation
Molybdenum	Tachycardia, tachypnea, altered mental status, visual changes, headache, nausea, vomiting
Selenium	Muscle weakness and pain, cardiomyopathy
Zinc	Dermatitis, hypogeusia, alopecia, diarrhea, apathy, depression

The **macrominerals** required by the human body are distinguished from **microminerals** only by the magnitude of their requirement. The daily requirements for these minerals range from about 0.3 g for sodium and magnesium to 2.0 g for potassium. Sodium and potassium are alkali metals, calcium and magnesium are alkali earths, and chloride and phosphate are accompanying anions. Water is not a mineral, but it is the oxide of hydrogen, which is in Group I in the periodic table. Water makes up about 60% of body weight and should be grouped with other substances that contribute to the matrix portion of the body. It has the highest requirement of any essential nutrient, namely 1 ml/kcal of energy expended, which for most persons exceeds 2 k/day.

states, protein-losing enteropathy, nephrotic syndrome, copper-free parenteral nutrition, and copper-deficient enteral nutrition.

Chromium deficiency is characterized by glucose intolerance but also may include neuropathy, increased free fatty acid concentrations, and a low respiratory quotient. Chromium deficiency has been identified in the setting of long-term, chromium-free parenteral nutrition.

Manganese deficiency has been reported only in association with chemically defined manganese-deficient oral diets. The symptoms include nausea, vomiting, dermatitis, color changes in hair, hypocholesterolemia, and growth retardation.

Selenium deficiency has been described in patients receiving long-term selenium-free total parenteral nutrition. Myopathy and abnormal **glutathione peroxidase** concentrations are most frequently observed.

Molybdenum deficiency in humans has rarely been observed. The presenting symptoms included tachycardia, tachypnea, headache, night blindness, nausea, vomiting, central scotomas, lethargy, disorientation, and ultimately coma.

Iodine deficiency may result in goiter formation. However, not everyone will experience goiter formation with an iodine-deficient diet. Iodine is needed for synthesis of the thyroid hormones thyroxine (T_4) and triodothyronine (T_3).

Patients with **iron deficiency** anemia present with fatigue, weakness, and pallor, and possibly also with glossitis, headache, dysphagia, fingernail changes, gastric atrophy, and paresthesias. Inadequate intake of iron, malabsorption, and blood loss from any origin are the principal causes of iron deficiency anemia.

FOODS THAT HELP*

Apples

- Low in calories and high in soluble fiber that help lower cholesterol

Apricots

- A rich source of beta carotene, iron, and potassium
- High in fiber, low in calories
- Dried apricots and apricot leather are nutritious, fat-free snack foods

Artichokes

- A good source of folate, vitamin C, and potassium
- Low in calories, high in fiber

Asparagus

- A good low-calorie source of folate and vitamins A and C
- Stalks high in fiber

Avocados

- A rich source of folate, vitamin A, and potassium
- Useful amounts of protein, iron, magnesium, and vitamins C, E, and B_6

Bananas

- A good source of potassium, folate, and vitamins C and B_6

Beans

- High in folate and vitamins A and C
- Mature (shelled) beans are high in protein and iron

Bean Sprouts

- Some are high in folate; others are fair to good sources of protein, vitamin C, B vitamins, and iron

---

* See Weiss, 1997; Gardner, 1982.

Beef and Veal

- Major sources of high-quality protein
- Contain a wide range of nutrients, especially vitamin B_{12}, iron, and zinc

Beer

- Lower in alcohol concentration than wine and hard liquor
- Contains modest amounts of niacin, folate, vitamin B_6, and some minerals

Beets

- A good source of folate and vitamin C
- The greens are a rich source of potassium, calcium, iron, beta carotene, and vitamin C
- Low in calories

Bioflavonoids

- Thought to function as antioxidants and also to enhance the antioxidant effects of vitamin C
- Believed to be instrumental in proper capillary function
- Some appear to be natural antibiotics and anticancer agents

Blackberries

- Low in calories and high in fiber
- A good source of vitamin C and bioflavonoids; also contain folate, vitamin E, iron, and calcium
- Contain anticancer chemicals
- Contain **salicylate**

Blueberries

- A good source of dietary fiber
- Provide some vitamin C and iron
- May protect against some intestinal upsets
- May help prevent some urinary tract infections

Bread

- A good source of protein and complex carbohydrates
- High in niacin, riboflavin, and other B complex vitamins
- Some provide good amounts of iron and calcium
- Whole-grain breads are high in fiber

Brussels Sprouts

- An excellent source of vitamin C
- A good source of protein, folate, vitamin A, iron, and potassium
- Contain bioflavonoids and other substances that protect against cancer

Buckwheat

- A good source of iron and magnesium
- High in starches, protein, and fiber

Butter and Margarine

- Add flavor to many foods
- Improve flavor, moistness, and texture of baked goods
- Good sources of vitamins A and D
- Margarine made with polyunsaturated oils has essential fatty acids

Cabbage

- An excellent source of vitamin C
- Low in calories and high in fiber
- May help prevent colon cancer and malignancies stimulated by estrogen
- Juice helps heal peptic ulcers

Caffeine

- Temporarily enhances mental alertness and concentration
- Can improve athletic performance by temporarily increasing muscle strength and endurance
- May abort an asthma attack by relaxing constricted bronchial muscles

Cakes, Cookies, and Pastries

- In small amounts, a good source of quick energy
- Delicious occasional snacks or desserts

Carrots

- An excellent source of β-carotene, the precursor of vitamin A
- A good source of dietary fiber and potassium
- Help prevent night blindness
- May help lower blood cholesterol levels and protect against cancer

Cauliflower

- An excellent source of vitamin C
- A good source of folate and potassium
- Low in calories and high in fiber

Celeriac

- A good source of fiber, including the soluble type that lowers elevated blood cholesterol levels
- A good source of potassium; also provides some vitamin C
- Low in calories

Celery

- Low in calories and high in fiber
- A good source of potassium
- May reduce inflammation and protect against cancer

Cereals

- High in complex carbohydrates
- Many are high in fiber
- Some are reasonable sources of protein; adding milk makes a bowl of cereal a high-protein dish
- Enriched cereals are high in iron, niacin, thiamine, and riboflavin, along with other B vitamins
- Iron-fortified infant cereals are ideal introductory solid foods

Cheese

- High in protein and calcium
- A good source of vitamin B_{12}
- Cheddar and other aged cheeses may fight tooth decay

Cherries

- A low-calorie fat-free snack or dessert
- A good source of vitamin C
- High in **pectin**, a soluble fiber that lowers cholesterol

Chestnuts

- Rich in folate and vitamins C and B_6
- A good source of iron, phosphorus, riboflavin, and thiamine
- Much lower in fat and calories than almost all other nuts

Chilies

- An excellent source of vitamins A and C
- May help relieve nasal congestion
- May help prevent blood clots that can lead to a heart attack or stroke

Chocolate and Candy

- Flavorful source of quick energy
- Eating chocolate elevates some people's moods

Coconuts

- A useful source of iron and fiber
- High in easy-to-digest fatty acids

Coffee

- Stimulates the central nervous system
- Can help one stay awake and alert

Corn

- A good source of folate and thiamine
- A fair amount of vitamins A and C, potassium, and iron
- Air-popped unbuttered popcorn is low in calories and very high in fiber

Cranberries

- A fair source of vitamin C and fiber
- Juice helps prevent or alleviate cystitis and urinary tract infections
- Contain bioflavonoids, thought to protect eyesight and help prevent cancer

Cucumbers

- Low in calories
- A good source of fiber
- A fair source of vitamin C and folate

Currants

- An excellent low-calorie source of vitamin C and potassium
- High in bioflavonoids

Dates

- An excellent source of potassium
- A good source of iron and calcium
- High in fiber

Eggplants

- Low in calories (unless cooked in fat)
- Meaty flavor and texture lends itself to vegetarian dishes

Eggs

- An excellent source of protein, vitamin B_{12}, and many other nutrients

Fennel

- An excellent source of vitamins A and C (especially the leaves)
- A good source of potassium, calcium, and iron
- High in fiber and low in calories

Fiber

- Helps prevent constipation
- Relieves the symptoms of diverticulosis and hemorrhoids
- May help reduce risk of colon cancer
- Plays a role in lowering elevated blood cholesterol levels
- Useful as a means of controlling weight

Figs

- A rich source of magnesium, potassium, calcium, and iron
- High in fiber

Fruits

- Excellent sources of vitamin C, beta carotene, and potassium; lesser amounts of other vitamins and minerals
- Contain bioflavonoids, which protect against cancer and other diseases
- High in fiber and low in calories
- A source of natural sugars that provide quick energy

Fish

- An excellent source of complete protein, iron, and other minerals
- Some are high in vitamin A
- Contains omega-3 fatty acids

Flours

- A concentrated source of starch
- Enriched flours are a good source of calcium, iron, and B vitamins

Garlic

- May help lower high blood pressure and elevated blood cholesterol
- May prevent or fight certain cancers
- Antiviral and antibacterial properties help prevent or fight infection
- May alleviate nasal congestion

Ginger

- May prevent motion sickness
- Can help to quell nausea
- Ginger wine may help to relieve menstrual cramps

Gooseberries

- A good source of vitamin C, potassium, and bioflavonoids
- Fair amounts of iron and vitamin A
- High in fiber, low in calories

Grains

- An excellent source of starchy carbohydrate and dietary fiber
- A good source of niacin, riboflavin, other B vitamins, iron, and calcium

Grapefruit

- High in vitamin C and potassium
- A good source of folate, iron, calcium, and other minerals
- Pink and red varieties are high in beta carotene, a precursor of vitamin A
- High in fiber, low in calories
- Contain bioflavonoids and other plant chemicals that protect against cancer and heart disease

Grapes

- High in pectin and bioflavonoids
- A fair source of iron, potassium, and vitamin C
- A low-calorie sweet snack and dessert
- Contains **salicylate**

Guavas

- An excellent source of vitamin C
- High in pectin and other types of soluble dietary fiber
- Good amounts of potassium and iron

Honey

- A source of quick energy
- Adds flavor to foods and beverages and improves the shelf life of baked goods
- Contains choline
- Memory enhancer and neuroprotectant

Ice Cream

- A good source of calcium
- Provides protein and digestible, high-calorie nutrition during an illness

Kale

- An excellent source of beta carotene and vitamins C and E
- A good source of folate, calcium, iron, and potassium
- Contains bioflavonoids and other substances that protect against cancer

Kiwi Fruits

- An excellent source of vitamin C
- A good source of potassium and fiber
- Can be used as a meat tenderizer

Kohlrabi

- High in vitamin C, potassium, and cancer-preventing antioxidants and bioflavonoids
- High in dietary fiber

Lamb

- A rich source of minerals, including iron, phosphorus, and calcium
- An excellent source of protein and B-complex vitamins

Leeks

- A good source of vitamin C, with lesser amounts of niacin and calcium

Legumes

- Contain more protein than any other plant-derived food
- A good source of starch, B-complex vitamins, iron, potassium, zinc, and other essential minerals
- Most are high in soluble fiber

Lemons

- An excellent source of vitamin C
- A low-calorie alternative to oil dressings
- May relieve dry mouth

Lettuce and Other Salad Greens

- Low in calories and high in fiber
- Some varieties are high in beta carotene, folate, vitamin C, calcium, iron, and potassium

Limes

- An excellent source of vitamin C
- Can be used to flavor and tenderize meat, poultry, and fish
- Peels contain **psoralens**, which increases sun sensitivity

Malted Milk, Malt Extracts

- Good food for convalescents
- Sugar provides quick energy
- Malt extract is a good source of niacin, iron, and potassium
- Warm malted milk promotes sleep

Mangoes

- An excellent source of beta carotene and vitamin C
- A good source of vitamin E and niacin

- High in potassium and iron
- Low in calories, high in fiber

Mayonnaise

- A good source of vitamin E, depending upon the type of oil used
- Contains small amounts of vitamin A and some minerals

Melons

- Sweet and flavorful, yet low in calories
- Yellow varieties are high in vitamin A
- Most are good sources of vitamin C and potassium
- Some are high in **pectin**, a soluble fiber that helps control blood cholesterol levels

Milk and Milk Products

- The best source of calcium
- A good source of vitamins A and D, riboflavin, phosphorus, and magnesium
- Low-fat dairy products are low in cholesterol and high in protein

Mushrooms and Truffles

- Fat-free and very low in calories
- Rich in minerals
- High glutamic acid content may boost immune function

Nectarines

- A fairly rich source of **beta carotene** and potassium
- Provide moderate amounts of vitamin C
- High in **pectin**, a soluble fiber
- The pits contain **cyanide** and should not be eaten

Nuts and Seeds

- Rich in vitamin E and potassium
- Most are high in minerals, including calcium, iron, magnesium, and zinc
- Some are good sources of folate, niacin, and other B vitamins
- A good source of protein, especially when combined with legumes
- Some are high in fiber
- **Cashew shells** contain **urushiol**, the same irritating oil that is in poison ivy; heating inactivates urushiol, so toasted cashews are safe to eat; the raw nuts, however, should never be eaten
- Molds that grow on nuts (especially peanuts) create **aflatoxins**, substances that cause liver cancer

Oils

- Provide essential fatty acids needed for hormone production
- Make possible the absorption of fat-soluble vitamins A, D, E, and K
- Oils may contain toxic compounds. For example, **Myristicin** is the compound that flavors nutmeg and mace; it is also found in black pepper and carrot, parsley, and celery seeds. Used in culinary quantities, myristicin is only a flavoring. In massive doses, it causes hallucination. **Thujone**, the anise-flavored oil in wormwood, caused an epidemic of brain disease in drinkers addicted to the now-banned liquor **absinthe**. **Sassafras** contains a toxic oil similar to thujone, which is why it is no longer used to make root beer. Very high doses of **menthol**, from peppermint, may cause dangerous irregularities in the heart's rhythm (Weiss, 1997).

Okra

- A good source of vitamins A and C, folate, and potassium
- High in dietary fiber and low in fat and calories

Olives

- High in monounsaturated fats, which benefit blood cholesterol levels
- A modest, low-calorie source of vitamin A, calcium, and iron

Onions

- The green tops are a good source of vitamin C and beta carotene
- May lower elevated blood cholesterol
- Reduce the ability of the blood to clot
- May help lower blood pressure
- Mild antibacterial effect may help prevent superficial infections

Oranges

- An excellent source of vitamin C
- A good source of **beta carotene**, folate, thiamine, and potassium

Organ Meats

- An inexpensive source of protein
- Liver and kidneys are excellent sources of vitamins A and B_{12}, folate, niacin, iron, and other minerals
- Most are good sources of potassium

Papayas

- An excellent source of vitamins A and C and potassium
- Papayas contain **papain**, an enzyme that is similar to the digestive juice **pepsin**. Because this enzyme breaks down protein, papain extract from papayas is marketed as a **meat tenderizer**.

Parsnips

- Low in calories and high in fiber and carbohydrates
- A flavorful alternative to potatoes
- A useful source of vitamin C, folate, and potassium

Pasta

- An excellent source of starches
- A useful source of protein, B vitamins, iron, and other minerals
- Low in fat and sodium
- Can be served in many different ways

Peaches

- A good source of vitamin A, with useful amounts of vitamin C and potassium
- A good source of dietary fiber

Pears

- A good source of vitamin C and folate
- A good source of dietary fiber

Peas and Pea Pods

- A good source of vitamins A and C, thiamine, **riboflavin**, and potassium
- High in **pectin** and other types of fiber
- Provide complete protein when served with grain products

Peppers

- An excellent low-calorie source of vitamins A and C

Pickles and Other Condiments

- Sauerkraut is a good source of vitamin C, iron, potassium, and other nutrients.
- Pickles are low-calorie snacks.
- A diet that is high in pickled or other salt-cured foods and condiments has been linked to an increased risk of stomach and esophageal cancers. This is thought to stem from their high levels of **nitrates**, which are converted to cancer-causing **nitrosamines** during digestion. Vitamins A and C, beta carotene, and other **antioxidants** are thought to inhibit the cancer-causing potential of nitrosamines; eating ample fresh vegetables and fruits may counteract any risk from pickled foods (Weiss, 1997).

Pineapples

- A good source of vitamin C, with useful amounts of vitamin B_6, folate, thiamine, iron, and magnesium

Plums

- A useful source of vitamin C, riboflavin and other B vitamins, and potassium

Pork

- Fresh, lean pork is a good source of high-quality protein and B vitamins.
- Ham and bacon have high levels of vitamin C, which is added as a preservative.
- Nitrites in cured pork products form cancer-promoting **nitrosamines**.

Poultry

- An excellent source of protein
- Lower in saturated fat than red meats
- A good source of vitamin A, the B vitamins, and minerals

Prunes

- A rich source of vitamin A
- High in B vitamins, vitamin E, potassium, and iron
- Help to relieve constipation

Pumpkins

- A rich source of beta carotene
- A good low-calorie source of vitamin C and potassium

- High in fiber
- The seeds are a good source of protein, iron, B vitamins, vitamin E, and fiber
- Can be stored for long periods of time

Quinces

- A good source of vitamin C, iron, and potassium
- High in **pectin**, a soluble fiber
- **Seeds contain a cyanide compound**

Quinoa

- An excellent source of iron, magnesium, potassium, phosphorus, zinc, and other minerals
- A good source of B-complex vitamins
- High in protein

Radishes

- A useful source of vitamin C
- Low in calories and high in fiber
- Contain **salicylate**

Raspberries

- A rich source of vitamin C
- Contain useful amounts of folate, iron, and potassium
- Provide bioflavonoids, which may protect against cancer
- High in fiber
- Contain **salicylate**
- Contain **oxalic acid**, which can aggravate kidney and bladder stones in susceptible persons

Rhubarb

- High in vitamin C and potassium
- Contains **oxalic acid**, which inhibits calcium and iron absorption
- **Leaves are highly poisonous**

Rice

- Enriched varieties provide B vitamins and iron
- Makes a complete protein when combined with beans and other legumes
- A starchy food that provides energy and contributes to protein synthesis

- **Gluten-free** and suitable for people with **celiac disease**
- Easy to digest and useful in restoring bowel function after a bout of diarrhea
- Rarely, if ever, causes food allergies
- Diets high in white rice may be deficient in **thiamine**
- A substance in brown rice can inhibit absorption of iron and calcium

Salad Dressings

- Add flavor and interest to lettuce and other salad greens
- A good source of vitamin E

Salt and Sodium

- Helps to maintain fluid balance, regulate blood pressure, and transmit nerve impulses
- Improves the flavor of many foods
- A useful food preservative
- Promotes fluid retention and may contribute to high blood pressure

Sauces and Gravies

- Used sparingly, sauces complement flavors and enhance appearance of other food.
- **Salsa-style garnishes** supply fiber and antioxidant vitamins, provided they are carefully prepared.
- **Pasta sauces** based on fresh vegetables and olive oil are good sources of vitamins, fiber, complex carbohydrates, and unsaturated fat.
- Traditional sauces made from butter, flour, cream, and egg yolks are very high in cholesterol.
- Asian-style sauces are high in salt and should be avoided by people on low-sodium diets.

Seaweed

- An excellent source of **iodine**
- Provides a wide spectrum of minerals, including calcium, copper, iron, magnesium, and potassium
- Some types rich in the B vitamins, vitamin C, and β-carotene
- Some are a good source of protein

Shellfish

- A low-fat source of high-quality protein
- A rich source of minerals, including calcium, fluoride, iodine, iron, and zinc
- A good source of B-group vitamins

Smoked, Cured, and Pickled Meats

- Preserved meats have more concentrated minerals (but generally lower vitamin content) than fresh.
- Used sparingly, they add flavor without excessive fat or calories.
- Cured meats have more concentrated fat than fresh meats.
- Nitrites in preserved meats may form cancer-causing **nitrosamines**.
- High sodium content makes most unsuitable for people on low-salt diets.
- Preserved meats must be carefully handled to prevent food poisoning.
- Sausages made with corn solids or syrup or cereal fillers may cause symptoms in people sensitive to these grains.
- Cured meats may contain high levels of **tyramine**, which triggers **migraine** in susceptible people and causes serious hypertension in patients taking a monoamine oxidase inhibitor such as **tranylcypromine**.

Snacks and Dips

- Snacks can help a person meet recommendations in the **Food Guide Pyramid**.
- Fruit and vegetable snacks are fat-free and high in vitamins and minerals.
- Well-timed low-calorie snacks can take the edge off hunger and help to prevent overeating at mealtimes.

Soft Drinks

- Carbonated drinks are refreshing and may provide a quick energy boost from their sugar or caffeine.
- Sipping ginger ale or cola can help to quell nausea and provide energy for people unable to take solid food.
- High phosphorus content may interfere with calcium absorption.
- Caffeine may cause health problems in adults or behavior and development problems in children.

Soups

- Can be highly nourishing
- An ideal food for convalescents
- Easy to make and economical
- Commercial varieties are often high in salt and fat

Spinach

- A rich source of vitamin A and folate
- High in vitamin C and potassium
- A vegetarian source of protein
- Contains **oxalic acid**, which reduces iron and calcium absorption, and can accelerate the formation of kidney and bladder stones

Squash

- Summer varieties provide some folate and vitamins A and C
- Winter varieties are extremely rich in vitamin A and are a good source of vitamin C, folate, and potassium

Strawberries

- An excellent source of vitamin C
- A good source of folate and potassium
- Low in calories and high in fiber
- Provide anticancer bioflavonoids
- Contain **oxalic acid**, which reduces mineral absorption and may aggravate kidney and bladder stones

Tangerines

- A good source of vitamin C, beta carotene, and potassium
- Contain **pectin**, a soluble fiber that helps control blood cholesterol

Tea

- Tea is a refreshing stimulant that is almost calorie-free if consumed plain.
- Tea contains **antioxidants** and **bioflavonoids**, which may lower the risk of cancer, heart disease, and stroke.
- Green tea is high in vitamin K.
- Tea contains **tannins** that may provide protection against **dental decay**.
- Herbal teas are caffeine-free.
- Tannins decrease iron absorption if tea is consumed with meals.
- Tea has a diuretic effect that increases urination.
- Tea may cause insomnia in caffeine-sensitive people.

Tomatoes

- A useful source of vitamins A and C, folate, and potassium
- A good source of **lycopene**, an antioxidant that protects against some cancers

Turnips

- A useful source of vitamin C, as well as some calcium and potassium
- A low-calorie source of fiber
- May protect against certain cancers
- Contain substances that interfere with the production of thyroid hormones

Vegetables

- Many are rich in vitamins A, C, and E, folate and other B vitamins, and potassium and other minerals
- High fiber content promotes regular bowel function
- Rich in bioflavonoids and other chemical compounds that help prevent disease
- **Goitrogens** in **cruciferous vegetables** may interfere with thyroid function

Vinegar

- Basis for a low-calorie salad dressing
- Can be used to preserve other foods

Watercress

- A good source of **beta carotene**, a precursor of vitamin A, and vitamin C
- A useful source of calcium, iron, and potassium
- Rich in **antioxidants**, which help prevent cancer and other diseases

Wine

- Moderate consumption may decrease the risk of heart disease and certain cancers.
- Wine increases the absorption of calcium, magnesium, phosphorus, and zinc.
- Pigments, such as **anthocyanins** and **tannins**, may protect against viruses and inhibit the formation of **dental plaque**.
- Wine promotes relaxation.

Yams and Sweet Potatoes

- A rich source of β-carotene
- A good source of vitamins C and B_6, folate, and potassium
- Naturally sweet and high in fiber

Yogurt

- An excellent source of calcium and phosphorus
- Provides useful amounts of vitamin A, several B vitamins, and zinc
- More digestible than milk for people with lactose intolerance
- Immunostimulant

Zucchini

- Low in calories
- A good source of vitamins A, C, and folate

Zwieback

- An ideal finger food for teething babies
- A nonperishable alternative to bread
- Provides some of the same nutrients found in bread

4

DIETARY ANTIOXIDANTS

- Reduce cardiovascular diseases
- Control diabetes mellitus
- Enhance the efficacy of antineoplastic agents

DIET IN CURING AND PREVENTING DISEASES

Diet is a strong factor in the control of **atherosclerosis** relating to general vascular disease, **coronary heart disease**, and **stroke**. The interrelated disorders in atherosclerosis of **hyperinsulinemia**, **hyperlipidemia**, and **hypertension** are strongly subject to dietary influence. The type of dietary protein, animal vs. plant, appears to be as important as the type of lipid, animal vs. plant, in atherosclerosis. Dietary protein type, with its differing amino acid ratios, appears to be a major secretagogue of insulin.

Diabetes mellitus, or Type II diabetes is a related disease where diet is a possible causal or at least a strong contributing factor. Diet is the beginning and continuing basis for the control of Type II diabetes. Interestingly, people with diabetes have a high incidence of atherosclerosis.

Renal failure has a long history of treatment with protein-restricted diets. Dietary plant protein is a possible therapy mechanism for the treatment of chronic and acute renal failure.

Patients with **rheumatic arthritis** appear to be helped by specialized dietary approaches. A few examples of the use of vegetarian diets look very promising. Individual arthritic sensitivities or reactivities to certain foods appear to warrant more study.

Osteoporosis is a disease or metabolic disturbance, particularly in postmenopausal women, that shows a need for very high dietary calcium intake. High calcium requirements appear to be related to the very high protein intake of the modern Western diet. A decreased protein intake, as can be obtained on a total vegetarian (vegan) diet, can allow for calcium balance, in a variety of age groups, from one third to one fifth the amount of daily calcium required on the Western diet.

Cancer of the breast, colon, and prostate appears to have a strong dietary relationship. The incidence of cancer is significantly greater on the modern Western diet than on a vegetarian or vegan diet. The greater antioxidant vitamin content, higher fiber level, higher complex carbohydrate, more unsaturated and less saturated fat, along with a variety of anticancer like compounds in vegetables, grains, legumes, nuts, and fruits, including tomatoes, are all dietary factors that appear to reduce the risk of cancer.

ANTIOXIDANT CAPACITY OF VEGETABLES AND FRUITS

Reactive oxygen species (ROS) could be important causative agents of a number of human diseases including coronary heart disease, cancer, and the degenerative processes associated with aging. It has been suggested that a high intake of fruit and vegetables, the main sources of antioxidants in the diet, could decrease the potential stress caused by ROS. The consumption of tomato and tomato products, for example, has been inversely related to the development of some types of cancer and to plasma lipid peroxidation. **Lycopene** has been hypothesized as being responsible for most of these effects, because of its high antioxidant activity and **singlet oxygen quenching capacity**. It has been shown that tomato intake significantly increases plasma lycopene and **β-carotene** concentrations.

It has been suggested that the total antioxidant capacity of plasma is due to the relative concentration in antioxidant compounds and to their synergism. It is possible that molecules that are not properly chain-breaking antioxidants, such as lycopene, may nevertheless maintain other antioxidant substances in the reduced form, thus indirectly determining an increased protection against oxidative stress. Evidence for the interaction between aqueous and lipophilic antioxidants (**vitamin C** and **vitamin E**, respectively) in defending **lipoproteins** against oxidative damage has been recently suggested in a human study. Tomato contains many antioxidant substances in addition to lycopene and β-carotene that could contribute to its protective effect.

ANTIOXIDANT ACTIVITY IN CEREAL GRAINS

The data from both experimental and epidemiological studies show that grains, vegetables, and fruits contain a large variety of substances called "plant chemicals" or "**phytochemicals**." The term *phytochemical* refers to every naturally occurring chemical substance present in plants, especially those that are biologically active. Major phytochemicals include **phenolic acids, flavonoids**, and **coumarin derivatives**, as well as many other **polyphenols**. The antioxidant activities in these phytochemicals range from extremely slight to very great. The natural antioxidants may have one or more of the following functions:

- Free-radical scavengers
- Reducing agents
- Potential complexers of prooxidant metals
- Quenchers of singlet oxygen

Consequently, phytochemical dietary components may actively contribute to the control of oxidative reactions and provide protection *in vivo*. Antioxidants play an important role in preventing undesirable changes in flavor and nutritional quality of foods. Cereal grains provide significant quantities of energy, protein, and selected micronutrients to the animal and human diet. The chemical composition and bioavailability of nutrients varies between species and varieties of grains and may be affected by forms of processing as feed and food. Cereal grains are rich in phenolic acids; the total amounts may approach 500 mg/kg of edible cereals. Other phytochemicals occurring in cereals are **phytosterines, saponins,** and **phytoestrogens**. In cereals, flavonoids are present in small quantities. Barley contains measurable amounts of **catechins** and some di- and trimer **procyanidins**. In contrast, cereals are the major source of dietary **lignans** in human nutrition. Lignans are potent antioxidants and exert anticancer effects. They decrease the production of reactive oxygen species by tumor cell types and cells of the immune system. Recent reports have described antioxidants and compounds with radical-scavenging activity present in **peabean, peanut, rice, buckwheat,** and **oat** (Zieliński and Kozlowska, 2000).

ANTIOXIDANT ACTIONS OF DU-ZHONG

Du-zhong (*Eucommia ulmoides* Oliv.) tea is commonly used in Japan in treatment of hypertension and is thought to be a functional health food. In Japan, statistics published by the Ministry of Health and Welfare show that billions of Japanese yen are spent on Du-zhong each year, ranking Japan at the top in the consumption of this specific health food. According to ancient records, roasted Du-zhong cortex was recommended as a folk medicine to reinforce muscle and lung, to lower blood pressure to prevent miscarriages, to improve the tone of the liver and kidney, and also to increase longevity.

Reactive oxygen species (ROS), such as **superoxide** ($O_2^{\cdot-}$) and **hydrogen peroxide** (H_2O_2), have been implicated both in aging and cancer. In particular, it is now recognized that the extremely reactive **hydroxyl radical** ($^{\cdot}OH$) derived from $O_2^{\cdot-}$ and H_2O_2 causes **DNA strand scission** in cellular damage. While in aerobes, the generation of ROS and the level of antioxidant defense systems are approximately balanced. Once the balance is tipped, oxidative stress occurs and diseases result. This oxidative stress may be decreased partially by increasing the dietary intake of the antioxidants. Therefore, much attention has been focused on the investigation of antioxidants, especially natural antioxidants, that can inhibit lipid peroxidation or protect biomolecules from damage caused by free radicals. Supplementation of antioxidants has been used successfully in some cases either to protect against physiological deterioration related to aging or to increase the average life span of some species. Natural antioxidants, especially phenolics and flavonoids, are those most often found in plants and are the most bioactive. **Phenolics** and **flavonoids** are major groups of nonessential dietary components that possess the bioactivity of an **antioxidant, antimutagenic, anti-inflammatory,** and **anticarcinogenic**, and can suppress **atherosclerosis** and **cancer** as **prophylactic agents**. The bioactivity of phenolics may be related to their ability to chelate metals, inhibit **lipoxygenase**, and scavenge free radicals. However, besides these beneficial effects, some phenolics can block the chain reaction of lipid peroxidation, whereas they cannot protect against oxidative damage to DNA, carbohydrates, and proteins when a transition metal ion exists.

Hsieh and Yen (2000) investigated the antioxidant effect of **water extracts of Du-zhong** (WEDZ) on oxidative damage in biomolecules such as deoxyribose, DNA, and 2′-deoxyguanosine (2′-dG) as induced by Fenton reaction. The WEDZ used included leaves, raw cortex, and roasted cortex. All of the WEDZ inhibited the oxidation of deoxyribose induced by Fe^{3+}-EDTA/H_2O_2/ascorbic acid in a concentration-dependent manner. At a concentration of 1.14 mg/ml, the inhibitory effect of the extracts of leaves, roasted cortex, and raw cortex was 85.2, 68.0, and 49.3%, respectively. The extract of leaves inhibited the strand-breaking of DNA induced by the Fenton reaction at concentrations of 5 and 10 µg/l. This inhibitory effect was similar to mannitol, whereas the extracts of raw cortex and roasted cortex had no inhibitory effect at all. WEDZ also inhibited the oxidation of 2′-dG to 8-OH-2′-dG induced by Fe^{3+}-EDTA/H_2O_2/ascorbic acid. **Gallic acid** had a prooxidant effect, but **trolox** and **mannitol** had an antioxidant effect. The leaf extract had a marked inhibitory effect on Fenton reaction-induced oxidative damage in biomolecules. The extract of roasted cortex exhibited modest inhibition while the extract of raw cortex had the least inhibitory effect on oxidative damage in biomolecules. This is in contrast to gallic acid in the same reaction system, whose higher reducing power and weaker chelating ability may contribute to its prooxidant effect. In the present study, leaf extract of Du-zhong had inhibitory effect on oxidative damage in biomolecules. Therefore, drinking of **Du-zhong tea** (leaf extract) over a long period of time may have anticancer potential (Hsieh and Yen, 2000).

Oxidative stress reduces the rate of cell proliferation, and that occurring during chemotherapy may interfere with the cytotoxic effects of antineoplastic drugs, which depend on rapid proliferation of cancer cells for optimal activity. Antioxidants detoxify ROS and may enhance the anticancer effects of chemotherapy. For some supplements, activities beyond their antioxidant properties, such as inhibition of **topoisomerase II** or **protein tyrosine kinases**, may also contribute. ROS cause or contribute to certain side effects that are common to many anticancer drugs, such as gastrointestinal toxicity and muagenesis. ROS also contribute to side effects that occur only with individual agents, such as **doxorubicin-induced cardiotoxicity**, **cisplatin-induced nephrotoxicity**, and **bleomycin-induced pulmonary fibrosis**. Antioxidants can reduce or prevent many of these side effects, and for some supplements the protective effect results from activities other than their antioxidant properties. Certain side effects, however, such as alopecia and myelosuppression, are not prevented by antioxidants, and agents that interfere with these effects may also interfere with the anticancer effects of chemotherapy (Conklin, 2000).

Chemotherapy has long been a cornerstone of cancer therapy. Although extensive research is done on the development of more effective and less toxic antineoplastic agents, much less attention has been paid to factors that may enhance the effectiveness of existing drugs. Nutritional factors may hold a key to enhancing the anticancer effects of chemotherapy and to reducing or preventing certain chemotherapy-induced side effects.

OXIDATIVE STRESS AND CHEMOTHERAPEUTIC EFFECTIVENESS

Free radicals and other ROS are essential for life, because they are involved in cell signaling and are used by phagocytes for their bacteriocidal action. In addition to these well-controlled and necessary functions, ROS are also produced in all respiring organisms as a consequence of mitochondrial respiration, which consumes oxygen in the process of generating ATP by the coupling of electron transport and oxidative phosphorylation. Nonessential production of ROS, i.e., oxidative stress, can also by induced by exogenous factors such as drugs and environmental toxins. Oxidative stress is potentially harmful to cells, and ROS are implicated in the etiology and progression of many disease processes including cancer. However, antioxidant mechanisms that scavenge ROS, by means of low-molecular-weight antioxidant enzyme systems, protect an organism from the damaging effects of oxidative stress. Under normal conditions, these antioxidant defense systems are able to detoxify ROS and prevent damage to cellular macromolecules and organelles. Under conditions of excessive oxidative stress, however, cellular antioxidants are depleted and ROS can damage cellular components and interfere with critical cellular activities (Conklin, 2000).

PRINCIPLE OF CANCER THERAPY

Malignant neoplastic diseases may be treated by various approaches: surgery, radiation therapy, immunotherapy, or chemotherapy, or a combination of these. The extent of a malignant disease (staging) should be ascertained to plan an effective therapeutic intervention.

CHEMOTHERAPY

Before discussing the specific pharmacokinetics and pharmacodynamics of each class of antineoplastic agents, several fundamental concepts and therapeutic objectives will be considered first.

Because a single cancerous cell is capable of multiplying rapidly and eventually causing the host's death, one of the therapeutic objectives is to **eradicate the last neoplastic cell**. Unlike normal cells, **cancerous cells multiply ceaselessly**, and, unless arrested, they will kill the host. In the early phase, cancerous cells grow exponentially. However, as the tumor grows in mass, the time needed for the number of cells to double also increases. The kinetics of cell multiplication are said to follow a **gompertzian growth curve**. Tumor growth may be divided into three phases: (1) the **subclinical phase**, in which 10×10^4 cells are present, (2) the **clinical phase**, in which 10×10^8 cells (1-cm^3 nodule) are present, and (3) the **fatal phase**, in which the number of cancerous cells equals or exceeds 10×10^{12}.

Most human cancerous cells evolve from the single clone of a malignant cell. As a tumor grows, significant mutation takes place, producing cells that exhibit diversified morphological and biochemical characteristics. During the subclinical phase, the rapidly growing cell population is uniform in character and thus highly sensitive to drug treatment; this is the reason for the importance of early diagnosis and treatment. During the clinical phase, the nondividing and slowly growing cells are nonuniform in character and less sensitive to drug treatment, thus necessitating the need for multiple-drug treatment.

Cell destruction with antineoplastic agents follows a **first-order kinetic**, indicating that the drugs kill a constant fraction of cells and not a constant number of cells. This concept is depicted mathematically in the following table.

No. of Treatments	% Killed	No. of Tumor Cells Killed Each Treatment	No. of Surviving Tumor Cells
Start	—	—	1,000,000
1	90	900,000	100,000
2	90	90,000	10,000
3	90	9,000	1,000
4	90	900	100
5	90	90	10
6	90	9	1

Cytotoxic drugs are not specific in their actions. Not only do they arrest cancerous cells, but also normal cells, especially those of the rapidly proliferating tissues such as the bone marrow, lymphoid system, oral and gastrointestinal epithelium, skin and hair follicles, and germinal epithelium of the gonads. Consequently, the therapeutic regimen must be carried out using **high-dose intermittent schedules** and not a low-dose continuous approach. Succeeding doses are given as soon as the patient has recovered from the previous treatment.

Antineoplastic agents may be **teratogenic, carcinogenic,** or **immunosuppressant,** and they exert their lethal effects on different phases of cell cycle by being either **cell-cycle specific** or **nonspecific** (Figure 4).

The management of cancer includes treatment with **alkylating agents** (nitrogen mustards and alkyl sulfonates), **antimetabolites** (methotrexate and purine analogues), **natural products** (vinca alkaloids and antibiotics), miscellaneous compounds (hydroxyurea, procarbazine, and *cis*-platinum), hormones (estrogens and corticosteroids), and radioactive isotopes.

ALKYLATING AGENTS

The alkylating agents exert their antineoplastic actions by generating highly reactive **carbonium ion intermediates** that form a covalent linkage with various nucleophilic components on both proteins and **DNA**. The 7 position of the purine base **guanine** is particularly susceptible to alkylation, resulting in miscoding, depurination, or ring cleavage. **Bifunctional alkylating agents** are able to cross-link either two nucleic acid molecules or one protein and one nucleic acid molecule. Although these agents are very active from a therapeutic perspective, they are also notorious for their tendency to cause carcinogenesis and mutagenesis. Alkylating agents that have a nonspecific effect on the cell-cycle phase are the most cytotoxic to rapidly proliferating tissues.

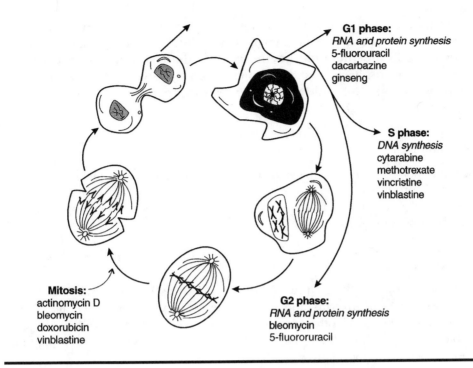

Figure 4   The actions of antineoplastic agents on different phases of the cell cycle.

Nitrogen Mustards

The activity of nitrogen mustards depends on the presence of a *bis*-(-2-chloroethyl) grouping:

$$\begin{array}{c} \text{CH}_2\text{-CH}_2\text{Cl} \\ \diagup \\ \text{N} \\ \diagdown \\ \text{CH}_2\text{-CH}_2\text{Cl} \end{array}$$

This is present in **mechlorethamine** (Mustargen), which is used in patients with **Hodgkin's disease** and other lymphomas, usually in combination with other drugs, such as in **MOPP therapy** (**mechlorethamine, Oncovin** [vincristine], **procarbazine**, and **prednisone**). It may cause bone marrow depression.

Chlorambucil

Chlorambucil (Leukeran) is the least-toxic nitrogen mustard, and is used as the drug of choice in the treatment of **chronic lymphocytic leukemia**. It is absorbed orally, is slow in its onset of action, and may cause bone marrow depression.

Cyclophosphamide

Cyclophosphamide (Cytoxan and Endoxan) is used in the treatment of **Hodgkin's disease**, **lymphosarcoma**, and other lymphomas. It is employed as a secondary drug in patients with acute leukemia and in combination with **doxorubicin** in women with breast cancer. A drug combination effective in the treatment of breast cancer is **cyclophosphamide**, **methotrexate**, **fluorouracil**, and **prednisone** (CMFP). Cyclophosphamide is also an immunosuppressive agent. The toxicity of cyclophosphamide causes alopecia, bone marrow depression, nausea and vomiting, and **hemorrhagic cystitis**.

Alkyl Sulfonate

The alkyl sulfonate busulfan (Myleran) is metabolized to an alkylating agent. Because it produces selective myelosuppression, it is used in cases of chronic myelocytic leukemia. It causes pronounced hyperuricemia stemming from the catabolism of purine.

Nitrosoureas

Carmustine (BCNU), lomustine (CCNU), and **semustine (methyl-CCNU)** generate **alkyl carbonium ions** and **isocyanate molecules** and hence are able to interact with DNA and other macromolecules. These agents, which are lipid soluble, cross the blood–brain barrier and are therefore effective in treating brain tumors. They are bone marrow depressants.

Triazenes

Dacarbazine (DTIC-Dome) is metabolized to an active alkylating substance. It is used in the treatment of **malignant melanoma** and causes myelosuppression.

Antimetabolites

Antimetabolites are structural analogues of naturally occurring compounds and function as fraudulent substances for vital biochemical reactions.

Folic Acid Analogues

Methotrexate (Amethopterin) is a folic acid antagonist that binds to **dihydrofolate reductase**, thus interfering with the synthesis of the active cofactor **tetrahydrofolic acid**, which is necessary for the synthesis of **thymidylate**, **purine nucleotides**, and the amino acids serine and methionine. Methotrexate is used for the following types of cancer:

- **Acute lymphoid leukemia.** During the initial phase, vincristine and prednisone are used. Methotrexate and mercaptopurine are used for maintenance therapy. In addition, methotrexate is given intrathecally, with or without radiotherapy, to prevent meningeal leukemia.
- **Diffuse histiocytic lymphoma.** Cyclophosphamide, vincristine, methotrexate, and cytarabine (COMA).
- **Mycosis fungoides**. Methotrexate.

- **Squamous cell**, **large-cell anaplastic**, **and adenocarcinoma**. Doxorubicin and cyclophosphamide, or methotrexate.
- **Head and neck squamous cell.** *Cis*-platinum and bleomycin, or methotrexate.
- **Choriocarcinoma.** Methotrexate.

Tumor cells acquire **resistance to methotrexate** as the result of several
factors:

- The deletion of a high-affinity, carrier-mediated transport system for
  reduced folates
- An increase in the concentration of dihydrofolate reductase
- The formation of a biochemically altered reductase with reduced affinity
  for methotrexate

To overcome this resistance, higher doses of methotrexate need to be administered.
 The effects of methotrexate may be reversed by the administration of **leucovorin**, the reduced folate. This leucovorin "rescue" prevents or reduces the toxicity
of methotrexate, which is expressed as mouth lesions (stomatitis), injury to the
gastrointestinal epithelium (diarrhea), leukopenia, and thrombocytopenia.

Pyrimidine Analogues

Fluorouracil and **fluorodeoxyuridine** (floxuridine) inhibit pyrimidine nucleotide biosynthesis and interfere with the synthesis and actions of nucleic acids.
To exert its effect, fluorouracil (5-FU) must first be converted to nucleotide
derivatives such as **5-fluorodeoxyuridylate** (5-FdUMP). Similarly, floxuridine
(FUdR) is also converted to FdUMP by the following reactions:

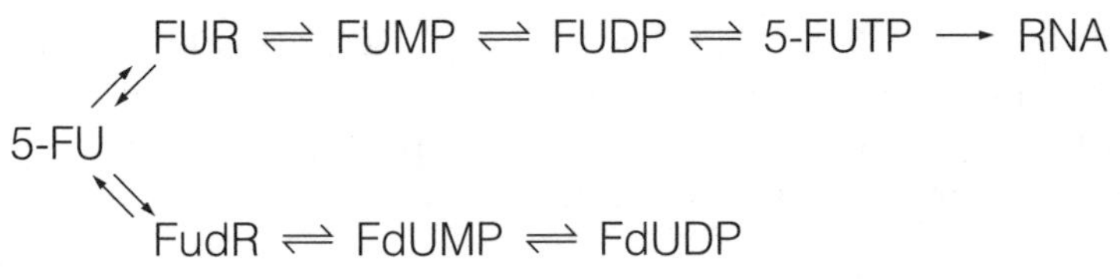

$$\text{FUR} \rightleftharpoons \text{FUMP} \rightleftharpoons \text{FUDP} \rightleftharpoons \text{5-FUTP} \longrightarrow \text{RNA}$$

5-FU

$$\text{FudR} \rightleftharpoons \text{FdUMP} \rightleftharpoons \text{FdUDP}$$

FdUMP inhibits **thymidylate synthetase**, and this is in turn inhibits the
essential formation of dTTP, one of the four precursors of DNA. In addition, 5-FU
is sequentially converted to 5-FUTP, which becomes incorporated into RNA, thus
inhibiting its processing and functioning. Fluorouracil is used for the following
types of cancer:

- **Breast carcinoma.** Cyclophosphamide, methotrexate, fluorouracil, and
  prednisone (CMP + P). The alternative drugs are doxorubicin and cyclophosphamide.
- **Colon carcinoma.** Fluorouracil.
- **Gastric adenocarcinoma.** Fluorouracil, doxorubicin (Adriamycin), and
  mitomycin (FAM), or fluorouracil and semustine.

- **Hepatocellular carcinoma.** Fluorouracil alone or in combination with lomustine.
- **Pancreatic adenocarcinoma.** Fluorouracil.

Resistance of 5-FU occurs as the result of one or a combination of the following factors:

- Deletion of uridine kinase
- Deletion of nucleoside phosphorylase
- Deletion of orotic acid phosphoribosyltransferase
- Increased thymidylate kinase

Because 5-FU is metabolized rapidly in the liver, it is administered intravenously and not orally. 5-FU causes myelosuppression and mucositis.

Deoxycytidine Analogues

Cytosine arabinoside (Cytarabine, Cytosar, and Ara-C) is an analogue of deoxycytidine, differing only in its substitution of sugar arabinose for deoxyribose. It is converted to Ara-CTP, and thereby inhibits DNA polymerase according to the following reactions:

$$\text{Ara-C} \xrightarrow{\text{Deoxycytidine kinase}} \text{Ara-CMP} \xrightarrow{\text{dCMP kinase}} \text{Ara-CDP}$$

$$\text{Ara-CDP} \xrightarrow{\text{NDP kinase}} \text{Ara-CTP}$$

$$\text{Deoxynucleotides} \xrightarrow[\text{DNA-polymerase}]{—/—/—} \text{DNA}$$

Cytosine arabinoside is used in the treatment of **acute granulocytic leukemia.** Doxorubicin; daunorubicin and cytarabine; cytarabine and thioguanine; or cytarabine, vincristine, and prednisone are the combinations of agents employed.

Resistance to cytosine arabinoside may stem from the following factors:

- The deletion of deoxycytidine kinase
- An increased intracellular pool of dCTP, a nucleotide that competes with Ara-CTP
- Increased cytidine deaminase activity, converting Ara-C to inactive Ara-U

The **toxic effects** of cytosine arabinoside are myelosuppression and injury to the gastrointestinal epithelium, which causes nausea, vomiting, and diarrhea.

PURINE ANTIMETABOLITES

6-Mercaptopurine (6MP) and 6-thioguanine (6TG) are analogues of the purines hypoxanthine and guanine, which must be activated by nucleotide formation, according to the following scheme:

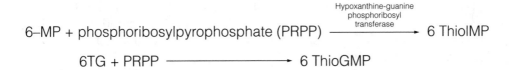

$$6\text{–MP} + \text{phosphoribosylpyrophosphate (PRPP)} \xrightarrow{\substack{\text{Hypoxanthine-guanine}\\\text{phosphoribosyl}\\\text{transferase}}} 6 \text{ ThioIMP}$$

$$6\text{TG} + \text{PRPP} \longrightarrow 6 \text{ ThioGMP}$$

ThioIMP and ThioGMP are **feedback inhibitors** of **phosphoribosylpyrophosphate amidotransferase**, which is the first and rate-limiting step in the synthesis of purine. In addition, these analogues inhibit the *de novo* biosynthesis of purine and block the conversion of inosinic acid to adenylic acid or guanylic acid. The triphosphate nucleotides are incorporated into DNA, and this results in delayed toxicity after several cell divisions.

6-Mercaptopurine is used in the treatment of acute lymphoid leukemia. Maintenance therapy makes use of both methotrexate and 6-mercaptopurine. Mercaptopurine is absorbed well from the gastrointestinal tract. It is metabolized through (1) methylation of the sulfhydryl group and subsequent oxidation and (2) conversion to **6-thiouric acid** with the aid of **xanthine oxidase**, which is inhibited by **allopurinol**. Mercaptopurine may cause hyperuricemia. Its chief toxicities are hepatic damage and bone marrow depression.

Thioguanine is used in patients with acute granulocytic leukemia, usually in combination with cytosine arabinoside and daunorubicin.

NATURAL PRODUCTS

Vinca Alkaloids

The **vinca alkaloids** (**vinblastine, vincristine,** and **vindesine**), which bind to tubulin, block mitosis with metaphase arrest. Vinca alkaloids are used for the following types of cancer:

- **Acute lymphoid leukemia.** In the induction phase, vincristine is used with prednisone.
- **Acute myelomonocytic or monocytic leukemia.** Cytarabine, vincristine, and prednisone.
- **Hodgkin's disease.** Mechlorethamine, Oncovin (vincristine), procarbazine, and prednisone (MOPP).
- **Nodular lymphoma.** Cyclophosphamide, Oncovin (vincristine), and prednisone (CVP).
- **Diffuse histiocytic lymphoma.** Cyclophosphamide, Adriamycin (doxorubicin), vincristine, and prednisone (CHOP); bleomycin, Adriamycin (doxorubicin), cyclophosphamide, Oncovin (vincristine), and prednisone (BACOP); or cyclophosphamide, Oncovin (vincristine), methotrexate, and cytarabine (COMA).
- **Wilms' tumor.** Dactinomycin and vincristine.
- **Ewing's sarcoma.** Cyclophosphamide, dactinomycin, or vincristine.
- **Embryonal rhabdomyosarcoma.** Cyclophosphamide, dactinomycin, or vincristine.
- **Bronchogenic carcinoma.** Doxorubicin, cyclophosphamide, and vincristine.

The chief toxicity associated with vinblastine use is bone marrow depression. The toxicity of vincristine consists of paresthesia, neuritic pain, muscle weakness, and visual disturbances. In addition, both vinblastine and vincristine may cause alopecia.

Dactinomycin

The antibiotics that bind to DNA are nonspecific to the cell-cycle phase. Dactinomycin (actinomycin D and Cosmegen) binds to double-stranded DNA and prevents RNA synthesis by inhibiting DNA-dependent RNA polymerase. It is administered intravenously in the treatment for pediatric solid tumors such as **Wilms' tumor** and **rhabdomyosarcoma** and for gestational **choriocarcinoma**. Dactinomycin causes skin reactions, gastrointestinal injury, and delayed bone marrow depression.

Mithramycin

The mechanism of action of mithramycin (Mithracin) is similar to that of dactinomycin. It is used in patients with advanced disseminated tumors of the testis and for the treatment of hypercalcemia associated with cancer. Mithramycin may cause gastrointestinal injury, bone marrow depression, hepatic and renal damage, and hemorrhagic tendency.

Daunorubicin and Doxorubicin

Daunorubicin (Daunomycin and Cerubidine) and doxorubicin (Adriamycin) bind to and cause the intercalation of the DNA molecule, thereby inhibiting DNA template function. They also provoke DNA chain scission and chromosomal damage. Daunorubicin is useful in treating patients with acute lymphocytic or acute granulocytic leukemia. Adriamycin is useful in cases of solid tumors such as sarcoma, metastatic breast cancer, and thyroid cancer. These agents cause stomatitis, alopecia, myelosuppression, and cardiac abnormalities ranging from arrhythmias to cardiomyopathy.

Bleomycin

Bleomycin (Blenoxane) causes chain scission and fragmentation of DNA. With the exception of the skin and lungs, most tissues can enzymatically inactivate bleomycin. Bleomycin is used in the management of **squamous cell carcinoma** of the head, neck, and esophagus in combination with other drugs in patients with testicular carcinoma, and in the treatment of Hodgkin's disease and other lymphomas. Bleomycin causes stomatitis, ulceration, hyperpigmentation, erythema, and pulmonary fibrosis.

MISCELLANEOUS ANTINEOPLASTIC AGENTS

Asparaginase

Normal cells are able to synthesize **asparagine**, but neoplastic tissues must obtain it from external sources. By metabolizing asparagine, **asparaginase** (Elspar) deprives the neoplastic tissues of asparagine, and in turn inhibits protein and nucleic acid synthesis. The resistant tumors are thought to possess higher-than-ordinary amounts of asparagine synthetase. Asparaginase, which is prepared from *Escherichia coli*, is used to induce remission of acute lymphocytic leukemia. Asparaginase causes malaise, anorexia, chills, fever, and hypersensitivity reactions. In general, it does not damage the bone marrow or other rapidly growing tissues as much as other antineoplastic agents do.

Hydroxyurea

Hydroxyurea suppresses DNA synthesis by inhibiting ribonucleoside diphosphate reductase, which catalyzes the reduction of ribonucleotides to deoxyribonucleotides. Hydroxyurea is used in chronic cases of granulocytic leukemia that are unresponsive to **busulfan**. In addition, it is used for acute lymphoblastic leukemia. Hydroxyurea may cause bone marrow depression.

Cis-Platinum

Cis-platinum (cisplatin) binds to intracellular DNA, causing both interstrand and intrastrand cross-linking. It is a cell-cycle phase nonspecific agent. Cis-platinum, which is ineffective orally, is used for testicular, bladder, and head and neck cancers. It precipitates nephrotoxicity, ototoxicity, and gastrointestinal injury.

Carboplatin

Carboplatin is a promising second-generation platinum agent. Because it is less reactive, it causes less nephrotoxicity, myelosuppression, and thrombocytopenia.

Procarbazine

Procarbazine (Matulane) inhibits DNA, RNA, and protein synthesis through the operation of an unknown mechanism. It is effective in patients with Hodgkin's disease when given in combination with mechlorethamine, vincristine, and prednisone (MOPP). Procarbazine causes neurotoxicity, bone marrow depression, and gastrointestinal injury.

Etoposide

Etoposide is used to combat several types of tumors, including testicular and small-cell lung cancers, lymphoma, leukemia, and Kaposi's sarcoma.

Cancer Chemotherapy–Induced Emesis

Nausea and vomiting are frequent side effects of radiotherapy and cancer chemotherapy. The incidence of this is relatively low for bleomycin, vincristine, and chlorambucil but is high for the remaining agents. Besides **prochlorperazine** and **metoclopramide (dopamine-receptor-blocking agents)**, **nabilone** (a **cannabinoid**), **batanopride**, **granisetron**, and **ondansetron** (all **serotonin-receptor-blocking agents**) have been shown to be effective in ameliorating these symptoms.

OXIDATIVE STRESS INTERFERES WITH THE CYTOTOXIC EFFECTS OF ANTINEOPLASTIC AGENTS

The major reactive oxygen species causing oxidative stress are:

Free radicals:

Hydroxyl radical	$HO^{-\bullet}$
Superoxide radical	$O_2^{\bullet-}$
Peroxyl	$ROO^{\bullet}$
Alkoxyl	$RO^{\bullet}$

Nonradical oxygen species:

Hydrogen peroxide	H_2O_2
Singlet oxygen	$^1O_2$

Considerable evidence supports the contention that excessive oxidative stress, i.e., the generation of greater-than-normal amounts of ROS, interferes with the cytotoxic effects of antineoplastic agents on cancer cells. Important to this concept is how oxidative stress influences the progression of a cell through the cell cycle and at which phases of the cell cycle antineoplastic agents are cytotoxic (see Figure 4).

The cell cycle consists of a presynthetic phase (G_1), which precedes DNA synthesis, the phase of DNA synthesis (S), an interval that follows DNA synthesis (G_2), and mitosis (M), during which the G_2 cell with a double complement of DNA divides into two daughter cells. Each of these daughter cells may then immediately reenter the division cycle at G_1 or pass into a nonproliferative state (G_0) for an indefinite period of time. As the length of the cell cycle increases, the duration of the G_1 phase increases, whereas the durations of the S, G_2, and M phases remain relatively constant.

The length of the cell cycle can be influenced by many factors, one of which is oxidative stress (Conklin, 2000). During oxidative stress, the excessive production of ROS results in lipid peroxidation. Because the rate of DNA synthesis and the rate of cell proliferation of cancer cells and normal cells are inversely related to the degree of lipid peroxidation, oxidative stress prolongs the G_1 phase or may result in cells entering the G_0 phase. This cell cycle prolongation by lipid peroxides is also illustrated by results of *in vivo* studies that demonstrate that lipid peroxidation, induced by feeding polyunsaturated fatty acids (PUFA), slows the growth rate of human breast carcinoma implanted into **athymic nude mice**. Reversal of the inhibitory effect of oxidative stress on cell proliferation by antioxidants that prevent lipid peroxidation supports the contention that the inhibitory effect is due to lipid peroxides.

Antineoplastic agents act by blocking the synthesis of DNA precursors, damaging the integrity of DNA, or interfering with DNA replication, separation of the two double helixes after replication, or the function of the mitotic spindle. Tumor cells in the nonproliferative G_0 state are little affected by anticancer drugs and can reenter the division cycle after chemotherapy is completed, resulting in recurrence of disease. Even drugs that do not require ongoing DNA synthesis for activity, e.g., the alkylating agents and platinum coordination complexes, which can damage the integrity of quiescent DNA, may have little effect on G_0 tumor cells, since repair mechanisms for DNA can repair damage before the cells reenter the division cycle. Tumor cells with prolonged G_1 phases may likewise be resistant to the cytotoxic effects of antineoplastic agents. Because anticancer drugs are cytotoxic only when tumor cells are proliferating rapidly, this provides a mechanism whereby oxidative stress, which slows or arrests cell growth, interferes with chemotherapeutic effectiveness. It also explains why slow-growing tumors, such as lung and colon carcinoma, which likely have many cells in the G_0 phase for prolonged periods of time, are relatively unresponsive to chemotherapy.

Cancer cells have highly evolved protective mechanisms to prevent lipid peroxidation so that rapid cell proliferation can occur. Several studies have demonstrated that lipid peroxidation is decreased significantly in tumor cells and tissues compared with that in corresponding normal cells and tissues. The primary mechanism whereby cancer cells prevent lipid peroxidation consists of a marked increase, compared with normal cells, in vitamin E relative to the peroxidizable moieties (methylene groups) of the PUFA in their biological membranes. Compared with normal cells, tumor cells also have relatively low levels of the components of the **NADPH-cytochrome P-450 electron transport chain**, which results in less favorable conditions for the initiating and propagation of lipid peroxidation. However, excessive oxidative stress may overcome even the highly evolved protective mechanisms of cancer cells, resulting in inhibition of cancer proliferation and interference with the cytotoxic activity of antineoplastic agents.

ANTIOXIDANT EFFECTS ON ANTINEOPLASTIC ACTIVITY AND CHEMOTHERAPY-INDUCED SIDE EFFECTS

Vitamin E

Vitamin E resides in the lipid domain of biological membranes and plasma lipoprotein, where it prevents lipid peroxidation of PUFA. Because vitamin E is the most important antioxidant for preventing lipid peroxidation, vitamin E adequacy may significantly impact the rate of cancer cell proliferation and the response to cancer chemotherapy. Vitamin E has been shown to effectively reverse the inhibitory effect of lipid peroxidation on cell proliferation.

Vitamin E, *in vitro,* has been shown to enhance the cytotoxic effect of several anticancer drugs, including **5-fluorouracil** (5-FU), **doxorubicin**, **vincristine**, **dacarbazine**, **cisplatin**, and **tamoxifen**.

Vitamin C

Vitamin C is a versatile water-soluble antioxidant. It protects against lipid peroxidation by scavenging ROS in the aqueous phase before they can initiate lipid peroxidation.

In vitro studies with several tumor cell lines have shown vitamin C to enhance the cytotoxic activity of **doxorubicin, cisplatin, paclitaxel, dacarbazine, 5-FU,** and **bleomycin.** Vitamin C has also been shown to increase drug accumulation and to partially reverse vincristine resistance of human non-small-cell lung cancer cells.

Coenzyme Q_{10} (Ubiquinone)

Coenzyme Q_{10} (CoQ_{10}) is an indispensable cofactor in the **electron transport chain of mitochondria,** functioning as an electron carrier between the enzyme complexes of the respiratory chain. CoQ_{10} is also a lipid-soluble antioxidant that scavenges lipid radicals within biological membranes. CoQ_{10} within mitochondria may play an important role in preventing lipid peroxidative damage of mitochondrial membranes.

A potentially important application of CoQ_{10} supplementation during chemotherapy is for the prevention of **doxorubicin-induced cardiotoxicity,** most importantly the chronic form, which is not prevented by other dietary antioxidants. The postulated mechanism of the dose-limiting chronic cardiotoxicity, a complication that occurs to a greater or lesser degree with all **anthracyclines,** involves the production of oxidizing agents through an iron-dependent process. Consistent with this hypothesis is the generation of free radicals by doxorubicin, which results in mitochondrial lipid peroxidation within myocardial cells. However, other effects of doxorubicin on the mitochondria of cardiac myocytes may be equally or more important. These effects of doxorubicin include (1) reduction of the CoQ_{10} content of mitochondrial membranes, (2) inhibition of mitochondrial biosynthesis of CoQ_{10}-dependent enzymes, which interferes with the aerobic generation of ATP, and (3) inhibition of mitochondrial biosynthesis of CoQ_{10}. These effects of doxorubicin on CoQ_{10} biosynthesis and function may explain the acute and chronic forms of doxorubicin-induced cardiotoxicity.

Acute doxorubicin-induced cardiotoxicity, a reversible side effect, is observed soon after a single dose of doxorubicin. It can be attributed to inhibition of respiratory enzymes by doxorubicin, which may result from competition between CoQ_{10} and doxorubicin for the enzymatic sites of the coenzyme, because both compounds contain a quinone group. Enzyme inhibition may also result from oxidation of CoQ_{10} by doxorubicin or doxorubicin-induced ROS, thus preventing CoQ_{10} from functioning as a cofactor for electron transport. Because the respiratory generation of ATP is essential to myocardial function, inhibition of this function by doxorubicin would interfere with the electrophysiological activity of the heart, resulting in electrocardiographic changes and reduction of ejection fraction, which are characteristic of acute doxorubicin-induced cardiotoxicity.

Chronic doxorubicin-induced cardiotoxicity, i.e., that after repeated doses of doxorubicin, may be attributable to depletion of mitochondrial CoQ_{10}, which disrupts electron transport and mitochondrial respiratory bioenergetics, may ultimately lead to loss of mitochondrial integrity and necrosis of cardiac myocytes. In support of this mechanism as the primary etiology are studies in rabbits that demonstrate that mitochondrial degeneration is the earliest and most prominent ultrastructural change associated with the chronic cardiomyopathy induced by doxorubicin. This mechanism is also consistent with the clinical observations that chronic doxorubicin-induced cardiotoxicity is not reversible, the associated congestive heart failure is not responsive to **digitalis**, and antioxidants such as vitamin E and vitamin C do not prevent it.

β-Carotene

β-Carotene is one of >600 **carotenoids** that are produced by microorganisms and plants. β-Carotene, a lipid-soluble antioxidant, occurs naturally as a mixture of *cis* and *trans* isomers. *cis* β-Carotene is a more effective antioxidant than is *trans* β-carotene.

β-Carotene has been shown to enhance the cytotoxicity of **melphalan** and **BCNU** on human **squamous carcinoma cells** and of **cisplatin** and **dacarbazine** on **melanoma cells**. In mice with transplanted mammary carcinoma, β-carotene enhanced the antitumor effect of cyclophosphamide, and in mice transplanted with FsaII fibrosarcoma or SCC VII carcinoma, β-carotene enhanced the antitumor effect of melphalan, BCNU, doxorubicin, and etoposide. β-Carotene (5 to 50 mg/kg) has been shown to reduce the genotoxicity of **cyclophosphamide** in mice and of **mitomycin C**, methyl methanesulfonate, and bleomycin in cultured cells. β-Carotene also reduced the rate of tumor induction in animals receiving chronic low doses of cyclophosphamide.

Glutathione (GSH) and Glutathione Esters

GSH, a tripeptide of glutamic acid, cysteine, and glycine (GluCysGly), is the major water-soluble antioxidant in the cytoplasm, nuclei, and mitochondria of cells. Many of the critical antioxidant functions of GSH require GSH peroxidase, which exists in several forms. Reduction of oxidized GSH (GSH disulfide), which is produced by reactions involving GSH peroxidase, requires GSH reductase.

GSH is not transported into cells. In order for circulating GSH to increase intracellular GSH concentrations, it must first be hydrolyzed to Glu and CysGly, which are subsequently transported into the cell and serve as substrates for GSH synthesis. Thus, GSH administered orally or parenterally and that produced by the liver and released into the circulation enhance tissue levels of GSH by providing a source of its constituent amino acids. In contrast, GSH monoesters, which are well absorbed after oral administration, as is GSH are readily transported to cells and then hydrolyzed to GSH and the corresponding alcohol. Thus, higher cellular levels of GSH result from oral administration of GSH monoesters than from oral administration of comparable doses of GSH.

GSH has been investigated for its protective effect against cisplatin-induced nephrotoxicity and peripheral neuropathy. Although oxidative damage most likely contributes to these toxicities, the protective effect of GSH can be accounted for by a chemical interaction between GSH and cisplatin instead of the antioxidant properties of GSH. GSH contains a thiol (sulfhydryl, an –SH moiety) group. Thiols are strongly nucleophilic and form stable covalent compounds with electrophilic compounds, such as the platinum coordination complexes **cisplatin** and **carboplatin**. Formation of the **thiol-platinum complex** inactivates the antineoplastic agent, which blocks not only its ROS-generating activity but also its cytotoxic effects. If inactivation occurs within the circulation before uptake of the drug by tumor cells, interference with the antineoplastic activity of cisplatin or carboplatin may occur. A similar concern exists if GSH is administered with alkylating agents, which are also strong electrophiles, since thiols can compete with DNA for alkylation and result in inhibition of antineoplastic activity. Thus caution should be employed when GSH or any thiol compound is administered with platinum coordination complexes or alkylating agents.

Studies in laboratory animals have shown that intravenous administration of a high dose of GSH (up to 500 mg/kg) within 30 minutes of cisplatin injection protects against **cisplatin-induced neurotoxicity and nephrotoxicity.** Subcutaneous injection of GSH or GSH monoisopropyl ester 2.5 hours before injection of cisplatin also protected mice against nephrotoxicity and the acute lethal toxicity of cisplatin, although the GSH ester was far more effective than GSH itself. In these studies, treatment with GSH or GSH ester did not interfere with the antitumor effectiveness of cisplatin, which can be explained by the characteristics of uptake of GSH and cisplatin. GSH and cisplatin are cleared rapidly from the circulation.

N-Acetylcysteine

N-Acetylcysteine (NAC) is well absorbed after oral administration and readily transported into cells, where it is deacetylated. Although NAC, a thiol compound, is a free radical scavenger, its more important antioxidant role is providing an intracellular source of cysteine, a substrate for GSH synthesis. However, because of its **nucleophilic thiol group**, NAC (like GSH) can form stable covalent compounds with electrophilic alkylating agents and platinum coordination complexes, which inactivate the anticancer drugs. In this regard, NAC has been shown to block the cytotoxicity and ROS-generating capability of cisplatin *in vitro*. Thus, as with GSH, one should exercise caution when administering NAC during chemotherapy with any electrophilic antineoplastic agent.

In animal studies, NAC has been shown to prevent **hemorrhagic cystitis** that results from administration of **cyclophosphamide** or its position isomer **ifosfamide**. Hemorrhagic cystitis results from the toxic effect of **acrolein**, a metabolic product of **cyclophosphamide** or its position isomer ifosfamide. The mechanism whereby NAC prevents this toxicity may be prevention of the intracellular depletion of antioxidants, such as GSH, by acrolein. Concomitant administration of NAC with cyclophosphamide or ifosfamide does not impair antineoplastic activity, because both anticancer drugs are inactive until they are metabolized by the liver to their **phosphoramide mustard metabolites**.

NAC (50 to 2000 mg/kg), by means of free radical scavenging or by enhancing intracellular levels of GSH, protects mice against acute doxorubicin-induced cardiotoxicity without interfering with the antitumor activity of the drug.

Glutamine

Glutamine is a conditionally essential amino acid. Although it does not possess antioxidant activity, it serves as a source of glutamate for GSH synthesis, thus supporting cellular antioxidant systems. Because glutamine is a primary fuel source for the rapidly proliferating enterocytes of the gastrointestinal (GI) tract, an important role of glutamine supplementation during chemotherapy is to reduce

GI injury that results from administration of antineoplastic drugs, especially the severe mucositis that results from treatment with antimetabolites such as **5-FU** and **methotrexate.**

Glutamine has been shown to enhance the antitumor effectiveness of **methotrexate** in laboratory animals, an effect that may be attributed to glutamine increasing the intracellular tumor concentration of methotrexate. Oral glutamine, but not intravenous glutamine, has also been shown to reduce the bacteremia and mucosal injury associated with methotrexate-induced enterocolitis of rats. Glutamine, administered by intragastric infusion, accelerates healing of the gut mucosa in rats receiving 5-FU.

Selenium

Although inorganic **selenium** does not have antioxidant properties, selenium has an important role in cellular antioxidant defenses as a necessary component of **selenoproteins.** Selenium is incorporated into **selenoproteins** as **selenocysteine.** The **glutathione (GSH) peroxidases** are the best-characterized selenoproteins, although other circulating selenoproteins also have antioxidant functions.

In laboratory animals, parenteral administration of organic and inorganic selenium (210 to 12,000 µg/kg) has been shown to protect against **cisplatin-induced nephrotoxicity.** Protection occurs without apparent inhibition of the antineoplastic activity of cisplatin, although this may be attributed to the fact that selenium administration allows for higher doses of cisplatin to be used. Additionally, selenium administration reduces cisplatin-induced myelosuppression. This raises a concern similar to that of administering cisplatin with thiol compounds, i.e., that the reduction of myelosuppression may indicate that selenium can also interfere with the antitumor activity of cisplatin. Selenium, with chemical properties similar to those of sulfur, can bind with platinum and inactivate the antineoplastic platinum coordination complexes. Thus, caution should be used when administering selenium during chemotherapy with **cisplatin** and **carboplatin.**

Parenteral administration of 27 to 60 µg/kg of selenium to laboratory animals has been shown to inhibit doxorubicin-induced decreases in myocardial vitamin E and GSH peroxidase levels and to reduce changes in myocardial function that are consistent with acute doxorubicin-induced cardiotoxicity. Oral supplementation of **sodium selenite** also protects against acute doxorubicin-induced cardiotoxicity in rabbits.

Genistein and Daidzein

Soybeans contain a number of **isoflavones**, including **genistein** and **daidzein**, which are the most extensively studied. Soybean isoflavones enhance cellular antioxidant status by scavenging ROS and by increasing the activity of antioxidant enzymes including GSH peroxidase, **GSH reductase**, and **superoxide dismutase**. Genistein has also been shown to abrogate chemical- and ligand-induced generation of ROS *in vivo* and *in vitro*, although these effects result, in part, from activities of genistein other than its antioxidant activities. In addition to its antioxidant properties, genistein has other activities that may enhance the antineoplastic effects of cancer chemotherapy. Although clinical studies are yet to be done to assess the impact of soy isoflavones on cancer chemotherapy, a substantial number of preclinical studies have been performed.

Genistein is an inhibitor of **topoisomerase I and II**. Inhibition of **topoisomerase II** by genistein has effects similar to those of doxorubicin and etoposide, i.e., enhanced **topoisomerase II-mediated DNA cleavage**, the formation of stable **covalent protein–DNA cleavage complexes**, and cell cycle arrest at the G_2–M junction. Genistein also inhibits the binding of ATP to its binding site on the enzyme, an activity not possessed by doxorubicin or etoposide. Thus administration of genistein during chemotherapy with either of these drugs may enhance antineoplastic activity. Enhanced antineoplastic activity may also result from inhibition of topoisomerase I by genistein, although this effect, which results in arrest of the cell cycle in the S phase, requires a higher concentration of genistein than is necessary to inhibit topoisomerase II. Genistein has been shown to induce apoptosis in a number of tumor cell lines. Although apoptosis can be explained by inhibition of topoisomerase II activity, inhibition by genistein of **protein tyrosine kinases** (PTKs) may also contribute to the process.

Quercetin

Quercetin, a ubiquitous polyphenolic flavonoid, is one of the most potent natural antioxidants. In addition to its antioxidant properties, quercetin exhibits other activities that may enhance antineoplastic activity. As with genistein, clinical studies are yet to be performed to determine the impact of quercetin on cancer chemotherapy.

Quercetin inhibits topoisomerase II activity. Like genistein, it enhances topoisomerase II-dependent DNA cleavage complexes. It also inhibits topoisomerase II-catalyzed ATP hydrolysis. Because quercetin is an intercalative compound, it may enhance DNA cleavage by a mechanism similar to that of doxorubicin. Quercetin also inhibits several PTKs, with IC_{50} values that are comparable to or lower than those of genistein.

Quercetin has been shown to inhibit the growth of several human and animal cancer cell lines *in vitro*. However, the sensitivity of different cell lines varies considerably. In human breast cancer cells the IC_{50} of quercetin for inhibition of growth was 23 μM (7 μg/ml), whereas in human gastric cancer cells the IC_{50} was 32 to 55 μM. A head and neck squamous cell carcinoma line is unaffected by quercetin at <110 μM. Because growth inhibition of sensitive cells results from arrest of cell cycle progression at the G_1–S boundary, this effect of quercetin is likely to be independent of its inhibition of topoisomerase II.

Quercetin has been shown to enhance the cytotoxicity of several antineoplastic agents. In multidrug-resistant cancer cells, quercetin markedly enhanced the growth-inhibitory effects of doxorubicin, although it did not affect the cytotoxicity of **doxorubicin** in drug-sensitive cells. In drug-sensitive cancer cells, quercetin has been shown to enhance the antiproliferative activity of **cisplatin, nitrogen mustard, busulfan**, and **cytosine arabinoside**. Quercetin also enhances the antitumor activity of cisplatin in athymic nude mice implanted with a human large-cell lung cancer.

Conclusions

Dietary supplementation with antioxidants may provide a safe and effective means of enhancing the response to cancer chemotherapy. **Vitamin E** may prove to be an important nutrient for enhancing antineoplastic activity because of its role in preventing lipid peroxidation, thus maintaining the rapid rate of proliferation of cancer cells. Other antioxidants may be important because of their antioxidant properties, as well as for activities such as inhibition of topoisomerase II and PTKs.

The quality of life of patients after chemotherapy may be improved by dietary supplementation with antioxidants that reduce or prevent chemotherapy-induced side effects. Although approved cytoprotectants are available, including **dexrazoxane** for **doxorubicin-induced cardiotoxicity**, **amifostine** for **cisplatin-induced nephrotoxicity**, and **mesna** for **ifosfamide-induced hemorrhagic cystitis**, these agents are not without adverse effects. For example, **dexrazoxane** can reduce the antineoplastic activity of doxorubicin. This most likely results from complex formation and inactivation of doxorubicin, which may also prevent doxorubicin from interfering with **coenzyme P_{10}** (CoP_{10}) biosynthesis and function, thus explaining the cardioprotectant effect of dexrazoxane. **Dexrazoxane** is also myelosuppressive and may increase the risk of developing secondary malignancies. **Amifostine** can induce hypotension, hypocalcemia, and nausea, and administration of mesna is associated with nausea, vomiting, and diarrhea. In contrast, certain dietary antioxidants, in doses that are without adverse effects, can ameliorate some side effects of cancer chemotherapy. In this regard, **CoQ_{10}** may prove to be an effective means of preventing cardiotoxicity without compromising antineoplastic activity when chemotherapy employs the versatile and highly effective doxorubicin. However, much more work is needed to establish a clear role for the use of dietary supplements as an adjunct to cancer chemotherapy (Conklin, 2000).

5

FOOD–DRUG
INTERACTIONS

"Poison and medicines are oftentimes the same substances given with different intents."

— Peter Mere Latham (1789–1875)

Grapefruit juice interacts with a variety of drugs, elevating their serum concentration.

INTRODUCTION

Grapefruit juice carries the American Heart Association's healthy "heart-check" food mark and contains compounds that may both reduce **atherosclerotic plaque formation** and inhibit **cancer cell proliferation**. However, unlike other citrus fruit juice, grapefruit juice interacts with a variety of prescription medications, raising the potential for concern. This is particularly worrying in that juice and medications are commonly consumed together at breakfast. This drug–food interaction seems to occur through inhibition by grapefruit juice of one of the intestinal cytochrome P-450 (CYP) enzyme systems, **cytochrome P-450 3A4** (CYP3A4). This enzyme system in the liver is well known for its involvement with drug–drug interactions. Most notably, **terfenadine**, **mibefradil**, and **cisapride** have been withdrawn from the U.S. market in recent years, in part because of deaths due to drug–drug interactions involving the hepatic CYP3A4 (Kane and Lipsky, 2000).

BIOTRANSFORMATION

Biotransformation may be defined as the enzyme-catalyzed alteration of drugs by the living organism. Although few drugs are eliminated unchanged, urinary excretion is a negligible means of terminating the action of most drugs or poisons in the body. As a matter of fact, the urinary excretion of a highly lipid-soluble substance such as **pentobarbital** would be so slow that it would take the body a century to rid itself of the effect of a single dose of the agent. Therefore, mammalian and other terrestrial animals have developed systems that allow the conversion of most lipid-soluble substances to water-soluble ones, so that they may be easily excreted by the kidney. In general, biotransformation may be divided into two forms of metabolism: hepatic and nonhepatic.

Hepatic Drug Metabolism

By far the major portion of biotransformation is carried out in the liver by **cytochrome P-450** (P-450), which is a collective term for a group of related enzymes or isoenzymes that are responsible for the oxidation of numerous drugs; endogenous substances such as fatty acids, prostaglandins, steroids, and ketones; and carcinogens such as **polycyclic aromatic hydrocarbons**, **nitrosamines**, **hydrazines**, and **arylamines**.

Nonhepatic Metabolism

Plasma

One of the drugs that is metabolized in the blood is **succinylcholine**, a muscle relaxant that is hydrolyzed by the **pseudocholinesterase** of liver and plasma to succinylmonocholine. The short duration of action of succinylcholine (5 minutes) is due to its rapid hydrolysis in plasma. Patients with atypical cholinesterase who cannot metabolize succinylcholine suffer pronounced apnea. **Procaine**, a local anesthetic, is also hydrolyzed by pseudocholinesterase.

Lung

The lung is involved in both the activation and inactivation of numerous physiological and pharmacological substances. For example, **angiotensin I** is converted to **angiotensin II** in the lung.

Intestinal Epithelium

The intestinal epithelium is capable of removing numerous agents.

MOLECULAR BIOLOGY OF MULTIPLE ISOENZYMES OF P-450

In recent years, an extensive number of complementary DNAs for the P-450 genes for humans have been isolated and sequenced. Genes that encode proteins that are less than 36% similar in their amino acid sequence belong to different families. Currently, eight different families have been identified in humans, each designated by a Roman numeral. The drug metabolizing P-450s belongs to family I, II, III, and IV. P-450s that are 70% or more similar are encoded by genes in the same subfamily. Finally, the individual gene is designated by an Arabic numeral.

During Phase I, most drugs are inactivated pharmacologically; some remain unaltered and some become more active and toxic. For example, **phenytoin** in the liver is first hydroxylated to **hydroxyphenytoin** (Phase I) and is then conjugated with glucuronic acid (Phase II) and excreted by the kidney as **phenytoin glucuronide conjugate**. During Phase I, besides introducing a polar group such as an –OH group, a potential polar group may also be unmasked from the

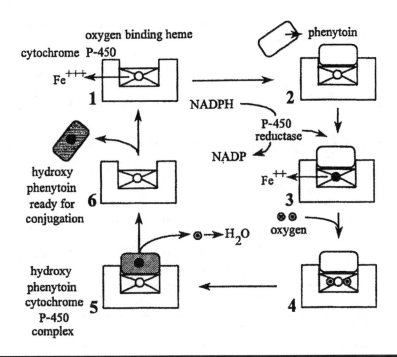

Figure 5 A simplified scheme for the mechanism of action of cytochrome P-450. NADP = nicotinamide adenine dinucleotide phosphate; NADPH = the reduced form of NADP.

drug to be metabolized. For example, compound R–OCH$_3$ is converted to compound R–OH by demethylation. **Codeine** becomes demethylated to **morphine**. The free or unmasked polar group is then conjugated with glucuronate, sulfate, glycine, or acetate. With the exception of **morphine 6-glucuronide**, almost all conjugates lack pharmacological activity.

SCHEME OF THE MIXED-FUNCTION OXIDATION REACTION PATHWAY

The hepatic endoplasmic reticulum possesses oxidative enzymes called mixed-function oxidases or monooxygenase with a specific requirement for both molecular oxygen and a reduced concentration of **nicotinamide adenine dinucleotide phosphate** (NADPH). Essential in the mixed-function oxidase system is P-450 (Figure 5). The primary electron donor is NADPH, whereas the electron transfer involves P-450, a **flavoprotein**. The presence of a heat-stable fraction is necessary for the operation of the system.

A drug substrate to be metabolized binds to **oxidized P-450**, which in turn is reduced by **P-450 reductase**. The drug-reduced P-450 complex then combines with molecular oxygen. A second electron and two hydrogen ions are acquired from the donor system, and the subsequent products are oxidized drug and water, with regeneration of the oxidized P-450. This process is summarized as follows:

1. NADPH + oxidized cytochrome P-450 + H$^+$ → reduced P-450 + NADP$^+$
2. Reduced cytochrome P-450 + O$_2$ → "active oxygen complex"

3. "Active oxygen complex" + drug substrate → oxidized drug + oxidized cytochrome P-450 + H_2O

$$NADPH + O_2 + \text{drug substrate} + H^+ \rightarrow NADP^+ + \text{oxidized drug} + H_2O$$

The overall scheme for the mechanism of action of P-450 is shown in Figure 5.

CONSEQUENCE OF BIOTRANSFORMATION REACTIONS

The process of biotransformation usually inactivates or detoxifies, or both, the administered drugs or the ingested poisons, but other reactions may also take place.

Precursor Activation. Occasionally, an inactive precursor such as **levodopa** is converted to an active metabolite such as **dopamine**.

Metabolic Activation of Drugs. Often an active drug is converted to another pharmacologically active substance. The following table lists a few examples of this.

Drug		Active Metabolite
Mephobarbital	is demethylated to	Phenobarbital
Primidone	is oxidized to	Phenobarbital
Imipramine	is demethylated to	Desmethylimipramine
Prednisone	is reduced to	Prednisolone

Conversion to Metabolites with Dissimilar Actions. In certain instances, the body converts a drug to several active metabolites possessing dissimilar pharmacological properties. For example, **phenylbutazone** undergoes aromatic hydroxylation to produce a metabolite that has sodium-retaining and antirheumatic activities, and also undergoes alkyl chain oxidation to produce a metabolite with a strong uricosuric property. Thus, phenylbutazone has both **uricosuric** and **antirheumatic effects**.

Conversion to More Active Products. The conversions of **cyclophosphamide** to **aldophosphamide** and **prednisone** to **prednisolone** are examples of active compounds that are converted to more active substances.

Lethal Synthesis. The metabolism of drugs and agents does not always lead to detoxification; occasionally, the metabolites are toxicologically more potent. Some examples of this are **sulfamethazine**, which is metabolized to N_4-**acetylsulfamethazine**, and **aminopyrine**, with **4-aminoantipyrine** as its metabolite.

DRUG METABOLITE KINETICS

In the majority of cases, drugs are converted to metabolites, which, in the more polar and water-soluble forms, are readily excreted. Often the concentration of a metabolite far exceeds the concentration of the drug. For example, orally administered **propranolol** is rapidly converted to **4-hydroxypropranolol**, which has a concentration that is severalfold higher than that of propranolol. Sometimes the metabolites are able to inhibit the further metabolism of the parent drug. For example, **phenytoin** becomes metabolized to **hydroxyphenytoin**. When given in higher than recommended individual doses, hydroxyphenytoin inhibits the hydroxylase system that metabolizes phenytoin, increasing its concentration in free form and its potential to produce toxicity.

The **enterohepatic circulation** may sometimes prolong the half-life of a drug. A drug that is absorbed from the gastrointestinal tract, excreted in the bile, and resorbed from the intestine is said to have undergone enterohepatic cycling. Drugs are delivered to the liver by both the portal vein and hepatic artery, and returned to the rest of the body by the hepatic vein. The difference between the concentration of drug transported to and removed from the liver accounts for the amount of drug metabolized or excreted, or both, in the bile. For example, if the liver has conjugated a drug containing glucuronic acid to its metabolite, the conjugated product may appear in the bile and finally be excreted in the small intestine. However, in the intestine, the β-glucuronidase originating from the resident flora may hydrolyze the glucuronide-drug conjugate back to the parent drug, thus allowing the parent drug to be resorbed. The continuous enterohepatic cycling will therefore increase the half-life of this agent in the body.

FACTORS THAT MODIFY THE METABOLISM OF DRUGS

Many environmental factors and pathophysiological conditions inhibit or stimulate the activity of drug-metabolizing enzymes, and hence may alter the outcome of a therapeutic regimen. **Pharmacogenetics**, the immaturity of drug-metabolizing enzyme systems, and drug–drug interactions are a few of the factors that have been shown to alter drug metabolism.

Pharmacogenetics

Pharmacogenetics represents the study of the **hereditary variation in the handling of drugs**. Pharmacogenetic abnormalities may be entirely innocuous, until the affected individual is challenged with particular drugs. The hyposensitivity and resistance of certain individuals to **coumarin** anticoagulants and the hypersensitivity of patients with **Down's syndrome** to atropine most probably stem from abnormalities in their respective receptor sites. **Acatalasia** and the decrease in the activities of **pseudocholinesterase**, **acetylase**, and **glucose 6-phosphate dehydrogenase** are a few examples of enzymatic deficiencies that can lead to adverse reactions that are mild to very severe.

Liver Disease

The liver is the principal metabolic organ, and hepatic disease or dysfunction may impair drug elimination. Any alteration in the serum albumin or bilirubin levels and in the prothrombin time indicates impaired liver function. Similarly, skin bruising and bleeding tendency indicate decreased production of clotting factors by the liver.

The Influence of Age

Drug metabolism is qualitatively and quantitatively very deficient in newborns. For example, chloramphenicol, when used injudiciously, may cause **gray syndrome**.

The mechanism of chloramphenicol toxicity is apparently the failure in the newborn to conjugate **chloramphenicol** with glucuronic acid, due to inadequate activity of hepatic **glucaronyl transferase**. This, in combination with inadequate renal excretion of the drug in the newborn, results in a higher-than-expected plasma level of chloramphenicol. Therefore, a newborn should receive doses of chloramphenicol not greater than 25 to 50 mg/kg of body weight.

Elderly people are also prone to toxicity from numerous drugs, including cardiac glycosides. A dose of **digitoxin**, which may be totally therapeutic and innocuous at the age of 60, may produce severe toxicity and even death at the age of 70. The abilities of the liver to metabolize drugs and of the kidney to excrete drug metabolites decline with aging.

Enzyme Induction and Inhibition

The activities of microsomal drug-metabolizing enzymes in humans can be enhanced by altering the levels of endogenous hormones such as androgens, estrogens, progestational steroids, glucocorticoids, anabolic steroids, norepinephrine, insulin, and thyroxine. This effect can also be elicited by the administration of exogenous substances such as drugs, food preservatives, insecticides, herbicides, and polycyclic aromatic hydrocarbons. This increase in the activities of drug-metabolizing enzymes appears to stem from an elevated rate of synthesis of the enzyme protein; hence, it is truly an enzyme-induction phenomenon.

Liver microsomal enzyme inducers that are lipid soluble at the physiological pH can be classified into two general groups. Some, like phenobarbital, tend to stimulate all enzymes; others, such as **3-methylcholanthrene**, tend to be selective. The administration of phenobarbital increases the amounts of **NADPH-cytochrome C reductase** and P-450, and the rate of P-450 reduction. In contrast, the administration of 3-methylcholanthrene increases the amount of P-450, but neither the activity of NADPH-cytochrome C reductase nor the rate of P-450 reduction.

Clinical Implications of Enzyme Induction and Inhibition

Patients are often given several drugs at the same time. The possibility that one drug may accelerate or inhibit the metabolism of another drug should always be kept in mind. When this phenomenon occurs, the removal of an enzyme inducer could be hazardous. The following examples reveal the consequence of enzyme induction. Phenylbutazone is an analgesic, antipyretic, uricosuric, and anti-inflammatory agent. Among its side effects are activation of peptic ulcer and gastrointestinal hemorrhage. If one gives a dog large amounts of phenylbutazone, side effects such as vomiting and diarrhea with bloody stool ensue. However, if phenylbutazone treatment is continued for several days, these side effects disappear. In this case, phenylbutazone "induces" its own hydroxylation, which results in a lower plasma level of the drug and ultimately the absence of the side effects. Long-term treatment with phenylbutazone and many other drugs should be expected to result in decreased effectiveness and toxicity.

Patients who are on anticoagulant therapy may suffer severe hemorrhage several days after discharge from the hospital. Often these patients are sedated with barbiturates during their hospitalization, which tends to stimulate the enzymes that metabolize dicumarol. The abrupt withdrawal of barbiturates after discharge tends to revert the activity of the drug-metabolizing enzymes to their prebarbiturate stage, which raises the free circulating level of the anticoagulant and results in hemorrhage. Obviously, treatment with phenobarbital should prompt altering the maintenance dosage of anticoagulants.

GRAPEFRUIT JUICE ACTION ON INTESTINAL CYTOCHROME P-450 (CYP) ENZYMES

The effects of some CYP3A4 inhibitors wane with repeated administration, as they cause induction of CYP3A4 through upregulation of CYP3A messenger RNA and protein over time. However, this is not the case with grapefruit juice. Recurrent ingestion of grapefruit juice leads to a selective decrease of both CYP3A4 and CYP3A5 protein expression in enterocytes, resulting in increased drug bioavailability. Messenger RNA expression is not reduced, which suggests that this decrease in activity is not transcriptionally mediated. The mechanism of the decrease in CYP3A4 protein most likely reflects either accelerated protein degradation or reduced messenger RNA translation. It would be reasonable to suppose that one or more components of grapefruit juice cause a rapid intracellular degradation of the intestinal CYP3A4 enzyme through irreversible "suicide" inhibition. This would explain the rapid and sustained onset of inhibition by grapefruit juice. A 47% reduction in intestinal CYP3A4 concentration occurs within 4 hours of the ingestion of grapefruit juice, and grapefruit juice maintains a bioavailability-enhancing effect for up to 24 hours (see Kane and Lipsky, 2000).

Grapefruit juice also inhibits the CYP1A2 enzyme system *in vitro* but not *in vivo*. This is consistent with the understanding that the effect of grapefruit juice occurs at the level of the intestinal wall where levels of CYP2A expression are low. These CYP2A substrates studied with grapefruit juice have included **caffeine**, **theophylline**, and **coumarin** (Table 3).

Table 3 Medications with Which Grapefruit Juice Should Not Be Considered in an Unsupervised Manner

Calcium Channel Blockers
   Felodipine
   Nimodipine
   Nisoldipine
   Nitrendipine
   Pranidipine
Immunosuppressants
   Cyclosporine
   Tacrolimus

Table 3 Medications with Which Grapefruit Juice Should Not Be Considered in an Unsupervised Manner (Continued)

HMG-CoA Reductase Inhibitors[a]
  Atorvastatin
  Cerivastatin
  Lovastatin
  Simvastatin
Antihistamines
  Ebastine
  Terfenadine
Psychiatric Medications
  Buspirone
  Carbamazepine
  Diazepam
  Midazolam
  Triazolam
Prokinetics
  Cisapride
Others
  Methadone
  Sildenafil

[a]HMG-CoA = 3-hydroxy-3-methylglutaryl coenzyme A.

6

HERB–DRUG INTERACTIONS

Herbs may mimic, magnify, or oppose the effects of many drugs.

INTRODUCTION

Concurrent use of herbs may mimic, magnify, or oppose the effect of drugs. Plausible cases of herb–drug interactions include:

- Bleeding when **warfarin** is combined with **ginkgo** (*Ginkgo biloba*), **garlic** (*Allium sativum*), **dong quai** (*Angelica sinensis*), or **danshen** (*Salvia miltiorrhiza*).
- **Mild serotonin syndrome** in patients who mix **St. John's wort** (*Hypericum perforatum*) with serotonin-reuptake inhibitors.
- Decreased bioavailability of **digoxin, theophylline, cyclosporin**, and **phenprocoumon** when these drugs are combined with **St. John's wort**.
- Induction of **mania** in depressed patients who mix **antidepressants** and **Panax ginseng**.
- Exacerbation of **extrapyramidal effects** with **neuroleptic drugs** and **betel nut** (*Areca catechu*).
- Increased risk of hypertension when **tricyclic antidepressants** are combined with **yohimbine** (*Pausinystalia yohimbe*).
- Potentiation of oral and topical effects of **corticosteroids** by **licorice** (*Glycyrrhiza glabra*).
- Decreased blood concentration of **prednisolone** when taken with the Chinese herbal product **xaio chai hu tang** (*sho-saiko-to*).
- Decreased concentrations of **phenytoin** when combined with the Ayurvedic syrup **shankhapushpi**.

Furthermore, **anthranoid-containing** plants, including **senna** (*Cassia senna*) and **cascara** (*Rhamnus purshiana*), and soluble fibers, including **guar gum** and **psyllium**, can decrease the absorption of drugs (Fugh-Berman, 2000; and Tables 4 and 5).

Table 4 Herbs Interacting with Drugs

Herbs and Drug(s)	Results of Interaction	Comments
Betal nut (*Areca catechu*)		
Flupenthixol and procyclidine	Rigidity, bradykinesia, jaw tremor	Betal contains **arecoline**, a cholinergic alkaloid
Fluphenazine	Tremor, stiffness, akithesia	
Prednisone and salbutamol	Inadequate control of asthma	Arecoline challenge causes dose-related bronchoconstriction in patients with asthma
Chili pepper (*Capsicum* spp.)		
ACE inhibitor	Cough	Capsaicin depletes substance P
Theophylline	Increased absorption and bioavailability	
Danshen (*Salvia miltiorrhiza*)		
Warfarin	Increased INR, prolonged PT/PTT	Danshen decreases elimination of warfarin
Devil's claw (*Harpagophytum procumbens*)		
Warfarin	Purpura	—
Dong quai (*Angelica sinensis*)		
Warfarin	Increased INR and widespread bruising	Dong quai contains coumarins
Eleuthero or **Siberian ginseng** (*Eleutherococcus senticocus*)		
Digoxin	Raises digoxin concentrations	Patients exhibit unchanged ECG despite digoxin concentration of 5×2 nmol/l.
Garlic (*Allium sativum*)		
Warfarin	Increased INR	Postoperative bleeding and spontaneous spinal epidural hematoma have been reported with garlic alone; garlic causes platelet dysfunction

Herb	Drug	Effect	Comment
Ginkgo (*Ginkgo biloba*)			
	Aspirin	Spontaneous hyphema	Ginkgolides are potent inhibitors of PAF
	Paracetamol and ergotamine/caffeine	Bilateral subdural hematoma	Subarachnoid hemorrhage and subdural hematoma have been reported with the use of ginkgo alone
	Warfarin	Intracerebral hemorrhage	This effect may be an unusual adverse reaction to the drug or herb; ginkgo alone has not been associated with hypertension.
	Thiazide diuretic	Hypertension	
Ginseng (*Panax* spp.)			
	Warfarin	Decreased INR	
	Phenelzine	Headache and tremor, mania	
	Alcohol	Increased alcohol clearance	Ginseng increases the activity of alcohol dehydrogenase and aldehyde dehydrogenase
Guar gum (*Cyamopsis tetragonolobus*)			
	Metformin, phenoxymethylpenicillin, glibenclamide	Slows absorption of digoxin, paracetamol, and bumetanide; decreases absorption of metformin, phenoxymethylpenicillin, and some formulations of glibenclamide	Guar gum prolongs gastric retention

ACE = angiotensin-converting enzyme; INR = international normalized ratio; PT = prothrombin time; PTT = partial thromboplastin time; ECG = electrocardiogram; PAF = platelet-activating factor; AUC = area under the concentration/time curve.

Table 5   Herbs Interacting with Drugs

Herbs and Drug(s)	Results of Interaction	Comments
Karela or **bitter melon** (*Momordica charantia*)		
Chlorpropamide	Less glycosuria	Karela decreases glucose concentrations in blood.
Licorice (*Glycyrrhiza glabra*)		
Prednisolone	Glycyrrhizin decreases plasma clearance, increases AUC, increases plasma concentrations of prednisolone	11β-dehydrogenase converts endogenous cortisol to cortisone; orally administered glycyrrhizin is metabolized mainly to glycyrrhetinic acid
Hydrocortisone	Glycyrrhetinic acid potentiates of cutaneous vasoconstrictor response	Glycyrrhetinic acid is a more potent inhibitor of 5α-, 5β-reductase and 11β-dehydrogenase than is glycyrrhizin
Oral contraceptives	Hypertension, edema, hypokalemia	Oral contraceptive use may increase sensitivity to glycyrrhizin acid; women are reportedly more sensitive than men to adverse effects of licorice
Papaya (*Carica papaya*)		
Warfarin	Increased INR	—
Psyllium (*Plantago ovata*)		
Lithium	Decreased lithium concentrations	Hydrophilic psyllium may prevent lithium from ionizing.
St. John's Wort (*Hypericum perforatum*)		
Paroxetine	Lethargy/incoherence	
Trazodone	Mild serotonin syndrome	A similar case is described with the use of St. John's wort alone
Sertraline	Mild serotonin syndrome	
Nefazodone	Mild serotonin syndrome	
Theophylline	Decreased theophylline concentrations	
Digoxin	Decreased AUC, decreased peak and trough concentrations	Most, but not all, studies indicate that St. John's wort is a potent inhibitor of cytochrome P-450 isoenzymes

Herb / Drug	Effect	Comment
Phenprocoumon	Decreased AUC	
Cyclosporin	Decreased concentrations in serum	
Combined oral contraceptive (ethinyloestradiol and desogestrel)	Breakthrough bleeding	
Saiboku-to (Asian herbal mixture)		
Prednisolone	Increased prednisolone AUC	Contains all the same herbs as sho-saiko-to, and *Porio cocos*, *Magnolia officinalis*, and *Perillae frutescens*
Shankhapushipi (Ayurvedic mixed-herb syrup)		
Phenytoin	Decreased phenytoin concentrations, loss of seizure control	Multiple coadministered doses (but not single doses) decreased plasma phenytoin concentrations; single doses decreased the antiepileptic effect of phenytoin; Shankhapushpi is used to treat seizures
Sho-saiko-to or Xiao chai hu tang (Asian herb mixture)		
Prednisolone	Decreased AUC for prednisolone	Contains licorice (*Glycyrrhiza glabra*), *Bupleurum falcatum*, *Pinellia ternata*, *Scutellaria baicalensis*, *Zizyphus vulgaris*, *Panax ginseng*, and *Zingiber officinale*
Tamarind (*Tamarindus indica*)		
Aspirin	Increased bioavailability of aspirin	Tamarind is used as a food and a medicine.
Valerian (*Valeriana officinalis*)		
Alcohol	A mixture of valepotriates reduces adverse effect of alcohol on concentration	—
Yohimbine (*Pausinystalia yohimbe*)		
Tricyclic antidepressants	Hypertension	Yohimbine alone can cause hypertension, but lower doses cause hypertension when combined with tricyclic antidepressants; effect is stronger in hypertensive than normotensive individuals

INR = international normalized ratio; AUC = area under the concentration/time curve.

Table 6 Stages in Hemostasis

Stage	Initiator	Response or Outcome
Vascular	Tissue injury	Vasoconstriction
Platelet	Adhesion and aggregation	Plug formation
Plasma	Fibrin generation	Coagulation

Hemostasis refers to a complex homeostatic mechanism within blood and on blood vessels that serves to maintain the patency of vessels after injury, while preserving the fluidity of blood.

POTENTIAL INTERACTIONS BETWEEN ALTERNATIVE THERAPIES AND WARFARIN

Alternative medicine therapies have become increasingly popular, and it has been estimated that one third of all Americans use herbal products. In 1997, herbal medicine sales increased nearly 59%, reaching an estimated total of $3.24 billion (see Heck et al., 2000). One particular safety concern is potential interactions of alternative medicine products with prescription medications. This issue is especially important with respect to drugs with narrow therapeutic indexes, such as warfarin. More food and drug interactions have been reported for warfarin than for any other prescription medication. Multiple pathways exist for interference with warfarin, and interactions may lead to either hemorrhage or thrombotic episodes by increasing or reducing the effect of this agent. Therefore, close monitoring of therapy and knowledge of potential interactions of herbs with warfarin are extremely important.

ANTICOAGULANTS AND THROMBOLYTIC AGENTS

The clotting of blood, which protects against hemorrhage, involves the sequential initiation, interaction, and completion of several stages in hemostasis (Table 6).

The **adhesion** and **aggregation** of platelets are mediated via the release of **adenosine diphosphate** (ADP). An extensive number of pharmacological agents such as **acetylsalicylic acid**, **indomethacin**, **phenylbutazone**, **sulfinpyrazone**, and **dipyridamole** inhibit both **platelet aggregation** and **thrombus formation**, and thus may be of value in the treatment of **thrombotic disorders**. The **formation of fibrin** itself takes place by means of a cascading group of reactions involving numerous blood-clotting factors and is accomplished in several stages (Table 7).

Hemorrhage may result from several causes:

- An abnormality or **deficiency of platelets** (thrombocytopenic purpuras)
- A **deficiency of clotting factors** (factors II, VII, IX, or X)
- **Vitamin K deficiency** (necessary for synthesizing clotting factors)
- **Liver diseases** that involve the synthesis of clotting factors

Increased clotting may occur in the presence of **thrombosis** (enhanced formation of fibrin), **stasis** and **phlebitis** (diminished circulation), or **embolism**

Table 7 Formation of Fibrin

Stage	Formation	Needed Factors or Precursors
One	Plasma thromboplastin	Hageman factor (XII), antihemophilic globulin (VIII), Christmas factor (IX), plasma thromboplastin antecedent (XI), calcium (IV), and platelet phospholipid
Two	Activated thromboplastin	Tissue thromboplastin (III), Stuart factor (X), proconvertin (VII), and calcium (IV)
Three	Thrombin	Prothrombin (II), proaccelerin (V), calcium (IV), and platelet
Four	Fibrin	Thrombin

The hemostatic process is initiated immediately after vascular injury and involves, in addition to contraction of the injured vessel, three sets of complicated, interrelated events: (1) Adhesion and accumulation of blood platelets at the site of injury. This process forms a platelet plug and is called primary hemostasis. (2) Activation of the coagulation enzyme cascade to generate a protein clot, of which the primary component is cross-linked fibrin. This process is called secondary hemostasis. (3) Activation of an enzymatic system to solubilize the fibrin clot and eventually restore blood flow. This process is called fibrinolysis.

(dislocation and lodging of blood clots). Although **heparin** is an extremely effective anticoagulant, it has certain limitations that are not shared by the newer thrombin inhibitors. As a result, these novel inhibitors may have advantages over heparin for use in certain clinical settings.

Thrombin activates platelets, converts fibrinogen to fibrin, activates factor XIII, which stabilizes fibrin, and activates factors V and VIII, which accelerate the generation of **prothrombinase**. Therefore, the inhibition of thrombin is essential in preventing and treating **thromboembolic disorders**.

AGGREGIN AND PLATELET AGGREGATION

Platelets circulate in blood without adhering to other platelets or to the endothelium. However, when the endothelial cells are perturbed, the platelets adhere and undergo a change in shape, and aggregate. **ADP** is known to induce the platelet shape change, aggregation, and exposure of fibrinogen binding sites.

The platelet surface contains **aggregin**, a membrane protein with a molecular weight of 100 kDa, which has physical and immunochemical properties that differ from those of platelet glycoprotein IIIa.

Binding to aggregin is required in order for epinephrine-induced platelet aggregation to take place. In turn, **epinephrine** increases the affinity of ADP for its receptor. **Thrombin** stimulates platelet aggregation independent of ADP, but, by raising the level of calcium in the cytoplasm, it activates platelet **calpain**, which in turn cleaves aggregin (Figure 6).

VITAMIN E AND PLATELET AGGREGATION

α-Tocopherol, a natural antioxidant, inhibits platelet aggregation and release. The effect of vitamin E is due to a reduction in **platelet cyclooxygenase activity**

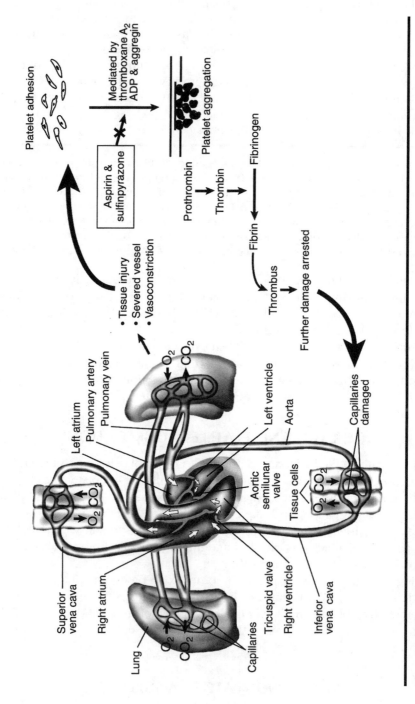

Figure 6   The role of platelets in arresting tissue injury. ADP = adenosine diphosphate.

and inhibition of **lipid peroxide formation**. It is believed that supplementing the diet with vitamin E could play a role in the treatment of thromboembolic disease, especially if it is given in conjunction with an inhibitor of platelet aggregation.

PLATELET-ACTIVATING FACTOR

A platelet-activating factor, **1-0-alkyl-2-(R)acetyl-sn-glyceryl-3-phosphocholine**, is released in the presence of shock and ischemia. The platelet-activating factor antagonist can protect the heart and brain against ischemic injury.

PLATELET-INHIBITING DRUGS

A combination of **acetylsalicylic acid** and **dipyridimole** has been found to be effective in preventing myocardial reinfarction and occlusion of aortocoronary grafts.

FIBRINOLYSIS

In fibrinolysis, **plasmin**, an endopeptidase that is converted from plasminogen by an activator, hydrolyzes fibrin, fibrinogen, factor V, and factor VIII to their inactive products. **Hageman factor** (factor XII) converts a proactivator to the active activator. Agents such as thrombin, **streptokinase**, and **urokinase** therefore enhance the formation of plasmin and hence have fibrinolytic properties. **Epsilon-aminocaproic acid** inhibits the activator-mediated formation of plasmin and hence may be used as an antidote to streptokinase-urokinase, or in a defibrination syndrome when bleeding from a mucous membrane occurs.

HIRUDIN AND THROMBIN

The saliva of the **medicinal leech** contains a battery of substances that interfere with the hemostatic mechanisms of the host. One of these compounds is **hirudin**, a potent anticoagulant, which maintains the fluidity of the ingested blood and which is the most potent inhibitor of thrombin. Upon binding to thrombin, the cleavage of fibrinogen and subsequent clot formation are prevented. The potency and specificity of hirudin suggest it as a useful antithrombin III-independent alternative to heparin for the control of thrombosis.

AGENTS THAT INTERFERE WITH COAGULATION

Therapeutic agents that interfere with blood coagulation fall into four classes:

1. **Anticoagulants**, which include **heparin** and the **coumadin-inanedione** oral anticoagulants
2. **Thrombolytic** agents such as **streptokinase, urokinase**, and **recombinant tissue-type plasminogen activator**
3. **Antiplatelet** drugs, which alter the aggregating ability of platelets
4. **Defibrinogenating** agents, which remove the fibrinogen from circulating blood

ORAL ANTICOAGULANTS

The **coumarin anticoagulants** include **dicumarol**, **warfarin sodium** (coumadin sodium), **warfarin potassium** (Athrombin-K), acenocoumarol (Sintrom), and **phenprocoumon** (Liquamar). The **inanedione derivatives** are **phenindione** (Hedulin), **diphenadione** (Dipaxin), and **anisindione** (Miradon). The pharmacological properties of oral anticoagulants are identical qualitatively, but their pharmacokinetic parameters and their toxicities vary. **Racemic warfarin sodium** is the most widely used anticoagulant. Antithrombotic drugs are used clinically either to prevent the formation of blood clots within the circulation (anticoagulant) or to dissolve a clot that has already formed (thrombolytic).

HERBS WITH COUMARIN, SALICYLATE, OR ANTIPLATELET DRUGS

Several natural products contain substances that have coumarin, salicylate, or exhibit antiplatelet properties. Therefore, a theoretical risk for potentiation of the pharmacological activity of warfarin exists when these herbs are taken with warfarin. Herbs thought to contain coumarin or coumarin derivatives include:

- **Angelica root**
- **Arnica flower**
- **Anise**
- **Asafoetida**
- **Celery**
- **Chamomile**
- **Fenugreek**
- **Horse chestnut**
- **Licorice root**
- **Lovage root**
- **Parsley**
- **Passionflower herb**
- **Quassia**
- **Red clover**
- **Rue**

Meadowsweet, **poplar**, and **willow bark** contain high concentrations of salicylates, while **bromelain**, **clove**, **onion**, and **turmeric** have been reported to exhibit antiplatelet activity. **Borage seed oil** contains **γ-linoleic acid**, which may increase coagulation time. **Bogbean** has been noted to demonstrate hemolytic activity, and **capsicum** has been reported to cause hypocoagulability. There have been no documented case reports of an interaction of warfarin with any of these herbs. However, patients taking any products containing these herbs concurrently with medications that have anticoagulant effects, such as warfarin, should be closely monitored for signs or symptoms of bleeding.

 Sweet clover also contains coumarin derivatives and therefore poses an increased risk of bleeding if given with warfarin. There have been no reports of an interaction between sweet clover and warfarin or hemorrhagic disease in humans. However, several cases of severe hemorrhage and death have been reported in cattle.

Feverfew

Feverfew (*Tanacetum parthenium*) is commonly used for the **treatment of migraine** headaches, arthritis, and various type of allergies. This herb is thought to exert its pharmacological activity by inhibiting serotonin release, histamine release, prostaglandin synthesis, and platelet release and aggregation. Several studies have shown feverfew to interfere with hemostasis and platelet aggregation by neutralizing platelet sulfhydryl groups, as well as preventing prostaglandin synthesis. **Parthenolide**, one of the many **sesquiterpene lactone** constituents of feverfew extract, has been shown to exert the greatest pharmacological activity.

Garlic

Garlic (*Allium sativum*) is thought to provide several cardiovascular benefits, such as blood pressure lowering, serum lipid lowering, and antithrombotic activity. **Garlic oil** has been reported to interrupt **thromboxane synthesis**, thereby inhibiting platelet function.

Ginger

Ginger (*Zingiber officinale*), promoted for use in motion sickness and arthritis, has been reported to reduce platelet aggregation through the inhibition of **thromboxane synthetase**. Ginger supplements, containing amounts of ginger much greater than regularly found in food products, may lead to an increased risk of bleeding when taken with warfarin.

Ginkgo

Ginkgo (*Ginkgo biloba*) is a common herbal product available in the United States and is advertised to improve **cognitive function**. Ginkgolide B, one component of ginkgo, inhibits platelet-activating factor by displacing it from its receptor-binding site, resulting in reduced platelet aggregation. Several cases of bleeding thought to be secondary to ginkgo ingestion have been reported.

Coenzyme Q_{10}

Coenzyme Q_{10} (also known as **ubiquinone** or **ubidecarenone**), while not an herb, is a provitamin found in the mitochondria of plant and animal cells. It is involved in electron transport and may act as a **free-radical scavenger**, an **antioxidant**, or a **membrane stabilizer**. Coenzyme Q_{10} supplementation is primarily promoted to treat a variety of cardiovascular disorders, including:

- Heart failure
- Hypertension
- Stable angina
- Ventricular arrhythmias

Many patients with these conditions may also be prescribed warfarin. Coenzyme Q_{10} is structurally related to **menaquinone** (vitamin K) and may have procoagulant effects.

Danshen

Although not commonly used in the United States, danshen (the root of *Salvia miltiorrhiza*), also known as **tan seng**, is a very popular herb recommended in the Chinese community for various cardiovascular diseases. The pharmacological effects of danshen have been described primarily *in vitro* and in animals and include:

- Hypotensive effects
- Positive inotropic effects
- Coronary artery vasodilation
- Inhibition of platelet aggregation

The available evidence contraindicates concurrent use of danshen and warfarin.

Devil's claw

Devil's claw (*Harpagophytum procumbens*) is an expensive herbal product that has been promoted for use as an analgesic in the treatment of arthritis, gout, and myalgia. Until more is known about this possible interaction, patients taking warfarin should be advised to avoid devil's claw.

Ginseng

Three ginseng species — **American ginseng** (*Panax quinquefolius*), **Oriental ginseng** (*Panax ginseng*), and **Siberian ginseng** (*Eleutherococcus senticosus*) — have been promoted as enhancing energy, reducing the effects of stress, and improving mood. The active components of ginseng are known as **ginsenosides**, more than 20 of which have been identified. The pharmacological activity of each ginsenoside appears to vary depending on where the plant grew and the extraction techniques used. Also, data suggest that the ginsenoside composition varies widely among commercially available ginseng products. This variability makes it difficult to evaluate the safety and efficacy of ginseng products. Although the exact pharmacological actions of ginsenosides in humans are not fully understood, studies suggest that these substances may increase adrenal hormone synthesis, decrease blood glucose concentrations, and promote immunomodulation. Oriental ginseng (Ginsana) may antagonize the anticoagulant effects of warfarin. The possible mechanism for this interaction has not been identified, and it is not known which ginsenoside or ginsenosides may be responsible.

Green tea

Green tea (*Camellia sinensis*), also known as **Chinese tea**, is a popular beverage purported to prevent various cancers, treat gastrointestinal disorders, and enhance cognition. Although dried green tea leaves have been found to contain substantial

Table 8 Potential and Documented Interactions of Herbs and Warfarin

Potential Increase in Risk of Bleeding

Angelica root	Ginkgo
Arnica flower	Horse chestnut
Anise	Licorice root
Asafoetida	Lovage root
Bogbean	Meadowsweet
Borage seed oil	Onion
Bromelain	Parsley
Capsicum	Passionflower herb
Celery	Poplar
Chamomile	Quassia
Clove	Red clover
Fenugreek	Rue
Feverfew	Sweet clover
Garlic	Turmeric
Ginger	Willow bark

Documented Reports of Possible Increase in Effects of Warfarin
Danshen
Devil's claw
Dong quai
Papain
Vitamin E

Documented Reports of Possible Decrease in Effects of Warfarin
Coenzyme Q_{10}
Ginseng
Green tea

The coumarin anticoagulants include dicumarol, warfarin sodium (coumadin sodium), warfarin potassium (Athrombin-K), aceno-coumarol (Sintrom), and phenprocoumon (Liquamar).

amounts of **vitamin K**, brewed green tea is generally not considered a significant source of the vitamin. However, large amounts of brewed green tea may potentially antagonize the effects of warfarin.

Papain

Papain is a mixture of proteolytic enzymes found in **extract of papaya**, the fruit of the **papaya tree** (*Carica papaya*). It is taken orally in the belief that it reduces edema, inflammation, herpes zoster symptoms, diarrhea, and psoriasis symptoms. The pharmacological mechanisms by which papain may affect coagulation are not known. Patients receiving warfarin should be advised to avoid papain supplementation until further information about this potential interaction becomes available (Table 8).

Vitamin E

Vitamin E has received much publicity as one of several antioxidants that may be useful in treating a variety of disorders, including cardiovascular disease. Vitamin E may inhibit the oxidation of reduced vitamin K. Vitamin K oxidation is necessary for carboxylation of vitamin K-dependent clotting factors, which must occur for these clotting factors to be fully functional. Increased prothrombin times induced by combined vitamin E and warfarin therapy may be managed by discontinuing vitamin E, and, if necessary, by administering vitamin K.

Because nearly 70% of patients who use alternative therapies do not inform their health-care providers about these products, pharmacists and other health-care professionals should question all patients about their use of alternative therapies. Health-care professionals should remain vigilant for potential interactions between alternative therapies and prescription medications, especially medications with a narrow therapeutic index, and should report suspected interactions to the **FDA MedWatch program**. The FDA recently established the Special Nutritionals Adverse Event Monitoring System, a searchable database including information about suspected adverse events associated with dietary supplements or nutritional products. This database includes reports that have been submitted to MedWatch and can be accessed via the Internet (http://vm.cfsan.fda.gov/~dms/aems.htm1). Continued efforts by health-care professionals to recognize and report suspected interactions between prescription medications and herbal and other alternative therapies should ultimately increase knowledge and awareness of interactions and improve the quality of patient care (see Heck et al., 2000).

7

NATURAL PRODUCTS AS A RESOURCE FOR ESTABLISHED AND NEW DRUGS

THE ROLE OF HERBAL MEDICINES IN HEALTH CARE

Natural products have served as a major source of drugs for centuries, and about half of the pharmaceuticals in use today are derived from natural products. **Quinine**, **theophylline**, **penicillin G**, **morphine**, **paclitaxel**, **digoxin**, **vincristine**, **doxorubicin**, **cyclosporin**, and **vitamin A** all share two important characteristics: they are cornerstones of modern pharmaceutical care and they are all natural products. The use of natural substances, particularly plants, to control diseases is a centuries-old practice that has led to the discovery of more than half of all "modern" pharmaceuticals.

Documentation of the use of natural substances for medicinal purposes can be found as far back as 78 A.D., when **Dioscorides** wrote "**De Materia Medica**," describing thousands of medicinal plants. This treatise included descriptions of many medicinal plants that remain important in modern medicine, not because they continue to be used as crude drug preparations, but because they serve as the source of important pure chemicals that have become mainstays of modern therapy. A few examples will be cited here.

The positive benefits of extracts of two species of **Digitalis purpurea** (foxglove and *lanata*) were recognized long before the active constituents were isolated and characterized structurally. The cardiac glycosides, which include **digoxin**, **digitoxin**, and **deslanoside**, exert a powerful and selective **positive inotropic action** on the cardiac muscle.

One of the most cited examples of important natural product-derived drugs is the neuromuscular blocker, **d-tubocurarine**, derived from the South American plant, **curare**, which was used by South American Indians as an **arrow poison**. Tubocurarine led to the development of **decamethonium**, which, although structurally dissimilar to **tubocurarine**, was nevertheless synthesized based on the presumption at the time that tubocurarine contained two quaternary nitrogens. Similarly, the synthetic local anesthetics such as **lidocaine**, **benzocaine**, and **dibucaine**, were synthesized to mimic the nerve-blocking activity of **cocaine**, a natural alkaloid obtained from the leaves of **Coca eroxylum**, but without the adverse side effects that have led to the abuse of cocaine.

The **opium alkaloids codeine** and **morphine** served as models for the synthesis of **naloxone**, an important analogue used to treat and diagnose opiate addicts and also led to the discovery of "endogenous opioids" (**enkephalins** and **endorphins**). Similarly, **Δ⁹-tetrahydrocannabinol** (THC), the component of **Cannibas sativa** responsible for the central nervous system (CNS) effect, has also been found to reduce nausea associated with cancer chemotherapy.

There can be no argument that the **antibiotics** are among the most important classes of therapeutic agents and have had enormous impact on both life expectancy and quality of life. With the discovery of the natural penicillins as secondary metabolites of species of the fungus **Penicillium**, the course of medical history was dramatically changed and the antibiotic era was introduced.

Another therapeutic area where natural products have had a major impact on longevity and quality of life is in the **chemotherapy of cancer**. In fact, most of the major anticancer drugs are natural products from plants or microorganisms. Examples include such important anticancer drugs as **bleomycin, doxorubicin, daunorubicin, vincristine, vinblastine, mitomycin, streptozocin**, and now include the recent addition of **paclitaxel** (Taxol™), **ironotecan** (a camptothecin derivative), and **etoposide** and **tenoposide** (podophyllotoxin derivatives).

Some of the most exciting natural products discovered in recent years are the **cholesterol-lowering agents** derived from fungi. These drugs act by inhibition of **3-hydroxy-3-methylglutaryl coenzyme A reductase (HMG-COA reductase)**, an enzyme critical in the biosynthesis of **cholesterol**. The first of the HMG-COA reductase inhibitors were isolated from **Pencillium sp**.

The word *healing* has its roots in the Greek, *holos*, the same word that has given us *whole* and *holistic*. The emotions through life and spiritual flow are as important to health as is the state of organs and tissues within the body. Whether we are concerned about being healthy, regaining health, or moving to greater health, the whole of the being, physical, mental, and spiritual, is involved in the process.

The therapeutic philosophy for plant use varies, but for thousands of years plants have demonstrated their efficiency as healing agents. We find them within the Indian **Ayurvedic system**, and in Chinese medicine alongside **acupuncture** and other techniques; they play a very important role in the spiritual healing ecology of the **Native North Americans**; and we see their constituents being utilized as a source of drugs in **"orthodox" medicine**.

DRUG THERAPY IN CHINESE TRADITIONAL MEDICINE

The origins and development of **Chinese traditional medicine** (CTM), attributed to **Shen Lon** are based in the accumulation of lifetimes of experience and practice of the Chinese people to maintain health and treat disease. Thus, it has a long history of more than 2000 years. Similarly, in ancient Europe, *materia medica* also have been employed for many centuries. The use of medicinal herbs was mentioned in the ancient Egyptian medical **papyri** that were discovered by **Ebers** in 1862 and translated into English in 1937 by Ebbell. More than 800 prescriptions in **Ebers Papyrus** emphasized that **medicines were originally foods**. Thus, humans from different civilizations discovered medicinal substances from food in

Table 9   Five-Element Theory in Chinese Traditional Medicine

Indicator	1	2	3	4	5
Taste	Sour	Bitter	Sweet	Pungent	Salty
Color	Blue (dark)	Red	Yellow	White	Black
Evils	Wind	Heat	Damp	Dry	Cold
Organs	Liver	Heart	Spleen	Lung	Kidney
Emotions	Anger	Joy	Worry	Grief	Fear
Senses	Eyes	Tongue	Mouth	Nose	Ear
Expressions	Shouting	Laughing	Singing	Wailing	Groaning
Orientations	East	South	Middle	West	North
Elements	Wood	Fire	Earth	Gold	Water

Each element has the correlative connection. For example, impaired function of the liver may produce problems in the eyes; excessive joy is harmful to the "heart," which included the functions of the brain; also, herbs with a black color would be likely to affect the function of the kidney.

the same way. An additional stimulus to the use of herbal medicines in the 16th century was the introduction of novel remedies from the East, such as **nutmeg** from the East Indies, **purgative rhubarb** from India, and **camphor** from Japan (Cheng, 2000).

Chinese people continue to believe in CTM, and herbal medications have remained very popular in Chinese societies around the world. One of the main reasons may be the relatively high incidence of side effects with Western medicine. Another reason that Chinese medicine is still used in therapeutics may be that it is used as a last resort, when Western medicine is too toxic or unable to provide the expected benefit.

The integrity of the human body and the relationship of the body with the environment are emphasized in CTM. The **yin–yang theory** and the **five-element theory** are the ancient cosmology to explain the nature, materialism, and dialectics in China. The five-element or phase theory is named ***Wun Shing*** in Chinese, which means five matters. As shown in Table 9, the five-element theory can be applied to widely different aspects of humans and the environment, including organs and emotions. The colors in the five-element theory can be applied to symptoms when making a diagnosis, and depending on the tastes and sometimes the colors, herbs were also classified by the five-element theory, which determined their usage in CTM. For example, an herb with a black color or salty taste would be assumed to affect the function of the kidney because they are in the same group of elements (see Table 9). Similarly, an herb with a red color or bitter taste would be useful to modify the function of the heart.

Another theory, that of **Zhang-Fu**, in CTM is mainly related to the pathophysiology of disease, and in the handling of diseases in CTM, the application of the theories of **Bian-Zheng (diagnosis)** and **Lun-Zhi (treatment)** is used. The use of herbal medications over the centuries has been refined by expert practitioners using their practical experience. One of the famous experts was **Chang Chon-Zhing**, who practiced during the **Jin** and **Yuan dynasties** (1271–1368).

Prescription of Multiple Herbs in Chinese Traditional Medicine

According to the experience of experts, several prescriptions were recorded in the ancient books of CTM. Therapeutic principles could be divided into eight points, which included the **diaphoretic method (Han Fah)**, the **emetic method (Tu Fah)**, the **purgative method (Xia Fah)**, the **regulatory method (Tee Fah)**, the **warming method (Wen Fah)**, the **heat-purging method (Ching Fah)**, the **resolving method (Xiao Fah)**, and the **tonic method (Bu Fah)**. Except for the emetic method, the other methods were widely used in the clinical practice of CTM. Sometimes, two or more methods were employed together to achieve the desired results. All the methods were applied with flexibility in accordance with individual cases (Cheng, 2000).

The prescription of herbs, named **Fang-Chi** in Chinese, may comprise one herb only or many herbs. The prescription is sometimes named depending on the number of herbs that are employed, such as **"7-component prescription" (Chi-Fang)** and **"10-component prescription" (Shi-Jie)**. The "large" prescriptions refer to the use of a large number of herbs to treat diseases due to virulent pathogens. However, experts in CTM can modify the old prescription by addition or deletion of herbs or modification of doses for the better treatment of the individual's diseases. This is called **Zhia-Gean Fang**. In practice, the experts in CTM revised the classic prescriptions depending on the symptoms of the patients. Also, the differences in the active principles in herbs due to variation from season or location could also be taken into account by this modification.

Herbs contained in the prescription are mainly classified according to four properties, depending on the presumed pharmacological activity. The major one is known as the "**chief**" or the king (**Zhing**) in the prescription. The minor ones are called "**adjuvant**" or "**minister**," which is named **Chen** in Chinese. The helper-like herbs are known as "**assistant**" and named **Zhou**, and the additive herbs or "**guider**" are called **Shi** in Chinese. Normally in most prescriptions of CTM, the four components of **Zhing**, **Chen**, **Zhou**, and **Shi** are included. However, each property may not be represented by only one herb. In any prescription, in practice two or more herbs may work as the major component or Zhing. Furthermore, in the "small" prescription or monoprescription, one herb may work as Zhing and Chen or more components. Also, an herb may work as Chen in one prescription, but it is used as Shi in another prescription. Therefore, the properties of herbs are seen as relative but not absolute, which is a key point of consideration in CTM. This principle could be explained by the drug interactions in the use of herbs (Cheng, 2000).

Pharmacological Classification of Herbs According to CTM

Classification of herbs in CTM is different from the modern pharmacological view. From the physical properties, either color or taste, each herb can be linked with the theory of five elements (Table 10), named **Woo-Wei** in Chinese. Also, the properties of herbs can be classified according to the principle named four-chi (**Su-Chi**) in Chinese. Four-chi means the four properties of herbs that include cold, heat, warm, and cool. Depending on the yin–yang theory, an herb with a cold property is suitable to treat a symptom, which is considered as "hot." For

Table 10   Pharmacological Classification of Chinese Herbs

Herbs for relieving exterior syndrome
Herbs for eliminating heat
Herbs for purgation (cathartics)
Herbs for eliminating wind dampness
Herbs for dispelling dampness
Diuretics
Herbs for warming the interior
Herbs for regulation of chi
Digestives
Anthelmintics
Hemostatics
Herbs for activating blood circulation and removing blood stasis
Expectorants, antitussives, and dyspnea-relieving herbs
Sedatives
Herbs for calming the liver and suppressing wind
Herbs for promoting resuscitation
Tonics
Astringents

Limited toxicological data are available on medicinal plants. However, there exists a considerable overlap between those herbs used for medicinal purposes and those used for cosmetic or culinary purposes, for which a significant body of information exists. For many culinary herbs used in herbal remedies, there is no reason to doubt their safety, providing the intended dose and route of administration is in line with their food use. When intended for use in larger therapeutic doses the safety of culinary herbs requires reevaluation.

example, the herb ***Rhizoma zingiberis*** belongs to the "hot" group and is therefore suitable for the treatment of diarrhea, which is considered a disorder of "cold."

Two other principles used in the practice of CTM are even more difficult to understand and reconcile with the practice of modern Western medicine. These are the **attributive meridians (*Qui Jing*)** and the direction of the herb's action. The direction of the herb is considered important in its application. **Ascending (*sheng*)**, **descending (*jiang*)**, **floating (*fu*)**, and **sinking (*chen*)** are the terms used to indicate the direction of herbs. According to the direction of herbs, practitioners can make the best use to reach the pathogenic factors. Herbs with a descending property are appropriate to treat vomiting and hiccups due to "abnormal ascent of the stomach." Also, herbs with a floating property can disperse and drive out pathogenic factors to treat the disorders on the surface of the human body. Moreover, the direction of herbs corresponds with the four-chi (**Su-Chi**) and the five-element (**Woo-Wei**) theories in the practice of CTM. Herbs with the tastes of acrid and sweet and the properties of warm and hot are mostly ascending and floating in action. Similarly, herbs with the tastes of bitter, sour, and salty and the properties of cool and cold are mostly sinking and descending in action. The flowers and leaves of plants and the herbs that are light in weight are mostly ascending and floating, while heavy herbs such as seeds and minerals are mostly sinking and descending in action.

Another principle applied in the use of herbs is the attributive meridians (**Qui Jing**) that are related with the theory of organs and meridians (**Zhang-Fu**). The meridians are considered to connect the interior and exterior parts of the human body. It will be more helpful if the herb can meet the right meridian. However, it does not mean that a disorder from a known meridian must be treated by an herb that can only enter this meridian.

In clinical application, herbs are classified mainly into 18 groups, as shown in Table 10. Some of these are similar to the therapeutic groups used in Western medicine, such as **digestives**, **anthelmintics**, **purgatives** (cathartics), **diuretics**, **expectorants**, and **antitussives**, which are widely used in herbal medication. However, some groups have no equivalent in the Western system, such as herbs for relieving exterior syndrome, herbs for eliminating heat, herbs for eliminating wind dampness, herbs for dispelling dampness, herbs for warming the interior, and the herbs for regulation of Chi. Also, hemostatics, dyspnea-relieving herbs, and the herbs used for activating blood circulation and removing blood stasis may have some similarity to the Western drugs used for effects on the cardiovascular system, but tonics and astringents are applied in a different way. Although the terms used in the classification may be similar, the main principle in the use of herbs is quite different from that of modern pharmacology.

The potential toxicity of herbs has been emphasized in CTM for a long time. The term **Du-Jing** in Chinese means the consideration of harmful effects of herbs. The degree of toxicity is generally defined as "**very poisonous**" and, "**slightly poisonous**" to promote the careful handling of certain herbs. It is important for the practitioner to obtain a full understanding of CTM, including the compatibility, indications, and contraindications of herbs (Cheng, 2000).

CLASSIC PLANT SUBSTANCES

Some plant-derived drugs come from lower organisms such as a fungi (e.g., the antihyperlipidemic agent **lovastatin** from *Aspergillus terreus* and the immunosuppressant **cyclosporin** from *Beauveria nivea*, whereas others have been obtained from higher plants (see Tables 11 and 12).

Table 11 Proposed Therapeutic Values of Some Herbal Medicines

Aloe Vera	Skin moisturizer; minor cuts; sunburn	Sometimes promoted as a laxative or for stomach disorders, but internal use is hazardous
Anise	Cough soother; digestive aid	Generally safe
Arnica	Anti-inflammatory painkiller for sprains, bruises; muscle aches	For external use only; causes cardiac toxicity if ingested; stop using if arnica liniment causes dermatitis
Chamomile	Tea: sedative, digestive aid, menstrual cramps; compress: arthritis, skin disorders	Generally safe in small amounts

Table 11   Proposed Therapeutic Values of Some Herbal Medicines (Continued)

Comfrey	Skin wounds, bruises, burns, and boils; reducing swelling of sprains	Should not be taken internally because of possible liver toxicity
Dandelion	Laxative; tonic; diuretic	Generally safe
Echinacea	Compress: skin sores and stings; tea: bladder infections, fever, headache	Generally safe in small doses
Eucalyptus	Nasal and sinus congestion; coughs	Oil irritating to eyes and mucous membranes
Evening primrose	Painkiller; sedative; autoimmune disorders	Can worsen bleeding disorders
Fenugreek	Expectorant; digestive tonic; laxative; poultice: soothing of boils and skin ulcers	Generally safe
Fennel	Digestive aid	Generally safe
Feverfew	Painkiller	Generally safe; may cause mouth sores when chewed
Garlic	Antiseptic; reducing of high blood pressure and high blood cholesterol	Generally safe, but raw garlic may irritate skin and mucous membranes; should not substitute for prescribed drugs
Ginseng	General tonic; blood thinner; aphrodisiac	Generally safe, but many claims of its healing power are unproved
Licorice	Irritated skin and mucous membranes; ulcers; digestive aid	Generally safe; excessive use raises blood pressure
Peppermint	Digestive aid	Generally safe
Uva-ursi	Diuretic; cystitis; kidney and bladder stones	Generally safe, but should not substitute for antibiotics
Valerian	Sleep aid; relieving of stress	Generally safe, but should not be used during pregnancy

Some culinary herbs contain potentially toxic constituents. The safe use of these herbs is ensured by limiting the level of constituent permitted in a food product to a level not considered to represent a health hazard. *Apiole* — The irritant principle present in the volatile oil of **parsley** is held to be responsible for the abortifacient action. Apiole is also hepatotoxic and liver damage has been documented as a result of excessive ingestion of parsley, far exceeding normal dietary consumption, over a prolonged period. *β-Asarone* — **Calamus rhizome oil** contains **β-asarone** as the major component, which has been shown to be carcinogenic in animal studies. Many other culinary herbs contain low levels of β-asarone in their volatile oils and therefore the level of β-asarone permitted in foods as a flavoring is restricted. *Estragole (Methylchavicol)* — Estragole is a constituent of many culinary herbs but is a major component of the oils of **tarragon, fennel, sweet basil**, and **chervil**. Estragole has been reported to be carcinogenic in animals. The level of estragole permitted in food products as a flavoring is restricted. *Safrole* — Animal studies involving safrole, the major component of **sassafras oil**, have shown it to be hepatotoxic. The permitted level of safrole in foods is 0.1 mg/kg.

Table 12 Examples of Classic Plant Drugs from Higher Plants with and without a Correlation between Clinical Action and Traditional Use

Drug Substance	Clinical Action/Use	Botanical Source
With Correlation		
Atropine	Anticholinergic	*Atropa belladonna*
Caffeine	CNS stimulant	*Camellia sinensis*
Camphor	Rubefacient	*Cinnamomum camphora*
Cocaine	Local anesthetic	*Erythroxylum coca*
Codeine	Analgesic/antitussive	*Papaver somniferum*
Colchicine	Antigout	*Colchicum autumnale*
Digitoxin	Cardiotonic	*Digitalis purpurea*
Digoxin	Cardiotonic	*Digitalis lanata*
Emetine	Amoebicide	*Cephaelis ipecacuanha*
Ephedrine	Sympathomimetic	*Ephedra sinica*
Gossypol	Male contraceptive	*Gossypium spp.*
Hyoscyamine	Anticholinergic	*Hyoscyamus niger*
Kawain	Tranquilizer	*Piper methysticum*
Methoxsalen	Psoriasis/vitiligo	*Ammi majus*
Morphine	Analgesic	*Papaver somniferum*
Noscapine	Antitussive	*Papaver somniferum*
Ouabain	Cardiotonic	*Strophanthus gratus*
Physostigmine	Cholinesterase inhibitor	*Physostigma venenosum*
Pilocarpine	Parasympathomimetic	*Pilocarpus jaborandi*
Podophyllotoxin	Topical treatment for condylomata acuminata	*Podophyllum peltatum*
Quinine	Antimalarial	*Cinchona ledgeriana*
Reserpine	Antihypertensive	*Rauwolfia serpentina*
Scopolamine (hyoscine)	Sedative	*Datura metel*
Sennosides A and B	Laxative	*Cassia spp.*
Theophylline	Bronchodilator	*Camellia sinensis*
Tubocurarine	Muscle relaxant	*Chondodendron tomentosum*
Yohimbine	Aphrodisiac	*Pausinystalia yohimbe*
Without Correlation		
Chymopapain	Chemonucleolysis	*Carica papaya*
Galanthamine	Cholinesterase inhibitor	*Lycoris squamigera*
Levodopa	Antiparkinsonian	*Mucuna deeringiana*
Menthol	Rubefacient	*Mentha spp.*
Methyl salicylate	Rubefacient	*Gaultheria procumbens*
Nordihydroguaiaretic acid	Antioxidant (lard)	*Larrea divaricata*
Quinidine	Antiarrhythmic	*Cinchona ledgeriana*
Tetrahydrocannabinol	Antiemetic	*Cannabis sativa*
Vinblastine	Anticancer	*Catharanthus roseus*
Vincristine	Anticancer	*Catharanthus roseus*

Table 12   Examples of Classic Plant Drugs from Higher Plants with and without a Correlation between Clinical Action and Traditional Use (Continued)

CNS = Central nervous system.

Quality assurance of herbal products may be ensured by control of the herbal ingredients and by means of good manufacturing practice. Some herbal products have many herbal ingredients with only small amounts of individual herbs being present. Chemical and chromatographic tests are useful for developing finished product specifications. Stability and shelf life of herbal products should be established by manufacturers. There should be no differences in standards set for the quality of dosage forms, such as tablets or capsules, of herbal remedies from those of other pharmaceutical preparations.

Over 120 of our present medicines are derived directly or indirectly from higher plants (Tables 11 & 12; De Smet, 1997).

Caffeine

Caffeine continues to be used therapeutically in analgesic combination preparations. It has superseded theophylline as the drug of choice for the treatment of neonatal apnea because of its ease of administration and its more predictable plasma concentrations.

Capsaicin

This pungent product is responsible for the intense irritant effects of topical *Capsicum* preparations. Repeated application of capsaicin can deplete and prevent reaccumulation of **substance P**, an endogenous mediator of pain impulses from the periphery to the CNS. Since the early 1990s, capsaicin cream has been available in the United States as an approved drug for relief of **postherpetic neuralgia** and pain due to **diabetic neuropathy** or **osteoarthritis**.

β-Carotene

In clinical trials, no preventive effect of β-carotene supplementation on cancer or cardiovascular disease could be demonstrated in a well-nourished population. This apparent lack of efficacy contrasts starkly with the promising results of numerous basic studies and epidemiological observations, which suggested that supplementation with this antioxidant was likely to have a protective effect.

Colchicine

This alkaloid still has a role in the management of **acute gout**.

Nordihydroguaiaretic Acid

Nordihydroguaiaretic acid occurs as the mesoform in the resinous exudate of the creosote bush (*Larrea divaricata* syn. *L. tridentata*). Under the generic name of **masoprocol**, it has become available in the United States as a 10% cream for topical treatment of **actinic keratoses**. Masoprocol is a potent inhibitor of **5-lipoxygenase** but the exact mechanism of its antiproliferative effect on keratinocytes remains to be established.

Pilocarpine

Oral pilocarpine is moderately effective for **xerostomia** after irradiation of head and neck cancer. It may also be beneficial in relieving refractory xerostomia associated with chronic graft-vs.-host disease.

Quinine

The alkaloid quinine occurs naturally in the bark of the **_Cinchona_** tree. Apart from its continued usefulness in the treatment of **malaria**, it can also be taken for the relief of **nocturnal leg cramps**.

Scopolamine (Hyoscine)

Transdermal scopolamine (hyoscine) was introduced in the 1980s as a convenient alternative for the prevention of **motion sickness**. It is also effective in the reduction of nausea and vomiting, after **ear surgery**, but it was not found to be useful for the prevention of **vasovagal syncope**.

Tetrahydrocannabinol

Δ^9-Tetrahydrocannabinol is the major psychoactive cannabinoid in **marijuana** (_Cannibis sativa_). Its synthetic form, **dronabinol**, became available in the United States in 1985 as an antiemetic for patients receiving **emetogenic chemotherapy**. However, it is seldom used as a first-line antiemetic because of its serious CNS adverse effects, and its use is usually limited to patients who have a low tolerance for or minimal response to other antiemetic drugs.

Theophylline

The mode of action of theophylline is different from that of β_2-adrenergic agonists and anticholinergic agents, so it still deserves a niche in the management of selected patients with **asthma** or **chronic obstructive pulmonary disease**.

ANTIMALARIAL AGENTS

After the introduction of synthetic antimalarial agents such as **chloroquine** and **mefloquine**, the use of the **_Cinchona_ alkaloid** quinine declined. However, because of the widespread emergence of chloroquine-resistant and multiple-drug-resistant strains of malarial parasites, its use has become firmly reestablished. Quinine is considered to be the drug of choice for severe chloroquine-resistant malaria due to _P. falciparum_. In the United States, the related alkaloid **quinidine** is recommended for this purpose because of its wide availability there in its use as an antiarrhythmic agent. In many clinics in the tropics, quinine is the only effective treatment for severe malaria, but unfortunately, decreasing sensitivity of *Plasmodium falciparum* to quinine has already been reported from Southeast Asia.

In 1972, a Chinese search for new antimalarial drugs from traditional sources resulted in the isolation of **artemisinin** (qinghaosu) from the leaves of ***Artemisia annua*** (qinghao). This herb has been described in Chinese *materia medica* texts as a treatment for febrile illnesses since the 4th century A.D. Artemisinin is a **sesquiterpene lactone peroxide** with marked therapeutic activity against malaria. It is effective against *P. vivax* and against chloroquine-sensitive and chloroquine-resistant strains of *P. falciparum*. Semisynthetic derivatives have even greater activity than the parent drug.

Artemisinin compounds clear the parasites from the blood more rapidly than any other antimalarial agent via a unique pharmacodynamic action. They are concentrated in parasitized erythrocytes, and structure–activity relations suggest that their endoperoxide bridge is essential for the antimalarial effect. A critical step in the mechanism of action seems to be a hemin-catalyzed reduction of the peroxide moiety, which results in free radicals and reactive aldehydes that subsequently kill the malaria parasites. The hemin-rich internal environment of the parasites is assumed to be responsible for the selective toxicity of artemisinin toward these organisms.

An interesting compound, **yingzhaosu**, has been isolated from the roots of ***Artabotrys uncinatus***. This botanical medicine has been used in southern China for malaria. A synthetic derivative of yingzhaosu A, **arteflene**, is currently under Western development for the treatment of malaria. *In vitro*, this compound has shown activity against drug-resistant strains of *P. falciparum* without signs of cross-resistance with existing antimalarials. Drug interaction studies have revealed an additive to synergistic effect with **chloroquine**, **mefloquine**, and **quinine** (De Smet, 1997).

CARDIOVASCULAR DRUGS

The positive benefits of extracts of two species of ***Digitalis* (foxglove)** and *lanata* were recognized long before the active constituents were isolated and characterized structurally. The cardiac glycosides, which include **digoxin, digitoxin**, and **deslanoside**, exert a powerful and selective positive inotropic action on the cardiac muscle. Digoxin is still produced by mass cultivation and extraction of a strain of foxglove (*D. purpurea*) that has been selected for maximum production of the bioactive glycosides. In addition to the cardiac glycosides, a number of naturally occurring alkaloids are important drugs in the control of various cardiovascular conditions. For example, **quinidine**, isolated from the bark of the ***Cinchona* tree**, is an important antiarrhythmic drug, and was also one of the earliest and most well known examples of the critical role of chirality in drug action. Its diastereoisomer, quinine, has virtually no cardiac activity, but was recognized as one of the first anti-infective agents by its efficacy against malaria. Other important naturally occurring alkaloids active as cardiovascular drugs include **reserpine**, once one of the most useful antihypertensive agents known, and **papaverine**, a non-narcotic peripheral vasodilator. In addition, **theophylline**, a xanthine alkaloid, is an important bronchodilator used to control asthma in children. **Ergotamine**, an ergot alkaloid obtained from a fungus that infects rye grass, is an important central vasoconstrictor that is used therapeutically to treat **migraine headaches**. One of the newest additions to this class of drugs is the dopamine **D₂ receptor agonist**, **cabergoline**, launched in Belgium in 1993 as an **antiprolactin**. This drug is also being evaluated as a potential therapy for **Parkinson's disease** and as an anticancer agent for the treatment of **breast cancer** (Clark, 1996).

The discovery of the pharmacological effects of components of the venom of the **pit viper** (*Bothrops jararaca*) ultimately led to the discovery of the role of **angiotensin converting enzyme (ACE)** in hypertension. As a result, **teprotide**, a specific ACE inhibitor, was discovered and a model system for the interaction of small peptides with the enzyme was developed. Ultimately, **captopril** was designed as a specific, orally effective ACE inhibitor and was introduced to the market in 1981 for the control of hypertension. The success and utility of captopril led to the design and synthesis of additional ACE inhibitors, such as **enapronil**. The ACE inhibitors constitute an important class of drugs that has played a major role in the management of cardiovascular disease (De Smet, 1997).

CNS DRUGS

One of the most cited examples of important natural product-derived drugs is the neuromuscular blocker, ***d*-tubocurarine**, derived from the South American plant, curare, which was used by South American Indians as an **arrow poison**. Tubocurarine led to the development of **decamethonium**, which, although structurally dissimilar to tubocurarine, was nevertheless synthesized based on the presumption at the time that tubocurarine contained two quaternary nitrogens. Similarly, the synthetic local anesthetics, such as **lidocaine**, **benzocaine**, and **dibucaine**, were synthesized to mimic the nerve-blocking activity of cocaine, a natural alkaloid obtained from the leaves of ***Coca eroxylum***, but without the adverse side effects that have led to the abuse of **cocaine**. The opium alkaloids **codeine** and **morphine** served as models for the synthesis of **naloxone**, an important analogue used to treat and diagnose opiate addicts and also led to the discovery of "**endogenous opioids**" (**enkephalins** and **endorphins**). Similarly, **Δ⁹-tetrahydrocannibinol** (THC), the component of ***Cannibas sativa*** responsible for the CNS effects, has also been found to reduce nausea associated with cancer chemotherapy. As a result, efforts to design semisynthetic or synthetic agents that mimic the desirable antiemetic effects, while reducing the CNS effects of THC, led to the development and introduction of nabilone as an antiemetic. Although **nabilone** retains some of the less desirable effects of THC, it is an effective antiemetic that is widely used.

Recently, the recognition of the possibility that a number of vastly different CNS and peripheral nervous system diseases may be therapeutically controlled by selective nicotinic acetylcholine receptor (nAChR) agonists has opened a new area of drug design based on the nicotine molecule. Disorders such as **Alzheimer's disease**, **Tourette's syndrome**, **Parkinson's disease**, as well as other **cognitive and attention disorders** may ultimately be more effectively treated if agonists specific for certain subtypes of nAChRs can be discovered or designed. The characterization and understanding of these receptors was based largely on studies using agonists or antagonists, most of which are natural products, such as **acetylcholine**, **arecoline**, **anabasine**, **lobeline**, and **methyllycaconitine**. Recently, reports that the nicotine analogue **epibatidine**, isolated from the skin of **poisonous frogs**, exhibits exceptional analgesic activity and is a very potent nAChR agonist have led to a number of synthetic studies on this molecule, including the synthesis of both pure enantiomers and the remarkable discovery that both enantiomers are very active. Although the toxicity of this compound probably precludes its development as a therapeutic agent, structure–activity relationship (SAR) studies are likely to be undertaken in an effort to separate its toxic from its analgesic effects.

Physostigmine, a naturally occurring alkaloid, and its carbamate ester, **neostigmine**, are also important acetylcholinesterase inhibitors. The cholinesterase inhibitors are used for the treatment of **myasthenia gravis** and as antagonists to neuromuscular blockade by nondepolarizing blocking agents. The acetylcholinesterase inhibitor **galanthamine** is an alkaloid that occurs in the bulbs of **daffodils**, and is currently being investigated as a possible therapy for cognitive impairment in **Alzheimer's disease** (De Smet, 1997).

ANTIBIOTICS

There can be no argument that the antibiotics are among the most important classes of therapeutic agents and have had enormous impact on both life expectancy and quality of life. With the discovery of the natural **penicillins** as secondary metabolites of species of the fungus **Penicillium**, the course of medical history was dramatically changed and the antibiotic era was introduced. Antibiotics are, by definition, natural products or derivatives of natural products. During the course of some 50 years that followed **Alexander Fleming's** observations and the subsequent isolation and characterization of the active constituent by **Howard Florey** and **Ernst Chain**, hundreds of antibiotics have been isolated from scores of microorganisms. On average, two or three new antibiotics are launched each year. Not only do these antibiotics serve as important drugs, but explorations into the mechanisms by which these natural products exert their action have led to an understanding of the biology of the target pathogens that would not likely have been possible without these important biochemical probes. With advances in molecular biology, similar advances in the utilization of natural products to probe specific molecular targets in the pathogens are also being made.

The discovery of important anti-infectives is not limited to antibacterial or antifungal antibiotics from microbial sources. Long before the discovery of penicillin, it was known that the bark of the South American "**fever tree**" **Cinchona succiruba** was effective in controlling malaria. **Quinine** was ultimately identified as the active antimalarial constituent of the **Cinchona bark**. When the natural source of quinine was threatened during World Wars I and II, massive programs to synthesize multitudes of quinoline derivatives based on the quinine prototype ensued. In fact, it could be said that this effort, if not responsible for the origination, was at the very least a major contributor to the early growth of the discipline of medicinal chemistry, which was founded in the synthesis of biologically active compounds modeled after natural products. From this intensive effort emerged the two drugs that remained the therapeutic standards for the treatment of malaria until the past decade: **primaquine** and **chloroquine**.

Today, new important anti-infectives are being discovered from microbial, plant, and animal sources. For example, the antimalarial agent, **artemisinin**, was isolated from the Chinese medicinal plant, *Artemisia annua*. Commonly known as **Qinghaosu**, this herbal remedy had been used in China for centuries for the treatment of malaria. In 1972, the active constituent was isolated and identified as the **sesquiterpene endoperoxide artemisinin**. This compound, in addition to having a structure very different from any of the known antimalarial agents (i.e., quinolines), also exhibits antimalarial and pharmacological profiles very different from the clinically useful agents. Specifically, with activity against strains of the parasite that had become resistant to conventional chloroquine therapy and the ability, due to its lipophilic structure, to cross the blood–brain barrier, it was particularly effective for the deadly **cerebral malaria**. For this reason, several major programs were undertaken to produce artemisinin derivatives with more desirable pharmaceutical properties, and much has been published on synthetic and semisynthetic studies, microbial transformations, biological evaluations, mechanism of action studies, and pharmacological profiles of artemisinin and related analogues. From these studies has emerged **artemether**, a derivative that is currently approved for the treatment of malaria in much of the world.

Another important class of anti-infective natural products to be introduced for human use in recent years are the **avermectins**, polyketide-derived macrolides that were originally isolated from several species of *Streptomyces*. The major drug of this class, **ivermectin**, was originally developed to treat and control nematodes and parasites of livestock. In recent years, however, the potential of ivermectin for the treatment of human disease has also been realized, and it is now used to treat **onchocerciasis (river blindness)**, a disease that afflicts 40 million people worldwide (De Smet, 1997).

ANTINEOPLASTIC AGENTS

Another therapeutic area where natural products have had a major impact on longevity and quality of life is in the chemotherapy of cancer. In fact, most of the major anticancer drugs are natural products from plants or microorganisms. Examples include such important anticancer drugs as **bleomycin**, **doxorubicin**, **daunorubicin**, **vincristine**, **vinblastine**, **mitomycin**, **streptozocin**, and now, the recent additions of **paclitaxel** (Taxol™), **ironotecan** (a camptothecin derivative), and **etoposide** and **tenoposide** (podophyllotoxin derivatives).

The observation that fractions of the **rosey periwinkle**, *Catharanthus rosea*, produced severe leukopenia, resulted in isolation and development of the two major anticancer drugs **vincristine** and **vinblastine**. These two complex, dimeric indole-indoline alkaloids are important therapies for the treatment of **acute childhood leukemia (vincristine)**, **Hodgkin's disease (vinblastine)**, and **metastatic testicular tumors (vinblastine)**, and continue to be manufactured today by mass cultivation and processing of the natural source.

Over 40 years ago, the National Cancer Institute initiated a program to explore higher plants as a source of anticancer agents. Although this program went through periods of low productivity and little support, the discovery of one of the most exciting new drugs in recent history, **paclitaxel (Taxol™)**, was a direct result of this effort. This compound, as well as most other anticancer drugs, was discovered using a system of screening large numbers of extracts of plants, microorganisms, and, more recently, marine organisms, for inhibition of cancer cells grown in culture. Although the total synthesis of paclitaxel has been reported, this drug is a complex natural product with many asymmetric centers and is not readily amenable to total synthesis on an economical basis. Although a semisynthetic derivative with improved water solubility, **Taxotere**™, is now available, the unaltered natural product is approved and used clinically. Studies on the mechanism of anticancer action by paclitaxel revealed that it blocks **depolymerization of microtubules**. Since Taxol is the only known compound to exhibit this activity, without its discovery and development this new target for anticancer drug discovery and development would likely not have been identified.

Another important plant-derived anticancer natural product isolated is **camptothecin**, an alkaloid from the Chinese tree, *Camptotheca acuminata* **Descne**. A semisynthetic water-soluble derivative of camptothecin known as **ironotecan (Topotecin**™, **Campto**™) was introduced in Japan in 1994 for the treatment of lung, ovarian, and cervical cancers. Unlike **Taxol**, camptothecin acts by inhibition of the enzyme **topoisomerase I**.

Two important recent additions to the cancer chemotherapeutic arsenal are **etoposide** and **tenoposide**. *Podophyllum peltatum*, used for years as a folk remedy, is the source of **podophyllin**, a crude resin used topically to treat **condylamata acuminata**. Podophyllin contains, among other things, the **lignan podophyllotoxin**. Studies directed at preparing a water-soluble derivative of podophyllotoxin ultimately led to the discovery of a minor, but very active, constituent of the podophyllin resin, **4′-desmethoxy-1-epipodophyllotoxin glucoside**. The ability to produce this minor, naturally occurring compound by semisynthetic modification of the more abundant podophyllotoxin was a breakthrough that allowed the preparation and evaluation of a number of analogues, some of which had extraordinary activity. Two of these analogues, **etoposide** and **tenoposide**, were introduced as anticancer drugs in 1983 and 1992, respectively. These compounds act by a mechanism (**topoisomerase II inhibitors**) different from that of podophyllotoxin (**spindle poison**), illustrating that structural similarity alone is not always a reliable predictor of similar biological effect.

Paclitaxel, camptothecin, vincristine, vinblastine, as well as other compounds currently under development as potential anticancer drugs (i.e., the **bryostatins**, isolated from marine **dinoflagellates**), were discovered as result of a broad-based screening program to identify, using a whole-cell inhibition assay, natural products that are active against a battery of representative cancer cell lines.

A final example of the importance of natural product drug discovery and development to advances in the chemotherapy of cancer is the growing body of evidence that the **retinoids**, derived from **vitamin A (retinol)**, may have potential utility in the treatment of cancer. Although several retinoids are already used clinically, i.e., **all-*trans*-retinoic acid (ATRA, Retin A**™**)** and **13-*cis*-retinoic acid (13-*cis* RA, Accutane**™**)** are used for the treatment of acne and the synthetic analogue, **etretinate (Tegison**™**)**, is used for the treatment of severe **psoriasis**, it has recently been recognized that these compounds may have significant potential in cancer chemotherapy. Although each of these drugs suffers from significant side effects that diminish its utility, it has been shown that the retinoids exert both their therapeutic and adverse effects through activation of **retinoid**

receptors (see Soprano and Soprano, 1995), for which there are several subtypes. These data, when coupled with synthetic accessibility, suggest that synthetic retinoids based on the naturally occurring prototypes may be prepared that are more selective for specific receptor subtypes, thus offering the hope of separating the beneficial effects from the undesirable effects. Increasing evidence supports the potential benefit of retinoids in the therapy of cancer and prevention of carcinogenesis, primarily as a result of their ability to regulate cellular growth and differentiation. **All-*trans*-retinoic acid**, **13-*cis*-retinoic acid**, **etritinate**, and **fenretinide** have undergone human clinical trials, with varying levels of response (De Smet, 1997).

CHOLESTEROL-LOWERING AGENTS (HYPOLIPAEMICS)

Some of the most exciting natural products discovered in recent years are the cholesterol-lowering agents derived from fungi. These drugs act by inhibition of **3-hydroxy-3-methylglutaryl coenzyme A reductase (HMG-CoA reductase)**, an enzyme critical in the biosynthesis of cholesterol. The first of the HMG-CoA reductase inhibitors were isolated from ***Penicillium* sp. Compactin**, from *P. brevicompactin*, was first reported as an antifungal agent. With the recognition of the mechanism of action of **compactin** came a search for other naturally occurring HMG-CoA reductase inhibitors that led to the discovery of **lovastatin**, a secondary metabolite of the fungus ***Aspergillus terreus***. Although lovastatin was introduced to the market in 1989, many studies were undertaken to prepare improved analogues and led to the development of **simvastatin**, **privastatin**, and **fluvastatin** (De Smet, 1997).

IMMUNOMODULATORS

The immunomodulator **cyclosporin** was originally isolated from a soil fungus, ***Trichoderma polysporum***. This compound was a major breakthrough for organ transplantation, since it suppressed immunological rejection of the transplanted organ. Recently, **Tacrolimus** (FK-506), a secondary metabolite of ***Streptomyces tsukabaensis***, was approved in 1994 for use as an immunosuppressant in organ transplantation (De Smet, 1997).

RECENTLY EVALUATED HERBAL PRODUCTS

The following are examples of the crude herbal preparations which have drawn the attention of clinical investigators in more recent years.

Aloe Vera Gel (*Aloe vera*)

Aloe vera gel is a mucilaginous preparation obtained from the leaves of *Aloe vera* (syn. *Aloe barbadensis*) once all the sap has drained away. It is widely used in cosmetic products for purported emollient and moisturizing effects. It has been claimed to have antibacterial, antifungal, and anti-inflammatory properties and has been proposed as a potentially useful agent for radiation and thermal wounds.

American Dwarf Palm (*Serenoa repens*)

Lipophilic extracts of **sabal fruit** (*Serenoa repens* syn. *Sabal seruulata*) are available for the treatment of benign **prostatic hyperplasia** (BPH). The hexane extract of the pulp and seed contains a complex mixture of free fatty acids and their esters, small quantities of **phytosterols** (such as β-sitosterol), **aliphatic alcohols**, and **polyprenic compounds**. The efficacy of this extract was recently compared for 6 months with that of the **5α-reductase inhibitor finasteride**, in a randomized, double-blind trial involving almost 1100 men with moderate BPH. Dwarf palm and finasteride reduced the International Prostate Symptom Score by 37 and 39% and increased urinary peak flow by 2.7 and 3.2 ml/s, respectively. In contrast to finasteride, dwarf palm had little effect on prostate volume.

Cranberry juice (*Vaccinium* spp.)

A randomized, double-blind, placebo-controlled trial suggests that the regular intake of cranberry juice reduces the frequency of bacteriuria and pyuria in elderly women.

Evening Primrose Oil (*Oenothera biennis*)

Evening primrose oil is rich in the essential fatty acid **linoleic acid** and its metabolite **γ-linolenic acid**. In the United Kingdom, evening primrose oil preparations have received approval as medicines for the relief of **atopic eczema** and for the symptomatic treatment of **breast pain**.

Feverfew Leaf (*Tanacetum parthenium*)

Double-blind withdrawal of feverfew leaf from migraine sufferers who were regular users led to a significant increase in the frequency and severity of headache. A prospective, randomized, double-blind, placebo-controlled trial subsequently found that feverfew was an effective prophylactic migraine agent. The **sesquiterpene lactone parthenolide** is considered to be an important constituent, but the possibility cannot be excluded that other constituents play a significant role.

Garlic (*Allium sativum*)

Garlic has been reported to have lipid- and blood pressure-lowering actions, as well as antiplatelet, antioxidant, and fibrinolytic effects. In animal models and human cell cultures, it has been shown to have antiatherosclerotic activity. The sulfur-containing component, **allicin**, is often considered to be the principal active ingredient of garlic, but several other bioactive ingredients have also been isolated.

Ginger Root (*Zingiber officinale*)

Several controlled clinical trials suggest that ginger root can relieve symptoms of motion sickness by a mechanism of action that differs from that of antihistamines. The responsible constituents are believed to be **gingerols** and **shoagols**.

Ginkgo Leaf (*Ginkgo biloba*)

Gingko leaf is recommended in Germany for **cerebral insufficiency** and for peripheral vascular diseases such as **intermittent claudication**. Although it may be of use to patients with mild symptoms of cerebral insufficiency, it is probably not useful in those with clear **dementia**. One of the major **terpenes** in gingko leaf, **ginkgolide B**, is a potent antagonist of **platelet-activating-factor** (PAF). Since PAF plays a role in various pathological processes, including asthma, shock, ischemia, anaphylaxis, graft rejection, and renal disease, ginkgolide B (BN-52021) has been submitted in recent years to clinical evaluations in such conditions.

Hawthorn (*Crataegus monogyna* and *C. laevigata*)

Among the reported activities of hawthorn are positive inotropic activity, coronary vasodilatation, slowing of heart rate, and slight lowering of blood pressure.

Horse Chestnut (*Aesculus hippocastanum*)

The edema-reducing effects of class II compression stockings in patients with chronic venous insufficiency have been compared with those of placebo and a dried horse-chestnut seed extract (HSCE, providing **aescin** 50 mg twice daily) in a randomized, partially blinded, parallel study. After12 weeks of therapy, HSCE and compression produced equivalent mean decreases in lower-leg volume (44 and 47 ml, respectively), whereas a mean increase of 10 ml was observed in the placebo group.

Ipecacuanha (*Cephaelis ipecacuanha* and *C. acuminata*)

In the United States, **syrup of ipecac** has long been valued as an emetic OTC agent in the treatment of poisonings, and it has been the primary form of decontamination in pediatric cases, both at home and in health-care facilities.

Ispaghula Husk (*Plantago ovato*)

Ispaghula husk (known as **psyllium husk** or **psyllium hydrophilic mucilloid** in the United States) is an effective bulk-forming laxative, which causes retention of fluid and an increase in fecal mass, resulting in stimulation of peristalsis. Ispaghula husk caused small reductions in total and low-density lipoprotein cholesterol (LDL-C) levels in double-blind, placebo-controlled studies involving patients with mild to moderate hypercholesterolemia. It should be regarded as an adjunct to dietary modification rather than as a primary intervention for hypercholesterolemia. There is also a study to suggest that the combination of ispaghula husk with a bile acid sequestrant resin at half its usual dose may be as effective as, and better tolerated than, the resin alone at its usual dose.

Kava-Kava (*Piper methysticum*)

Several clinical studies suggest that extracts from the rhizome of kava-kava may alleviate mild symptoms of anxiety.

Marijuana (*Cannabis sativa*)

Among the herbs that have recently drawn substantial attention from Anglo-Saxon countries to their potential as prescription drugs, is marijuana (*C. sativa*). The beneficial effects of marijuana are as follows:

- Symptomatic relief in nausea and vomiting associated with chemotherapy (but with a high incidence of central adverse effects)
- Alleviation of muscle spasms, tremor, ataxia, and bladder dysfunction in multiple sclerosis and after spinal cord injury
- Anticonvulsant and muscle relaxant in spastic disorders
- Reduction of intraocular pressure in glaucoma
- Appetite stimulant in the wasting syndrome of HIV infection
- Relief of phantom pain, menstrual cramps, and other types of chronic pain (see De Smet, 1997)

St. John's Wort (*Hypericum perforatum*)

According to a recent meta-analysis, St. John's wort is more effective than placebo for the treatment of mild to moderately severe depressive disorders.

Tea Tree Oil (*Melaleuca alternifolia*)

Tea tree oil had an ameliorating effect on mild to moderate acne in a single-blind, randomized trial in which benzoyl peroxide served as comparator. It has also shown promise in **onychomycosis** in a comparative trial with **clotrimazole**. Among the antimicrobial constituents are **terpinen-4-ol**, **linalool**, and **α-terpineol**.

Wheat Bran (*Triticum* spp.)

There is substantial evidence that bran is useful in the treatment of simple constipation. If used in sufficient quantities, it softens stools, eliminates scybala, and prevents straining. However, a recent study has challenged its clinical merits in the management of the **irritable bowel syndrome**.

NEW USES FOR "OLD" DRUGS

It should also be continually recognized and emphasized that as we acquire additional experience with known drugs, it often occurs that new uses for existing agents are identified. For example, it has been shown in recent years that the combination of the antibacterial antibiotic **clindamycin** (used clinically for years for its relatively narrow spectrum of activity against important Gram-positive bacteria and anaerobes) and the antimalarial **8-aminoquinoline primaquine** (synthesized after the quinine model) is effective in the treatment of the important **AIDS-related opportunistic infection *Pneumocystis carinii* pneumonia**. This combination therapy is currently undergoing clinical trials for the treatment of AIDS-related PCP, and demonstrates the importance of continuing to explore older, known drugs for new therapeutic applications. The original therapeutic utility of an agent may be expanded as new disease states are recognized and as experience with an agent affords more empirical observations upon which to base recommendations for such new therapeutic uses.

8

HERBAL PRODUCTS FOR THE CHEMOTHERAPY OF HUMAN IMMUNODEFICIENCY VIRUS (HIV) INFECTION

- Blue-green alga
- Seaweed
- Licorice root
- Malaysian tree

INTRODUCTION

The **acquired immune deficiency syndrome** (AIDS) was first recognized in the United States in the summer of 1981, when the Centers for Disease Control and Prevention (CDC) reported the unexplained occurrence of ***Pneumocystis carinii*** pneumonia in five previously healthy homosexual men in Los Angeles and of **Kaposi's sarcoma** in 26 previously healthy homosexual men in New York and Los Angeles. Within months, the disease became recognized in male and female injection drug users (IDUs), and, soon thereafter, in **recipients of blood transfusions** and in **hemophiliacs.**

In 1983, human immunodeficiency virus (HIV) was isolated from a patient with **lymphadenopathy**, and by 1984 it was demonstrated clearly to be the causative agent of AIDS. The virus, which belongs to the **lentivirus subfamily** of the large family of **retroviruses** (Retroviridae), was formerly called ***lymphadenopathy associated virus*** (LAV), ***human T-lymphotropic virus*** (HTLV) ***type III***, and ***AIDS-associated retrovirus*** (ARV).

At the present time, there are at least 14 compounds that have been formally approved for the treatment of **human immunodeficiency virus (HIV) infections**. There are six **nucleoside reverse transcriptase inhibitors** (NRTIs), that, after their intracellular conversion to the 5'-triphosphate form, are able to interfere as competitive inhibitors of the normal substrates (dNTPs): these are **zidovudine** (AZT), **didanosine** (ddI), **zalcitabine** (ddC), **stavudine** (d4T), **lamivudine** (3TC), and **abacavir** (ABC). These are three **non-nucleoside reverse transcriptase inhibitors** (NNRTIs) that, as such, directly interact with the reverse transcriptase at a nonsubstrate binding, allosteric site (**nevirapine, delavirdine, efavirenz**). There are five **HIV protease inhibitors** (PIs: **saquinavir, ritonavir, indinavir, nelfinavir, amprenavir**) that block the cleavage of precursor to mature HIV proteins, thus impairing the infectivity of the virus particles produced in the presence of these inhibitors.

A number of products exist that are targeted at any of the successive events implicated in the replicative cycle of HIV: **virus entry, viral adsorption, virus–cell fusion, reverse (RNA→DNA) transcription, proviral DNA integration, viral (DNA→RNA) transcription** (transactivation), **viral (mRNA→protein) translation, virus release, viral assembly, budding,** and **maturation** (see Fauci et al., 1998; De Clercq, 2000).

HERBAL PRODUCTS FOR THE CHEMOTHERAPY OF HUMAN IMMUNODEFICIENCY VIRUS (HIV) INFECTION

- **Blue-green algae: Cyanovirin-N,** an 11-kDA protein from *Cyanobacterium* (blue-green alga) irreversibly inactivates HIV and also aborts cell-to-cell fusion and transmission of HIV, due to its high-affinity interaction with gp120.
- **Seaweed:** Various sulfated polysaccharides extracted from seaweed (i.e., *Nothogenia fastigiata, Aghardhiella tenera*) inhibit the virus adsorption process.
- **Ingenol derivatives:** Ingenol derivatives may inhibit virus absorption at least in part through downregulation of CD4 molecules on the host cells.
- **Flavonoids:** Inhibition of virus adsorption by flavonoids such as (–) **epicatechin** and its 3-o-gallate has been attributed to an irreversible interaction with gp120 (although these compounds are also known as reverse transcriptase inhibitors).
- **Licorice root:** For the triterpene **glycyrrhizin** (extracted from the licorice root *Glycyrrhiza radix*) the mode of anti-HIV action may at least in part be attributed to interference with virus–cell binding.
- **Lectins:** The mannose-specific plant lectins from *Galanthus, Hippeastrum, Narcissus, Epipactis helleborine,* and *Listera ovata,* and the *N*-acetylglucosamine-specific lectin from *Urtica dioica* would primarily be targeted at the virus–cell fusion process.
- **Streptomyces spp:** The **siamycins, siamycin I** (BMY-29304), **siamycin II** (RP71955, BMY 29303), and NP-06 (FR901742), which are tricyclic 21-amino acid peptides have been isolated from *Streptomyces* spp.
- **Horseshoe crabs:** The **betulinic acid derivative** RPR 103611, and the peptides **tachyplesin** and **polyphemusin,** which are highly abundant in hemocyte debris of the horseshoe crabs *Tachypleus tridentatus* and *Limulus polyphemus.*
- **Tropical rainforest tree and Malaysian tree:** A number of natural products have been reported to interact with reverse transcriptase, i.e., **baicalin, avarol, avarone, psychotrine, phloroglucinol derivatives,**

and, in particular, **calanolides** (from the tropical rainforest tree, *Calophyllum lanigerum*) and **inophyllums** (from the Malaysian tree, *Calophyllum inophyllum*).

■ **Marine products:** The natural marine substance **illimaquinone** would be targeted at the RNase H function of the reverse transcriptase.

■ **Tumeric: Curcumin** (**diferuloylmethane**, from tumeric, the roots/rhizomes of Curcuma spp).

■ **Fungal products: Dicaffeoyiquinic** and **dicaffeoyltaric acid**, L-**chicoric acid**, and a number of fungal metabolites (**equisetin, phomasetin, oteromycin**, and **integric acid**) have all been proposed as HIV-1 **integrase inhibitors** (see Clercq, 2000).

ETIOLOGICAL AGENT

The etiological agent of AIDS is HIV, which belongs to the family of human **retroviruses** and the subfamily of **lentiviruses**. Non-oncogenic lentiviruses cause disease in other animal species, including sheep, horses, goats, cattle, cats, and monkeys. The four recognized human retroviruses belong to two distinct groups: the **human T lymphotropic viruses**, HTLV-I and HTLV-II, which are transforming retroviruses; and the **human immunodeficiency viruses**, HIV-1 and HIV-2, which are cytopathic viruses. The most common cause of HIV disease throughout the world, and certainly in the United States, is HIV-1.

LIFE CYCLE OF HIV

HIV is an RNA virus whose hallmark is the reverse transcription of its genomic RNA to DNA by the enzyme *reverse transcriptase.* The life cycle of HIV begins with the high-affinity binding of the gp120 protein via a portion of its V1 region near the *N* terminus to its receptor on the host cell surface, the **CD4 molecule** (Figure 7).

The CD4 molecule is a **55-kDa protein** found predominantly on a subset of **T lymphocytes** that are responsible for helper or inducer function in the immune system. It is also expressed on the surface of monocytes/macrophages and dendritic/Langerhans cells. It has recently been demonstrated that the coreceptor that must be present together with the CD4 molecule for fusion and entry of T-cell-tropic strains of HIV-1 is a molecule termed **CXCR4**, whereas the coreceptor for macrophage-tropic strains of HIV-1 is the **α-chemokine receptor CCR5**. Both receptors belong to the family of seven-transmembrane-domain G protein-coupled cellular receptors. Following binding, fusion with the host cell membrane occurs via the gp41 molecule, and the HIV genomic RNA is uncoated and internalized into the target cell. The **reverse transcriptase enzyme**, which is contained in the infecting virion, then catalyzes the reverse transcription of the genomic RNA into double-stranded DNA. The DNA translocates to the nucleus, where it is integrated randomly into the host cell chromosomes through the action of another virally encoded enzyme, *integrase*. This provirus may remain transcriptionally inactive (latent), or it may manifest varying levels of gene expression, up to active production of virus.

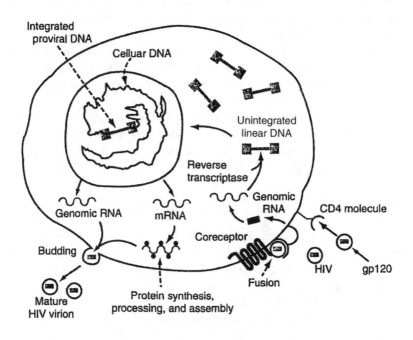

Figure 7   The life cycle of HIV.

Cellular activation plays an important role in the life cycle of HIV and is critical to the pathogenesis of HIV disease. Following initial binding and internalization of virions into the target cell, incompletely reverse-transcribed DNA intermediates are labile in quiescent cells and will not integrate efficiently into the host cell genome unless cellular activation occurs shortly after infection. Furthermore, activation of the host cell is required for the initiation of transcription of the integrated proviral DNA into either genomic RNA or mRNA. In this regard, activation of HIV expression from the latent state depends on the interaction of a number of cellular and viral factors. Following transcription, HIV mRNA is translated into proteins that undergo modification through **glycosylation**, **myristylation**, **phosphorylation**, and **cleavage**. The viral core is formed by the assembly of HIV proteins, enzymes, and genomic RNA at the plasma membrane of the cells. Budding of the progeny virion occurs through the host cell membrane, where the core acquires its external envelope. Each point in the life cycle of HIV is a real or potential target for therapeutic intervention. Thus far, the reverse transcriptase and protease enzymes have proved to be susceptible to pharmacological disruption (Fauci et al., 1998).

IMMUNOPHARMACOLOGY

The immune system defends the body against invading organisms, foreign antigens, and host cells that have become neoplastic. In addition, the immune system is an active participant in autoimmune diseases, hypersensitivity reactions, and transplant tissue rejections.

The ability of the immune system to discriminate between foreign molecules and antigenic sites of foreign cells, normal endogenous molecules, and cells of the host results in elimination of most diseases before an overt pathological condition is established. Moreover, after an infection or neoplasm becomes established and chemotherapy initiated, drugs kill only a fraction of the invading organisms or neoplastic cells, and a functional immune system is needed to eradicate the remaining organisms or cells. This ability to recognize foreign antigens allows destruction and removal of invading organisms by various effector mechanisms of the immune system. However, in autoimmune diseases, inappropriate immune responses against host cells (**self-antigens**) may occur. In addition, an overt response to an antigen may result in tissue-damage reactions known as **hypersensitivity responses**.

IMMUNE SYSTEM COMPONENTS

The cells of the immune system are formed from **pluripotent stem cells** produced in bone marrow. These stem cells undergo a sequence of cellular differentiations, to form B lymphocytes, T lymphocytes, erythrocytes, polymorphonuclear leukocytes, monocytes, macrophages, and mast cells.

Lymphocytes are one of the primary cell types involved in the immune response. There are two general types of lymphocytes, B and T. Both are derived from bone marrow lymphoid stem cells, but T cells go through an additional maturation process in the thymus. Although the morphology of T cells and B cells is similar, the functions of these two types are distinct. After antigen exposure, B cells develop into **antibody-producing plasma cells**, whereas T cells are divided into functional subtypes that possess distinct cell surface antigens.

The other major participant in the immune response is the **bone marrow-derived macrophage**. When circulating, it is referred to as a monocyte; however, when it migrates into the extravascular spaces, it is known as a macrophage. In comparison to unactivated lymphocytes, macrophages are larger and possess a greater cytoplasmic-to-nuclear ratio. The macrophage contains lysosomes filled with various catabolic enzymes. The macrophage membrane possesses digestive enzymes and receptors for binding complement components and the constant or Fc region of antibodies (**immunoglobulins**). The size and cytoplasmic contents are determined by the activation state. The primary functions of macrophages are **phagocytosis**, **antigen presentation**, and **cytokine production**. Phagocytic macrophages are distributed throughout many tissues and collectively are called the ***reticuloendothelial system.*** The generation of superoxide free radicals and digestive enzymes allows the macrophage to destroy other phagocytized organisms or molecules. This phagocytic process is also important in altering and cleaving large antigen molecules before presentation to T lymphocytes (**antigen processing**).

Cells That Constitute the Reticuloendothelial Systems

Tissues	Cell Types
Liver	Kupffer cells
Brain	Microglial cells
Spleen	Macrophages
Lymph node	Macrophages
Lung	Aveolar macrophages

Table 13   The Targets and Compounds for Anti-HIV Therapy

Targets (within the HIV replicative cycle)	Compounds (interacting with the identified target)
Virus particles, infectivity, entry	Canovirin-N (Hypericin)
Virus adsorption	Sulfated polysaccharides
	Ingenol derivatives
	Flavanoids
	Glycyrrhizin
	(Mannose-specific plant lectins)
Virus–cell fusion	Mannose-specific plant lectins
	Siamycins
	Betulinic acids
	Polyphemusins
Reverse transcription	Baicalin
	Avarol, Avarone, Psychotrine, Phloroglucinol derivatives
Reverse transcription	Calanolides, Inophyliums
	Illimaquinone
	(Flavanoids)
Integration	Curcumin, dicaffeoylquinic acid, L-chicoric acid
	Equisetin
	(Hypericin)
	(Cyclosporins)
Transcription, transactivation	Flavonoids
	Melittin
	EM2487
Transiation	Trichosantin, MAP30, GAP31, DAP30, DAP32
Assembly, secondary viral spread	Bellenamine
	Hypericin, Pseudohyperion
	Cyclosporins

AIDS is caused by HIV, which impairs both cellular and humoral immune functions, and this results in increased susceptibility to opportunistic infection and certain malignancies. The medical aspects of HIV infection include AIDS-related complex (weight loss, chronic diarrhea, fever, thrush, herpes zoster, fatigue); opportunistic infections and cancer; end-stage renal disease; blindness (cytomegalovirus); HIV encephalopathy and dementia. The psychological aspects of HIV infection include major depression, regression, and suicidal impulses; delirium; substance abuse; antisocial personality; bereavement.

The targets of anti-HIV therapy are shown in Table 13.

CYANOVIRIN-N

Cyanovirin-N is an 11-kDa protein that has been isolated from the **cyanobacterium (blue-green alga)** *Nostoc ellipsosporum*. At low nanomolar concentrations, cyanovirin-N irreversibly inactivates both laboratory strains and primary isolates of HIV-1 and HIV-2. In addition, cyanovirin-N aborts cell-to-cell fusion and transmission of HIV-1 infection. The antiviral activity of cyanovirin-N has been attributed to a tight binding of the compound to the viral envelope glyco-protein gp120, which must result in a reduced infectivity of the virus as well as reduced capacity of virus-infected cells to fuse with uninfected cells.

SULFATED POLYSACCHARIDES

Sulfated polysaccharides, such as **dextran sulfate**, **pentosan polysulfate**, and **heparin**, as well as sulfated polysaccharides extracted from **sea algae**, have long been recognized as effective anti-HIV agents. The sulfated galactan from the **red seaweed *Aghardhiella tenera*** inhibits HIV-1 and HIV-2 infection at a concentration that is tenfold higher than the IC_{50} (50% inhibitory concentration) required for a dextran sulfate to inhibit these viruses (IC_{50}: 0.5 and 0.05 µg/ml for HIV-1 and HIV-2, respectively). The **sulfated xylomannan** from the **red seaweed *Nothogenia fastigiata*** is only weakly active against HIV-1 and HIV-2 (IC_{50}: ~10 µg/ml), but much more so against other enveloped viruses, such as **herpes simplex virus** (HSV), **human cytomegalovirus** (HCMV), **respiratory syncytial virus** (RSV), and **influenza A virus** (IC_{50}: 0.2 µg/ml for the last). Antivirally active polysaccharides have also been isolated from the **marine brown alga *Fucus vesiculosus***. The sulfated polysaccharides extracted from the marine *Pseudomonas* and the marine plant **Dinoflagellata** inhibit HIV-1 and HIV-2 infection at a similar level (IC_{50}: ~1 µg/ml) as dextran sulfate, and again, inhibit other enveloped viruses (HSV-1, influenza A and B virus, RSV, and measles virus, as well).

Sulfated polysaccharides are notorious for their anticoagulant properties, but, apparently, the anti-HIV activity of sulfated polysaccharides can be dissociated from their antithrombin activity. They invariably show activity against various enveloped viruses other than HIV, particularly HSV, HCMV, influenza A, RSV, and **vesicular stomatitis virus** (VSV). They block the binding (adsorption) of these viruses to the cells, which, in the case of HIV, is due to a direct interaction of the sulfated polysaccharides with the V3 loop of the viral envelope gp120. As a consequence, the sulfated polysaccharides also block **syncytium formation** (fusion) between HIV-infected cells (expressing the **gp120 glycoprotein** on their surface) and uninfected cells (expressing the **CD4 receptor** for gp120). All retro- and myxoviruses that are sensitive to the sulfated polysaccharides share a tripeptide segment (Phe-Leu-Gly), which may be involved either directly or indirectly in the inhibitory effects of the compounds on virus–cell binding and fusion.

Sulfated polysaccharides are very poorly absorbed following oral administration, and efficacy upon parenteral administration has not been demonstrated. Their greatest potential may well reside in their topical application, i.e., as a gel formulation in the prevention of **sexual HIV**, and HSV, transmission.

INGENOL DERIVATIVES

Ingenol derivatives extracted from **kansui**, the **dried roots of *Euphorbia kansui***, have various biological activities, including anti-HIV activity, due to inhibition of virus adsorption to the host cells. Among the ingenol derivatives, **13-hydroxy-ingenol-3-(2,3-dimethylbutanoate)-13-dodecanoate (RD4-2138)** proved to be exquisitely potent (IC_{50}: 0.07 nM) against HIV-1 (III_B strain) in MT-4 cells, and selective (50% cytotoxic concentration, or CC_{50}: 8.7 µM). The mechanism of anti-HIV action of RD4-2138 could at least in part be attributed to downregulation of the CD4 receptor for gp120, thus accounting for the inhibitory activity of this class of compounds on virus adsorption.

Although ingenol derivatives are highly potent inhibitors of HIV-1 replication in acutely infected cells, it was recently shown that some ingenol derivatives enhance HIV-1 replication in chronically infected cells, whether or not through activation of the nuclear factor kB (NF-kB) and protein kinase C (PKC).

FLAVONOIDS

Of a series of flavonoids that were evaluated for their anti-HIV potential, **(–)-epicatechin** and **(–) epicatechin-3-o-gallate** (from ***Detarium microcarpum***) were found to block HIV infection through an irreversible interaction with the glycoprotein gp120: IC_{50} for HIV replication in C8166 cells: 2 and 1 µg/ml, respectively, as compared to a CC_{50} of >100 µg/ml. (–)-Epicatechin gallate and (–)-epigallocatechin gallate, two components from the **tea plant *Camelliasinensis***, were previously described as inhibitors of reverse transcriptase and cellular RNA and DNA polymerases. They may thus be able to interact at multiple targets of the HIV replicative cycle.

The flavonoid **crysin** and **benzothiophenes** have been shown to inhibit **casein kinase II** (CKII), a cellular protein that may regulate HIV-1 transcription by phosphorylating (other) cellular proteins involved in the HIV-1 transcription transactivation process. This mechanism of action is independent of the **nuclear factor kB-driven transcription pathway**. Thus, flavonoids interfere with HIV-1 transcription, and hence prevent HIV expression in latently infected cells. Their specificity and usefulness as HIV transcription inhibitors remain to be assessed.

GLYCYRRHIZIN

Glycyrrhizin from **licorice root** (*Glycyrrhiza radix*), has been known for some time as an antiviral agent; its IC_{50} for HIV-1(III$_B$) in MT-4 cells is 0.15 mM. The mechanism of action of glycyrrhizin may at least partially be attributed to an interference with virus–cell binding, although the site of interaction of glycyrrhizin (at the envelope glycoprotein) has not been further characterized.

MANNOSE-SPECIFIC PLANT LECTINS

A number of **mannose-specific agglutinines (lectins)** from *Galanthus nivalis, Hippeastrum hybrid, Narcissus pseudonarcissus, Listera ovata, Cymbidium* **hybrid**, *Epipactis helleborine,* and the **N-acetylglucosamine-specific** *lectin* from *Urtica dioica* have been found to inhibit HIV-1 and HIV-2 infection at similar concentrations as dextran sulfate (IC_{50}: 0.2 to 0.6 µg/ml), or even lower (IC_{50}: 0.04 to 0.08 µg/ml. Akin to sulfated polysaccharides, the plant lectins also exhibit activity against various enveloped viruses other than HIV, i.e., HSV-1, HSV-2, CMV, RSV, and influenza virus. Plant lectins would primarily interfere with the virus–cell fusion process. Their precise mode of action remains to be resolved.

SIAMYCINS

Three varieties of siamycin, from **Streptomyces**, have been described: **siamycin II** (RP 71955, BMY 29303), NP-06 (FR 901724), and **siamycin I** (BMY-29304). They differ from one another only at position 4 (Val or Ile) or 17 (again, Val or Ile), which should not make much of a difference in terms of conformational or functional properties. The siamycins have been found to inhibit HIV infection *in vitro*; they exert a strong inhibitory effect on syncytium formation while only weakly inhibiting virus–cell binding. The most likely target for the mode of action of siamycins is the HIV envelope glycoprotein gp41. In fact, 33% sequence identity was observed between siamycin II and residues 608 to 628 of gp41 (amino acid numbering for the gp160 precursor). This sequence is part of the **ectodomain of gp41** (residues 527 to 655). Siamycins can be assumed to interfere with the **fusogenic activity of gp41**. They may do so by various mechanisms: for example, through direct binding to the hydrophobic gp41 fusion domain, through inhibition of the anchoring of gp41 in the cellular plasma membrane, or through induction of conformational alterations in gp41. The exact mechanism of action of siamycins remains to be resolved, and so are their clinical potential and pharmacokinetics.

BETULINIC ACIDS

Betulinic acid and **platanic acid** are **triterpenoids** isolated from **Syzygium claviflorum**. They exhibit inhibitory activity against HIV-1 replication in H9 lymphocyte cells at an IC_{50} of 1.4 and 6.5 μM, respectively (selectivity index: 9.3 and 14, respectively). Hydrogenation of betulinic acid yielded **dihydrobetulinic acid**, which showed an IC_{50} of 0.9 μM and a selectivity index of 14. Introduction of a **3,3-dimethylsuccinyl group** at the C3 hydroxyl group of betulinic acid and dihydrobetulinic acid significantly increased the anti-HIV activity of betulinic acid and dihydrobetulinic acid (IC_{50}: < 0.35 nM; selectivity index >20,000 and >14,000, respectively). This is a remarkably high potency and selectivity, which leaves one wondering about the mechanism underlying the anti-HIV activity of 3-0-(3′,3′-dimethylsuccinyl)betulinic acid and –dihydrobetulinic acid.

The betulinic acid derivative RPR 103611 blocks HIV infection at an IC_{50} of about 10 nM, through inhibition of a postbinding, envelope-dependent step involved in the fusion of the virus with the cell membrane. The target for the anti-HIV action of RPR 103611 is the HIV-1 glycoprotein gp41. HIV resistance to RPR 103611 is associated with amino acid substitutions in positions 22 (Arg ? Ala) and 84 (Ile ? Ser) of gp41. RPR 103611 has so far remained the only small-molecular-weight, nonpeptidic agent blocking HIV-1 entry by affecting gp41. It can be considered an attractive lead compound for the design and development of HIV–cell fusion inhibitors.

POLYPHEMUSINS

The polypeptides **tachyplesins** and **polyphemusins** are highly abundant in hemocyte debris of the **horseshoe crabs Tachypleus tridentatus** and **Limulus polyphemus.** An analogue of one of these polypeptides, T22 or [Tyr-5,12,Lys-

7]polyphemusin II, a synthetic peptide that consists of 18 amino acid residues was found to inhibit HIV infection at an IC_{50} of 8 ng/ml (CC_{50}: 54 µg/ml). A postbinding process, probably corresponding to virus–cell fusion, has been suggested as the target for the anti-HIV action of T22. Structure–activity relationship studies have pointed to the importance of the arginine residues in the anti-HIV activity of T22, and in further studies the compound was shown to act as a specific antagonist of CXCR4, the coreceptor used by T-tropic (X4) HIV-1 strains to enter the cells. This antagonism toward the CXCR4 coreceptor can be attributed to the polycationic (polyargininic) character of T22. New T22 derivatives with lower cytotoxicity have been synthesized, and one of these T22 derivatives, T134, was shown to block HIV-1 infection with a higher selectivity index (33,889 as compared with 645) than T22.

BAICALIN

Baicalin monohydrate (TJN-151) is a flavonoid **(5,6,7-trihydroxyflavone-7-O-β-_D_-glucopyranoside uronic acid monohydrate)**, from *Scutellariae radix*, that has been reported to inhibit HIV-1 replication in peripheral blood mononuclear cells (PBMC) at an IC_{50} of 0.5 µg/ml. Baicalin also inhibits **HIV-1 reverse transcriptase**, but not human **DNA polymerases α and γ** (DNA polymerase β being slightly inhibited), so that it was tempting to attribute the anti-HIV-1 effect of baicalin, even partly, to inhibition of HIV-1 reverse transcriptase. No inhibition of virus adsorption was noted, which contrasts to the anti-HIV activity of other flavonoids that have been claimed to inhibit HIV adsorption through an (irreversible) interaction with the glycoprotein gp120.

AVAROL, AVARONE, PSYCHOTRINE, AND PHLOROGLUCINOL DERIVATIVES (I.E., MALLOTOJAPONIN)

Avarol and avarone derivatives (from the **Red Sea sponge _Dysidea cinerea_**), the alkaloids **psychotrine** and **O-methylpsychotrine** (from **ipecac**, the dried rhizome and root of **_Cephaelis ipecacuanha_**), and phloroglucinol derivatives such as **mallotojaponin**, from the pericarps of **_Mallotus japonicus_**, have all been reported to inhibit the reverse transcriptase activity of HIV-1, noncompetitively with respect to the natural substrate (dNTP). In neither case was the anti-HIV-1 activity determined in cell culture, so it is not clear whether any of these compounds is really an effective inhibitor of HIV replication, and, if so, whether they inhibit HIV replication through an inhibitory effect on HIV-1 reverse transcriptase.

CALANOLIDES AND INOPHYLLUMS

The calanolides and inophyllums represent novel HIV-inhibitory coumarin derivatives, isolated from the **tropical rainforest tree, _Calophyllum lanigerum_**, and the **Malaysian tree, _Calophyllum inophyllum_**, respectively. Both compounds were found to inhibit HIV-1 replication at an IC_{50} of approximately 0.1 and 1 µ*M*.

These compounds can be considered NNRTIs, as they are primarily active against HIV-1 reverse transcriptase, but differ from the classical (synthetic) NNRTIs in their HIV sensitivity/resistance profile. Several structural analogues of (+)-calanolide A have been prepared. Two isomers of **calanolide A**, **(−)-calanolide B (costatolide)** and **(−)-dihydrocalanolide B (dihydrocostatolide)**, possess antiviral properties similar to those of calanolide A. The calanolide analogues exhibit tenfold enhanced antiviral activity against drug-resistant viruses that bear one of the most prevalent NNRTI-resistance reverse transcriptase mutations (Y181C); in turn, the calanolide analogues lead to the selection of drug-resistant virus strains carrying the T1391, L1001, Y188H, or L187F mutations in their reverse transcriptase. The calanolide isomers represent a novel and distinct subgroup of the NNRTI family; they should be further evaluated for their therapeutic potential in combination with other anti-HIV agents.

ILLIMAQUINONE

Illimaquinone, like **avarone** and avarol isolated from a Red Sea sponge, i.e., *Smenospongia*, has been reported to inhibit the **RNase H activity**, associated with the HIV-1 reverse transcriptase, at a concentration of 5 to 10 µg/ml, while not being active against the **RNA-dependent DNA polymerase** (RDDP) and **DNA-dependent DNA polymerase** (DDDP) activities of the enzyme at a concentration of 50 µg/ml.

CURCUMIN, CURCUMIN ANALOGUES, DICAFFEOYLQUINIC ACID, AND L-CHICORIC ACID

Curcumin corresponds to the yellow pigment in **turmeric (*Curcuma longa*)** that is widely used as a spice, food coloring (**curry**), and preservative. Curcumin has been reported to inhibit **HIV-1 integrase** at an IC_{50} (for strand transfer) of 40 µ*M*. **Dicaffeoylquinic acids** inhibit HIV-1 integrase at submicromolecular concentrations: molecular modeling of these ligands with the core catalytic domain of integrase indicate that the most potent inhibitors fill a groove within the catalytic site and interact with the integrase in an energetically favorable fashion. The **dicaffeoylquinic acids** and **L-chicoric acid** were reported to inhibit HIV-1 integrase at concentrations ranging from 0.06 to 0.66 µg/ml; and HIV-1 replication in cell cultures at 1 to 4 µg/ml. Within the HIV-1 integrase different activities can be discerned: i.e., 3′-end processing and strand transfer; curcumin analogues such as **dicaffeoylmethane** and **rosmarinic acid** inhibit both activities of integrase with IC_{50} values below 10 µ*M*.

In follow-up studies, dicaffeoylquinic acids (DCQAs) and dicaffeoyltartaric acids (DCTAs) were found to inhibit HIV-1 integrase at concentrations between 0.15 and 0.84 µ*M*, and HIV replication at concentrations between 2 and 12 µ*M*; no inhibition of gp120 binding to CD4 was noted at concentrations up to 80 µ*M*. Similarly, no inhibition of reverse transcription or RNase H was noted, and it was concluded that the DCQAs and DCTAs act as specific integrase inhibitors, and that their activity against integrase is consistent with their observed anti-HIV activity in cell cultures. That the integrase would be an "excellent" target for combination chemotherapy of HIV infection was further ascertained by combination experiments where L-chicoric acid, the putative integrase inhibitor, was combined with a protease inhibitor (AG1350) and zidovudine.

EQUISETIN

The screening of natural product extracts has recently yielded a few more HIV integrase inhibitors: equisetin, from the **fungus *Fusarium heterosporum***, the enantiomeric homologue of equisetin from Phoma species, and two additional analogues, a novel decalin derivative, **integric acid**, and **oteromycin**. Equisetin inhibits 3′-end processing and strand transfer at an IC_{50} of approximately 10 μM. Equisetin and related compounds also inhibit disintegration catalyzed by either full-length enzyme or the truncated integrase core domain (amino acids 50 to 212). These compounds also inhibit strand transfer reactions catalyzed by stable complexes assembled *in vitro* and integration reactions catalyzed by preintegration complexes isolated from HIV-1-infected cells, but whether these fungal metabolites also inhibit HIV-1 replication within the cells, whether or not due to inhibition of integration, remains to be determined.

MELITTIN

Melittin, a 26-amino acid **amphipathic α-helical peptide** that occurs in **bee venom**, has recently been found capable of suppressing HIV-1 gene expression. Melittin inhibits HIV-1 infection in both acutely and persistently infected t-lymphoma (KE37/1) and fibroblastoid (LC_5) cells at an IC_{50} of 0.5 to 1.5 μM; this effect is apparently mediated by a direct suppressive action on the HIV long terminal repeat (LTR). Antimicrobial peptides such as **melittin (honeybees)**, **cecropin (moths)**, and **magainin (frogs)** may thus inhibit cell-associated HIV-1 production at the transcription level.

TRICHOSANTIN, MAP30, GAP31, DAP31, AND DAP32

A series of proteins, i.e., **α-trichosanthin** (α-TCS) isolated from the **root tubers of *Trichosanthes kirilowii***, MAP30 isolated from the seeds and fruits of the **bitter melon *Momordica charantia***, GAP31 from **Euphorbiaceae himalaya seeds (*Gelonium multiflorum*)**, and DAP30 and DAP32 from **carnation leaves (*Dianthus caryophyllus*)**, have been described to inhibit HIV infection, allegedly, at noncytotoxic concentrations. These compounds (also referred to as RIPs for "**ribosome-inactivating proteins**") are known to abrogate eukaryotic protein synthesis through inactivation of the ribosomes. RIPs inactivate ribosomes by cleaving the adenine *N*-glycosidic bond at position 4324 of 28S rRNA. In addition to this RNA *N*-glycosidase activity, RIPs, such as MAP30, also act as **DNA glycosylase/apurinic (ap)lyases**, thus explaining the apparent ability of MAP to inhibit HIV-1 integrase and irreversibly relax supercoiled DNA.

BELLENAMINE

Bellenamine, (R)-3,6-diamino-N-(aminomethyl)hexanamide is a (very) small-molecular-weight (MW: 174) antibiotic produced by **Streptomyces nashvillensis**. It has only weak antibacterial activity but inhibits HIV-1 infection at an IC_{50} of 0.62 µg/ml (3.6 µM). With a CC_{50} of >2 mg/ml (11.5 mM), the selectivity index of bellenamine can be estimated at well above 2000. Similarly, to the well-established glycosylation inhibitors **castanospermine** and **1-deoxynojirimycin**, bellenamine inhibits the secondary spread of HIV, although, unlike the glycosylation inhibitors, bellenamine had no apparent inhibitory effect on the glycosylation process. Therefore, its antiviral mechanism remains to be elucidated.

HYPERICIN AND PSEUDOHYPERICIN

The aromatic polycyclic diones hypericin and pseudohypericin, which are present in plants of the family **Hypericum** (**St. John's wort**), have since more than a decade been accredited with anti-HIV activity. They possess the capacity to inactivate HIV virions directly as well as to interfere with their assembly or processing. Hypericin has received continued interest as a potential therapeutic agent and was, more recently, found to interact with preintegration complexes (PICs) and thus affect the **proviral DNA integration process** (Clercq, 2000).

9

ALOE AND
COLON FUNCTION

INTRODUCTION

Aloe is one of the few medicinal plants that has maintained its popularity for a long period of time. **Aloe latex** is used for its **laxative effect** and should be distinguished from **aloe gel**, used both in cosmetics and in ointments for skin ailments. **Aloe whole leaf** is another preparation used internally as a drink in a wide range of human diseases including **cancer**, **AIDS**, **ulcerative colitis**, and **other disturbances**. In this chapter, only the laxative properties of aloe that have been proved scientifically will be described in detail.

The genus *Aloe* comprises about 600 species, most of which are indigenous to South Africa (*A. ferox* **Miller**, *A. africana* **Miller**, *A. spicata* **Baker**); some have been introduced in Asia (*A. chinensis* **Bak**), Barbados Islands in Central America (*A. barbadensis* **Miller**), and Europe (*A. arborescens* **Miller**). The genus *Aloe* includes trees (*A. ferox*), shrubs and herbs (*A. barbadensis*); they are xerophytic plants with large, fleshly leaves, carrying spines at the margin and they resemble the agave or century plant (*Agave americana* **Linne**).

The word *Aloe* in pharmacopoeias and formularies means a drug derived from the dried leaf juice. This has always created confusion because the leaves of the genus *Aloe* are the source of two products that are quite different in their chemical composition and therapeutic properties: **aloe latex** and **aloe gel**. These two products are obtained from two different specialized cells, **latex** from **pericyclic cells** and **gel** from **parenchymatous cells**. Therefore, the term *juice* must be avoided, as it could mean either the latex from the pericyclic cells or the gel after extraction from the leaf. However, to add even more confusion for people interested in medicinal herbs, there is also a preparation obtained from the whole leaf (total extract) and another obtained from the aloe wood, so-called **lignaloe** or **aloe of the Bible**, a fragrant wood obtained from an entirely different plant that was once used as an "**incense**" (Capasso et al., 1998).

ALOE PREPARATIONS

Aloe latex is prepared by cutting transversely the leaf near the base and taking it inclined so that the latex contained in the specialized pericyclic cells and

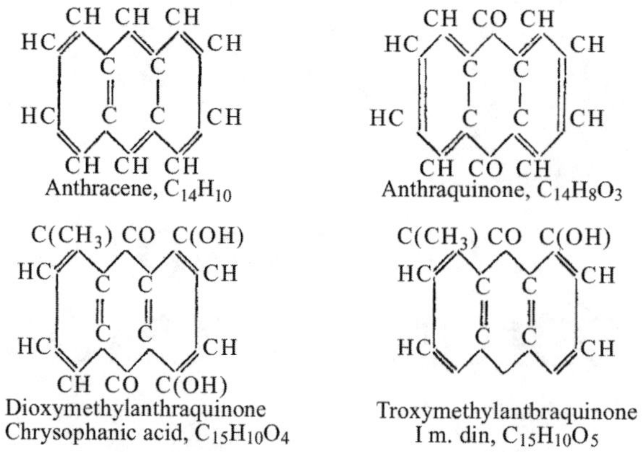

Anthracene, $C_{14}H_{10}$

Anthraquinone, $C_{14}H_8O_3$

Dioxymethylanthraquinone
Chrysophanic acid, $C_{15}H_{10}O_4$

Troxymethylantbraquinone
I m. din, $C_{15}H_{10}O_5$

Figure 8 Emodin is probably the most important of these principles. It is present in the amount of 0.8 to 1% in senna; 2.6% in frangula; 0.6% in cascara; 0.8% in Cape aloes; and 1.5% in rhubarb (Tschirch and Hiepe).

sometimes in the adjacent parenchyma flows out in about 6 hours. No pressure must be applied or the product will be contaminated with the mucilage present in the inner part of the leaves. The latex obtained is bitter and yellow; condensed to dryness it becomes a shiny mass, like broken glass, of a yellow greenish to red-black color. Slow evaporation, carried out at an inappropriate temperature, yields an opaque mass with a waxlike fracture. The taste is nauseating and bitter and the odor sour, recalling that of **rhubarb**, **apple-tart**, or **iodoform** (Sollmann, 1944).

Aloe latex contains **anthraquinones** (see Figure 8) and is totally different from aloe gel, a colorless gelatin obtained from the central portion of aloe leaf. The mucilaginous parenchymous tissue is excised from fresh leaves and immediately utilized for pharmaceutical preparations or lyophilized and kept dry until use. During extraction of the gel, it is practically impossible to prevent contamination by the leaf latex as the leaves are cut. On the other hand, in intact leaves, anthraquinones may diffuse into gel from the bundle sheath cells. To reduce such contamination, the starting material must be from varieties of aloe with a reduced anthraquinone content.

Scientific research has proved in recent years that the rind and the outer leaf, normally thrown away, contain much more active therapeutic ingredients. On the other hand, the gel is sensitive to heat and light and can quickly deteriorate at high temperature. For these reasons, the gel is now available as a cold-processed extract obtained from aloe whole leaf. The fresh whole leaf is cut into small pieces and whipped to obtain a homogeneous yellowish or reddish material. The product (10% aloe leaf and 90% water), concentrated, and in some cases deprived partially of **aloins**, is bitter.

CHEMICAL COMPOSITION OF ALOE PLANTS

Aloe plant produces **anthraquinone glycosides** (10 to 30%) like aloins (A and B), mucilage (30%), a resinous material (16 to 63%), sugars (about 25%), mucopolysaccharides like **acemannan** and **betamannan**, fatty acids (cholesterol,

campesterol, β-sitosterol), glycoproteins (aloctins A and B), enzymes (including **cyclooxygenase** and **bradykinase**), together with other compounds such as **lupeol, salicylic acid**, urea nitrogen, **cinnamic acid**, phenol, sulfur, magnesium lactate, **prostanoids**, and fiber.

- **Aloins** have laxative properties.
- **Aloctins, campesterol, β-sitosterol**, and **acemannan** have anti-inflammatory properties.
- **Acemannan** has immunostimulant properties.
- **Lupeol, salicyclic acid, phenol**, and **sulfur** have antiseptic properties.

THE THERAPEUTIC USES OF ALOE

Aloe and Cancer

Several studies have been carried out utilizing total leaf extracts internally, which therefore combine aloe gel with aloe latex. These extracts contain, among other substances, immunostimulatory and mild anti-inflammatory mucopolysaccharides of which acemannan is the most notable. **Mucopolysaccharides** are normally produced by the human body until puberty, after which time these substances must be introduced from outside sources. Their deficiency could produce drastic degenerative diseases. It has been shown that administration of acemannan stimulates macrophages to release **tumor necrosis factor** (TNF), **interleukin**, and **interferon** and it is able to abrogate viral infection in both animals and humans (Figure 9). Aloe extract also contains **aloctins** — substances that possess many biological activities, such as mitogenic activity for lymphocytes, binding of human α_2-**macroglobulin**, and complement activation via the alternative pathway. Recent studies also show that **aloctin A** is a promising candidate as an immunomodulator. Anti-inflammatory and immunomodulating agents are also **anthraquinones**. The presence of all these substances might be enough to explain the prophylactic, and possibly therapeutic, effect of aloe extract and its antitumor activity against leukopenia caused by exposure to cobalt-60, sarcoma-180, and Ehrich ascites (Janeway and Travers, 1994).

Experimental studies have also reported **antimetastatic activity** of aloe juice in rats and mice and inhibitory effect of aloe extract on induction of **preneoplastic hepatocellular lesions**. Antitumor effects of aloe extract may also depend on the ability to augment tumor-specific immunity. These findings have encouraged cancer treatment in humans with a preparation as follows: aloe (5 years old) fresh leaf 300 g, honey 500 g, two tablespoons of gin (or vodka or whisky); the mixture can be left for 10 days in a jar, filtered, and taken (1 tablespoon 1 or 2 times a day for 14 days) or mixed in a blender and then taken as above. Honey increases the palatability of the preparation and could enhance the effect of aloe because its content of **caffeic acid phenethyl ester**, a potent chemopreventive agent, is useful in combating diseases with a strong inflammatory component, i.e., various types of cancer. However, well-controlled clinical trials are insufficient to allow conclusions.

Aloe extract, used in association with **vitamin E** and **squalene**, is also able to cause regression of any tumor already formed in a mouse skin model.

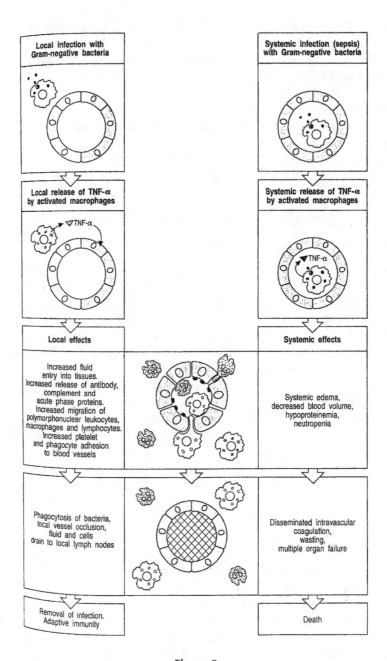

Figure 9

Aloe and AIDS

Aloe extract is also considered a possible therapy for acquired immunodeficiency syndrome (AIDS), in association with an antiviral agent (**asidothymidine**).

Aloe and Asthma

There is also evidence to show the efficacy of aloe extract in patients with chronic bronchial asthma. The effect of aloe extract seems due to formation of some

prostanoids during dark storage of aloe extract at 4 to 30°, for a period of 3 to 10 days.

Aloe and Diabetes

Hypoglycemic actions have also been referred to for aloe extract in five patients with non-insulin-dependent diabetes.

Aloe Gel and Skin Ailments

Aloe gel is reported to contain **glycoproteins, polysaccharides**, and other constituents (enzymes, etc.) and is essentially used for the treatment of various skin conditions (burns, abrasions, bruises, cuts, psoriasis, herpes simplex, etc.). Therefore, it is incorporated in ointments, creams, and lotions and other preparations for external use. Several studies have shown that the activity of aloe gel varies and that it is attributed to changes in the gel composition during growth or after harvest. On the other hand, there is a general agreement that aloe gel used in a fresh state, not on storage, is more able to show anti-inflammatory properties. This variation is due in part to the instability of the active ingredients, **aloctin A** and **aloctin B**. Acemannan has been recently indicated as another important active ingredient in aloe gel. The anti-inflammatory activity of aloe gel has been postulated to be due to inhibition of **eicosanoids**. Other suggested alternative mechanisms are inhibition of histamine formation by magnesium lactate, destruction of **bradykininase**, or inhibition of **bradykinin** activity. Many pharmacologists believe that the presence of small amounts of **anthraquinones** may enhance the beneficial effect of the gel when used in the treatment of inflammation. However, the literature on the efficacy of aloe in skin ailment management is still incomplete, and further studies are required to clarify the capacity of **acemannan** to interact with **integrins**, heterodimeric cell surface

---

Figure 9 (Opposite) The release of TNF-α, by macrophages induces local protective effects, but TNF-α can have damaging effects when released systemically. The left-hand panels show the causes and consequences of local release of TNF-α, the right-hand panels show the causes and consequences of systemic release. The central panels illustrate the common effects of TNF-α, which acts on blood vessels, especially venules, to increase blood flow, vascular permeability to fluid, proteins, and cells, and increased adhesiveness for white blood cells and platelets. Local release thus allows an influx into the infected tissue of fluid, cells, and proteins that participate in host defense. The small vessels later clot, preventing spread of the infection to the blood, and the fluid drains to regional lymph nodes where the adaptive immune response is initiated. When there is a systemic infection, or sepsis, with bacteria that elicit TNF-α production, then TNF-α acts in similar ways on all small blood vessels. The result is shock, disseminated intravascular coagulation with depletion of clotting factors and consequent bleeding, multiple organ failure, and death. Antibodies to TNF-α promote the dissemination of local infections, but prevent septic shock; they are used clinically only in systemic infection, where the effects of TNF-α are mainly deleterious. (From Janeway, C.A., Jr. and Travers, P., Immunobiology: The Immune Systems in Health and Disease, Current Biology Ltd, London, 1994. With permission.)

receptors. Integrins play a crucial role in inflammation, permitting inflammatory cells to leave the bloodstream and enter the damaged tissues (Capasso et al., 1998).

Aloe gel is also present in many cosmetics, for its emollient properties and antiaging effects on the skin. Although aloe is considered hypoallergenic even in large doses, allergic **contact dermatitis**, due to the topical application of gel preparations, has been described.

Aloe and Constipation

Aloe latex is an active laxative. Its effect is due to **anthraquinone glycosides**, **aloin A** and **B** (formerly designated **barbaloin**). **Glycosides** are probably chemically stable in the stomach (under conditions of pH 1 to 3) and the sugar moiety prevents their absorption in the upper part of the gastrointestinal tract. Once it has reached the large intestine, aloin is hydrolyzed by the bacterial flora to form **aloe-emodin-9-anthrone**, the active metabolite. **Hydroxyanthracene derivative** induces the secretion of fluid and electrolytes into the lumen through an inhibition of the Na^+-K^+-ATPase activity in the epithelial cells of the intestine, increases mucosal permeability of the endothelial cells of the colon, raises the level of cAMP in the enterocytes, and alters the motility of the large intestine by releasing endogenous substances. Taken in doses of 0.25 g, aloe causes laxation after 6 to 12 hours. Excessive doses or a prolonged use may cause nephritis, gastritis, vomiting, and diarrhea with elimination of mucus and blood. These side effects, occurring after chronic use or overdosage of aloe, have reduced its use as a laxative in recent years in some countries while in others it is still widely used alone or in association with **choleretic** and **cholagogue** drugs in constipation (Capasso et al., 1998).

MOVEMENTS OF THE COLON

The functions of the colon are (1) absorption of water and electrolytes from the chyme and (2) storage of fecal matter until it can be expelled. The proximal half of the colon, illustrated in Figure 9, is concerned principally with absorption, and the distal half with storage. Since intense movements are not required for these functions, the movements of the colon are normally sluggish. Yet in a sluggish manner, the movements still have characteristics similar to those of the small intestine and can be divided into **mixing movements** and **propulsive movements**.

Mixing Movements — Haustrations

In the same manner that segmentation movements occur in the small intestine, large circular constrictions also occur in the large intestine. At each of these constriction points, about 2.5 cm of the circular muscle contracts, sometimes constricting the lumen of the colon to almost complete occlusion. At the same time, the longitudinal muscle of the colon, which is aggregated into three longitudinal strips called the ***tineae coli***, contract. These combined contractions of the circular and longitudinal smooth muscle cause the unstimulated portion of the large intestine to bulge outward into baglike sacs called ***haustrations***. The haustral contractions, once initiated, usually reach peak intensity in about 30 seconds and then disappear during the next 60 seconds. They also at times move slowly analward during their period of contraction, especially in the cecum and ascending colon. After another few minutes, new haustral contractions occur in other areas nearby. Therefore, the fecal material in the large intestine is slowly dug into and rolled over in much the same manner that one spades the earth. In this way, all the fecal material is gradually exposed to the surface of the large intestine, and fluid is progressively absorbed until only 80 to 150 ml of the 750 ml daily load of chyme is lost in the feces.

Propulsive Movements — "Mass Movements"

Peristaltic waves of the type seen in the small intestine only rarely occur in most parts of the colon. Instead, most propulsion occurs by (1) the **haustral contractions** discussed above and (2) **mass movements**.

Most of the propulsion in the cecum and ascending colon results from the slow but persistent haustral contractions, requiring as many as 8 to 15 hours to move the chyme only from the ileocecal valve to the transverse colon, while the chyme itself becomes fecal in quality and also becomes a semisolid slush instead of a semifluid.

From the transverse colon to the sigmoid, mass movements take over the propulsive role. These movements usually occur only a few times each day, most abundantly for about 15 minutes during the first hour after eating breakfast.

A mass movement is characterized by the following sequence of events: First, a constrictive ring occurs at a distended or irritated point in the colon, usually in the transverse colon, and then rapidly thereafter the 20 cm or more of colon distal to the constriction contract almost as a unit, forcing the fecal material in this segment *en masse* down the colon. During this process, the haustrations disappear completely. The initiation of contraction is complete in about 30 seconds, and relaxation then occurs during the next 2 to 3 minutes before another mass movement occurs. But the whole series of mass movements will usually persist for only 10 minutes to half an hour, and they will then return perhaps a half day or even a day later.

Mass movements can occur in any part of the colon, although most often they occur in the transverse or descending colon. When they have forced a mass of feces into the rectum, the desire for defecation is felt.

Initiation of Mass Movements by the Gastrocolic and Duodenocolic Reflexes. The appearance of mass movements after meals is caused at least partially by gastrocolic and duodenocolic reflexes. These reflexes result from distension of the stomach and duodenum. They can take place, although with decreased intensity, when the autonomic nerves are removed; therefore, it is probable that the reflexes are basically transmitted through the **myenteric plexus**, although reflexes conducted through the autonomic nervous system probably reinforce this direct route of transmission.

It is likely that the hormone **gastrin**, which is secreted by the stomach antral mucosa in response to distension, also plays some role in this effect, because gastrin has an excitatory effect on the colon and an inhibitory effect on the ileocecal valve, thus allowing rapid emptying of ileal contents into the cecum. This, in turn, elicits increased colonic activity.

Irritation in the colon can also initiate intense mass movements. For example, a person who has an ulcerated condition of the colon (**ulcerative colitis**) frequently has mass movements that persist almost all of the time.

Mass movements are also initiated by intense stimulation of the parasympathetic nervous system or simply by overdistension of a segment of the colon.

DEFECATION

Most of the time, the rectum is empty of feces. This results partly from the fact that a weak functional sphincter exists approximately 20 cm from the anus at the juncture between the sigmoid and the rectum. However, when a mass movement forces feces into the rectum, the desire for defecation is normally initiated, including reflex contraction of the rectum and relaxation of the anal sphincters.

Continual dribble of fecal matter through the anus is prevented by (1) tonic constriction of the **internal anal sphincter**, a circular mass of smooth muscle that lies immediately inside the anus, and (2) the **external anal sphincter**, composed of striated voluntary muscle that both surrounds the internal sphincter and also extends distal to it; the external sphincter is controlled by the somatic nervous system and therefore is under voluntary control.

The Defecation Reflexes. Ordinarily, defecation results from the **defecation reflexes**, which can be described as follows: When the feces enter the rectum, distension of the rectal wall initiates afferent signals that spread through the myenteric plexus to initiate peristaltic waves in the descending colon, sigmoid, and rectum, forcing feces toward the anus. As the peristaltic wave approaches the anus, the internal anal sphincter is inhibited by the usual phenomenon of **receptive relaxation**, and if the external anal sphincter is relaxed, defecation will occur. This overall effect is the **intrinsic defecation reflex** of the colon itself.

However, the intrinsic defecation reflex itself is usually weak, and to be effective in causing defecation it must be fortified by another type of defecation reflex, a **parasympathetic defecation reflex** that involves the sacral segments of the spinal cord, as illustrated in Figures 10 and 11.

When the afferent fibers in the rectum are stimulated, signals are transmitted into the spinal cord and thence, reflexly, back to the descending colon, sigmoid, rectum, and anus by way of parasympathetic nerve fibers in the **nervi erigentes**. These parasympathetic signals greatly intensify the peristaltic waves and convert the intrinsic defecation reflex from an ineffectual weak movement into a powerful process of defecation that is sometimes effective in emptying the large bowel in one movement all the way from the splenic flexure to the anus. Also, the afferent signals entering the spinal cord initiate other effects, such as taking a deep breath, closure of the glottis, and contraction of the abdominal muscles to force downward on the fecal contents of the colon while at the same time causing the pelvic floor to pull outward and upward on the anus to evaginate the feces downward.

However, despite the defecation reflexes other effects are also necessary before actual defecation occurs because relaxation of the internal sphincter and forward movement of feces toward the anus normally initiates an instantaneous contraction of the external sphincter, which still temporarily prevents defecation. Except in babies and mentally inept persons, the conscious mind then takes over voluntary

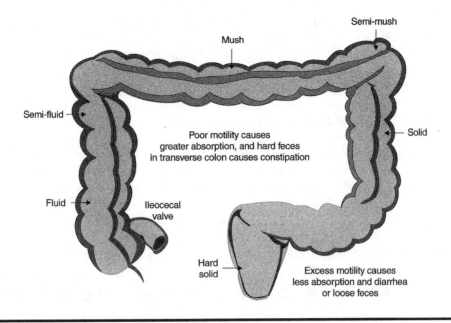

Figure 10 Absorptive and storage functions of the large intestine.

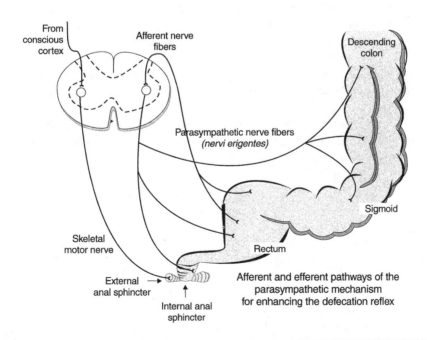

Figure 11 The afferent and efferent pathways of the parasympathetic mechanism for enhancing the defecation reflex.

control of the external sphincter and either inhibits it to allow defecation to occur or further contracts it if the moment is not socially acceptable for defecation. When the contraction is maintained, the defecation reflexes die out after a few minutes and usually will not return until an additional amount of feces enters the rectum, which may not occur until several hours thereafter.

When it becomes convenient for the person to defecate, the defecation reflexes can sometimes be excited by taking a deep breath to move the diaphragm downward and then contracting the abdominal muscles to increase the pressure in the abdomen, thus forcing fecal contents into the rectum to elicit new reflexes. Unfortunately, reflexes initiated in this way are never as effective as those that arise naturally, for which reason people who too often inhibit their natural reflexes become severely constipated.

In the newborn baby and in some persons with transacted spinal cords, the defecation reflexes cause automatic emptying of the lower bowel without the normal control exercised through contraction of the external anal sphincter.

OTHER AUTONOMIC REFLEXES AFFECTING BOWEL ACTIVITY

Aside from the duodenocolic, gastrocolic, gastroileal, enterogastric, and defecation reflexes several other important nervous reflexes can affect the overall degree of bowel activity. These are the **intestino-intestinal reflex**, **peritoneo-intestinal reflex**, **reno-intestinal reflex**, **vesico-intestinal reflex**, and **somato-intestinal reflex**. All these reflexes are initiated by sensory signals that pass to the spinal cord and then are transmitted through the sympathetic nervous system back to the gut. And they all inhibit gastrointestinal activity. Thus, the **intestino-intestinal reflex** occurs when one part of the intestine becomes overdistended or its mucosa becomes excessively irritated; this blocks activity in other parts of the intestine while the local distension or irritability increases activity in the localized region and moves the intestinal contents away from the distended or irritated area.

The **peritoneo-intestinal reflex** is very much like the intestino-intestinal reflex, except that it results from irritation of the peritoneum; it causes intestinal paralysis. The **reno-intestinal** and **vesico-intestinal reflexes** inhibit intestinal activity as a result of kidney or bladder irritation. Finally, the **somato-intestinal reflex** causes intestinal inhibition when the skin over the abdomen is irritatingly stimulated (Guyton, 1981).

CONSTIPATION

Definition

Constipation is a complaint commonly seen in clinical practice. Because of the wide range of normal bowel habits, constipation is difficult to define precisely. Most persons have at least three bowel movements per week, and constipation has been defined as a frequency of defecation of less than three times per week. However, stool frequency alone is not a sufficient criterion to use, because many constipated patients describe a normal frequency of defecation but subjective complaints of excessive straining, hard stools, lower abdominal fullness, and a sense of incomplete evacuation. Thus, a combination of objective and subjective criteria must be used to define constipation.

Causes

Pathophysiologically, constipation generally results from **disordered colonic transit** or **anorectal function** as a result of a primary motility disturbance, certain drugs, or in association with a large number of systemic diseases that affect the gastrointestinal tract. Constipation from any cause may be exacerbated by chronic illnesses that lead to physical or mental impairment and result in inactivity or physical immobility. Additional contributing factors may include a lack of fiber in the diet, generalized muscle weakness, and possibly stress and anxiety.

In the patient presenting with a recent onset of constipation, an **obstructing lesion** of the colon should be sought. In addition to a colonic neoplasm, other causes of colonic obstruction include **strictures** due to colonic ischemia, diverticular disease, or inflammatory bowel disease; foreign bodies; or **anal strictures**. Anal sphincter spasm due to painful **hemorrhoids** or **fissures** also may inhibit the desire to evacuate.

In the absence of an obstructing lesion of the colon, **disturbed colonic motility** may mimic colonic obstruction. Disruption of parasympathetic innervation to the colon as a result of injury or lesions of the lumbosacral spine or sacral nerves may produce constipation with hypomotility, colonic dilatation, decreased rectal tone and sensation, and impaired defecation. In patients with **multiple sclerosis**, constipation may be associated with neurogenic dysfunction of other organs. Similarly, constipation may be associated with **lesions of the central nervous system** caused by parkinsonism or a cerebrovascular accident. In South America, the parasitic infection **Chagas' disease** may result in constipation because of damage to myenteric plexus ganglion cells. **Hirschsprung's disease**, or **aganglionosis**, is characterized by absence of myenteric neurons in a segment of distal colon just proximal to the anal sphincter. This results in a segment of contracted bowel that produces obstruction and proximal dilatation. In addition, an absent **rectosphincteric inhibitory reflex** results in the failure of the internal anal sphincter to relax following rectal distention. Most patients with Hirschsprung's disease are diagnosed by 6 months of age, but in occasional cases symptoms are mild enough that the diagnosis may be delayed into adulthood.

Drugs that may lead to constipation include those with **anticholinergic properties**, such as **antidepressants** and **antipsychotics**, **codeine** and other narcotic analgesics, **aluminum-** or **calcium-containing antacids**, **sucralfate**, **iron supplements**, and **calcium channel blockers**. In patients with certain endocrinopathies such as **hypothyroidism** and **diabetes mellitus**, constipation is generally mild and responsive to therapy. Rarely, life-threatening megacolon occurs in patients with **myxedema**. Constipation is common during **pregnancy**, presumably as a result of altered progesterone and estrogen levels, which decrease intestinal transit. **Collagen vascular diseases** may be associated with constipation, which may be a particularly prominent feature of **progressive systemic sclerosis**, in which delayed intestinal transit results from atrophy and fibrosis of colonic smooth muscle (Fauci et al., 1998).

Cathartics

Cathartics are drugs that are employed in medicine to cause evacuation of the bowels. They may act by irritating the intestines or by augmenting the fluid of the feces. In the first instance, intestinal peristalsis is increased directly, and in the second instance it is increased indirectly by distention of the bowel. Cathartics are drugs that promote defecation. They are probably used more than any other class of medicinal compounds. This fact is not an index of their usefulness or value but rather a measure of the misconceptions that exist concerning their worth and the indications for their use.

The cathartics are so numerous that they require classification. This has been done in a variety of ways. The drugs may be classified according to the intensity of their action, and thus can be designated as laxatives, purgatives, and drastics, in order of increasing potency. The group can also be delineated according to the site of action, or by the interval of time before they are effective. The most useful classification, however, is that which takes into account both their mechanism of action and their chemical nature (Table 14).

Cathartics act by one of three fundamental mechanisms. The so-called **irritant cathartics** are drugs that are presumed to irritate the intestinal tract and thus to increase its motor activity. This leads to a rapid propulsion of the intestinal contents and the passage of a stool that may vary in consistency from one that is formed to one that is liquid, depending upon the degree of irritation. The **bulk cathartics** increase the contents of the intestinal tract and are therefore more physiological in action. They are **hydrophilic colloids**, which resist destruction the gastrointestinal tract, or **indigestible fiber**, or inorganic salts, which are slowly absorbed from the intestinal tract and thus hold water at that site due to the osmotic pressure which they exert. The **emollient cathartics** lubricate the intestinal tract and thereby facilitate the passage of feces.

EMODIN OR ANTHRAQUINONE CATHARTICS

Senna, rhubarb, cascara, aloes, and related drugs contain glucosidal compounds that are themselves inactive. In the alkaline intestine they are hydrolyzed or oxidized, yielding various **oxymethyl-anthraquinones**, especially **emodin** and chrysophanic acid, which produce a cathartic effect chiefly by acting on the large intestines as **laxatives**, **aperients**, or **ecproctics**, with easier defecation of formed stools. They are often more effective when taken in divided doses (three times daily) than as a single dose. With larger doses and with the more active members, the action extends upward into the small intestines, with griping and forced evacuation of liquid stools; however, it does not readily progress to inflammation. The effects are fairly easily controlled by grading the dosage, and by repeating the smaller doses, as the action tends to be cumulative. The dosage may be reduced or omitted when the desired degree is attained. These drugs are therefore used especially in chronic constipation, in addition to or in alternation with mineral oil. Prolonged use tends to decrease the response of the intestines, so that the effect occurs only with larger doses, which are apt to set up chronic inflammation of the mucosa. **Cascara** is the mildest of these drugs and the most easily controlled. **Senna** is the most active and generally causes griping and semiliquid stools. **Aloe** and **aloin** are intermediate, but can be pushed to drastic action and this may result in abortion. Rhubarb catharsis is apt to be followed by constipation, which has been attributed to its **tannic acid**. **Phenolphthalein** acts similarly to cascara and aloes, but the response is variable. Its chemical structure is somewhat different (see Figure 8).

Table 14   Classification of Cathartics

"Irritant" Cathartics
Emodin cathartics
 Cascara sagrada
 Senna
 Rhubarb
 Aloe
Resinous cathartics
 Jalap
 Colocynth
 Elaterin
 Podophyllum
Irritant oil cathartics
 Castor oil
 Croton oil
Miscellaneous cathartics
 Phenolphthalein
 Mercurous chloride

Bulk Cathartics
Saline cathartics
 Magnesium sulfate
 Magnesium oxide
 Magnesium citrate
 Magnesium carbonate
 Sodium sulfate
 Sodium phosphate
 Potassium sodium tartrate
Hydrophilic colloids and indigestible fiber
 Bran
 Agar
 Tragacanth
 Psyllium seed
Emollient cathartics
 Liquid petrolatum
 Vegetable oils

Constipation may be defined as the passage of excessively dry stools, infrequent stools, or stools of insufficient size. Constipation is a symptom and not a disease. It may be of brief duration (e.g., when one's living habits or diet change abruptly) or it may be a life-long problem, as occurs in congenital aganglionosis of the colon (Hirschsprung's disease).

The pure substances would be too irritant, but their action is graded by their slow liberation, and by the presence of colloid extractives. The most common of these substances are **emodin** and **chrysophanic acid**. Numerous isomers of these are possible (15 for emodin alone). The special character of the different drugs is probably due to differences in these isomers, in the stability of their glucosidal combination, and to the presence of other associated substances (**tannin in rhubarb**).

Fate. Emodin is partly absorbed and is excreted into the urine and milk (laxative effects on sucklings). The drugs containing emodin color the urine yellowish-brown when it is acid, reddish or violet when alkaline. It may be advisable to acquaint the patient with this fact.

Other drugs that change the color of the urine are as follows. **Logwood (hematoxylon)** does not color acid urine, but produces a reddish or violet color in alkaline urine. **Santonin** gives a yellowish color to acid urine, with a yellow foam; if the urine is made alkaline it gives a very pronounced pink. **Picric acid** gives reddish-brown color in both acid and alkaline urine. The various coal-tar products give a brownish-black color. **Methylene blue** imparts a green color.

CASCARA SAGRADA

This is used chiefly in habitual constipation. It softens the stools without noticeable irritation. The susceptibility is not lost with use; on the contrary, the establishment of regular habits usually permits it to be gradually withdrawn. By its bitter taste, it acts also as stomachic. It may be given as the fluid extract, 15 drops three times daily, or in a single dose, a half teaspoonful, in the evening. This causes a single soft stool in the morning. In the **aromatic fluid-extract**, which is much more pleasant, some of the bitter and cathartic principles have been destroyed by an alkali (magnesia). It may be given in twice the dose.

SENNA

The effects are more extensive and more irritant than with the other drugs of the group, so that it is better suited for acute than for chronic constipation. There is considerable griping, which may be corrected by **carminatives**, or by the previous extraction of the resins with alcohol (as in the fluid extract).

RHUBARB

This contains a considerable quantity of **tannic acid** as well as cathartic principles. The astringent action predominates with smaller doses (0.05 to 0.01 g) and these are used as astringent bitters in gastric catarrh, and in diarrhea. Larger doses (1 to 5 g) are laxative with little colic. They may be employed in chronic constipation, but the astringent action makes it inferior to cascara. It may cause skin eruptions.

ALOE

As described in previous sections, the inspissated juice of the leaves of various species of *Aloe* from the West Indies, Africa, etc. furnishes the commercial varieties of aloes, which differ somewhat in composition and action. They contain **anthraquinone** derivatives (see Figure 8), especially **aloins** (10 to 16%). These are glucosides, which are probably inactive themselves, but which yield emodin and other active anthraquinone derivatives on cleavage. This cleavage occurs in the intestine; it is favored by alkalis and perhaps also by bile and by iron salts. Aloin may be used as such. Aloe is intensely bitter, so that it must be administered in pills. Very small doses may be used as **stomachic**. Moderate doses are simply laxative, but somewhat apt to gripe, and are therefore commonly combined with **belladonna** or **carminatives (myrrh**, etc.). They do not lose their efficiency on continued use and would be suited to habitual constipation, but have no advantage over cascara. Aloes are supposed to produce considerable congestion of the pelvic organs and are therefore contraindicated in hemorrhoids, menstruation, and pregnancy. On the other hand, they may be used as **emmenagogues**. They are especially prescribed for correcting the constipative action of iron medication.

CASTOR OIL

This consists mainly of the triglyceride of an unsaturated fatty acid, **ricinoleic acid**. The neutral fat is not active, but becomes so when it is saponified in the intestine. The cathartic effect is due mainly to motor stimulation of the small intestine. The intestinal secretion is not increased, the fluid character of the stools being due to the quicker passage of the feces.

With the usual dose, ½ to 1 ounce, the response is certain and complete, resulting in a few hours in one or two extensive semifluid stools, with but slight hyperemia or irritation, there being little or no colic. It is especially suited to the treatment of acute constipation or acute diarrhea from dietetic errors, ptomaine and other poisoning, etc. It is not used continuously, since this may produce gastric disturbance. There is some pelvic congestion, so that it is avoided during menstruation and pregnancy.

RICIN

This toxic protein is contained in **caster seeds**, but does not pass into the oil. Similar **phytotoxins** occur in **croton seeds (Crotin)**; and in **jequirity seeds (Abrin)**; in the **bark of the locust tree**, **Robinia pseudo-acacia (Robin)**; and in the seeds of some **leguminous plants (Phasin)**. The last is but weakly toxic. The ricin is responsible for the toxic effects on eating the castor seeds; 5 or 6 of these are fatal to a child, 20 to adults, and 3 or 4 seeds may cause violent gastroenteritis, with nausea, headache, persistent vomiting, colic, sometimes bloody diarrhea, thirst, emaciation, and great debility. The symptoms usually do not set in until after several days. More severe intoxications cause small frequent pulse, cold sweat, icterus, and convulsions. Death occurs in 6 to 8 days, from the convulsions or from exhaustion. The fatality is about 6%. This small fatality rate is due to the destruction of the poison in the alimentary canal. The treatment would be evacuant and symptomatic. Usually, 3 to 10 days are required to complete recovery.

ABRIN

Abrin, from **jequirity beans (*Abrus precatorius*)**, resembles ricin so closely in its action that the difference was established only when it was noted that immunity against one did not constitute immunity against the other.

CROTON OIL

This contains, besides ordinary fats, about 10% of "**croton-resin**," the active component producing the local and systemic effects of the oil. It is destroyed by boiling with alkalis. Croton oil is the most violent of all cathartics, ½ to 1 drop producing burning in the mouth and stomach, often vomiting, and after ½ to 3 hours, several extensive fluid evacuations, with much colic and tenesmus. It was formerly used in extreme cases of constipation, and in unconscious patients (**uremic** and **apoplectic coma**), but it should never be employed. It may produce severe local effects on the mouth, as well as gastroenteritis. Toxic doses cause collapse. A dose of 20 drops has been fatal, but recovery is said to have followed ½ ounce — presumably because of vomiting. The treatment would be evacuant, demulcent, opiates, and symptomatic (directed against the enteritis and collapse).

RESIN OF PODOPHYLLUM

This acts very slowly, generally only after 12 to 24 hours or longer, producing several soft stools. Larger doses (20 mg, ⅓ grain, or more) act as a **hydragogue cathartic**, but it is chiefly employed in small doses (5 to 10 mg, ⅒ or ⅙ grain) as a laxative in chronic constipation, often combined with aloes or calomel; 0.3 to 0.5 g has been fatal.

Podophyllum Basin. U.S.P. (**Podophyllin**) is prepared from the dried rhizome and roots of *Podophyllum peltatum* (may apple, mandrake), a plant used by native peoples. An amorphous greenish-yellow powder, of slight peculiar odor and faintly bitter taste. Active constituents, resin and a crystalline principle, **podophyllotoxin**. Dose of the resin, 0.01 to 0.065 g, ⅙ to 1 grain. Very irritating to mucous membranes, especially of the eyes.

Other roots and rhizomes related to this group include **Leptandra (Culver's root)**, from *Veronica virginica*. Contains an amorphous bitter principle, and not a crystalline glucoside, as formerly reported. "**Leptandrin**" is a resinoid, pharmacologically inert, but was supposed to be cholagogic without producing intestinal irritation.

JALAP, SCAMMONIUM, IPOMEA

Jalap is perhaps the mildest member of the group. It produces watery stools within 3 or 4 hours, with little intestinal irritation, without gastric disturbance or much prostration. It has little taste and is therefore easily administered. It has been used in ascites, cerebral congestion, hypertension, and as adjuvant to anthelmintics. **Scammony** and its equivalent **Ipomea** are closely related to jalap.

COLOCYNTH, ELATERIN, AND BRYONIA

A number of plants of the cucumber family yield these drastic drugs. They are of complex composition, each containing several active principles, resinous and alkaloidal. They produce continuous watery stools, with intense irritation and hyperemia (Sollmann, 1944).

10

ALKALOIDS

An extensive number of important drugs are naturally occurring alkaloids:

- Belladonna
- Muscarine
- Oxotremorine
- Pilocarpine
- Rauwolfia
- Vinca alkaloids

INTRODUCTION

One of the most common definitions states that an **alkaloid** is "a nitrogenous base of plant origin having marked physiological action." This is quite in error since some of the alkaloids are not necessarily basic in reaction and many of them possess little or no physiological action. All alkaloids contain one or more nitrogen atoms, which is also true of the proteins. Although it is true that most alkaloids are derived from plants, there are substances of animal origin that from any chemical reasoning certainly ought to be included with the alkaloids. A case in point includes **ephedrine** (of plant origin) and **epinephrine** (of animal origin), both of which are quite similar in chemical constitution. The best that can be said is that alkaloids are nitrogenous, are usually of plant origin, frequently basic in character, and often have a definite physiological action.

In spite of the difficulty in definitely characterizing alkaloids by definition, they do show a surprising number of physical and chemical properties in common. For the most part the alkaloids are insoluble or sparingly so in water, but form salts (by metathesis or addition) that are usually freely soluble. The free alkaloids are usually soluble in ether or chloroform or other immiscible solvents, in which, however, the alkaloidal salts are insoluble. This permits a ready means for the isolation and purification of the alkaloids as well as for their quantitative estimation. Most of the alkaloids are crystalline solids, although a few are amorphous and an additional few (**coniine**, **nicotine**, **sparteine**) are liquids. It is interesting to note that the liquid alkaloids have no oxygen in their molecules. Alkaloidal salts are invariably crystalline and their crystal form and habit is often a useful means of their rapid microscopical identification (Sollmann, 1944).

The alkaloids, like other amines, form double salts with compounds of mercury, gold, platinum, and other heavy metals. These double salts are usually obtained as precipitates and many of them are microcrystallographically characteristic. Among the common alkaloidal reagents are **Wagner's** (iodine in potassium iodide), **Mayer's** (potassium mercuric iodide), **Dragendorff's** (potassium bismuth iodide), **Hager's** (picric acid), **Marme's** (potassium cadmium iodide), **Sonnen-schein's** (phosphomolybdic acid), gold chloride, mercuric chloride, tannic acid, and many others. The alkaloids usually possess a bitter taste.

The alkaloids appear to have a restricted distribution in the vegetable kingdom. Among the **angiosperms** the **Apocynaceae, Leguminosae, Papaveraceae, Ranunculaceae, Rubiaceae, Solanaceae,** and **Fumariaceae** are outstanding for plants yielding alkaloids. The **Labiatae** and **Rosaceae** are quite free and the **monocotyledons** and **gymnosperms** only rarely contain them. A specific alkaloid is usually confined to a specific plant family, although **caffeine** and **berberine** are exceptions to this rule. Alkaloids may occur in various plant parts: in seeds and fruits (**nux vomica** and **black pepper**), in leaves (the **solanaceous leaves**), in rhizomes and roots (**aconite, corydalis, hydrastis**), and in barks (**cinchona, pomegranate**).

The names of the alkaloids are obtained in various ways: (1) from the generic and specific names of the plants yielding them (**colchicine, belladonnine**); (2) from the name of the drug yielding them (**ergotoxine**); and (3) from their physiological activity (**emetine, morphine**) and occasionally from the discoverer (**pelletierine**). Sometimes a prefix or suffix is added to the name of a principal alkaloid to designate another alkaloid from the same source (**quinine, quinidine, hydroquinine**). By agreement, chemical rules designate that the names of all alkaloids should end in "ine."

Various schemes for the classification of alkaloids based upon their constitution have been suggested. Because of the wide variation in the constitution of alkaloids, however, none of these is quite satisfactory.

Alkaloids usually contain one nitrogen atom, although some, like **ergotoxine**, may contain up to five. The nitrogen may exist as a primary amine ($R-NH_2$), a secondary amine such as (R_2NH) or cyclic, a tertiary amine such as (R_3N) or cyclic, or a quaternary ammonium hydroxide such as (R_4N-OH) or cyclic. These forms of nitrogen linkage are basic and account for the common basic nature of the alkaloids. **Acid amides** (neutral) and **acid imides** (acid) may also be found in the alkaloids.

CLASSIFICATION

A. The Group Containing Aliphatic Bases with an Aromatic Nucleus

Including such compounds as **tyramine** (*p*-hydroxyphenylethylamine) from ergot, **ephedrine** from ephedra, and **colchicine** from colchicum (Figure 12).

B. The Pyrrolidine Group

Pyrrolidine occurs free in small quantities in **tobacco** and **opium** and is related to its mother substance pyrrole (Figure 13). To this group belong such alkaloids

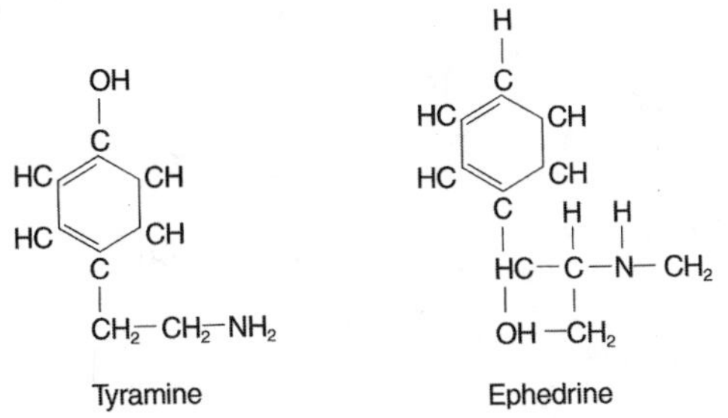

Figure 12   The structures of tyramine and ephedrine.

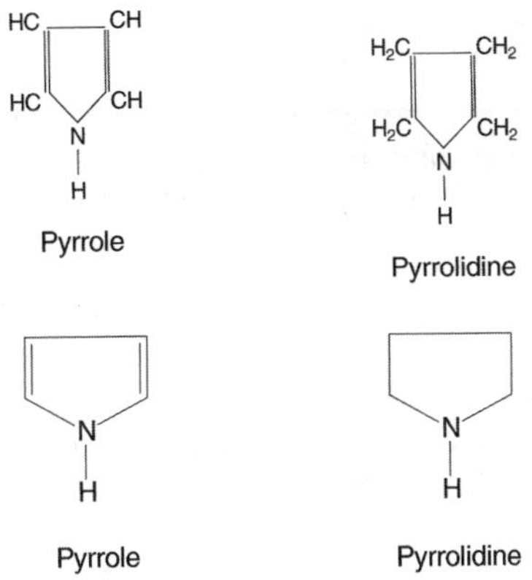

Figure 13   The structures of pyrrole and pyrrolidine.

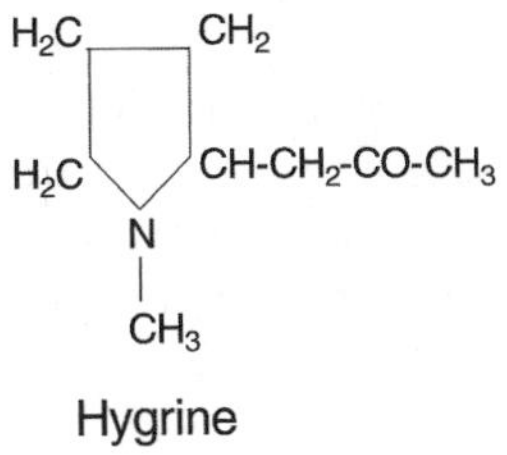

Figure 14   The structure of hygrine.

as **hygrine** from ***Erythroxylon coca*** and **stachydrine** from ***Stachys tuberifera*** (Figure 14).

C. The Pyridine Group

Upon reduction the tertiary base **pyridine** is converted into the secondary base **piperidine.** These two nuclei form the basis of this group (Figure 15). The group is sometimes divided into four subgroups: (1) derivatives of piperidine including

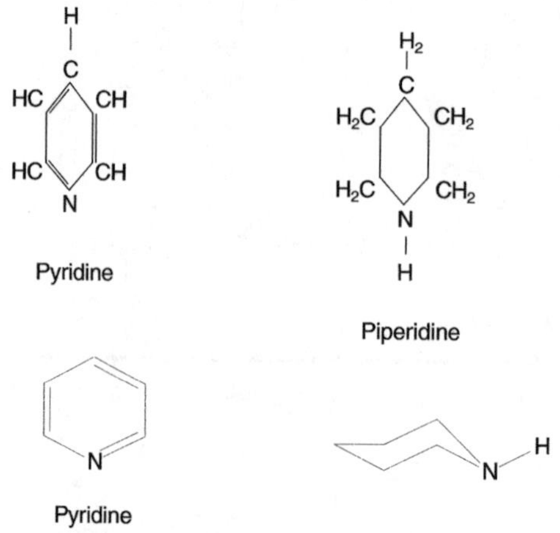

Figure 15 The structures of pyridine and piperidine.

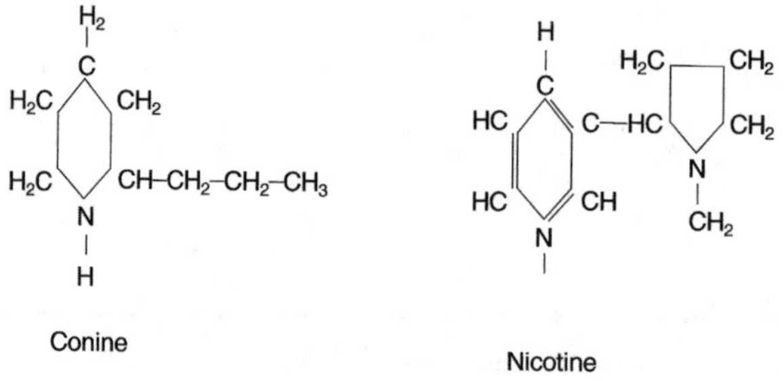

Figure 16 The structures of coniine and nicotine.

piperidine from **black pepper** and methysticin from **kava**: (2) derivatives of **α-propylpiperidine** including coniine from **conium**; (3) derivatives of nicotinic acid including **arecoline** from areca; and (4) derivatives of both pyridine and pyrrolidine including **nicotine** from tobacco (Figure 16).

D. The Glyoxaline Group

Imidazole (glyoxaline) is the principal nucleus in **histamine** from **ergot** and **pilocarpine** from pilocarpus (Figure 17).

E. Alkaloids with Condensed Pyrrolidine and Piperidine Rings

Tropane is formed when **pyrrolidine** and **piperidine** are condensed. Closely related to tropane are tropine, the principal nucleus of the solanaceous alkaloids, **atropine**, **hyoscyamine**, **hyoscine** and **belladonnine**, and **ecgonine**, the nucleus of **cocaine** (Figure 18).

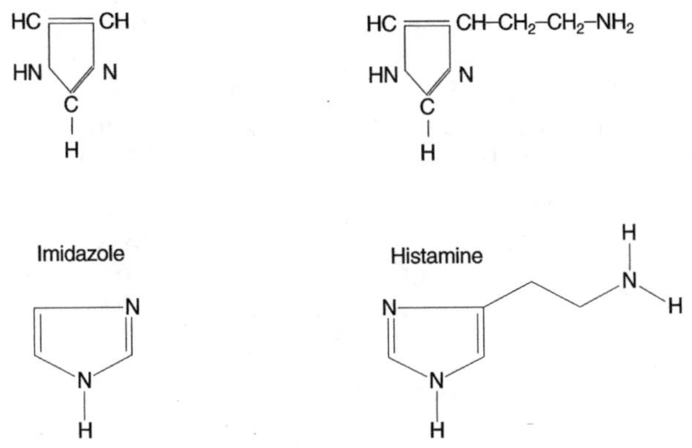

Figure 17 The structures of imidazole and histamine.

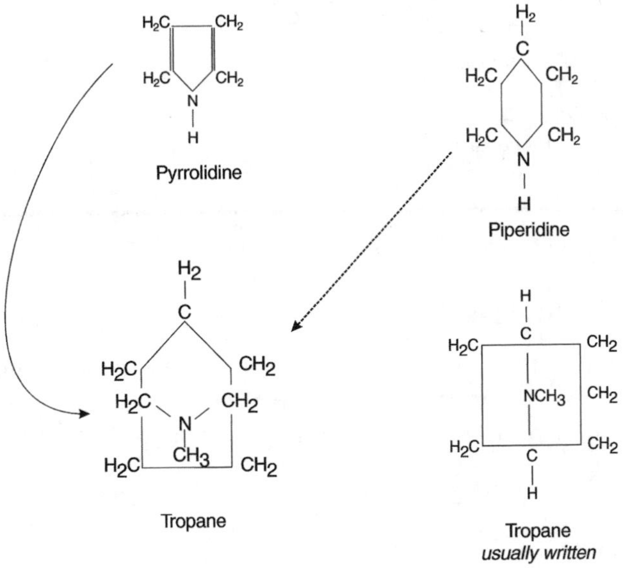

Figure 18 The structures of pyrrolidine, piperidine, and tropane.

F. Derivatives of Quinoline

Alkaloids containing **quinoline** as the principal nucleus include **anemonine** from *Anemone thalictroides*, **galipine** from **Angostura bark** (*Galipea officinalis*), and the **cinchona** alkaloids, **quinine**, **quinidine**, **cinchonine**, and **cinchonidine** (Figure 19).

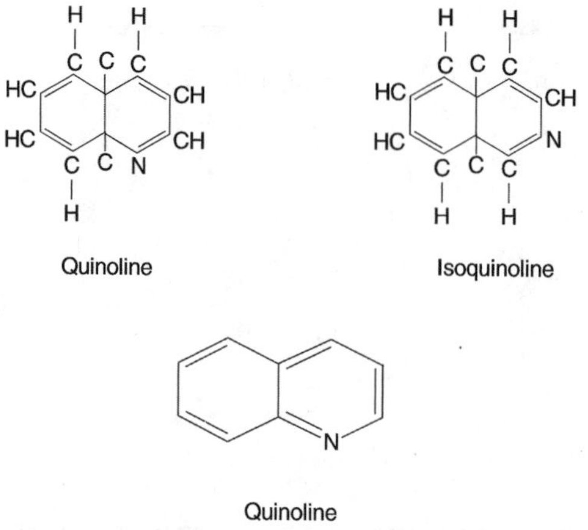

Figure 19 The structures of quinoline and isoquinoline.

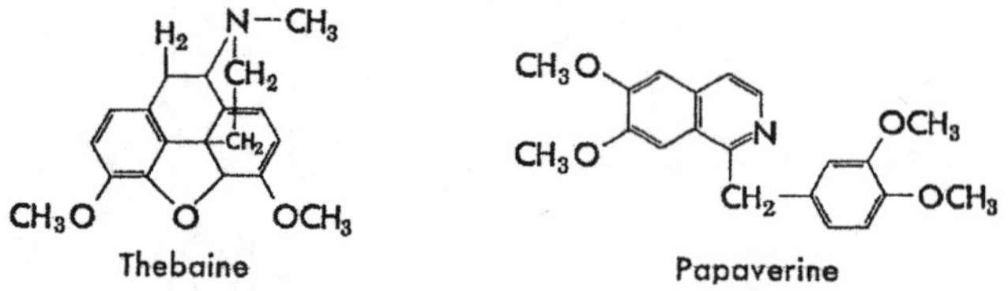

Figure 20 The structures of thebaine and papaverine.

G. Derivatives of Isoquinoline

Alkaloids containing the isoquinoline nucleus may be divided into two groups: (1) **papaverine**, **narcotine**, **cotarnine**, **narcine**, and certain others of the opium group, and (2) the hydrastis group including **hydrastine**, **berberine**, and **canadine** (Figure 20).

H. Phenanthrene

The opium alkaloids **morphine**, **codeine**, and **thebaine** have a phenanthrene nucleus (Figure 21).

I. Indole Derivatives

Certain alkaloids upon distillation with zinc dust yield indole. It is natural, therefore, to assume that they possess an indole ring as part of their structure. **Strychnine** and **brucine** from **nux vomica** and **physostigmine** from **physostigma** belong to this group. Strychnine and brucine contain in addition a quinoline nucleus and many authors classify them in the quinoline group. The formula for strychnine shown in

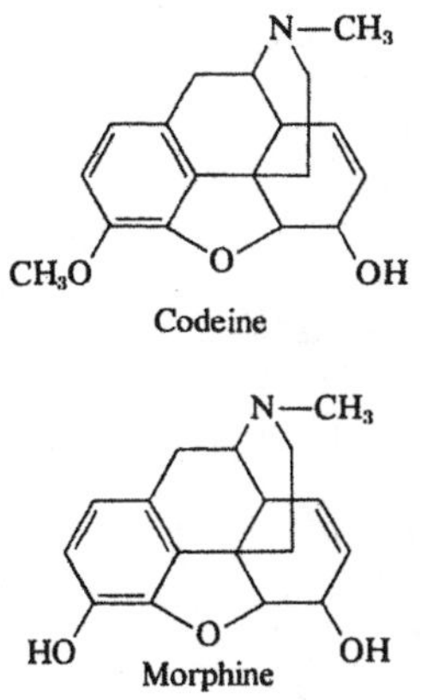

Figure 21   The structures of codeine and morphine.

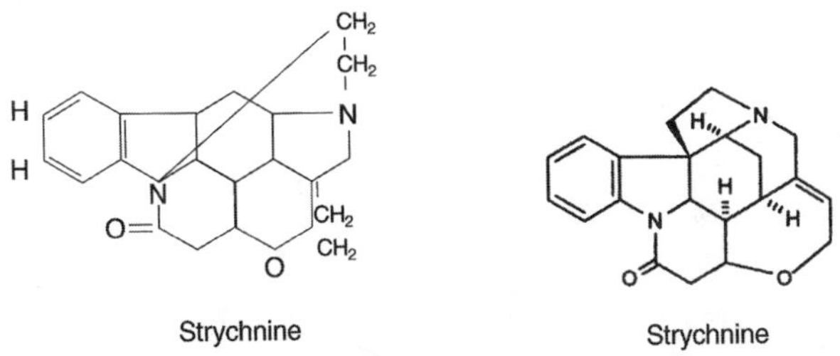

Figure 22   The structure of strychnine.

Figure 22 is now generally accepted. Brucine is **dimethyoxystrychnine**, in which the methoxy groups replace the two hydrogen atoms indicated in the formula.

J. The Purine Bases

The purine nucleus is constructed by a "condensation" of the pyrimidine and imidazole (glyoxaline) nuclei (Figure 23). **Xanthine** is 2,6-dioxypurine; caffeine is 1,3,7-trimethylxanthine, **theophylline** is 1,3- and **theobromine** 3,7-dimethylxanthine. These alkaloids are found in **coffee, tea, cacao, kola, maté,** and **guarana** (Figure 24).

K. Alkaloids of Unknown Constitution

The constitution of a number of alkaloids has not as yet been definitely established. Among these are **aconitine** from **aconite, delphinine** from **larkspur, jervine** and **protojervine** from **veratrum viride, cevadine** from **sabadilla seed, lobeline** from **lobelia,** the curare alkaloids from *Chondrodendron tomentosum,* **sanguinarine** from **sanguinaria,** and many others.

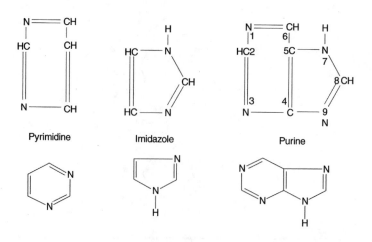

Figure 23 The structures of pyrimidine, imidazole, and purine.

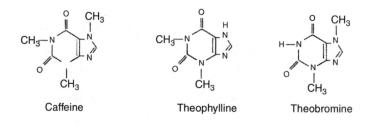

Figure 24 Pharmacology of important xanthines.

As with most attempts at chemical classification, some alkaloids do not fall satisfactorily into any of the above groups and others can be classified in more than one group.

11

MANUKA AND FUNGAL DISEASES

TEA TREE OILS FROM AUSTRALIA AND NEW ZEALAND

Both Australia and New Zealand have indigenous "**tea trees**" in the family **Myrtaceae**, which were reputedly used for brewing tea by **Captain Cook**. There is, however, no resemblance between real tea *Camellia sinensis*, **Camelliaceae**, and the taste or odor of these species. The Australian tea tree oil from *Melaleuca alternifolia* and other *Melaleuca* species has strong antimicrobial potential (see Lis-Balchin et al., 2000, for a review and references).

The whole-plant extract had been used originally by aboriginals and the essential oils themselves had been used during the World War II as a general antimicrobial agent and insect repellent, and provided in the first-aid kits of serving Australian soldiers. The essential oil is today used as a strong antimicrobial and antifungal agent in creams, soaps, toothpastes, and other preparations and has been used both externally and internally by both **herbalists** and **aromatherapists** for some years.

There is scant evidence that **Manuka (*Leptospermum scoparium*)** and **Kanuka (*Kunzea ericoides*)** have such potential, but as the essential oils are said to have remarkable powers of healing, based on folk medicinal usage, these oils are being used by some aromatherapists, although there have been no safety/toxicological evaluations performed on them.

Many of the folk-medicinal uses of the New Zealand tea tree oils are related to both species, e.g., the leaves of Manuka and Kanuka were used as vapor baths for colds; an infusion was very astringent and various uses were found for concoctions including urinary complaints and as a febrifuge. Kanuka was applied to scalds and burns, used to stop coughing and as a sedative; it was also used against dysentery. The decoction of boiled leaves and bark was used to treat stiff backs etc. Seed capsules were boiled to yield a decoction to apply externally to treat inflammation or to drink for diarrhea; the capsules or leaves were also chewed for dysentery. The water from boiled bark was used for treating inflamed breasts and also to treat mouth, throat, and eye problems.

The antibacterial effect of honeys derived from Kanuka and Manuka blossom against ***Staphylococcus aureus*** was shown and more recently Manuka honey was shown to be active against *Helicobacter pylori*. *Leptospermum scoparium* contains **leptospermone**, which has antihelminthic properties and is closely related to compounds having similar properties in male ferns; **leptospermone** has insecticidal properties, and is similar in structure to the insecticide **valone**.

Lis-Bulchin et al. (2000) studied three different species of **Myrtaceae** growing in Australia and New Zealand all known as "tea tree": the Australian tea tree (*Melaleuca alternifolia*), the New Zealand Manuka (*Leptospermum scoparium*), and Kanuka (*Kunzea ericoides*). All three essential oils are used by aromatherapists, although only *Melaleuca* has been tested for toxicity and its antimicrobial effects studied. The pharmacology and antimicrobial activity of the three tea tree oils was determined using guinea pig ileum, skeletal muscle (chick biventer muscle and the rat phrenic nerve diaphragm), and also rat uterus *in vitro*. Differences were shown between the three essential oils in their action on smooth muscle: Manuka had a spasmolytic action, while Kanuka and *Melaleuca* had an initial spasmogenic action. Using the diaphragm, Manuka and *Melaleuca* decreased the tension and caused a delayed contracture; Kanuka had no activity at the same concentration. The action on chick biventer muscle was, however, similar for all three oils, as was the action on the uterus, where they caused a decrease in the force of the spontaneous contractions. The latter action suggests caution in the use of these essential oils during childbirth, as cessation of contractions could put the baby, and mother, at risk. The comparative antimicrobial activity showed greater differences between different samples of Manuka and Kanuka than *Melaleuca* samples. The antifungal activity of Kanuka was inversely proportional to its strong antibacterial activity, while Manuka displayed a stronger antifungal effect, although not as potent as *Melaleuca*. The antioxidant activity of Manuka samples was more consistent than that of Kanuka, whereas *Melaleuca* showed no activity. The variability in the Manuka and Kanuka essential oils suggests caution in their usage, as does the fact that the oils have not been tested for toxicity.

FUNGAL DISEASES

The fungal diseases are **histoplasmosis**, **coccidioidomycosis**, **blastomycosis**, **paracoccidioidomycosis**, **cryptococcosis**, **sporotrichosis**, **candidiasis**, **aspergillosis**, and **mucormycosis**.

ANTIFUNGAL AGENTS

The antifungal agents consist of either topical or systemic medications. The systemic antifungal agents include:

■ Amphotericin B
■ Flucytosine
■ Ketoconazole
■ Griseofulvin
■ Nystatin

The topical antifungal agents comprise:

- Clotrimazole
- Econazole
- Miconazole
- Terconazole and butoconazole
- Ciclopirox olamine
- Haloprogin
- Totnaftate
- Naftifine
- Nystatin
- Undecylenic acid
- Benzoic acid

SYSTEMIC ANTIFUNGAL AGENTS

Amphotericin B

Amphotericin B (Fungizone), which is ineffective in ridding infections caused by bacteria, *Rickettsia*, or viruses, is either fungicidal or fungistatic, depending on the drug concentration used or the sensitivity of the particular fungus. Numerous pathogenic yeasts (***Cryptococcus neoformans***), pathogenic yeastlike organisms (***Monilia***), dimorphic fungi (***Blastomyces***), filamentous fungi (***Cladosporosium***), and other fungi are highly sensitive to amphotericin B. Furthermore, the antifungal actions of amphotericin B are enhanced by **flucytosine**, **minocycline**, or **rifampin**, agents otherwise devoid of antifungal activity.

Pharmacodynamics

Amphotericin B imposes its antifungal effects by binding to the sterol moiety of the membrane and damaging its structural and functional integrity (Figure 25).

Pharmacokinetics and Proper Usage

Amphotericin B is available in the form of sterile lyophilized powder. Because it is insoluble in water, it is marketed with sodium deoxycholate for dispersal in sterile water and 5% dextrose. The **polyene antibiotics**, **amphotericin B**, **nystatin**, and **candicidin**, are all poorly absorbed from the gastrointestinal tract. In the plasma, amphotericin B binds to lipoproteins including cholesterol. It is extensively metabolized and the inactive metabolite, or metabolites, are slowly excreted in the urine.

Amphotericin B is the only polyene antibiotic given parenterally. When the intravenous route is contemplated, amphotericin B is dispersed fresh, as discussed, and infused slowly. Amphotericin B should not be administered rapidly because this causes cardiac toxicity. **Heparin** (1000 units) is often added to the infusion suspension to avert the risk of **thrombophlebitis**. Amphotericin B can also precipitate normocytic or **normochromic anemia**, **leukopenia**, and **thrombocytopenia**.

During the infusion of amphotericin B, the patient's temperature will rise, which may or may not be accompanied by hypotension and delirium. Often

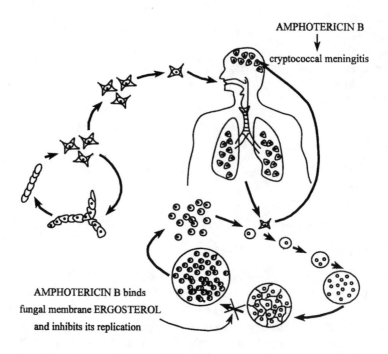

Figure 25   The mechanism of action of amphotericin B.

hydrocortisone sodium succinate is added to the infusion during the initial but not the succeeding alternate-day treatment with amphotericin B.

Amphotericin B is nephrotoxic in most patients, and often causes a permanent reduction in the glomerular filtration rate. Furthermore, hypokalemia may occur, requiring the oral administration of potassium chloride.

Amphotericin B has been used intrathecally in patients with coccidioidal or **cryptococcal meningitis**. The side effects associated with this route of administration are headache, paraesthesia, nerve palsy, and visual impairment. To treat coccidioidal arthritis, amphotericin B may be injected intraarticularly (Figure 26).

Flucytosine

Flucytosine (Ancobon) possesses clinically useful activity against *Cryptococcus neoformans*, *Candida* species, *Torulopsis glabrata*, and the agents of chromomycosis. Susceptible fungi deaminate flucytosine to **5-fluorouracil**, which becomes an antimetabolite. Flucytosine, which is excreted by the kidney, should be used cautiously in the setting of renal impairment. Flucytosine is a bone marrow depressant. Flucytosine is used in combination with **amphotericin B**.

Ketoconazole

Ketoconazole (Nizoral) has a broad therapeutic potential for a number of superficial and systemic fungal infections. Ketoconazole dissolves in an acidic media; therefore, antacids or histamine₂-receptor-blocking agents reduce its effectiveness (Figure 27). Ketoconazole can cause gastrointestinal disturbances.

AMPHOTERICIN B injected intra-articularly to treat

COCCIDIOIDAL ARTHRITIS

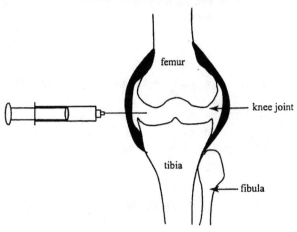

Figure 26 Amphotericin B is used to treat coccidioidal arthritis.

KETOCONAZOLE dissolves in acid media

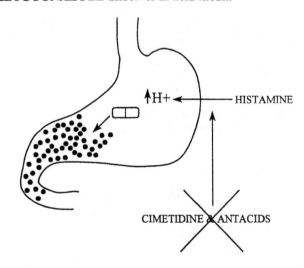

Figure 27 The action of ketoconazole.

Fluconazole

Fluconazole, which is absorbed from the stomach, has been shown to be effective in the treatment of the oral and esophageal forms of candidiasis in patients with acquired immunodeficiency syndrome.

Griseofulvin

Griseofulvin (Fulvicin and Grisactin) is a fungistatic effective against various dermatophytes, including *Microsporum*, *Epidermophyton*, and *Trichophyton*, that produce diseases of the skin, hair, and nails. It exerts its effect by inhibiting fungal mitosis.

Nystatin

Nystatin (Mycostatin) is poorly absorbed from the gastrointestinal tract. It is both fungistatic and fungicidal but has no effect on bacteria, viruses, or protozoa. It exerts its effect by binding to the sterol moiety and hence damaging the fungal membrane. It is used primarily as a topical agent to treat **candidal infections** of the skin and mucous membrane (**paronychia**, **vaginitis**, and **stomatitis**), and so causes no major toxicities.

TOPICAL ANTIFUNGAL AGENTS

Clotrimazole

Clotrimazole, which is used only topically, is available as a 1% cream, lotion, or solution (**Lotrimin** and **Mycelex**), as a 1% vaginal cream, as 100- or 500-mg vaginal tablets (**Gyne-Lotrimin**, **Mycelex-G**), and as 10-mg troches (**Mycelex**).

Econazole

Econazole nitrate (**Spectazole**) is administered only topically and comes as a water-miscible cream (1%), which is applied twice a day.

Miconazole

Miconazole nitrate, applied topically, is available as a 2% dermatological cream, spray, powder, or lotion (**Micatin** and **Monistat-Derm**).

Terconazole and Butoconazole

Terconazole and **butoconazole nitrate** (**Femstat**) are available as a 2% vaginal cream. They are used at bedtime for 3 days in nonpregnant females. There is a slower response during pregnancy, which requires a 6-day course of treatment.

Ciclopirox Olamine

Ciclopirox olamine is available as a 1% cream and lotion applied topically for the treatment of cutaneous candidiasis and tinea corporis, cruris, pedis, and versicolor.

Haloprogin

Haloprogin (Halotex) is available as a 1% cream or solution, which is applied topically. It is applied twice a day for 2 to 4 weeks. It is used principally in the treatment of tinea pedis.

Tolnaftate

Tolnaftate (Aftate, Tinactin, and others) is available in a 1% concentration as a cream, gel, powder, aerosol powder, and topical solution, or as a topical aerosol liquid.

Naftifine

Naftifine hydrochloride (Naftin) is available as a 1% cream. It is effective for the topical treatment of tinea cruris and corporis.

Undecylenic Acid

Undecylenic acid (Desenex) is available as a foam, ointment, cream, powder, soap, and liquid, all administered topically.

Benzoic Acid and Salicylic Acid

Benzoic and **salicylic acid ointment** is known as **Whitfield's ointment**. It combines the fungistatic action of benzoate with the keratolytic action of salicylate. It contains benzoic acid and salicylic acid in a ratio of 2:1 (usually 6%:3%). It is used mainly in the treatment of tinea pedis.

AROMATHERAPY CAUTIONS

There is a vogue for many scientifically nonqualified aromatherapists to practice clinical aromatherapy, where they prescribe the internal usage of essential oils. Internal prescribing involves oral, rectal, and vaginal intake: however, the use of tampons soaked in various potentially toxic essential oils, like the various tea tree oils with variable biological potential could have a possible harmful effect on the delicate internal mucosal membranes. The possibility of misdiagnosis of a urino-genital condition by medically unqualified aromatherapists or by the patients themselves could also result in serious consequences.

The problem with many Myrtarcae is that of their genetic variation or the production of spontaneous chemotypes, giving many different essential oil compositions with differing bioactivities. The quality of commercial Australian tea tree has now been generally stabilized, mainly by the selection of clones and the blending of different essential oils from different species of *Melaleuca* to conform with the Australian Standards, but the problem of diversity remains in other parts of the world where *Melaleuca* species grow, e.g., Zimbabwe, New Zealand, Indonesia, etc. The diversity of Manuka and Kanuka oils is even more pronounced and the use of these New Zealand oils, especially in aromatherapy, may therefore be premature, unless their quality can be assured and also toxicological studies (which are available for all essential oils used in the food and cosmetics industry) are undertaken (Lis-Balchin et al., 2000).

12

BANISTERINE, SELEGILINE, AND PARKINSON'S DISEASE

- **Banisterine** from *Banisteria caapi* and *Nicotiana tabacum* is a monoamine oxidase inhibitor that is similar pharmacologically to **selegiline** (deprenyl), which is used in the treatment of Parkinson's disease.
- Plant-based drugs for Parkinson's disease include **L-dopa**, which comes from sources such as the seeds of *Mucuna pruriens* and *Vincia faba*. L-Dopa is the chemical compound that has been used for Parkinson's disease for the last 30 years.
- Seeds of *Datura stramonium* have anticholinergic effects, similar to those seen with trihexyphenidyl and benztropine.

INTRODUCTION

Parkinson's disease consists of a severe reduction in the dopamine content in all components of the basal ganglia. Four separate groups of symptoms constitute the symptom complex that makes up parkinsonism. These consist of tremor at rest, akinesia or bradykinesia, rigidity, and loss of postural reflexes.

The parkinsonian tremors have been referred to as "pill-rolling," "cigarette rolling," and "to and fro" movements. These tremors are present during rest but often disappear during purposeful movement or sleep. Stress or anxiety-provoking situations aggravate the tremors and the initiation of movement becomes increasingly difficult, extremely fatiguing, and ponderously inefficient.

The akinesia or bradykinesia is characterized by a poverty of spontaneous movements and slowness in initiation of movements. Rigidity or increased muscle tone occurs in response to passive movements. The loss of normal postural reflexes is a disorder of postural fixation and equilibrium.

THE ACTIONS OF ACETYLCHOLINE AND DOPAMINE

Neurochemically, parkinsonism is considered as a **striatal dopamine-deficiency syndrome**, and the main extrapyramidal symptoms — tremor, akinesia, and rigidity — correlate positively with the degree of this deficiency. Although eight separate neurotransmitters interact in the nigro-striato-nigral loop, the basic therapeutic problem in parkinsonism has been to find suitable compounds that (1) increase the concentration of dopamine, (2) stimulate the dopamine receptor sites directly, or (3) suppress the activity at cholinergic receptor sites. These drugs fall into two main categories: those that increase dopaminergic function and those that inhibit cholinergic hyperactivity.

HISTORY

Physicians in ancient India first used ***Mucuna* seeds** in the treatment of Parkinson's disease over 4500 years ago. The Indian medical system is called ***Ayurveda***, which is the world's oldest system of medicine based on scientific principles. Ayurveda is founded on scientific principles. It has a long history of use of herbal remedies and has documented data on mechanism of action, specific action, short-term and long-term toxic effects, drug–drug interactions, and drug–diet interactions with a long history of use in humans. As per historical evidence, **Parkinson's disease** existed in ancient India and was called ***Kampavata***. This was over 4500 years ago even though the disease acquired its present name from **James Parkinson** who redescribed the disease in A.D. 1817. In the Ayurvedic system, powder of *Mucuna* is used for treating Parkinson's disease and is subjected to special processing. The English name "**cowage**" plant (***Mucuna pruriens***) is derived from **Hindi Kiwach**. In Sanskrit, it is called **Atmagupta**. *Mucuna* is a twiner with trifoliate leaves, purple flowers, and turgid S-shaped pods covered with hairs that cause intense itching on contact with the skin. The plant belongs to the family **Leguminosae**, which is indigenous to India and has long been used in Ayurveda since ancient times. Overdose effects of *Mucuna* were also recognized in Ayurveda. These included headache, dystonia, fatigue, tremor, syncope, and thirst. Many of these could also occur from synthetic L-dopa.

In 1929, Louis Lewin reported the use of a hallucinogenic compound prepared from the South American vine, ***Banisteria caapi***, to treat Parkinson's disease (see Lewin and Schuster, 1929; Holmstedt, 1979). This psychoactive compound, named banisterine, proved to be identical to **harmine**. The first reports of the use of banisterine to treat postencephelic parkinsonism in 1929 created a stir in the popular press and banisterine was hailed as a "**magic drug**." **Banisterine** is a reversible **monoamine oxidase inhibitor** (see Sanchez-Ramos, 1991, for a review), which in turn resulted in the discovery of **selegiline** and **rasagiline** that are now used in the treatment of Parkinson's disease.

SELEGILINE

Selegiline is metabolized in the liver to (–)-**desmethylselegiline** and (–)-**methamphetamine**, which is further metabolized to **amphetamine**. Desmethylselegiline is also an irreversible inhibitor of monoamine oxidase B in human. **Rasagiline** is not metabolized to amphetamine-like derivatives. No gender- or age-related differences in the pharmacokinetics of selegiline have been reported. **Selegiline** may be used safely with **citalopram** or with **cabergoline**.

Table 15   Pharmacokinetic Parameters of Selegiline

Parameters Tested	Values Reported
C_{max} (µg/l)	220 ± 1.17
t_{max} (h)	0.90 ± 0.38
$t_{1/2}$ (h)	1.20 ± 0.56
CL/F (l/min)	59.4 ± 43.7
V/F (l)	1854 ± 824
Dose excreted in urine (%)	0.01 ± 0.005
Absolute bioavailability (%)	NR

Abbreviations: CL/F = apparent total body clearance; C_{max} = maximum plasma drug concentration; NR = not reported; t_{max} = time to reach peak drug concentration; $t_{1/2}$ = half-life; V/F = apparent volume of distribution.

- **Selegiline** inhibits the activity of monoamine B, enhances the release of dopamine, blocks the uptake of dopamine, is a calmodulin antagonist, and enhances the level of cyclic AMP, which in turn protects dopaminergic neurons.
- **Selegiline** protects against 1-methyl-4-phenyl-1,2,3,5-tetrahydro-pyridine (MPTP)-induced neurotoxicity.
- **Selegiline** protects against neurotoxic factor(s) present in the spinal fluid of patients with Parkinson's disease.
- **Selegiline** protects dopaminergic neurons from glutamate-induced exatatoxicity.
- **Selegiline** possesses cognition-enhancing function, rejuvenates serum insulin-like growth factor I in aging humans, and enhances life expectancy.
- **Selegiline** possesses neurotrophic-like actions, and rescues axotomized motorneurons independent of monoamine oxidase B (MAO-B) inhibition.
- **Selegiline** increases the striatal superoxide dismutase, protects against peroxynitrite- and nitric oxide-induced apoptosis, and guards dopaminergic neurons from toxicity induced by glutathione depletion.
- **Selegiline** stimulates the biosynthesis of interleukin-1β and interleukin-6, is an immunoenhancing substance, possesses antiapoptotic action, and is a neuroprotectant in nature.
- Finally, like banisterine, **selegiline** has been shown to be efficacious in Parkinson's disease.

CLINICAL PHARMACOKINETICS OF SELEGILINE

The pharmacokinetic parameters of selegiline in healthy subjects have been studied and the data are shown in Table 15. One study stated that the plasma selegiline concentration–time data are best described by a biexponential model with first-order absorption and a lag time in three volunteers, whereas a triexponential model with first-order absorption and a lag time was fitted to plasma selegiline concentration–time data in the remaining two volunteers. Selegiline in two dosage forms exhibited a lag time, the absorption was rapid, and the mean time to reach peak drug concentration (t_{max}) was observed between 30 and 45 minutes (t_{max} = 36 ± 8.2 minutes) for the solution and 30 to 90 minutes (mean t_{max} = 54 ± 23 minutes) for the tablets. The mean C_{max} was 2.2 ± 1.2 µg/l for tablet and 4.06 ± 2.6 µg/l for solution. The apparent volume of distribution (V/F) was 1854 ± 824 l for the tablet and 1446 ± 1315 l for the solution. The $t_{1/2}$ was approximately 1.20 ± 0.65 hours following the administration of the 10-mg tablet. The oral clearance was 59 ± 44 l/min and 50 ± 43 l/min, respectively, following the administration of tablet and solution. This very high oral clearance (almost 30-fold higher than the haptic blood flow) indicates that selegiline is extensively metabolized not only in the liver but also through extrahepatic metabolism (see Table 15).

No gender-related differences in the pharmacokinetics of selegiline were observed following a single oral dose of selegiline 10 mg to six elderly males and six elderly females between 60 and 85 years old.

A comparison of the pharmacokinetics of selegiline in 14 male patients (18 to 30 years of age) with 6 elderly male patients (age >60 years) indicated that the C_{max} and the AUC remains unaltered between the young and the elderly patients. The $t_{1/2}$ of selegiline in the elderly patients was almost twofold longer than in the young patients (3.2 hours in the elderly patients vs. 1.5 hours in the young); however, because of the small sample size and high variability in $t_{1/2}$ β, this difference may not be a real difference and may not have any clinical impact.

Three selegiline metabolites have been detected in human plasma. Selegiline is metabolized in the liver to **(–)-desmethylselegiline** and **(–)-methamphetamine**, which is further metabolized to **amphetamine**. Furthermore, both methamphetamine and amphetamine can be metabolized into the corresponding parahydroxy derivatives, which are pharmacologically inactive after being conjugated with **glucoronic acid**. Selegiline can also be metabolized in the lung to form methamphetamine or desmethylselegiline. In the kidneys, only methamphetamine is formed, but the degree of metabolism in these tissues is minimal when compared with that in the liver (Figure 28).

Desmethylselegiline is also an irreversible inhibitor of monoamine oxidase B in humans. There is evidence that the l-stereoisomers of l-amphetamine and l-methamphetamine may have some qualitatively different actions than their d-isomer counterparts, which might result in beneficial clinical effects and could complement any beneficial clinical actions of selegiline itself. Food has no effect on the pharmacokinetics of desmethylselegiline, methamphetamine, and amphetamine. At a dose of 10 mg/day, selegiline is devoid of the "**cheese effect**," that is, it does not cause hypertension when taken with tyramine-containing foods such as cheese.

LACK OF ADVERSE INTERACTION BETWEEN SELEGILINE AND CITALOPRAM OR BETWEEN SELEGILINE AND CABERGOLINE

The concomitant use of serotonergic antidepressive drugs with MAO inhibitors has been associated with serious adverse drug interactions manifesting as hyperexcitation of the central nervous system, known as the **serotonin syndrome**. The serotonin syndrome is caused by excessive activation of central serotonergic receptors, resulting in typical clinical symptoms that may be lethal. These symptoms include confusion,

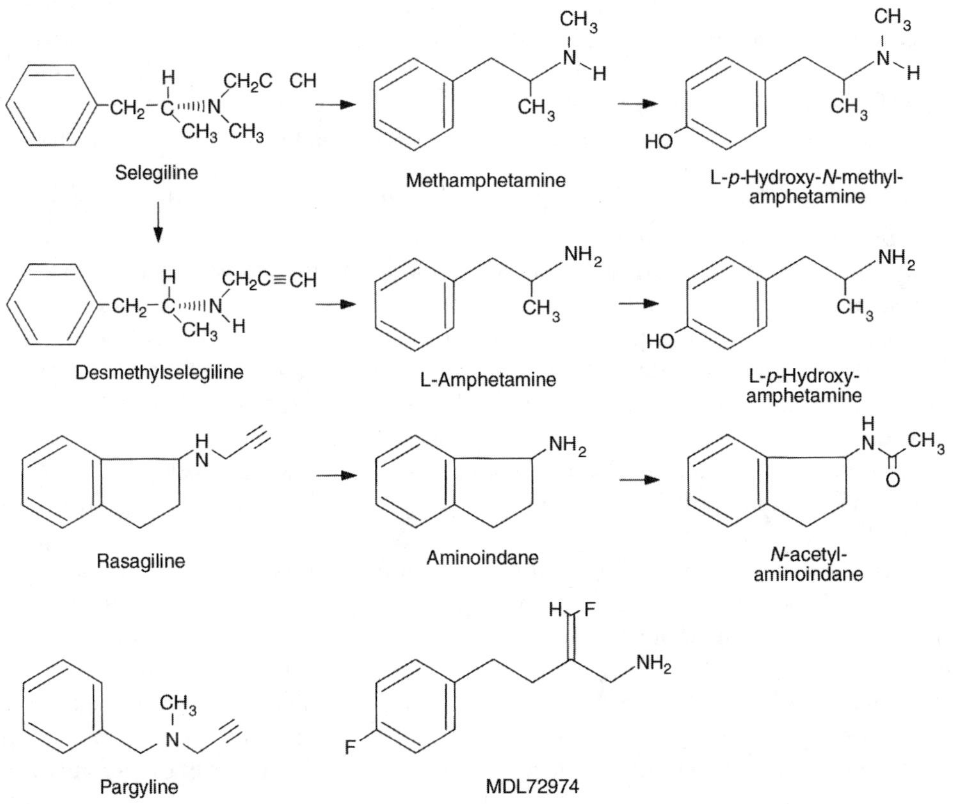

Figure 28 Metabolism of selegiline and rasagiline.

hypomania, restlessness, myoclonus, hyperreflexia, diaphoresis, shivering, tremor, and, in more severe cases, elevated blood pressure, tachycardia, and hyperthermia. Several case reports indicate that the concomitant use of selegiline with the serotonin reuptake inhibitor **fluoxetine** may lead to adverse drug interactions. The risk for pharmacokinetic or pharmacodynamic interactions of concomitantly administered **selegiline**, a selective monoamine oxidase type B inhibitor, and **citalopram**, a widely used selective serotonin uptake inhibitor antidepressant, have been studied.

Two parallel groups of healthy volunteers received 20 mg of citalopram ($n = 12$) or placebo ($n = 6$) once daily for 10 days in a randomized, double-blind fashion, followed by concomitant selegiline, 10 mg once daily for 4 days. The safety of this drug combination was assessed by measurements of blood pressure, heart rate, body temperature, and inquiries for adverse events. Blood samples were taken for the analysis of serum concentrations of selegiline, citalopram, and their metabolites. In addition, plasma was obtained to measure prolactin, epinephrine, norepinephrine, and **3,4-dihydroxyphanolglycol (DHPG)**, the urinary excretion of norepinephrine and 5-hydroindoleacetic acid (5-HIAA), the urinary metabolite of serotonin.

After a 5-week washout, the 12 subjects who took citalopram in the first part of the study received 10 mg of selegiline once daily for 4 days to compare the pharmacokinetics of selegiline with and without concomitant citalopram. The safety analysis showed no significant differences in vital signs or the frequency of adverse events between the study groups. Plasma prolactin concentrations were increased by 40% after 10 days of treatment with citalopram ($p = 0.03$) and this effect was not potentiated by concomitantly administered selegiline. The comparison of plasma concentrations of norepinephrine, epinephrine, and DHPG and the amount of serotonin and 5-HIAA excreted into urine between the study groups indicated no signs of subclinical pharmacodynamic interaction between selegiline and citalopram. The relative bioavailability of selegiline was slightly reduced (by 29%; $p = 0.008$) when citalopram was coadministered compared with selegiline alone. However, no indication of a pharmacokinetic interaction was found in the analysis of serum concentrations of the three main metabolites of selegiline. The pharmacokinetics of citalopram remained unaffected by concomitant selegiline. The present results indicate lack of clinically relevant pharmacodynamic or pharmacokinetic interactions between selegiline and citalopram. The interaction between cabergoline and selegiline has been studied.

Since selegiline is hardly detectable in plasma, the plasma levels of its metabolites amphetamine, methamphetamine, and desmethylselegiline were used to assess the effect of cabergoline coadministration. Plasma levels of the selegiline metabolites were determined first after selegiline administration (10 mg/day) for 8 days, and then after administration of both drugs for 22 additional days (day 30). Cabergoline plasma levels were measured on day 30, and then after administration of cabergoline (1 mg/day) alone an additional 22 days. No statistical difference was found between the values of cabergoline and of the selegiline metabolites when the two drugs were given alone or in combination, indicating the absence of pharmacokinetic interaction between **cabergoline** and **selegiline**.

DISTRIBUTION OF MONOAMINE OXIDASES A AND B mRNAS IN THE RAT BRAIN BY *IN SITU* HYBRIDIZATION

MAO-A and MAO-B are the major catabolic isoenzymes of catecholamines and serotonin in the mammalian brain. The two isoenzymes have 70% amino acid identity, identical exon–intron organization, and are encoded by two different genes that reside closely between bands Xp11.23 to Xp22.1 of the X chromosome.

MAO-B-containing structures in the nucleus accumbens, using MAO-A-deficient transgenic mice and MAO enzyme histochemistry have been studied. Among the striatal structures, the **nucleus accumbens**, and in particular its dorsal shell, showed the strongest MAO-B activity. MAO-B-active cell bodies were embedded in a dense MAO-B-active fiber plexus. MAO-B-positive terminals formed axo-dendritic synapses which were exclusively of the asymmetric type. It is suggested that dopamine in the nucleus accumbens shell is transported into MAO-B-positive fibers where it is degraded by MAO-B.

By applying *in situ* hybridization histochemistry and by using 35-S-labeled **oligodeoxynucleotide probe** complementary to cloned human sequences, MAO-A and MAO-B mRNAs in the monkey brain stem have been studied. Furthermore, hybridization studies to visualize MAO-A and MAO-B mRNAs in the rat brain by using specific cDNA and oligonucleotide probes have been carried out. These studies have shown that MAO-A mRNA was present in major monoamine groups of the brain, such as the dorsal vagal complex, the C1/A1 groups, the locus ceruleus, the raphe nuclei, substantia nigra, and the ventral tegmental area. It was also present in important brain structures of the forebrain, such as the cerebral cortex, the hippocampus, the medial habenula, the thalamus, and the hypothalamus; and these are major projection sites of the monoamine groups of the midbrain and hind brain. Thus, MAO-A mRNA was found in serotonergic and noradrenergic cell groups as well as in dopaminergic groups. The presence of MAO-A mRNA in these groups is consistent with a role for MAO-A in oxidizing serotonin, norepinephrine, and dopamine in the rat.

In contrast to MAO-A mRNA, the distribution of MAO-B mRNA in the brain was very circumscribed. There were only three brain regions in which the amount of MAO-B mRNA exceeded the limits of detection; the **area postrema**, the **dorsal raphe**, and the **subfornical organ**. The area postrema and the subfornical organ are both circumventricular organs that contain a wide variety of neurotransmitters, including monoamines. MAO-B expression, therefore, may play a role in the oxidation of local neurotransmitters. The area postrema and the subfornical organs also lack blood–brain barriers and are exposed to blood-borne substances; they are functionally important as detectors of humoral factors and toxic substances in the blood. It is possible that MAO-B may aid in the degradation of blood-borne factors in these circumventricular organs.

Although MAO-B mRNA was observed in the serotonergic cell group of the dorsal raphe, MAO-B was not present in the median raphe or in any major catecholaminergic cell group, such as the C1/Al groups, the nucleus of the solitary tract, the locus ceruleus, the substantia nigra, or the ventral tegmental area. MAO-B mRNA was also not observed in projection sites of the monoaminergic groups in the forebrain.

Although there is a considerable amount of confusion regarding the substrate specificity of MAO-A and MAO-B toward dopamine, it is widely accepted that dopamine is metabolized by both forms of the enzyme in the rat brain and it is deaminated solely by MAO-B in the human brain. The activity of MAO-B increases with age and the age-related changes may lead to an overproduction of neurotoxic agents. The inhibition of the enzyme activity may play a preventive role against neurodegradation brain disorders.

INHIBITION OF MONOAMINE OXIDASE B BY BANISTERINE AND SELEGILINE AND THE INITIAL RATIONALE OF USING THEM IN PARKINSON'S DISEASE

Amine oxidases catalyze the oxidation of amines, diamines, and polyamines. According to their ability to recognize one of those substrates preferentially, amine oxidases may be divided into **monoamine oxidases**, **diamine oxidases**, and **polyamine oxidases**, respectively. Several different enzymes fall into the amine oxidase class and the classification of some of them still remains ambiguous. The term "monoamine oxidase" (flavin-containing, EC 1.4.3.4) was introduced to contrast with copper-containing amine oxidases (EC 1.4.3.6).

In the case of MAO, oxidation of the substrates is coupled to the reduction of **flavin adenine dinucleotide (FAD) cofactor**. The product of the reaction is the amine of the substrate, which hydrolyzes spontaneously to yield the corresponding aldehyde and ammonia. Reoxidization of the cofactor by molecular oxygen produces H_2O_2, according to the following reactions:

$$R\text{--}CH_2NH_2 + E\text{-}FAD \rightarrow R\text{--}CH = NH + FADH_2$$

$$R\text{--}CH = NH + H_2O \rightarrow + R\text{--}CHO + NH_3$$

$$FADH_2 + O_2 \rightarrow E\text{-}FAD + H_2O_2$$

Flavin-containing mitochondrial MAO-A and MAO-B catalyze the oxidative deamination of neurotransmitters, such as dopamine, serotonin, and norepinephrine in the central nervous system and peripheral tissues. The enzymes share 73% sequence homology and follow the same kinetic and chemical mechanism but have different substrate and inhibitor specifities. Chemical modification experiments provide evidence that a **histidine residue** is essential for the catalysis. There is also strong evidence that two **cysteine residues** are present in the active site of MAO.

Both MAO-A and MAO-B contain a redox-active disulfide at the catalytic center. The results imply that MAO may be a novel type of **disulfide oxidoreductase** and may open the way to characterizing the catalytic and chemical mechanism of the enzyme.

When an amine is oxidized, electrons pass from the amine to the disulfide and then to the flavin. The formation in MAO of a **carboxyl–imidazole–disulfide triad**, similar to that demonstrated for **disulfide oxidoreductases**, would result in an increased positive charge on the disulfide which would facilitate electron transfer from the amine substrate. Similarities with the disulfide oxidoreductase family of enzymes, which includes **glutathione oxidoreductase**, **lipoamide dehydrogenase**, and **thioredoxin reductase**, may be useful in elucidating the mechanism. However, although there are structural analogies between the N-terminal region of MAO and the nucleotide-binding region of lipoamide dehydrogenase, the **dithiol sequence motif** common to the disulfide oxidoreductase family is not found in MAO.

MAO exists in different forms with different substrate specificities and inhibitor sensitivities (Table 16). These studies paved the way for utilization of selegiline in Parkinson's disease and for the discovery of a new generation of MAO inhibitors.

Placenta expresses mainly MAO-A. Platelets, lymphocytes, and chromaffin cells (Table 17) express mainly MAO-B. MAO-A and MAO-B are present in diversified cell population of the central nervous system (Table 18).

RASAGILINE PROTECTS DOPAMINERGIC NEURONS

The success of selegiline as an adjunct to L-dopa treatment of Parkinson's disease has prompted the development of other selective inhibitors of MAO-B. The antiparkinsonian effect of selegiline is thought to comprise both symptomatic and neuroprotective components, although the extent of each of these elements in attaining clinical antiparkinsonian action is unclear. The former may depend on an increase in synaptic dopamine and **phenylethylamine** when selegiline is administered alone or together with L-dopa, and both its inhibition of MAO and the amphetamine-like properties of the molecule may play a part in achieving

Table 16   Classification of Monoamine Oxidases (MAO); Substrate Specificities and Inhibitor Sensitivities

	Substrates for MAO-A	Substrates for MAO-B	Substrates for both MAO-A and MAO-B
	Serotonin	Dopamine	Tyramine
	Epinephrine	2-Phenylethalamine	Inhibitors for both
	Norepinephrine	Tryptamine	MAO-A and MAO-B
	Octopamine	MPTP	
	Inhibitors for MAO-A	Inhibitors for MAO-B	
Irreversible	Clorgyline	Selegiline	Phenelzine
		Pargyline	Tranylcypromine
Reversible	Harmine		
	Harmaline		
	α-Ethyl-tryptamine		
	Quipazine		

MPTP = 1-methyl-4-phenyl-1,2,3,6-tetrahydropyridine.

this effect; the latter has been demonstrated definitively only in isolated cell or intact animal experiments.

An indane derivative, AGN-1135 (N-propargyl-l-aminoindane), which produces selective inhibition of MAO-B without the amphetamine-like properties of selegiline, has been described. It has been shown that R(+)-N-propargyl-l-aminoindane (**rasagiline**) possessed nearly all of the enzyme-inhibitory activity of selegiline. Furthermore, rasagiline is not metabolized to amphetamine-like derivatives. Like selegiline, when given in a dose selective for inhibition of MAO-B, it did not affect striatal extracellular fluid dopamine levels, but when administered chronically (21 days) it increased striatal microdialysate dopamine without reduction in deaminated metabolites. Similar to selegiline, rasagiline ($10^{-6}M$) increased the percentage of **tyrosine hydroxylase-positive cells** in a primary culture of **fetal mesencephalic cells** (6 days in culture). Rasagiline, but not selegiline, also increased the number of neurons per field in this organotypic culture. The production of accumulation of neurotoxin can be blocked by drugs such as **selegiline** that inhibit the activity of MAO-B. Moreover, monkeys treated simultaneously with MPTP and selegiline do not develop Parkinson's syndrome (Figure 29).

SELEGILINE PROTECTS AGAINST MPTP-INDUCED NEUROTOXICITY

Endogenously produced neurotoxins have long been suspected of being involved in the pathogenesis of Parkinson's disease; however, little mechanistic evidence existed to support this concept until the neurotoxic effects of **MPTP** were found to induce a Parkinson-like syndrome. The discovery that MPTP produces a central nervous system pathology very similar to that observed with Parkinson's disease has strengthened the endogenous neurotoxin hypothesis and provided a heuristic

Table 17   Tissue Distribution of MAO in Humans

Tissues	MAO-A	MAO-B
Placenta	++(+)	±
Fibroblasts	++	+
Platelets	−	++
Lymphocytes	−	++
Erythrocytes	−	−
Brain	++	++(+)
Intestine	+++	−
Ileum	++	++
Colon	++	++
Lung	++(+)	+
Liver	+++	+++
Kidney	++(+)	++(+)
Adrenals	++(+)	++(+)
Heart	++(+)	++(+)
Pancreas	++(+)	+
Skeletal muscle	+/+++	+/+++
Stomach	+	+
Spleen	+	+
Tongue	++	+
Esophagus	++	+
Skin	+	+
Diaphragm	+	++
Blood vessels		
Aorta	+/+++	+/+++
Cerebral	+	= /−

− = no activity; ± = very low to negligible activity; + = low activity; ++ = intermediate activity; +++ = high enzyme activity.

model for investigating the pathological process of Parkinson's disease. The first stage in the mechanism of action of MPTP appears to be its deamination by MAO-B, possibly in glial cells, which results in the formation of 1-methyl-4-phenylpyridinium ion (**MPP+**). The MPP+ is then selectively accumulated in dopamine nerve terminals by way of the high-affinity dopamine reuptake system. Neurons lacking dopamine transporter remain unaffected. Once inside the nerve terminals, MPP+ seems to act in a manner similar to **6-hydroxydopamine** by generating **hydrogen peroxide** and free radicals that interfere with mitochondrial respiration. MPP+ is concentrated in mitochondria, where it impairs mitochondrial respiration by inhibiting **complex I** of the electron transfer complex and consequently causing death of neurons. The activity of complex I is reduced in the brains of patients with Parkinson's disease. It has been suggested that the neuromelanin present in dopaminergic neurons (by binding and storing the MPP+) acts as a storage site for MPP+ or other neurotoxins. Dopaminergic cell death by MPTP (Figure 30) is caused by oxidative stress followed by lipid peroxidation caused by inhibition of mitochondrial

Table 18 The Distribution of MAO Isoenzymes in the Human CNS

Brain Areas	MAO-A	MAO-B
Cortices:		
Frontal cortex	+(+)	+(+)
Cingulate gyrus	+	++(+)
Insula cortex	+	++(+)
Temporal cortex	+	+(+)
Parietal cortex	+	+(+)
Occipital cortex	+	+(+)
Prefrontal cortex	+	+(+)
Cerebellum	+	+(+)
Basal ganglia		
Caudate	+(+)	++
Putamen	+	++
Globus pallidus	+	++(+)
Thalamus	+	++
Limbic areas		
Nucleus accumbens	+++	
Amygdala	++	++
Hippocampus	+(+)	+(+)
Dentate gyrus	++	++
Hypothalamus	+++	+++
Mammilary complex	+++	+
Interpeduncular nucleus	++	+++
Entorhinal cortex	++	++
Brain stem		
Locus coeruleus	+++	+(+)
Raphe nuclei	++	+++
Pons	++	++
Medulla	++	++
Substantia innominata		+++
Substantia nigra	++(+)	+++
Reticular formation	+(+)	++
Others		
Choroid plexus	+	+++
Nucleus ruber	++	++
White matter		+
Periaqueductal grey	++	+++
Superior colliculus		+++

+ = low activity; ++ = intermediate activity; +++ = high activity.

enzymes participating in ATP synthesis. The production and accumulation of neurotoxin can be blocked by drugs such as **selegiline** that inhibit the activity of MAO-B. Moreover, monkeys treated simultaneously with MPTP and selegiline do not develop Parkinson's syndrome (Figure 30).

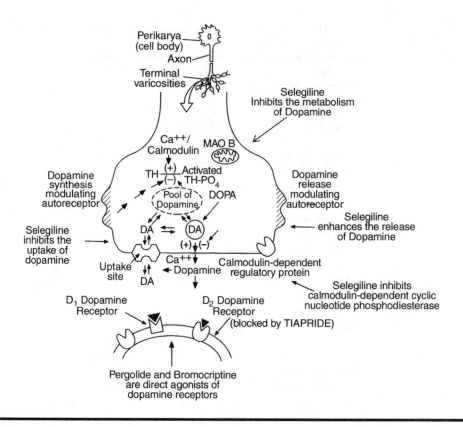

Figure 29 Selegiline inhibits the metabolism of dopamine, stimulates its release, and inhibits its uptake. In addition to possessing autoreceptors capable of modulating the release of dopamine, the corpus striatum also contains γ-aminobutyric acid (GABA)-producing feedback pathways, whose activation leads to the release of GABA and subsequent inhibition of dopaminergic neurons. Moreover, Ca^{2+}-dependent release of dopamine exists that itself is attenuated. The bottom portion of the figure provides a schematic representation of a Ca^{2+}-dependent release of dopamine, which itself is attenuated by release-modulating autoreceptors via a coupling mechanism that may involve protein carboxymethylation whose substrate may be calmodulin or a calmodulin-dependent enzyme. In addition, synthesis modulating autoreceptors exist, which through a Ca^{2+}-calmodulin-dependent mechanism may lead to phosphorylation of tyrosine hydroxylase (TH), producing phosphorylated tyrosine hydroxylase (TH-PO4), which possesses a greater affinity for tetrahydrobiopterin cofactor, and subsequently enhances the synthesis of dopa and, in turn, dopamine.

SELEGILINE PROTECTS DOPAMINERGIC NEURONS IN CULTURE FROM TOXIC FACTORS PRESENT IN THE CEREBROSPINAL FLUID OF PATIENTS WITH PARKINSON'S DISEASE

In addition to MPTP, other endogenously produced neurotoxins, namely, the monoamine-derived 1,2,3,4-**tetrahydroisoquinolines** and 6,7-dihydroxy-1,2,3,4-tetrahydroisoquinolines, have been proposed as factors accelerating dopamine cell death. N-methylated isoquinolines were found to be oxidized by MAO, and hydroxyl radicals were found to be produced by this reaction. In addition, by incubation with the N-methylated isoquinolines, ATP was depleted

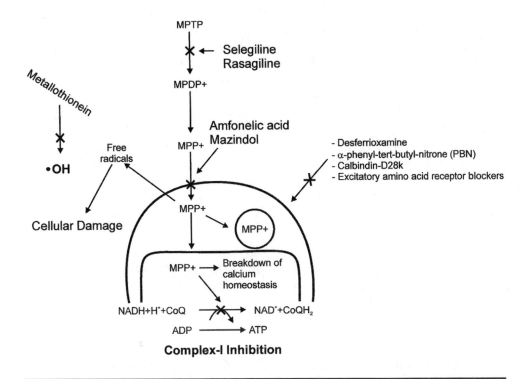

Figure 30 Oxidation of MPTP to 1-methyl-4-phenyl-2,3-dihydropyridium ion (MPDP$^+$), and finally to MPP$^+$, which generates free radicals and causes parkinsonism in humans. A deficiency of NADH Co (complex I) also causes striatal cell death. A deficiency of complex I may signify that an MPTP-like neurotoxin is generated endogenously, enhancing the vulnerability of striatum to oxidative stress reactions.

from a dopaminergic cell model. Pretreatment of the cells with MAO inhibitors such as **selegiline** could protect against ATP depletion. These results suggest that oxidation of neurotoxic isoquinolines is directly involved in the oxidative stress to induce the cell death of dopamine neurons. On the other hand, 1-methyl-1,2,3,4-tetrahydroisoquinoline and 1-methyl-6,7-dihydroxy-1,2,3,4-tetrahydroiso-quinoline were found to inhibit the activity of MAO, indicating that they may be neuroprotective agents in the brain.

The cerebrospinal fluid of patients with Parkinson's disease contains substances that inhibit the growth and function of dopaminergic neurons in culture. Moreover, selegiline, a MAO-B inhibitor (0.125 to 0.250 μM) enhances the number of tyrosine hydroxylase (TH)-positive neurons, augments the high-affinity uptake of dopamine, and averts the neurotoxic effects of the cerebrospinal fluid of patients with Parkinson's disease on rat mesencephalic neurons in culture.

Selegiline in protecting dopaminergic neurons from toxic factors present in the cerebrospinal fluid of patients with Parkinson's disease acts in a manner similar to those of **neurotrophic factors**, capable of partially protecting neurons against the neurotoxic effects of MPP+. In addition to blocking the neurotoxic effects of factors present in the cerebrospinal fluid of patients with Parkinson's disease, selegiline stimulated the uptake of dopamine, enhanced the activity of TH, and increased the number of TH-positive neurons. Selegiline stimulates the uptake of [^{3}H]dopamine at a lower concentration (0.125 μM) and inhibits it at a higher concentration (2.5 μM). This effect may be related to the higher level of dopamine, which, in turn, causes an attenuation in carrier-mediated functions through a feedback regulatory mechanism (see Figure 29).

Both **selegiline** and **rasagiline** (at a concentration of 1 to 10 μM) increase survival of mesencephalic dopaminergic neurons that had been primed with 10% serum. Rasagiline, but not selegiline, also increases total neuronal survival. Under serum-free conditions, rasagiline, but not selegiline, retains its neuroprotective action on dopaminergic neurons; GABAergic neurons are not affected by either selegiline or rasagiline. **Chlorgyline**, an MAO-A inhibitor, does not exert any of these effects. The protective action of rasagiline on dopaminergic neurons, even under stringent serum-free conditions, is striking, and warrants a role in the treatment of Parkinson's disease.

SELEGILINE PROTECTS DOPAMINERGIC NEURONS FROM GLUTAMATE-INDUCED EXCITATOTOXICITY

The excessive activity of excitatory amino acids, such as L-**glutamate** and L-**aspartate**, followed by elevation of intracellular free Ca^{2+} concentration and accumulation of free radicals has been postulated to underlie the neurodegeneration that occurs after ischemic insults and trauma. Additionally, an excitotoxic component has been shown to play an important role in the pathogenesis of chronic neurodegenerative disorders, such as **Alzheimer's disease**, **Parkinson's disease**, **amyotrophic lateral sclerosis**, and **Huntington's disease**, which are characterized by progressive loss of neuronal elements.

Activation of glutamate receptors (in particular, the NMDA type) leads to an intracellular accumulation of Ca^{2+}, which initiates a cascade of alterations resulting in the formation of free oxygen radicals. Glutamate perfusion of the striatum enhances dopamine release, a fact that also might contribute to increases in the formation of **hydroxyl radicals**. Alternatively, free radical formation can induce a release of excitatory amino acids, as demonstrated in hippocampal slices (Figure 31).

There is evidence from experimental Parkinson's disease models that excessive activation of NMDA receptors might be an important factor for induction of Parkinson's disease. However, the involvement of NMDA receptors has not been shown in all parkinsonian models.

The effect of selegiline on cultured mesencephalic dopamine neurons subjected to daily changes of feeding medium, an experimental paradigm that causes neuronal death associated with activation of the N-methyl-D aspartate (NMDA) subtype of glutamate receptors, has been studied. Both selegiline (0.5 to 50 μM) and the **NMDA receptor blocker MK-801** (10 μM) protected dopamine neurons from damage caused by medium changes. The nonselective MAO inhibitor **pargyline** (0.5 to 50 μM) was not protective, indicating that protection by selegiline was not due to MAO inhibition. Selegiline (50 μM) also protected dopamine neurons from delayed neurotoxicity caused by exposure to NMDA. Because selegiline had no inhibitory effect on NMDA receptor binding, it is likely that selegiline protects from events occurring downstream from activation of glutamate receptors. As excitotoxic injury has been implicated in neurodegeneration, it is

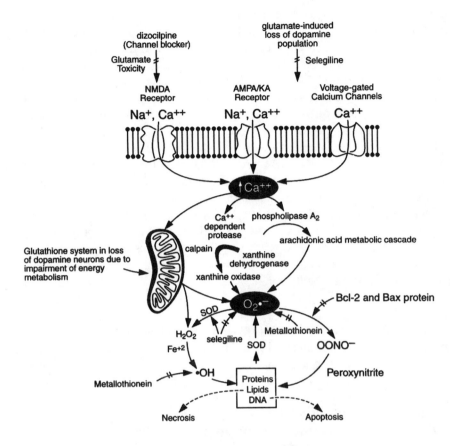

Oxidative Stress

Figure 31 Growing evidence points to the involvement of free radicals in mediating nigral death in Parkinson's disease. Circumstantial evidence suggests that activation of glutamate receptors and excitotoxicity play a role in Parkinson's disease and other neurodegenerative disorders. Calcium-activated protease and phospholipase A_2 generate superoxide anions ($O_2^{\bullet-}$) and peroxynitrite (OONO⁻) capable of damaging proteins, lipids, and DNA and producing either necrosis or apoptosis. Glutathione deficiency, as seen in the substantia nigra of Parkinson's patients, causes conversion of H_2O_2 to excess hydroxyl radicals (•OH). Selegiline (which enhances the level of superoxide dismutase), metallothionein isoforms (which scavenge •OH radicals and $O_2^{\bullet-}$ anions), or dizocilpine (an NMDA receptor channel blocker) are postulated to avert oxidative stress in Parkinson's disease.

possible that selegiline exerts its beneficial effects in Parkinson's disease by suppressing **excitotoxic damage**.

Selegiline might exert its neuroprotection by (1) preventing free radical formation, (2) increasing antioxidant cell defenses, (3) stimulating synthesis of growth factors, or (4) inhibiting apoptosis by interfering with the apoptotic pathways (see Figure 31).

Glutamate has been applied into the striatum by microdialysis, and the formation of **hydroxyl radicals** has been measured in the presence and absence of selegiline. In this study, microdialysis probes were implanted into the striatum 1 day before measurement of levels of hydroxyl radicals. The next day, the probes were first perfused for 120 minutes with a modified Ringer's solution containing 5 mM **salicylic acid**, to obtain stable baselines. Afterward, the perfusion solution was switched to another solution that, in addition, contained 50 mM glutamate to stimulate radical formation. After 20 minutes, **α-phenyl-tert-butylnitrone** (PBN; 100 mg/kg), **selegiline** (10 mg/kg), or saline was administered intraperitoneally. The glutamate perfusion produced marked 2- to 2.5-fold increases in 2,3-DHBA content. Treatment with PBN, a spin-trapping compound, significantly antagonized the rise of 2,3-DHBA level, indicating that PBN is a direct radical scavenger not only *in vitro* but also *in vivo*. Acute treatment with selegiline failed to reduce significantly the glutamate-induced radical formation.

Selegiline, in combination with the MAO-A inhibitor **chlorgyline**, inhibits the enhanced formation of hydroxyl radicals induced by the 2′-methyl analogue of MPTP. Similarly, pretreatment with selegiline administered into the striatum decreases the formation of hydroxyl radicals elicited by intrastriatal injection of MPP+. The contrasting results might be due to the different neurotoxic agents used (glutamate vs. MPP+ or its derivative) or to different routes of administration of selegiline.

Under treatment with the MAO-B inhibitor selegiline, the degradation of **putrescine** might be diminished, which, in turn, would suppress polyamine synthesis. Hence, the reported neuroprotective effect of selegiline might also receive a contribution from the diminished potentiation of the NMDA receptor by the polyamine-binding site. On the other hand, regarding inhibition of MAO-B, as is the case in patients with Parkinson's disease on selegiline, it is also likely that N^1-**acetylated spermine** and **spermidine** will be present in the brain in increased concentrations. Therefore, it seems possible that selegiline will exert a neuroprotective effect via an antagonistic modulation of the polyamine-binding site of the NMDA receptor (see Figure 31).

SELEGILINE IS A CALMODULIN ANTAGONIST AND THE INCREASED CYCLIC AMP PROTECTS DOPAMINERGIC NEURONS

The second messenger molecules Ca^{2+} and cyclic AMP (cAMP) provide major routes for controlling cellular functions. In many instances, calcium (Ca^{2+}) achieves its intracellular effects by binding to the receptor protein **calmodulin**. Calmodulin has the ability to associate with and modulate different proteins in a Ca^{2+}-dependent and reversible manner. **Calmodulin-dependent cyclic nucleotide phosphodiesterase** (CaMPDE, EC 3.1.4.17) is one of the key enzymes involved in the complex interactions that occur between the cyclic-nucleotide and Ca^{2+} second messenger systems (see Figure 29). CaMPDE exists in different isozymic forms, which exhibit distinct molecular and catalytic properties. The differential expression and regulation of individual phosphodiesterase (PDE) isoenzymes in different tissues relates to their function in the body.

Dopamine activates adenylate cyclase and **phospholipase C** (PLC) via a D_1 receptor and inhibits through a D_2 receptor, thereby regulating the production of intracellular second messengers, cAMP, Ca^{2+}, and **1,2-diacylglycerol**. D_1 and D_2 receptors are decreased in the striatum of patients with dementia. There is considerable evidence that suggests that intracellular levels of cAMP have a protective role for dopaminergic neurons. Intracellular concentrations of cyclic nucleotides are regulated by **cyclic nucleotide phosphodiesterases** and CaMPDE, one of the most intensely studied and best-characterized phosphodiesterases.

The heterogeneity of dopaminergic neurons in the central nervous system may be judged by the fact that they may vary not only in terms of their projection areas, but also in terms of the type of feedback pathways that they receive. Moreover, dopaminergic neurons and their receptors undergo dramatic developmental alterations that may foster a unique circumstance to produce dissimilar or opposite pathological states even with the same drugs. For example, in aged animals treated with **6-hydroxydopamine**, the loss of nigrostriatal dopamine is accommodated by a compensatory increase in the activity of the remaining neurons, whereas the **tuberinfundibular dopaminergic systems** are unable to compensate in a similar manner.

The heterogeneity of dopaminergic neurons may also be judged by the fact that the cotransmitter systems involving dopamine and peptides are varied in the central nervous system. For example, in the corpus striatum, in addition to dopamine, acetylcholine, γ-aminobutyric acid, serotonin, glutamate, and aspartate, one also finds peptides such as enkephalin, substance P, somatostatin, neuropeptide Y, cholecystokinin, neurotensin, and vasoactive intestinal peptide. Although many neuroleptics block dopamine receptors, they may have selective effects on the peptides and other parts of the brain. A few examples will be cited.

A high degree of coexistence of **cholecystokinin-like** and **tyrosine hydroxylase-like immunoreactivities** has been observed in the substantia nigra pars compacta. Moreover, ventral mesencephalic cholecystokinin projections encompass the full range of the well-known dopaminergic mesolimbic, mesostriatal, and mesocortical projections. Therefore, it should not be surprising that the intrastriatally injected cholecystokinin is able to stimulate dopamine-mediated transmission and to elevate the density of brain D_2-dopamine receptors.

The chronic injection of **haloperidol**, but not **clozapine**, increases selectively the concentration of enkephalins in the striatum. Protracted blockade of dopamine receptors by haloperidol causes a reduction in nigral content of substance P-like immunoreactivity, and of **substance P** and **substance K** mRNAs. Moreover, the effects of haloperidol on substance P is nonuniform in various areas of brain.

Subchronic oral administration of **lithium** causes a time-dependent increase in the substance P level in the striatum, which is prevented by coadministration of **haloperidol**. In PC12 pheochromocytoma cells, lithium dramatically increases the intracellular levels of the neuropeptide **neurotensin** and the mRNA encoding it. An extensive overlap between specific and high-affinity neurotensin binding sites and dopamine perikarya and dendrites has been shown to occur in the mesocorticolimbic and nigrostriatal projection systems. Consistent with this observation are the results of observations showing that **cocaine**, an indirect sympathomimetic agent that enhances the extrapyramidal dopaminergic activity, increases dramatically the striatal content of **neurotensin-like immunoreactivity**.

The effects of selegiline on bovine brain CaMPDE isoenzymes have been investigated. The findings indicated that selegiline inhibited brain 60-kDa isozyme; however, the inhibition for brain 63-kDa CaMPDE was observed to a lesser extent. The inhibition of brain 60-kDa CaMPDE was overcome by increasing the concentration of calmodulin, suggesting that selegiline may be a calmodulin antagonist or act specifically and reversibly on the action of calmodulin. The 60-kDa CaMPDE isozyme is predominantly expressed in brain and its inhibition can result in increased intracellular levels of cAMP. The increased intracellular levels of cAMP play a protective role for dopaminergic neurons. The cAMP may produce long-lasting physiological effects on synaptic membranes through a mechanism comprised of activation of cAMP-dependent protein kinase and consequent catalysis of phosphorylation of protein constituents of synaptic membrane. Intracellular signaling systems are complicated, but the multiple sites for pharmacological intervention may provide a basis for their mechanism of action and development of new drugs.

SELEGILINE INDUCES DOPAMINE RELEASE THROUGH AN ATP-SENSITIVE POTASSIUM CHANNEL

Selegiline is metabolized to methamphetamine (see Figure 28) and, by possessing amphetamine-like effects, is thought to release dopamine. By using an *in vivo* brain microdialysis technique, the chronic effects (21 days) and low doses of selegiline (0.25 mg/kg) on striatal release of dopamine have been investigated (see Figure 29). To determine whether amphetamine-like metabolites were responsible for the effects of selegiline, its action was compared with that of another selective propargyl MAO-B inhibitor, TVP-1012[R(+)-*N*-propargyl-*l*-aminoindan mesylate], which does not yield amphetamine-like metabolites. The action of **chlorgyline** was also studied, to compare the effects of MAO-A and MAO-B inhibition. In addition to measuring the effects of chronic treatment with MAO inhibitors on basal dopamine release, the potassium-induced release of dopamine was also measured. The results of these studies showed that dopamine metabolism was reduced only by chlorgyline, whereas neuronal release of dopamine was enhanced by both MAO-A and MAO-B inhibitors on chronic administration. The enhanced dopamine release by chronic MAO-B inhibition does not appear to be dependent on production of amphetamine-like metabolites of the inhibitor. Possible mechanisms for the release-enhancing effect of the MAO-B inhibitors include elevation in levels of endogenous β-phenylethylamine, or an inhibition of dopamine reuptake, which develops only on chronic administration because selegiline has a very weak effect on amine uptake in acute experiments.

Superfusion chambers have been used to investigate the role of ATP-sensitive potassium (K_{ATP}) channels in dopamine release elicited by the MAO inhibitor selegiline in the caudate-putamen *in vitro*. Selegiline induced increases in the extracellular concentrations of dopamine, with a maximal increase to 185% in comparison to basal outflow at 0.1 mM selegiline. **Butanedione** (0.1 mM), a specific K_{ATP} channel blocker, also significantly enhanced extracellular dopamine levels in the caudate-putamen to approximately 260%. Selegiline only led to an additional increase of dopamine outflow when added to submaximal concentrations of **butanedione** or **tolbutamide**, implying that selegiline was acting on identical sites. When the K_{ATP} channel opener **cromakalim** was added to the incubation medium, basal as well as butanedione-enhanced dopamine levels markedly decreased to about 40% when compared to baseline values. Selegiline-activated dopamine release was also antagonized by cromakalim. The selegiline effect was modulated neither by preincubation with the dopamine uptake inhibitor

nomifensine nor by the dopamine receptor agonist **quinpirole** and the dopamine receptor antagonist **sulpiride**. These results suggest that selegiline is able to modulate K_{ATP} channels in the caudate-putamen resulting in an enhancement of striatal dopamine release.

SELEGILINE POSSESSES NEUROTROPHIC-LIKE ACTION AND RESCUES AXOTOMIZED MOTORNEURONS INDEPENDENT OF MONOAMINE OXIDASE INHIBITION

Selegiline possesses "trophic-like" action and may be of value for treating acute nervous system damage, particularly damage caused by trauma. Neuronal death is difficult to establish conclusively. Immunocytochemistry for marker proteins is essential to identify specific neuronal populations, but this phenomenon suffers from the limitation that damaged nerve cells, like motorneurons, often decrease or even halt the synthesis of some proteins such as **choline acetyltransferase**, while increasing the synthesis of cytoskeletal proteins. Hence, a loss of immunoreactive somata may indicate neuronal death or a reactive decrease in protein synthesis. Furthermore, motoneuronal somata can be premarked by the retrograde transport of an agent from their terminal areas, and the number of remaining premarked somata may be used to estimate neuronal death. The technique carries the limitation that the entire neuronal population may not be able to transport the marking substance, either due to incomplete delivery of the agent to the target tissues or due to inadequate uptake or retrograde transport by some of the neurons. If the retrograde transport of a premarker is combined with immunocytochemistry for a marker protein, it can then be used to determine if a loss of immunoreactivity has contributed to a reduction in the counts of immunoreactive somata. However, this cannot offer absolute values for neuronal death unless the entire neuronal population is shown to have transported the premarker.

The role of MAO-B and the influence of selegiline on the rescue of facial motoneurons axotomized at postnatal day 14 using the (+)- and (−)enantiomers of selegiline [S(+)selegiline and R(−)selegiline] have been investigated. Motoneuronal survival was measured 21 days after unilateral facial nerve transaction. Doses of R(−)selegiline of 0.005, 0.01, and 10.0 mg/kg/2 days increased the surviving facial motoneuron to 38, 51, and 48%, respectively. Selegiline increased motoneuronal survival without significant inhibition of brain stem MAO-B activity. The dosage relationships for motoneuronal rescue vs. MAO-A and MAO-B inhibition and the marked difference in the stereospecificity of MAO-B inhibition vs. that of motoneuronal rescue show that the increased survival is unlikely to be dependent on the interaction of selegiline with the FAD site of MAO-A or MAO-B.

The finding that the rescue of the axotomized motoneurons appears independent of an interaction with MAO-B is in accord with work showing that selegiline was able to promote a marked increase in the survival of murine substantia nigra compacta neurons, after they had sustained otherwise lethal damage by MPTP, at dosages too low to cause any MAO inhibition. Similarly, the finding that the rescue of axotomized motoneurons occurs with doses too small to cause MAO-A inhibition is in accord with the finding that **chlorgyline**, a selective MAO-A inhibitor, did not rescue MPTP-damaged substantia nigra neurons. **Pargyline**, which inhibits

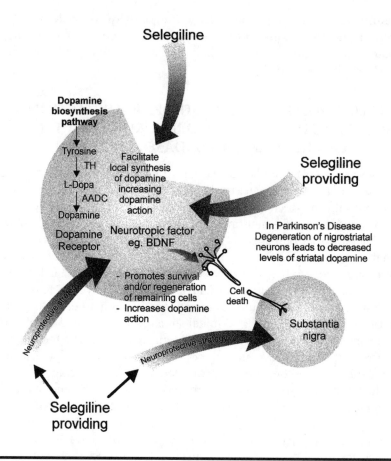

Figure 32 Selegiline enhances the release of dopamine and inhibits its metabolism. In addition, it has neurotrophic-like action capable of repairing damaged neurons.

both MAO-A and MAO-B and is structurally similar to selegiline, increases the survival of the axotomized immature motoneuron but is considerably less effective on an equal-molar basis. Methamphetamine and amphetamine, the major metabolites of selegiline, do not increase the survival of the motoneuron. It would seem plausible that selegiline induces compensation for loss of target-derived trophic support. This does not imply that selegiline itself acts as a trophic factor but rather that it interferes with or triggers some event that supports the survival of motoneurons after the withdrawal of the trophic support. Recent research has shown that neurotrophic factors, **basic fibroblast growth factor (BFGF)**, **brain-derived neurotrophin factor (BDNF)**, and **ciliary neurotrophic factor (CNTF)**, reduce axotomy-induced death of facial or hypoglossal motoneurons in neonatal or immature rats in experiments similar to selegiline (Figure 32 and Table 19).

Neurotropic factors of dopaminergic neurons may represent a potential neuroprotective therapy for Parkinson's disease. Apart from **nerve growth factor** (NGF), the family of the "**neurotrophins**" subsumes at the present time BDNF, **neurotrophin-3** (NT-3), and **neurotrophin-4/5** (NT-4/5). However, the concept of neurotrophic factors as specific, target-derived molecules, each acting on distinct neuronal types, has to be modified because of their high degree of pleiotropism

Table 19 The Actions of Major Cytokines

Family	Members	Major Activities and Features
Interleukins	IL-1α, IL-1β, IL-Ira, and IL-2-IL-15	Multiple tissue and immunoregulatory activities; no similarity of activity is implied by membership of this family
Chemokines	IL-8/NAP-1, NAP-2, MIP-1α and β, MCAF/MCP-1, MGSA and RANTES	Leukocyte chemotaxis and cellular activation
Tumor necrosis factors	TNF-α and TNF-β	Similar to IL-1, in addition to tumor cytotoxicity
Interferons	IFN-α, β, and γ	Inhibition of intracellular viral replication and cell growth regulation; IFN-γ is primarily immunoregulatory
Colony-stimulating factors	G-CSF, M-CSF, GM-CSF, IL-3, and some of the other ILs	Colony cell formation in the bone marrow and activation of mature leukocyte functions
Growth factors	EGF, FGF, PDGF, TGFα, TGFβ, and ECGF	Cell growth and differentiation
Neurotrophins	BDNF, NGF, NT-3-NT-6, and GDNF	Growth and differentiation of neurons
Neuropoietins	LIF, CNTF, OM, IL-6	Cytokines acting on the nervous system, and acting via a related receptor complex

BDNF, brain-derived neurotrophin factor; CNTF, ciliary neurotrophic factor; EGF, epidermal growth factor; ECGF, endothelial cell growth factor; FGF, fibroblast growth factor; GDNF, glial-derived neurotrophic factor; G, M, and GM-CSF, granulocyte, macrophage, and granulocyte/macrophage colony-stimulating factor(s); IFN, interferon; IL, interleukin; IL-Ira, interleukin 1 receptor antagonist; MCAF, monocyte chemotactic and activating factor; MCP, monocyte chemotactic protein; MGSA melanoma growth stimulatory activity; MIP, macrophage inflammatory protein; NGF, nerve growth factor; NT, neurotrophin; NAP, neutrophil activating protein; OM, oncostatin M; PDGF, platelet-derived growth factor; RANTES, regulated upon activation normal expressed and secreted; TGF, transforming growth factor; TNF, tumor necrosis factor.

and a considerable overlap in biological activities. In addition, a number of other non-target-derived molecules exert trophic actions on certain neurons and non-neuronal cells. These factors include CNTF, **fibroblast growth factors 1 and 2 (FGF-1/FGF-2)**, **insulin-like growth factors 1 and 2** (IGF-1/IGF-2), **muscle-derived differentiation factor** (MDF), and members of the **transforming growth factor β-superfamily** such as **glial cell line-derived neurotrophic factor** (GDNF) (see Figure 32).

Glial cell line-derived neurotrophic factor (GDNF) has been purified and cloned. Furthermore, it has been identified by its promotion of survival and morphological differentiation of dopaminergic neurons with an increase of their high-affinity dopamine uptake in embryonic midbrain cultures. In 6-hydroxy-dopamine (6-OHDA)-induced parkinsonism, neurochemical and behavioral improvements following intranigral administration of GDNF were demonstrated. Following hemitransection of the medial forebrain bundle, a significant protection of tyrosine hydroxylase (TH-iR) nigral neurons by GDNF was reported. Daily injections of 5 μg GDNF for 14 days above the substantia nigra led to a survival of 84% of TH-iR nigral cells, whereas in control animals only 50% of these cells survived. In the **MPTP**-mouse model of Parkinson's disease, the protective effects of GDNF were investigated in respect to the time point of its administration. Striatal injections of GDNF (10 μg on two consecutive days), 24 hours before MPTP exposure, significantly reduced the normally observed TH-iR cell loss of 30% to about 15%. In addition, this effect was accompanied by a protection of the density of dopaminergic nerve terminals and dopamine levels in striatum as well as by increased motor behavior. When GDNF was administered 7 days after MPTP exposure, there was still some recovery of tyrosine hydroxylase immunoreactive (TH-iR) cells, increased locomotor behavior, and partial restoration of dopamine levels.

Brain-derived neurotrophic factor (BDNF) exhibits distinct effects on survival, morphological differentiation, neuritic growth, protection against MPP+ cytotoxicity, and dopamine uptake of fetal mesencephalic dopaminergic neurons. This neurotrophin was shown to be retrogradely transported to substantia nigra after injection into the striatum, indicating that there are functional receptors for BDNF on adult dopaminergic nigrostriatal neurons. Additionally, BDNF has been reported to prevent cell death of axotomized spinal motor neurons *in vivo*.

The effects of **metallothionein**, **neurotrophins**, and **selegiline** in providing neuroprotection in Parkinson's disease have been studied. The finding that MPTP elicits parkinsonism in humans suggests that endogenous or xenobiotic neurotoxic compounds may be involved in the etiology of Parkinson's disease. The studies have shown that the cerebrospinal fluid of newly diagnosed and drug-untreated patients with Parkinson's disease contains a low-molecular-weight substance that inhibits the growth and function of dopaminergic neurons in culture. In addition, selegiline in a dosage below the level that inhibits MAO-B, was able to protect dopaminergic neurons in culture against toxic factors present in the cerebrospinal fluid of patients with Parkinson's disease, and the said effect was mediated via elaboration of BDNF (see Figures 30 and 31).

That 6-hydroxydopamine (6-OHDA) or MPTP causes parkinsonism by generating **free radicals** and that inducers of **metallothionein** isoforms avert the neurotoxicity led to the investigation of whether metallothionein isoforms were capable of scavenging free radicals. By employing **electron spin resonance spectroscopy** (ESR), the free radical scavenging effects of metallothionein-I (MT-I) and metallothionein-II (MT-II) isoforms on four types of free radicals were examined for the first time. Solutions of 0.15 mM of MT-I and 0.3 mM of MT-II scavenged the **1,1-diphenyl-2-picrylhydrazyl radicals** completely. Further, they were able to scavenge hydroxyl radicals generated in a Fenton reaction. Moreover, MT-I scavenged almost 90% of the superoxide generated by the **hypoxanthine** and **xanthine oxidase system**, while MT-II could only scavenge 40%. By using **2,2,6,6-tetramethyl-4-piperidone** as a "spin trap" for the reactive oxygen species (containing **singlet oxygen**, superoxide, and hydroxyl radicals) generated by photosensitized oxidation of riboflavin, and by measuring the relative signal intensities of the resulting stable nitroxide adduct, **2,2,6,6-tetramethyl-4-piperidone-1-oxyl**, it was observed that MT-II could scavenge 92%, and MT-I could completely scavenge all the reactive species generated (see Figure 30).

The results of this investigation are interpreted to suggest that selegiline, by preventing the generation of free radicals, metallothionein isoforms, by scavenging free radicals, and neurotrophins, by rescuing dopaminergic neurons, are capable of attenuating oxidative stress and of providing neuroprotection in Parkinson's disease.

Epidermal growth factor (EGF) was shown to support survival of both embryonic dopaminergic midbrain and cholinergic forebrain neurons. The trophic actions of EGF required the presence of glial cells, proposing an indirect mode of action of EGF on dopaminergic neurons. In this context, it was reported that EGF acts also on glial cells promoting their ability to proliferate and to differentiate. *In vivo* studies revealed that intraventricular administration of EGF in 5 weeks after hemitransection of the medial forebrain bundle restores about 20% of tyrosine hydroxylase nigral neurons in comparison to vehicle-treated animals. Moreover, intraventricular infusion of EGF also accelerates recovery of striatal dopaminergic parameters, i.e., the dopamine content and tyrosine hydroxlase activity, in the MPTP-mouse model of Parkinson's disease.

Fibroblast growth factor-2 (FGF-2) exhibits trophic effects on central embryonic dopaminergic and GABAergic neurons in culture that appear to be mediated via glial cells. Transient increases in the amounts of FGF-2 have been described in distinct lesion paradigms of the central nervous system. The same was suggested to occur in a MPTP-induced lesion of the nitrostriatal dopaminergic system in mice. This may indicate a possible role of FGF-2 in neuronal regeneration, for example, in form of induction of synthesis of NGF or of other trophic molecules in astrocytes. In 6-OHDA-lesioned animals, intrastriatal FGF-2 infusions neither prevent striatal dopamine depletions nor diminish behavioral deficits. In the MPTP-mouse model, intrastriatal application of FGF-2 via gel foam partially attenuates the toxin-induced damage. However, this effect was only observed if FGF-2 was applied simultaneously or 3 days after the intraperitoneal MPTP injection, whereas a delay of FGF-2 administration for 7 days after MPTP injection aborted restoration of transmitters and tyrosine hydroxylase levels. Apart from these findings, FGF-2 ameliorates rotational behavior of substantia nigra-transplanted animals with lesions of the nigrostriatal dopaminergic system.

Muscle-derived differentiation factor (MDF) induces tyrosine hydroxylase expression in a variety of central nervous system neurons, including those of striatum, cerebellum, and cortex. Normally, i.e., without MDF, these neurons do not express this enzyme of catecholamine synthesis. Further *in vitro* studies revealed that MDF enhances TH-mRNA 40-fold in fetal mesencephalic neurons. *In vivo* studies, employing infusion of partially isolated MDF, reported this molecule to enhance tyrosine hydroxylase activity in dopamine-depleted striata of 6-OHDA-lesioned animals. Furthermore, an increase of striatal dopamine concentrations and a partial compensation of rotational asymmetry were observed. In contrast, dopaminergic parameters were not affected by administration of MDF in control animals, suggesting that adult dopaminergic neurons may regain sensitivity toward differentiation factors after lesion.

Transforming growth factor-β (TGF-β1, β2, β3) exert a survival-promoting activity on cultured dopaminergic neurons of the developing substantia nigra. In addition, TGF-β2 and TGF-β3 mRNAs were detected in developing striatum and substantia nigra. However, TGF-β3 did not prevent delayed degeneration of nigral dopaminergic neurons following intrastriatal 6-OHDA lesion.

The effect of chronic treatment with selegiline in an animal model of Parkinson's disease induced by unilateral knife transection of the medial forebrain bundle (MFB) in adult rats has been studied. The experimental conditions included a 3-week pretreatment with selegiline before stereotaxic transection of the MFB. Following surgery, selegiline treatment was maintained for 3 weeks. Neurochemical and immunohistochemical procedures were used to study the dopaminergic system and reactive astrocytes in the nigrostriatal system. Selegiline treatment failed to counteract the axotomy-induced degenerative changes of the nigrostriatal dopaminergic system. However, it was effective in increasing the density of reactive astrocytes in terms of glial fibrillary acidic protein (GFAP) immunoreactivity in the intact contralateral substantia nigra and also in further enhancing the axotomy-induced increase of GFAP-immunolabeled astrocytes in the lesioned substantia nigra. This selegiline-induced effect on GFAP immunoreactivity was confined to substantia nigra without effect in striatum. In addition, the investigators also found a medial-to-lateral gradient decrease in the distribution pattern of GFAP-immunolabeled astrocytes. Axotomy increased the number of reactive astrocytes in both striatal areas examined, but the preferential distribution pattern of reactive astrocytes in striatum was still evident.

SELEGILINE RESTORES SERUM INSULIN-LIKE GROWTH FACTOR-I IN AGED RATS

It has been reported that selegiline improves the functions of meso-limbo-cortical dopamine neurons that are known to be associated with cognitive processes. Furthermore, selegiline protects against the aging processes, increases the life span of some animals, and slows the age-related decline of performance in behavioral tests.

Selegiline is able to protect the substantia nigra in aging animals against oxidative stress, an effect that is mediated via dopaminergic neurons. Furthermore, it has been shown that moderate exposure to MPTP alters cognitive function in mice. Furthermore, it has been shown that a defect in norepinephrine release may contribute to the impaired accuracy of attention in patients with severe Parkinson's disease, and dopamine may be important in maintaining rapid motor responding. The hypothalamic dopaminergic neurons have been reported to participate in the control of gonadotropin secretion and prolactin, and they also affect the secretion of thyrotropin (TSH), adrenocorticotropin, and growth hormone (GH) from the anterior pituitary. Aging and a decline in bodily functions is associated with an alteration in many of the aforementioned neurohormones. For example, the concentration of dopamine in the median eminence of an older person is lower than that in a young person. A decrease of norepinephrine concentration, with some signs of neurodegeneration of serotonin (5-HT) fibers, in hypothalamus of aged animals has also been reported. All these results support the idea that the decrease of some important hormones such as GH and TSH, which are known to be reduced in aging, could be due to the decrease in catecholamines in hypothalamic areas. Administration of drugs that increase hypothalamic catecholamine activity has been shown to inhibit or reverse many effects of physical decline due to aging.

The effects of treatment with selegiline, an MAO-B inhibitor, on plasma levels of insulin-like growth factor I (IGF-I) (as indicator of GH secretion), levels of monoamines and their metabolites, and the activity and content of tyrosine hydroxylase — the rate-limiting enzyme in the biosynthesis of catecholamines — in the hypothalamus and hypophysis of old animals have been studied. It is believed that the antiaging effects of selegiline are due to restoration of hypothalamic hormones.

Insulin-like growth factors (IGFs) IGF-I and IGF-II are single-chain polypeptides that have structural homology with proinsulin. IGF-I has insulin-like short-term metabolic effects and long-term effects similar to those of growth factor on cell proliferation and differentiation of various cell types. The mitogenic activity of IGF-I is mediated through binding mainly to the IGF-I receptor, also known as the type I IGF receptor, located on the cell surface. The receptor binds IGF-I and IGF-II with high affinity and insulin with low affinity. The IGF-I receptor may also form a hybrid with the insulin receptor. Animals carrying null mutations of the genes encoding IGFs or IGF receptors demonstrate embryonic and postnatal growth retardation. Distinct from other growth factors, IGFs are known to associate with specific binding proteins (IGFBPs) in plasma and tissue.

IGF-I may act as a survival factor via stimulation of the **Bcl-2 family of proteins**. Several signaling pathways such as tyrosine kinase, P13 kinase, and MAP kinase have also been suggested to mediate the **antiapoptotic effects of IGF-I** (Figure 33).

SELEGILINE PROTECTS DOPAMINERGIC NEURONS FROM TOXICITY INDUCED BY GLUTATHIONE DEPLETION

Oxygen radicals are created in the brain during oxidative stress and at other times. These oxygen radicals must be neutralized to prevent neuronal death. **Glutathione** (GSH) is one of the most important antioxidants used in detoxifying these oxygen radicals. Brain glutathione levels range from 1 to 3 mM. Astrocytes and nerve terminals have high concentrations of GSH, whereas neuronal somas have limited amounts of GSH. Therefore, it may be important for neuron cell bodies to increase the synthesis of GSH during oxidative stress to protect themselves.

GSH is synthesized by two enzymes. **γ-glutamyl cysteine** is formed from cysteine and glutamate by **γ-glutamyl synthetase**. GSH is formed from γ-glutamyl cysteine and glycine by glutathione synthetase. Both enzymes are found in the brain (Figure 34).

The turnover rates of GSH and the activity of GSH synthetase during oxidative stress induced by **butylhydroperoxide** in young and mature animals have been measured. These studies have found that defense mechanisms against oxidative stress in the brain differ with age. Young people can increase the cellular availability of GSH, whereas older people can increase GSH synthetase activity during oxidative stress. These differences make older people more susceptible to brain oxidative damage.

It has been shown that incubation with L-dopa increased the level of GSH in neurons. The pure neuronal cultures were destroyed by incubation with L-dopa,

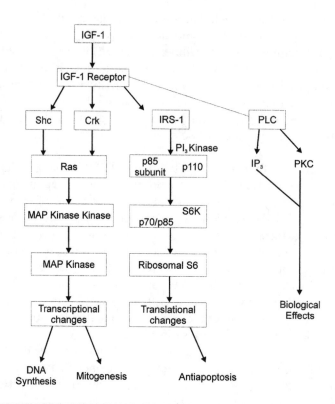

Figure 33 Overview of IGF-1 signaling. IGF-1 receptors exist in a number of tissues. The IGF-1 receptor is a heterotetrameric protein with intrinsic tyrosine kinase activity. Various domains in the IGF-1 receptor share 40 to 84% amino acid sequence homology with the insulin receptor. One of the earliest steps in signal transduction by IGF-1 receptor is the phosphorylation of insulin receptor substrate (IRS). Tyrosine-phosphorylated IRS-1 and IRS-2 interacts with specific cytoplasmic proteins containing SH2 (src homology 2) domains, leading to the transduction of downstream signals. The IGF-1 signaling cascades are activated by the association of growth receptor binding protein 2-Son of Sevenless (Grb-2-SOS) with phosphorylated Crk and Shr, resulting in the activation of Ras, and the sequential activation of serine/threonine kinase Raf (A-Raf, B-Raf, and c-Raf) and MAP kinase kinase (MEK, also designated as MKK1 and MKK2). MEK is just upstream of and very specific for MAP kinases (ERK-1 and ERK-2), which it activates by phosphorylation on specific residues. MAP kinases activate a number of transcription factors and there is evidence that they are involved in mediating IGF-1 stimulation of DNA synthesis and mitogenesis. The p85 regulatory subunit of PI3-kinase is another important SH2 domain-containing protein, the binding of which to tyrosine-phosphorylated IRS-1 activates the catalytic function of the 110-kDa subunit of PI3-kinase. PI3-kinase is essential for the transduction of metabolic growth and functional effects of IGF-1 and insulin, including stimulation of glucose transport, antilipolysis, protein and glycogen synthesis, and inhibition of apoptosis.

whereas the addition of ascorbic acid or superoxide dismutase protected the cells. These results show that the upregulation of cellular GSH evoked by auto-oxidizable agents is associated with significant protection of cells. Glia play an essential role in the response of mesencephalic cell cultures. An ability to upregulate GSH may serve a protective role.

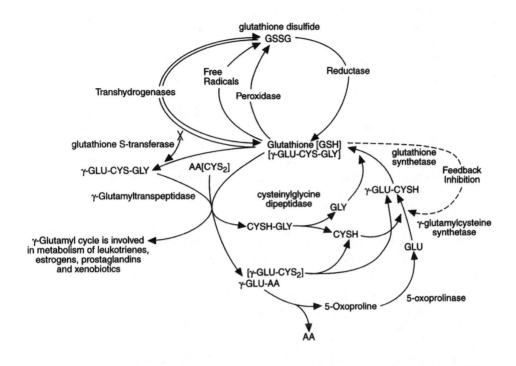

Figure 34 Glutathione (GSH), through γ-glutamyl pathway, recycles glutamate, cysteine, and glycine. GSH is synthesized by the consecutive actions of the ATP-dependent enzymes γ-glutamylcysteine synthetase and glutathione synthetase. Levels of GSH are regulated in part by feedback inhibition of γ-glutamylcysteine synthetase by GSH. In the presence of a suitable amino acid (AA) acceptor, GSH is catabolized by the action of γ-glutamyltranspeptidase to yield a γ-glutamyl amino acid (γ-GLU-AA) and cysteinylglycine (CYSH-GLY). The γ-GLU-AA is converted to free amino acid and 5-oxoproline by the action of γ-glutamylcyclotransferase. 5-Oxoproline is converted back to glutamate by the ATP-dependent 5-oxoprolinase reaction. The glutamate released in this process can then be reused in the synthesis of GSH, thus completing the cycle with the glutamate component of GSH. The cysteinylglycine released in the γ-glutamyltranspeptidase reaction is hydrolyzed by the action of a dipeptidase to glycine and cysteine, completing the cycle with the glycine and cysteine components of GSH.

GSH protects human neuronal cells from **dopamine-induced apoptosis**. The role of GSH and other antioxidants in dopamine-induced apoptosis in cultures of the human neuronal cell line has been studied. Apoptosis, induced by 0.1 to 0.3 mM dopamine, was blocked by GSH in a dose- and time-dependent manner. This was observed by monitoring cell morphology, cell viability, and the release of the cytosolic enzyme **lactate dehydrogenase** into the culture medium. L-Cysteine and N-acetylcysteine had a similar effect in protecting against dopamine neurotoxicity, but at lower concentrations than GSH. The dopamine-induced alteration in the cell cycle profile, detected by flow cytometry, and **intranucleosomal DNA fragmentation**, were both blocked by GSH. Treatment of cells with **buthionine sulfoximine**, an irreversible inhibitor of **γ-glutamylcysteine synthetase**, increased the neurotoxic effect of dopamine, suggesting that endogenous

GSH participates in reducing dopamine neurotoxicity. The relationship between GSH and dopamine was further investigated by testing the effect of dopamine on the endogenous GSH level. Dopamine decreased GSH levels within 16 to 24 hours; however, this effect was preceded by a transient increase in the level of the tripeptide within the first 0.5 to 7 hours. Two other types of endogenous antioxidants, (+)-α-tocopherol (**vitamin E**) and ascorbic acid (**vitamin C**), were tested; vitamin E (at 1 to 100 µg/ml) was inactive against dopamine toxicity, whereas vitamin C had no effect at 0.05 to 0.2 mM, but increased dopamine toxicity at 0.5 to 2 mM. The results indicate that GSH has a selective role in protecting human neural cells from the toxic effect of dopamine. This study may contribute to a better understanding of the mechanisms underlying the excessive loss of dopaminergic neurons in neurodegenerative diseases, such as parkinsonism, and in the aging process.

A reduction in GSH has been detected in the substantia nigra pars compacta of patients with Parkinson's disease. Furthermore, a similar reduction in GSH has been found in substantia nigra pars compacta of cases with incidental **Lewy body diseases**, who are thought to have a preclinical form of Parkinson's disease. These findings suggest that a reduction in GSH, with the ensuing oxidative stress, may be a critical factor in the pathogenesis of Parkinson's disease.

Studies have been designed to learn whether or not selegiline and its desmethyl metabolite could protect mesencephalic neurons in culture from death brought about by depletion of GSH by ʟ-**buthionine-(S,R)-sulfoximine** (BSO). The results of these studies showed that BSO (10 µM) caused extensive cell death after 48 hours as demonstrated by disruption of cellular integrity and release of lactate dehydrogenase into the culture medium. Both selegiline and desmethyl-selegiline, at concentrations of 5 and 50 µM, significantly protected dopaminergic neurons from toxicity without preventing the BSO-induced loss of GSH.

Protection was not associated with MAO-B inhibition in that pargyline, a potent MAO inhibitor, was ineffective and pretreatment with pargyline did not prevent the protective effects of selegiline. Protection was not associated with inhibition of dopamine uptake by selegiline because the dopamine uptake inhibitor **mazindol** did not diminish BSO toxicity. Antioxidant ascorbic acid (200 µM) also protected BSO-induced cell death, suggesting that oxidative events were involved. This study demonstrates that selegiline and its desmethyl metabolite can diminish cell death associated with depletion of GSH.

SELEGILINE INCREASES THE STRIATAL SUPEROXIDE DISMUTASE

Reactive oxygen species are suspected of contributing to the etiology of Parkinson's disease with increasing frequency. **Superoxide dismutase** (**SOD**: EC 1.15.1) is a metalloenzyme system that exists in cystolic (Cu,Zn-SOD) and mitochondrial (Mn-SOD) forms. SOD confers protection against oxidative cellular injury by conversion of superoxide radical ($O_2^{\bullet-}$) to hydrogen peroxide (H_2O_2) (see Figure 31). Observations of increased Cu,Zn-SOD and Mn-SOD activities in substantia nigra in Parkinson's disease are consistent with a protective effect. Elevated SOD activity in the parkinsonian brain could be an adaptive response to sustained oxidative stress induced by either defective **mitochondrial electron transport** or exposure to an MPTP-like toxicant. However, excessive SOD may produce cell injury if H_2O_2 is not effectively metabolized by **catalase** and **GSH perioxidase**, because the highly reactive **hydroxyl radical** (·OH) will be generated from H_2O_2 by a Fenton reaction with free iron, which is abundant in substantia nigra.

Several studies have explored SOD activity in peripheral tissues of patients with Parkinson's disease as potential markers of oxidative stress, and have yielded inconsistent results. Elevations of total (combined Cu,Zn-SOD and Mn-SOD) SOD activity and GSH peroxidase activity in serum of patients with Parkinson's disease have been reported. Furthermore, elevations of immunoreactive Mn-SOD in cerebrospinal fluid of patients with Parkinson's disease but normal levels of Cu,Zn-SOD have been detected. Erythrocyte SOD activity, which is exclusively due to Cu,Zn-SOD, is unaltered in Parkinson's disease. Moreover, no differences in mean activities of total SOD, GSH peroxidase, **GSH reductase**, or catalase have been detected in either the lymphocytes, granulocytes, or cerebrospinal fluid of patients with Parkinson's disease.

The levels of Cu,Zn-SOD and Mn-SOD activities in peripheral lymphocytes of 43 newly diagnosed idiopathic Parkinson's disease cases and 62 age- and sex-matched controls free of neurodegenerative disorders have been measured. Significant excesses of both SOD forms were found among Parkinson's disease cases compared with controls; however, the excesses were found exclusively among patients treated with the Parkinson's disease MAO inhibitor selegiline (L-deprenyl). Enzyme-linked immunosorbent assays (ELISAs) confirmed that the activity excesses were due to increased protein rather than more highly reactive enzymes in lymphocytes of patients with Parkinson's disease. These findings clearly indicate the importance of selegiline on measured Cu,Zn-SOD and Mn-SOD activity in peripheral lymphocytes. Characterizing a possible therapeutic value of SOD will require longitudinal assessments of SOD in relation to the progression of Parkinson's disease.

L-Selegiline induces a dose-dependent inhibition of oxygen (O_2) consumption during ATP synthesis in the presence of complex I (pyruvate and malate) and complex II (succinate) substrates in fresh mitochondrial preparations. D-Selegiline produces a similar inhibitory profile, whereas MDL 72974, a selective MAO-B inhibitor, is less effective. Administration of D-selegiline or MDL 72974 to animals results in an increase in both striatal CuZn-SOD and Mn-SOD activities. This observation is consistent with previous reports in which lower L-selegiline concentrations were shown to be sufficient to induce SOD activity in the striatum. Selegiline is not a direct modulator of SOD activity and has no effects on purified SOD activity *in vitro* with concentrations as high as 4 mM. Striatal CuZn-SOD and Mn-SOD activities remain elevated following L-selegiline withdrawal even though MAO-B activity had recovered by 50%. This observation suggests that the modulation of striatal SODs by L-selegiline is not readily reversible and appears to be independent, at least in part, from the inhibition of MAO-B.

Several studies have shown that SOD is upregulated in animals exposed to O_2 radicals. Accordingly, the mechanism by which L-selegiline modulates SOD activities within the nigrostriatal pathway and the cerebellum may represent an adaptive mechanism reacting to the endogenous formation of O_2 radical. Numerous potential sources of O_2 exist in the cell, including the electron transport chain and the autooxidation of catecholamines. Therefore, selegiline could directly modulate free radical production by altering respiratory function within mitochondria.

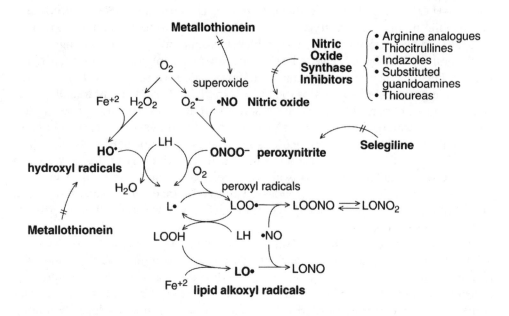

Figure 35 Nitric oxide regulates critical lipid membrane and lipoprotein oxidation events by (1) contributing to the formation of more potent secondary oxidants from $O_2^{\cdot-}$ (i.e., ONOO⁻), (2) catalyzing the redirection of O_2^- and H_2O_2-mediated cytotoxic reactions to other oxidative pathways, and (3) termination of lipid radicals to possibly less reactive secondary nitrogen-containing products. Moreover, metallothionein isoforms I and II protect against neurotoxic effects of superoxide anions and hydroxyl radicals and selegiline protects against peroxynitrite- and nitric oxide-induced apoptosis.

L-Selegiline alters the redox state of ubiquinone, suggesting that the flow of electrons is impaired in the respiratory chain. Furthermore, a decrease in **ubiquinone** levels has been observed, whereas **ubiquinol** (reduced ubiquinone) concentrations are increased in the striatum. Ubiquinol levels have been shown to augment as a result of impaired mitochondrial respiration. For example, ubiquinol concentrations were demonstrated to increase in tubular kidney cells exposed to complex IV inhibitors and in disease states with defects in respiratory chain components. These results are also consistent with the hypothesis that L-selegiline enhances O_2 formation by altering the rate of electron transfer within the respiratory chain leading to increases in SOD activities in the mouse striatum.

SELEGILINE PROTECTS AGAINST PEROXYNITRITE- AND NITRIC OXIDE-INDUCED APOPTOSIS

Apoptosis is induced by various intra- and extracellular stimuli, and recently nitric oxide was reported to induce apoptosis in cultured cerebellar granule cells and cultured cortical neurons. The toxicity of nitric oxide is mainly ascribed to **peroxynitrite**, a reaction product of **nitric oxide** with **superoxide** (Figure 35). Cells producing an increased amount of SOD (superoxide, superoxide oxidoreductase; EC 1.15.1.1) are resistant to nitric oxide-mediated apoptosis. In contrast, superoxide levels which have been increased by downregulation of Cu,Zn-SOD lead to apoptotic cell death in PC12 cells, which required the reaction with nitric oxide to generate peroxynitrite. Peroxynitrite itself was found to induce apoptosis in PC12 cells and in cultured cortical neurons.

The pathological characteristic of Parkinson's disease is the selective degeneration of dopamine neurons in the pars compacta of the substantia nigra. The mechanism for the loss of neurons remains to be elucidated, and recently **apoptosis** has been proposed as a death process in Parkinson's disease. For example, the level of a product of the oxidative stress, 4-hydroxy-2-nonenal protein adduct, was found to increase in the nigral neurons of parkinsonian brains. **Peroxynitrite** (see Figure 35) has been proposed to be involved in the neuronal cell death in some neurodegenerative diseases, such as **amyotrophic lateral sclerosis**.

The induction of apoptosis by nitric oxide and peroxynitrite in human dopaminergic neuroblastoma SH-SY5Y cells and the antiapoptotic activity of selegiline have been reported. After the cells were treated with **1-methyl-4-phenyl-1,2,3,6-tetrahydropyridine** (NOR-4), a nitric oxide donor, or **N-morpholino-sydnonimine** (SIN 1), a peroxynitrite donor, DNA damage has been studied using a single-cell gel electrophoresis (comet) assay. NOR-4 and SIN-I induce DNA damage dose dependently. **Cycloheximide** and alkaline treatment of the cells prevented the DNA damage, indicating that the damage is apoptotic in nature and depends on a signal transduction mechanism. SOD and the antioxidants reduced GSH and α-tocopherol protected the cells from the DNA damage. Selegiline protected the cells from the DNA damage induced by nitric oxide or peroxynitrite almost completely. The protection by selegiline was significant even after it is washed from the cells, indicating that selegiline may activate the intracellular system against apoptosis. These results suggest that selegiline or related compounds may be neuroprotective to dopamine neurons through its antiapoptotic activity.

The neuroprotection of SH-SY54 cells from apoptosis caused by NOR-4 and SIN-1 is not related to inhibition of MAO-B, since SH-SY54 contains only MAO-A. Selegiline is able to protect dopaminergic neurons from the toxicity of MPTP and also of MPP+, an oxidized product of MPTP by MAO-B. This indicates that the neuroprotection by selegiline does not require blockage of the conversion of MPTP to MPP+.

The administration of selegiline increases the activity of SOD and catalase in the rat striatum. Selegiline reduces apoptosis in PC12 cells induced by withdrawal of serum and growth factor and that transcription induction was required for its antiapoptotic effect. Furthermore, selegiline increases the expression of Cu,Zn-SOD, Mn-SOD, GSH peroxidase, tyrosine hydroxylase bcl-2, bcl-x, and c-fos and decreased that of bax and c-jun. Moreover, selegiline is able to attenuate **glutamate receptor-mediated toxicity**, which is induced by activation of neuronal nitric oxide synthase and increased nitric oxide formation (see Figure 31). In addition, selegiline protects cells against endogenously produced toxins. These results suggest that selegiline may be neuroprotective in general through the inhibition of **"death" signal transduction**, induced by endogenous and environmental factors, causing Parkinson's disease or perhaps other neurodegenerative disorders.

SELEGILINE POSSESSES ANTIAPOPTOTIC ACTIONS IN A VARIETY OF NEURONS

Selegiline reduces neuronal death even after neurons have sustained seemingly lethal damage at concentrations too small to inhibit MAO-B and the rescuing action is related to antiapoptotic action of selegiline.

Initially, it was thought that selegiline only reduces the death of nigrostriatal dopaminergic neurons by inhibiting MAO-B and thereby blocking the conversion of MPTP to its active radical MPP+. Subsequently, it was shown that selegiline maintains dopamine levels and dopamine uptake in mesencephalic explants after MPP+ treatment, thereby establishing that selegiline could reduce dopaminergic neuronal damage by a mechanism other than the blockade of the conversion of MPTP to MPP+. Selegiline was shown to reduce the death of murine dopaminergic nigrostriatal neurons even when first administered 72 hours after MPTP treatment when MPTP was fully converted to MPP+ and maximal striatal damage had occurred. It was then shown using the same delayed administration that selegiline at doses of 0.01 mg/kg every 2 days induces a maximal increase in the survival of the dopaminergic nigrostriatal neurons. The 0.01 mg/kg dose was insufficient to inhibit MAO-A or MAO-B even after 20 days of administration. It has also been shown that selegiline increases the survival of dopaminergic mesencephalic neurons in culture, either after trophic withdrawal or MPP+ treatment. Selegiline is also shown to reduce **N-(2-chloroethyl)-N-ethyl-2-bromobenzylamine** (DSP-4) toxicity, which does not involve MAO-A or MAO-B.

Selegiline has been shown to reduce neuronal death in a variety of noncatecholaminergic neuronal types and after a variety of insults, including immature facial neurons after axotomy, adult murine facial motoneurons crush, rat CA1 hippocampal neurons after ischemia/hypoxia, and rat cerebellar Purkinje cells and granule cells with aging. Studies with axotomized immature rat facial motoneurons have shown that the maximum effectiveness of intraperitoneal (–)-selegiline for increasing the survival of the motoneurons was greater than that seen with intrathecal delivery of ciliary neurotrophic factor (CNTF) and it did not induce the cachexia caused by CNTF. Furthermore, it has been shown that a dose of selegiline insufficient to inhibit MAO-B or MAO-A induces maximum increases in the survival of the immature facial motoneuron whereas doses of selegiline sufficient to cause 80% MAO-B inhibition did not increase motoneuron survival.

With the exception of selegiline and pargyline, most MAO-A or MAO-B inhibitors including **iproniazid**, **phenelizine**, **semicarbazide**, **tranylcypramine**, **nialamide**, MDL 72974A, RO-16-6491, **chlorgyline**, and **brofaromine** did not increase the survival of the axotomized immature motoneurons. All in all, these results show that the neuronal "rescue" action of selegiline did not involve MAO-A or MAO-B inhibition and is dependent on a structural entity shared by (–)-selegiline and pargyline but not (+)-selegiline or other structurally similar compounds like chlorgyline.

SELEGILINE STIMULATES BIOSYNTHESIS OF CYTOKINES INTERLEUKIN-1β AND INTERLEUKIN-6

Cytokines are a heterogenous group of polypeptide mediators that have been associated with activation of numerous functions, including the immune system and inflammatory responses. The cytokine families include, but are not limited to, **interleukins**, **chemokines**, **tumor necrosis factors**, **interferons** (INF-α, -β, and -γ), **colony-stimulating factors**, growth factors, **neuropoietins**, and **neurotrophins** (see Table 19).

The neurotrophins represent a family of survival and differentiation factors that exert profound effects in the central and peripheral nervous systems. The neurotrophins are currently under investigation as therapeutic agents for the treatment of neurodegenerative disorders and nerve injury either individually or in combination with other trophic factors such as ciliary neurotrophic factor (CNTF) or fibroblast growth factor (FGF).

Responsiveness of neurons to a given neurotrophin is governed by the expression of two classes of cell surface receptor. For nerve growth factor (NGF), these are $p75^{NTR}$ (p75) and $pl40^{trk}$ (referred to as trk or trkA), which bind both BDNF and neurotropin (NT)-4/5, and trkC receptor, which binds only NT-3. After binding ligand, the neurotrophin–receptor complex is internalized and retrogradely transported in the axon to the soma. Both receptors undergo ligand-induced dimerization, which activates multiple signal transduction pathways. These include the ras-dependent pathway utilized by trk to mediate neurotrophin effects such as survival and differentiation. Indeed, cellular diversity in the nervous system evolves from the concerted processes of cell proliferation, differentiation, migration, survival, and synapse formation. Neural adhesion and extracellular matrix molecules have been shown to play crucial roles in axonal migration, guidance, and grown cone targeting.

Proinflammatory cytokines, released by activated macrophages and monocytes during infection, can act on neural targets that control thermogenesis, behavior, and mood. In addition to induction of fever, cytokines induce other biological functions associated with the acute-phase response, including hypophagia and sleep. Cytokine production has been detected within the central nervous system as a result of brain injury, following stab wound to the brain, during viral and bacterial infections (AIDS and meningitis), and in neurodegenerative processes (multiple sclerosis and Alzheimer's disease). Novel cytokine therapies, such as anticytokine antibodies or specific receptor antagonists acting on the cytokine network may provide an optimistic feature for treatment of multiple sclerosis and other diseases in which cytokines have been implicated.

An immune-mediated pathophysiology has been proposed for Parkinson's disease and Alzheimer's disease. The MAO-B inhibitor selegiline appears to slow the progression of neurological deficits in patients with Parkinson's disease and with Alzheimer's disease, but its mechanism of action is still under discussion. *In vitro* and animal trials show that the putative neuroprotective effects of selegiline may be due to reduction of oxidative stress or altered expression of a number of mRNAs and proteins in nerve and glial cells. The results of a recent animal trial suggest that selegiline might also influence the immune system, because it increases the survival rate of immunosuppressed animals. Stimulation of the immune response to bacterial or viral infection and to chronic inflammatory processes, and biosynthesis of neurotrophic factors, are managed by an increased synthesis of the cytokines interleukin-1β (IL-1β) and subsequently interleukin-6 (IL-6).

Interleukin 1

Activated macrophages and microglia are likely cellular sources of IL-1 in the central nervous system. IL-1α and β, both 17-kDa proteins, are the products of two distinct genes and produce many of the same effects that TNFα has on glial cells. IL-1 upregulates cytokine production, includes cell surface molecules, activates nitric oxide, and stimulates proliferation. When used alone, IL-1 and TNFα both stimulate nitric oxide production in C6 cells. However, in human fetal astrocyte cultures, IL-1 is a better nitric oxide inducer when used in combination with IFNγ.

IL-1 has been demonstrated to affect the production of other cytokines or growth factors. Treatment of astrocytes with IL-1 increases basic fibroblast growth factor concentrations *in vitro*. Following administration of intraventricular IL-1β, nerve growth factor-like immunoreactivity is evident in hippocampal astrocytes as well as in neurons. *In vitro* production of nerve growth factor by astrocytes is most evident following combined stimulation with IL-1β and TNFα. Transforming growth factor-β-I (TGFβ1), as measured by secreted protein and immunohistochemical presence in cells, is induced by IL-1α in astrocytes and microglia. Oligodendrocytes, in response to IL-1α, demonstrate increased TGFβ by immunohistochemistry, but functional TGFβ1 protein is not significantly secreted. An IL-1β-stimulated human astroglioma produces increased TNFα mRNA and protein concentrations. In contrast, TNFα is not induced or elevated after stimulating first-trimester human microglia with IL-1α or rat astrocytes with IL-lβ. When combined with IFNγ, IL-1β induces astrocytes to produce TNFα.

Regulation of major histocompatibility complex (MHC) expression by IL-1 may be cell and species dependent. IFNγ-induced Class Ia expression on murine astrocytes is inhibited by IL-1β treatment. IL-1β is ineffective in modulating IFNγ-induced Class II expression on murine microglia and Class I expression on astrocytes. Additionally, Class II expression on human fetal and adult astrocytes does not appear to be modulated by IL-1.

In mixed glial cultures, IL-1 indirectly increases the proliferation of microglia, astrocytes, and oligodendrocytes but not glial precursors. The combination of IL-1 and TNFα leads to microglial and astrocytic *in vitro* proliferation reminiscent of microgliosis and astrogliosis *in vivo*. IL-1 and TNFα induce astrocytes to lose their flat polygonal shape and appear as cells with long, fine processes and small compact bodies.

Interleukin 6

Interleukin 6 (IL-6) is a protein of approximately 23 to 30 kDa with heterogeneity in size depending on differential glycosylation. It is produced by monocytes, T cells, B cells, fibroblasts, mesangial cells, keratinocytes, and endothelial cells. IL-6 has a major role in B-cell proliferation and differentiation. In the nervous system, it is produced by endothelial cells, microglia, and astrocytes. IL-6 belongs to a family of neuropoietic/hematopoietic factors, which, while biochemically different, are structurally and functionally related. The neuropoietic family includes **cholinergic differentiation factor** (also known as **leukemia inhibitory factor**), ciliary neurotrophic factor (CNTF), and oncostatin M. The hematopoietic cytokines are IL-6, G-CSF, and myelomonocytic factor.

IL-6 induces nerve growth factor production and augments bFGF production in astrocytes. IL-6, induced *in vivo* by IL-1, may contribute to acute-phase response and fever through induction of prostaglandin E (PGE). IL-6 is present in elevated levels in several central nervous system diseases such as AIDS, multiple sclerosis, and Alzheimer's disease, of which all three pathologies may involve activated microglia and astrocytes. Interestingly, IL-6 has been shown to enhance HIV-1 expression in transfected primary rodent astrocytes but not human malignant astrocytoma cell lines.

The IL-6/LIF family members are also critical for long-term and short-term survival, but not proliferation of mature oligodendrocytes and their precursors.

The influence of selegiline on the synthesis of IL-1β and IL-6 in peripheral blood mononuclear cells (PBMC) from healthy blood donors cultured with or without selegiline (10^{-8}, 10^{-9}, or 10^{-10} M) in a humidified atmosphere (7% CO_2) has been studied. Treatment of cultured PBMC with selegiline significantly increased synthesis of both cytokines. The effect of selegiline on cytokine biosynthesis may contribute to its putative neuroprotective properties. The results of this study indicate immunomodulatory properties of selegiline. The elevated levels of IL-6 may be explained by direct stimulatory activity of selegiline. An indirect effect caused by IL-1β also has to be considered because IL-1β induces release of IL-6. The mechanism of action of selegiline on PBMC has remained unclear. Recent findings, however, demonstrate that selegiline can alter gene expression/protein synthesis in nerve and glial cells. These results suggest that immunostimulation by selegiline may be caused in a similar way, by selective action on transcription. It remains to be established whether selegiline also increases release of IL-1β and IL-6 from microglial cells and astrocytes in the brain.

Because of the growing evidence for a bidirectional interaction between the peripheral immune system and the brain, it has been proposed that even peripheral circulating cytokines such as IL-1β or IL-6 may influence regenerative and degenerative processes in the brain, acting, for example, via the organum vasculosum lamina termalis, where the blood–brain barrier is weak. IL-1β stimulates the proliferation of astrocytes and the release of IL-6 and attenuates excitatory amino acid-induced neurodegeneration *in vitro*. IL-6 induces acute-phase protein synthesis and differentiation of neuronal cells and improves catecholaminergic and cholinergic cell survival.

Immune-related phenomena have been demonstrated in studies of the pathogenesis of Parkinson's and Alzheimer's diseases. Alterations in IL-1β and IL-6 levels in brain parenchyma, serum, and cerebrospinal fluid of patients with Parkinson's and Alzheimer's diseases have been found. A recent study showed elevated IL-1β and IL-6 levels in the cerebrospinal fluid of untreated patients with so-called *"de-novo"* Parkinson's and Alzheimer's diseases, which may reflect, especially in the case of Parkinson's disease, the original condition at the probable beginning of the disease. From the results of this study and evidence of the influence of IL-1β and IL-6 on regenerative processes in the brain, one might speculate that this cytokine status represents the regenerative potential of the brain in the fight against the neurodegenerative process.

BANISTERINE AND SELEGILINE FOR THE TREATMENT OF PARKINSON'S DISEASE

Parkinson's disease is characterized by tremor at rest, rigidity, hypokinesia, and postural abnormality. The 4- to 6-Hz tremor of parkinsonism is characteristically most conspicuous at rest; it increases at times of emotional stress and often improves during voluntary activity. It commonly begins in the hand or foot, where it takes the form of rhythmic flexion–extension of the fingers or of the hand or foot or of rhythmic pronation–supination of the forearm. It frequently involves the face in the area of the mouth as well. Although it may ultimately be present in all of the limbs, it is not uncommon for the tremor to be confined to one limb — or to the two limbs on one side — for months or years before it becomes more generalized.

Rigidity. Rigidity, or increased tone — i.e., increased resistance to passive movement — is a characteristic clinical feature of parkinsonism. The disturbance in tone is responsible for the flexed posture of many patients with parkinsonism. The resistance is typically uniform throughout the range of movement at a particular joint and affects agonist and antagonist muscles alike — in contrast to the findings in spasticity, where the increase in tone is often greatest at the beginning of the passive movement (clasp-knife phenomenon) and more marked in some muscles than in others. In some instances, the rigidity in parkinsonism is described as *cogwheel rigidity* because of rachetlike interruptions of passive movements that may be due, in part, to the presence of tremor.

Hypokinesia. The most disabling feature of this disorder is hypokinesia (sometimes called bradykinesia or akinesia) — a slowness of voluntary movement and a reduction in automatic movement, such as swinging the arms while walking. The patient's face is relatively immobile (*masklike facies*), with widened palpebral fissures, infrequent blinking, with certain fixity of facial expression, and a smile that develops and fades slowly. The voice is of low volume (*hypophonia*) and tends to be poorly modulated. Fine or rapidly alternating movements are impaired, but power is not diminished if time is allowed for it to develop. The handwriting is small, tremulous, and difficult to read.

Abnormal Gait and Posture. The patient generally finds it difficult to get up from bed or an easy chair and tends to adopt a flexed posture on standing. It is often difficult to start walking, so that the patient may lean farther and farther forward while walking in place before being able to advance. The gait itself is characterized by small, shuffling steps and absence of the arm swing that normally accompanies locomotion; there is generally some unsteadiness on turning, and there may be difficulty in stopping. In advanced cases, the patient tends to walk with increasing speed to prevent a fall (*festinating gait*) because of the altered center of gravity that results from the abnormal posture.

Other Clinical Features. There is often mild **blepharoclonus** (fluttering of the closed eyelids) and occasionally **blepharospasm** (involuntary closure of the eyelids). The patient may drool, perhaps because of the impairment of swallowing. There is typically no alteration in the tendon reflexes, and the plantar responses are flexor. Repetitive tapping (about twice per second) over the bridge of the nose produces a sustained blink response (*Myerson's sign*); the response is not sustained in normal subjects. Cognitive decline sometimes occurs but is usually mild.

The etiology of progressive death of dopaminergic neurons in substantia nigra of Parkinson's disease brains remains unclear. Dopamine deficiency in Parkinson's disease is commonly treated with L-dopa and **carbidopa**, a **periphera dopa decarboxylase inhibitor** (**Sinemet**). Since its introduction, L-dopa has been shown to be effective in treating Parkinson's disease. However, high concentrations of L-dopa produce side effects such as psychosis, "**on–off**" effects, abnormal involuntary movements, and akinetic crisis.

By using microdialysis of substantia nigra it has been shown that L-dopa treatment produced a dose-dependent increase in output of nigral hydroxyl ($\cdot$OH) radicals, which could be increased further with acute inhibition of mitochondrial complex I activity. Selegiline is able to attenuate the generation of hydroxyl radicals caused by enhanced turnover in the synthesis of striatal dopamine according to the following reactions.

Unfortunately, Sinemet (L-dopa–carbidopa) does not arrest nigrostriatal degeneration or prevent the progressive disabling course of the disease. Moreover, the therapeutic effect is not sustained during long-term use. The reasons for the decline in efficacy of selegiline and increasing response fluctuations that characterize long-term L-dopa therapy are not fully understood, but may reflect the inevitable progression of the disease. Other contributing factors may include deficits in storage of dopamine, changes in dopamine receptor function, competition with other amino acids, variation in gastric emptying time, and other poorly understood pharmacokinetic changes.

Irreversible MAO-B inhibitors, namely, **banisterine** and **selegiline**, are efficacious in the treatment of Parkinson's disease, and selegiline prolongs the action of L-dopa and enhances the life expectancy of patients suffering from Parkinson's disease.

Physicians have studied 54 patients who were randomly assigned to selegiline (10 mg/day) or placebo treatment groups and followed until L-dopa therapy was indicated or until the patient had been in the study for 3 years. Analysis of Kaplan–Meier survival curves for each group showed that selegiline delayed the need for L-dopa therapy; the average time until L-dopa was needed was 312.1 days for patients in the placebo group and 548.9 days for patients in the selegiline group. Disease progression, as monitored by five different assessment scales, was slowed (by 40 to 83% per year) in the selegiline group compared with placebo. Therefore, early selegiline therapy delays the requirement for antiparkinsonian medications, possibly by slowing progression of the disease.

A large multicenter trial was initiated in 1987 to examine the efficacy of selegiline (10 mg/day) and α-tocopherol (2000 U/day) in delaying the onset of disability sufficiently to require no, or reduced, L-dopa therapy. The **Deprenyl and Tocopherol Antioxidative Therapy of Parkinsonism** (**DATATOP**) study involved 800 patients with early, untreated Parkinson's disease who had been randomized in 2×2 factorial design to receive L-deprenyl (selegiline) alone, tocopherol alone, selegiline plus tocopherol, or double placebo. Of 399 patients who received active selegiline, 302 were able to function without L-dopa, compared with 225 of the 401 patients who did not receive selegiline. These striking and very significant differences favoring selegiline treatment also translated into a significant delay until patients ceased full-time employment due to emerging disability. The project concluded that the use of selegiline (10 mg/day) delayed the onset of disability associated with early, otherwise-untreated cases of Parkinson's disease.

The field of neurology seems to have more than its share of untreatable and progressive diseases. For many years, virtually all of the neurodegenerative diseases of aging (**Parkinson's disease**, **Alzheimer's disease**, and **amyotrophic lateral sclerosis**) fell into this category. Parkinson's disease became the exception with the advent of L-dopa therapy (given as carbidopa–levodopa) in the late 1960s; but like these other diseases, it has remained an inexorably progressive process. This is a major therapeutic problem, because progression of disease almost certainly contributes to the increasing management difficulties encountered in the later stages of the illness. Indeed, it could be argued that the management of patients with advanced Parkinson's disease represents one of the most-challenging demands in the day-to-day practice of neurology.

For these reasons, the need for an alternative strategy for the treatment of Parkinson's disease has become increasingly obvious. Stated in another way, while L-dopa therapy remains a central pillar in the management of Parkinson's disease, it is not the "answer" that many had originally hoped. The advent of the dopaminergic agonists has provided another therapeutic alternative when L-dopa begins to fail, but these drugs, too, represent only a temporary station along the way to increasing disability and/or side effects of therapy. In this setting, it is not difficult to see why a new therapeutic strategy aimed at slowing the progress of Parkinson's disease might be warmly welcomed.

In discussing the possible utility of selegiline, physicians have concluded that selegiline deserves consideration for symptomatic use as the first measure against L-dopa-related motor fluctuations that do not respond to manipulation of the L-dopa regimen. Furthermore, a clear benefit of L-dopa with selegiline in altering the course of Parkinson's disease compared to L-dopa alone has been shown. By studying selegiline as an adjunct to L-dopa therapy in 200 patients, physicians have reported on the limitation of selegiline as a long-term antiparkinsonian adjunct to conventional L-dopa therapy. Selegiline does improve parkinsonism during the initial 6 months to 12 to 24 months of combined therapy in a third to almost half of patients with an end-of-dose type of response to long-term L-dopa therapy. But even this particular class of patients is unable to maintain such improvement by 36 months, much less by 48 months from the start of the trial. About a quarter of poor responders to L-dopa and those with random deterioration show improvement in their parkinsonian status in the first 6 months of the selegiline trial, but quickly deteriorate by 1 year.

Selegiline is able to ameliorate depression associated with Parkinson's disease. However, selegiline is not effective in parkinsonian bradyphrenia.

The efficacy of selegiline on the progression of disability in early Parkinson's disease has been studied and the following conclusions have been reached: selegiline (10 mg/day) delays the onset of disability associated with early, otherwise-untreated Parkinson's disease. It remains unclear whether these benefits derive from mechanisms that are symptomatic (dopaminergic), protective (antineurotoxic), or both.

By studying 50 *de novo* patients with Parkinson's disease, physicians have reported that selegiline was effective in decreasing the hypokinesis, while the rigidity was improved less by this drug. The reduction of parkinsonian symptoms developed slowly and was independent of the severity of Parkinson's disease. The maintenance of selegiline monotherapy showed a high individual variation but the average period was about 1 year.

The beneficial effects of selegiline are symptomatic in nature. Furthermore, it has been postulated that there are several mechanisms in which such a response might be achieved:

- **Banisterine** and **selegiline** could increase the concentration of dopamine in the striatal synaptic cleft by simply impeding degradation of the neurotransmitter;
- **Banisterine** and **selegiline** promote accumulation of phenylethylamine, which potentiates the action of dopamine;
- **Banisterine** and **selegiline** could enhance the release of dopamine and block its reuptake;
- **Banisterine** and **selegiline** might elicit an antidepressant action by inhibiting MAO.

The comparative effectiveness of L-dopa and carbidopa and that of L-dopa, carbidopa, and selegiline has been studied, and the results have shown that the major usefulness of selegiline is to modify the fluctuating therapeutic response seen with L-dopa–carbidopa.

SUMMARY AND CONCLUSIONS

Banisterine from *Banisteria caapi* and *Nicotiana tabacum* was the first MAO inhibitor that has shown efficacy in the treatment of Parkinson's disease. Then, selegiline, a selective inhibitor of MAO-B, was one of the first adjunct therapies in clinical use. A retrospective analysis of data from long-term patients found a significant increase in survival in those treated with selegiline plus L-dopa compared with L-dopa alone. These studies, coupled with evidence that selegiline blocked parkinsonism induced by MPTP in animals, suggested that MAO-B inhibition might protect against neurodegeneration caused by an endogenously occurring neurotoxin. The findings triggered widespread interest in the potential neuroprotective effects of selegiline in humans.

Extensive data, gathered by using tissue and experimental animals, suggest that selegiline possesses a plethora of actions that may be outlined under at least three separate major categories:

1. Enhancement of dopaminergic transmission
   - Selegiline inhibits oxidation of MPTP to MPP+ (Figure 30).
   - Selegiline produces amphetamine-like effect, enhances release of dopamine, blocks reuptake of dopamine (Figure 29).
   - Selegiline increases striatal phenylethylamine levels, enhances release of dopamine, activates dopamine receptors (see Figure 29).
   - Selegiline stimulates gene expression of L-amino-acid decarboxylase in PC12 cells (see Figure 29).

2. Neuronal protection
   - Selegiline reduces the production of oxidative radicals (see Figures 31).
   - Selegiline upregulates SOD and catalase (see Figure 31).
   - Selegiline suppresses nonenzymatic, iron-catalyzed autooxidation of dopamine.

3. Neuronal rescue
   - Selegiline compensates for loss of target-derived trophic support (stereospecific).
   - Selegiline enhances glial activation (see Figure 32).
   - Selegiline induces NT-3trikC receptor (see Figure 32).
   - Selegiline upregulates ciliary neurotrophic factor (CNF) gene expression in astroglial cell culture.
   - Selegiline delays apoptosis in serum-deprived PC12 cells.
   - Selegiline blocks apoptosis-related fall in mitochondrial membrane potential.

Most of the aforementioned properties occur independent of the efficacy of selegiline to inhibit MAO-B. No firm clinical data exist at this time to either accept or reject the contentions that these diversified therapeutic possibilities are seen in banisterine- or selegiline-treated patients with neurological disorders.

When used as a monotherapy in the early stage of Parkinson's disease, selegiline provides a mild antiparkinsonian effect. When given concomitantly with L-dopa, it potentiates the effects of L-dopa and allows a reduction in daily L-dopa dosage. In patients with fluctuating responses, selegiline decreases their severity, particularly the "wearing off," but less so the "off–on" phenomenon.

Long-term double-blind studies have both accepted and rejected the contentions that monotherapy with selegiline is able to slow the progression of Parkinson's disease and hence delay the initiation of therapy with L-dopa. However, general agreement exists that selegiline is a relatively safe drug, since the incidences of fatigue, vertigo, constipation, insomnia, dry mouth, depression, nausea, sweating, orthostatic dizziness, involuntary movements, anxiety, and palpitation are identical to those caused by placebo.

Some of the beneficial effects attributed to high (20 to 40 mg/day) and not low (5 to 10 mg/day) doses of selegiline, such as being useful in the treatments of depression or narcolepsy, may be due to inhibition of MAO-A by selegiline altering adrenergic and serotonergic signals.

It is certain that today's selegiline and rasagiline will be replaced by even more selective and efficacious MAOs of tomorrow, possessing undisputed **neurotrophic**, **immunoenhancing**, **neuron rescuing**, **antiapoptotic**, and **neuroprotectant** properties capable of averting the deleterious effects of MPTP-like compounds produced endogenously, of protecting the striatal neurons, and of alleviating the physical disability and mental suffering of patients with Parkinson's disease.

13

BELLADONNA

- Atropine
- Hyoscyamine
- Hyoscine

SOLANACEAE OR NIGHTSHADE FAMILY

The plants are of varied form, most abundant in tropical regions, and the family comprises 85 genera and about 1800 species. The leaves are usually alternate; the flowers are mostly regular, except in *Hyoscyamus*; the anthers are connivent, with the pollen sacs apically or longitudinally dehiscent; and the fruits are superior berries or capsules. The plants are usually malodorous. They furnish a number of important economic products, including potato, tomato, eggplant, and tobacco, as well as a number of drugs.

Occurrence

The atropine series contains a number of very closely allied alkaloids of which the chief are **atropine, hyoscyamine** and **hyoscine** or **scopolamine**. They are found in the roots and leaves of many plants of the **Solanaceae**, notably **belladonna** (*Atropa belladonna*), henbane (*Hyoscyamus niger*), the **thorn apple** or **jimson weed** (*Datura stramonium*), and some members of the *Duboisia* and *Scopolia* species. These plants were used during the Middle Ages to form the "sorcerer's drugs" and have been smoked, chewed, or imbibed in the form of decoctions by primitive people for the hallucinations and frenzy they produce (Figure 36).

History

Galenical preparations of belladonna have been employed in medicine for many centuries and were known to the ancient Hindus. The **professional poisoners** of the Middle Ages frequently used the deadly nightshade plant to produce a type of intoxication that was often prolonged and obscure. This prompted Linné to name the shrub *Atropa belladonna*, after **Atropos**, the oldest of the Three Fates who severs the thread of life. The word *belladonna* itself is a reminder of

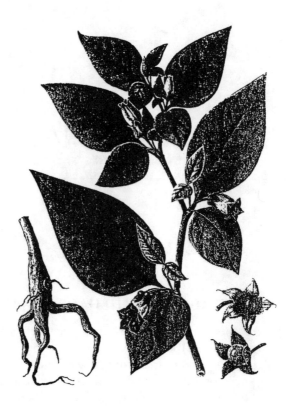

Figure 36 Belladonna (*Atropa belladonna*) from which atropine and the belladonna alkaloids are obtained.

the antiquity of the drug in that it signifies "beautiful lady," because the women of long ago were wont to instill a decoction of the plant in their eyes to produce dilated pupils, a sign of comeliness. The resulting disturbance of vision evidently was a small price to pay for beauty. The mydriatic action of belladonna, so widely used today, was thus well understood hundreds of years before the invention of the retinoscope.

Chemistry

The alkaloids of this group are derived from a combination of a piperidine and a pyrrolidine ring, designated as **tropane** (Figure 37). The 3-hydroxy derivative of **tropane** is known as **tropine** and is the basic component of atropine. When atropine is hydrolyzed, it forms **tropine** and **tropic acid** (α-phenyl-β-hydroxy-propionic acid). Atropine is the tropic acid ester of tropine. It has been prepared synthetically. Tropic acid contains an asymmetric carbon atom. The racemic compound (atropine) as obtained naturally or as synthesized may be resolved into its optically active components, **d-** and **l-hyoscyamine**. Atropine is racemic hyoscyamine; that is, it consists of equal parts of *l*-hyscyamine and *d*-hyoscyamine, but, as the latter is only feebly active in the body, the action of atropine is practically that of its *l*-hyoscyamine moiety. *l*-Hyoscyamine is formed in the plants, but is

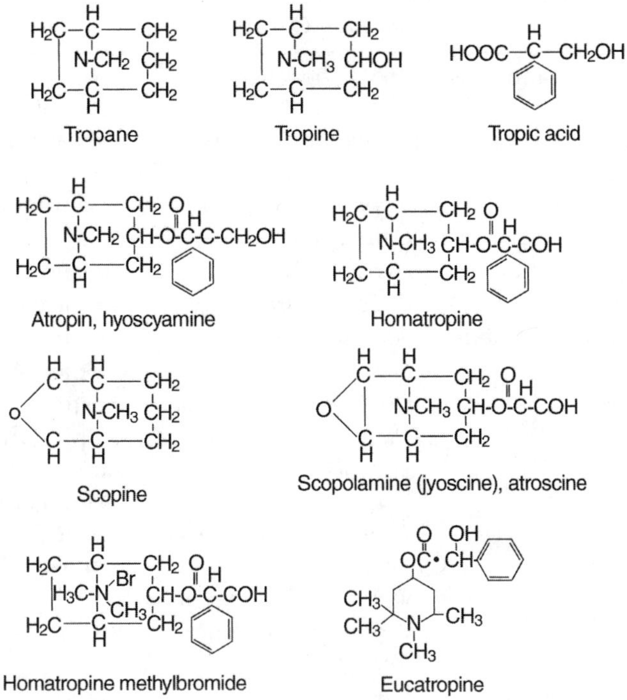

Figure 37   Structural formula of atropine and its related anticholinergic drugs.

readily changed to atropine in the plant cells and also in the process of extraction, so that the relative proportion of the isomers in the plants and in the preparations varies. However, atropine itself does exist in small amount as such in the plants although most of it is formed from the *l*-hyoscyamine in the process of extraction.

As noted from its formula (see Figure 37), **scopolamine** differs from atropine chemically in having an oxygen atom attached to the end carbon atoms of the **scopine** part of the molecule. Scopolamine is levorotatory but is readily racemized to the *d,l*-form designated as **atroscine**.

A number of related compounds are also present in the **solanaceous** plants and in some *Duboisia* and *Scopolia* **species**. **Apoatropine (atropamine)**, which is found in belladonna root, may also be prepared synthetically by dehydrating atropine. It is an ester of tropine and **atropic acid (α-phenylacrylic acid)**. **Belladonnine**, an isomer of apoatropine, also is present in belladonna root. **Norhyoscyamine (pseudohyoscyamine)** present in *Duboisia, Scopolia,* and *Datura* consists of tropic acid and **nortropine**, a tropine containing an NH group in place of NCH_3.

Botany of Solanaceae

Herbs (potato, tomato, horsenettle), **shrubs** (*Lycium* spp.), **vines** (*Solanum dulcamara,* or bittersweet), rarely trees as in some tropical *Datura*. Stems and

Table 20   Drugs Obtained from Nightshade Family

Drugs	Parts Used	Botanical Origins	Habitats
Belladonna Folium	Leaves and tops	*Atropa belladonna*	Central and Southern Europe
Belladonna Radix	Root	*Atropa belladonna*	Asia Minor and Persia
Atropina	Alkaloid	*Atropa belladonna*	Asia Minor and Persia
Stramonium	Leaves	*Datura stramonium*	Asia and Tropical America
Hyoscyamus	Leaves and tops	*Hyoscyamus niger*	Europe, Asia
Capsicum	Fruit	*Capsicum frutescens*	Tropical America
Dulcamara	Twigs and stems	*Solanum dulcamara*	Europe and Asia
Manaca	Root	*Brunfelsia hopeana*	Tropical America
Duboisia	Leaves	*Duboisia myoporoides*	Australia
Tabacum	Leaves	*Nicotiana tabacum*	Tropical America
Scopola	Rhizome	*Scopola carniolica*	Alps and Carpathian Mountains
Paprika	Fruit	*Capsicum annuum*	Tropical America; cultivated
		Variety of *C. longum*	
Pimiento	Fruit	Variety of *Capsicum annuum*	Spain
Miré	Roots, stems, and leaves	*Brunfelsia, hydrangaeformis*	South America
Solanum	Ripe fruit	*Solanum carolinense*	United States

leaves possess bicollateral bundles. Many of the plants, as **belladonna**, **stramonium**, **hyoscyamus**, and **scopola**, contain **mydriatic alkaloids**. Leaves alternate, exstipulate, entire or more-or-less lobed, rarely compound; often glandular-hairy. Flowers in cymes or solitary; regular and rarely irregular (*Petunia, Tobacco* spp.), pentamerous, perfect, synphyllous; sepals green (rarely petaloid), rotate to tubular, usually persistent and accrescent; petals rotate (*Solanum*), to tubular (*Atropa*), to funnel-shaped (*Tobacco*), and so (1) open to all comers, or (2) to bees or wasps, or (3) to butterflies, moths; color, greenish-yellow, or greenish-white, to white, to pink, crimson, purple, rarely blue; stamens five, epipetalous, hypogynous, along with style usually forming nectar glands. Filaments short to long, anthers dehiscing longitudinally or by apical pores; pistil bicarpellate, syncarpous, with or without nectar girdle; superior ovary, two-celled with central placentation, ovules numerous, style more or less elongate with bilobed or bifid stigma. Fruit, a capsule dehiscing longitudinally (*Tobacco, Stramonium*) or transversely (henbane); or a berry (potato, eggplant, tomato, red pepper) (Table 20).

PHARMACOGNOSY OF BELLADONNA

The *Atropa* belladonna is an herbaceous, perennial plant, of the nat. Ord. **Solanaceae**, having dark-purple, bell-shaped flowers, and glossy, purplish-black berries about the size of cherries. It is indigenous in the mountainous districts of central and southern Europe and Asia, and is cultivated in Europe and the United States. It contains the official alkaloid **atropine**, $C_{17}H_{23}NO_3$, which may be decomposed into **tropine** and **tropic acid**, also the alkaloids **belladonnine, hyoscyamine, hyoscine**, and **atropamine**, in varying quantity, all existing as malates in the plant. It also contains the usual vegetable constituents, as albumin, gums, etc., and a coloring principle named **Atrosin**. The historical titles are as follows:

- **Belladonnae Folia**, Belladonna leaves, ovate and tapering, brownish-green above, grayish-green below, of slight odor and bitter, disagreeable taste. Stramonium leaves are more wrinkled; hyoscyamus leaves are more hairy.
- **Belladonnae Radix**, Belladonna root, occurs in cylindrical, tapering, wrinkled pieces, ½ to 1 inch thick, nearly odorless, taste is bitter and acrid.

Allied plants are **hyoscyamus, stramonium, duboisia**, and **scopola**, containing alkaloids that are closely allied to atropine, both chemically and physiologically.

Preparation of the Leaves

- **Extractum Belladonnae Foliorum**. Extract of belladonna leaves
- **Tinctura Belladonnae Foliorum**. Tincture of belladonna leaves, 10%
- **Emplastrum Belladonnae**. Belladonna Plaster, has the above extract 30%, mixed with adhesive plaster
- **Unguentum Belladonnae**. Belladonna ointment, has the above extract 10%, diluted alcohol 5, benzoinated lard 65, and hydrous wool-fat, 20.

Preparations of the Root

- **Fluidextractum Belladonnae Radicis**. Fluid extract of the belladonna root
- **Linimentum Belladonnae**. Belladonna liniment, has a camphor, dissolved in fluid extract of belladonna root

Atropine and Its Derivatives

- **Atropina**. Atropine, $C_{17}H_{23}NO_3$, white, acicular crystals, odorless, of bitter taste and alkaline reaction; very soluble in alcohol and in chloroform, also soluble in 130 of water at 59°F. It is decomposed by prolonged contact with caustic alkalies and is resolvable into tropine and tropic acid.
- **Atropinae Sulphas**. Atropine sulfate, $(C_{17}H_{23}NO_3)_2H_2SO_4$, a white powder of bitter taste and neutral reaction, soluble in 0.4 of water and 6.2 of alcohol at 59°F.
- **Oleatum Atropinae**. Oleate of atropine, a 2% solution of the alkaloid in alcohol 2, oleic acid 50, and olive oil to 100.

Figure 38 The **henbane** (*Hyoscyamus niger* Linné), a member of the nightshade family, whose leaves, with or without the tops, constitute the official drug Hyoscyamus and are a source of the valuable medicinal alkaloids **hyoscyamine** and **scopolamine**.

- **Homatropinae Hydrobromidum**. Homatropine Hydrobromide, $C_{16}H_{21}NO_3HBr$, is the hydrobromide of an alkaloid obtained by the condensation of tropine and **mandelic acid**. It is soluble in 6 of water, 33 of alcohol, insoluble in ether (Figure 38).

BELLADONNA ALKALOIDS AND THE AUTONOMIC NERVOUS SYSTEM

The nervous and endocrine systems control an extensive number of functions in the body. The nervous system is divided into the central nervous system and the peripheral nervous system. The peripheral nervous system is further divided into the somatic nervous system (a voluntary system innervating skeletal muscles) and the autonomic nervous system (an involuntary system innervating smooth muscle, cardiac muscle, and glands).

COMPARISON OF THE SOMATIC AND AUTONOMIC NERVOUS SYSTEMS

The somatic and autonomic nervous systems differ anatomically and physiologically (Table 21).

Table 21   Comparison of Somatic Nervous System and Autonomic Nervous System

Somatic Nervous System	Autonomic Nervous System
Target tissues are skeletal muscles	Target tissues are smooth muscle, cardiac muscle, and glands
Cell bodies of somatic efferent neurons are found at all levels of the spinal cord	Cell bodies are absent in cervical, lower lumbar, and coccygeal levels of the spinal cord
The neurons have voluntary and involuntary (reflex) regulation	The neurons have involuntary regulation only
Target tissues (effectors) receive only one efferent neuron	Target tissues receive two (sympathetic and parasympathetic) efferent neurons
Only one myelinated neuron is interposed between the CNS and the effector organ.	The neurons between the CNS and the organ are myelinated preganglionic fibers and unmyelinated postganglionic fibers
Acetylcholine is the transmitter	Acetylcholine and norepinephrine are the transmitters
Transmitter is necessary for skeletal muscle contraction, which is rapid	Separate transmitters are necessary for the contraction and relaxation of smooth muscles, which are slow
Denervation causes paralysis of skeletal muscle	Smooth muscle shows autoregulation
Only drugs causing skeletal muscle relaxation are used for clinical purposes	Both stimulants and depressants of smooth muscles are used for clinical purposes

The autonomic nervous system (ANS) regulates the functions of internal viscera such as the heart, blood vessels, digestive organs, and reproductive organs.

NEUROCHEMICAL BASIS OF CHOLINERGIC TRANSMISSION

Acetylcholine, an ester of choline and acetic acid, is synthesized in cholinergic neurons according to the following scheme:

$$\text{Acetyl CoA} + \text{choline} \xrightarrow{\substack{\text{Choline} \\ \text{acetyltransferase}}} \text{acetylcholine}$$

The acetylcholine, in turn, is hydrolyzed by both **acetylcholinesterase** and plasma **butyrylcholinesterase**. Choline is actively transported into nerve terminals (synaptosomes) by a high-affinity uptake mechanism. Furthermore, the availability of choline regulates the synthesis of acetylcholine (Figure 39).

Hemicholinium blocks the transport of choline into synaptosomes, whereas botulinum toxin blocks the calcium-mediated release of acetylcholine. The released acetylcholine is hydrolyzed rapidly by acetylcholinesterase to choline and acetate.

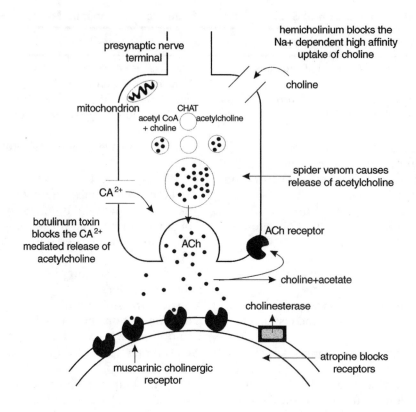

Figure 39 The metabolism and actions of acetylcholine (ACh). CoA = coenzyme A; CHAT = choline acetylase.

DIVISION AND FUNCTIONS OF THE AUTONOMIC NERVOUS SYSTEM

The autonomic nervous system consists of central connections, visceral afferent fibers, and visceral efferent fibers. The hypothalamus is where the principal integration of the entire autonomic nervous system takes place, and it is involved in the regulation of blood pressure, body temperature, water balance, metabolism of carbohydrates and lipids, sleep, emotion, and sexual reflexes. To a lesser extent, the medulla oblongata, limbic system, and cerebral cortex also integrate, coordinate, and adjust the functions of the autonomic nervous system (Figure 40).

Most blood vessels, the sweat glands, and the spleen are innervated only by one division of the autonomic nervous system. In the salivary glands, the two divisions of the autonomic nervous system supplement one another. In the bladder, bronchi, gastrointestinal tract, heart, pupil, and sex organs, the two divisions of the autonomic nervous system have opposing effects (see Figure 40).

CLASSIFICATION OF CHOLINERGIC RECEPTORS

Acetylcholine receptors are classified as either **muscarinic cholinergic receptors** or **nicotinic cholinergic receptors**. The alkaloid **muscarine** mimics the

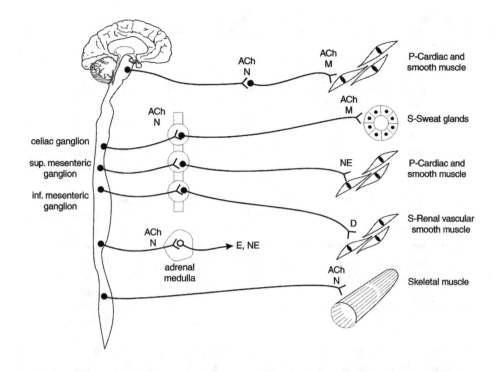

Figure 40 The autonomic nervous system innervates smooth muscle, cardiac muscles, and gland. ACh = acetylcholine, N = nicotinic cholinergic receptors; M = muscarinic cholinergic receptors; S = sympathetic chain; P = parasympathetic chain; E = epinephrine; D = dopamine; NE = norepinephrine; sup. = superior; inf. = inferior.

effects produced by stimulation of the parasympathetic system. These effects are postganglionic and are exerted on exocrine glands, cardiac muscle, and smooth muscle. The alkaloid **nicotine** mimics the actions of acetylcholine, which include stimulation of all autonomic ganglia, stimulation of the adrenal medulla, and contraction of skeletal muscle.

Dimethylphenylpiperazinium stimulates the autonomic ganglia, **tetraethyl-ammonium** and **hexamethonium** block the autonomic ganglia, **phenyltri-methylammonium** stimulates skeletal motor muscle end plates, **decamethonium** produces neuromuscular blockade, and ***d*-tubocurarine** blocks both the autonomic ganglia and the motor fiber end plates.

Among the agents cited, only *d*-tubocurarine is useful as a drug (skeletal muscle relaxant); the rest are useful only as research tools. Cholinesterase, found in liver and plasma, can hydrolyze other esters such as **succinylcholine** (a skeletal muscle relaxant). Cholinergic peripheral receptors are located on (1) postganglionic parasympathetic fibers, (2) postganglionic sympathetic fibers, (3) all autonomic ganglia, and (4) skeletal end plates.

CHOLINERGIC RECEPTOR BLOCKING AGENTS

Atropine and scopolamine, which are obtained from belladonna alkaloids, as well as other synthetic anticholinergic drugs, inhibit the actions of acetylcholine and

Table 22 The Pharmacological Actions of Atropine

Doses, mg	Effects
0.5	Slight cardiac slowing; some dryness of mouth; inhibition of sweating
1.0	Definite dryness of mouth; thirst; acceleration of heart, sometimes preceded by slowing; mild dilatation of pupil
2.0	Rapid heart rate; palpitation; marked dryness of mouth; dilated pupils; some blurring of near vision
5.0	All the above symptoms marked; difficulty in speaking and swallowing; restlessness and fatigue; headache; dry, hot skin; difficulty in micturition; reduced intestinal peristalsis
10.0 and more	Above symptoms more marked; pulse rapid and weak; iris practically obliterated; vision very blurred; skin flushed, hot, dry, and scarlet; ataxia, restlessness, and excitement; hallucinations and delirium; coma

The best-known antimuscarinic or anticholinergic drugs are the belladonna alkaloids. The major drugs in this class are atropine, hyoscyamine, and scopolamine. A number of plants belonging to the potato family (*Solanaceae*) contain similar alkaloids. *Atropa belladonna* (deadly nightshade), *Hyoscyamus niger* (henbane), *Datura stramonium* (jimsonweed or thorn apple), and several species of *Scopolia* also contain belladonna alkaloids.

cholinomimetic drugs at muscarinic receptors in smooth muscles, heart, and exocrine glands. In addition to these peripheral effects, anticholinergic drugs, by blocking the acetylcholine receptor sites in the central nervous system, have pronounced effects such as restlessness, irritability, excitement, and hallucinations. Scopolamine, on the other hand, depresses the central nervous system and, in therapeutic doses, produces fatigue, hypnosis, and amnesia. Therefore, it is used extensively in numerous medications, often in combination with antihistamines.

DOSE-DEPENDENT EFFECTS OF ATROPINE

The pharmacological effects of atropine in general are dose dependent (Table 22).

It produces mydriasis (blockade of the iris sphincter muscle), cycloplegia (blockade of the ciliary muscle), and cardiovascular effects characterized by transient bradycardia (central vagal stimulation) and tachycardia (vagal blockade at the sinoatrial node). Lacking any significant effects on circulation, atropine is often used as a preanesthetic medication to depress bronchial secretion and prevent pronounced bradycardia during abdominal surgical procedures. In still larger doses, it depresses the tone and motility of the gastrointestinal tract, the tone of the urinary bladder, and gastric secretion. Therefore, the effective doses for use in acid-pepsin diseases are preceded by numerous side effects.

Atropine is absorbed orally and crosses the placental barrier, whereupon it causes fetal tachycardia. Atropine has been used to examine the functional integrity of the placenta.

THE HISTORICAL USES OF BELLADONNA ALKALOIDS

Belladonna was one of the most valuable agents in the materia medica, ranking high in its efficacy and its wide range of usefulness. It was employed to **relieve**

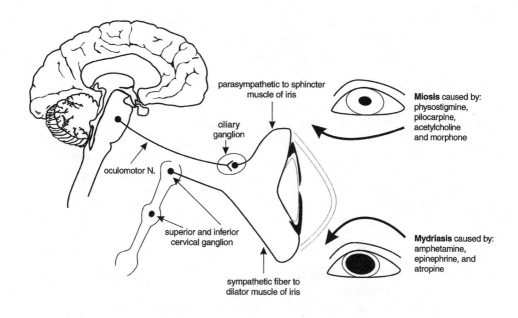

parasympathetic to sphincter
muscle of iris

ciliary
ganglion

oculomotor N.

superior and inferior
cervical ganglion

sympathetic fiber to
dilator muscle of iris

Miosis caused by:
physostigmine,
pilocarpine,
acetylcholine
and morphone

Mydriasis caused by:
amphetamine,
epinephrine, and
atropine

Figure 41　Miosis and mydriasis caused by autonomic drugs. N = nerve.

pain, **relax spasm**, **stimulate the circulation**, **decrease secretion**, and **check local inflammation**. Atropine was used for the same purposes, as well as to antagonize the effects of certain poisons, to dilate the pupils, and to paralyze the accommodation of the eye. These agents were administered in **rheumatic torticollis**, **lead colic**, **spasmodic colic**, **spasmodic dysmenorrhea**, **larynigismus stridulus**, **whooping cough**, **asthma**, **constipation**, **irritability of the bladder**, and **many other spasmodic affections**.

THE MODERN USES OF SYNTHETIC MUSCARINIC RECEPTOR ANTAGONISTS

The various applications of synthetic muscarinic receptor antagonists are listed below:

Ophthalmology

Homatropine. Produces **mydriasis** and **cycloplegia**
Eucatropine. Produces only mydriasis (Figure 41)

Preoperative Uses

Atropine. To prevent excess salivation and bradycardia
Scopolamine. In obstetrics, to produce sedation and amnesia

Cardiology

Atropine:
- To reduce severe bradycardia in **hyperactive carotid sinus reflex**
- Diagnostically in **Wolff–Parkinson–White syndrome** to restore the PRS complex to normal duration

Gastroenterology

Propantheline, **oxyphenonium**, **pirenzepine**:
- To diminish vagally mediated secretion of gastric juices and slow gastric emptying in peptic ulcer
- To reduce diarrhea associated with dysenteries and diverticulitis
- To reduce excess salivation associated with heavy metal poisoning or parkinsonism

Neurology

Trihexyphenidyl, **benztropine**. In parkinsonism and drug-induced pseudoparkinsonism

Scopolamine. In vestibular disorders such as motion sickness

Methantheline and **propantheline** are synthetic derivatives that, besides their antimuscarinic effects, are ganglionic blocking agents and block the skeletal neuromuscular junction. Propantheline and **oxyphenonium** reduce gastric secretion, while **pirenzepine**, in addition to reducing gastric secretion, also reduces gastric motility.

CONTRAINDICATIONS TO ANTICHOLINERGIC AGENTS

Conditions that are contraindications to the use of atropine and related drugs are **glaucoma** and **prostatic hypertrophy**, in which they cause urinary retention.

Atropine toxicity is characterized by dry mouth, burning sensation in the mouth, rapid pulse, mydriasis, blurred vision, photophobia, dry and flushed skin, restlessness, and excitement.

Physostigmine, given intravenously, counteracts both the peripheral and central side effects of atropine and other anticholinergic drugs such as **thioridazine** (neuroleptic), **imipramine** (antidepressant), and **benztropine** (antiparkinsonian medication).

HYOSCYAMINE

Hyoscyamine, which, as already noted, is the *l*-isomer of racemic atropine, exerts the same action as the latter but is approximately twice as potent. It is, however, rarely obtainable in pure form, as it is almost always mixed with atropine, into which it changes when kept in solution. The action of atropine, as has been stated, is compounded by that of natural or levorotary hyoscyamine with that of its dextrorotary isomer. The latter does not exist free in nature and possesses little or no parasympatholytic action.

SCOPOLAMINE

Scopolamine or hyoscine resembles atropine closely in its peripheral action, except that it passes off more quickly. Following the administration of therapeutic doses, the pulse remains unaltered in rate or may be slower, owing to the hypnotic action. Applied to the conjunctiva it produces mydriasis and loss of accommodation more quickly than atropine, but for a much shorter time. The effects on the central nervous system present the greatest divergencies from those described under atropine, for the characteristic stimulation is absent in the great majority of cases. As a general rule, scopolamine produces a marked sensation of fatigue. In drowsiness, the patient moves about less and speaks less, and a condition in no way dissimilar to natural sleep follows. In many cases, however, a short stage of excitement with giddiness, uncertain movements, and difficult and indistinct speech precedes sleep, and occasionally symptoms exactly resembling those produced by atropine follow the administration of scopolamine, especially if large doses are employed. Sleep generally lasts from 5 to 8 hours, and the patient may then remain quiet for several hours longer. As a general rule, after small doses, no confusion is complained of on awakening, but dryness of the throat and thirst are often present. Larger doses do not cause deeper sleep but give rise to delirium and excitement resembling those following atropine. Rarely, collapse has been observed after scopolamine. The respiratory center does not seem to be stimulated as by atropine, the respiration generally becoming slower from the beginning.

In lower mammals, scopolamine reduces the excitability of the motor areas as tested by electric shocks, while the reflex excitability in the frog is not increased as by atropine. Scopolamine appears to be excreted or destroyed in the tissues much more rapidly than atropine, for its effects last a shorter time.

The action of scopolamine, then, seems to correspond with that of atropine, except that the central nervous system is here depressed, while the action on the peripheral structures is of shorter duration. It depresses the brain in very small quantities; 0.5 mg is generally sufficient to induce quiet. The therapeutic dose is well below the fatal dose, but medicinal doses occasionally produce toxic symptoms, apparently a form of idiosyncrasy. A certain degree of tolerance is produced after repeated use, so that the dose has to be increased after a week or two.

Scopolamine is much less reliable as a hypnotic than morphine or the members of the chloral hydrate group. It is most effective when sleep is prevented by motor excitement, and the sleep seems to arise from the relief of this condition and not from depression of the consciousness.

Scopolamine is levorotary to polarized light; the racemic form, **atroscine**, which is often present in commercial scopolamine, acts only one half as strongly on the peripheral organs, because in it the levorotary is mixed with the dextrorotary isomer, which is almost inactive. The cerebral action is equal, however, in the two forms.

SYNTHETIC AND SEMISYNTHETIC SUBSTITUTES FOR BELLADONNA ALKALOIDS

Quaternary Ammonium Muscarinic Receptor Antagonists

There are several differences in the pharmacological properties of the quaternary muscarinic receptor antagonists as compared to the belladonna alkaloids described above. Compounds with a quaternary ammonium structure are poorly and unreliable absorbed after oral administration. Penetration of the conjunctiva also is poor, so that most quaternary ammonium compounds are of little value in ophthalmology. Central effects are generally lacking, because these agents do not readily cross the blood–brain barrier. The quaternary ammonium compounds usually have a somewhat more prolonged action than the belladonna alkaloids; little is known of the fate and excretion of most of these agents. Since the ratio of ganglionic-blocking (nicotinic) to muscarinic-blocking activity is greater for quaternary ammonium compounds than for tertiary amines, side effects attributable to ganglionic blockade, such as impotence and postural hypotension, can occur. Poisoning with quaternary ammonium muscarinic receptor antagonists also may cause a curariform neuromuscular block, leading to respiratory paralysis.

There is a clinical impression that the quaternary ammonium compounds have a relatively greater effect on gastrointestinal activity and that the doses necessary to treat gastrointestinal disorders are, consequently, somewhat more readily tolerated than are other agents of this type; this effect has been attributed to the additional element of ganglionic block. Nevertheless, most of these drugs, like atropine, generally control gastric secretion or gastrointestinal motility only at doses that also cause significant side effects due to muscarinic blockade at other sites.

Ipratropium

Ipratropium bromide (Atrovent) is a quaternary ammonium compound formed by the introduction of an isopropyl group to the N atom of atropine. A similar agent, **oxitropium bromide**, also is available in Europe; it is a quaternary ammonium derivative of scopolamine, formed by the introduction of an ethyl group. The most recently developed and bronchoselective member of this family is **tiotropium bromide**, which has a longer duration of action and currently is in clinical trial. Ipratropium bromide produces effects that are similar to those of atropine when each agent is administered parenterally. These include bronchodilatation, tachycardia, and inhibition of salivary secretion. Although somewhat more potent than atropine in these actions, ipratropium lacks appreciable effect on the central nervous system and has greater inhibitory effects on ganglionic transmission, similar to other quaternary ammonium muscarinic receptor antagonists. One unexpected and therapeutically important property of ipratropium is the relative lack of effect on mucociliary clearance, in contrast to the inhibitory effect of atropine; this difference is evident upon either local or parenteral administration and remains unexplained. Hence, the use of ipratropium in patients with airway disease avoids the increased accumulation of lower airway secretions and the antagonism of β-adrenergic agonist-induced enhancement of mucociliary clearance encountered with atropine.

When solutions are inhaled, the actions of ipratropium are confined almost exclusively to the mouth and airways. Even when administered in amounts many times the recommended dosage, little or no change occurs in heart rate, blood pressure, bladder function, intraocular pressure, or pupillary diameter. This selectivity results from the very inefficient absorption of the drug from the lungs or the gastrointestinal tract. The degree of bronchodilatation produced by ipratropium is thought to reflect the level of parasympathetic tone, supplemented by reflex activation of cholinergic pathways brought about by various stimuli. In normal individuals, the inhalation of ipratropium can provide virtually complete protection against the bronchoconstriction produced by the subsequent inhalation of such substances as sulfur dioxide, ozone, nebulized citric acid, or cigarette smoke. However, patients with asthma or with demonstrable bronchial hyperresponsiveness are less well protected. Although ipratropium causes a marked reduction in sensitivity to **methacholine** in asthmatic subjects, more modest inhibition of responses to challenge with **histamine**, **bradykinin**, or **prostaglandin $F_{2\alpha}$** is achieved, and little protection is provided against the bronchoconstriction induced by serotonin (5-HT) or the **leukotrienes**. The principal clinical use of ipratropium is in the treatment of chronic obstructive pulmonary disease; it is less effective in most patients with asthma.

Methoscopolamine

Methoscopolamine bromide (Pamine) is a quaternary ammonium derivative of scopolamine and therefore lacks the central actions of scopolamine. It is less potent than atropine and is poorly absorbed; however, its action is more prolonged, the usual oral dose (2.5 mg) acting for 6 to 8 hours. Its use has been limited chiefly to gastrointestinal diseases.

Homatropine Methylbromide

Homatropine methylbromide is the quaternary derivative of homatropine. It is less potent than atropine in antimuscarinic activity, but it is four times more potent as a ganglionic blocking agent. It is available in some combination products intended for relief of gastrointestinal spasm.

Methantheline

Methantheline bromide (Banthine) differs from atropine in having a particularly high ratio of ganglionic-blocking activity to antimuscarinic activity. High doses may cause impotence, an effect rarely produced by purely antimuscarinic drugs and indicative of ganglionic block. Toxic doses may paralyze respiration by neuromuscular block. Restlessness, euphoria, fatigue, or, very rarely, acute psychotic episodes may be seen in some patients despite the relatively poor penetration of the drug into the central nervous system. The duration of methantheline action is slightly more prolonged than that of atropine, the effects of a therapeutic dose (50 to 100 mg) lasting 6 hours. An additional toxic manifestation unrelated to the blocking actions is the occasional appearance of skin rashes, including exfoliative dermatitis.

Propantheline

Propantheline bromide (Pro-banthine) resembles methantheline chemically (isopropyl groups replace the ethyl substituents on the quaternary N atom). Its pharmacological properties are also similar to those of methantheline, but it is two to five times more potent. It is one of the more widely used of the synthetic muscarinic receptor antagonists. Very high doses block the skeletal neuromuscular junction. The usual clinical dose (15 mg) acts for about 6 hours.

Other Compounds

Other drugs in this category include **anisotropine methylbromide** (Valpin), **clidinium bromide** (**Quarzan**; also in combination with chlordiazepoxide as **Librax** and others), **glycopyrrolate** (Robinul; also used parenterally in conjunction with anesthetics), **isopropamide iodide** (Darbid), **mepenzolate bromide** (Cantil), **tridihexethyl chloride** (Pathilon), and **hexocyclium methylsulfate** (Tral).

Tertiary-Amine Muscarinic Receptor Antagonists

Certain of these agents are particularly useful in ophthalmology; included in this category are **homatropine hydrobromide** (Isopto homatropine) (a semisynthetic derivative of atropine); **cyclopentolate hydrochloride** (Cyclogyl), and **tropicamide** (Mydriacyl). These agents are preferred to atropine or scopolamine because of their shorter duration of action.

Tertiary-amine muscarinic receptor antagonists gain access to the central nervous system and are therefore the anticholinergic drugs used to treat parkinsonism and the extrapyramidal side effects of antipsychotic drugs. Specific agents used primarily for these conditions include **benztropine mesylate** (Cogentin) and **trihexyphenidyl hydrochloride** (Artane, others).

Tertiary amines used for their antispasmodic properties are **dicyclomine hydrochloride** (Bentyl, others), **oxyphencyclimine hydrochloride** (daricon), **flavoxate hydrochloride** (Uripas), and **oxyburynin chloride** (Ditropan). The latter two are indicated specifically for urological disorders. These agents appear to exert some nonspecific direct relaxant effect on smooth muscle. In therapeutic doses they decrease spasm of the gastrointestinal tract, biliary tract, ureter, and uterus; characteristic atropine-like effects on the salivary glands and the eye also are seen with oxybutynin.

Pirenzepine is a tricyclic drug, similar in structure to imipramine. Pirenzepine has selectivity for M_1-, relative to M_2-, and M_3-muscarinic receptors. **Telenzepine** is an analogue or pirenzepine that has higher potency and similar selectivity for M_1-muscarinic receptors. Both drugs are used in the treatment of peptic ulcer. At therapeutic doses, the incidence of dry mouth and blurred vision is relatively low with pirenzepine. Central effects are not seen because of the low lipid solubility of the drug and its limited penetration into the central nervous system. Some studies also have shown pirenzepine and telenzepine to be of therapeutic value in **chronic obstructive bronchitis**, presumably as a result of blockade of **vagally mediated bronchoconstriction**. In both the gastrointestinal tract and airway, the site of M_1 receptor antagonism is believed to be on receptors in ganglia.

14

BOTULINUM TOXIN

BOTULINUM TOXIN TYPE A

(Oculinum)

Botulinum, a neurotoxin with muscle relaxant properties, is used in the treatment of strabismus.

BOTULISM ANTITOXIN, TRIVALENT (ABE) EQUINE

Botulism antitoxin, which binds and neutralizes toxin, is used in the treatment of botulism.

USES OF BOTULINUM TOXIN INJECTION IN MEDICINE TODAY

Botulinum neurotoxin is produced by the anaerobic bacterium *Clostridium botulinum*. It is the most poisonous substance known. Very small amounts of botulinum toxin can lead to botulism, a descending paralysis with prominent bulbar symptoms and often affecting the autonomic nervous system (see Chapter 12). Botulism can occur in two ways. It can result from infection with bacterial spores that produce and release the toxin in the body, as in **enteric infectious botulism**, when the bacteria grow in the intestine, and in **wound botulism**, when the wound becomes infected. Alternatively, botulism occurs after ingestion of the toxin (**food-borne botulism**).

- Botulinum toxin inhibits release of **acetylcholine** (Ach) at the neuromuscular junction and in cholinergic sympathetic and parasympathetic neurons.
- Local injections of toxin weaken overactive muscles and control hypersecretion of glands supplied by cholinergic neurons.
- Botulinum toxin injections have an established role in some disorders of ocular motility.
- Botulinum toxin is the treatment of choice for **focal dystonias** such as **torticollis** and **writer's cramp** and for **hemifacial spasm** and may complement the management of **spasticity**.

- Local injections have also been shown to be beneficial in many other conditions including **achalasia**, **chronic anal fissure**, and **hyperhidrosis**.
- Treatment is usually well tolerated, the main side effect being weakness in adjacent muscles (Münchau and Bhatia, 2000).

Serotype A is the only one commercially available for clinical use, although experience is emerging with serotypes B, C, and F. Two preparations exist: **Dysport**, which is most widely used in the United Kingdom, and **Botox**, which is used in the United States and elsewhere.

Strains of **C. botulinum** produce seven antigenetically distinct neurotoxins designated as serotypes A through G. All seven serotypes have a similar structure and molecular weight, consisting of a heavy (H) chain and a light (L) chain joined by a disulfide bond. They all interfere with neural transmission by blocking the release of ACh (see Figure 39), which is the principal neurotransmitter at the neuromuscular junction.

After synaptic transmission is blocked by botulinum toxin, the muscles become clinically weak and atrophic. The affected nerve terminals do not degenerate, but the blockage of neurotransmitter release is irreversible. Function can be recovered by the sprouting of nerve terminals and formation of new synaptic contacts; this usually takes 2 to 3 months.

Estimates of the ACh content of synaptic vesicles range from 1000 to over 50,000 molecules per vesicle, and it has been calculated that a single motor-nerve terminal contains 300,000 or more vesicles. In addition, an uncertain but possibly significant amount of ACh is present in the extravesicular cytoplasm. The release of ACh and other neurotransmitters by exocytosis through the prejunctional membrane is inhibited by toxins from *Clostridium*. Some of the most potent toxins known are produced by this microbe. *Clostridium* toxins inhibit **synaptobrevin** and related proteins in the nerve terminals. Specificity of these toxins for particular nerves is due to the larger subunit of the toxin heterodimer, which binds to surface proteins on the nerve prior to internalization. Internalization, in turn, enables the small subunit to proteolyze synaptobrevin or other target proteins. Botulinum toxin A binds to cholinergic motor nerve terminals, giving rise to flaccid paralysis. **Tetanus toxin**, also from *Clostridium*, selectively binds to and enters spinal neurons where it blocks **glycine** release and causes spastic paralysis. Once internalized, the larger subunit functions as a selective Zn^{2+}-dependent protease and hydrolyzes the target protein. The toxin from **black widow spider venoms** (**α-latrotoxin**) promotes massive vesicle release, presumably by binding to the neurexins on the neuronal membrane (see Figure 39).

RATIONALE FOR TREATMENT WITH BOTULINUM NEUROTOXIN

Botulinum toxin induces weakness of striated muscles by inhibiting transmission of α motor neurons at the neuromuscular junction. This has led to its use in conditions with muscular overactivity, such as **dystonia**. Transmission is also inhibited at γ neurons in muscle spindles, which may alter **reflex overactivity**.

The toxin also inhibits release of Ach in all parasympathetic and cholinergic postganglionic sympathetic neurons. This has fueled interest in its use as a treatment for overactive smooth muscles (for example, in **achalasia**) or abnormal

Table 23 Disorders Caused by Overactivity of Muscles for Which Treatment with Botulinum Toxin A Is Established

Ophthalmological Disorders

Concomitant misalignment

  Primary or secondary esotropia or exotropia

Nonconcomitant misalignment

  Paralytic strabismus (III, IV, VI nerve palsy, internuclear ophthalmoplegia, skew deviation)

  Duane's syndrome

  Restrictive or myogenic strabismus

Movement Disorders

Idiopathic focal dystonias

  Craniocervical (torticollis and isolated head tremor, blepharospasm, oromandibular dystonia, lingual dystonia, laryngeal dystonia)

  Other focal dystonias (writer's cramp, occupational cramps such as musician's cramp)

Tardive dystonia

Hemifacial spasm/postfacial nerve palsy synkinesis

Many drugs and toxins block neuromuscular transmission by other mechanisms, such as interference with the synthesis or release of acetylcholine, but most of these agents are not employed clinically for this purpose. One exception is **botulinus toxin**, which has been administered locally into muscles of the orbit in the management of **blepharospasm** and **strabismus**. This treatment produces a long-lasting interruption of neuromuscular transmission and reduction of spasmodic ocular movements. Another exception is *dantrolene*, which blocks release of Ca^{2+} from the sarcoplasmic reticulum and is used in the treatment of **malignant hyperthermia**.

activity of glands (for example, **hyperhidrosis**). Over the past 15 years, botulinum toxin has been shown to be useful in many conditions, especially strabismus and various movement disorders (Table 23).

Encouraging clinical reports have generated an abundance of ideas for other uses. Following are examples of overactive muscle conditions and other disorders for which treatment with botulinum toxin A has been tried (Münchau and Bhatia, 2000).

Ophthalmic Disorders

- Disorders of ocular motility (nystagmus and oscillopsia)
- Thyroid disease (upper eyelid retraction, glabellar furrowing)
- Therapeutic ptosis for corneal protection

Movement Disorders

- Secondary dystonia
- Tic disorders (simple tics, **Tourette's syndrome**, dystonic tics)

- Tremor (essential, primary writing, palatal, cerebellar)
- Painful spinal myoclonus
- Parkinson's disease (freezing of gait, off-period dystonia, severe constipation)
- Cephalic tetanus, **stiff man syndrome**, neuromyotonia
- Muscle stiffness, cramps, spasms

Spasticity

- Multiple sclerosis
- Stroke
- Traumatic brain injury
- Cerebral palsy
- Spinal cord injury

Neuromuscular Disorders

- Myokymia
- Neurogenic tibialis anterior hypertrophy with myalgia
- Benign cramp-fasciculation syndrome

Pain

- Headache (tension type, migraine, cervicogenic)
- Backache (neck, lower back)
- Myofascial pain
- Tennis elbow

Ear, Nose, and Throat Disorders

- Oromandibular disorders (**bruxism**, Masseter hypertrophy, temporomandibular joint dysfunction)
- Pharyngeal disorders (cricopharyngeal dysphagia, closure of larynx in chronic aspiration)
- Laryngeal disorders (vocal fold granuloma, ventricular dysphonia, mutational dysphonia)
- Stuttering with glottal blocks

Disorders of Pelvic Floor

- Anismus
- Vaginismus
- Anal fissures
- Detrusor-sphincter dyssynergia

Cosmetic Applications

- Wrinkles, frown lines
- Rejuvenation of aging neck

Overactivity of Smooth Muscles

- Esophageal disorders (achalasia, diffuse esophageal spasm, esophageal diverticulosis)
- Sustained sphincter of Oddi hypertension
- Gastric pyloric spasms

Hypersecretion of Glands Supplied by Cholinergic Sympathetic or Parasympathetic Neurons

- Ptyalism
- Increased tearing
- Hyperhidrosis (axillary, palmar, gustatory)
- Intrinsic rhinitis (see Münchau and Bhatia, 2000)

15

BRUSSELS SPROUTS

Cancer prevention by naturally occurring substances in the diet is an area of great scientific interest. Fruits and vegetables such as brussels sprouts, cabbage, leeks, citrus, herbs and spices, and food ingredients such as antioxidant vitamins, flavonoids, glucosinolates, and organo-sulfur compounds may have antimutagenic or anticarcinogenic potential.

EFFECTS OF BRUSSELS SPROUTS ON OXIDATIVE DNA DAMAGE

The hypothesis that certain components of plant foods are protective against cancers at many sites has gained interest in the past two decades and is supported in many studies. A large number of potentially anticarcinogenic agents have been suggested in fruits and vegetables so far, e.g., fiber, vitamins C and E, **carotenoids, flavonoids, phenols, phytoestrogens, diallylsulfides, limonene**, and hydrolysis products of **glucosinolates**. The mechanisms by which these agents may act include dilution and binding of carcinogens in the digestive tract (fiber), antioxidant effects, inhibition of **nitrosamine formation**, inhibition of activation of promutagens/procarcinogens, induction of detoxification enzymes, alteration of hormone metabolism, and others.

One group of vegetables that has been widely regarded as potentially cancer protective are vegetables of the Cruciferae family. **Cruciferous vegetables** are the major source of glucosinolates in the diet which distinguishes them from other vegetables. **Brassica vegetables**, including all cabbage-like vegetables, are a genus of the family Cruciferae and contribute most to our intake of glucosinolates.

In the 1960s interest emerged in the possibility that certain aromatic and indolic glucosinolate hydrolysis products might influence carcinogenesis. In *in vivo* experiments animals were fed glucosinolate hydrolysis products together with a carcinogen and it was found that fewer animals developed tumors in comparison with control animals not receiving the glucosinolate hydrolysis products. From that time on, many studies have been carried out to examine the possible anticarcinogenic effect of brassica vegetables and glucosinolate hydrolysis products.

Cancer prevention by naturally occurring substances in the diet is an area of great scientific interest. Fruits and vegetables such as cabbage, leeks, citrus, herbs, and spices and food ingredients such as antioxidant vitamins, flavonoids, glucosinolates, organo-sulfur compounds may have antimutagenic or anticarcinogenic potential. However, such claims are often based on results obtained in short-term genotoxicity assays *in vitro* or studies with experimental animals. Direct proof whether antimutagenic or anticarcinogenic effects also occur in humans under normal dietary conditions can only be obtained from studies with humans. Indeed, epidemiological studies indicate that consumption of fruits and vegetables is inversely associated with the risk of cancer. In particular, **brassica vegetables** like **cabbage**, **brussels sprouts**, and **broccoli** have been attributed with beneficial health effects in humans.

Oxidative DNA damage is considered an important mechanism underlying many cancers. Therefore, by reducing the amount of oxidative DNA damage, one could reduce the risk of developing cancer. The DNA adduct most abundantly formed by reactive oxygen species is **8-oxo-7,9-dihydro-2′-deoxyguanosine** (8-oxodG). *In vivo* this mutagenic lesion induced in DNA is repaired by excision. The resulting product, 8-oxodG, is excreted unchanged and independently of diet into the urine. Thus, the rate of this excretion serves as a biomarker of the integrated rate of oxidative DNA damage in the whole body.

Verhagen et al. (1997) investigated the effect of consumption of brussels sprouts on the excretion of 8-oxodG into human urine. Ten nonsmoking volunteers (five males and five females) were randomly assigned to two groups. Five volunteers started on a diet of 300 g of **glucosinolate-free vegetables**, whereas the other five consumed 300 g of brussels sprouts per day. After 1 week, the dietary regimes were reversed. Levels of 8-oxodG in 24-hour urine samples were measured by high-performance liquid chromatography. In four of five males a reduction in 8-oxodG was found, whereas in the fifth male the 8-oxodG excretion was high in the control period and was even much higher in the sprouts period. In females, no effect of consumption of brussels sprouts on excretion of 8-oxodG was found. These findings support the results of epidemiological studies that consumption of brassica vegetables may diminish cancer risk.

Brussels sprouts has putative antimutagenic and anticarcinogenic properties in humans by induction of **glutathione-*S*-transferase** and by decreasing the rate of **oxidative DNA damage**. These findings lend further support to the notion that brassica vegetables have health-promoting effects under normal dietary conditions and apparently without side effects. In general, the beneficial effects of fruits and vegetables have been related to their content of antioxidant vitamins, such as **vitamins C** and **β-carotene**, as well as other antioxidants. However, brassica vegetables differ in particular from other vegetables by the presence of **glucosinolates**. Apparently, the specific reducing effect of brussels sprouts on oxidative DNA damage may seem to be attributable to one or more of these **phytochemicals**. In addition to an induction of the detoxification enzyme glutathione-*S*-transferase the protection against oxidative DNA damage possibly is a second mechanism underlying cancer prevention by **phytochemicals** in brassica vegetables. The urinary excretion of 8-oxodG as a biomarker for integrated oxidative DNA damage in the body allows for determination of dose–effect relationships with vegetable intake and identification of the preventive principle(s) of fruits and vegetables, such as brussels sprouts.

The **indolyl isothiocyanates** liberated from myrosinase catalyzed breakdown of, for example, **glucobrassicin** and **neoglucobrassicin**, are highly unstable and ultimately condense to products such as **indole-3-carbazole**, which are potent inducers of **cytochrome P-450 enzymes**, particularly CYP1A1 and CYP1A2 in the liver and intestine. However, in cell cultures responsive to CYP1A1 inducers, indole-3-carbazole increased 8-oxoGua formation along with the enzyme activity.

Cooked brussels sprouts and in particular an extract from these vegetables decrease the spontaneous rate of oxidative DNA damage as shown by reduced 8-oxodG excretion. Moreover, the extract prevented **2-nitropropaine** (2-NP)-induced 8-oxodG formation as shown by the reduced urinary excretion of 8-oxodG and the lowering of the induced nuclear levels of 8-oxodG in liver, kidney, and bone marrow. In contrast, raw brussels sprouts, noncruciferous vegetables, or isolated intact or myrosinase-degraded indolyl glucosinolates had no effect. This indicates that cooked brussels sprouts contain bioactive substances with a potential for reducing the physiological as well as oxidative stress-induced oxidative DNA modification. Remarkably, this applies to the overall DNA modification measured by urinary 8-oxodG excretion and the organ-specific changes measured by the nuclear DNA level of oxidized guanine.

In a previous study in humans, cooked brussels sprouts decreased 8-oxodG excretion whereas noncruciferous vegetables had no effect. The data by Verhagen et al. (1997) and Deng et al. (1998) confirm and extend this observation indicating that the protective effect may be specific for cruciferous vegetables that are characterized by their content of glucosinolates. This large family of compounds and their transformation products have various potent pharmacological and toxicological effects, among which some could be responsible for the beneficial effect on oxidative DNA damage. Moreover, cooking was essential for the protective effect. Antioxidants, including **vitamin E**, **ellagic acid**, and **tea extracts** have been shown to protect the liver from 8-oxoG induction by 2-NP although other signs of heptotoxicity were unaffected.

The brassica vegetables are rich in **glucosinolates**, which are hydrolyzed by **myrosinase** contained in the plants and liberated by processing and storage. Many of the resulting **isothiocyanate** breakdown products have potent pharmacological and toxicological effects. The indolyl derivatives are unstable and their further breakdown and condensation products are potent inducers of **cytochrome P-450 (CYP) enzymes**, particularly the CYP1A subfamily, and some phase II enzymes. Other resulting **isothiocyanates** with aryl, alkyl, and/or sulfinyl side chains, such as **sulforaphane** isolated from broccoli, are potent inducers of particular phase II enzymes and may inhibit **chemical carcinogenesis**.

A specific cancer-preventive mechanism of cruciferous vegetables could be related to reduction of oxidative DNA damage, particularly guanine oxidation, which is the most abundant and mutagenic oxidative modification known. Similarly, **green tea** and its extracts can reduce oxidative DNA damage, measured by 8-oxodG formation, induced by the **hepatocarcinogen 2-NP**. Indeed, in an intervention study in humans 300 g of brussels sprouts as compared with noncruciferous vegetables reduced the urine excretion of 8-oxodG, a repair product of guanine oxidation and biomarker of the rate of oxidative DNA damage. The effects of cruciferous vegetables on the rate of DNA damage induced by oxidative stress and the levels of 8-oxodG in the relevant organs or surrogate tissue have not been investigated (Deng et al., 1998).

16

CANNABIS, OR MARIJUANA

- **Cannabis**, whether ingested or smoked, has a long history of reportedly safe and effective use in the treatment and prophylaxis of migraine.
- **Cannabis** has a mild but definite analgesic effect in its own right.
- **Cannabis** seems to affect nociceptive processes in the brain, and may interact with serotonergic and other pathways implicated in migraine.
- **Cannabis** is an effective antiemetic, which is a useful property in migraine treatment.
- **Cannabis**, even when abused, has mild addiction potential, and seems to be safe in moderate doses, particularly under the supervision of a physician.
- **Cannabis's** primary problem as a medicine lies in its possible pulmonary effects, which seem to be minimal in occasional and intermittent use.
- **Cannabis**, when inhaled, is rapidly active, obviates the need for gastrointestinal absorption (impaired markedly in migraine), and may be titrated to the medical requirement of the patient for symptomatic relief.
- **Cannabis**, delivered by **pyrolysis**, in the form of a marijuana cigarette, or "joint," presents the hypothetical potential for quick and effective parenteral treatment of acute migraine.

There can be no doubt that a plant that has been in partnership with humans since the beginnings of agricultural efforts, that has served humans in so many ways, and that, under the searchlight of modern chemical study, has yielded many new and interesting compounds will continue to be a part of our economy. It would be a luxury that we could ill afford if we allowed prejudices, resulting from the abuse of *Cannabis*, to deter scientists from learning as much as possible about this ancient and mysterious plant.

CANNABIS, OR MARIJUANA

The hemp plant, ***Cannabis sativa*** (Figure 42), when cultivated in warm climates, develops products that induce derangement of the central nervous system, and has been used as an intoxicant in Asiatic countries and Africa from time immemorial, under the names of **hashish**, **bhang**, **ganja**, **charas**, or **churrus**. Some

261

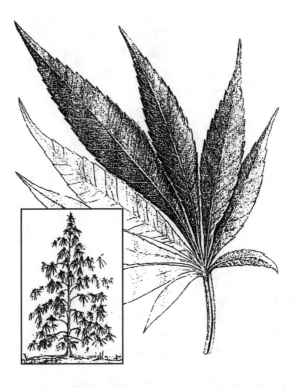

Figure 42 A marijuana leaf (*Cannabis sativa*) used in the preparation of marijuana cigarettes.

of the preparations are smoked either alone or mixed with tobacco; others form an intoxicating drink, while in others the plant is mixed with sugar or honey and taken as a confection.

Marijuana is the flowering tops and leaves of the hemp plant known as *C. sativa* (**grass**, **pot**, **weed**, **Mary Jane**). The dried leaves are most commonly smoked as cigarettes (**joint, reefer, doobie, number**) or in a pipe. The drug may be ingested orally in a prepared food (brownies) or beverage, but onset is slow (45 to 60 minutes) and absorption incomplete. Although all parts of the male and female plant contain psychoactive substances, the highest concentrations are found in the flowering tops. The dried resinous exudate of the tops is called **hashish**. More than 60 cannabinoids have been isolated from marijuana. The euphoric and psychoactive effects may be due to a combination of several of them. However, **THC** is the agent believed to produce most of the characteristic effects. Hundreds of other compounds are produced by pyrolysis when marijuana is smoked and may be important in the long-term toxic effects of chronic use.

Another form of the drug, "**hashoil**," is extracted from the plant with organic solvents and may contain a concentration of up to 50 to 60% THC. However, much of the "pure THC" or hashoil that is found on the street is often **phenyl-cyclidine (PCP)**. The major effects of marijuana are seen in the central nervous system and range from a mild relaxation and sedation to euphoria. Intoxication with marijuana is known as **being stoned** or **high**. Users commonly describe an increased sense of well-being, giddiness, alteration of time and space perception, and subjective enhancement in senses of touch, taste, smell, and sound. Users experience unusual hunger and craving for food referred to as the **munchies**. Some users claim that the drug increases their creativity and awareness of their surroundings. Many of the subjective effects appear to depend on the expectations of the user and the circumstances under which the drug is used. Other physiological effects are a mild increase in heart rate, dry mouth and throat, dryness of the eyelids, and conjunctival reddening. Performance of simple motor tasks and reaction times are unimpaired in small doses but are significantly affected by doses equivalent to one or two cigarettes. Driving performance is clearly impaired for 4 to 8 hours, well beyond the time that the user thinks the subjective effects of the drug have worn off. Impairment produced by marijuana is additive to that of alcohol if the two are used concurrently. The effects of smoking marijuana occur rapidly within 5 to 10 minutes, may last 2 to 3 hours, depending on the dose, and are occasionally followed by a period of drowsiness.

Relative to opiates, cocaine, depressants, and hallucinogens, emergency treatment of marijuana intoxication is rare. Occasionally, a first-time user may experience a syndrome known as a **panic attack**, consisting of anxiety, disorientation, and paranoid feelings. Previous users of the milder forms of marijuana may also experience this reaction if exposed to more potent forms of cannabis THC. Treatment for this reaction is primarily reassuring and talking down the fear and panic exhibited. Placing the patient in quiet surroundings and avoiding stimulants are also beneficial. Panic attacks seldom last more than 2 to 3 hours and may be relieved by small doses of **benzodiazepines** if "**talk-down therapy**" is not fully effective. Patients with combative or aggressive behavior may have been using a product known as "**super weed**" that has been laced with **PCP**.

Tolerance to the effects of marijuana clearly exist even though chronic users have described a "**reversed tolerance**" and claim that smaller doses of the drug are necessary to produce the desired effects. This effect is probably related to the manner of use and the expectations of the user. Chronic, high-dose cannabis users may experience an abstinence or **withdrawal syndrome** on abrupt discontinuation of use. Signs and symptoms include irritability, restlessness, nervousness, weight loss, insomnia, and rapid eye movement (REM) rebound. Onset of this syndrome is several hours after the last dose, and it lasts 4 to 5 days. Since withdrawal is not life-threatening, treatment involves little more than supportive therapy with short-term, low doses of benzodiazepines.

In addition to the slowed psychomotor response while under the influence of the drug, there is a decrease in **short-term memory** that is reversible on discontinuing the drug. Several reported physiological effects produce concern. The respiratory effects of long-term marijuana use produce chronic cough, laryngitis, bronchitis, and pathological changes like those in chronic tobacco smokers. Chronic smoking of one marijuana cigarette daily will decrease vital lung capacity equal to that seen after chronic smoking of one pack of tobacco cigarettes daily. In addition, the smoke from a marijuana cigarette contains high concentrations of several key carcinogenic agents found in tobacco smoke. Another concern is the possibility of **pulmonary fibrosis** from smoking marijuana that has been sprayed with **paraquat**, other herbicides, or other insecticides. However, the latest reports indicate that smoking destroys paraquat by pyrolysis.

Other effects of marijuana are decreased **testosterone** concentrations and a **reversible decreased spermatogenesis** in males. Females experience decreased **follicle-stimulating hormone (FSH)** and **lutenizing hormone (LH)** concentrations in the normal ovulatory cycle. Although THC crosses the placenta, there are no reports in humans of teratogenic effects directly attributable to marijuana alone. However, avoiding any unnecessary drug during pregnancy is advisable. Children born to mothers who were chronic marijuana smokers during pregnancy may be at increased risk for learning disabilities such as **attention-deficit disorders**. *Salmonella* and *Aspergillus* **contaminants** have been reported to produce infections.

Chronic use of marijuana has been implicated in producing psychological changes and production of what has been called an "**amotivational syndrome**," characterized by diminished drive, ambition, and motivation; loss of effectiveness; impairment of judgment, concentration, memory, and communication skills; and inability to set goals or manage stress. Cessation may lead to gradual improvement over several weeks but may require months. It is difficult to determine whether these effects were present prior to and not caused by marijuana use, but they are observed very commonly. Several studies have implicated marijuana with structural brain damage, but much controversy exists about these. Seizures have occurred in some epileptics.

There has been much recent information on the dangers of passive tobacco smoke. Concern could also be expressed over potential problems of **passive marijuana smoke**, although intoxication is rarely reported from passive inhalation of marijuana. However, there are numerous reports of cannabis metabolites detected in the urine of the passive inhaler. THC metabolites may be present in the urine up to 2 months after heavy chronic use.

Marijuana is considered, along with alcohol and tobacco products, as one of the major "**gateway drugs**." Although this does not imply that users of marijuana will all progress to more dangerous drugs (e.g., **cocaine**), use of one or more of these commonly available agents is part of the history of most drug dependencies.

MARIJUANA IN HISTORY

Cannabis, Indian hemp, American hemp (listed in the U.S.P. from 1873 to 1942) consists of the dried flowering tops of the pistillate plants of *Cannabis sativa* **Linné**. The plant is an annual herb indigenous to central and western Asia, and is cultivated in India and other tropical and temperate regions for the fiber and seed. *Cannabis* is the ancient Greek name for hemp.

Cannabis was used in China and India, spreading slowly through Persia to the Arabs, and it probably was introduced into European and American materia medica about the time of Napoleon.

The amount of resin found in the pistillate flowering tops of *C. sativa* markedly decreases as the plants are grown in more temperate climates. Thus, Indian cannabis yields 20% or more of resin; Mexican cannabis 15% or less; Kentucky hemp 8% or less; Wisconsin hemp 6% or less. The active principles are found in the resin in about the same, or even smaller, ratio as that indicated above. The hemp leaves contain a small amount of the resin.

Indian cannabis is prepared rather carefully from the pistillate flowerheads only, with relatively few leaves, but Mexican and American cannabis consist of the whole upper portion of the stalk of the pistillate plant. Indian cannabis may have an activity about ten times as great as that of a poor quality of American cannabis.

The *U.S. Pharmacopoeia* recognized Cannabis Americana in a separate monograph in the 1870 and the 1880 revisions; then the monograph was deleted because of poor-quality drug. During World War I (1914 to 1918) the importation of Indian cannabis ceased, and as a war measure, American cannabis was permitted in the cannabis monograph in the U.S.P. of 1916. It remained in the monograph until the drug was deleted in 1942.

The pleasurable sensations induced in humans by the controlled use of cannabis are obtained from inhaling the smoke of burning cannabis, more promptly and with less drug, than by oral dosage. Some decades ago unscrupulous persons began the importation into the United States of rather crude Mexican cannabis (marijuana) cigarettes for sale at high prices. As the demand for these cigarettes increased and the habit of smoking them spread even to school children, federal and state narcotics agents started a campaign to stamp out their sale. The importation of Indian cannabis was prohibited and even large areas of growing American hemp were destroyed. This campaign has been successful; however, the official substitution of a poor-quality drug for the high-quality Indian cannabis, along with the marijuana campaign, has cost the materia medica a valuable drug, for the medicinal use of cannabis in the United States has been discontinued. The therapeutic potential of marijuana as an analgesic, antimigraine medication, and antiemetic is being rediscovered.

BOTANICAL PROPERTIES OF *CANNABIS INDICA* OR INDIAN CANNABIS (INDIAN HEMP)

Hemp is an annual plant, from 4 to 8 feet or more in height, with an erect, branching, angular stem. The leaves are alternate or opposite, on long, lax footstalks, roughish, and digitate, with linear-lanceolate, serrated segments. The stipules are subulate. The flowers are axillary; the male hemp in long, branched, drooping racemes; the female hemp in erect, simple spikes. The stamens are five, with long pendulous anthers; the pistils two, with long, filiform, glandular stigmas. The fruit is ovate and one-seeded. The whole plant is covered with a fine pubescence, scarcely visible to the naked eye, and somewhat viscid to the touch. The hemp plant of India, from which the drug is derived, has been considered by some as a distinct species, and named *C. indica*; but the most observant botanists, upon comparing it with our cultivated plant, have been unable to discover any specific difference. It is now, therefore, regarded merely as a variety, and is distinguished by the epithet *indica*.

The seeds have been used in medicine. They are about ⅛ inch long, roundish-ovate, somewhat compressed, of a shining ash-gray color, and of a disagreeable, oily, sweetish taste. They yield by expression about 20% of a fixed oil, which has a drying property, and is used in the arts. They contain also uncrystallizable sugar and albumen, and when rubbed with water form an emulsion, which may be used advantageously in inflammations of the mucous membranes, although without narcotic properties. The seeds are much used as food for birds, as they are fond of them. They are generally believed to be in no degree poisonous.

In Hindostan, Persia, and other parts of the East, hemp has long been habitually employed as an intoxicating agent. The parts used are the tops of the plant, and a resinous product obtained from them. *Ganja* or *gunjah* is at the top of cultivated female plants, cut while unfertilized directly after flowering, and formed into bundles from 2 to 4 feet long by 3 inches in diameter. The utmost care is taken to prevent fertilization, it being affirmed that a single male plant will spoil a whole field. When hemp is cultivated in India for its fiber or seed, male and female plants are grown together.

The name *bang* is given to a mixture of the larger leaves and capsules, without the stems, of wild plants, male and female. There is on the surface of the plant a resinous exudation, to which it owes its clammy feel. Men clothed in leather run through the hemp fields, brushing forcibly against the plants, and thus separating the resin, which is subsequently scraped from their dress and formed into balls. These balls and also masses formed out of resin mechanically separated from **gunjah bundles** are called *churrus*. In these different states of preparation the hemp is smoked like tobacco, with which it is said to be frequently mixed. *Momea* or *mimea* is a hemp preparation said to be made in Tibet with human fat. An infusion or decoction of the plant is also sometimes used as an "**exhilarating drink**."

Fresh hemp has a peculiar narcotic odor, which is said to be capable of producing vertigo, headache, and a species of intoxication. It is much less in the dried tops, which have a feeble bitterish taste. *Churrus* is pure when of a blackish-gray, blackish-green, or dirty olive color, with a fragrant and narcotic odor and a slightly warm, bitterish, and acrid taste. The Indian hemp is officially described as branching, compressed, brittle, about 5 cm or more long, with a few digitate leaves, having linear-lanceolate leaflets, and numerous sheathing, pointed bracts, each containing two small, pistillate flowers, sometimes with the nearly ripe fruit, the whole more or less agglutinated with a resinous exudation.

MARIJUANA-INDUCED INEBRIATION

The effects of marijuana in humans are identical for oral administration (about 2 grains of the solid extract) and for smoking. The inebriation sets in rather abruptly, after a latent period of about an hour, in which there may be temporary digestive phenomena, anorexia, dryness or burning thirst. The phenomena differ in kind as well as in degree according to the individual personality and disposition, the surroundings, and the mental preparation. The *emotional sphere* develops euphoria, expansiveness, and gaiety, less commonly anxiety. Waves of exaltation sweep over the consciousness and may alternate precipitously with fear and terror. The subject is likely to be quarrelsome and violent if crossed. In the subjective sphere, a feeling of unreality is among the first effects, with a consciousness of double personality. Objects and sensations appear exaggerated in size and in degree, the limbs lengthened, the head swollen, the body alternately light and heavy, time enormously lengthened. *Motor excitement* and restlessness vary in different persons; tremors of the tongue and fingers, and muscular rigidity may occur. The ***pupils*** become widely dilated and barely react to light. The eyes appear glazed, the conjunctiva congested, the lids reddened. The pulse is generally rapid, with marked palpitation. ***Mental confusion*** may deepen to visual illusions and hallucinations. The imagery is apt to become terrifying in the dark. ***Aphrodisiac desire***, rather than impulse, occurs in some and not in others. The ***after effects*** are generally light. The exaltation is succeeded by irritability and somnolence, and after some hours by sleep, from which the subject awakens hungry and thirsty, with a clear recollection of the episodes. ***Toxic doses*** cause vertigo and collapse, with soft arrhythmic pulse, and blood concentration sometimes to 50%. Serious poisoning is rare because the margin between the effective and fatal dose is wide; the intravenous fatal dose for dogs is a hundred times the narcotic dose.

Addiction to cannabis is usually casual, since psychopathic individuals tend to drop it for a more potent narcotic. It generally takes the form of an occasional hedonistic indulgence, whose chief harm lies in the introduction to more serious addictions. No relation to criminal offenders, sexual or other, has been established. Physical consequences are no more significant than with tobacco. Digestive disturbances and marasmius are not prominent. Conjunctival congestion is common. There is apparently no material tolerance, either in humans or in animals, nor are there withdrawal symptoms, except in extreme cases. Insanity is a rare sequel, and in these the cannabis is probably only a provocative agent that unveils the latent psychopathy. **Psychopathic patients** react to cannabis like normal individuals, but usually become more self-absorbed and less communicative, in contrast to their response to cocaine and amytal (Sollmann, 1944).

In summary, subjective effects from smoking marijuana include usually elation and a tendency to be amused easily. Occasionally, there may be feelings of depersonalization with detachment from reality but overt behavior is usually limited to giggling, singing, and dancing. The conjunctiva are reddened and the upper lids become ptotic giving an appearance of sleepiness. The appetite is enhanced. In susceptible individuals, toxic psychoses may occur with acute mania and convulsions. No great degree of tolerance to the drug develops and no abstinence syndrome follows its withdrawal.

Death from acute poisoning is extremely rare, and recovery has occurred after enormous doses. The continued abuse of hashish in the East sometimes leads to mania and dementia.

CANNABINOID RECEPTOR SUBTYPES

Δ^9-Tetrahydrocannabinol (Δ^9-THC) is considered to be the predominant compound in preparations of *C. sativa* (marijuana, hashish, bhang) responsible for the central nervous system effects in humans. The recognized central nervous system responses to these preparations include alterations in cognition and memory, euphoria, and sedation. Potential therapeutic applications of *Cannabis* preparations that are of either historical or contemporary interest include analgesia, attenuation of the nausea and vomiting of cancer chemotherapy, appetite stimulation, decreased intestinal motility of diarrhea, decreased bronchial constriction of asthma, decreased intraocular pressure of glaucoma, antirheumatic

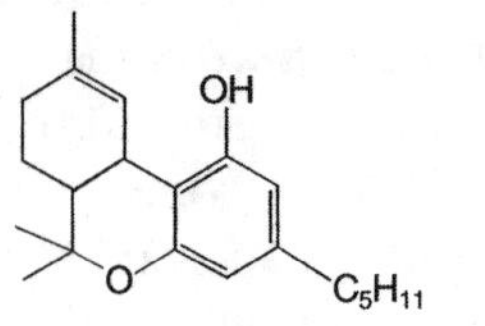

(-)Delta-9-THC

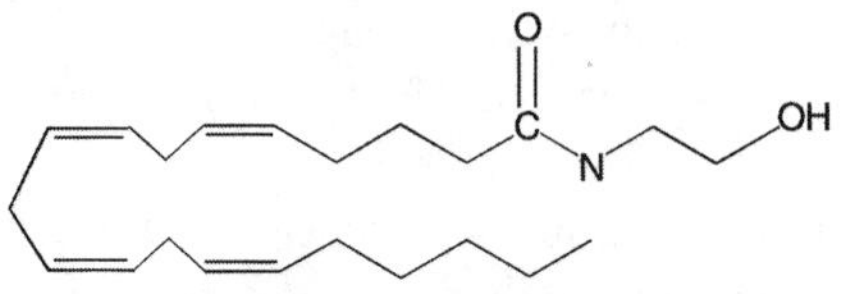

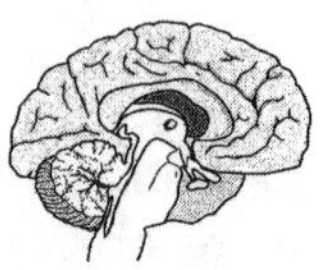

Anandamide Arachidonylethanolamide

Figure 43 Chemical structures of Δ^9-tetrahydrocannabinol (Δ^9-THC) derived from marijuana, and the endogenous cannabimimetic eicosanoid, anadamide (arachidonylethanolamide), identified in human brain.

and antipyretic actions, and treatment of convulsant disorders. These effects have been reviewed recently (Howlett, 1995).

Two subtypes of **cannabinoid receptors**, CB_1 and CB_2, have been described to date, although future investigations may elucidate other receptors. Actions of cannabimimetic agents via CB_1 receptors in brain are mediated by $G_{I/O}$ to inhibit adenylate cyclase and Ca^{2+} channels. Little is known about signal transduction mechanisms utilized by CB_2 receptors. Three classes of agonist ligands regulate cannabinoid receptors: **cannabinoid**, **aminoalkylindole**, and **eicosanoid derivatives**. Cannabinoid receptors produce analgesia and modify cognition, memory, locomotor activity, and endocrine functions in mammals (Figures 43 and 44).

CB_1

The first identified cannabinoid receptor subtype, CB_1, was cloned and demonstrated to have an amino acid sequence consistent with a tertiary structure typical of the seven transmembrane-spanning proteins that are coupled to G proteins. In addition to being found in the central nervous system, mRNA for CB_1 has also been identified in testes. The central nervous system responses to cannabinoid compounds are believed to be mediated exclusively by CB_1, inasmuch as CB_2 transcripts could not be found in brain tissue by either Northern analysis or *in situ* hybridization studies. CB_1 transduces signals in response to central nervous system–active constituents of *C. sativa* as well as synthetic bicyclic and tricyclic cannabinoid analogues, **aminoalkylindole**, and **eicosanoid cannabimimetic compounds**. CB_1 is coupled to G_1 to inhibit adenylate cyclase activity and to a pertussis-sensitive G protein to regulate Ca^{2+} currents.

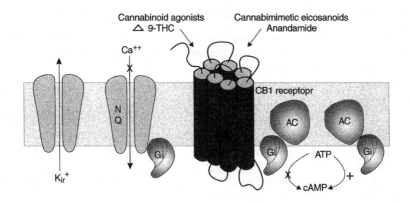

Figure 44 Δ^9-THC and anandamide bind to the CB_1 cannabinoid receptor and stimulate receptor coupling to heterotrimeric guanine nucleotide binding proteins Gi and Gs and subsequent activation or inhibition of signal transduction effectors. The receptor is a single subunit polypeptide that spans the plasma membrane seven times. The CB_1 receptor couples to the inhibition of adenylate cyclase and Q-type voltage-dependent calcium channels through Gi, and to the stimulation of adenylate cyclase through Gs. CB_1 receptor-mediated stimulation of inwardly rectifying potassium channels may be a consequence of cAMP formation and independent of G protein coupling.

CB_2

The second cannabinoid-binding seven transmembrane–spanning receptor, CB_2, exhibits 68% identity to CB_1 within the helical regions, and 44% identity throughout the total protein. The CB_2 clone was derived from a **human promyelocytic leukemia cell** HL60 cDNA library. The rat or human CB_2 clones were able to hybridize with mRNA in undifferentiated HL60 cells and in HL60 cells that had been differentiated into granulocytes or macrophages. Probes also hybridized with a splenic macrophage/monocyte preparation but not to splenic T cells. Northern blots detected CB_2 mRNA from spleen, but not from mature blood neutrophils, thymus, liver, brain, lung, or kidney, indicating that the distribution is distinct from that of CB_1. Probes derived from CB_1 have identified a signal in splenocytes and mononucleocytes after reverse transcription–polymerase chain reaction amplification. At present unclear are which cell types express CB_1 or CB_2, and the relative abundance of each subtype within immune or other tissues.

CANNABINOID RECEPTORS
AND THEIR ENDOGENOUS AGONIST, ANANDAMIDE

Analogous to the discovery of endogenous opiates, isolation of cannabinoid receptors provided the appropriate tool to isolate an endogenous **cannabimimetic eicosanoid, anandamide**, from porcine brain. Recent studies indicate that **anandamide** is a member of a family of fatty acid ethanolamides that may represent a novel class of lipid neurotransmitters (see Figures 43 and 44; Axelrod and Felder, 1998).

Table 24   Evidence for an Interaction between the Cannabinoid-Anandamide System and Other Neuroreceptor Systems

Δ^9-Tetrahydrocannabinol (THC) and anandamide enhance γ-aminobutyric acid (GABA)ergic neurotransmission

The cannabinoid receptor gene is under the negative control of dopamine receptor-mediated events

THC regulates substance P and enkephalin messenger RNA (mRNA) levels in the caudate putamen

Dizocilpine (MK-801; a noncompetitive antagonist of N-methyl-D-aspartate receptors) ameliorates THC-induced catalepsy

The mRNA for the cannabinoid receptor in the caudate putamen of adrenalectomized rats is 50% higher than that in control animals

Anandamide and THC significantly increase serum levels of corticotrophin (ACTH) and corticosterone in a dose-dependent manner

---

The cannabis plant has been cultivated for centuries both for the production of hemp fiber and for its presumed medicinal and psychoactive properties. The smoke from burning cannabis contains many chemicals, including 61 different cannabinoids that have been identified. One of these, Δ^9-tetrahydrocannabinol (Δ^9-THC), produces most of the characteristic pharmacological effects of smoked marijuana.

The cannabinoid–anandamide system interacts with several neuroreceptor systems (Table 24).

CENTRAL NERVOUS SYSTEM CANNABINOID RECEPTORS

Cannabis sativa was one of the first plants to be used by humans to treat a wide range of medical conditions. In the early 19th century, O'Shaughnessy demonstrated that many of the therapeutic claims for cannabis that had arisen as a result of its use in Indian medicine (such as treatment of cramps, convulsions in children, migraine, neuralgia, and tetanus) were well founded. His reports made "**Indian hemp**" an accepted therapeutic agent, first in England and later in other European countries and North America. Indeed, Reynolds, a noted English physician, stated that "Indian hemp when pure and administered carefully is one of the most valuable medicines we possess" (Mechoulam et al., 1994).

Around the turn of the century, the medical use of cannabis declined because reproducible clinical effects could not be obtained. At that time, the active constituent in cannabis had not been isolated in pure form. Instead, plant extracts were generally used and these were known to deteriorate rapidly, thus accounting for the nonreproducible effects.

Work on the medicinal use of cannabis restarted when the major, and essentially only, psychoactive constituent of cannabis, **THC** (see Figure 43), was isolated in pure form. The structure of THC was elucidated in 1964. Simple synthetic routes became available and the pharmacological activity of numerous cannabinoids was examined in animals and humans.

A long list of potential therapeutic effects was recorded for THC, including **analgesic**, **bronchodilatory**, **antiemetic**, **anticonvulsant**, and **anti-inflammatory action**, **reduction of intraocular pressure**, and **alleviation of some**

neurological conditions (such as seizure disorders, spasticity associated with spinal cord injuries, and multiple sclerosis) (Mechoulam et al., 1994).

THERAPEUTIC POTENTIAL OF CANNABINOID LIGANDS

Movement Disorders

Autoradiographic localization of brain cannabinoid receptors using a labeled probe has revealed a unique distribution. Binding is most dense in the outflow nuclei of the basal ganglia, in the substantia nigra pars reticulata, the globus pallidus, the hippocampus, and cerebellum.

The high levels recorded in the basal ganglia and cerebellum, regions that coordinate motor functions, strongly suggest that the **cannabinoid–anandamide system** is involved in this activity. This is supported by the wide spectrum of neurological effects of cannabis or THC in laboratory animals (impaired coordination and balance, decreased muscle tone and reflexes, and, at higher doses, tremor, myoclonic jerks, muscle spasms, and convulsions). In humans, uncoordinated movements, muscle weakness, and tremulousness are frequently observed after cannabis use.

At present, there is no approved medical use for cannabis in patients with neurological disorders. However, it is illegally used for **spasticity** and ataxia in patients with **multiple sclerosis** and **spinal cord injury**, and for the treatment of **trigeminal nerve pain** and, to a lesser extent, **attention deficit hyperactivity disorder**. Individuals with spinal cord injury have reported a reduction in spasticity after cannabis use.

Multiple Sclerosis

The effect of THC in patients with multiple sclerosis has been investigated in a placebo-controlled, double-blind, crossover trial involving nine patients who were orally administered placebo or THC 5 to10 mg/day. Clinical and electromyographic measurements indicated a significant reduction in spasticity, with negligible adverse effects. In another trial, patients with multiple sclerosis reported subjective improvement of tremor and sense of well-being. Smoking cannabis was reported to result in symptomatic improvement, as judged by clinical rating, electromyographic investigation of leg flexor reflexes, and recording of hand action tremor, in a single patient with multiple sclerosis.

The ability of THC to suppress the effects of **experimental autoimmune encephalomyelitis (EAE)**, the laboratory model of multiple sclerosis, has been investigated in Lewis rats and guinea pigs. All animals treated with placebo developed severe clinical EAE 10 to 12 days postinjection of myelin and more than 98% died. THC-treated animnals had either no clinical signs or mild signs with delayed onset, with survival greater than 95%. Examination of central nervous system tissues revealed a marked reduction of inflammation in the THC-treated animals. Multiple sclerosis is considered to be an autoimmune disease. Because the cannabinoid CB_2 receptor has been found in the spleen (an organ involved in the immune system), it is possible that the effect of THC on multiple sclerosis is mediated through both CB_1 and CB_2 receptors.

Huntington's Chorea

A major effect of **Huntington's chorea** is the loss of neostriatal neurons, accompanied by the loss of dopamine D_1 receptors. A group of investigators has found that this loss of neurons also causes a 97.5% reduction of CB_1 receptors in the substantia nigra pars reticulata. On the basis of these data the authors have suggested a role for cannabinoids in the control of movement and possibly in the therapeutics or symptomology of **hyperkinetic** and **dystonic disorders**, such as Huntington's disease, as well is in the treatment of Parkinson's disease.

Neuroprotection

Cannabinoids may also cause effects via mechanisms distinct from the cannabinoid receptor pathways. The most extensively investigated compound is (+)-HU 211, a synthetic cannabinoid with a stereochemistry opposite to that present in the naturally occurring compounds. It does not produce THC-type effects in animals and shows insignificant binding to the CB_1 receptor. However, HU 211 blocks **N-methyl-D-aspartate (NMDA) receptors** and calcium uptake through the NMDA-receptor-ion channel in primary cell cultures. HU 211 is a potent blocker of NMDA-induced tremor, seizures, and lethality in mice. It may therefore prove useful as a nonpsychoactive drug that protects against NMDA-receptor-mediated neurotoxicity. This is supported by the potent attenuation of NMDA-receptor-mediator neurotoxicity in cell cultures by HU 211.

Cerebroprotective effects of HU 211 have been demonstrated after experimental closed head trauma in rats. The drug is very effective in improving recovery of motor function. The drug was given 1 or 2 hours after the trauma and the clinical status was evaluated 24 and 48 hours later. The percentage of rats able to perform a beam-walking task increased significantly. The drug was also effective in reducing by more than four-fold the blood–brain barrier breakdown that is observed during head trauma. It also attenuates cerebral edema.

Cannabis for Migraine

Cannabis, or marijuana, has been used for centuries for both symptomatic and prophylactic treatment of **migraine**. It was highly esteemed as a headache remedy by the most prominent physicians of the age between 1874 and 1942, remaining part of the Western Pharmacopoeia for this indication even into the mid-20th century. Current ethnobotanical and anecdotal references continue to refer to its efficacy for this malady, while biochemical studies of THC and anandamide have provided a scientific basis for such treatment (Russo, 1998).

ANTIEMETIC EFFECTS OF CANNABIS

The Physiology of Vomiting

The physiological purpose of nausea is to discourage food intake, and vomiting is meant to expel food or other toxic substances present in the upper part of the gastrointestinal tract. Protracted vomiting may not only cause electrolyte imbalance, dehydration, or a malnutrition syndrome, but may also lead to mucosal laceration and upper gastrointestinal hemorrhage (**Mallory–Weiss syndrome**).

Nausea and vomiting may occur when the stomach is overly irritated, stimulated, or distended (from overeating). In addition, nausea and vomiting may occur when the **chemoreceptor trigger zone for emesis** or the vomiting center, or both, are directly stimulated.

Pharmacological agents such as aspirin and l-dopa may cause vomiting by directly irritating the stomach. Agents such as aminophylline, isoniazid, reserpine, anti-inflammatory steroids, and caffeine may also elicit vomiting in susceptible individuals by causing the release of hydrochloric acid. This drug-induced emesis may be avoided by having patients take the drugs with meals. Antiemetics are not effective in rectifying these conditions and their use is not justified.

The Central Control of Vomiting

In addition to agents that stimulate or irritate the stomach, many other factors may be responsible for inducing emesis centrally. The central control of vomiting is vested in two areas:

1. The **vomiting center**, which is located in the lateral reticular formation in the midst of a group of cells governing such activities as salivation and respiration
2. The **chemoreceptor trigger zone**, which is a narrow strip along the floor of the fourth ventricle located close to the vomiting center

The functions of these two areas are distinct but interdependent.

The vomiting center is activated by impulses that originate from the gastrointestinal tract and other peripheral structures. In addition, there are unidentified tracts that extend from the cerebral cortex to the vomiting center, such that emotional trauma and unpleasant olfactory and visual stimuli may cause nausea and vomiting.

Stimulation of the vestibular apparatus that responds to movements of the head, neck, and eye muscles may also cause nausea and vomiting by stimulating the vomiting center. On the other hand, circulating chemicals, toxins, virus, and ions may provoke nausea and vomiting by first stimulating the chemoreceptor zone for emesis, which in turn stimulates the vomiting center (Figure 45).

Drugs to Control the Vomiting Caused by Cancer Therapy

Several currently available antiemetic drugs can prevent vomiting caused by cancer chemotherapy. Anticancer drugs that cause vomiting are listed in Table 25.

Ondansetron

Ondansetron antagonizes the action of serotonin at receptors in the upper gastrointestinal tract and in the chemoreceptor trigger zone of the central nervous system. Given intravenously (IV), it prevents vomiting completely for 24 hours in 40% of patients treated with high doses (>100 mg/m^2) of **cisplatin** and in 70 to

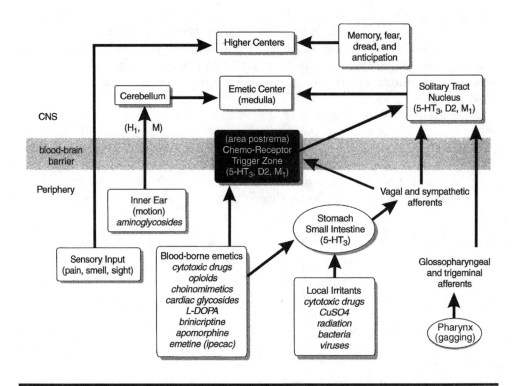

Figure 45 Pharmacologist's view of emetic stimuli. Myriad signaling pathways lead from the periphery to the emetic center. Stimulants of these pathways are noted in *italics*. These pathways involve specific neurotransmitters and their receptors (**bold** text). Receptors are shown for: dopamine, D; acetylcholine (muscarinic), M; histamine, H; and 5-hydroxy-tryptamine, 5-HT. Some of these receptor types also may mediate signaling in the emetic center. This knowledge offers a rationale for current antiemetic therapy.

80% of patients treated with 50 to 100 mg/m^2 of cisplatin or a combination of **cyclophosphamide** and **doxorubicin**.

Ondansetron is also available in an oral formulation, which can be used for mildly or moderately emetogenic chemotherapy. In one double-blind trial in 318 patients taking combinations of cyclophosphamide with methotrexate and/or doxorubicin, oral ondansetron 1, 4, or 8 mg t.i.d. for 3 days was completely effective in preventing emesis in 57, 65, and 66% of patients, respectively, compared with 19% who did not vomit with placebo.

Adverse Effects. Ondansetron generally does not cause severe toxicity. Headache and constipation are the most frequent adverse effects. Light-headedness, dizziness, and transient increases in serum aminotransferase activity can occur. Extrapyramidal effects have occurred rarely, and anaphylactoid reactions have been reported.

Dexamethasone

How **corticosteroids** prevent vomiting is unclear, but many studies have confirmed the effectiveness of a single dose of dexamethasone against a variety of anticancer agents. A randomized trial found low doses of dexamethasone (Decadron, and

Table 25   Vomiting with Anticancer Drugs[a]

Severe	Moderate	Mild
Cisplatin (Platinol)	Carboplatin (Paraplatin)	Bleomycin (Blenoxane)
Cyclophosphamide (Cytoxan, others) — high dose	Carmustine (BiCNU)	Chlorambucil (Leukeran)
Cytarabine (Cytosar-U, others) — high dose	Cyclophosphamide (Cytoxan)	Cytarabine (Cytosar-U)
Dacarbazine (DTIC-Dome, others)	Dactinomycin (Cosmegen)	Etoposide (VePesid)
Mechlorethamine (Mustargen)	Daunorubicin (Cerubidine)	Fluorouracil (Adrucil, others)
Streptozocin (Zanosar, others)	Doxorubicin (Adriamycin, others)	Hydroxyurea (Hydrea)
	Idarubicin (Idamycin)	Melphalan (Alkeran)
	Ifosfamide (Ifex)	Methotrexate (Folex, others)
	Lomustine (CeeNu)	Paclitaxel (Taxol)
	Mitomycin (Mutamycin)	Plicamycin (Mithracin)
	Mitoxantrone (Novantrone)	Thioguanine
	Pentostatin (Nipent)	Vinblastine (Velban, others)
	Procarbazine (Matulane)	Vincristine (Oncovin, others)

[a]Antineoplastics are often given in combination. Dose, route, and schedule of administration also affect the incidence and intensity of nausea and vomiting.

others) alone equivalent to low doses of ondansetron alone for control of acute emesis in patients receiving moderately emetogenic chemotherapy that included cyclophosphamide and/or doxorubicin. A double-blind crossover trial found that IV ondansetron given with dexamethasone to patients treated with cisplatin ($\geq$50 mg/m^2) prevented vomiting for 24 hours in 81 (91%) of 89 patients, compared to 57 (64%) who did not vomit with ondansetron alone.

Adverse Effects. Short courses of dexamethasone generally cause no adverse effects, although rapid injection can cause intense genital or perineal pain. Mild insomnia and epigastric discomfort can occur.

Metoclopramide

Metoclopramide (Reglan, and others) is both a serotonin- and dopamine-receptor antagonist. Given alone, it is less effective than IV ondansetron. Given concurrently with dexamethasone and either the antianxiety agent lorazepam (Ativan, and others) or the antihistamine **diphenhydramine** (Benadryl, and others), it can control emesis due to high-dose cisplatin (120 mg/m^2) in older adults. One study in older patients, however, found metoclopramide, dexamethasone, and diphenhydramine less effective than IV ondansetron plus dexamethasone in patients taking ≥50 mg/m^2 of cisplatin. Another double-blind study found metoclopramide combined with dexamethasone at least as effective as oral ondansetron in women receiving cyclophosphamide-containing regimens.

Adverse Effects. Metoclopramide can cause sedation, **akathisia** (motor restlessness), involuntary movements, diarrhea, and dizziness. The extrapyramidal reactions, which are more common in patients ≤30 years old, can be relieved by intravenous or oral **diphenhydramine** or **benztropine** (Cogentin).

PHENOTHIAZINES

Prochlorperazine (Compazine, and others) can be effective for prevention of vomiting due to cancer chemotherapy, but is generally less so than dexamethasone or metoclopramide. Phenothiazines can cause orthostatic hypotension, sedation, dystonic reactions, and akathisia.

Cannabinoids

Dronabinol (**Δ^9-tetrahydrocannabinol**; THC; **Marinol**) is FDA approved for treatment of nausea and vomiting associated with cancer chemotherapy in patients who have failed to respond adequately to conventional antiemetics. With mildly or moderately emetogenic chemotherapy, it is more effective than placebo and equivalent or superior to oral prochlorperazine. Dry mouth, sedation, orthostatic hypotension, ataxia, dizziness, and dysphoria occur frequently, particularly in middle-aged and older patients.

Benzodiazepines

Lorazepam (Ativan, and others) is used as an adjunct to antiemetic regimens, particularly in patients with anticipatory vomiting. **Alprazolam** (Xanax) has also been used as an adjunct. Benzodiazepines can cause somnolence and amnesia lasting for several hours, which may be beneficial.

Anticipatory Nausea and Vomiting

If chemotherapy-associated nausea and vomiting are not well controlled, some patients develop nausea and vomiting in anticipation of their next chemotherapy treatment. This conditioned response, once it occurs, is often difficult to treat. Adequate early antiemetic treatment, particularly with regimens that include a benzodiazepine, may prevent this reaction.

Delayed Emesis

Even after an effective regimen for prophylaxis, nausea or vomiting can begin again or persist 24 hours or more after chemotherapy, particularly with cisplatin. Concurrent use of oral dexamethasone (8 mg twice daily for 2 days, then 4 mg

Table 26   Dosage and Cost of Antiemetic Drugs

Drug	Dosage[a]	Cost[b]
Ondansetron — Zofran	0.15 mg/kg IV q2–4h × 3 or 32 mg IV once[c]	$207.50[d]
	8 mg PO tid[e]	53.50
Dexamethasone	20 mg IV once	
Generic — average price		3.65
Decadron		27.89
Metoclopramide	1–3 mg/kg IV q2h × 2–6	
Generic — average price		30.60
Reglan		36.32
Prochlorperazine	10 mg IV q4–6h × 4	
Generic — average price		10.60
Compazine		25.44
Dronabinol — *Marinol*	10 mg PO q3–4h × 4	47.84
Lorazepam — *Ativan*	1–2 mg IV once	6.01

[a] Beginning ½ hour before chemotherapy.
[b] Cost to the pharmacist for the lowest recommended dosage for a 70-kg patient, according to average wholesale price.
[c] An 8-mg IV dose may be equally effective in some patients.
[d] Based on cost of a 20-ml vial containing 40 mg.
[e] 4 mg t.i.d. may be effective in some patients.

twice daily for 2 days) and oral metoclopramide (0.5 mg/kg four times daily for 4 days) has been effective for this condition. Ondansetron alone has not been effective for treatment of delayed emesis following high doses of cisplatin.

Multiple-Day Cisplatin Chemotherapy

Many chemotherapeutic regimens include cisplatin for 4 or 5 days. Concurrent intravenous ondansetron and dexamethasone in two studies prevented emesis in 58 and 66% of patients for the entire 5-day period (see Table 26).

Summary

Intravenous ondansetron plus dexamethasone, with or without lorazepam, is the most effective treatment available for prevention of severe vomiting due to cancer chemotherapy and causes little toxicity. Metoclopramide is less effective than ondansetron and can cause extrapyramidal effects, but it is much less expensive and, used together with dexamethasone and lorazepam, is often effective, particularly for older patients and those taking less emetogenic drugs.

17

CINCHONA TREE

■ **Quinine**, an antimalarial agent

INTRODUCTION

The medicinal use of **quinine** dates back over 350 years. Quinine is the chief alkaloid of **cinchona**, the bark of the South American **cinchona tree**, otherwise known as **Peruvian bark**, **Jesuit's bark**, or **Cardinal's bark**. In 1633, an Augustinian monk named **Calancha of Lima, Peru**, first wrote that a powder of cinchona "given as a beverage, cures the fevers and tertians." By 1640, cinchona was used to treat fevers in Europe, a fact first mentioned in the European medical literature in 1643. The Jesuit fathers were the main importers and distributors of cinchona in Europe, hence the name Jesuit's bark. Cinchona also was called Cardinal's bark because it was sponsored in Rome by the eminent philosopher, **Cardinal de Lugo**.

CINCHONA, A PERUVIAN BARK

The cinchona tree belongs to the nat. ord. **Rubiaceae** and is a native of the eastern slope of the Andes, but has been largely planted in India, Ceylon, Java, and Burma, with the result of improving the quinine-yielding value of many species by cultivation. There are two official species:

1. **Cinchona**, the dried bark of ***Cinchona ledgeriana, C. calisaya, C. officinalis***, and of hybrids of these and of other species of *Cinchona*, yielding, when assayed by a prescribed process, not less than 5% of total alkaloids.
2. ***Cinchona rubra, red cinchona***, is the bark of *C. Succirubra* or of its hybrids, containing not less than 5% of cinchona alkaloids. From it is prepared the compound tincture of cinchona.

Figure 46 *Cinchona succirubra.*

VARIETIES OF CINCHONA

The principal varieties of the suborder Cinchoneae, the barks of which are found in commerce and are used by manufacturers of the alkaloids, are ***Cinchona calisaya***, *C. flava*, **yellow bark**, from Peru, Bolivia, and India; *C. succirubra* (Figure 46), **red bark**, from Ecuador, Java, and Ceylon; *C. condaminea*, pale bark, from Ecuador and Peru; *C. pitayensis*, **pitaya bark**, from New Granada; *C. micrantha*, **gray bark**, from Peru and Bolivia. Altogether, there are some 31 species acknowledged by botanists, and the list is constantly increasing, from the tendency of different trees to hybridize. Several trees formerly acknowledged as cinchonas are now placed in the **genus *Cascarilla***, but their barks are to be found on the market. Cuprea bark is from trees of the **genus *Remijia***, growing in Colombia; it contains quinone and a peculiar alkaloid, ***cinchonamine***, but no cinchonidine (Potter, 1910).

COMPOSITION OF CINCHONA

Cinchona bark contains 21 natural alkaloids, 3 of which are official, 8 artificial alkaloids, 2 simple acids, 2 tannic acids, a resinoid, and a coloring matter, as follows:

- **Quinine**, $C_{20}H_{24}N_2O_2$, a strong base, fluorescent, the most valuable of all the alkaloids; heated with glycerin to 374°F, it is converted into the isomeric base, *quinicine*.
- **Quinidine**, $C_{20}H_{24}N_2O_2$, isomeric with quinine, fluorescent, probably the most powerful as an antiperiodic, but existing in very small quantity.
- **Cinchonine**, $C_{19}H_{22}N_2O$, the least active of the official three, having about half the therapeutic power of quinine. Not fluorescent.
- **Cinchonidine**, $C_{19}H_{22}N_2O$, isomeric with cinchonine, not fluorescent, one of the most powerful of the alkaloids.

The other alkaloids are of no interest medicinally.

- **Kinic and kinovic acids** are combined in the bark with the alkaloids. The former is used to make a kinate of quinine, and the latter occurs in the nonofficial pharmacy as **Kinovate of Lime**, an ingredient in **Deloudre's Extract**, which is used in Europe and India for dysentery.
- **Kino-tannic and kinovo-tannic acids** give to bark its peculiar and powerful astringent qualities. They have not been fully studied.
- **Kinovin** is a bitter, amorphous resinoid, which is resolvable into **kinovic acid** and sugar. It is soluble in alcohol, but not in water.
- **Cinchona red** is a reddish-brown, insipid, inodorous substance.

QUININE FOR TREATMENT OF MALARIA

The causative organisms of malaria are protozoa of the genus *Plasmodium*, with four species known to infect the humans only:

- *Plasmodium falciparum*
- *Plasmodium vivax*
- *Plasmodium malariae*
- *Plasmodium ovale*

The major antimalarial agents are the **4-aminoquinoline derivative** (e.g., **chloroquine**), **8-aminoquinoline derivative** (e.g., **primaquine**), **folic acid antagonist** (e.g., **pyrimethamine**), and alkaloid (e.g., **quinine**). These antimalarial agents can be classified according to their effectiveness in interfering with various stages of parasitization. These are the primary tissue schizonticides. An agent such as primaquine destroys the primary tissue schizonts in the liver soon after infection has taken place (Figure 47).

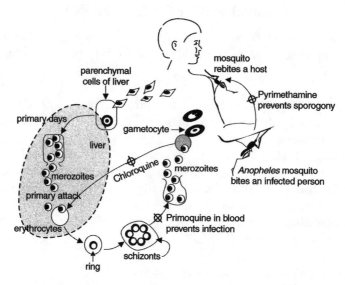

Antimalarial agents are effective against either erythrocytic or exoerythrocytic stages of *Plasmodium* protozoa

Figure 47 The actions of quinine, chloroquine, and primaquine.

BLOOD SCHIZONTICIDES

Agents such as **quinine**, **quinacrine**, **amodiaquine**, and **chloroquine** suppress the symptoms of malaria by destroying the schizonts and merozoites in the erythrocytes.

- **Gametocides**. Agents such as primaquine, by destroying the gametocytes in the blood, prevent infection caused by bites of the *Anopheles* mosquito.
- **Sporonticides**. Agents such as **chlorguanide** and **pyrimethamine** help to eradicate the disease by preventing sporogony and multiplication of the aforementioned parasites.
- **Secondary Tissue Schizonticides**. Agents such as primaquine destroy exoerythrocytic tissue schizonts such as those developing in the liver.

QUININE SULFATE, A CLASSIC DRUG

Quinine sulfate, the classic drug, the salt of an alkaloid obtained from cinchona bark, has been superseded by newer drugs in most parts of the world. It is only a fairly good suppressant, even in toxic dosage, and is so rapidly eliminated that it must be given at very frequent intervals to maintain its effects. When used to check an established attack, it achieves the control of **parasitemia** in 96 hours that chloroquine accomplishes in little more than 72 hours, and its effects on fever and most other symptoms also lag (Figure 48).

Absorption from the upper intestinal tract is rapid and reliable, and less than 10% is excreted as such in the urine; excretion begins in about 15 minutes, reaches

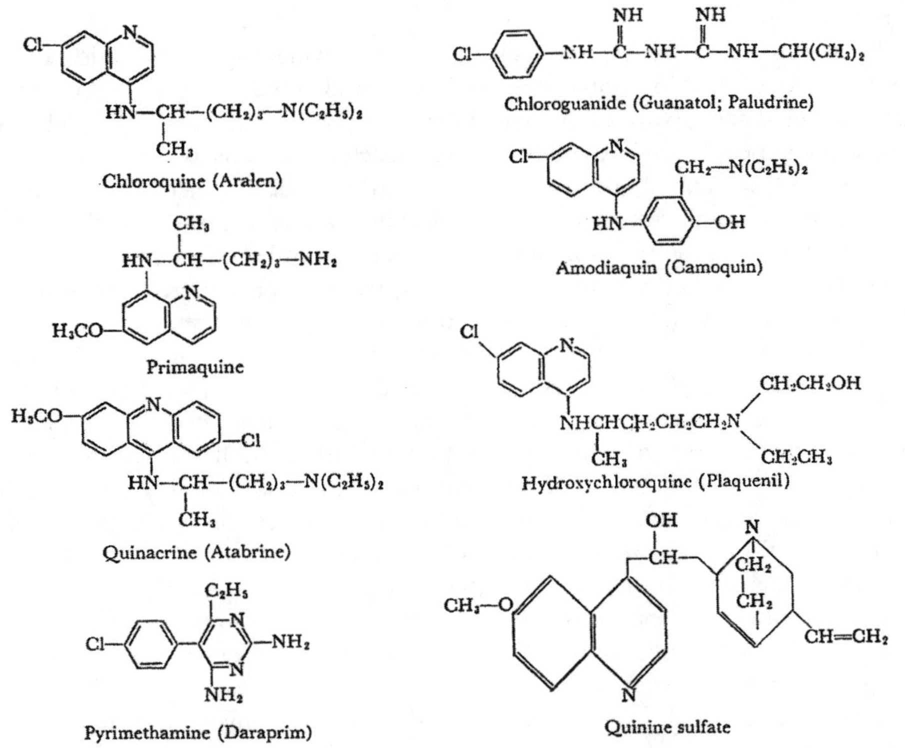

Figure 48 The structure of quinine and other antimalarial agents.

its maximum in 4 hours, gradually declines, and is complete in 24 hours. The other 90% or more of the drug is rapidly destroyed in liver, kidneys, and muscles, some of it being oxidized to the **carbostyril derivative** that is only about one third as active as quinine in experimental infections. Quinine does not accumulate in the body even though taken daily for a long period. Full action is obtainable with a plasma concentration of 5 mg/l, but frequent administration is required to maintain this level. Adult dosage of quinine sulfate (or bisulfate, which is more soluble) in tablets or capsules is usually 1 g (15 grains) three times daily (preferably some time before the taking of food) for 2 days, followed by 0.6 g (10 grains) at the same intervals for an additional 5 days. However, in Holland and Indonesia it has been customary to use only 1.5 g (22 grains) daily dosage in *P. falciparum,* and 1 g (15 grains) in *P. vivax* cases, which probably at least partially accounts for the greater satisfaction with the drug in those regions. As good results are claimed as with higher dosage, and there must certainly be a lower incidence of disturbing reactions. On the higher dosage scale used in this part of the world, infants under 1 year are given 30 mg (½ grain) three times daily; 1 year, 60 mg (1 grain); 2 years, 120 mg (2 grains); 3 and 4 years, 200 mg (3 grains); 5 and 6 years, 250 mg (4 grains); 7 and 8 years, 300 mg (5 grains), etc. In serious *P. falciparum* cases in some tropical practices a continuous intravenous drip has been employed, delivering 30 to 40 drops per minute of a solution containing

0.5 to 0.6 g (7½ to 10 grains) per 1000 ml, with the aim of administering 2 g (30 grains) in 24 hours.

Quinine in usual dosage frequently makes the patient uncomfortable because one or more of the following symptoms of **cinchonism** appear: nervousness, headache, giddiness, palpitation, tremors, nausea, and disturbances of sight and hearing. Some people also experience allergic reactions such as skin rashes, pseudo-asthmatic and anginal phenomena, and hemorrhagic disturbances (the action of vitamin K is apparently antagonized). **Quinine blindness**, which is usually a reflection of retinal spasm, is probably an allergic reaction in the beginning.

All the new synthetic antimalarials may apparently be administered safely to the pregnant woman, but it is a firmly entrenched clinical impression that quinine in full dosage is contraindicated in the mother and may cause serious visual and oral disturbances in the infant. **Black-water fever**, a little understood hemolytic syndrome sometimes complicating *P. falciparum* cases, has been associated with quinine administration in a high proportion of instances. If quinine is injected rapidly intravenously, it causes severe hypotension that is probably compounded of myocardial depressant and peripheral vasodilator actions, with the possible addition of a flocculent "nitritoid" type of occurrence.

Quinine has certain minor pharmacological actions quite distinct from its antimalarial role: analgesic (in prohibitively high dosage), striated muscle relaxant, and antipyretic. Other uses of quinine are as a sclerosing agent, a bowel evacuant, and in dermatological practice.

The structures of the newer antimalarial agents are given in Figure 48 and their pharmacological properties are outlined in Table 27.

Other pharmacological agents are sometimes used in combination with the antimalarial agents for greater effect. These include:

- **Sulfonamides** — Sulfadoxine or sulfadiazine is used with pyrimethamine.
- **Sulfones** — Dapsone (DDS) is used in place of or in addition to the sulfonamides and pyrimethamine.
- **Acridines** — Quinacrine (Atabrine) has an action similar to that of chloroquine.
- **Biguanides** — Clorguanide (Proguanil and Paludrine) has suppressive as well as prophylactic actions.

OTHER ACTIONS OF QUININE

The bark of the cinchona tree (Peruvian or Jesuit's bark or Cardinal's bark) has a very selective effect on malaria, so that a single adequate dose suppresses the plasmodia in the blood. This action is due chiefly to the alkaloid **quinine**, but is shared more or less by other cinchona alkaloids, **quinidine**, **cinchonine**, **cinchonidine**, etc., and by most of the numerous chemical derivatives of quinine and also by certain benzene derivatives, notably **atabrine** and **plasmoquine**, and to a minor degree by several other drugs. It is especially toxic to ameboid cells, and relatively nontoxic to higher organisms. Quinine is a fairly effective antipyretic, increasing heat loss by hydremia and by dilatation of the skin vessels. It is useful as an analgetic in colds; and in the form of **quinine-urea hydrochloride** as a local anesthetic. Through its bitterness, it acts as a **stomachic**

Table 27   Pharmacological Properties of Antimalarial Agents

Antimalarial Drug	Properties
Chloroquine (Aralen) **Amodiaquine** (Camoquin, as an alternate drug)	Chloroquine destroys schizonts in erythrocytes by interfering with DNA synthesis. The phosphate salts are active orally, whereas the hydrochloride salt is used for intravenous purposes. It accumulates in normal and parasitized erythrocytes. Overdosage has caused reversible corneal damage and permanent retinal damage. In toxic doses, chloroquine causes visual disturbances, hyperexcitability, convulsions, and heart block. It is an antimalarial of choice in all cases except chloroquine-resistant *Plasmodium falciparum*. In addition, it has a certain degree of effectiveness in amebiasis and in the late stages of rheumatoid arthritis.
Primaquine	Primaquine attacks plasmodia in the exoerythrocytic stages. It is effective for preventing relapse and as a prophylactic measure when staying in an infested area. Primaquine may cause hemolytic anemia, especially in patients who are deficient in glucose 6-phosphate dehydrogenase.
Quinine	Quinine is a naturally occurring alkaloid obtained from *Cinchona* bark, with a mechanism of action similar to that of chloroquine. Quinine is very useful in treating chloroquine-resistant *Plasmodium falciparum*. In toxic doses, it may cause cinchonism characterized by tinnitus, headache, nausea, and visual disturbances.
Pyrimethamine (Daraprim)	Pyrimethamine is a folic acid antagonist (antifol) with pharmacological actions similar to chlorguanide, methotrexate, and trimethoprim. Pyrimethamine may be used in combination with sulfadoxine for suppression and sulfadiazine for treatment of chloroquine-resistant *Plasmodium falciparum*.

The main objective in the clinical management of patients suffering from an acute malaria attack is the prompt elimination of the parasite form responsible for the symptoms, that is, the asexual erythrocytic form. Drugs that are particularly effective in this regard are called *schizontocidal* or *suppressive* agents. They include such compounds as amodiaquine, chloroguanide, chloroquine, hydroxychloroquine, pyrimethamine, quinine, and tetracycline.

tonic. It may stimulate the uterus. Full doses produce "**cinchonism**," characterized especially by auditory and visual disturbances, which may assume grave proportions with excessive doses. These also cause renal irritation. Very large quantities have sometimes been consumed without serious poisoning, chiefly because the absorption of the drug is limited. Hypersusceptibility and other idiosyncrasies are quite common. Cinchona also contains considerable **tannic acid**.

Colloids and Surface Phenomena

Quinine, like **saponin** and proteins, tends to concentrate at the surface of solutions, to form rather rigid films. These interfere with the condensation of other substances at the surface, and therefore hinder catalytic phenomena, inorganic as well as organic. This may explain the decreased metabolism. The surface action is also shown by arrest of "**Brownian movement**," and there is a tendency to precipitate colloidal solutions, such as the proteins. Quinine may therefore produce **anaphylactoid phenomena**. The quinine film presumably diminishes the permeability of the cell, and would thereby lead to a narcotic action, such as is seen on local application to nerves and muscles. Quinine tends to arrest cell movements, for example, of the white blood cells, infusoria, cilia, spermatozoa, etc., presumably through the rigidity of its film. All these render quinine a "**general protoplasmic poison**," toxic to all kinds of cells. However, its toxicity to vertebrates is relatively low, probably because it is largely bound and deposited in insoluble and harmless form.

Local Actions

Quinine produces some local irritation and anesthesia. When given hypodermically, it occasions severe pain, and may lead to abscess formation. When given by the stomach, large doses cause gastralgia, nausea, vomiting, and diarrhea. It also retards the absorption of salts, and probably of food. Its excretion through the kidneys may give rise, with large doses, to albuminuria and hemoglobinuria. The hemoglobinuria of severe malaria (**black-water fever**) may be partly due to increased sensitiveness to quinine.

Ulcers

The sulfate has been used as antiseptic, styptic, and stimulant dusting powder on ulcers.

Sunburn Prophylaxis

Quinine sulfate is protective through its fluorescence, which scatters the actinic rays.

Quinine as Anesthetic

Quinine and its derivatives, when brought in contact with sensory nerves, produce local anesthesia. This is used clinically in the form of the more soluble and less irritant double salts, especially quinine urea hydrochloride. The anesthesia is more lasting than with other local anesthetics; it may persist for several days. It is indeed due to necrosis of the axis cylinders and sheaths, with subsequent regeneration. However, the concentrations above ¼% sometimes produce considerable irritation, edema, and fibrous indurations and above 1% there may be sloughing. Its employment has therefore greatly diminished, although some clinicians consider these fears as exaggerated.

Antiseptic Action

Bacteria and yeasts are killed by quinine, but require a higher concentration (2 to 8:1000) than do the protozoa. The efficiency is not materially hindered by serum or pus.

Effect on Digestion

Small doses of quinine (0.05 g, 1 grain), taken before meals, act as a stomachic and tonic; but they are probably inferior to other bitters. Larger doses diminish gastric secretion.

Antipyretic Action

The suppression of malarial fever by quinine is etiological. However, quinine also reduces other febrile temperatures to normal, analogous to the other antipyretic drugs, i.e., by adjustment of the temperature-regulating centers, primarily through increased heat loss, assisted somewhat by diminished heat production.

Use of Quinine in Colds, Headaches, and Neuralgias

Like other antipyretics, quinine has an analgetic effect in these conditions. It is used in doses of 0.05 to 0.2 g, 1 to 3 grains.

Skeletal Muscle

Quinine, quinidine, cinchonine, and cinchonidine have similar actions. Dilute solutions cause direct depression with prolonged refractory phase. Concentrated solutions produce rigor.

Cinchonism

The larger therapeutic doses of quinine produce headache, ringing in the ears, and disturbed vision, a complex of symptoms grouped under the name of cinchonism. These are probably due to a selective action on the vessels of the eye and ear, which can also be demonstrated experimentally on animals. Toxic doses produce photophobia, deafness, and blindness, at first partial, later complete. These are probably partly central. Similar phenomena occur with salicylates and related drugs, as indeed also with fever itself, and with "colds." Cinchonism is more easily produced in some individuals than in others. It is said that the minor symptoms may be somewhat diminished by bromides. Hyperthyroid patients generally have marked tolerance for quinine, and absence of cinchonism has some diagnostic significance. With very high doses of quinine, the symptoms of cinchonism are accompanied by difficulty of speech, confusion of ideas and somnolence, followed by loss of consciousness, alternating with delirium, coma, and at times convulsions. The treatment is by evacuation and is symptomatic.

Skin Eruptions and Idiosyncrasy

Scarlatinal or urticarial rashes are not rare after therapeutic doses of quinine, and susceptible individuals may develop dangerous edemas of the mucous membranes and serious collapse (Sollmann, 1944).

18

CAPSICUM, ROSEMARY, AND TURMERIC

Several spices possess antioxidative properties — a **turmeric extract (curcuminoid)**, a **hexane extract** of **rosemary**, and the **α-tocopherol-supplemented capsicum pigment** exhibit their antioxidative effects *in vivo* by dietary supplementation. Turmeric extract has demonstrated the ability to reduce liver triacylglycerol deposition as well as cholesterol.

Many spices have been extensively used as natural food additives for flavoring, seasoning, coloring, and antiseptic properties. Several spices are known to exhibit antioxidative activities. In recent years, the antioxidative property of food constituents has been seriously noted by medical and nutritional experts, since the reactive oxygen species-mediated oxidation of biological molecules has been proposed to induce a variety of pathological events such as atherogenesis, carcinogenesis, and even aging.

Although many *in vitro* studies on the antioxidative property of food constituents have been reported, little is known about the biological functions of dietary antioxidants *in vivo*, except for several well-known antioxidants such as **tocopherols**, **β-carotene**, and **ascorbic acid**. Since the bioavailability of food constituents is limited by their digestibility and metabolic fate, an oral administration trial of a dietary antioxidant is favored to evaluate its biological function.

Asai et al. (1999) examined extracts of **turmeric (*Curcuma longa* L.)**, **rosemary (*Rosmarinus officinalis* L.)**, and **capsicum** (red pepper, *Capsicum annum* L.) as antioxidative food supplements. These spices contain types of antioxidants different from each other. The rhizome of **turmeric** has been used as a traditional remedy for treating sprains and inflammation in several Asian countries. **Curcuminoids (curcumin, demethoxycurcumin, and bis-demethoxycurcumin)**, the phenolic yellowish pigments of turmeric, have been suggested to have antioxidative, anticarcinogenic, anti-inflammatory, and hypo-cholesterolemic activities. Several **phenolic diterpenes** isolated from rosemary have been reported to prevent lipid peroxidation in bulk, emulsified, liposomal, and microsomal systems. The main constituents of capsicum pigment are **hydroxylated carotenoids (xanthophylls)**. Both hydroxylated and nonhydroxylated carotenoids are expected to act as potential membrane antioxidants because of their reactivity with **singlet molecular oxygen** and **peroxyl radicals**.

Asai et al. (1999) determined that **phospholipid hydroperoxides** (PLOOH) are key products for oxidative injury in membranous phospholipid layers in the plasma, red blood cells (RBC), and liver of mice. The formation and accumulation of PLOOH have been confirmed in several cellular disorders, diseases, and in aging. A lower PLOOH level was found in RBC of the spice extract-fed mice (65 to 74% of the nonsupplemented control mice). The liver lipid peroxidizability induced with Fe^{2+}/ascorbic acid was effectively suppressed by dietary supplementation with the turmeric and capsicum extracts to mice. Although no difference in the plasma lipids was observed, the liver triacylglycerol concentration of the turmeric extract-fed mice was markedly reduced to one half of the level in the control mice. These findings suggest that these spice extracts could act antioxidatively *in vivo* by food supplementation, and that the turmeric extract has the ability to prevent the deposition of triacylglycerols in the liver.

Liu and Ng (2000) studied the antioxidative and superoxide- and hydroxyl radical-scavenging activities and pro-oxidant effect of 12 selected medicinal herbs. The aqueous extracts of ***Coplis chinensis***, ***Paeonia suffruitcosa***, ***Prunella vulgaris***, and ***Senecio scandens*** exhibited the highest potency in inhibiting rat erythrocyte hemolysis and lipid peroxidation in rat kidney and brain homogenates. The aforementioned four herbs also demonstrated strong superoxide- and hydroxyl radical-scavenging activity.

19

CAROTENOIDS

Carotenoids reduce:

- Risk of prostate cancer
- Breast cancer
- Head and neck cancers
- Cardiovascular diseases
- Age-related macular degeneration

INTRODUCTION

Carotenoids, a class of yellow to deep-red pigments present in many commonly eaten **fruits** and **vegetables**, have been hypothesized to play a role in the prevention of chronic diseases such as cancer and heart disease. Most research to date has focused on the relationship between β-carotene and lung cancer. Lung cancer is one of the most common cancers in the United States, and β-carotene was initially the most thoroughly studied carotenoid because of its role as a **vitamin A precursor** (Cooper et al., 1999).

Vitamin A is an essential nutrient for humans and other vertebrates. Dietary sources of vitamin A are provided either by **retinol esters**, which are present in foods of animal origin and are hydrolyzed in the intestine to form **retinol**, or by **plant carotenoids**. More than 600 carotenoids have been identified in nature, of which 50 to 60 possess provitamin A properties and ~10 have nutritional relevance (De Flora et al., 1999).

Provitamin A carotenoids may be converted by central cleavage through a **15,15′-dioxygenase** present in the intestinal mucosa to yield one or two molecules of **retinaldehyde (retinal)**, which is then reduced by an **aldehyde reductase** to **retinol**. **Retinoic acid** may be formed by eccentric cleavage of such carotenoids as β-carotene. A central feature of vitamin A physiology is the strong mechanism of homeostatic control of the circulating concentrations of retinol in blood across a broad range of intake of preformed vitamin A and provitamin A. Circulating retinol concentrations remain fairly constant until liver reserves fall to very low levels. This is in contrast to concentrations of carotenoids in blood, which are mainly determined by the levels of carotenoids in the diet.

The **antigenotoxic** and **anticarcinogenic** effects of carotenoids or vitamin A have been studied extensively. The bulk of available data support the view that carotenoids and vitamin A do not induce genotoxic effects per se. Even in the absence of any genotoxic agent, these nutrients appear, on the contrary, to display some mechanisms that play protective roles in tumor promotion and progression, such as inhibition of **N-myc gene expression** resulting in antiproliferative effects, upregulation of **cell-to-cell communication**, an increase in **connexin 43 gene expression**, a decrease in the "spontaneous" **cell transformation frequency**, and induction of **differentiation** *in vitro*. A large number of studies investigated the modulation by carotenoids and vitamin A of genotoxic and related effects produced by 69 genotoxicants, including biological agents, physical agents, chemical compounds, and complex mixtures. In spite of some discrepant data, the general trend was that both carotenoids and vitamin A are poorly effective in acting as **nucleophiles**, nor do they appear to interfere substantially with the induction or repair of DNA damage produced by direct-acting agents. In contrast, vitamin A and carotenoids, irrespective of their provitamin A role, in most studies inhibited those genotoxicants that require metabolic activation to **electrophilic derivatives** in either bacterial or mammalian cells. Coupled with biochemical data, the distinctive patterns observed with genotoxic agents belonging to different chemical classes suggest a complex modulation of both phase I and phase II enzymes involved in the metabolism of **xenobiotics**. Furthermore, carotenoids and vitamin A shared other protective mechanisms, such as scavenging of **genotoxic oxygen species**, modulation of signal transduction pathways, inhibition of cell transformation induced by physical and chemical agents, and facilitation of intercellular communication inhibited by genotoxic compounds. Therefore, carotenoids and vitamin A appear to work via multiple mechanisms, which would support a potential protective role in cancer initiation and in the pathogenesis of other mutation-related diseases. These conclusions are consistent with the recognized cancer-preventive activity of these nutrients in certain animal models and with the evidence provided by observational epidemiological studies, which suggested cancer-protective effects at many sites as related to their dietary intake or plasma levels (De Flora et al., 1999).

PROSTATE CANCER

Prostate cancer is the most common cancer in U.S. males, and there is evidence that **dietary fat**, **soy proteins**, **vitamin E**, and **selenium** affect risk for this disease. Furthermore, it has been shown that consumption of **lycopene** and lycopene-containing foods (tomato and tomato paste) is associated with reduced risk of prostate cancer.

BREAST CANCER

Breast cancer is the most common form of cancer among women in developed countries. Approximately one in eight U.S. women will be diagnosed with breast cancer at some point in her life. A number of dietary factors have been investigated for their role in breast cancer etiology, including caloric intake, dietary fat, protein, fiber, vitamins A, C, and E, carotenoids, and fruits and vegetables. Higher β-carotene consumption has been reported to be associated with a lower risk of breast cancer in some epidemiological studies; however, the evidence is insufficient to reach a firm conclusion.

CANCER OF THE HEAD AND NECK

Cancers of the head and neck include cancers of the oral cavity, pharynx, and larynx. In the United States, approximately 42,000 of these squamous cell cancers are diagnosed each year. Worldwide, head and neck cancers are the sixth most common cancers. Tobacco and alcohol use are the most significant risk factors for development of head and neck cancers. Epidemiological and experimental data have suggested that **retinoids** and **carotenoids** may have chemopreventive activity causing regression of **oral leukoplakia**, a premalignant lesion, thus preventing its progression to oral cancer.

CARDIOVASCULAR DISEASE

Research on carotenoids and cardiovascular disease (CVD) stems from the discovery that the etiology of this disease involves oxidative processes that may be slowed by exogenous antioxidants. One of the best-understood processes contributing to development of CVD is the oxidation of **low-density lipoprotein** (LDL). When LDL becomes oxidized, it is readily taken up by **foam cells** in the vascular endothelium where it contributes to the development of **atherosclerotic lesion**. Enhancement of the oxidative stability of LDL may also prevent other oxidative steps involved in clinical expression of coronary disease (e.g., **myocardial infarction**) and possibly steps not related to LDL oxidation. There is optimism about the potential role of β-carotene in prevention of CVD because β-carotene is carried on LDL and has the capacity to trap **peroxyl radicals** and quench **singlet oxygen** *in vitro*.

AGE-RELATED MACULAR DEGENERATION

Age-related macular degeneration (AMD) is the leading cause of legal blindness in people over the age of 65 in the United States. This disease is characterized by degenerative changes in the central region of the retina, the macula, that eventually lead to loss of central detail vision. The cause of AMD is unknown; however, age, smoking, and genetics are risk factors. In addition, many medical, lifestyle, and environmental factors have also been associated positively or negatively with risk of AMD.

The role of carotenoids in the pathogenesis of AMD has gained considerable attention because the macula contains the carotenoids **lutein** and **zeaxanthin** in concentrations high enough to give it a yellow color. It has been hypothesized that these carotenoids reduce the risk of AMD by preventing damage to the retina by absorbing high-energy blue light or by antioxidant activity. The retina is a highly oxidizing environment because of light exposure, a high rate of photoreceptor catabolism and oxidative metabolism, and a high content of polyunsaturated fatty acids. Risk factors for atherosclerosis have also been associated with increased risk of AMD in some studies. Atherosclerotic disease of the retina could play a role in AMD by limiting the flow of blood, oxygen, and nutrients to and from the area. If carotenoids inhibited the atherosclerotic process, this could provide another mechanism for preventing AMD. At the present time, no significant association seems to exist between intake or serum concentrations of carotenoids and risk of AMD (Cooper et al., 1999).

20

COCA

The leaves of Erythroxylon coca, the divine plant of the Incas, contain a local anesthetic and a psychostimulant.

INTRODUCTION

Erythroxylaceae is a very small family, represented by two genera, the more important of which is **Erythroxylon**. They are mostly tropical shrubs with entire leaves and 5-merous flowers, and the fruit is one-seeded, reddish drupe resembling that of dogwood. The anatomy of the plants of this family closely resembles that of the **Linaceae**. Of special interest is the development of *papillae* on the dorsal surface of the leaves. This is found in most species of *Erythroxylon*.

COCA

Coca. Cocaina. Cocaine, *Erythroxylon coca,* Lamarck and other species. An alkaloid obtained from the leaves.

Habitat

Peru, Bolivia, Ecuador — eastern slopes of the Andes (Colombia, Brazil, India, Ceylon, Java); cultivated.

Syn.

Erythroxylon, Spadic, Coca leaves, Cuca, Hayo, Ipado, Coca Folia; Fr. Feuilles de Coca; Ger. Kokablätter; Cocain, Methyl-benzoyl-ecgonine; Br. Cocaina; Fr. Cocaine; Ger. Cocainum, Kokaina.

Er-y-throx'y-lon L.

See Etymology, above, of Erythroxylaceae.

Co'ca

Sp. from the native name, meaning tree or plant, *par excellence.*

Figure 49 *Erythoxylon coca*. Right, Coca leaf; under side, usual size.

HISTORY

Coca leaves were highly valued by the natives long before the Spanish conquest, the tree being known as "**The Divine Plant of the Incas**." Monardes published an extensive article on the drug in 1569. The natives chew the leaf, either as such or mixed with lime, and are thus able to travel great distances, often with heavy loads, without experiencing fatigue and without any but the most meager food rations. Cocaine was first isolated in 1860, but until 1884, coca was considered as but an inferior substitute for tea. In that year **Koller** discovered its local anesthetic properties.

PLANT

The plants are shrubs or small trees that attain a height of about 2 m; they are indigenous to Peru (**E. truxillense**) and Bolivia (**E. coca**) and are cultivated not only in these countries but also in Java (*E. truxillense*), and to some extent in Ceylon. *Erythroxylon* is from two Greek words meaning red and wood, alluding to the color of the plants; *coca* is the Spanish name for the tree; and *truxillense* is from Truxillo, a coastal city in northern Peru.

Huanuco coca leaves are oval, obovate, or elliptical, 3 to 7 cm in length, 2 to 3 cm in breadth (Figure 49), with an acute, slightly mucronate apex, an acute base, and an entire, somewhat revolute margin. The upper surface is dark green, glabrous, and the under surface is yellowish green and distinctly undulate with numerous minute papillae. A parallel line about 4 mm from the midrib on either side and extending from the base to the summit is often noted. The texture is somewhat coriaceous, the odor distinct, and the taste bitter, producing a sensation of numbness.

Truxillo coca leaves are usually more broken, a little larger, and the curved line on either side of the midrib is usually wanting.

Javanese coca leaves resemble Truxillo coca and are employed for the manufacture of cocaine in Holland.

Commercial Properties

Coca, although not introduced into England until 1870, was used in South America prior to the Spanish conquest, 1569, by the aborigines, who extolled it as a **God-given plant** ("**The Divine Plant of the Incas**") that satisfied hunger, strengthened the weak, and banished man's misfortunes. However, the invaders, intolerant of such homage, forbade its use and cultivation until they observed that it enabled the conquered to perform better work and service. Previous to 1884, the alleged properties were thought legendary and imaginative, and coca was considered simply a mild stimulant like tea. Then, in 1884, **Koller** proclaimed its local anesthetic power (see Culbreth, 1927).

Species differ when wild and under cultivation, and after collection soon degenerate and show marked changes in leaf characteristics. Cocoa is cultivated extensively in the Andes on terraced plantations, *cocales*, cleared from the forests on the warm declivities, thriving best in a moist atmosphere amid scattered trees, but not deep shade, which, as well as low elevation, develops bitterness; consequently, any variation in this or in the prescribed soil, exposure, and curing may affect quality. Propagation is similar to peach, yielding leaves the second year and continuing for 50 years thereafter. Leaves when bright green above and yellow-green below are picked carefully to avoid breaking or injuring young leaf-buds that form the next crop; the leaves are removed in baskets, spread on unroofed floors, and dried quickly for a few hours in the sun: if dried too rapidly, they lose odor and green color; if too slowly, they acquire a disagreeable odor and taste. After remaining 2 to 3 days in the coca house, in loose piles, they are exposed again for a short time to the sun, to drive off developed sweat, and then compressed into bales (*cestos*), 25 to 50 pounds (11.6 to 23.3 kg) or, better, packed in tin-lined boxes that prevent likely deterioration in shipping through fermentative decomposition. Irrespective of the care in drying and storing, impairment begins at once, and the cocaine decreases materially, especially in dampness; the leaves should be discarded after a few months. Although in some localities collection is almost continuous, there are at least two to three yearly harvests; the September harvest is the best and the April harvest next, each yielding when dried 60 to 80 pounds (27 to 37 kg) per acre. The annual production is about 80,000,000 pounds (37,383,177 kg), exported largely from Huanuco, Lima, Truxillo, etc.

There are two varieties:

1. *Huanuco* (*Cuzco-E. Coca*, short-styled), named after cities of southern Peru, has best aroma, most cocaine, and less isatropyl-cocaine; grown mostly in Bolivia and southern Peru, thriving and yielding maximum product at 1050 to 1800 *M* (3500 to 6000°), in 18° south, and inferior grades at lower elevations.
2. The true Bolivian (*E. bolivia'num*, long-styled) is prized most highly.

These two distinct varieties or, perhaps, species of Erythroxylon coca, one the Bolivian form, the other the Peruvian, are distinguished by much larger fruits in the Bolivian variety and much larger leaves in the Peruvian. The Bolivian drug is the more highly esteemed in Peru. The coca, which grows wild in various parts of South America, is cultivated not only in its native country, but also in British India, in Ceylon, and in Java. The coca plant, which is propagated from seed in nurseries, begins to yield in 18 months, and continues productive for half a century. The leaves, when mature, are carefully picked by hand so as to avoid breaking them or injuring the young buds, are slowly dried in the sun, and are then packed in bags (**cestos**) holding from 25 to 150 pounds each. They were in general use among the natives of Peru at the time of the conquest (Wood and Bache, 1894).

Botanical Properties

The leaves resemble in size and shape those of tea; they are oval-oblong, pointed, 2 inches or more in length by somewhat over an inch in their greatest breadth, and furnished with short delicate footstalks. However, unlike tea-leaves, they are not dentate, and are distinguished from most other leaves by a slightly curved line on each side of the midrib, running from the base to the apex. These lines are not ribs, but curves, which have been produced by the peculiar folding of the leaf in the bud. Good specimens are perfectly flat, of a fine green color; brown leaves should always be rejected as inferior. They have an agreeable odor resembling that of tea, and a peculiar taste, which, in decoction, becomes bitter and astringent. Leaves vary between ovate, lanceolate, and obovate–oblong, and are from 2 to 5 or 7 cm in length; they are short-petiolate, entire, rather obtuse or emarginate at the apex, slightly reticulate on both sides, with a prominent midrib, and on each side, a curved line runs from base to apex; their odor is slight and tealike; taste is somewhat aromatic and bitter. When chewed, it temporarily numbs the lips and tongue (Culbreth, 1927).

Chemical Composition

In 1853, it was demonstrated that coca leaves contained **tannic acid** and a peculiar bitter principle, resin, **tannin**, an aromatic principle, extractive, **chlorophyll**, a substance analogous to **theine**, and **salts of lime**. In 1855, the crystalline alkaloid **erythroxyline** was discovered in coca leaves, which was called **cocaine**. The earlier methods for isolating cocaine were as follows:

1. The leaves are exhausted with 85% alcohol acidulated with 2% of sulfuric acid; the tincture is treated with milk of lime and filtered; the filtrate is neutralized with sulfuric acid, and the alcohol distilled off. The syrupy residue is treated with water to separate resin, and then precipitated by sodium carbonate. The deposited matter is exhausted by ether, and the etheral solution, after most of the ether has been distilled, is allowed to evaporate spontaneously. The cocaine is thus obtained in colorless crystals, mixed with a yellowish-brown matter of a disagreeable odor, which is separated by washing with cold alcohol.

2. Coarsely ground coca leaves are repercolated with an aqueous 5% solution of sulfuric acid, and a very dense, slightly acid percolate is obtained, which is thoroughly agitated with pure coal oil and an excess of sodium carbonate. The liberated alkaloid is retained by the coal oil, and is nearly free from coloring matter. The oily solution is then agitated with acidulated water, and again precipitated by sodium carbonate in the presence of ether. The ethereal solution of cocaine is treated with diluted hydrochloric acid fractionally, and the nearly colorless solutions of cocaine hydrochlorate are cautiously evaporated in shallow porcelain pans almost to dryness. The product is in the form of a white, crystalline, granular powder, and is a nearly pure anhydrous salt.

3. Pure cocaine is in colorless, transparent prisms, inodorous, of a bitterish taste, soluble in 704 parts of cold water, more soluble in alcohol, and freely so in ether. The solution has an alkaline reaction and a bitterish taste, leaving a peculiar numbness on the tongue. The alkaloid melts at 97.7°C (208°F), and on cooling congeals into a transparent mass, which gradually becomes crystalline. Heated above this point it changes color, and is decomposed. It is inflammable, burning with a bright flame, and leaving charcoal. With the acids it forms soluble and crystallizable salts, which are more bitter than the alkaloid itself. The formula is $C_{17}H_{21}NO_4$.

Cocaine as a Local Anesthetic

When applied locally to nerve tissue in appropriate concentrations, local anesthetics (Figure 50) reversibly block the action potentials responsible for nerve conduction. They act on any part of the nervous system and on every type of nerve fiber. Thus, a local anesthetic in contact with a nerve trunk can cause both sensory and motor paralysis in the area innervated. The necessary practical advantage of the local anesthetics is that their action is reversible at clinically relevant concentrations; their use is followed by complete recovery in nerve function with no evidence of damage to nerve fibers or cells.

GENERAL PHARMACOLOGY OF LOCAL ANESTHESIA

When a local anesthetic is injected near a nerve, it blocks the flow of electrons along the axons and eliminates the pain without loss of consciousness. These effects are reversible. When administering a local anesthetic, one must remember that the larger the diameter of the nerve fiber, the more anesthetic is needed to produce anesthesia.

All local anesthetics contain a **lipophilic group**, an **amino derivative**, and an **intermediate chain** (Figure 51). The intermediate chain may be either an **ester**, as seen in compounds such as **cocaine**, **procaine** (Novocain), and **tetracaine** (Pontocaine), or an **amide**, as seen in compounds such as **lidocaine** (Xylocaine), **mepivacaine** (Carbocaine), **bupivacaine** (Marcaine), and **dibucaine** (Nupercaine). **Cross-sensitivity** occurs among drugs in the same group (e.g., cocaine, procaine, and tetracaine), but not between compounds containing ester and amide linkages.

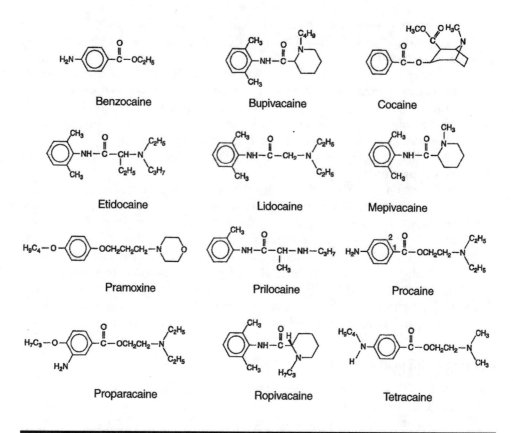

Figure 50 Structural formulas of selected local anesthetics.

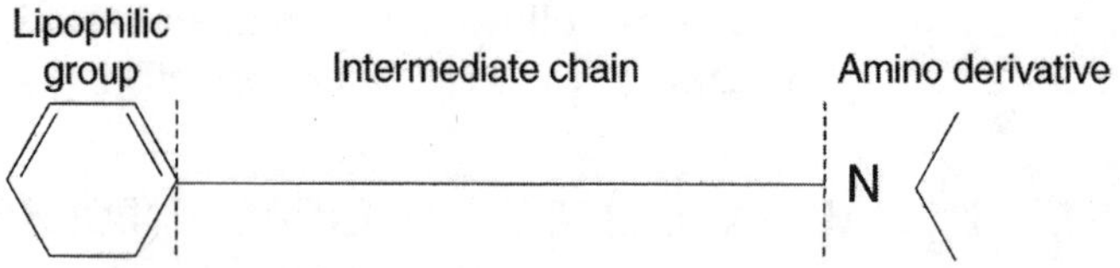

Amides	Potency	Duration of Action
Lidocaine	4	Medium
Medpicacaine	2	Medium
Prilocaine	3	Medium
Bupicacaine	16	Long
Etidocaine	16	Long
Esters		
Cocaine	2	Medium
Procaine	1	Short
Tetracaine	16	Long

Figure 51 Duration of action of various local anesthetics.

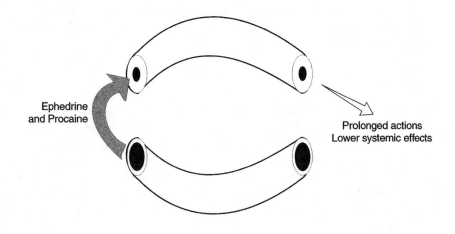

Ephedrine
and Procaine

Prolonged actions
Lower systemic effects

Figure 52 Epinephrine prolongs the duration of action of local anesthetics.

Local anesthetics are unstable and insoluble in **basic solution**. If a local anesthetic lacks an amine, it is insoluble in water and is only used topically.

VASCULAR SUPPLY AT THE SITE OF INJECTION

Epinephrine is used in combination with a local anesthetic to reduce its uptake, prolong its duration of action, produce a bloodless field of operation, and protect against systemic effects (Figure 52). Local anesthetic solutions containing epinephrine should not be used in areas supplied by end arteries such as in the digits, ear, nose, and penis, because of the threat of ischemia and subsequent gangrene. Furthermore, under no circumstances should anesthetic solutions containing epinephrine be used intravenously in patients with cardiac arrhythmias. In general, solutions designed for multiple doses should not be used for **spinal** or **epidural anesthesia**.

SYSTEMIC REACTIONS

Cardiovascular Effects

Local anesthetics block the sodium channels, are cardiac depressants, and bring about a ventricular conduction defect and block that may progress to cardiac and ventilatory arrest if toxic doses are given. In addition, these agents produce arteriolar dilation. Circulatory failure may be treated with vasopressors such as **ephedrine**, **metaraminol** (Aramine), or **mephentermine** (Wyamine). Artificial respiration and cardiac massage may also become necessary. Among the local anesthetics, only cocaine blocks the uptake of norepinephrine, causes vasoconstriction, and may precipitate cardiac arrhythmias.

Central Nervous System Effects

An overdosage of local anesthetics can produce dose-dependent central nervous system (CNS) side effects such as insomnia, visual and auditory disturbances, nystagmus, shivering, tonic–clonic convulsions, and finally fatal CNS depression. The initial CNS excitation and convulsions may be brought under control by **diazepam** or **thiopental**.

Allergic Reactions

The **ester-containing local anesthetics** become metabolized to *p*-**aminobenzoic acid derivatives**, which have a potential for causing hypersensitivity reactions. Allergic reactions to amide are extremely rare. In general, patients who have shown hypersensitivity reactions to ester compounds are treated with amide compounds. If a patient proves sensitive to both ester and amide compounds, then the hypersensitivity reaction is treated with certain antihistaminics such as **diphenhydramine**, which also possesses local anesthetic properties.

Effect of Cocaine

Cocaine, which blocks the uptake of catecholamines, produces dose-dependent effects, initially causing euphoria, vasoconstriction, and tachycardia, and, in toxic doses, convulsions, myocardial depression, ventricular fibrillation, medullary depression, and death. Cocaine is able to block nerve conduction and currently is used only for topical anesthesia.

MECHANISM OF LOCAL ANESTHETIC ACTION OF COCAINE

Cocaine and other local anesthetics prevent the generation and the conduction of the nerve impulse. Their primary site of action is the cell membrane. Local anesthetics block conduction by decreasing or preventing the large transient increase in the permeability of excitable membranes to Na^+ that normally is produced by a slight depolarization of the membrane. This action of local anesthetics is due to their direct interaction with **voltage-gated Na^+ channels**. As the anesthetic action progressively develops in a nerve, the threshold for electrical excitability gradually increases, the rate of rise of the action potential declines, impulse conduction slows, and the safety factor for conduction decreases, these factors decrease the probability of propagation of the action potential, and nerve conduction fails (Figure 53).

DETAILED PHARMACOLOGY OF COCAINE AS A LOCAL ANESTHETIC

Many substances of widely different chemical structure abolish the excitability of nerve fibers on local application, in concentrations that do not cause permanent injury, and that may not affect other tissues. Sensory nerve fibers are most susceptible, so that these agents produce a selective sensory paralysis, which is utilized especially to suppress the pain of surgical operation. This property was first discovered in **cocaine**, but because of its toxicity and addiction liability, this has been largely displaced by synthetic chemicals. The oldest of these, **procaine** (novocain), is still the most widely used. Its relatively low toxicity renders it especially useful for injections but it is not readily absorbed from intact mucous

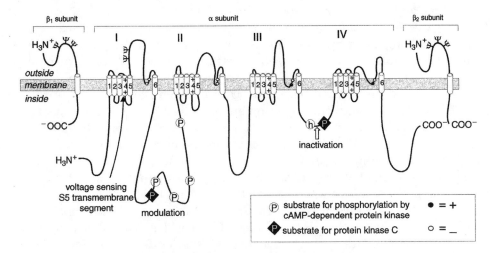

Structure and function of voltage-gated Na$^+$ channels

Figure 53 Structure and function of voltage-gated Na$^+$ channels. A two-dimensional representation of the α (center), β_1 (left), and β_2 (right) subunits of the voltage-gated Na$^+$ channel from mammalian brain. The polypeptide chains are represented by continuous lines with length approximately proportional to the actual length of each segment of the channel protein. Cylinders represent regions of transmembrane, α helices. ψ indicates sites of demonstrated N-linked glycosylation. Note the repeated structure of the four homologous domains (I through IV) of the α subunit. **Voltage Sensing.** The S4 transmembrane segments in each homologous domain of the α subunit serve as voltage sensors. (+) represents the positively charged amino acid residues at every third position within these segments. Electrical field (negative inside) exerts a force on these charged amino acid residues, pulling them toward the intracellular side of the membrane. **Pore.** The S5 and S6 transmembrane segments and the short membrane-associated loops between them (segments SS1 and SS2) form the walls of the pore in the center of an approximately symmetrical square array of the four homologous domains. The amino acid residues indicated by circles in segment SS2 are critical for determining the conductance and ion selectivity of the Na$^+$ channel and its ability to bind the extracellular pore-blocking toxins tetrodotoxin and saxitoxin. **Inactivation.** The short intracellular loop connecting homologous domains III and IV serves as the inactivation gate of the Na$^+$ channel. It presumably folds into the intracellular mouth of the pore and occludes it within a few milliseconds after the channel opens. Three hydrophobic residues (isoleucine-phenylalanine-methionine, IFM) at the position marked h appear to serve as an inactivation particle, entering the intracellular mouth of the pore and binding to an inactivation gate receptor there. **Modulation.** The gating of the Na$^+$ channel can be modulated by protein phosphorylation. Phosphorylation of the inactivation gate between homologous domains III and IV by protein kinase C slows inactivation. Phosphorylation of sites in the intracellular loop between homologous domains I and II by either protein kinase C or cyclic AMP-dependent protein kinase (p) reduces Na$^+$ channel activation.

membranes and is therefore not very effective for them. Many of its chemical derivatives are also used. They differ in penetration, toxicity, irritation, and local injury as well as in duration of action and potency. Absolute potency is not so important for practical use as is its balance with the other qualities. If cocaine is absorbed in sufficient quantity, it produces complex systemic actions, involving stimulation and paralysis of various parts of the CNS. These are mainly of toxicological and scientific interest. Its continued use leads to the formation of a habit, resembling **morphinism**. This is not the case with the other local anesthetics. Direct contact with cocaine or procaine paralyzes all forms of nervous tissue, without preceding stimulation. The susceptibility of the various nerve fibers presents marked and characteristic quantitative differences. Sensory fibers are especially easily attacked, and, by using appropriate dilutions, the paralysis is as complete as if the nerve fibers had been severed with a knife. If the drug is washed away, or absorbed, the nerve recovers its functions promptly and completely (but very strong solutions of cocaine may produce neuritis and permanent paralysis). Since these effects are strictly local, it follows that the anesthetic agent must be applied in such a way that an effective concentration reaches the nerve supply of the part which it is desired to affect. This may be accomplished by painting a solution on mucous membranes (from which cocaine is very readily absorbed); or by injecting it into or under the skin (**infiltration and subcutaneous anesthesia**); or around or into the trunk (**peri- or intraneural anesthesia**); or around the spinal nerve roots (**subdural or spinal anesthesia**). The intact skin is practically impermeable even to cocaine, although this penetrates slightly if the skin has been macerated.

Selective Action on Motor and Sensory Fiber

When a ½ to 2% solution of cocaine is injected into a mixed nerve or into the subdural canal, the selective action is very marked, so that there is complete anesthesia, without motor impairment. If the contact is prolonged, by stopping the circulation, or if stronger solutions are employed, the motor fibers become paralyzed, so that the difference is merely quantitative. A similar difference in the susceptibility of motor and sensory structures exists also for the alcohol group, **aconitine**, **phenol**, **hydrocyanic acid**, and for pressure, and even for the centrally acting narcotics, ether, etc. It is therefore a characteristic of nervous tissue rather than of cocaine. It is now generally accepted that the selective action in peripheral nerves is due chiefly if not entirely to differences in the size of the myelinated nerve fibers, for it follows this order closely, the smallest fibers being blocked first, the largest last. The selective action may therefore depend on penetration of the axis cylinder fibrillae.

Peripheral Paralysis of Special Sense Nerves

Cocaine and other local anesthetics abolish not only the sensation of pain, but other special sensations, if they are suitably applied. Here also there is some selection. *In the skin*, they paralyze first the vasoconstrictor reaction, then progressively the sensation of cold, warmth, touch, tickling, pressure, pain, "joint sense." *In the nose*, they abolish the olfactory sense. *On the tongue*, they destroy the taste for bitter substances, but have less effect on sweet and sour taste, and none on salty taste. When cocaine is applied to the appropriate nerves, it is found that the **centrifugal *vagus fibers*** are paralyzed before the centripetal; vasoconstrictor fibers before vasodilator; bronchial constrictors before the dilators, etc. (Sollmann, 1944).

Site of Action

All local anesthetics including cocaine paralyze the nerve fibers anywhere in their course, wherever they are brought into contact with them. When applied to mucous membranes or hypodermically, they select the portions peripheral to the main trunks, the thinner sheath of the terminal fibrils facilitating its penetration. It is therefore unnecessary to assume selective action on the histological sensory endings.

The **oxygen consumption** of nerve is decreased by cocaine, procaine, and urethane, approximately in the ratio of their anesthetic concentrations.

Combination of Anesthetics

The efficiency of mixtures of the local anesthetics corresponds to more or less complete summations, generally without a potentiation: If solutions of different anesthetics are diluted until they are "just effective" — then a mixture of any two of these solutions will also be "just effective" — no more and no less.

Potassium Synergism

Potassium salts may considerably potentiate other anesthetics. However, this seems to hold mainly for motor fibers and not for sensory. A 1% potassium chloride solution could be used with some advantage in place of sodium chloride for making up isotonic anesthetic solutions.

Alkalinization

The addition of sodium bicarbonate increases the potency of the local anesthetics two to four times, for direct application or injection into nerve trunks, and, probably for subdural injection and on application to mucous surfaces. This is due to the easier penetration of the free anesthetic bases, as compared with their salts. For these purposes the usual solutions of the anesthetic salts may be mixed with an equal volume of 0.5% sodium bicarbonate solution, without loss of efficiency, and with a saving of one half of the anesthetic, and correspondingly smaller danger of accidents. The mixtures, however, do not keep well, and must be made just before use. Alkalinization or buffering has *no advantage for hypodermic or intradermic injections*, since these do not require much penetration. Procaine base dissolved by the aid of carbon dioxide is also more potent than the hydrochloride when applied to the cornea, but not for intramuscular injection.

The **effects of temperature on local anesthesia** of frog's nerve differ for various types of anesthetics; **cocaine**, **salicylamide**, and **monacetine** are more effective on cooled nerves, whereas alcohol and chloral are more effective on heated nerves. This is probably connected with differences in solubility and the "partition coefficient."

Other Synergisms

Additions of ⅒ to 2½% of urea to dilute procaine solutions increase materially their potency, to ten times, on direct application to nerve trunks, presumably by favoring penetration. The effect is not due to alkalinization. The action of local anesthetics is reported to be increased by intravenous injection of **methylene blue**, by the local application of **caffeine** or **theophylline**, by morphine, and by the antipyretic analgesics. Cocaine anesthesia is said to be ineffective in tissue impregnated with **oxalic acid.** Inflamed tissues are less susceptible to local anesthesia, probably because of their difficult penetration.

Epinephrine Anesthetic Combinations

The anesthetic effect of cocaine injections is greatly improved by the addition of epinephrine (about 1:50,000). The duration of the anesthesia is prolonged for over an hour. The effective concentration can therefore be considerably lowered, and the danger of poisoning greatly reduced. These results are due to the vasoconstriction, which practically arrests absorption into the circulation. On the other hand, the nervousness induced by epinephrine in hypersensitive individuals may make them more susceptible to cocaine collapse. The greatest field for the epinephrine combination is in connection with operations involving bleeding. It is useless on the intact cornea. Epinephrine is similarly efficient with procaine and most other local anesthetics. Its effect is not as pronounced with **stovaine**, and is practically absent with **tropacocaine**, since ½% of this destroys the vasoconstriction effect of epinephrine. **β-eucaine** also counteracts epinephrine.

Epinephrine is not directly anesthetic either alone or in combination with cocaine, for it does not increase the anesthetic effect when the circulation has previously been arrested. It also does not lower the threshold concentration in wheals, although the duration of the anesthesia is greater for a given concentration.

Local Vascular Action

Cocaine produces vasoconstriction (blanching), probably by sensitizing to sympathetic stimulation. This action is lacking in the other local anesthetics. Procaine has practically no vascular effect; **alypin, eucaine**, and **stovaine** cause some dilatation. Cocaine decreases capillary hemorrhage; **procaine, apothesine**, and other synthetic anesthetics tend, rather, to increase bleeding.

Practical Application of Local Anesthetics

This involves considerable art, which must be learned in the clinic. It falls into two divisions: (1) surface application to the mucous membranes, especially of the eye, nose, throat, and urethra; and (2) injections about nerves, in different parts of their course and distribution, from their spinal roots to their ultimate fibrils. The advantages and disadvantages in comparison with general anesthesia and the selection of the local anesthetic agent also depend on clinical discrimination. Nervous, fearful, and excitable patients often suffer severely from apprehension, which also disposes toward accidents. They may be at least somewhat quieted by sedatives, morphine (0.015 g hypodermically) half an hour before the operation, or by barbiturates. The latter also tend to prevent convulsions.

The onset of the action depends on the penetration into the nerve fibers, and the duration on the sojourn of the drug. In general, the sensation becomes blunter in a few minutes but the anesthesia does not reach its maximum for 10 to 20 minutes and lasts 10 to 30 minutes, after a single application. The duration depends more on the concentration than on the total amount of the agent, so that it is more economical to use small quantities in high concentration, repeating if necessary.

Local Ischemia

As the anesthetic agent is absorbed and thus removed from the site of application, its local action ceases and its systemic and toxic effects start. Since most of these drugs, especially cocaine, are rapidly destroyed in the body, the systemic toxicity increases with the rapidity of absorption. It is therefore desirable and often necessary to delay the absorption. This may be done by restricting the local circulation. Cocaine itself tends to do this by producing a local vasoconstriction, an action that is not shared by its substitutes. This vasoconstriction should be reinforced by the addition of epinephrine. More dilute solution may thus be used, and the anesthetic effect is much more prolonged. With intracutaneous infiltration, the pressure and edema also result in ischemia. In suitable situations, the circulation may be slowed by bandages, or arrested by temporarily clamping the arterial blood supply.

Surface Anesthesia

Anesthesia from intact surfaces requires that the drug must be absorbable. This excludes the intact skin. **Cocaine**, **butyn**, **metycaine**, and **diothane** are readily absorbed from mucous membranes, and are therefore efficient, but also correspondingly toxic. Procaine and **apothesine** are relatively inabsorbable and therefore less applicable. **Alypin** and **β-eucaine** are intermediate, but the order of absorbability varies for different mucosae.

Ophthalmic Anesthesia

Cocaine, 4%, or butyn, 2%, is applied to the conjunctiva for deep penetration, as in cataract operations; **phenacaine (holocaine)**, 2%, or **butyn**, 2%, for surface action (foreign bodies), etc. Toxic manifestations are rare and slight because only small quantities are used. For infiltration, 2% procaine is advised. Several of the anesthetic agents produce some conjunctival irritation, smarting, and congestion. Cocaine blanches the conjunctiva, but cloudiness and even ulcers of the cornea sometimes follow its application; they are partly explainable by the drying and by the irritation of dust and other foreign matters that are not perceived on account of the anesthesia, and by the abolition of the winking reflex. The protoplasmic toxicity may play a part in their production.

Actions of Cocaine on the Eye

When cocaine is applied to the conjunctiva, it produces local anesthesia (which may not extend to the iris); local anemia (which extends to the iris, but not to the retina); and submaximal dilatation of the pupil by peripheral sympathetic stimulation. Mydriasis occurs also on systemic administration. The accommodation is impaired, but the light reflex is preserved; there is some **exophthalmos**. The intraocular pressure is usually lowered, but may be increased. The mydriasis and its associated phenomena are not produced by most of the cocaine substitutes.

Insoluble Anesthetics for Open Wounds and Ulcers

The soluble anesthetics are absorbed too rapidly to be of material use in persistent pain. The limited and slow solubility of some of the free synthetic bases renders them more adapted to these conditions. However, their practical usefulness is not great. They are employed for painful wounds, ulcers, etc., of the skin and accessible mucous membranes, for example, after dental operations. **Ethyl aminobenzoate (anesthesine)** may be applied diluted with 5 to 20 parts of talcum or boric acid, or lozenges of 0.01 to 0.02 g, but it has been reported that it delays the healing of X-ray burns and increases their depth and extent. Insoluble anesthetics have also been used against the pain of gastric ulcer or cancer and other gastralgias, but their efficiency is doubtful. Anesthesine, **cycloform, orthoform**, and **propaesin** are about equally effective through intact mucous membranes. They are practically nonirritant and nontoxic.

Conduction and Infiltration Anesthesia

This consists in the injection of the local anesthetic into or around the nerve trunk or in the area of its distribution, so as to block off sensory impulses from the operative field. Since fatal effects may arise from the absorption of the anesthetic, the smallest amount of the least-toxic agent that is effective should be employed, under conditions that minimize absorption. Procaine with the addition of epinephrine (1:100,000) is generally preferred. A well-planned technique is important. It is not necessary to flood the entire field of operation, as in the earliest methods, nor even to infiltrate the whole line of incision, as in "infiltration anesthesia." It is now aimed at confining the anesthetic mainly to the nerves, by placing it where the nerves chiefly run, or injecting it into the nerves themselves.

Spastic Rigidity of Skeletal Muscles

The local injection of 0.1 to 2% procaine or other anesthetics blocks the **centripetal proprioceptive impulses**, and thereby relaxes muscular tonus, normal and abnormal, such as **spasmodic torticollis**. It effects almost instantaneous relief of the pain, stiffness, malposition, and incapacity of **fibrositis, lumbago,** and acute sprains and fractures. The site of greatest tenderness may be infiltrated with 10 to 30 cc of 1 or 2% procaine hydrochloride. Injected systemically, it relaxes traumatic tetanus and removes decerebrate rigidity, so that spontaneous movements of the limbs and of the respiration return. Its curare action may also be concerned in this effect. It relaxes parkinsonian, but not myotonic, rigidity.

Spinal Anesthesia

Spinal anesthesia is really nerve-root anesthesia, produced by lumbar or sacral subdural injection of procaine or other local anesthetics, which anesthetizes the sensory nerve roots at their emergence from the spinal cord, and abolishes sensation in their entire peripheral distribution, beginning in 2 to 10 minutes, and lasting ½ to 2 hours. There is little if any interference with motor functions, and consciousness is, of course, preserved. The anesthesia extends to the level of the nerve roots reached by the anesthetic, with the aim of confining it to the lower half of the body. If it extends to the fourth ventricle, it paralyzes respiration. Alarming symptoms and fatalities from this cause are not infrequent. Other side actions are a drop in blood pressure, severe headache which may sometimes last for days or weeks, and cord injury. Bronchial pneumonia and atelectasis are fairly common. The technique requires some practice, is disagreeable to the patient, and the anesthesia is not always completely successful. Spinal anesthesia causes more psychic shock than general anesthesia, but usually less somatic disturbance. It is especially useful in pulmonary disease, arteriosclerosis, bladder and rectal operations, impending uremia, and in patients with diabetes. It is more difficult and more dangerous for operations above the costal margin. If the respiration stops, artificial respiration is the most effective treatment. **Strychnine** and caffeine have little or no value. Epinephrine or ephedrine is injected if the blood pressure falls.

Sympathetic Stimulation

The phenomena of systemic cocaine poisoning are largely those of sympathetic stimulation, but not as consistently as with epinephrine. The sympathetic stimulation is mainly central (midbrain) but partly peripheral. The chief manifestations of sympathetic stimulation are (1) sensitization to epinephrine (but antagonization to ephedrine) by peripheral action; (2) mydriasis and slight exophthalmos by central and peripheral action; and (3) cardiac acceleration (chiefly central). Other sympathetic symptoms are constriction of the blood vessels, erection of hair, and relaxation of the intestines. High concentrations of cocaine paralyze all smooth muscle. Procaine also produces sympathetic stimulation on systemic administration, dilating the pupils, augmenting the heart, inhibiting the intestines, increasing the pressor action of tyramine.

Circulation

The effects of cocaine on the general circulation are partly central, and partly peripheral. They vary according to the dose, and are also influenced by individual susceptibility. **The systemic actions of cocaine** are rather variable and complex, depending largely on the dose.

Cerebral Stimulation

The first systemic effects of cocaine consist in physical stimulation, somewhat resembling caffeine. This differs considerably in degree and in kind with individuals, and may be manifested as alertness, loquacity, elation, wakefulness, excitement, sometimes markedly sexual; or anxiety and flutter. The stimulation is greatest in excitable individuals and may interfere seriously with operations. Patients should be forewarned of the anxiety complex.

Incoordination, Narcosis, Convulsions

The stage of stimulation may be very short or even absent. With somewhat larger doses it is usually succeeded by depression, first of the coordinating functions. The movements lose their purposive type and become choreic. This is followed by general narcosis, and then by epileptiform convulsions, located chiefly in the midbrain. If the paralysis is rapid, the convulsive stage may not appear.

The **respiration** is at first accelerated. During the spasms it is irregular. The volume then diminishes. It may assume the Cheyne–Stokes type. Respiratory paralysis is the usual cause of death. This is also the first center to fail when the cocaine is applied locally to the fourth ventricle.

Emesis

The vomiting that frequently occurs in cocaine poisoning is perhaps due to the medullary stimulation, but its mechanism has not been fully investigated.

Temperature

Cocaine produces hyperpyrexia in mammals, due to increased heat production, by the motor excitement and rise of basal metabolic rate, reinforced by the convulsions. It probably acts directly on the midbrain temperature centers. Cocaine lowers the temperature of birds. Procaine lowers the temperature of mammals by increasing heat loss. It is antagonized by **thyroxin** and by **dinitrophenol**.

EXAMPLES OF CURRENTLY USED LOCAL ANESTHETICS

Procaine

Procaine, which has a pKα of 8.9, is highly ionized at physiological pH and has a short duration of action. Because it causes vasodilation, a vasoconstricting substance is added to the procaine solution to delay systemic absorption. Procaine may prolong the effect of **succinylcholine**, because both drugs are metabolized by the same enzyme. **Cholinesterase inhibitors** alter metabolism of procaine.

Chloroprocaine

Chloroprocaine has an onset of action of 6 to 12 minutes and its pharmacological properties resemble those of procaine.

Tetracaine

Tetracaine, which is metabolized slowly by **pseudocholinesterase**, has a relatively long duration of action.

Lidocaine

Lidocaine is effective in producing analgesia when administered by infiltration, or by nerve, epidural, caudal, and spinal block. In addition, it is effective when applied topically, with an onset of action of 5 minutes and a duration of action of 15 to 30 minutes. Lidocaine (1.5 μg/ml) is the agent of choice for the acute suppression of most ventricular arrhythmias.

Mepivacaine

Mepivacaine, which is ineffective topically, is used for infiltration plus nerve, epidural, and caudal block. Its potency and speed of action are similar to those of lidocaine.

Prilocaine

Prilocaine, which is equal in potency to lidocaine, has a longer duration of action. It is metabolized to **o-toluidine**, which in toxic doses may cause **methemoglobinemia**.

Bupivacaine

Bupivacaine, which is more potent than lidocaine, has a prolonged duration of action.

Etidocaine

Etidocaine is a long-acting derivative of lidocaine but is far more potent. It is effective for infiltration anesthesia, peripheral nerve block, and epidural and caudal blockade.

CLINICAL USES OF LOCAL ANESTHETICS

Local anesthesia is the loss of sensation in a body part without the loss of consciousness or the impairment of central control of vital functions. It offers two major advantages. The first is that the physiological perturbations associated with general anesthesia are avoided; the second is that neurophysiological reponses to pain and stress can be modified beneficially.

Infiltration Anesthesia

Infiltration anesthesia is the injection of local anesthetic directly into tissue without taking into consideration the course of cutaneous nerves. Infiltration anesthesia can be so superficial as to include only the skin. It also can include deeper structures, including intra-abdominal organs when these, too, are infiltrated. The advantage of infiltration anesthesia and other regional anesthetic techniques is that it is possible to provide satisfactory anesthesia without disruption of normal bodily functions.

Field Block Anesthesia

Field block anesthesia is produced by subcutaneous injection of a solution of local anesthetic such that the region distal to the injection is anesthetized.

Nerve Block Anesthesia

Injection of a solution of a local anesthetic into or about individual peripheral nerves or nerve plexuses produces even greater areas of anesthesia than do the techniques described above. Blockade of mixed peripheral nerves and nerve plexuses also usually anesthetizes somatic motor nerves, producing skeletal muscle relaxation, which is useful for some surgical procedures.

Intravenous Regional Anesthesia (Bier's Block)

This technique relies on using the vasculature to bring the local anesthetic solution to the nerve trunks and endings. In this technique, an extremity is exsanguinated with an Esmarch (elastic) bandage, and a proximally located tourniquet is inflated to 100 to 150 mmHg above the systolic blood pressure. The Esmarch bandage is removed, and the local anesthetic is injected into a previously cannulated vein. Typically, complete anesthesia of the limb ensues within 5 to 10 minutes.

Spinal Anesthesia

Anesthesia of the lower extremities and abdomen may be induced by the introduction of anesthetic drugs into the **subarachnoid space** (Figure 54). The drug most often used for this purpose is **bupivacaine**. The latency period plus the duration of the maximal cephalad level for both **plain and hyperbaric bupivacaine** lasts from 10 to 60 minutes. A bupivacaine solution is made hyperbaric by the addition of 5 to 8% glucose. The distribution of bupivacaine in the cerebrospinal fluid (CSF) is affected by gravity and is therefore influenced by the patient's position. With a dose of 15 mg of plain 0.5% bupivacaine, a half-life of about 3 hours is achieved. The addition of epinephrine to bupivacaine prolongs the duration of block.

The advantages of spinal anesthesia are the ease of administration, rapid onset of anesthesia, and good muscular relaxation, plus it allows patients to remain awake.

The disadvantages of spinal anesthesia are hypotension (**ephedrine** and **methoxamine** may prevent this), nausea and vomiting (avoided by thiopental), respiratory depression (treated by artificial respiration), and postoperative headache (treated by increasing the CSF pressure).

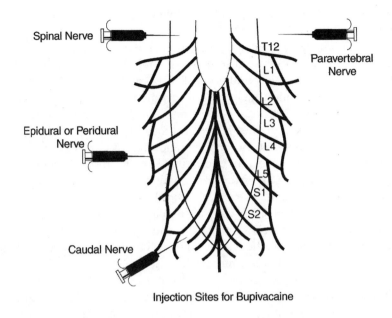

Injection Sites for Bupivacaine

Figure 54 Administration of anesthetics into subarachnoid space. T_{12} = 12th thoracic nerves; L_1, L_2, L_3, L_4, L_5 = first, second, third, fourth, and fifth lumbar nerves; S_1, S_2 = first and second sacral nerves.

Epidural Anesthesia

Epidural anesthesia is administered by injecting local anesthetic into the epidural space — the space bounded by the ligamentum flavum posteriorly, the spinal periosteum laterally, and dura anteriorly. Epidural anesthesia can be performed in the sacral hiatus (caudal anesthesia) or in the lumbar, thoracic, or cervical regions of the spine.

Epidural and Intrathecal Opiate Analgesia

Small quantities of opiate injected intrathecally or epidurally produce segmental analgesia. This observation led to the clinical use of spinal and epidural opiates during surgical procedures and for the relief of postoperative and chronic pain. As with local anesthesia, analgesia is confined to sensory nerves that enter the spinal cord dorsal horn in the vicinity of the injection. Presynaptic opioid receptors inhibit the release of **substance P** and other neurotransmitters from primary afferents, whereas postsynaptic opioid receptors decrease the activity of certain dorsal horn neurons in the spinothalamic tracts.

SAFE AND TOXIC DOSAGE OF LOCAL ANESTHETICS

Acute poisoning by local anesthetics is fairly common, especially with cocaine, and is apt to be fatal. Some accidents are due to excessive doses, through mistakes or faults of technique; but the susceptibility varies greatly, owing to differences of absorption and of destruction, and to varying response to the actions. Fright plays a considerable part.

Probably all commonly used local anesthetics have produced fatal accidents. These depend not only upon the absolute dose, but also on the rate of absorption, on idiosyncrasy, and on other conditions. The Local Anesthetic Committee of the American Medical Association has made recommendations that should minimize the accidents. The following are the most important: Procaine appears the safest of the more widely used local anesthetics and may be employed for subcutaneous and submucosal injections; but the concentration should not exceed 1%. Cocaine and butyn should not be injected under the skin or mucous membranes, but restricted to surface application. The total quantity of cocaine should not exceed 0.06 to 0.1 g (1 to 1½ grains). The patient should be recumbent, if the operation permits. With nervous patients, it is advisable to inject morphine 15 minutes before the local anesthetic, and to delay the start of the operation until 20 minutes after the injection of the local anesthetic. Urethral injections are especially dangerous, and should be avoided if there is trauma or stricture.

Strict attention should be given to prevent the confusion of procaine and cocaine. The solutions should be kept in different kinds of bottles. Cocaine solution may be distinguished by tinting with sodium fluorescein, 1 mg/g of cocaine. This would also show the concentration of the solution.

The **treatment of poisoning by local anesthetics** should begin with prevention and the selection, dosage, and technique of the administration; gross errors and carelessness have caused many deaths. The previous administration of a sedative, especially of the barbituric series, diminishes the risk, by suppressing the convulsions and their interference with respiration, so that animals survive one and a half to four times the ordinary fatal dose of cocaine or procaine, if administered hypodermically.

Clinically, 0.2 g of phenobarbital or 0.6 g of sodium barbital may be administered an hour before operation. Barbiturates do not prevent the direct circulatory collapse and depression of respiration that occur on intravenous injection of procaine, and are useless or harmful for either prophylaxis or treatment if the anesthetic agent is rapidly absorbed. However, if the symptoms develop slowly, the suppression of the convulsions is at least helpful.

If an overdose of the local anesthetic has been taken by mouth, evacuation and chemical antidotes are indicated. If the anesthetic was injected, absorption should be blocked by ligation, if possible. Whatever delays death is likely to save the life in view of the rapid destruction of these drugs. If the symptoms have developed, the head should be lowered, the convulsions arrested by intravenous injection of **sodium pentobarbital** or amytal, and artificial respiration instituted if necessary, with cardiac massage if the circulation fails. The use of epinephrine or morphine is not advised.

DESTRUCTION OF COCAINE *IN VITRO*; STERILIZATION

Long-continued boiling decomposes cocaine into **benzoyl-ecgonine** and methyl alcohol. It was therefore believed that solutions could not be sterilized by boiling. In fact, however, the decomposition, on boiling half an hour, is insignificant;

longer boiling only decreases the activity, because the decomposition products are merely inactive and not toxic. Any decomposition on boiling is probably due to alkalinity from the glass.

COCAINE, A "NERVE STIMULANT"

As a nerve stimulant, coca leaves have been used immemorially by the Peruvian and Bolivian natives. In 1853, it was stated (Wood and Bache, 1894):

> Coca leaves produces a gently excitant effect, with an indisposition to sleep, in these respects resembling tea and coffee; also that it will support the strength for a considerable time in the absence of food, but does not supply the place of nutriment, and probably in this respect also acts like the two substances referred to. The Indians, while chewing it, mixed with some alkaline substance, as the ashes of certain plants, or lime, pass whole days in traveling or working without food. It is, however, clearly proved that these leaves do not take the place of nutriment, but simply put off the sense of fatigue and hunger, the Indian making up at his evening meal for the day's abstinence. It is probable that they prevent hunger simply by their local benumbing influence upon the nerves of the stomach. Their moderate habitual use does not seem to be injurious, but the habit is said readily to grow upon the person, and finally the inveterate excessive coca-chewer can be recognized by his uncertain step, general apathy, sunken eyes surrounded by deep purple aureoles, trembling lips, green and crusted teeth, and excessively fetid breath, with peculiar blackness about the corners of the mouth. An incurable insomnia is apt to be developed, emaciation becomes extreme, dropsy appears, and even death results in a condition of general marasmus. When coca is taken in a single large dose it produces a condition of peculiar physical beatitude and calm, followed by a sensation of excessive power, which is affirmed to be accompanied by a real increase of physical ability.

In one case, where 900 grains of coca leaf were used by a physician named Mantegazza the following symptoms appeared:

> The ingestion of coca leaf caused great increase in the number of heart-beats, and a condition of intoxication resembling that produced by hasheesh. He was possessed by a feeling of intense joyousness, while a succession of visions and phantasmagoria, most brilliant in form and color, trooped rapidly before his eyes. He then passed into a condition of delirious excitement, which was succeeded by a deep sleep lasting three hours. (Culbreth, 1927)

Coca leaf can cause poisoning; in mild cases the ordinary symptoms have been great restlessness and nervous excitement, but no sense of beatitude, rather a condition of fear and terror. With this state come usually distinctly accelerated pulse, increased frequency of respiration, and, perhaps, muscular twitching or even mild convulsions. In the more severe cases of poisoning the symptoms vary; sometimes there have been nausea, vomiting, rapid almost imperceptible pulse, great perspiration, collapse with or without loss of consciousness; in other cases the pulse has been slow and feeble, and sometimes pronounced cyanosis, with slow or almost arrested respiration, has been the most alarming manifestation. The pupils are usually dilated, but are reported in some cases as "contracted." After very large doses, convulsions usually occur; they are often violent and epileptiform; and in many cases **opisthotonos** has been pronounced. Consciousness is usually lost, but sometimes it merges into a mania with hallucinations and delusions; the mania may become violent and even homicidal. Poisoning has followed both the internal administration and the local uses of the alkaloid. The occasional over-effects of small doses are quite remarkable; thus, four drops of a 2% solution in the eye produced in an old woman intoxication that persisted 4 days; eight drops of a 10% solution in the eye of a girl of 12 years caused violent poisoning; and even one drop of a 1% solution in the eye of a child 14 years old is said to have been followed by violent symptoms. A number of cases are on record in which one grain of the alkaloid given hypodermically has caused very severe fainting. Death is reported in several cases from the local use of the remedy, and 12 drops of a 4% solution given hypodermically to a girl of 11 caused death in 40 seconds. On the other hand, large doses have been recovered from: 22 grains by the mouth, 10 grains hypodermically, 5 grains hypodermically, and 6 grains hypodermically.

Cocaine Addiction

The elation and euphoria that cocaine induces in susceptible individuals tend to render it a "**habit drug**"; but it is more apt to be used for occasional "**jags**," by criminals, rather than daily, as is morphine. The treatment is therefore easier and more likely to be successful. However, it may be used continuously, as it was by the South American natives. (In East India it is combined with the chewing of **betel leaf**). Cocaine inebriation is active and sociable, in contrast to the dreamy state induced by morphine. It may induce a more or less maniacal condition, in which the individual becomes dangerous to others. The immediate effects include the phenomena of sympathetic stimulation, resembling hyperthyroidism: tachycardia, polypnea, mydriasis, exophthalmos, and fine tremors. Some individuals, especially women, react with marked erotic excitement. The stimulation is succeeded by depression, tremors, pallor, and sunken and unsteady eyes.

The chronic effects show rapid emaciation and severe psychic disturbances, insomnia, hallucinations, apathy, melancholia, and suicidal mania. The pupils are inconstant. Acne is common. It is claimed that considerable tolerance is acquired, so that the daily hypodermic consumption may reach 2.5 g, or even 10 g. Morphinists are relatively tolerant to cocaine. Sudden withdrawal leads to abstinence symptoms similar to morphine. With cocaine snuffing, where smaller amounts are used in intermittent debauches, the chronic effects, the craving, and the abstinence symptoms are proportionately less marked. Such patients develop atrophic rhinitis, with characteristic ulceration of the nasal fossae, also present in people who snuff heroin.

Uses of Cocaine in 18th Century

In melancholia, hysteria, epilepsy, spinal paralysis, insanity, diabetes, headache, typhoid state, opium habit, uterine inertia, vomiting of pregnancy, gastric irritability, cholera morbus, spermatorrhea, debility, poisoning by chloral hydrate, opium, or bromides. Locally, to burns, painful ulcers, fissures of anus, hay fever, sore throat, laryngitis, hemorrhoids, bronchitis, coryza, and in surgical operations; hypodermically in fingers, toes, small tumors — for amputation; for spinal anesthesia not as safe as novocaine, stovaine, eucaine; no more than ¾ g (0.045 g) should be applied at once.

Cocaine, a Psychostimulant

Not all users of cocaine become addicts. A key factor is the widespread availability of relatively inexpensive cocaine in the alkaloidal (free base, "**crack**") form suitable for smoking and the hydrochloride powder form suitable for nasal or intravenous use. Drug abuse in males occurs about twice as frequently as in females. However, smoked cocaine use is particularly common in young women of childbearing age, who may use cocaine in this manner as commonly as do males.

The half-life of cocaine in plasma is about 50 minutes, but inhalant (crack) users typically desire more cocaine after 10 to 30 minutes. Intranasal and intravenous uses also result in a high of shorter duration than would be predicted by plasma cocaine levels, suggesting that a declining plasma concentration is associated with termination of the high and resumption of cocaine seeking. This theory is supported by positron emission tomography (PET) imaging studies using C^{11}-labeled cocaine, which show that the time course of subjective euphoria parallels the uptake and displacement of the drug in the corpus striatum.

Cocaine Blocks Dopamine Transporter

The reinforcing effects of cocaine and cocaine analogues correlate best with their effectiveness in blocking the **dopamine transporter**, which leads to increased dopaminergic stimulation at critical brain sites. However, cocaine also blocks both norepinephrine (NE) and serotonin (5-HT) reuptake, and chronic use of cocaine produces changes in these neurotransmitter systems as measured by reductions in the neurotransmitter metabolites MHPG (**3-methoxy-4 hydroxyphenethyleneglycol**) and 5-HIAA (**5-hydroxyindoleacetic acid**).

Biosynthesis of Dopamine

Dopamine, a catecholamine, is synthesized in the terminals of dopaminergic neurons from tyrosine, which is transported across the blood–brain barrier by an active process (Figure 55). The rate-limiting step in the synthesis of dopamine is the conversion of L-tyrosine to L-dihydroxy-phenytalanine (L-dopa), catalyzed by the enzyme tyrosine hydroxylase, which is present within catecholaminergic neurons.

The major route for cocaine metabolism involves hydrolysis of each of its two ester groups. **Benzoylecgonine**, produced upon loss of the methyl group, represents the major urinary metabolite and can be found in the urine for 2 to 5 days

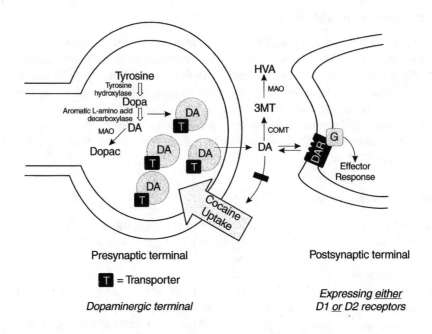

Figure 55 Dopamine (DA) is synthesized within neuronal terminals from the precursor tyrosine by the sequential actions of the enzymes tyrosine hydroxylase, producing the intermediary L-dihydroxyphenylalanine (Dopa), and aromatic L-amino acid decarboxylase. In the terminal, dopamine is transported into storage vesicles by a transporter protein (T) associated with the vesicular membrane. Release, triggered by depolarization and entry of Ca^{2+}, allows dopamine to act on postsynaptic dopamine receptors (DAR). Several distinct types of dopamine receptors are present in the brain, and the differential actions of dopamine on postsynaptic targets bearing different types of dopamine receptors have important implications for the function of neural circuits. The actions of dopamine are terminated by the sequential actions of the enzymes catechol-*O*-methyl-transferase (COMT) and monoamine oxidase (MAO), or by reuptake of dopamine into the terminal.

after a binge. As a result, benzoylecgonine tests are useful for detecting cocaine use; heavy users have been found to have detectable amounts of the metabolite in urine for up to 10 days following a binge.

Cocaine frequently is used in combination with other drugs such as heroin or alcohol. An important metabolic interaction occurs when cocaine and alcohol are taken concurrently. Some cocaine is transesterified to **cocaethylene**, which is equipotent to cocaine in blocking dopamine reuptake. Other risks of cocaine use, beyond the potential for addiction, involve cardiac arrhythmias, myocardial ischemia, myocarditis, aortic dissection, cerebral vasoconstriction, and seizures. Death from trauma also is associated with cocaine use. Pregnant cocaine users may experience premature labor and **abruptio placentae**. Reports of developmental abnormalities in infants born to cocaine-using women are confounded by prematurity, multiple drug exposure, and poor pre- and postnatal care. Cocaine has been reported to produce a prolonged and intense orgasm if taken prior to intercourse, and its use is associated with often compulsive and promiscuous sexual activity. Long-term cocaine use, however, often results in reduced sexual drive; complaints of sexual problems are common among cocaine users presenting for treatment. Psychiatric disorders, including anxiety, depression, and psychosis, are common in cocaine users who request treatment. Although some of these psychiatric disorders undoubtedly existed prior to the stimulant use, many develop during the course of the drug abuse.

Experienced users require more cocaine over time to obtain euphoria. Since cocaine typically is used intermittently, even heavy users go through frequent periods of withdrawal or "**crash**." The symptoms of withdrawal are seen in users admitted to the hospital.

Cocaine Withdrawal Symptoms and Signs

- Dysphoria, depression
- Sleepiness, fatigue
- Cocaine craving
- Bradycardia

At present, there is general agreement that no medication is yet available that can be used reliably in the treatment of cocaine addiction.

21

CHOCOLATE

Chocolate has antioxidant properties for low-density lipoproteins and hence could prevent heart disease. Foods and beverages derived from cocoa beans have been consumed by humans since A.D. 460. Cocoa pods from the cocoa tree (***Theobroma cacao***) are harvested and the beans removed and fermented. Dried and roasted beans contain about 300 chemicals including **caffeine**, **theobromine**, and **phenethylamine**. Chocolate liquor is prepared by finely grinding the nib of the cocoa bean and is the basis for all chocolate products. Cocoa powder is made by removing part of the cocoa butter from the liquor. **Bittersweet chocolate**, sometimes called **dark chocolate**, contains at least 15% chocolate liquor but may contain as much as 60% with the remainder **cocoa butter**, sugar, and other additives. **Milk chocolate** is the predominant form of chocolate consumed in the United States and typically contains 10 to 12% **chocolate liquor**.

The appeal of chocolate is universal, but the pleasures of eating chocolate products may perhaps be tempered by their fat and sugar content. However, in a series of human feeding studies it has been shown that the high proportion of **stearic acid** in the cocoa butter of chocolate does not adversely affect plasma lipids. Two recent reports of antioxidant activity have increased interest in the health aspects of chocolate: an *in vitro* low-density liposprotein (LDL) oxidation study and a short-term *in vivo* study. **Epicatechin**, the major monomeric polyphenol antioxidant in chocolate and an extract of chocolate liquor were both found to stimulate **cellular immune response** *in vitro*.

Polyphenol consumption as **flavonoids** has been shown to decrease the risk of heart disease in a cross-cultural epidemiological study. Most recently, an epidemiology study found that Harvard male graduates who ate a "moderate" amount of chocolate and other candy had a 36% lower risk of death compared with noncandy eaters. The authors speculate that it is the antioxidants present in the chocolate that provide a health benefit.

Vinson et al. (1999) investigated in detail the antioxidant properties of polyphenols in cocoa and two types of chocolate using an *in vitro* model of heart disease. These authors found that **epicatechin** and **catechin** were found in all of the six samples analyzed by high-performance liquid chromatography (HPLC). These two monomeric polyphenols were minor components of the chocolates, ranging from 1.97 to 2.76 mol% of the phenols in the milk chocolate, from 2.98 to 5.48 mol%

in the dark chocolate, and from 5.45 to 6.11 mol% in the cocoa powder. Assuming a 35% content of fat and moisture, the concentrations of the catechins are 15 to 16 mg/100 g in milk chocolate, 48 to 137 mg/100 g in dark chocolate, and 296 to 327 mg/100 .g in cocoa powder. A recent HPLC analysis of single samples of European chocolate found 16 mg/100 g in milk chocolate and 54 mg/100 g in dark chocolate. Another report found 300 mg/100 g in cocoa powder, which also agrees well with these results. The total concentration of the two catechins in chocolate correlates well with the total polyphenols as measured by the Folin method, with a Pearson correlation coefficient of 0.9326, $p < 0.01$.

The quality of the phenol antioxidants was assessed by Vinson et al. (1999) using the IC_{50} for LDL + VLDL (very low density lipoprotein) oxidation, with smaller values indicating a higher quality. Quality of the antioxidants was due to free radical–scavenging activity and not chelation as the concentration of polyphenols for 50% inhibition was $< 1 \mu M$ and cupric ion 25 μM in the oxidation medium. There was less percent variation within the groups for this parameter than for the total polyphenol content. The quality order was dark chocolate > cocoa > milk chocolate. The dark chocolate and cocoa were significantly different from the milk chocolate ($p < 0.05$), but not from each other. There was no correlation between the quantity of phenolic antioxidants in the chocolate and the quality as measured by IC_{50}, $p > 0.05$. There was also no correlation between the amount of **epicatechin** and **catechin** in the chocolates and the IC_{50}. **Theobromine** and **caffeine**, two major ingredients in chocolate, were neither pro-oxidants nor antioxidants.

Some interesting comparisons can be made between the antioxidant properties of chocolate and other foods and beverages. The average total polyphenol content of dry, defatted chocolate was 133.9 μmol/g. The foods with the highest dry weight total phenols were **beet**, 53.4 μmol/g, apple 6.4 μmol/g, and **black tea** 12.0 mol/ml. Chocolate has a higher polyphenol content than the 23 vegetables and the several fruits and beverages studied. The IC_{50} average for chocolate products is 0.33 μM and is superior to the average for vegetables, 0.69 μM. The **phenol antioxidant index** (PAOXI) for chocolate averaged 378×10^3, whereas the highest vegetable, asparagus, was 144×10^3 and the highest commonly consumed beverage was black tea, 31.5×10^3.

An important antioxidant parameter with respect to head disease is the **lipoprotein-bound antioxidant activity**. The value for epicatechin, a polyphenol component of chocolate, was 63% greater than the control average, increase in lag time for the three representative chocolate products was 41%. All beverages, such as wine and tea, that had this antioxidant activity were also found to produce an increase in lag time after ingestion.

Thus, chocolates contain both a high quantity and a high quality of phenol antioxidants. When expressed as catechin equivalents on a fresh-weight basis, the average chocolate contains 28.7 mg/g, assuming a 35% content of fat and moisture. A serving (40 g of chocolate is the reference amount commonly consumed per eating occasion) of milk chocolate, the most popular type of chocolate, provides 394 mg of polyphenol antioxidants and a serving of dark chocolate 951 mg. For comparison, the average black tea contains 943 mg/240 ml serving and red wine

431 mg/240 ml. The average hot cocoa mix made according instructions provides 45 mg of polyphenols in a 240-ml serving. Using 5 g of the average cocoa powder (35% fat and moisture), a homemade serving of hot cocoa has 211 mg of polyphenols as catechin. Therefore, polyphenols from chocolate may provide additional antioxidant protection for LDL and thus be beneficial for preventing heart disease (Vinson et al., 1999).

22

COLCHICINE AND GOUT

Colchicine (*Colchicum autumnale*) is useful in:

- Acute attacks of gout
- Familial Mediterranean fever
- Amyloidosis
- Psoriasis
- Behcet's syndrome

The focus of this chpater is on effects on gout.

HISTORY

Colchicine is an alkaloid found in the corm and seeds of **Colchicum autumnale** (**autumn crocus**, **meadow saffron**), a plant so named because it grew in Colchis in Asia Minor (Figure 56). It is a bulbous liliaceous plant that blooms during the fall in the wet meadows of central and southern Europe. Although the poisonous action of colchicum was known to **Dioscorides**, preparations of the plant were first recommended for pain of articular origin by **Alexander of Tralles** in the 6th century A.D. Colchicum was known to Byzantine physicians as hermodactyl (**finger of Hermes**); it was also called *articulorum* (**soul of the joints**). Colchicum was introduced for the therapy of acute gout by **Baron Anton Von Storck** in 1763, and its specificity for this syndrome soon resulted in its incorporation in a number of "**gout mixtures**" popularized by charlatans. **Benjamin Franklin**, himself a sufferer from gout, is reputed to have introduced colchicum therapy in the United States. The alkaloid colchicine was isolated from **colchicum** in 1820 by Pelletier and Caventou (Goodman and Gilman, 1955).

GOUT AND HYPERURICEMIA

Gout is a hyperuricemic state (>6 mg/dl) that is effectively diagnosed through the detection of **monosodium urate crystals** in the synovial fluid of the involved joint. Conditions causing hyperuricemia include the excessive synthesis of uric acid, the excessive synthesis of purine-precursor to uric acid, a high dietary intake of purine

Figure 56 *Colchicum autumnale.*

(shellfish, organ meat, anchovies, and wild game), diminished renal excretion of uric acid, and tissue destruction following injury or therapeutic irradiation.

Numerous agents, when used in therapeutic doses, can also cause hyperuricemia. This includes an analgesic dose of aspirin, thiazide diuretics, nicotinic acid, chronic consumption of alcohol, and antineoplastic agents.

If left untreated, the hyperuricemic state may precipitate an acute attack of gout, which first appears in metatarsal phalangeal joints. Ultimately, tophaceous deposits form in the joints and soft tissues such as the kidneys. The hyperuricemic state may be corrected either by inhibiting the synthesis of uric acid by **allopurinol** or by enhancing the elimination of uric acid by uricosuric agents.

ALLOPURINOL

Allopurinol (Zyloprim) reduces the synthesis of uric acid by inhibiting the activity of xanthine oxidase, according to the scheme shown in Figure 57. The reduction in the uric acid pool occurs slowly. Because **xanthine** and **hypoxanthine** are more soluble than uric acid, they are easily excreted.

Allopurinol is used not only in treating the hyperuricemia associated with gout, but also in the secondary hyperuricemia associated with the use of antineoplastic

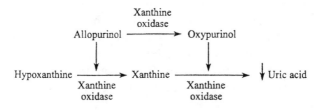

Figure 57 Allopurinol reduces the pool of uric acid.

agents. However, allopurinol may interfere with the metabolism of antineoplastic agents such as **azathioprine** and **6-mercaptopurine**.

URICOSURIC AGENTS

The most commonly used uricosuric agents are **probenecid** (Benemid) and **sulfinpyrazone** (Anturane). In low doses, these agents block tubular secretion, but, at higher doses, they also block the tubular resorption of uric acid. Because the solubility of uric acid is increased in alkaline urine, the administration of sodium bicarbonate may at times be advantageous for offsetting this condition. In addition, probenecid and sulfinpyrazone inhibit the excretion of agents such as aspirin, penicillin, ampicillin, and indomethacin. Although probenecid and sulfinpyrazone may be coadministered, neither should be given with aspirin, as their uricosuric effects will then be nullified.

TREATMENT OF COLCHICINE FOR GOUT

Colchicine (an alkaloid obtained from **meadow saffron** or autumn crocus) may be used both diagnostically to ascertain the presence of gout and prophylactically to prevent its further occurrence. Usually, 0.5-mg oral doses of colchicine are given hourly until either the therapeutic effects appear or the side effects develop. In addition to colchicine, phenylbutazone, indomethacin, adrenocorticotropic hormone (ACTH), and steroidal anti-inflammatory agents may be used to treat the acute attack of gout.

Pharmacological Properties

The anti-inflammatory effect of colchicine in acute gouty arthritis is relatively selective for this disorder. Colchicine is only occasionally effective in other types of arthritis; it is not an analgesic and does not provide relief of other types of pain.

Colchicine is an **antimitotic agent** and is widely employed as an experimental tool in the study of cell division and function.

Effect in Gout

Colchicine does not influence the renal excretion of uric acid or its concentration in blood. By virtue of its ability to bind to tubulin, colchicine interferes with the function of the mitotic spindles and causes depolymerization and disappearance of the fibrillar microtubules in granulocytes and other motile cells. This action is apparently the basis for the beneficial effect of colchicine, namely, the inhibition of the migration of granulocytes into the inflamed area and a decreased metabolic and phagocytic activity of granulocytes. This reduces the release of lactic acid and proinflammatory enzymes that occurs during phagocytosis and breaks the cycle that leads to the inflammatory response.

Neutrophils exposed to urate crystals ingest them and produce a glycoprotein, which may be the causative agent of **acute gouty arthritis**. Injected into joints, this substance produces a profound arthritis that is histologically indistinguishable from that caused by direct injection of urate crystals. Colchicine appears to prevent the elaboration by leukocytes of this glycoprotein.

Effect on Cell Division

Colchicine can arrest plant and animal cell division *in vitro* and *in vivo*. Mitosis is arrested in metaphase, due to failure of spindle formation. Cells with the highest rates of division are affected earliest. High concentrations may completely prevent cells from entering mitosis, and they often die. The action also is characteristic of the **vinca alkaloids** (**vincristine** and **vinblastine**), **podophyllotoxin**, and **griseofulvin**.

Other Effects

Colchicine inhibits the release of histamine-containing granules from mast cells, the secretion of **insulin** from β-cells of pancreatic islets, and the movement of **melanin** granules in melanophores. Although it is questionable whether these effects occur at clinically achieved concentrations of colchicine, all of these processes may involve the translocation of granules by the microtubular system.

Colchicine also exhibits a variety of other pharmacological effects. It lowers body temperature, increases the sensitivity to central depressants, depresses the respiratory center, enhances the response to sympathomimetic agents, constricts blood vessels, and induces hypertension by central vasomotor stimulation. It enhances gastrointestinal activity by neurogenic stimulation but depresses it by a direct effect, and alters neuromuscular function.

Pharmacokinetics and Metabolism

Colchicine is rapidly absorbed after oral administration, and peak concentrations occur in plasma by 0.5 to 2 hours. Large amounts of the drug and metabolites enter the intestinal tract in the bile and intestinal secretions, and this fact, plus the rapid turnover of intestinal epithelium, probably explains the prominence of intestinal manifestations in colchicine poisoning. The kidney, liver, and spleen also contain high concentrations of colchicine, but it is apparently largely excluded from heart, skeletal muscle, and brain. The drug can be detected in leukocytes and in the urine for at least 9 days after a single intravenous dose.

Colchicine is metabolized to a mixture of compounds *in vitro*. Most of the drug is excreted in the feces; however, in normal individuals, 10 to 20% of the drug is excreted in the urine. In patients with liver disease, hepatic uptake and elimination are reduced and a greater fraction of the drug is excreted in the urine.

Toxic Effects

The most common side effects reflect the action of colchicine on the rapidly proliferating epithelial cells in the gastrointestinal tract, especially in the jejunum. Nausea, vomiting, diarrhea, and abdominal pain are the most common and earliest untoward effects of colchicine overdosage. To avoid more serious toxicity, administration of the drug should be discontinued as soon as these symptoms occur. There is a latent period of several hours or more between the administration of the drug and the onset of symptoms. This interval is not altered by dosage or route of administration. For this reason, and because of individual variation, adverse effects may be unavoidable during an initial course of medication with colchicine. However, since patients often remain relatively consistent in their response to a given dose of the drug, toxicity can be reduced or avoided during subsequent courses of therapy by reducing the dose. The drug is equally effective when given intravenously; the onset of the therapeutic effect may be faster, and the gastrointestinal side effects may be almost completely avoided.

In acute poisoning with colchicine, there is hemorrhagic gastroenteritis, extensive vascular damage, nephrotoxicity, muscular depression, and an ascending paralysis of the central nervous system.

Colchicine produces a temporary leukopenia that is soon replaced by a leukocytosis, sometimes due to a striking increase in the number of basophilic granulocytes. The site of action is apparently directly on the bone marrow. Myopathy and neuropathy also have been noted with colchicine treatment, especially in patients with decreased renal function. Long-term administration of colchicine entails some risk of agranulocytosis, aplastic anemia, myopathy, and alopecia; azoospermia has also been described.

Therapeutic Uses

Colchicine provides dramatic relief from acute attacks of gout. The effect is sufficiently selective that the drug has been used for diagnostic purposes, but the test is not infallible. Colchicine also has an established role to prevent and to abort acute attacks of gout. However, its toxicity and the availability of alternative agents that are less toxic have substantially lessened its usefulness.

Acute Attacks

When colchicine is given promptly within the first few hours of an attack, fewer than 5% of patients fail to obtain relief. Pain, swelling, and redness abate within 12 hours and are completely gone in 48 to 72 hours. Although for many years colchicine was administered orally, current practice is to administer the drug intravenously. Although a number of regimens have been used, a single dose of 2 mg, diluted in 10 to 20 ml of 0.9% sodium chloride solution, usually is adequate; a total dose of 4 mg should not be exceeded. To avoid cumulative toxicity, treatment with colchicine should not be repeated within 7 days.

Great care should be exercised in prescribing colchicine for elderly patients, and for those with cardiac, renal, hepatic, or gastrointestinal disease. In these patients and in those who do not tolerate or respond to colchicine, indomethacin or another nonsteroidal anti-inflammatory agents (NSAID) is preferred.

Prophylactic Uses

For patients with chronic gout, colchicine has established value as a prophylactic agent, especially when there is frequent recurrence of attacks. Prophylactic medication also is indicated upon initiation of long-term medication with allopurinol or the uricosuric agents, since acute attacks often increase in frequency during the early months of such therapy.

The prophylactic dose of colchicine depends upon the frequency and severity of prior attacks. As small an oral dose as 0.5 mg two to four times a week may suffice; as much as 1.8 mg per day may be required by some patients. Colchicine should be taken in larger abortive doses immediately upon the first twinge of articular pain or the appearance of any prodrome of an acute attack. Before and after surgery in patients with gout, colchicine should be given for 3 days (0.5 or 0.6 mg, three times a day); this greatly reduces the very high incidence of acute attacks of gouty arthritis precipitated by operative procedures.

OTHER USES OF COLCHICINE

Daily administration of colchicine is useful for the prevention of attacks of **familial Mediterranean fever (familial paroxysmal polyserositis)** and for prevention and treatment of **amyloidosis** in such patients. Colchicine appears to benefit patients with primary **biliary cirrhosis** in terms of improvement of liver function tests and perhaps of survival. Colchicine also has been employed to treat a variety of skin disorders, including **psoriasis** and **Behcet's syndrome**.

23

COMPOSITAE

Compositae (Mexican Asteraceae) possess numerous properties and exhibit numerous actions, which include but are not limited to:

- Antimigraine efficacy
- Antiulcer efficacy
- Anti-inflammatory property
- Antitumor action
- Antiepileptic property
- Antihelminthic property
- Antiseptic action
- Antitussive property
- Antidiabetic action
- Cardiotonic
- Diuretic
- Diaphoretic
- Analgesic

INTRODUCTION

There are 380 genera with 3000 species recognized in the **Asteraceae** in Mexico. The most important source of knowledge about indigenous plant use in historical Mexico is the **Florentine Codex** compiled by the Franciscan monk **Fray Bernardino de Sahagun**. Also important are the work of the Spanish physician **Francisco Hernandez** and an herbal guide written in Nahuatl by the Aztec healer **Martin de la Cruz** from Tezcoco that was translated into Latin by **Juan Badiano** and sent to the King of Spain, Philip II, in 1552 (see Heinrich et al., 1998).

SESQUITERPENE LACTONES

Sesquiterpene lactones (SQLs), one of the major classes of chemical compounds reported from the Asteraceae, are a large and diverse group of biologically active plant constituents that have been reported from ten families of flowering plants; however, the greatest number are derived from the Asteraceae, with over 3000 reported structures. Some of the important medicinal plants from this family, such

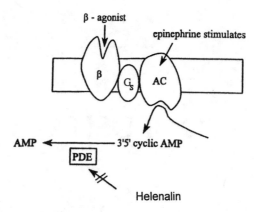

Figure 58 Amrinone exerts positive inotropic and vasodilating actions. AMP = adenosine monophosphate; AC = adenylate cyclase; PDE = phosphodiesterase; G_s = stimulatory guanine nucleotide-binding regulatory protein.

as *Arnica montana, Artemisia annua,* and *Tanacetum parthenium,* contain SQLs as the active constituents. SQLs are commonly present in large quantities (>2% dry wt of leaves); these SQLs are usually comparatively nontoxic but have an intensely bitter taste that presumably deters herbivores. A few SQLs are highly toxic (e.g., **repin, helenalin**), but many have anti-inflammatory activity and relax smooth muscle *in vitro*. SQLs have antitumor activity (e.g., **Parthenin**), antiulcer activity (e.g., *Artemisia douglasiana*), and cardiotonic activity (e.g., *Helenium autumnale*) as they are ionotropic in nature. Helenalin inhibits the activity of phosphodiesterase, increases the concentration of cyclic adenosine monophosphate, causes Ca^{2+} influx, and subsequently enhances the contractility of myocardium (Figure 58).

SESQUITERPENE-INDUCED NEUROTOXICITY

Ingesting a large amount of **Centaurea repens**, a member of Asteraceae containing a copious amount of sesquiterpene by horses causes "**chewing disease.**" Neuropathological examination of the brain from an intoxicated horse revealed bilateral necrosis of the anterior globus pallidus and zona reticulata of the substantia nigra.

TANACETUM PARTHENIUM (FEVERFEW)

Feverfew is widely consumed in England as a remedy of arthritis and migraine. Feverfew contains **Parthenolide**, which is a member of sesquiterpene. Parthenolide inhibits the activity of **prostaglandin synthetase**. It also inhibits **platelet aggregation** and alter **serotonin release** (Figure 59).

Compositae behave like aspirin that has an anti-inflammatory action as well as antirheumatic and antiarthritic effects, and may therefore be used in the treatment of rheumatic fever. However, aspirin cannot alter the cardiac lesion and other visceral effects of the disease. Aspirin is extremely effective in managing

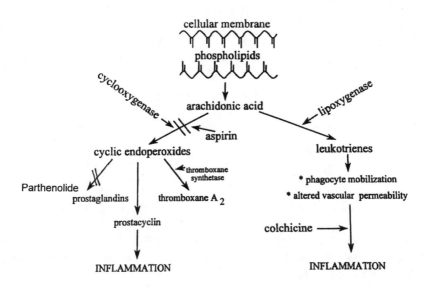

Figure 59 Sites of action of aspirin, colchicine, and parthenolide.

rheumatoid arthritis and allied diseases involving the joints, such as **ankylosing spondylitis** and **osteoarthritis**. It is thought that aspirin and indomethacin exert their anti-inflammatory effects by inhibiting prostaglandin synthesis through the inhibition of **cyclooxygenase**. The presynthesized prostaglandins are released during a tissue injury that fosters inflammation and pain. Furthermore, aspirin reduces the formation of prostaglandin in the platelets and leukocytes, which is responsible for the reported hematological effects associated with aspirin (see Figure 59).

Parthenolide may have a similar mechanism of action as seen following administration of **sumatriptan**, which is an effective analgesic and antimigraine drug. Sumatriptan have a selective agonist for serotonin receptor subtype, causing vasoconstriction of cranial arteries.

The introduction of **sumatriptan** in the therapy of migraine led to significant progress in preclinical and clinical research on migraine. At the scientific level, the selective pharmacological effects of sumatriptan at **serotonin** (5-HT$_1$) receptors have led to new insights into the pathophysiology of migraine. At the clinical level, sumatriptan is an effective acute antimigraine agent. Its ability to decrease, rather than exacerbate, the nausea and vomiting of migraine appears to be an important advance in the treatment of the condition.

The significance of one or more 5-HT$_1$ receptors in migraine pathophysiology relates to the observation that both 5-HT$_{1B}$ and 5-HT$_{1D}$ receptors serve as "autoreceptors," presynaptic receptors that modulate neurotransmitter release from neuronal terminals. Conceivably, 5-HT$_1$ agonists may block the release of proinflammatory neurotransmitters at the level of the nerve terminal in the perivascular space. It has been shown that the antimigraine drugs such as **ergotamine**, **dihydroergotamine**, and **sumatriptan** are able to block the development of **neurogenic plasma extravasation** in dura mater that follows depolarization of perivascular axons following **capsaicin** injection or unilateral electrical stimulation of the trigeminal nerve. The ability of potent 5-HT$_1$ agonist such as sumatriptan or maybe parthenolide to inhibit endogenous transmitter release in the perivascular space could account for their efficacy in the acute treatment of migraine.

OTHER ASTERACEAES

The therapeutic efficacies of other asteraceaes containing SQLs are hereby summarized.

Achillea

The Achillea genus is distributed in temperature regions of the Northern Hemisphere, especially in Europe and Asia. *Achillea millefolium* is a species frequently used in Europe and America alike. Many species of this genus are rich in essential oil, SQLs, and flavonoids. It is used internally as an **antitussive agent** and externally as an **analgesic**.

- *Achillea fragantissima.* **Cirsiliol**, a flavone isolated from *A. fragantissima,* causes a concentration-dependent relaxation of isolated ileum, interferes with calcium channels, and thus inhibits Ca^{2+} influx from the extracellular compartment. Its therapeutic efficacy is not known.

Artemisia

The genus *Artemisia* contains more than 200 species that have been reported from many parts of the world and has been selected for remedies by cultures throughout the world for its medicinal properties. The reported active constituents of the genus include all the types of common natural products of the Asteraceae: SQLs, polyacetylenes, flavonoids, essential oils (monoterpenes), **diterpenes**, **triterpenes**, and **coumarins**. Their therapeutic efficacies are not known.

- *Artemisia abrotanum.* Four flavonols isolated from *A. abrotanum* reportedly possess spasmolytic activity. These flavonols show a dose-dependent relaxing effect on the **carbacholine**-induced contraction of trachea. Its therapeutic efficacy is not known.
- *Artemisia herba-alba. Artemisia herba-alba* is widely used for the treatment of diabetes mellitus.
- *Artemisia verlotorum.* Water-alcoholic extract showed a protective effect against experimental convulsions elicited by various agents, as well as analgesic and hypothermic actions. *Artemisia verlotorum* is used as an anticonvulsant and analgesic.
- *Artemisia monosperma.* The relaxant effects of **7-O-methyleriodictyol**, a flavone isolated from the aerial parts of *A. monosperma,* on smooth muscle contraction has been studied. The inhibition of contraction induced by known therapeutic agents such as acetylcholine and **oxytocin** are consistent with the use of this plant in the treatment of certain gastrointestinal disorders.

- *Artemisia tridentata.* *Artemisia tridentata* is frequently used as an **anthelminthic**, **antiseptic**, and analgesic. Many of the compounds identified in this species have notable pharmacological activities, particularly the monoterpenes, SQLs, coumarins, and flavonoids.

Baccharis

Baccharis is a large, strictly American genus with approximately 350 species. Only a few are important medicinal plants, and their uses seem to be rather diverse. Some of the principal medicinal species are discussed below.

- *Baccharis conferta.* In the region of Veracruz, Mexico, *B. conferta* is used to treat stomachaches.
- *Baccharis glutinosa.* **Axixtlacotl** was used as a diuretic, against fevers, as a diaphoretic, and to remove blotches from the face. Tea made from the leaves is taken to lose weight and as a contraceptive. The tea is also used to stop blood loss after giving birth. The leaves, heated over coals, are placed on the head as a remedy for headache or on a sore area of the body.
- *Baccharis salicifolia.* The green leaves/leafy branches of *B. salicifolia* are applied externally as a remedy for inflammation, diarrhea, and dysentery.
- *Baccharis saathroides.* A tea made by boiling the twigs of *B. saathroides* is taken as a remedy for colds. The same tea is rubbed on sore muscles for relief.
- *Baccharis serraefolia.* *Baccharis serraefolia* is a popular form of treatment for various gastrointestinal illnesses.
- *Baccharis multiflora, Baccharis* **spp.** An infusion prepared from the leaves of *B. multiflora, Baccharis* spp. is said to be effective against catarrhs and was used for urinary problems in 19th century Mexico. Two unidentified species of this genus — **quappatli** and **malinalli** — are listed in the **Florentine Codex** as treatments for blotched face (applied internally) and for worms in the eyes and crab lice in the eyelids (espoecially applied externally), respectively.
- *Baccharis trinervis.* *Baccharis trinervis* is used by the Huastec in the treatment of high fever, edema, sores, and muscle cramps. It is also applied in the case of dizziness and lack of blood. The Huastec regard it as effective if one feels sleepy and if there is insufficient milk production. In Veracruz, a preparation made from the leaves is used for the treatment of typhoid fever.
- *Baccharis vaccinoides.* *Baccharis vaccinoides* is one of the most popular remedies from Chiapas for gastrointestinal disorders.

Elephantopus

This tropical genus is particularly well known because one species, ***Elephantopus scaber***, is now a widely distributed, invasive herb in many countries. There are approximately one dozen species of these herbaceous perennials in North and South America.

■ *Elephantopus spicatus* [syn.: *Pseudelephantopus spicatus* **C.F. Baker**, **Vernonieae**]. The leaves of *E. spicatus* are used for the treatment of cough and headache. Applied topically, they are employed as an antipyretic, for the treatment of erysipelas, skin infections, and measles. A preparation made from the roots is taken as a remedy for colic; the whole plant helps against diarrhea. It is one of the most popular cough remedies of middle America.

Helenium

This American genus has yielded several ornamental plants (e.g., *Helenium autumnale*) but is also known for its toxic effects on herbivores. *Helenium* genus contains about 40 species of annual and perennial herbs. *Helenium mexicanum* is taken orally if a person is very sick or to treat a stomachache, faint heartbeat, throbbing temples, quivering nerves, fever, or pus in the genital organ.

■ *Helenium quadridentatum.* Is used for treatment of fever, catarrh, and testicular inflammations and as a diuretic and insecticidal.

Montanoa

■ *Montanoa tomentosa.* Most Asteraceae are herbs or little shrubs. The trees of the genus *Montanoa* (**tree daisy**) therefore stand out as very unusual plants. They also are very showy during the flowering season. But the genus also includes one of the most important medicinal plants of Mexico.

Packera

■ *Packera candissima.* The whole plant *P. candissima* is used as a tea for kidney ailments; as a general medicine; and to cure sores, ulcers, and vaginal ailments. Similar uses are reported for **P. bellidifolia**.

Parthenium

Parthenium is a small genus of approximately 16 species of shrubs, herbaceous perennials, and annuals. A well-known species of this genus is **guayule** (*P. argentatum*), which is considered an alternative source for rubber and which was an emergency rubber during World War II. All species of this genus but especially **P. hysterophorus** are known to cause severe contact dermatitis.

■ *Parthenium Hysterophorus.* Is used medicinally by many indigenous and mestizo groups of Mexico as a remedy for malaria, for neuralgia, and as a vermifuge.

Tithonia

Tithonia is a small genus of fewer than ten species. The Mexican and Central American genus *Tithonia* has yielded some ornamental plants (e.g., ***T. rotundi-folia***), and at least one species, ***T. diversifolia***, is now a pantropically distributed bad weed. It is used to treat malaria (see Heinrich et al., 1998).

24

CURARE AND NEUROMUSCULAR BLOCKADE

Curare is a South American arrow poison causing muscular relaxation and used in anesthesiology.

Curare, Woorara, is a vegetable extract obtained from various members of the **Strychnos family**, also from *Paullinia curare* and other plants. It is used in South America as an arrow poison under the names **Caroval** and **Vao**. Its active principle is the very poisonous alkaloid *Curarine*, $C_{18}H_{35}N$.

CURARE

Curare, an arrow poison used by the native people of South America, was first brought to the attention of Elizabeth of England by **Sir Walter Raleigh**. The remarkable ability of the drug to induce paralysis of the muscles led to its early use by **Claude Bernard**, **Kölliker**, **Langley**, and subsequent physiologists to study the mechanism of neuromuscular activity. However, the uncertain supplies of the drug, its variability, and the effects of its contaminants prevented its therapeutic application.

Curare is obtained from various members of the **Menisperm** or **moonseed** family, particularly from the vines of *Chondodendron tomentosum* and other members of the *Chondodendron* genus as well as from various species of *Strychnos* (Figure 60).

The earliest available preparations, prepared as infusions and concentrated to a syrup by the native people, were designated as **calabash** (gourd), tubo-(bamboo), or pot (clay pot) curare depending upon the containers in which the drug was packaged. Curare is obtained from the upper regions of the Amazon River, the Orinoco basin, and the eastern slopes of the Ecuadorian plateau. The term *curare* is derived from the Indian name (**woorari**, **urari**) for poison (Grollman, 1962).

Curare or **South American arrow poison** varies in composition among Indian tribes. However, the bark of one or more species of *Strychnos* apparently is always used in its preparation. *Strychnos castelnaei* Weddell, *S. toxifera* Bentham, *S. Crevauxii* G. Planchon, and *Chondodendron tomentosum* Ruiz Pavon are commonly employed.

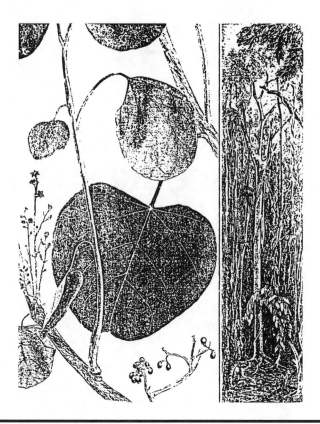

Figure 60 Left, *Chondodendron tomentosum* from which curare is derived; right, the rain forest of the Amazon, the habitat of the plant.

The young bark is scraped off the plants, mixed with other items, boiled in water, and strained, or extracted by crude percolation with water, then evaporated to a paste over the fire or in the sun. The imported article is a brownish or black, shiny, resinoid, very bitter mass; readily soluble in cold water and in dilute alcohol.

The drug contains several alkaloids: **curarine** and **protocurarine** give the typical effects. Curine is devoid of muscle-nerve action, but first stimulates, then depresses the heart action. Curare has a definite paralyzing action on voluntary muscle-nerve endings. Hypodermic injection of a suitable dose causes a total loss of motion, first of the voluntary, then of the respiratory muscles, and death results from respiratory failure. Therapeutically, the drug is a powerful antispasmodic, used in the convulsions of strychnine poisoning and of tetanus; as an adjunct to shock therapy in neuropsychiatry, and to promote muscular relaxation in surgical anesthesia. It is an important laboratory item for paralyzing the skeletal muscles of test animals.

SKELETAL MUSCLE RELAXANTS

Neuromuscular blocking agents may be used to diagnose myasthenia gravis, facilitate endotracheal intubation, relieve laryngeal spasm, provide relaxation during brief diagnostic and surgical procedures, prevent bone fracture in electroconvulsive therapy, produce apnea and controlled ventilation during thoracic surgery and neurosurgery, reduce muscular spasticity in neurological diseases (multiple sclerosis, cerebral palsy, or tetanus), and reduce the muscular spasm and pain resulting from sprains, arthritis, myositis, and fibrositis. Skeletal muscle relaxants can be classified into four categories:

1. **Depolarizing agents**
   Succinylcholine chloride (Anectine, Quelicin, Suxcert, Sucostrin)
2. **Nondepolarizing or competitive-blocking agents**
   Tubocurarine chloride (Tubarine)
   Pancuronium (Pavulon)
   Metocurine
   Alcuronium
   Fazadinium
   Atracurium
   Vecuronium
   Pipecuronium
   Doxacurium
   Mivacurium
3. **Direct-acting relaxants**
   Dantrolene sodium (Dantrium)
4. **Centrally acting muscle relaxants**
   Chlordiazepoxide (Librium)
   Diazepam (Valium)
   Baclofen (Lioresal)

THE NEUROMUSCULAR JUNCTION

The neuromuscular junction consists of a motor nerve terminal and skeletal muscle motor end plate, separated by a synaptic cleft that is filled with extracellular fluid (Figure 61).

NICOTINIC CHOLINERGIC RECEPTORS

Acetylcholine and agents acting at the autonomic ganglia or the neuromuscular junctions interact with nicotinic cholinergic receptors to initiate the end plate potential in muscle or an excitatory postsynaptical potential in nerve. The nicotinic receptor in skeletal muscle is a pentamer composed of four distinct subunits.

MECHANISM OF ACTION
OF NEUROMUSCULAR BLOCKING AGENTS

Neuromuscular agents are classified as either depolarizing agents (e.g., succinylcholine) or nondepolarizing agents (e.g., tubocurarine). Succinylcholine has dual modes of action in that it possesses two phases of blocking action: depolarization and desensitization.

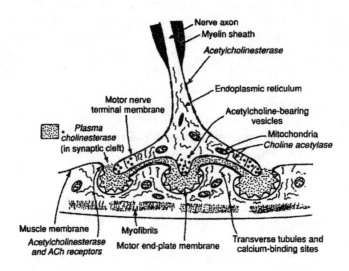

Nerve axon
Myelin sheath
Acetylcholinesterase
Endoplasmic reticulum
Motor nerve
terminal membrane
Acetylcholine-bearing
vesicles
Plasma
cholinesterase
(in synaptic cleft)
Mitochondria
Choline acetylase
Muscle membrane
Acetylcholinesterase
and ACh receptors
Myofibrils
Motor end-plate membrane
Transverse tubules and
calcium-binding sites

Figure 61 *Botulinum* poisoning causes progressive parasympathetic and motor paralysis. The toxin exerts its effects by inhibiting the release of acetylcholine.

Phase I. Blockade (Depolarization): Succinylcholine, like acetylcholine, interacts with the cholinergic receptors at the end plate region of the muscle, resulting in depolarization of the chemically excitable membrane. This, in turn, creates local action potentials, spreading them to and depolarizing the adjacent excitable membranes, finally culminating in a muscle contraction, or fasciculation, which is an uncoordinated muscle contraction. However, unlike acetylcholine, succinylcholine is not metabolized by acetylcholinesterase, and hence causes persistent depolarization of the end plate. The continuous presence of succinylcholine leads to inexcitability of the membrane adjacent to the end plate, resulting in neuromuscular blockade, which is not reversed by the administration of cholinesterase inhibitors. In fact, agents such as **neostigmine** may even prolong neuromuscular blockade.

Phase II. Blockade (Desensitization): In time and as the result of repeated administrations of succinylcholine, depolarization no longer promotes the formation of the neuromuscular blockade (desensitization). The blockade is then changed in character and becomes antagonized by neostigmine through the operation of an ill-defined mechanism that is very distinct from competitive blockade.

Agents such as tubocurarine and pancuronium compete with acetylcholine for the cholinergic receptors at the end plate (Figure 62). They combine with the receptors but do not activate them. Competitive or nondepolarizing agents are antagonized by neostigmine. Nondepolarizing muscle relaxants can also be classified according to their duration of action, as shown in Table 28.

Dantrolene, a hydantoin derivative, reduces the contraction of skeletal muscle directly on the muscle and not at the neuromuscular junction. It is thought to reduce the amount of calcium released and hence prevent excitation–contraction

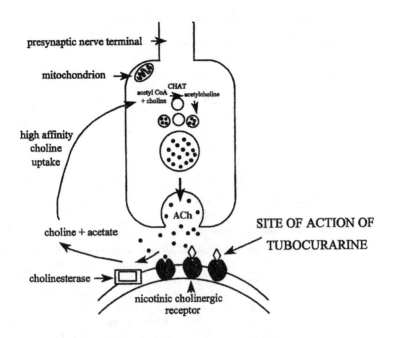

presynaptic nerve terminal

mitochondrion

CHAT
acetyl CoA → acetylcholine
+ choline

high affinity
choline
uptake

ACh

choline + acetate

cholinesterase

nicotinic cholinergic
receptor

SITE OF ACTION OF
TUBOCURARINE

Figure 62 Tubocurarine competes with acetylcholine (ACh) at cholinergic receptor. CHAT = choline acetylase; CoA = coenzyme A

coupling (Figure 63). Its usefulness in the treatment of anesthetic-induced malignant hyperthermia may be due to its calcium-related uncoupling actions.

Chlordiazepoxide and **diazepam** augment presynaptic inhibition by enhancing the release of γ-aminobutyric acid, an inhibitory transmitter (Figure 64). **Baclofen** interferes with the release of excitatory transmitters.

SEQUENCE OF AND ONSET OF NEUROMUSCULAR BLOCKADE

Neuromuscular blockade takes place in the following sequence: rapidly contracting muscles (eye, fingers, and toes) followed by slowly contracting muscles (diaphragm, limbs, and trunk). The onset and duration of action of succinylcholine are 1 and 5 minutes, respectively. The onset and duration of action of tubocurarine are 5 and 20 minutes, respectively.

FACTORS AFFECTING NEUROMUSCULAR BLOCKADE

Drug Interactions

Antibiotics such as neomycin, streptomycin, kanamycin, gentamycin, polymyxin A, polymyxin B, colistin, lincomycin, and tetracycline have all been implicated as augmenting the action of the nondepolarizing agents.

Substances that inhibit plasma cholinesterase prolong the response of succinylcholine. Succinylcholine is metabolized according to the following scheme:

Table 28   Duration of Action of Neuromuscular Blocking Agents

Neuromuscular Blocking Agent	Duration of Action (min)	Potency Relative to Tubocurarine
Atracurium	20–30	1.5
Doxacurium	>30	6
Metocurine	>30	4
Mivacurium	10–20	4
Pancuronium	>30	6
Pipecuronium	>30	6
Tubocurarine	>30	1
Vecuronium	20–30	6

Several drugs have as their major action the interruption or mimicry of transmission of the nerve impulse at the neuromuscular junction of skeletal muscle or autonomic ganglia. These agents can be classified together, since they interact with a common family of receptors; these receptors are called *nicotinic cholinergic*, since they are stimulated by both the natural transmitter acetylcholine (ACh) and the alkaloid nicotine. Distinct subtypes of nicotinic receptors exist at the neuromuscular junction and the ganglia, and the pharmacological agents that act at these receptors discriminate between them. Neuromuscular blocking agents are distinguished by whether or not they cause depolarization of the motor end plate and, for this reason, are classified either as *competitive (stabilizing)* agents, of which curare is the classical example, or as *depolarizing* agents, such as succinylcholine. The depolarizing and competitive agents are used widely to achieve muscle relaxation during anesthesia. Ganglionic agents usually act by stimulating or blocking nicotinic receptors on the postganglionic neuron.

$$\begin{array}{c} \text{Succinylcholine} \\ \text{Succinylmonocholine} \end{array} \xrightarrow[\text{Liver}]{\text{Plasma cholinesterase}} \begin{array}{c} \text{Succinylmonocholine} \\ \text{Succinate and choline} \end{array}$$

Because cholinesterase is synthesized in the liver, the duration of action of succinylcholine is elevated in the presence of liver disease. Cholinesterase inhibitors dramatically increase the duration of action of succinylcholine. In patients with atypical cholinesterase, the intensity and duration of the effects of succinylcholine are enhanced.

Antiarrhythmic agents such as **quinidine, procainamide,** and **propranolol** have all been shown to augment *d*-tubocurarine-induced blockade. Quinidine has also been reported to unmask or worsen the symptoms of **myasthenia gravis** and to cause postoperative respiratory depression after the use of muscle relaxants.

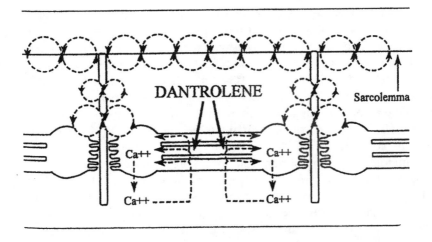

Figure 63 Dantrolene reduces the release of calcium.

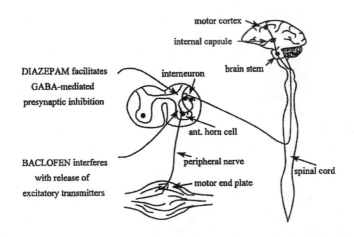

Figure 64 Diazepam facilitates GABAergic transmission. GABA = γ-aminobutyric acid.

Diuretics such as thiazides, **ethacrynic acid**, and **furosemide** intensify the effects of nonpolarizing muscle relaxants, possibly because of a diuresis-induced reduction in the volume of distribution and an associated electrolyte imbalance, such as hypokalemia.

The local anesthetics **procaine** and **lidocaine** enhance the neuromuscular block produced by nondepolarizing and depolarizing muscle relaxants.

Phenytoin has been shown to interfere with neuromuscular transmission, and the drug has been reported to exacerbate myasthenia gravis. **Lithium** augments the effects of both depolarizing and nondepolarizing muscle relaxants and also reportedly unmasks myasthenia gravis.

Chlorpromazine has been shown to potentiate nondepolarizing relaxants. The administration of steroids may lead to a transient worsening of symptoms in patients with myasthenia gravis, but the mechanism by which they interfere with neuromuscular transmission is unknown. The antagonism of **pancuronium-induced blockade** by hydrocortisone has also been reported. **d-Penicillamine**, which is used in the treatment of **Wilson's disease**, may cause a myasthenia gravis-like syndrome. These patients have elevated serum levels of antibody to acetylcholine receptors, suggesting than an immunological mechanism is involved in this drug-induced syndrome. **Azathioprine** antagonizes nondepolarizing neuromuscular blockade, possibly by inhibiting phosphodiesterase.

Calcium ions play an important role in the presynaptic release of acetylcholine, and prolonged neuromuscular blockade has been reported after calcium antagonist administration during anesthesia that includes concurrent nondepolarizing neuromuscular blockade. **Ketamine** potentiates neuromuscular blockade produced by tubocurarine and **atracurium**, but not that produced by pancuronium or succinylcholine.

All of the inhalational anesthetic agents augment both the degree and duration of the neuromuscular blockade induced by the nondepolarizing muscle relaxants.

Electrolytes

The generation of action potentials by muscle and nerve results from changes in the conductance of their membranes to sodium and potassium, and normal neuromuscular function depends on the maintenance of the correct ratio between intracellular and extracellular ionic concentrations.

An acute decrease in the extracellular potassium concentration tends to elevate the end plate transmembrane potential, causing hyperpolarization and an increased resistance to depolarization, together with a greater sensitivity to the nondepolarizing muscle relaxants. Conversely, an increased extracellular potassium concentration lowers the resting end plate transmembrane potential and thereby partially depolarizes the membrane, which should augment the effects of the depolarizing agents and oppose the action of the nondepolarizing drugs. Diuretic-induced chronic hypokalemia reduces the pancuronium requirements for neuromuscular blockade and thus more neostigmine is required to achieve antagonism.

The release of acetylcholine from the motor nerve terminal is also affected by calcium and magnesium ion concentrations, which have opposing effects. Calcium increases the quantal release of acetylcholine from the nerve terminal, decreases the sensitivity of the postjunctional membrane to transmitter, and enhances the excitation–contraction coupling mechanisms of muscle. In contrast, magnesium decreases acetylcholine release and reduces the sensitivity of the postjunctional membrane to acetylcholine. Consequently, the action of the nondepolarizing muscle relaxants can be accentuated by low calcium and high magnesium levels. In addition, magnesium augments the block produced by depolarizing relaxants. Therefore, the dose of a muscle relaxant should be reduced in patients who have toxemia associated with pregnancy and are undergoing **magnesium replacement therapy**.

Acid–Base Balance

Respiratory acidosis enhances d-tubocurarine- and pancuronium-induced neuromuscular block and opposes reversal by neostigmine.

Hypothermia

Hypothermia prolongs the neuromuscular blockade produced by *d*-tubocurarine and pancuronium.

Disease States

The plasma concentrations of *d*-tubocurarine and pancuronium are increased in patients with impaired liver functions because liver disease interferes with the metabolism of pancuronium.

Age

Neonates are more sensitive to nondepolarizing muscle relaxants and the response of the small infant to some extent resembles that of an adult patient with myasthenia gravis.

SIDE EFFECTS
OF NEUROMUSCULAR BLOCKING AGENTS

Cardiovascular Effects

Although the main site of action of the neuromuscular blocking agents is the nicotinic receptor of striated muscle, they may act at other cholinergic receptor sites throughout the body, such as the nicotinic receptors in the autonomic ganglia and the muscarinic receptors in the heart.

Succinylcholine may cause tachycardia, cardiac arrhythmias, and hypertension, which is brought about by stimulation of the sympathetic ganglia. It may also provoke bradycardia, caused by stimulation of muscarinic receptor sites in the sinus node of the heart. This effect is more pronounced following a second dose of succinylcholine. The bradycardia may be blocked by thiopental, atropine, and ganglionic blocking agents.

Ocular and Muscular Effects

Succinylcholine increases intraocular pressure transiently. It can also cause muscle pain, which may be due to fasciculation and uncoordinated muscle contraction. The prior administration of a competitive blocking agent may prevent both fasciculation and pain.

Patients with **myotonia congenita** and **myotonia dystrophica** respond differently to succinylcholine, in that their muscles are contracted rather than relaxed.

Histamine Release

Tubocurarine, **metocurine**, and **succinylcholine** have all been shown to elicit histamine release in humans. However, histamine release is less common with **pancuronium** and **alcuronium**. **Vecuronium** does not cause histamine release.

ERYTHRINA ALKALOIDS

The beanlike seeds of the trees and shrubs of the genus *Erythrina*, a member of the legume family, contain substances that possess curare-like activity. The plants are widely distributed in the tropical and subtropical areas of the American continent, Asia, Africa, and Australia, but apparently they are not used by the natives in the preparation of arrow poisons. Of 105 known species, the seeds from more than 50 have been tested and all were found to contain alkaloids with curariform properties. *Erythroidine*, from *E. americana*, was the first crystalline alkaloid of the group to be isolated. It consists of at least two isomeric alkaloids, α and **β-erythroidine**; both are dextrorotatory. Most experimental and clinical study has centered on the β form because it is more readily obtainable in pure state. β-Erythroidine is a tertiary nitrogenous base. Several hydrogenated derivatives of β-erythroidine have been prepared; of these, **dihydro-β-erythroidine** has been studied most carefully and subjected to clinical trial. Conversion of β-erythroidine into the quaternary metho salt (**β-erythroidine methiodide**) does not enhance but rather almost entirely abolishes its curariform activity; this constitutes a notable exception to the rule that conversion of many alkaloids into quaternary metho salts results in the appearance of curare-like action.

The pharmacological properties of β-erythroidine and its dihydro derivative are very similar to those of *d*-tubocurarine and therefore need not be described in any detail. The two compounds differ from curare in three important respects, namely, less potent paralytic action on neuromuscular junctions, briefer duration of paralysis, and oral efficacy. Indeed, gastrointestinal absorption of the alkaloids is so rapid and complete that the difference between effective oral and subcutaneous doses is rather small. Dihydro-β-erythroidine is longer acting than β-erythroidine and about six times as active. Like curare, β-erythroidine and its dihydro derivative are antagonized at the neuromyal junction by anticholinesterases such as **neostigmine**.

METHONIUM COMPOUNDS

In the search for synthetic agents capable of blocking neuromuscular and autonomic ganglionic transmission, that is, for substances blocking the nicotinic effects of acetylcholine, thousands of interesting compounds have been made and scores of new chemical series explored. One of the most fruitful discoveries was that of the British investigators who introduced the **polymethylene bis-trimethylammonium** series, referred to herein by the generic term "**methonium compounds**," a name approved by the British Pharmacopoeia Commission. Not only has the methonium series revealed an entirely new field of investigation for structure–activity relationship and provided valuable laboratory tools for autonomic pharmacology, but it has also yielded at least three clinically useful drugs, one causing neuromuscular blockade in skeletal muscle (**decamethonium**) and two causing blockade in autonomic ganglia (**pentamethonium** and **hexamethonium**) (Goodman and Gilman, 1955).

25

DAFFODIL AND ALZHEIMER'S DISEASE

Daffodil contains Galanthamine used in mood disorders and in Alzheimer's disease.

> When daffodils begin to peer,
> With heigh! The doxy over the dale,
> Why, then comes in the sweet o' the year.

> — Autolycus, *Winter's Tale*, iv.3

INTRODUCTION

In Greek mythology, ***Narcissus*** was a beautiful youth who was loved by the goddess **Echo**. His beauty, however, was so grand that he could love no one but himself after catching sight of his beauty in a pool. He became so focused on his beauty that the gods felt pity for him and turned him into a **daffodil**, which has thus become known as *Narcissus*. It is said that the scent is so strong that it was used by **Pluto**, the god of the underworld to lure the beautiful **Proserpine** to hell and by the gods in their coronas to adorn their brows. Finally, the Greeks have used the flower to ward off the Furies and Pluto and to subdue the dark spirits.

HISTORY AND GENERAL EFFECTS

Since ancient times, it has been known that daffodil bulbs exert toxic effects upon ingestion. Even the name of the daffodil genus, *Narcissus*, implies that daffodils have physiological effects. The word *Narcissus* is derived from the Greek term *narkao*, "to benumb," because of the narcotic properties attributed to the daffodil bulb. Similarly, the narcotic effects of the daffodil inspired **Socrates** to proclaim the daffodil "**Chaplet of the infernal Gods**." Daffodil bulbs contain powerful emetics; the flowers themselves are slightly poisonous (Figure 65).

Ancient uses of daffodil components included systemic administration for "hysterical affections," emesis, dysentery, chronic bronchial congestion, ague, and spasms. In addition, a plaster made of the roots was applied topically to dissolve hard swellings. An extract of the bulbs was also applied to open wounds,

Figure 65   Narcissus Pseudo-Narcissus. Daffodil. Narcisse des prés Porillon, Fr. Gelbe Narcisse, G. (Nat. ord. Amaryllidaceae). Both the bulb and the bright yellow flowers of this common garden plant have been used in medicine. The flowers have a feeble peculiar odor, and both have a bitter mucilaginous taste. They are an uncertain emetic. It is probable that the flowers of the wild European plant are more powerful than those of the cultivated.

presumably as a crude anti-inflammatory agent, although wound treatment with *Narcissus* bulb extract could produce "staggering, numbness of the whole nervous system, and paralysis of the heart."

COMPOUNDS FOUND IN DAFFODILS

Many compounds are found in daffodil bulbs: **lycorine, narcitine, pseudolycorine, suisemine, tazettine, galanthamine, narzettine, nartazine, haemanthanmine, homolycorine, fiancine, hippeastrine pluviine, galanthine, panceratine, narcissidine** (Figure 66).

Lycorine (formerly Narcissine)

Although narcissine was isolated as early as 1578, much of the initial information that exists concerning narcissine dates to the early 1900s. In 1910 narcissine was purified from both resting bulbs and flowering bulbs, at a yield of 0.2 and 0.1%, respectively. In the same year it was demonstrated that narcissine was a very stable alkaloid, not easily decomposed. Other scientists from this era noted that cats treated with narcissine were inflicted with vomiting, salivation, and purgation. In 1920, it was demonstrated that narcissine is the same compound as the alkaloid lycorine previously isolated from ***Lycoris radiata***, and the name narcissine was dropped.

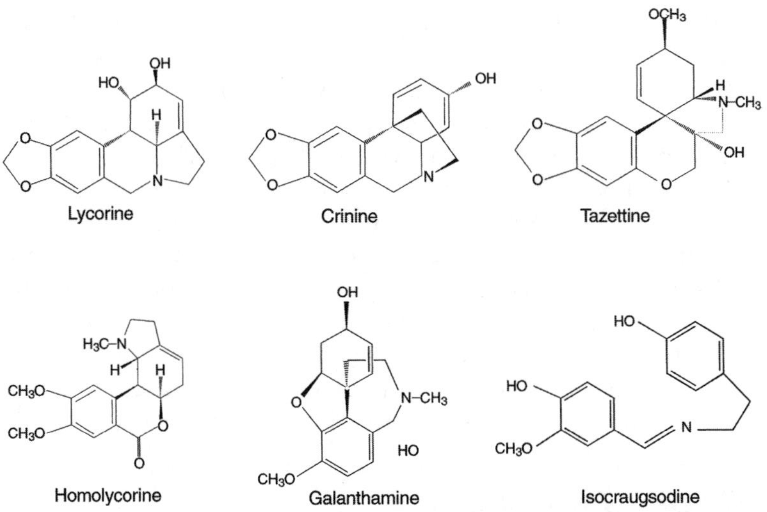

Figure 66   Structures of some drugs found in daffodils.

Lycorine is an **isoquinolone (phenathridine derivative)** alkaloid of molecular weight 287.32; its structure is shown in Figure 66. Although further details on the pharmacokinetics and pharmacological effects of lycorine are not available, the results of treating cats with lycorine suggest that it has **cholinomimetic effects**.

Galanthamine

A second compound isolated from daffodils, more extensively characterized than lycorine, is **galanthamine** (or **galantamine**) (see Figure 66). It has been suggested that galanthamine was the active compound in the potion that protected Odysseus from poisoning in **Homer's *Odyssey***. Known effects of galanthamine include antinociceptive effects and cholinomimetic effects. Current clinical trials are assessing the use of galanthamine in **Alzheimer's disease**, which is a neurodegenerative disease, not a mood disorder. However, successful treatment of the mood disorder mania with galanthamine has also been reported.

Chemistry of Galanthamine

Galanthamine is a phenanthrene-derivative alkaloid of molecular weight 368.43; its structure is shown in Figure 66. Galanthamine can be purified from the bulbs of different strains of *Narcissus*; the quantity recovered per 100 mg of dried bulb ranges from 3.9 to 78.7 mg depending on the specific daffodil cultivar used.

Pharmacological Mechanisms of Galanthamine Activity

Galanthamine is a long-lasting reversible **acetylcholinesterase (AChE) inhibitor**; since it is a tertiary amine, it rapidly crosses the blood–brain barrier and has effects on the central nervous system. Galanthamine is relatively selective for

AChE over **butyrylcholinesterase**, with IC$_{50}$ values of 0.35×10^{-7} for erythrocyte AChE and 18.6×10^{-7} for plasma butyrylcholinesterase. The inhibition of AChE by galanthamine results in slower degradation of **acetylcholine** (ACh) and thus generates longer responses to endogenous ACh.

Galanthamine also increases the frequency of opening of **nicotinic receptor channels** and potentiates agonist-activated receptors. This activity suggests that galanthamine can modulate the activity of nicotinic currents. Galanthamine binds to an allosteric site on the α-subunits of nicotinic ACh receptors. This result might be related to the ability of galanthamine to increase responses to the nonhydrolyzable **carbachol**, since theoretically galanthamine cannot increase the duration of response to a nondegradable cholinomimetic.

Analysis of the cholinergic effects of galanthamine in a mouse model suggests that galanthamine does not affect **choline acetyltransferase**, the choline carrier, or agonist binding to the active site of either muscarinic or nicotinic ACh receptors.

Pharmacokinetics of Galanthamine

Single doses of 10 mg are tolerated by humans without any unwanted side effects. Maximal plasma concentrations are reached 2 hours after either oral or subcutaneous ingestion. Elimination half-life is 5.3 (standard deviation 4.2) hours. Both oral and subcutaneous application show similar drug absorption parameters. Galanthamine shows first-order pharmacokinetics and good oral bioavailability (estimates range from 65 to 100%). Galanthamine is absorbed in all segments of the gastrointestinal tract, including the stomach, duodenum, colon, and rectum (16%, 54 to 85%, 43%, 76%). Studies in mice suggest that galanthamine accumulates in the kidney, liver, and parencyhmous tissues like the brain (drug levels were tenfold, fivefold, and twofold increased over plasma levels).

Galanthamine is metabolized in the liver; small amounts of the active metabolites **epigalanthamine** and **galanthaminone** can be measured in the plasma. These metabolites are 130-fold less potent than galanthamine. Negligible quantities of these two metabolites can also be detected in urine. Galanthamine is cleared from the body by excretion in the kidneys; renal clearance has been estimated at 0.084 l/kg/hour.

In one study, 65% inhibition of AChE was observed within 2 minutes of bolus injection of 10 mg ganathamine; enzyme activity returned to baseline within 24 hours. The median maximal value of AChE inhibition in this study was 53%. Studies have also undertaken to analyze the effects of galanthamine on AChE inhibition in human brain tissue. IC$_{50}$ values for galanthamine for AChE inhibition in postmortem human frontal cortex and hippocampus were 3.2×10^{-6} M and $2.8 \times 10^{-6} M$, respectively. Interestingly, the IC$_{50}$ was tenfold higher for galanthamine treatment of human erythrocytes. In a mouse model of Alzheimer's disease (lesion of the **nucleus basalis magnocellularis**), the IC$_{50}$ for AChE inhibition was 4.1×10^{-7} M.

Use of Galanthamine to Treat Depression

The potential role of galanthamine in depression is complicated: manic and depressive symptoms are thought to be mediated by hyperadrenergic and hypercholinergic

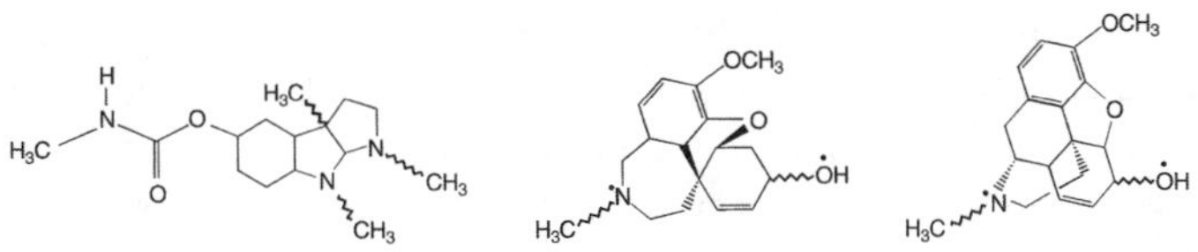

Figure 67   Chemical structures of physostigmine, galanthamine, and codeine.

stimuli, respectively. In this model, a cholinomimetic is appropriate for restoring the adrenergic/cholinergic balance in mania. However, galanthamine also has antinociceptive (analgesic) effects; perhaps these effects inspired the ancients to promote daffodil use in depression.

The one published case report using galanthamine hydrobromide to treat mania appeared in the *Lancet* in 1991. A 74-year-old woman was admitted to the hospital with lithium-resistant manic depression. Within 2 hours of treatment with 10 mg galanthamine hydrobromide she exhibited considerable reduction in manic symptoms. As cessation of galanthamine therapy led to a relapse in her manic symptoms, she was maintained on galanthamine hydrobromide, 10 mg three times a day as a long-term therapy (Snorrason and Stefansson, 1991). Interestingly, galanthamine is useful in the cholinergic REM test, which is used to diagnose depression on the basis of REM onset.

Specific Studies Showing Antinociceptive Effects of Galanthamine

Galanthamine has antinociceptive effects. This may partially be explained by the similarity in the structure of galanthamine to the structure of codeine (Figure 67). In the rat hot plate test, the antinociceptive effect could be partially blocked by **naloxone**, suggesting that galanthamine may have opioid-like activity. Galanthamine also provided analgesia in the mouse acetic acid writhing test, similar to that induced by morphine. In a separate study, intrathecal administration of galanthamine was shown to inhibit bradykinin-induced bioelectric activity in the rat spinal ventrolateral tracts.

Summary

In summary, although bulbs of the *Narcissus* genus contain many alkaloid compounds, the most-studied alkaloid component of daffodils is galanthamine. Galanthamine is a long-lasting AChE inhibitor, which also affects the nicotinic ACh receptor directly by binding an allosteric site. Galanthamine has cholinomimetic as well as antinociceptive effects. Because of its cholinomimetic properties, galanthamine is used in clinical trials to treat Alzheimer's disease. In addition, galanthamine has been used successfully to treat mania.

THE TREATMENT OF COGNITIVE IMPAIRMENT IN ALZHEIMER'S DISEASE

Alzheimer's disease is characterized by a progressive dementia of insidious onset leading to a gradual deterioration of intellectual abilities, neuropsychological deficits, and personality changes. The disease process eventually results in **anomia**, **agnosia**, and **apraxia** and in **complete loss of memory** and **learning abilities**. The disease is occasionally accompanied by sleep disturbances, agitation, anxiety, depressive states, or psychosis. In the final stages of the disease some patients are mute, unable to stand or walk, bedridden, and incontinent.

Generally, but not invariably, postmortem macroscopic examination of the brain reveals **cerebral atrophy** with narrowed convolutions, widened sulci, and enlarged lateral and third ventricles. On microscopic examination brain specimens from patients with a clinical diagnosis of Alzheimer's disease are characterized by widely spread cortical **senile plaques**, **neurofibrillary tangles**, and **granulovascular degeneration**.

Histochemically, Alzheimer's disease is associated with multiple deficits in various neurotransmitters or their associated markers such as **acetylcholine**, **noradrenaline**, **somatostatin**, and **others**.

THERAPEUTIC OBJECTIVES IN ALZHEIMER'S DISEASE

Although the disease is accompanied by various psychiatric noncognitive symptoms, the progressive deterioration represents the core phenomenon of Alzheimer's disease and the major target of current therapeutic efforts. As in any other disease characterized by progressive deterioration, the therapeutic intent can vary from wanting to alleviate some symptoms of the disease to slowing or arresting the deterioration, or if possible, reversing the damage completely. The neurotransmitter-based approaches described below offer modest palliation by augmenting deficient neurotransmission. Recent advances in the understanding of the biology of Alzheimer's disease may permit the development of strategies aimed at retarding or halting the progression of the illness. These strategies based on various pathophysiological models of the illness attempt to interfere with putative pathogenetic mechanisms involved, such as **glutamate's neurotoxicity**, **free radical production**, **central nervous system (CNS) inflammation**, **amyloid production**, and **aluminum accumulation**. Regenerative treatments, however, aimed at restoring incurred damage are not yet in sight.

THE CHOLINERGIC HYPOTHESIS OF ALZHEIMER'S DISEASE AND CHOLINERGIC ENHANCEMENT STRATEGIES

The cholinergic hypothesis of Alzheimer's disease postulates that:

- There is significant cerebral cholinergic neurotransmission deficit.
- This deficit causes some of the cognitive disturbances observed in the patient with Alzheimer's disease.
- Enhancing cholinergic neurotransmission will produce some amelioration in cognitive function in this population (see Figure 39).

Several lines of evidence support this hypothesis. Centrally active anticholinergic drugs have been shown to induce dose-related cognitive deficits in humans while cholinergic neurotransmission has been shown to be specifically involved in

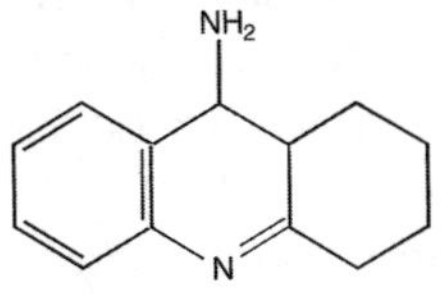

Tacrine
Molecular structure of THA

Figure 68   Molecular structure of THA.

memory and learning. The hypothesis is further supported by studies demonstrating that chemical, surgical, or pharmacological lesions to cerebral cholinergic systems impair learning and memory in animals and that cholinomimetic agents can reverse lesion-induced behavioral disturbances. Finally, brains of patients with Alzheimer's disease exhibit consistent cholinergic cell loss in the **septum** and **nucleus basalis of Meynert (NBM)**, a decrease in the cholinergic markers choline-acetyl transferase (CAT) and acetylcholinesterase (AChE), and a correlation between these neurochemical changes and the degree of cognitive impairment in Alzheimer's disease.

Analogous to the dopaminergic treatment approach used in Parkinson's disease, therapeutic trials in Alzheimer's disease have been aimed at augmenting cerebral cholinergic neurotransmission. The various cholinomimetic agents that have been assessed in Alzheimer's disease can be classified according to their particular cholinomimetic mechanism of action into (1) AChE inhibitors, (2) cholinergic agonists, (3) acetylcholine (ACh) precursors, (4) ACh-releasing agents, and (5) agents with various other cholinomimetic effects.

ACETYLCHOLINESTERASE INHIBITORS

Tetrahydroaminoacridine

1,2,3,4-Tetrahydro-9-acridinamine (THA), known as **tacrine**, is a synthetic aminoacridine, which was initially synthesized more than 40 years ago. Tacrine and physostigmine are the two AChE inhibitors that have been most extensively evaluated in the treatment of Alzheimer's disease. Tacrine was the first drug ever to be approved by the FDA (available as of 1993) for the treatment of cognitive impairment in Alzheimer's disease in the United States (Figure 68).

Pharmacology

Cholinesterase Inhibition. **Tacrine is thought to be a noncompetitive reversible inhibitor of AChE**. The drug has been shown to be a more potent inhibitor of **butyrylcholinesterase** than of AChE. The AChE inhibitory activity of Tacrine is thought to be mediated by its binding to a hydrophobic area close to the active site.

Effects on Choline Metabolism. THA has been reported to increase ACh brain levels.

Effects on Cholinergic Receptors. THA has been shown to bind to muscarinic and nicotinic receptor sites.

Inhibition of Monoamine Oxidase (MAO). The enzymatic activity of monoamine oxidases of type A (MAO-A) and type B (MAO-B) have been shown to be reduced by THA. MAO-A seems to be inhibited to a larger degree than MAO-B. At therapeutic concentrations, THA is believed to produce a significant decrease in MAO activity leading to an enhanced monoaminergic activity. Given that some cognitive deficits in Alzheimer's disease are believed to be due to a monoaminergic deficit, this monoaminergic enhancement might also contribute to the therapeutic effect of THA in the illness.

Monoamine Release and Uptake. Tacrine has been shown to induce monoamine release and to inhibit monoamine uptake, leading to an increase in several monoamine neurotransmitters including dopamine, serotonin, and norepinephrine.

Pharmacokinetics

Cognex®, the THA brand currently available on the U.S. market, is rapidly absorbed and reaches maximal plasma concentration 2 hours after oral administration.

Clinical Applications

Cognex is available in 10, 20, 30, and 40 mg capsules. It is recommended that treatment should be started with 10 mg orally four times a day (po qid) for 6 weeks. Additional dosage increments of 10 mg po qid up to a maximum of 160 mg po/day should be initiated every 6 weeks. Weekly transaminase monitoring should be performed for the first 18 weeks of treatment and for at least 6 weeks after each dosage increment. Thereafter, checks at 3-month intervals are recommended.

Side Effects

Reversible **liver toxicity** was common in THA-treated patients.

Drug–Drug Interactions

Cimetidine was found to increase the concentration and rate of absorption of THA by 50%. THA administration may double the elimination half-life and plasma concentrations of *theophylline*.

Overdosage

THA overdosage can result in a **cholinergic crisis** with severe nausea, vomiting, salivation, sweating, bradycardia, hypotension, convulsions, etc. Progressive muscle weakness may result in death from asphyxiation. Treatment with anticholinergics (such as intravenous atropine sulfate) and general supportive treatment are recommended.

Physostigmine and Its Derivatives

Physostigmine (PHS) is a natural alkaloid, first isolated in 1864 (Figure 69). **Heptylphysostigmine** was well tolerated in healthy volunteers and up to 48 mg

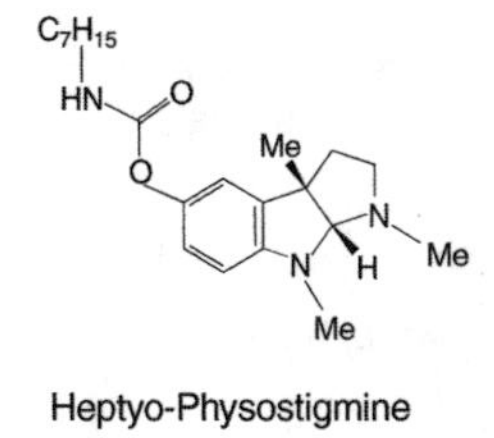

Heptyo-Physostigmine

Figure 69 Molecular structure of heptyl-physostigmine.

tid in patients with Alzheimer's disease. Unfortunately, heptylphysostigmine trials in humans have been abandoned because of **neutropenia** observed with this agent.

Limitations to AChE-Inhibitors Therapy in Alzheimer's Disease

Adverse effects of AChE inhibitors are due to the enhancement of peripheral cholinergic tonus and are primarily gastrointestinal (nausea, vomiting, abdominal cramping, and diarrhea) but may include **diaphoresis** and light-headedness. However, some nausea is undoubtedly of central origin and could be related to the rate of rise of the drug in plasma. **Bradycardia** and **hypotension** are anticipated risks, but appear infrequently. Depression and agitation have also been encountered. Individual susceptibility to such side effects varies widely. Considering the inverted U-shape form of the dose–response curve to cholinergic enhancement in Alzheimer's disease, AChE inhibitor levels that are too high or too low may easily be outside the therapeutic window. Nonoptimal levels may easily occur with fluctuations in drug levels due to uncertain CNS availability of the drug. Under such circumstances, fixed-dose treatment studies can be of limited benefit. Thus, studies should include individual titrations and confirmation of CNS availability using biological markers such as AChE activity in cerebrospinal fluid (CSF) or endocrinological parameters. Ultimately, some patients might present with such a severe degree of presynaptic cholinergic cell loss that any degree of AChE inhibition would fail to induce a clinically significant enhancement of the cholinergic transmission. Overall, it appears that short-term administration of AChE inhibitors can induce a mild to moderate improvement of memory in Alzheimer's disease, and that long-term AChE treatment should be evaluated for additional benefits.

CHOLINERGIC AGONISTS

Bethanechol, Arecoline, Oxotremorine, and **Nicotine** have been tried and had no beneficial effect on memory.

ACETYLCHOLINE PRECURSORS

Phosphatidylserine, L-α-**Glyceryl-Phosphorylcholine**, and **Cytidine Diphosphate Choline** have shown positive effects.

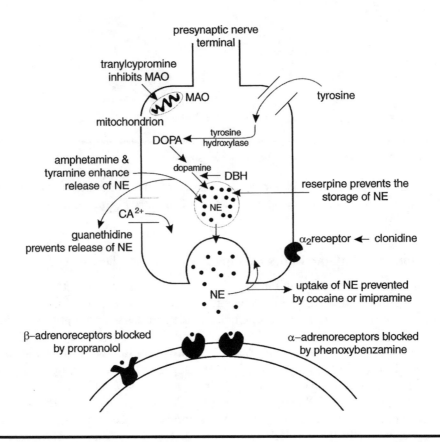

Figure 70   The synthesis and actions of norepinephrine. MAO = monamine oxidase; NE = norepinephrine; DBH = dopamine β-hydroxylase.

ACETYLCHOLINE-RELEASING AGENT

4-Aminopyridine or **linopirdine** provides some improvement.

NONCHOLINERGIC NEUROTRANSMITTER REPLACEMENT STRATEGIES

In addition to the cholinergic deficit, Alzheimer's disease has also been shown to be characterized by marked deficits in the monoamine neurotransmitters **norepinephrine** and **serotonin**, as well as in **glutamate** and some neuropeptide neurotransmitters (Figure 70).

Monoaminergic Drugs

L-Deprenyl (**selegiline**), a monoamine oxidase B inhibitor, **clonidine** and **guanfacine**, α_2-adrenoreceptor agonists, and **levodopa** (L-dopa) have been reported to improve cognitive function in some subjects. **Zimeldine**, **citalopram**, and **alaproclate**, selective serotonin uptake blockers, have no beneficial effects.

Neuropeptides

Adrenocorticotropic hormone (ACTH), **vasopressin** (VP), **somatostatin**, and **cholecystokinin** (CCK8) look promising.

Drugs Affecting Glutamate Receptor Sites

Milacemide and **D-cycloserine** look promising.

Nootropics

Piracetam and derivatives, such as **pramiracetam, oxiracetam, aniracetam**, or **acetyl-1-carnitine** showed no significant improvement.

Antioxidants

Several studies have demonstrated increased free radical production in aging in general and in Alzheimer's disease in particular. Increased **superoxide dismutase**-derived hydrogen peroxide fluxes, metal ions, and damaged mitochondria can contribute to cell damage mediated by free radicals. Free radical production in Alzheimer's disease may also be caused by amyloid β-protein and glutamate. These findings suggest that antioxidants could have beneficial effects in Alzheimer's disease by interfering with free radical production and preventing consequent cell injury. **L-deprenyl** is thought to possess antioxidant properties, and is believed to have neuroprotective effects in Alzheimer's disease as well as in Parkinson's disease by acting as scavenger of free radicals. **Vitamin E** and **idebenone** are also potential antioxidant treatments for Alzheimer's disease since they prevent cell death caused by glutamate and amyloid β-protein. Clinical trials with L-deprenyl and vitamin E are currently evaluating their ability to slow the progression of Alzheimer's disease.

Desferrioxamine, an Iron-Chelating Agent

Several findings suggest an association between aluminum and Alzheimer's disease. Epidemiological studies have reported an association between the **aluminum** concentration in drinking water and the occurrence of Alzheimer's disease. In addition, aluminum administration has been shown to be neurotoxic to the cholinergic system. These findings have led clinical trials with **desferrioxamine mesylate** in Alzheimer's disease, which has a particularly high affinity for aluminum and has been used to treat iron and aluminum overload. In a 2-year-long double-blind controlled trial, desferrioxamine administered intramuscularly was compared to placebo. Patients who received desferrioxamine treatment showed less decline in daily living skills when compared to the placebo group. The therapeutic effects observed with this agent may not necessarily be due to its chelating action because it has been shown to inhibit **free radical formation** and **inflammation** as well. The required intramuscular administration and toxic side effects of this compound, however, might limit its clinical utility. Replication studies are necessary to confirm the efficacy of this approach.

Nerve Growth Factor

Another conceivable therapeutic approach for Alzheimer's disease could be based on the administration of **neurotrophic factors**. **Nerve growth factor** (NGF) is a 118 amino acid polypeptide with no blood–brain barrier penetrance. Other substances with neurotrophic activity such as **epidermal growth factor, brain-derived neurotrophic factor, gangliosides**, and the **β1-28 peptide** of the β-amyloid protein might also have a therapeutic potential. Intracerebroventricular (ICV) administration of NGF has been shown to partially reverse lesion-induced deficits of cortical AChE and choline acetyltransferase (CHAT) activities to promote survival of septal cholinergic neurons after fimbrial transection in adult rats and to reverse behavioral deterioration in rats with such lesions.

Anti-Inflammatory Agents

The involvement of the immune system and inflammation in the pathophysiology of Alzheimer's disease has been suggested by several findings, such as the presence of inflammation markers in the brain and the colocalization of some acute-phase and complement proteins in senile plaques. Furthermore, increased numbers of **reactive glia** and **microglia** (believed to be related to macrophages) have been observed. **Activated T lymphocytes, a hallmark of the cell-mediated response** observed in chronic inflammatory states, have also been observed in postmortem studies of Alzheimer's disease cases. The use of anti-inflammatory agents in the treatment of Alzheimer's disease is further supported by the low prevalence of Alzheimer's disease found in patients with rheumatoid arthritis and in elderly patients with leprosy. These findings could potentially be explained by the long-standing exposure of these patients to chronic anti-inflammatory therapy. Further-more, preliminary observations suggest that **steroids** and **nonsteroidal anit-inflammatory agents (NSAIDs)** such as **indomethacin** and aspirin may have protective effects against Alzheimer's disease. In a recent 6-month-long double-blind placebo-controlled study in Alzheimer's disease, indomethacin 100 to 150 mg/po/day was found to improve cognitive function and potentially delay further deterioration. These results are promising; however, the toxicity of indomethacin might limit its use.

Corticosteroids offer a logical anti-inflammatory therapy for Alzheimer's disease since these agents are widely used and efficacious for several inflammatory diseases in the CNS, including **lupus cerebritis** and **multiple sclerosis**. Unfor-tunately, the systemic toxicity of steroids limits the use of high doses or long-term treatments with these agents. Animal studies suggest that prolonged exposure to high doses of glucocorticoids is toxic to hippocampal neurons. Low-dose steroid therapy may be the safest strategy because this dose is well tolerated and effective in patients with rheumatoid arthritis.

Colchicine is another possible candidate for the treatment of Alzheimer's disease. This drug effectively treats **familial Mediterranean fever**, a condition in which recurrent inflammation and **renal amyloidosis** occur. Although the amyloid constituents in familial Mediterranean fever and Alzheimer's disease differ, both illnesses involve chronic inflammation, elevated acute-phase proteins, and abnormal processing of a precursor protein leading to deposition of insoluble amyloid fragments. These similarities suggest the potential therapeutic efficacy of colchicine for patients with Alzheimer's disease.

Hydroxychloroquine is historically an antimalarial agent that has been adopted as a safe and effective second-line agent for the treatment of **rheumatoid arthritis** and **lupus erythematosus**. The efficacy of hydroxychloroquine is thought to be related to its effects on the immune response and lysosomal functioning. This agent suppresses cytokine and acute-phase reactant levels in these illnesses. Hydroxychloroquine also interferes with lysosomal enzymatic activity by increasing the pH in these organelles and by stabilizing lysosomal membranes. The safe clinical profile of hydroxychloroquine as a chronic treatment for rheumatoid arthritis supports it as a possible candidate for the treatment of Alzheimer's disease.

26

EPHEDRINE, OR MA HUANG, AS A DECONGESTANT AND VASOPRESSOR

THE ACTIONS OF EPHEDRINE

Ephedrine Sulfate, *Kondon's Nasal*, *Pretz-D*, or *Vicks Vatronol* is classified as a decongestant and vasopressor.

Ephedrine stimulates both α- and β-receptors, causing increased heart rate, unchanged or augmented stroke volume, enhanced cardiac output, and increased blood pressure. It causes relaxation of smooth muscle of bronchi and gastrointestinal tract, stimulation of cerebral cortex, and pupil dilation.

Ephedrine, given im/iv/sc, is indicated for the treatment of acute hypotensive states; treatment of Adams–Stokes syndrome with complete heart block; stimulation of the central nervous system (CNS) to combat **narcolepsy** and depressive states; treatment of acute bronchospasm; treatment of enuresis; treatment of **myasthenia gravis**. When given in nasal form, ephedrine is used in the treatment of nasal congestion; promotion of nasal or sinus drainage; or relief of **eustachian tube congestion**.

The contraindications to ephedrine are angle-closure glaucoma; patients anesthetized with cyclopropane or halothane; cases in which vasopressor drugs are contraindicated (e.g., thyrotoxicosis, diabetes mellitus, hypertension of pregnancy); or treatment with monoamine oxidase (MAO) inhibitor therapy.

Ephedrine may negate the antihypertensive effects of **guanethidine**, and may potentiate the pressor effects of MAO inhibitors causing hypertensive crisis and intracranial hemorrhage.

HISTORY

Ephedrine is an alkaloid occurring in various plants belonging to the genus **Ephedra**, a gymnosperm related to the pines and firs. The genus is indigenous to the temperate and subtropical latitudes of Europe, Asia, and America, and grows especially in northern and western China, northern India, and Spain. In the United States, ephedra plants grow along the Rocky Mountains but most of the naturally occurring types in this country do not yield the ephedrine alkaloid. The stems of the plant contain up to about 1% of the alkaloid and its isomers. The content increases through the spring and summer and becomes maximal just before the frosts; the old Chinese custom of collecting the plant for medicinal use in the autumn is thus vindicated.

The Chinese herb containing ephedrine and called *ma huang* (yellow astringent) has been employed empirically by native physicians for over 5000 years. The **Emperor Shen Nung** (2760 B.C.) is supposed to have tasted *ma huang*, as well as all other drugs in the pharmacopeia, and placed it in the "medium class." The Chinese Dispensatory (*Pentsao Kang Mu*) written in 1569 by **Shih-Cheng Li** mentions that the drug is valuable as an antipyretic, diaphoretic, circulatory stimulant, and sedative for cough. Ephedrine came to be an ingredient of several favorite native prescriptions. The plant was also widely employed in Russia for many centuries, particularly for respiratory disorders and rheumatism. The Native Americans and Spaniards of the southwestern United States used ephedrine plants for various medicinal purposes, especially venereal diseases.

An active principle was first isolated from *ma huang* by Yamanashi in 1885. In 1887, Nagai obtained the alkaloid in pure form and named it **ephedrine**. Pharmacological investigation by Japanese workers indicated that the drug was toxic, but large doses were used in the study. However, its mydriatic properties were noted by Miura (1887), and as early as 1888 Takahashi and Miura concluded that the drug caused dilatation of the pupil through stimulation of the sympathetic nerves (see Sollmann, 1944). Ephedrine then enjoyed a passing vogue as a mydriatic, several years before the isolation of epinephrine. Japanese scientists continued their investigations of the alkaloid, observed its sympathomimetic properties, and employed it in the treatment of asthma.

Interest in ephedrine in Western medicine was created by the classical investigations of Chen and Schmidt, which began in 1923 as a result of a Chinese druggist's assurance that *ma huang* was really a potent drug. These workers reported the cardiovascular effects of the alkaloid, its similarity to epinephrine, and its absorption from the intestinal tract. Numerous clinical and experimental studies soon followed, and the use of ephedrine spread so rapidly that several tons of the alkaloid are now consumed yearly. Synthetic ephedrine (racemic) was first prepared in 1927 and marketed under the name *Ephetonin* (Goodman and Gilman, 1955).

EPHEDRA

Ephedra or *ma huang* is the entire plant, or the overground portion of *Ephedra sinica* grown in China. In Chinese characters "**ma**" means astringent and "**huang**" means yellow, probably referring to the taste and color of the drug. It has been used as a medicine in China for more than 5000 years. Its use in modern medicine began with the recent discovery of the valuable properties of ephedrine. The plant is found near the seacoast in southern China and the drug is exported from Canton.

The plant is a low, dioecious, practically leafless shrub, 60 to 90 cm high. The stem, green in color, is slender, erect, small ribbed and channeled, 1.5 mm in diameter, and usually terminates in a sharp point. Nodes are 4 to 6 cm apart, at which the leaves appear as whitish triangular scarious sheaths. Small blossoms appear in the summer.

Various species of *Ephedra* grow in different parts of the world. It is believed that the Chinese species contain the official levorotatory ephedrine, while the same species and other varieties from Europe yield only dextrorotatory **pseudoephedrine**. East Indian species range from rich to none. Ephedrine exists in the seed of another, totally unrelated Indian plant, **Sida cordifolia**, which was also used in folk medicine. *Ma huang* also contains another alkaloid, **ephedine**, which lowers blood pressure and has other complex actions.

EPHEDRINE

This **alkaloid** was first isolated from **Ephedra equisetina**, a plant (*ma huang*) used as medicine by the Chinese since antiquity. Most of the present supply is probably synthetic. Its chemical structure is closely related to **epinephrine** and **tyramine**, and differs from epinephrine chiefly by the absence of the two phenolic hydroxyls. Its effects on the circulation, intestines, bronchi, iris, etc., are superficially similar to those of epinephrine. It requires that larger doses be given but they are more lasting, due probably to ephedrine's much greater stability and resistance to oxidation. The effects can be produced by oral administration. Unlike epinephrine, it is not sensitized by **cocaine** or by denervation. From this, it has been argued that its point of attack is not sympathomimetic but muscular. It also stimulates the CNS. A number of isomers with similar actions are known. Ephedrine is used therapeutically in hay fever and asthma, where it is less effective than epinephrine, but more convenient; as a mydriatic with minimal effect on intraocular tension; and as a stimulant to respiration, especially against barbiturate poisoning in conjunction with **metrazol** or **picrotoxin** (Sollmann, 1944).

Ephedrine is an alkaloid obtained from **E. equisetina**, **E. sinica**, and other species of *Ephedra,* or produced synthetically.

Ephedrine occurs in white, rosette, or needle crystals, or as an unctuous mass. It is soluble in water, alcohol, chloroform, ether, and in liquid petrolatum, the latter solution being turbid if the ephedrine is not dry. Ephedrine melts between 34 and 40°C, depending upon the amount of water it contains; it contains not more than 0.1% of ash; its solutions are alkaline to litmus; it readily forms salts with acids; and it responds to the usual tests for alkaloids. Ephedrine excites the sympathetic nervous system, depressing smooth and cardiac muscle action, and produces effects similar to those of epinephrine. It produces a rather long-lasting rise of blood pressure and mydriasis and diminishes hyperemia. The alkaloid may be used in 0.5 to 2% oil spray.

Ephedrine hydrochloride ($C_{10}H_{15}ON \cdot HCl$), when dried over sulfuric acid for 24 hours, contains not less than 80% and not more than 82.5% of anhydrous ephedrine, $C_{10}H_{15}ON$.

Ephedrine sulfate ($(C_{10}H_{15}ON)_2H_2SO_4$), when dried over sulfuric acid for 24 hours contains not less than 75.5% and not more than 77.3% of anhydrous ephedrine $C_{10}H_{15}ON$.

Both of these salts are readily soluble in water and in hot alcohol, but not in ether. They have the same pharmacological properties as ephedrine and are used orally in an average dose of 0.25 g. They are also used, in aqueous solution, intramuscularly and intravenously.

PHARMACOLOGICAL PROPERTIES OF EPHEDRINE

The change from the phenolic structure of epinephrine to the phenyl structure of ephedrine results in a marked difference in action. Unlike epinephrine, ephedrine is effective orally, has a prolonged action, exhibits tachyphylaxis, and is a potent corticomedullary stimulant. The oral effectiveness and prolonged action of ephedrine are apparently due to the presence of the methyl group on the α carbon atom, a configuration that renders the molecule refractory to deamination by the amine oxidase of the liver.

At first, ephedrine was assumed to have a mechanism of action similar to that of epinephrine, acting through the same receptive mechanism. Later, since it was found that not all the effects of epinephrine could be duplicated by ephedrine, it was conjectured that some of its effects were due to direct excitation of smooth muscle.

EFFECT ON ORGAN SYSTEMS

Cardiovascular

In experimental animals the effect of ephedrine administered intravenously is similar to that of epinephrine. The arterial pressure — systolic, diastolic, and mean pressure — rises and vagal slowing occurs. Compared with epinephrine, the pressor response to ephedrine occurs somewhat more slowly, and lasts about ten times longer. Furthermore, it requires more ephedrine than epinephrine to obtain an equivalent pressor response. How much more depends on the species tested, type and degree of anesthesia, dose level, and individual variability of the test animal. It is, therefore, almost impossible to give a definite figure for the relative potency of ephedrine and epinephrine. It is commonly accepted that it requires about 250 times more ephedrine than epinephrine to achieve equipressor responses.

If a second dose of ephedrine is administered too soon, its pressor response proves weaker than that of the first dose. This phenomenon, known as ***tachyphylaxis***, occurs with many adrenergic agents, and is related to the duration of action of the drugs. The longer acting the adrenergic agent, the more marked the tachyphylaxis.

The shorter the time interval between doses, the smaller the pressor response to each subsequent dose. Tachyphylaxis probably represents a dynamic blockade of the adrenergic receptors. Ephedrine, for example, activates the receptor, and this activation persists until the drug is completely eliminated. Before the drug is eliminated, however, the arterial pressure returns toward the control level because of compensatory cardiovascular changes, involving only an apparent loss of activity rather than a real one. When another dose is then administered, it acts on receptors still activated so that the pressor response obtained is less than expected.

The pressor response to ephedrine is due in part to peripheral constriction and in part to myocardial stimulation. Vasoconstriction can be demonstrated by intra-arterial injection, but compared to epinephrine, ephedrine is only about one thousandth as active. This would imply that the cardiac effect is predominant in increasing the arterial pressure. This, however, is difficult to demonstrate. In perfused hearts, ephedrine produces only minor stimulation, and if the drug is repeated, cardiac depression appears.

In humans, ephedrine increases the arterial pressure both by peripheral vasoconstriction and by cardiac stimulation. The heart rate is usually increased, as is the pulse pressure, both suggesting an increased cardiac output. The cardiovascular response to ephedrine is, however, quite variable. In some cases the arterial pressure is not elevated; in others bradycardia occurs at the height of the pressure rise. The peripheral resistance has been reported as increasing, decreasing, or remaining unchanged. However, the hypotension that commonly occurs during surgery under spinal anesthesia is practically always prevented by ephedrine. The vasoconstrictive action of ephedrine is demonstrated by the decongestive effect of solutions applied topically to the mucous membranes of the nose. Like epinephrine, ephedrine often produces a secondary congestive response.

In summary, based on the cardiovascular responses that can be observed, it can be stated that ephedrine activates the same adrenergic receptors as does epinephrine but is less potent and has a longer duration of action.

Bronchi

The smooth muscle of the bronchial tree is relaxed by ephedrine. Compared with epinephrine, the action of ephedrine is slow in onset, becoming complete only an hour or more after administration. Ephedrine prevents **histamine-induced bronchoconstriction** in patients with asthma.

Eye

Ephedrine produces mydriasis when applied locally to the conjunctiva, as well as upon systemic absorption. In humans, there is a striking disparity between the mydriatic effects of ephedrine in Caucasians and in Chinese or Blacks. It is most active in the first, less active in the second, and almost completely inactive in the last. The reason for this differential effect on irides of different color has not yet been completely explained.

Other Smooth Muscle

In general, ephedrine produces the same effects on smooth muscle as epinephrine. Inhibition of the intact gastrointestinal musculature and contraction of the splenic capsule and of pilomotor muscles are produced. Ephedrine has the same myometrial and urinary bladder actions as does epinephrine.

Glands

The effects of ephedrine are similar to those of epinephrine. However, ephedrine is much less potent, and with routes of administration other than intravenous, the glandular responses do not always occur. Ephedrine does produce hyperglycemia and eosinopenia.

Central Nervous System

All the adrenergic agents possessing the unsubstituted phenyl ring and the methyl group on the α carbon atom are corticomedullary stimulants. Depending on the dose, this stimulant action in humans results in feelings of anxiety, tremor, insomnia and mental alertness, and increased respiration. When ephedrine is used for its adrenergic effects, the central stimulation produced by it may be considered as an undesirable side action. When it is used as a central stimulant, the adrenergic effect becomes the side action.

At one time, ephedrine was advocated as a useful central stimulant in **narcolepsy** and depressant poisoning. Although it still can be used if necessary, other agents such as **amphetamine** and **methamphetamine** are more often used today.

Skeletal Muscle

Although it has long been known that ephedrine occasionally influences the muscular weakness of **myasthenia gravis** favorably, the exact mechanism of this action is unknown.

ROUTES OF ADMINISTRATION

Ephedrine can be administered by almost any route. Most frequently it is given orally. It is also injected subcutaneously, intramuscularly, or intravenously. For local action, as in the eye or on the nasal mucosa, ephedrine solutions are applied directly by drops, on a tampon, or as a spray.

ABSORPTION, FATE, AND EXCRETION

Ephedrine is readily and completely absorbed after oral or parenteral administration. As it is less active than epinephrine, it does not produce enough local vasoconstriction to hinder absorption after subcutaneous or intramuscular injection. As has been indicated, ephedrine is resistant to amine oxidase, but it is deaminated to some extent in the liver, probably by the **ascorbic–dehydroascorbic acid system**. Conjugation also occurs. In addition, up to 40% of the ephedrine administered may be excreted unchanged in the urine. Inactivation and excretion are so slow that the action of ephedrine may persist for several hours.

Urinary acidifiers may increase elimination of ephedrine, and **urinary alkalinizers** may decrease the elimination of ephedrine. Ephedrine is chemically incompatible with sodium bicarbonate and the two should not be used together.

THERAPEUTIC USES OF EPHEDRINE

Bronchial Asthma

In this condition, ephedrine is used primarily as a chronic medication for mild or only moderately acute cases, especially in children. In severe asthma, the response to ephedrine is usually poor. Compared with epinephrine, ephedrine is less reliable, is slower in action and longer in duration, and probably more often produces undesirable side effects. The average dose is 25 to 50 mg, orally, repeated three or four times a day. Resistance often develops; it may often be controlled by discontinuing the drug for a few days. In many patients, ephedrine produces anxiety, nervousness, and insomnia, so a barbiturate is often administered at the same time. Capsules containing either 15 mg of **amobarbital**, 25 mg of **pento-barbital sodium**, or 30 mg of **phenobarbital** in addition to 25 mg of ephedrine sulfate have been used in the past.

Allergic Disorders

One of the principal uses of ephedrine and other adrenergic activators is to relieve nasal congestion when applied locally to the mucous membrane. This decongestive effect, brought about by vasoconstriction, is used in acute coryza, vasomotor rhinitis, acute sinusitis, hay fever, and other allergic reactions. To be most useful, a decongestive agent should have the following characteristics:

- It should have a prompt, reliable action.
- A secondary or "rebound" congestion should not occur.
- Side effects due to systemic absorption should be minimal.
- Tachyphylaxis should not occur.
- The solution or preparation applied should not in itself be irritative or harmful to cilia.
- The duration of action should be fairly long so that frequent application does not become necessary.

Ephedrine rates as a fairly good decongestive; its principal disadvantages are that secondary congestion often occurs, tachyphylaxis is common, and symptoms of central stimulation usually appear. Ephedrine is used in concentrations ranging from 0.5 to 3% in aqueous or oily solution. The oil solutions should probably not be used, especially in children, because of the danger of producing lipoid pneumonia. Ephedrine administered orally very often produces a satisfactory decongestive effect. In many cases of allergy this route should be used. Added to cough mixtures, ephedrine acts as an adjuvant to the expectorants by controlling edema in the respiratory tract.

Spinal Anesthesia

Hypotension frequently occurs during surgery under spinal anesthesia. Ephedrine has long been used to prevent or abate this possibly undesirable phenomenon. The skin and subcutaneous area over the spinal puncture site are usually anesthetized with procaine solution as a preliminary step. Adding 50 mg of ephedrine sulfate to this solution, and injecting it partly at the puncture site and partly into the interspinous muscles or ligaments, serves as a convenient way to administer the pressor agent. Some anesthesiologists add this amount of ephedrine sulfate to the anesthetic–spinal fluid mixture. The purpose here is to prolong the anesthesia as described above for epinephrine. The usual dose of ephedrine employed to treat, rather than prevent, the hypotension is 50 mg intramuscularly or 15 mg intravenously. The central stimulant action of ephedrine, which may be objectionable, is controlled by adequate preanesthetic sedation or the concomitant use of a short-acting barbiturate.

Heart Block

In complete heart block with Stokes–Adams syncope, ephedrine may prove of value in a manner similar to epinephrine. In the attempt to increase ventricular rate and prevent ventricular asystole, an initial dose of about 8 mg of ephedrine sulfate orally may be tried. Later, the dosage may be increased to 25 mg three or four times daily. Syncope due to ventricular tachycardia can also be prevented in some cases with ephedrine.

Mydriasis

Many of the adrenergic agents produce a transient, incomplete mydriasis when applied to the eye. **Cycloplegia** does not occur. Ephedrine sulfate in a 3% concentration has been used for this purpose. As has been indicated, the effect is most positive in Caucasians with light-colored irides. Ephedrine is not considered to be a good mydriatic, since it is none too reliable and its solutions are quite irritating to the cornea. **Phenylephrine** should probably be the adrenergic mydriatic of choice.

Enuresis

Medical treatment of this condition consists of either diminishing cholinergic activation of the bladder by means of atropine-like drugs, or increasing adrenergic control by means of drugs like ephedrine. For the latter, ephedrine sulfate in fairly large doses, 60 to 120 mg given orally at bedtime, can be tried. Since ephedrine does tend to prevent micturition, however, urinary retention may occur in any patient being treated with this drug. This usually occurs in elderly male patients and may in some cases be considered a definite contraindication.

Myasthenia Gravis

This serious condition is usually treated with **neostigmine**. Before the value of this drug had been discovered, ephedrine was the only drug available that in any measure would benefit this disease. How it works is not known. Its action may be related to the observation that in certain nerve–muscle preparations epinephrine exhibits an anticurare effect. Compared to neostigmine, ephedrine is much less effective and should not be considered the drug of choice. Small doses of 8 to 25 mg several times a day should be used, since large doses may aggravate the disease.

MISCELLANEOUS ACTIONS OF EPHEDRINE

The effect of ephedrine on secretions is not prominent. Sweating occurs after ephedrine in humans but is not demonstrable in animals. Secretions of the gastro-intestinal tract are usually inhibited. Ephedrine decreases the output of pancreatic juice, and the suggestion has been made that the drug may be serviceable in the management of patients with pancreatic fistula. The metabolic effects of ephedrine are generally similar to those of epinephrine. Thus, the drug causes hyperglycemia, but the action is neither as prominent nor as constant as after epinephrine; indeed, in animals, ephedrine has been shown to block epinephrine hyperglycemia. The metabolic rate is increased in humans, but the response is not marked. Changes observed in the formed elements of the blood after administration of ephedrine are generally similar to those noted after epinephrine and are mainly the result of contraction of the spleen and other blood depots.

Ephedrine exerts interesting but as yet inadequately explained actions on skeletal muscle. Isolated skeletal muscle is contracted by ephedrine; this indicates that the observed effect is independent of actions of the drug on the vascular system or CNS. Numerous investigations have established the ability of ephedrine, epinephrine, and related substances to improve neuromuscular transmission depressed by curare, and to enhance the tension response of both normal and denervated fatigued skeletal muscle. **Tyramine** was twice as effective and **dihydroxyphenylethylamine** 20 times as active as ephedrine in this respect, and the influence on skeletal muscle far outlasted the vasopressor action. In myasthenia gravis, ephedrine produces a real but modest increase in motor power. It will be recalled that epinephrine may potentiate the effects of acetylcholine, anticholinesterases, and cholinergic nerve impulses in skeletal muscle; ephedrine intensifies these effects of epinephrine. The exact mechanism by which ephedrine influences skeletal muscle contraction remains unknown. The effect of ephedrine on autonomic ganglia has been studied, and it has been shown that doses larger than those affecting peripheral autonomic receptor cells cause depression of ganglionic transmission.

The relative potency of ephedrine with other sympathomimetic agents is shown in Table 29.

SIGNS AND SYMPTOMS OF EPHEDRINE TOXICITY

Toxic doses of ephedrine may cause:

- Convulsions
- Nausea
- Vomiting
- Chills
- Cyanosis
- Irritability
- Nervousness
- Fever
- Suicidal behavior
- Tachycardia
- Dilated pupils
- Blurred vision
- Opisthotonos
- Spasms
- Pulmonary edema

Table 29 Relative Potency and Duration of Pressor Activity of Epinephrine Congeners

			CH	CH	NH	Activity of Epinephrine = 1.0	
						Pressor Activity	*Duration of Pressor Action*
β-Phenylethylamine	H	H	H	H	H	0.005–0.0125	3–4
Tyramine	OH	H	H	H	H	0.01–0.05	2
Synephrine	OH	H	OH	H	CH₃	0.01–0.04	4
Phenylephrine	H	OH	OH	H	CH₃	0.2	2
Levarterenol	OH	OH	OH	H	H	1.5	2
Epinephrine	OH	OH	OH	H	CH₃	1.0	1.0
Amphetamine	H	H	H	CH₃	H	0.002–0.01	5–10
Phenylpropanolamine	H	H	OH	CH₃	H	0.003–0.017	7
Ephedrine	H	H	OH	CH₃	CH₃	0.005–0.01	7–10
Paredrinol	OH	H	H	CH₃	CH₃	0.033	10

The *adrenomimetic drugs* mimic the effects of adrenergic sympathetic nerve stimulation on sympathetic effectors. Thus, these drugs are also referred to as *sympathomimetic agents*. The adrenergic transmitter norepinephrine and the adrenal medullary hormone epinephrine also are included under this broad heading. The adrenomimetic drugs are an important group of therapeutic agents, which may be used to maintain blood pressure or to relieve a life-threatening attack of acute bronchial asthma. These drugs are also present in many over-the-counter preparations, in which advantage is taken of their ability to constrict mucosal blood vessels and thus relieve nasal congestion.

- Gasping respirations
- Coma
- Respiratory failure
- Personality changes

27

ERGOT, ITS ALKALOIDS, AND HEADACHE

Ergot alkaloids:

- Cause abortion
- Treat headache
- As ergoloid mesylate, are psychotherapeutic agents

ERGOT AND THE ERGOT ALKALOIDS

Source

Ergot (***Claviceps purpurea***) is a fungus that grows on rye and other grains. The parasitic nature of ergot was not appreciated until the 19th century; before that time it was looked upon as being merely diseased rye. The parasite is especially prevalent during moist, warm weather and can be found in the grain fields of North America and Europe.

The spores of ergot are carried by insects or the wind to the ovaries of young rye, where they germinate into hyphal filaments. The filaments secrete an enzyme that enables the fungus to penetrate the basal portions of the ovary. The decomposition product resulting from this enzymatic action is a yellow mucoid material, called **honeydew**, which at one time was erroneously thought to be the active substance of ergot. The honeydew attracts insects and in this way the fungus is further disseminated. As the hyphal filaments penetrate deep into the ovary of the rye, a dense tissue forms. This tissue gradually consumes the entire substance of the grain and hardens into a purple, curved body called the **sclerotium**. This sclerotium is the commercial source of ergot. It is either picked by hand or separated from the grain by a special threshing process (Figure 71).

Constituents of Ergot

Ergot has been termed a "veritable treasure house of pharmacological constituents." The substances isolated from ergot have been divided into two main groups. In the first group are those products peculiar to ergot and not obtainable from any other source. The second group consists of a large number of compounds

Figure 71   Ergot. (A) head of rye with ergo (a) growing in it; (B) young ergot, showing the growing sclerotium carrying upon its top the old diseased ovary.

that can be isolated from ergot but are also obtainable elsewhere. The important members of the first group are the ergot alkaloids. In addition, there are certain pigments peculiar to ergot but without pharmacological significance. A heterogeneous collection of compounds forms the second group. Included are inorganic constituents, carbohydrates, glycerides, sterols, acids, amino acids, amines, and quaternary ammonium bases. The amines and bases are of pharmacological importance. Ergot contains amines that result from the decarboxylation of six amino acids: leucine, tyrosine, histidine, lysine, arginine, and ornithine. Bacteria are capable of effect in the decarboxylation of amino acids, and it is thought that a similar ability of fungi may account for the presence of amines in ergot. At least three amines are of pharmacological importance — histamine, tyramine, and isoamylamine. The pharmacologically active quaternary ammonium bases found in ergot are choline and acetylcholine.

History

The contamination of an edible grain by a poisonous, parasitic fungus spread death and destruction for centuries. As early as 600 B.C., an Assyrian tablet alluded to a "**noxious pustule in the ear of grain**," and in one of the sacred books of the **Parsees** (400 to 300 B.C.) the following pertinent passage occurs: "Among the evil things created by Angro Maynes are noxious grasses that cause pregnant women to drop the womb and die in childbed." It was fortunate for the ancient Greeks that they objected to the "**black malodorous product of Thrace and Macedonia**," and therefore did not eat rye. Rye was also comparatively unknown to the early Romans for it was not introduced into southwestern Europe until after the beginning of the Christian era. Consequently, there is no undisputed reference to ergot poisoning in the early Greek and Roman literature.

It was not until the Middle Ages that written descriptions of ergot poisoning first appeared, although it is probable that the disease was prevalent long before this time. Strange epidemics were described in which the characteristic symptom was gangrene of the feet, legs, hands, and arms. In severe cases, the tissue became dry and black and the mummified limbs separated off without loss of blood. Limbs were said to be consumed by the **Holy Fire** and blackened like charcoal. Mention was also made of agonizing burning sensations in the extremities. The disease was called Holy Fire or **St. Anthony's fire**; the latter name was in honor of the saint at whose shrine relief was said to be obtained. The relief that followed migration to the shrine of St. Anthony was probably real, for the sufferers received a diet free of contaminated grain during their sojourn at the shrine. The symptoms of ergot poisoning were not restricted to the limbs. Indeed, a frequent complication of ergot poisoning was abortion. A convulsive type of ergotism was also known. There still is no proved explanation of why, in certain instances, ergotism was associated with symptoms referable to the central nervous system. It has been suggested that the convulsive form of ergotism occurred in persons with vitamin deficiency.

It was not until 1670 that ergot was proved to be the cause of the destructive epidemics that, for centuries, had raged uncontrolled. At present, knowledge of the etiology of ergot poisoning makes its prevention quite simple. Yet, outbreaks of ergot poisoning have occurred in the present century: in Russia in 1926, in Ireland in 1929, and in France in 1953. A fungus growing on corn and causing "**corn smut**" (*Ustilago maydis*) is analogous to that found on rye and has been the cause of small outbreaks of maize poisoning especially in the northern Mediterranean countries. The condition is known as *ustilaginism* and closely resembles **ergotism**.

Ergot was known as an obstetrical herb before it was identified as the cause of St. Anthony's fire. It was mentioned to cause "**pain in the womb**." It was used by midwives long before it was recognized by the medical profession (Goodman and Gilman, 1955).

Chemistry

The ergot alkaloids can all be considered to be derivatives of the tetracyclic compound **6-methylergoline**. The naturally occurring alkaloids contain a substituent in the β configuration at position 8 and a double bond in ring D. The natural alkaloids of therapeutic interest are amide derivatives of *d-lysergic acid*; these compounds contain a double bond between C9 and C10 and thus belong to the family of **9-ergolene compounds**. Many alkaloids, containing either a methyl or a hydroxymethyl group at position 8, are present in ergot in small quantities. These have been called **clavine alkaloids** and consist principally of both **9-ergolenes** (e.g., **lysergol**) and **8-ergolenes** (e.g., **elymoclavine**, the **8-ergolene isomer of lysergol**). A crystalline, pharmacologically active preparation first isolated from ergot was called **ergotoxine**. It is now known to be a mixture of four alkaloids: **ergocomine, ergocristine, α-ergocryptine**, and **β-ergocryptine**. The first pure ergot alkaloid, **ergotamine**, was obtained. The discovery of the "water soluble uterotonic principle of ergot" was subsequently determined to be ergonovine (also designated **ergometrine**) (Peroutka, 1996). A classification of the natural alkaloids of ergot and their dextrorotatory isomers follows.

	Natural Alkaloids	Dextrorotatory Isomers	Formula
A. Ergotamine group	Ergotamine	Ergotaminine	$C_{33}H_{35}O_5N_5$
	Ergosine	Ergosinine	$C_{30}H_{37}O_5N_5$
B. Ergotoxine group	Ergocristine	Ergocristinine	$C_{35}H_{39}O_5N_5$
	Ergocryptine	Ergocryptinine	$C_{32}H_{41}O_5N_5$
	Ergocornine	Ergocorninine	$C_{31}H_{39}O_5N_5$
C. Ergobasine group	Ergonovine (Ergobasine, Ergometrine)	Ergonovinine (Ergobasinine, Ergometrinine)	$C_{19}H_{23}O_2N_3$

THE EFFICACY OF ERGOTAMINE IN HEADACHE

The following drugs are used in migraine and cluster headache treatment:

I. Migraine
    A. Abortive Therapy
       - Acetaminophen
       - Aspirin
       - **Ergotamine preparations**
       - NSAIDs (nonsteroidal anti-inflammatory agents)
       - Midrin
       - Sumatriptan
       - Others (chlorpromazine, prochlorperazine, metoclopramide)
    B. Prophylactic Therapy
       - Antidepressants
       - Aspirin
       - β-Blockers
       - Calcium-channel blockers
       - **Ergotamine preparations**
       - Methysergide
       - NSAIDs
       - Others (cyproheptadine, clonidine, anticonvulsants, steroids, calcitonin)

II. Cluster Headaches
  A. Abortive Therapy
    ■ **Ergotamine preparations**
    ■ Local anesthetic agents
    ■ Oxygen
    ■ Sumatriptan
  B. Prophylactic Therapy
    ■ Calcium-channel blockers
    ■ Corticosteroids
    ■ **Ergotamine preparations**
    ■ Lithium
    ■ Methysergide
    ■ Others (capsaicin, leuprolide)

Introduction

Headache, one of the most common symptoms, can be precipitated by a great variety of stimuli: emotional stress; fatigue; sensitivity to certain foods and beverages, including alcohol; medications, and acute illness. There may be no apparent underlying cause. In some individuals, headaches occur frequently but irregularly; however, they are usually acute and short-lived and can be relieved by over-the-counter preparations containing aspirin, acetaminophen, or ibuprofen. This type of headache is usually not debilitating and does not require physician consultation.

In contrast, chronic recurrent headache, for which patients most often consult physicians, is associated with various medical, neurological, or psychogenic disorders. Appropriate therapy depends on an accurate diagnosis of the type of headache.

The Headache Classification Committee of the International Headache Society (1988*) has developed diagnostic criteria for classification of headache disorders, cranial neuralgias, and facial pain; the criteria include painful and nonpainful disorders of the entire head and are based on the diagnosis rather than on the underlying pain mechanisms.

A major type of headache that must be considered in differential diagnosis is that caused by underlying disease: intracranial disturbances (e.g., vascular anomalies, infections, tumors, trauma); diseases involving the head and neck but not the brain (e.g., cervical osteoarthritis; disorders of eye, ear, nose, sinuses, and throat; cranial neuralgias); and systemic diseases (e.g., sudden and severe hypertension, hyperthyroidism). These headaches usually can be relieved by specific therapy for the underlying disorder (e.g., surgical correction of tumors, antibiotics for infections, antiarthritic drugs for osteoarthritis).

Considerations in the management of recurrent migraine cluster, or tension-type headache include the following:

---

* Headache Classification Committee of the International Headache Society, Classification and diagnostic criteria for headache disorders, cranial neuralgias and facial pain, *Cephalalgia,* 8 (Suppl. 7):1–96, 1988.

Table 30   Classification of Headaches

I. Primary Headache Disorders
    A. Muscle contraction headaches
    B. Vascular headaches
       1. Migraine
          a. Classic migraine
          b. Common migraine
          c. Complicated migraine
       2. Cluster
          a. Episodic
          b. Chronic
             i. Primary
             ii. Secondary
          c. Chronic paroxysmal hemicrania
II. Psychogenic Headache Disorders
III. Secondary Headache Disorders

Most headaches are dull, deeply located, and aching. Superimposed on such nondescript pain may be other types of pain that have greater diagnostic value. It is useful to identify all types of pain that have been experienced by the patient, regardless of their frequency or intensity. A throbbing quality and tight muscles about the head, neck, and shoulder girdle are common nonspecific accompaniments of headache. It was formerly believed that tight "hatband" headaches indicated stress, anxiety, or depression, but investigations have not supported this view. Jabbing, brief, sharp cephalic pain, often occurring multifocally (ice pick-like pain), is the signature of a benign disorder.

1. No single therapy is effective in all patients with the same type of headache, which serves to underscore the uncertainty about the pathophysiology of these disorders and the variability among individuals. Therefore, drug therapy must be individualized, and trials of different therapies and drugs may be required to establish an effective regimen.
2. In addition to treatment, some patients should receive prophylactic therapy; these individuals should be monitored closely and adjustments made in choice of therapy or dosage when necessary.
3. Many patients with chronic headaches have received drugs that may cause drug dependence (e.g., barbiturates, antianxiety agents, ergotamine, narcotics, analgesic and caffeine mixtures), and their withdrawal along with instruction regarding their proper use is necessary.

This last consideration may be a primary obstacles in the long-term relief of headache.

Classification of Headaches

Headaches are classified into three major categories: primary, psychogenic, and secondary (Table 30). This section focuses on the characteristics, pathophysiology, management, and prevention of only **migraine headache**.

Precipitating Factors of Headache Symptoms

Many factors are known to precipitate attacks of headache (Table 31). Stress, anxiety, excitement, depression, fatigue, anger, exertion, hunger, loss of sleep, and poor diet are identified as the most common precipitating factors of nonorganic headaches. Patients may experience headaches precipitated by particularly stressful events, or by sudden relaxation after a period of stress (i.e., during weekends or the first few days of a holiday). Other factors include being subjected to environmental extremes (high altitudes, high humidity, rapid changes in the weather, temperature extremes), slight trauma to the head, and ingestion of vasoactive substances, including some types of beer, grain alcohol, red wines, or certain foods (e.g., those high in tyramine content, those using preservatives or flavor-enhancing substances such as nitrites or monosodium glutamate, chocolate, hard cheeses, and citrus fruits). Cigarette smoking precipitates cluster attacks in some patients. Many drugs may precipitate headaches.

MIGRAINE: THE INVOLVEMENT OF SEROTONIN AND THE EFFECTIVENESS OF SUMATRIPTAN IN MIGRAINE HEADACHE

Metabolism of Serotonin

5-Hydroxytryptamine (serotonin, 5HT) is metabolized according to the following scheme:

$$
\begin{array}{l}
\text{Tryptophan} \\
\quad\downarrow\ \ \text{Tryptophan 5-hydroxylase} \\
\text{5-Hydroxytryptophan} \\
\quad\downarrow\ \ \text{Aromatic L-amino acid decarboxylase} \\
\text{5-Hydroxytryptamine (serotonin)} \\
\quad\downarrow\ \ \text{Monoamine oxidase} \\
\text{5-Hydroxyindoleacetaldehyde} \\
\quad\downarrow\ \ \text{Aldehyde dehydrogenase} \\
\text{5-Hydroxyindoleacetic acid}
\end{array}
$$

Actions of Serotonin

Serotonin possesses many actions:

- Serotonin is involved in the neural network that regulates intestinal mobility.
- Serotonin is released by platelets (also ADP) during aggregation.
- Serotonin causes vasoconstriction by stimulating $5HT_2$ receptors, and this effect is blocked by **ketanserine**.
- Serotonin causes vasodilation by stimulating $5HT_1$ receptors.
- Serotonin causes positive inotropic and chronotropic effects by interacting with both $5HT_1$ and $5HT_3$ receptors.

Table 31   Precipitating Factors of Headache Symptoms

Physiological factors: stress, nervous tension, anxiety
Physical exertion, fatigue
Alcohol
Hormonal changes
Environmental: temperature extremes of rapid changes, glare
Foods containing: nitrites, glutamates, tyramine
Drugs: nitroglycerine, hydralazine, oral contraceptives, reserpine, theophylline,
  digitalis, estrogens, ergotamine, corticosteroids, or caffeine withdrawal

Pain is most commonly due to tissue injury resulting in stimulation of peripheral nociceptors in an intact nervous system, as in the pain of scalded skin or appendicitis. Pain can also result from damage to or anomalous activation of pain-sensitive pathways of the peripheral or central nervous system. Headache may originate from either mechanism. The following cranial structures are sensitive to mechanical stimulation: the scalp and aponeurotica, middle meningeal artery, dural sinuses, flax cerebri, and the proximal segments of the large pial arteries. The ventricular ependyma, choroid plexus, pial veins, and much of the brain parenchyma are pain insensitive. Electrical stimulation of the midbrain in the region of the dorsal raphe has resulted in migraine-like headaches. Thus, whereas most of the brain is insensitive to electrode probing, a site in the midbrain represents a possible source of headache generation.

- Serotonin increases the motility of the stomach as well as small and large intestines.
- Serotonin causes uterine contractions.
- Serotonin causes bronchial contractions.

Serotonergic Abnormalities. Serotonin (5-hydroxytryptamine; 5HT) is a biogenic amine neurotransmitter that has been implicated in the pathogenesis of migraine. Biochemical studies have documented abnormalities of serotonergic systems in migraine. For example, plasma and platelet levels of 5HT have been reported to vary during different phases of the migraine attack. At the same time, increased amounts of 5HT and its metabolites are excreted in the urine during most headache attacks. The observation that migraine may be precipitated by drugs that cause release of this biogenic amine from tissue stores, such as **reserpine** and **fenfluramine**, also supports a role for 5HT in the disorder.

Other Hypotheses. Many theories have been proposed to explain migraine pathogenesis. Alterations in neurotransmitter systems (e.g., glutamate, nitric oxide, opioids), anatomical structures (e.g., the raphe system, vasculature), and the autonomic nervous system may be either primary or secondary factors in the evolution of a migraine attack.

Serotonin Receptors and Their Subtypes

Extensive ligand-binding studies and molecular biologic examination of membrane preparations have revealed that there are many types of serotonin receptors, including: $5HT_{1A}$, $5HT_{1B}$, $5HT_{1C}$, $5HT_{1D}$, $5HT_2$, $5HT_3$, and $5HT_4$.

Serotonin Receptor Antagonists

Following are the serotonin receptor antagonists:

Ketaserine, a $5HT_2$ and alpha$_1$-adrenergic receptor antagonist, lowers blood pressure.

Methysergide, a $5HT_{1C}$ antagonist, has been used for the prophylactic treatment of migraine and other vascular headache, including **Horton's syndrome**.

Cyproheptadine, a serotonin and histamine receptor- and muscarinic cholinergic receptor-blocking agent, has been used in the treatment of **postgastrectomy dumping syndrome** and the intestinal hypermotility seen with carcinoid.

Sumatriptan, antagonist of the $5HT_1$-like receptor, is highly effective in the treatment of migraine.

Ondansetron, **granisetron**, **topisetron**, and **batanopride** are antagonists of the $5HT_3$ receptor, and are considered effective in controlling cancer chemotherapy-induced emesis.

Clozapine, an effective antipsychotic agent (neuroleptics) with little or no extrapyramidal side effects blocks the $5HT_2$ receptor.

Clinical Efficacy of Sumatriptan

When administered during the headache phase of migraine attack, sumatriptan has a rapid effect on all the symptoms: headache, nausea, vomiting, photophobia, and phonophobia. This effect applies to both migraine attacks preceded by an aura (classic migraine) and those without an aura (common migraine).

Side Effects of Sumatriptan

The acute side effects of sumatriptan are generally minor and transient and seldom lead to serious complications. Following subcutaneous administration, short-lasting pain and redness at the injection site are the most frequently reported adverse events. Other adverse effects reported include a tingling feeling in the chest and limbs, hot flashes, nausea and vomiting, dizziness, and painfulness in the limbs, fatigue, drowsiness, and painfully stiff neck. It should be noted, however, that some of these symptoms are part of the recovery phase of a migraine attack and may reflect migraine relief rather than a direct drug-related effect. Some patients may experience a frightening feeling of severe pressure on the chest that mimics cardiac ischemia.

Mechanisms of Antimigraine Action of Sumatriptan

The pathology of unilateral migraine is based on the anatomy of the trigeminal nerve. Cranial blood vessels, including the large cerebral arteries and vessels in the dura mater, are innervated by a dense perivascular network of sensory nerves. Vasodilation of these vessels may cause profound headache, mediated by nociceptive impulse transmission via the **trigeminal (fifth cranial) nerve**. As a result,

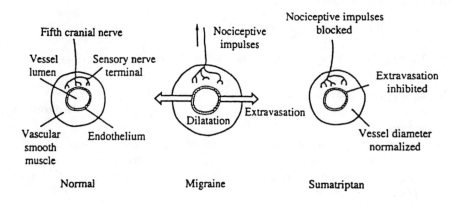

Figure 72 Summary of the vascular effects of sumatriptan. A schematic representation of the action of the sumatriptan in migraine therapy is shown. Vasoconstriction leads to reduced extravasation, blockage of sterile inflammatory response, and diminished head pain.

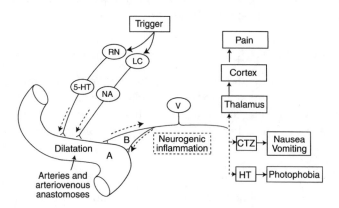

Figure 73 Putative sites of action of sumatriptan. Changes in the activity of raphe nuclei (RN), with their 5HT-containing neurons, and the locus ceruleus (LC) with its noradrenaline-containing (NA) efferents, might induce dilation, of arteries and arteriovenous anastomoses in cephalic (dural and scalp) circulation. This, in turn, may stimulate perivascular sensory afferents of the fifth cranial nerve (V) to cause headache and possibly nausea, vomiting, and photophobia through putative neurons to the chemoreceptor trigger zone (CTZ) and hypothalamus (HT). In addition, neurogenic inflammation through retrograde release of vasoactive neuropeptides and local ischemia caused by arteriovenous shunting may accentuate pain sensation. Sumatriptan appears to abort migraine attacks by constricting the dilated cephalic vessels (A). Sumatriptan may also have inhibitory influence at the perivascular nerve terminals (B) and CTZ, outside the blood–brain barrier.

vasoactive neuropeptides, such as **substance P** and **calcitonin gene-related peptide**, are released, causing increased vascular permeability in the meninges and leading to protein extravasation. Sumatriptan stimulates vascular 5HT receptors that constrict the affected vessels and block the inflammatory response, reducing extravasation, and consequently relieving pain (Figures 72 and 73).

Table 32 Summary Pharmacology of β-Adrenergic Receptor-Blocking Agents

Drug Name	ISA	β-1 Selectivity	α-Blocking Activity	Lipophilicity	Bioavailability,[a] %
Propranolol	–	–	–	+++	30
Timolol	–	–	–	++	75
Metoprolol	–	+	–	++	45
Nadolol	–	–	–	–	30
Atenolol	–	+	–	–	55
Acebutolol	±	+	–	+	40
Labetolol	–	–	+	++	30
Pindolol	+++	–	–	+	100
Penbutolol	±	–	–	+	100

The adrenergic blocking agents, presumably because of their structural similarity to the adrenergic agonists, also have an affinity for the adrenoreceptors. The antagonists, however, have only limited or no capacity to activate the receptors; that is, they have little or negligible intrinsic activity. The blocking drugs compete with the adrenomimetic substances for access to the receptors. Thus, these agents reduce the effects produced by both sympathetic nerve stimulation and by exogenously administered adrenomimetics. This action forms the basis for their therapeutic and investigational use.

PREVENTION AND TREATMENT OF MIGRAINE

The β-Adrenergic Blockers

The most important drugs for the prevention of migraine are the β-adrenergic blocking agents (Table 32). Their value in chronic daily headache and mixed forms is also recognized. Nadolol (Corgard) and propranolol (Inderal) are the most widely administered and tested agents, but others may be of similar value.

Methysergide

Methysergide (Sansert), the oldest of the migraine prophylactic agents, is marred by a history of adverse consequences, but most authorities believe that selective and carefully monitored usage is both appropriate and necessary in difficult-to-manage cases. The drug is given as 2 mg three to four times a day in equally divided dosages. Treatment should be continued for no more than 6 months (Figure 74).

CALCIUM AND CALCIUM-CHANNEL BLOCKERS

Calcium Antagonists

Of the available calcium antagonists — **verapamil** (Isoptin, Calan), **diltiazem** (Cardizem), and **nifedipine** (Procardia) — verapamil and diltiazem seem to have the greatest effectiveness for treatment of headache. Calcium antagonists do not appear to be as effective as the β-blockers for prophylaxis of migraine but may be very effective in certain instances. They are agents of choice in patients who cannot take β-adrenergic blockers, such as patients with asthma. Verapamil is

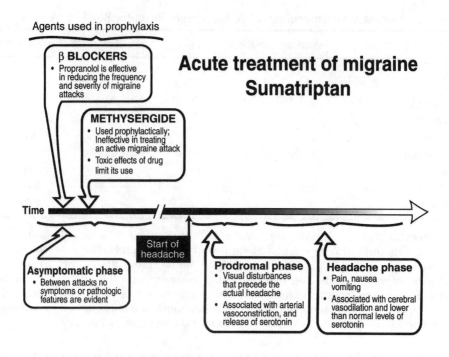

Figure 74 Drugs useful in the treatment and prophylaxis of migraine headaches.

given in a dose range of 40 to 160 mg, two to three times per day; diltiazem is administered in a dose range of 30 to 90 mg, two to three times per day. Many patients administered nifedipine report increased headache, at least initially, although successful results are occasionally evident.

Other Preventative Agents

Clonidine hydrochloride (Catapres), **carbamazepine** (Tegretol), and **methylphenidate** (Ritalin) are occasionally useful in intractable cases of migraine. **Cyproheptadine** (Periactin) may be effective in adults with migraine; it is of considerable importance in the treatment of childhood migraine, and many consider it to be the drug of first choice. Dosages range from 4 to 3 mg, three to four times a day, in adults and 4 mg, two to three times a day, in children.

CRANIAL NEURALGIAS

A variety of drugs with antiepileptic, antispastic, or antidepressant activity are used to manage cranial neuralgias. These compounds are usually more useful than agents with general analgesic properties.

TRIGEMINAL AND GLOSSOPHARYNGEAL NEURALGIAS

Trigeminal neuralgia, an episodic facial pain syndrome, occurs most often in elderly patients. Unilateral paroxysms of severe shooting pain in one or more divisions of the trigeminal nerve are followed by a period of relief and then by another paroxysm of severe pain. This pain is characteristically triggered by tactile stimuli on the face and by using the jaws or the mouth (e.g., chewing, talking, brushing teeth).

Carbamazepine (Epitol, Tegretol) is the drug of choice. If attacks are not controlled completely by carbamazepine or the patient cannot tolerate this agent, the addition of **baclofen** (Lioresal) has been effective. **Phenytoin** (Dilantin) also may be administered when results are suboptimal with carbamazepine. Both drugs may be given alone, with carbamazepine, or with each other, but phenytoin is less effective than baclofen.

A recent limited but controlled study indicated that **pimozide** (Orap) was more effective than carbamazepine. However, the potential for severe extrapyramidal adverse effects with pimozide is considerable, and it is not a drug of choice.

Clonazepam (Klonopin), a benzodiazepine anticonvulsant, has been useful in a limited number of patients, but controlled studies comparing this drug with carbamazepine and baclofen have never been performed.

Glossopharyngeal neuralgia is much less common than trigeminal neuralgia and is characterized by similar paroxysms of lancinating pain in the pharynx, tonsils, and ear. Pain is often triggered by swallowing. It may be accompanied by syncope that is thought to be due to vagal activity and usually is responsive to atropine. Carbamazepine is the drug of choice, but the other agents employed in trigeminal neuralgia may be substituted or used concomitantly. If pain persists, **cocainization of the pharynx** may provide temporary relief. When glossopharyngeal neuralgia is refractory to drug therapy, **microvascular decompression of the glossopharyngeal nerve** or surgical section of the ninth and tenth cranial nerve roots is recommended.

ERGOLOID MESYLATES (DIHYDROGENATED ERGOT ALKALOIDS; DIHYDROERGOTOXINE)

Ergoloid is a psychotherapeutic agent that is indicated in the treatment of age-related decline in mental capacity, primary progressive dementia, Alzheimer's dementia, multi-infarct dementia, and senile onset. Ergoloid is thought to increase cerebral blood flow.

28

FOLATE
IN FRUIT

Both **vitamin B$_{12}$** and **folic acid** are essential for the synthesis of DNA, and this process is impaired in patients with megaloblastic anemia. In the absence of adequate DNA synthesis, cells cannot divide but continue to grow.

There exists an inverse association between folate and cardiovascular disease. **Folic acid deficiency** may result from:

- Nutritional deficiency
- Malabsorption syndrome
- Reduced folate-binding protein
- Folic acid antagonists (e.g., **methotrexate**)
- Drugs reducing the level of folic acid (anticonvulsants and **pyrimethamine**)
- Agents blocking purine synthesis (e.g., **mercaptopurine**, thioguanine) or pyrimidine synthesis (**5-fluorouracil**)
- Hemolytic diseases (accelerated hematopoiesis)
- Proliferative diseases and other conditions

Folic acid is administered orally, and should not be used in the treatment of pernicious anemia.

FUNCTIONS OF VITAMIN B$_{12}$

Vitamin B$_{12}$ is a biologically active corrinoid, a group of cobalt-containing compounds with macrocyclic pyrrol rings. Vitamin B$_{12}$ functions as a cofactor for two enzymes, methionine synthase and L-methylmalonyl coenzyme A (CoA) mutase. Methionine synthase requires methylcobalamin for the methyl transfer from methyltetrahydrofolate to homocysteine to form methionine tetrahydrofolate. L-Methylmalonyl-CoA mutase requires adenosylcobalamin to convert L-methylmalonyl-CoA to succinyl-CoA in an isomerization reaction. An inadequate supply of vitamin B$_{12}$ results in neuropathy, megaloblastic anemia, and gastrointestinal symptoms (Baik and Russell, 1999).

BIOAVAILABILITY OF VITAMIN B$_{12}$ FROM DIFFERENT FOOD SOURCES

In healthy adults, the percentage of vitamin B$_{12}$ absorbed from eggs is 24 to 36%, from trout 25 to 47%, and from chicken, mutton, and liver 60, 65, and 9%, respectively. The bioavailability of vitamin B$_{12}$ from liver is low because its content of vitamin B$_{12}$ is high.

FOOD SOURCES OF VITAMIN B$_{12}$ FOR ELDERLY PEOPLE

Animal-origin food is the only natural food source of vitamin B$_{12}$. Plant foods do not provide it unless the plant is exposed to vitamin B$_{12}$-producing bacteria, contaminated with vitamin B$_{12}$-containing substances (soil, insect parts, etc.), or fortified with vitamin B$_{12}$ (e.g., fortified ready-to-eat breakfast cereals). Foods high in vitamin B$_{12}$ are dairy products, meat, liver, fish, eggs, and shellfish. Because atrophic gastritis with decreased acid pepsin production is prevalent in elderly people, absorption of food-bound vitamin B$_{12}$ is lower in older than in younger, healthier people. Milk is the most important source of vitamin B$_{12}$ for lactovegetarians because it contains 0.4 μg/100 ml (0.9 μg/cup).

Factors Contributing to Declining Vitamin B$_{12}$ Status with Aging

Pernicious anemia associated with gastric atrophy is the most common cause of clinically apparent vitamin B$_{12}$ deficiency in North American and European populations.

Prevalence of Vitamin B$_{12}$ Deficiency in Elderly People

Serum vitamin B$_{12}$ levels decrease with age, and serum methylmalonic acid concentrations increase with age. These findings reflect a decline in vitamin B$_{12}$ status in elderly people.

EFFECTS OF DEFICIENCY

Neurological Effects of Deficiency

In the past, neurological complications were thought to occur at a later stage of vitamin B$_{12}$ deficiency than hematological changes, but recent reports indicate that neurological changes can occur in the absence of any hematological abnormalities.

Hematological Effects of Deficiency

Megaloblastic anemia is a classical finding of vitamin B$_{12}$ deficiency. However, recent studies have demonstrated that subjects with vitamin B$_{12}$ deficiency often lack anemia and macrocytosis, and that there is a dissociation between the neurological and the hematological manifestations.

Gastrointestinal Effects of Deficiency

Gastrointestinal signs and symptoms of vitamin B$_{12}$ deficiency occur in 26% of cases. These include sore tongue, stomatitis, mucosal ulceration, appetite loss, flatulence, and constipation or diarrhea. Appetite loss, excess gas, and diarrhea are probably related to the underlying gastric disorder (i.e., gastric atrophy) in pernicious anemia. Gastrointestinal symptoms may occur in the absence of symptomatic anemia or macrocytosis.

Indicators of Hematological Status

Hematological indices are the simplest way to diagnose megaloblastic anemia, a classical finding of vitamin B_{12} deficiency.

Replacement Therapy

In pernicious anemia, vitamin B_{12} should be given as intramuscular injection or high-dose oral supplements. Intramuscular injections of 100 to 1000 µg of cyanocobalamin for 5 days and 100 to 1000 µg of cyanocobalamin each month thereafter is a sufficient protocol for treating pernicious anemia (see Baik and Russell, 1999).

FUNCTION OF FOLIC ACID IN FESTATION

More than 30 years ago, it was suggested that maternal intake of certain vitamins during pregnancy affected the incidence of serious fetal malformations. Subsequent research has revealed that folate (folic acid), a B vitamin, plays a crucial role in the development of the central nervous system during the early weeks of gestation, which is generally before the pregnancy is confirmed. In a significant number of embryos, an inadequate supply of folate at this time leads to a failure of the primitive neural tube to close and differentiate normally and results in neural tube birth defects (NTD). Numerous studies have confirmed the importance of an adequate intake of folate during the weeks just before and after conception. Overall, the data predict that if women consume multivitamin supplements containing folic acid during the periconceptional period, the number of children born with serious malformations (such as spina bifida and anenephaly) could be reduced by half. Although programs to increase dietary folate intake of potential mothers may be effective in reducing NTD, the only proven and practical preventive measure currently available is to take oral multivitamin supplements containing folic acid. Multivitamin supplementation has also been associated with reduced incidence of other congenital malformations (see Butterworth, 1996).

FUNCTION IN CARDIOVASCULAR DISEASE

The **atherosclerotic lesions** develop in a complex, chronic process. The first detectable lesion is the so-called "fatty streak," an aggregation of lipid-laden macrophage foam cells. A further stage of development is the formation of plaques consisting of a core of lipid and necrotic cell debris covered by a layer of connective tissue and smooth muscle cells. These plaques hinder arterial blood flow and may precipitate clinical events by plaque rupture and thrombus formation. Platelets from the thrombi, activated macrophages, and smooth muscle cells release growth factors and cytokines resulting in an inflammatory–fibroproliferative response that leads to the advanced lesions of atherosclerosis.

Evidence that a diet rich in fruit and vegetables may protect against coronary heart disease is accumulating. It is unclear exactly which substances in fruit and vegetables are responsible for the observed inverse association with cardiovascular disease. The inverse association may be attributed to **folate, antioxidant vitamins**, or other constituents such as **fiber, potassium, flavonoids**, or other **phytochemicals**. The protective effect of folate may be attributed to its role as a cosubstrate in **homocysteine metabolism** (Eichholzer et al., 2001).

ANTIOXIDANT VITAMINS

Fruits and vegetables are rich sources of antioxidant vitamins. Oxidation of low-density lipoproteins (LDL) by free radicals facilitates the uptake of LDL by macrophages to form the cholesterol-laden **foam cells** involved in the development of atherosclerotic lesions. Oxidative processes may also play a role in lesion maturation and the precipitation of clinical events. The antioxidant vitamins **α-tocopherol** and **β-carotene** are both carried within LDL particles. **Vitamin E** prevents the oxidation of unsaturated fatty acids to peroxides and β-carotene has singlet oxygen scavenging ability. **Vitamin C** may regenerate vitamin E and scavenge free radicals in the cytoplasm. Vitamin E may also have an additional role in that it may inhibit the proliferation of smooth-muscle cells and reduce the adhesiveness of platelets to collagen.

The observed associations between folate, antioxidant vitamins, and cardiovascular disease may be confounded by other substances in fruits and vegetables, as the following examples of studies show: **Flavonoids** are naturally occurring, water-soluble antioxidants found widely distributed in **vegetables, fruits, tea**, and **wine**. There exists an inverse relationship between flavonoids and decreased risk of coronary heart disease. **Lycopene**, the key antioxidant in **tomatoes**, shows an inverse association with myocardial infarctions. There exists an inverse association between folate and cardiovascular disease.

VITAMIN B₁₂, FOLIC ACID, AND THE TREATMENT OF MEGALOBLASTIC ANEMIAS

As described in previous sections, vitamin B_{12} and folic acid are dietary essentials. A deficiency of either vitamin results in defective synthesis of DNA in any cell in which chromosomal replication and division are taking place. Since tissues with the greatest rate of cell turnover show the most dramatic changes, the hematopoietic system is especially sensitive to deficiencies of these vitamins. An early sign of deficiency is a megaloblastic anemia. Abnormal macrocytic red blood cells are produced, and the patient becomes severely anemic. Recognition of this pattern of abnormal hematopoiesis, more than 100 years ago, permitted the initial diagnostic classification of such patients as having "**pernicious anemia**" and spurred investigations that subsequently led to the discovery of **vitamin B₁₂** and **folic acid**. Even today, the characteristic abnormality in red blood cell morphology is important for diagnosis and as a therapeutic guide following administration of the vitamins.

RELATIONSHIPS BETWEEN VITAMIN B₁₂ AND FOLIC ACID

Intracellular vitamin B_{12} is maintained as two active coenzymes: **methylcobalamin** and **deoxyadenosylcobalamin**. Deoxyadenosylcobalamin (deoxyadenosy B_{12}) is a cofactor for the mitochondrial mutase enzyme that catalyzes the isomerization of L-**methylmalonyl-CoA** to **succinyl CoA**, an important reaction in both carbohydrate and lipid metabolism. This reaction has no direct relationship to the metabolic pathways that involve folate. In contrast, methylcobalamin (CH_3B_{12}) supports the methionine synthetase reaction, which is essential for normal metabolism of folate. Methyl groups contributed by methyltetrahydrofolate ($CH_3H_4PteGlu_1$) are used to form methylcobalamin, which then acts as a methyl group donor for the conversion of **homocysteine** to **methionine**. This folate–cobalamin interaction is pivotal for normal synthesis of purines and pyrimidines and, therefore, of DNA. The methionine synthetase reaction is largely responsible for the control of the recycling of folate cofactors; the maintenance of intracellular concentrations of **folylpolyglutamates**; and, through the synthesis of methionine and its product, **S-adenosyl-methionine**, the maintenance of a number of methylation reactions.

29

FLAVONOIDS

INTRODUCTION

The flavonoids are found in fruits, vegetables, nuts, seeds, herbs, spices, stems, flowers, as well as tea and red wine. They are prominent components of citrus fruits. Flavonoids are effective in:

- Chronic inflammation
- Allergic diseases
- Coronary artery disease
- Breast cancer

The flavonoids protect cells against oxidative stress.

Flavonoids may have existed in nature for over 1 billion years and thus have interacted with evolving organisms over the eons. Clearly, the flavonoids possess some important purposes in nature, having survived in vascular plants throughout evolution. The very long association of plant flavonoids with various animal species and other organisms throughout evolution may account for the extraordinary range of biochemical and pharmacological activities of these chemicals in mammalian and other biological systems. Unique examples are:

- The inhibition of gamete membrane fusion in sea urchins caused by **quercetin** during egg fertilization
- Modulation of mammalian sperm motility by quercetin
- Influencing sexual differentiation

Over 4000 structurally unique flavonoids have been identified in plant sources (Middleton et al., 2000; Table 33). They are primarily recognized as the pigments responsible for the autumnal burst of hues and the many shades of yellow, orange, and red in flowers and food.

Flavonoids, low-molecular-weight substances found in all vascular plants, are **phenylbenzo-pyrones (phenylchromones)** with an assortment of structures based on a common three-ring nucleus. They are usually subdivided according to their substituents into **flavanols, anthocyanidins, flavones, flavanones**, and **chalcones**.

On average, the daily U.S. diet was estimated to contain approximately 1 g of mixed flavonoids expressed as glycosides. The flavonoid consumed most was

Table 33 Some Examples of Subclasses of Naturally Occurring Flavonoids

Classes	Flavonoids
Flavan-3-ols	(+)-Catechin
Anthocyanidins	Cyanidin
	Pelargonidin
Flavones	Apigenin
	Diosmin
	Luteolin
Flavanones	Naringenin
	Naringin
	Hesperetin
	Hesperedin
Chalcones	Phloretin
	Phloridzin
Flavon-3-ols	Quercetin
	Kaempferol
	Myricetin
	Fisetin
	Morin

Flavonoids are among the most widely distributed natural product compounds in plants with over 2000 different compounds reported occurring both in the free state and as glycosides. Their chemical structures are based upon a C_6–C_3–C_6 carbon skeleton with a chroman ring bearing a second aromatic ring in position 2, 3, or 4. The major general structural categories are flavones, flavanones, flavonols, anthocyanidins, and isoflavones. In some cases, the six-membrane heterocyclic ring is replaced by a five-membered ring (aurones) or exists in an open-chain isomeric form (chalcones). In addition to glycosylated derivatives, methylated, acylated, prenylated, or sulfated derivatives also occur. A variety of flavonoids have been assigned different roles in nature as antimicrobial compounds, stress metabolites, or signaling molecules.

quercetin, and the richest sources of flavonoids consumed in general were **tea** (48% of total), **onions**, and **apples**.

The amount consumed could be considerably higher in the Mediterranean diet, which is rich in olive oil, citrus fruits, and greens. These quantities could provide pharmacologically significant concentrations in body fluids and tissues. Nevertheless, flavonoid dietary intake far exceeds that of **vitamin E**, a monophenolic antioxidant, and that of **β-carotene** on a milligram per day basis.

- The flavonoids have long been recognized to possess anti-inflammatory, antioxidant, antiallergic, hepatoprotective, antithrombotic, antiviral, and anticarcinogenic activities.
- The flavonoids are typical phenolic compounds and, therefore, act as potent metal chelators and free radical scavengers.
- The flavonoids are potent chain-breaking antioxidants.

THE PHARMACODYNAMICS OF FLAVONOIDS

Flavonoids Inhibit Protein Kinase C

Protein kinase C (PKC), the ubiquitous, largely Ca^{2+}- and phospholipid-dependent, multifunctional serine- and threonine-phosphorylating enzyme, is involved in a wide range of cellular activities, including tumor promotion, mitogenesis, secretory processes, inflammatory cell function, and T-lymphocyte function. PKC has been shown to be inhibitable *in vitro* by certain flavonoids. It has been demonstrated that **quercetin** inhibited the phosphorylating activity of the Rous sarcoma virus transforming gene product both *in vitro* and *in vivo*. In addition, quercetin was competitive toward the nucleotide substrates ATP and GTP. Mitogen-activated protein (MAP) kinase in human epidermal carcinoma cells was strongly inhibited by quercetin (30 μM). **Fisetin**, **quercetin**, and **luteolin** are the most active flavonoid inhibitors of protein kinase C.

Flavonoids Inhibit Phospholipase A_2

Phospholipase A_2 (PLA_2), an enzyme involved in many cell activation processes, catalyzes the hydrolysis of phospholipids esterified at the second carbon in the glycerol backbone. Arachidonic acid is commonly esterified in this position, and the action of PLA_2 releases arachidonic acid for subsequent metabolism via the **cyclooxygenase** (CO) and **lipoxygenase** (LO) pathways. PLA_2 is likely an important intra- and extracellular mediator of inflammation. **Quercetin** was found to be an effective inhibitor of PLA_2 from human leukocytes. **Quercetagetin**, **kaempferol-3-O-galactoside**, and **scutellarein** inhibited human recombinant synovial PAL_2 with IC_{50} values ranging from 12.2 to 17.6 μM.

Flavonoids Inhibit Na$^+$ and K$^+$-ATPases

Flavonoids can affect the function of plasma membrane transport Na$^+$- and K$^+$-ATPase, mitochondrial ATPase, and Ca^{2+}-ATPase. The Mg^{2+}-ectoATPase of human leukocytes is inhibited by **quercetin**, which acts as a competitor of ATP binding to the enzyme. The sarcoplasmic reticulum Ca^{2+}-ATPase of muscle is effectively inhibited by several flavonoids that were also active inhibitors of antigen-induced mast cell histamine release.

Flavonoids Inhibit Lipoxygenases and Cyclooxygenases

Arachidonic acid released from membrane phospholipids or other sources is metabolized by the LO pathway to the smooth muscle contractile and vasoactive leukotrienes (LT), LTC_4, LTD_4, as well as to the potent chemoattractant, LTB_4. These molecules are intimately involved in inflammation, asthma, and allergy, as well as in multiple other physiological and pathological processes. For example, **cirsiliol** (3′,4′,5-trihydroxy-6,7-dimethoxyflavone) proved to be a potent inhibitor of 5-LO (IC_{50}, 0.1 μM) derived from basophilic leukemia cells and peritoneal polymorphonuclear leukocytes.

Flavonoids Inhibit Cyclic Nucleotide Phosphodiesterase

The cyclic nucleotides cAMP and cGMP mediate many biological processes through their ability to stimulate cyclic nucleotide-dependent protein kinases, which in turn phosphorylate cellular protein substrates and evoke specific responses. cAMP and cGMP are formed from ATP and GTP by the catalytic activity of adenylate and guanylate cyclases stimulated by various agonists. Their activity is terminated by the cyclic nucleotide phosphodiesterases (PDE). The cyclic nucleotides are involved in regulation of many cellular processes, such as cell division, smooth muscle contractility, secretary functions, immunological processes, and platelet aggregation, to name a few. Flavonoid inhibition of PDEs from many cellular sources has been described. The minimal structural requirements for PDE inhibitor activity include a **flavone**, **flavonol**, or **flavylium** skeleton.

Flavonoids Inhibit Reverse Transcriptase

Selected naturally occurring flavonoids have been shown to inhibit three reverse transcriptases (RT) (avian myeloblastosis RT, Rous-associated virus-2 RT, and Moloney murine leukemia virus, or MMLV, RT) when poly(rA)oligo(dT) 12–18 or rabbit globin mRNA was used as a template. **Amentoflavone**, **scutellarein**, and **quercetin** were the most active compounds, and their effect was concentration dependent. The enzymes exhibited differential sensitivity to the inhibitory effects of the flavonoids. These flavonoids also inhibited rabbit globin mRNA-directed MMLV RT-catalyzed DNA synthesis. **Amentoflavone** and **scutellarein** inhibited ongoing new DNA synthesis catalyzed by Rous-associated virus-2 RT.

Flavonoids Inhibit HIV-1 Proteinase

This enzyme is a necessary component for the processing and replication of HIV-1. It has been suggested that certain flavones may be potential nonpeptidic inhibitors of the enzyme. **Gardenin A**, **myricetin**, **morin**, **quercetin**, and **fisetin** exhibited activity with IC_{50} values in the 10 to 50 M range. Lineweaver–Burk analysis indicated competitive inhibition for fisetin and quercetin. Yet another enzyme involved in HIV replication could be inhibited by quercetin, namely, the **integrase**. This inhibition required at least one ortho pair of phenolic hydroxyl groups and at least one or two additional hydroxyl groups.

Flavonoids Inhibit Ornithine Decarboxylase

Ornithine decarboxylase catalyzes the transformation of ornithine to the polycationic bases, **putresine**, **spermine**, and **spermidine**. These compounds exert regulatory effects on cell growth. It has been shown that quercetin (10 to 30 μmol/mouse) markedly suppressed the stimulatory effect of the transporters associated with antigen processing (TPA) on ornithine decarboxylase (ODC) activity and on skin tumor formation in mice initiated with **dimethylbenzan-thracene**. Such inhibition may be related to the activation of the catalytic site, which is under nonconventional regulation by small molecules. Also, the synthetic flavonoid **flavone acetic acid** was shown to inhibit the activity of ODC in stimulated human peripheral blood lymphocytes and human colonic lamina propria lymphocytes.

Flavonoids Inhibit Topoisomerase

DNA topoisomerases are enzymes that introduce transient breaks in linear DNA sequences. They participate in several genetically related processes, including replication, transcription, recombination, integration, and transposition. DNA topoisomerase II is an important cellular target for several antineoplastic DNA intercalators and nonintercalators. Flavonoids have different effects on these enzymes. It has been shown that **genistein** inhibited mammalian DNA topoisomerase II as well as **protein tyrosine kinase**. Two flavones, **fisetin** and **quercetin**, also showed the same activity. Genistein selectively suppressed the growth of the ras-transformed NIH 3T3 cells, but not the normal NIH 3T3 cells, and inhibited topoisomerase II-catalyzed ATP hydrolysis. In contrast, **baicalein**, **quercetin**, **quercetagetin**, and **myricetin**, known inhibitors of RT, unwound DNA and appeared to promote mammalian DNA topoisomerase-mediated site-specific DNA cleavage.

Flavonoids Inhibit Xanthine Oxidase

Xanthine oxidase catalyzes the formation of urate and superoxide anion from xanthine. Quercetin and baicalein exhibit an inhibitory action on both xanthine oxidase and xanthine dehydrogenase. The level of xanthine oxidase is high in patients with hepatitis and brain tumor and select flavonoids might be useful in treating this disorder.

Flavonoids Inhibit Aromatase

The conversion of **androstenedione** to **estrone** is catalyzed by **aromatase**. Inhibition of aromatase (human estrogen synthetase) by several naturally occurring flavonoids, including **quercetin**, **chrysin**, and **apigenin**, has been described. The synthetic **flavone 7,8-benzoflavone** was most active. Aromatization of androstenedione was affected by several flavonoids, of which **7-hydroxyflavone** and **7,4-dihydroxyflavone** were the most potent.

Inhibition by 7-hydroxyflavone was competitive with respect to the substrate androstenedione. Flavonoids of the 5,7-dihydroxyflavone series could bind to the active site human cytochrome P-450 aromatase with affinity. The flavonoid **kaempferol** inhibited aromatase enzyme activity competitively in a human Glyoxalase cell culture system. Such results suggest that diets rich in these compounds could contribute to the control of estrogen-dependent conditions, such as breast cancer.

Flavonoids Inhibit Aldose Reductase

Lens aldose reductase has been implicated in the pathogenesis of cataracts in diabetic and galactosemic animals. The enzyme catalyzes the reduction of glucose and galactose to their polyols, which accumulate in large quantities in the lens and ultimately lead to mature lens opacities. Several key bioflavones have activity against this enzyme. Oral administration of **quercitrin** decreased the accumulation of sobital in the lens. Therefore, the accumulation of lens opacities could be partially abrogated by certain flavonoids. In a study of 30 flavones, 4 isoflavones, and 13 coumarins, many potent inhibitors were found, but **5,7,3′,4′-tetrahydroxy-3,6-dimethoxyflavone** and **6,3′,4′-trihydroxy-5,7,8-trimethoxyflavone** were especially active.

Flavonoids Inhibit FAD-Containing Monoamine Oxidase

Flavones, coumarins (neoflavonoids), and other oxygen-containing compounds were found to inhibit monoamine oxidases A and B in a reversible and time-independent manner.

Flavonoids Inhibit Carbonyl Reductases (Aldo-Keto-Reductase Family of Enzymes)

Carbonyl reduction is a metabolic pathway widely distributed in nature. Many endogenous substances, such as prostaglandins, biogenic amines, and steroids, together with xenobiotic chemicals of several varieties, are transformed to the corresponding alcohols before further metabolism and elimination. Carbonyl reduction in several continuous cell lines was investigated using metyrapone as a substrate ketone. Quercitrin was reported to inhibit carbonyl reductase.

Flavonoids Inhibit Hyaluronidase

Hyalurondiases, which depolymerize **hyaluronic acid** to **oligosaccharides** by breaking glucosaminidic bonds, have been referred to as a "**spreading factor**," and are possibly involved in tumor cell invasiveness. The most effective flavonoids are **kaempferol** and **silybin**.

Flavonoids Inhibit Histidine Decarboxylase and DOPA Decarboxylase

It has been shown that histidine decarboxylase is inhibited by flavonoids such as **quercetin** and **(+)-catechin**, whereas the flavonoid glycosides were inactive. Histamine stimulates gastric acid secretion, making the reported inhibition of histamine-induced gastric secretion by the synthetic flavone-6-carboxylic acid of interest (Figure 75). It has been shown that the flavan derivative **3-methoxy-5,7,3′,4′-tetrahydroxyflavan**, a compound that appears to be a specific histidine decarboxylase inhibitor, is as effective as **cimetidine** in reducing gastric acid secretion. This flavan also reduced gastric tissue histamine content. **Naringenin**, the aglycone of **naringin**, is a weak inhibitor of **histidine decarboxylase** and also exhibits some gastric antiulcer activity. **Orobol** and **3′,4′,5,7-tetrahydroxy-**

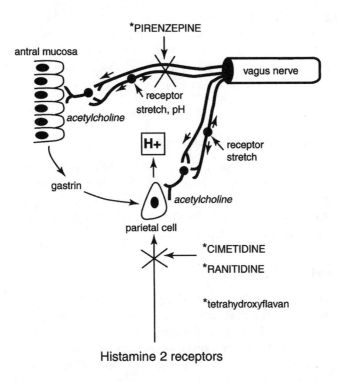

*PIRENZEPINE

antral mucosa

vagus nerve

acetylcholine

receptor
stretch, pH

H+

receptor
stretch

gastrin

acetylcholine

parietal cell

*CIMETIDINE

*RANITIDINE

*tetrahydroxyflavan

Histamine 2 receptors

Figure 75 Flavonoids inhibit and decrease gastric secretion.

8-methoxy isoflavone from culture filtrates of fungi and streptomyces were effective inhibitors of **DOPA decarboxylase**, and **orobol** had a significant hypotensive effect in spontaneously hypertensive animals.

Flavonoids Inhibit Aldehyde and Alcohol Dehydrogenases

An extract of ***Radix puerariae***, an herb long-used in traditional Chinese medicine for alcohol addiction and intoxication, suppresses the free-choice ethanol intake of ethanol. The isoflavonoids **daidzein** (4.,7-dihydroxy-isoflavone) and **daidzin** (7-glucoside of daidzein) isolated from the extract were shown to account for this effect by inhibiting human alcohol dehydrogenase. Daidzin and daidzein, at doses that suppressed ethanol intake, exhibited no effect on overall acetaldehyde and ethanol metabolism in hamsters, although they inhibited human mitochondrial aldehyde dehydrogenase and γ-γ alcohol dehydrogenase *in vitro*. These observations clearly distinguish the action(s) of these isoflavones from those of the classic, broadly acting inhibitors of aldehyde dehydrogenase and of class I alcohol dehydrogenase enzymes. Consequently, daidzin and daidzein represent a new class of compounds offering promise as safe and effective therapeutic agents for alcohol abuse.

Flavonoids Inhibit RNA and DNA Polymerases

Studies have shown that **quercetin**, **kaempferol**, and **fisetin** inhibited transcription with RNA polymerase II in permeabilized normal human fibroblasts (WI-38 cells); **flavone** and **chrysin** exhibited weak activity. Addition of quercetin to an ongoing transcription reaction arrested it promptly, suggesting that quercetin was inhibiting the elongation step. The effects of several flavonoids (**quercetin**, **quercetagetin**, **myricetin**, and **baicalein**) exhibited complex interactions with DNA and RNA polymerases, depending on the particular flavonoid and the enzyme species.

Flavonoids Inhibit Human DNA Ligase I

The effects of several natural products for their ability to disrupt the function of human DNA ligase I, which catalyzes the covalent joining of single-stranded breaks in double-stranded DNA, have been tested. Interestingly, a flavonoxanthone glucoside, **swertifrancheside** (isolated from *Swerua franchetiana*), inhibited enzyme function with IC_{50} of 11 μM.

Flavonoids Inhibit Sialidase

Sialidase (neuraminidase) catalyzes the hydrolysis of sialic acid residues from sialoglyco conjugates and may have an effect on biological functions such as antigen presentation and receptor function. Mouse liver sialidase was noncompetitively inhibited by **isoscutellarein-8-O-glucuronide** (IC_{50}, 40 μM), whereas influenza virus sialidase was only weakly inhibited. **Flavanone** and **chalcone** structures essentially lacked activity against the liver enzyme. In studies of influenza sialidase, the effect of other flavonoids derived from *Scutellaria baicalensis*. **5,7,4'-Trihydroxy-8-methoxyflavone** proved to be a moderately active compound among 103 tested. Because binding of influenza virus to target cells takes place via sialic acid residues in the viral envelope glycoprotein, it is of interest that **5,7,4'-trihydroxy-8-methoxyflavone** also inhibited infection by influenza virus.

ANTIOXIDANT PROPERTIES OF FLAVONOIDS

Flavonoids are a family of antioxidants found in fruits and vegetables as well as in popular beverages such as red wine and tea. Although the physiological benefits of flavonoids have been largely attributed to their antioxidant properties in plasma, flavonoids may also protect cells from various insults. Nerve cell death from oxidative stress has been implicated in a variety of pathologies, including stroke, trauma, and diseases such as Alzheimer's disease and Parkinson's disease.

To determine the potential protective mechanisms of flavonoids in cell death, Ishige et al. (2001) used the mouse hippocampal cell line HT-22, a model system for oxidative stress. In this system, exogenous glutamate inhibits cystine uptake and depletes intracellular glutathione (GSH), leading to the accumulation of reactive oxygen (ROS) and an increase in Ca^{2+} influx, which ultimately causes neuronal death. Many, but not all, flavonoids protect HT-22 cells and primary neurons from glutamate toxicity as well as from five other oxidative injuries. Three structural requirements of flavonoids for protection from glutamate are the hydroxylated C3, an unsaturated C ring, and hydrophobicity. These authors also found three distinct mechanisms of protection. These include increasing intracellular GSH, directly lowering levels of ROS, and preventing the influx of Ca^{2+} despite high levels of ROS. These data show that the mechanism of protection from oxidative insults by flavonoids is highly specific for each compound.

MUTAGENICITY AND GENOTOXICITY OF EXCESSIVE FLAVONOID INTAKE

Plant **flavonoids** are **diphenylpropane derivatives** that exert a wide range of biochemical and pharmacological effects. Their antioxidant properties, cytostatic effects in tumorigenesis, and ability to inhibit a broad spectrum of enzymes, such as protein kinase C, tyrosine protein kinase, and topoisomerase II, have led researchers to regard these compounds as potential **anticarcinogens** and **cardioprotective** agents. These findings have contributed to the dramatic increase in the consumption and use of dietary supplements containing high concentrations of plant flavonoids by some health-conscious individuals. Marketing strategies by manufacturers of flavonoid concentrates and herbal remedies advertise and often exaggerate their nontoxic therapeutic effects, most of which are not substantiated by regulated clinical trials. Furthermore, manufacturers' recommended doses of these products might far exceed the flavonoid dose one could attain from a typical vegetarian diet. This, coupled with the common misconception that, if a little of something is good, then more is better, may result in individuals ingesting extremely high levels of these compounds (Skibola and Smith, 2000).

Despite the apparently beneficial health effects of flavonoids, several studies indicate their mutagenicity and genotoxicity in mammalian systems. This may be due to their activity as pro-oxidants in generating free radicals that damage DNA or their inhibition of DNA-associated enzymes such as topoisomerase. Unrepaired or misrepaired oxidative DNA damage can result in DNA strand breaks and mutations that may lead to irreversible preneoplastic lesions. Furthermore, high intakes of these compounds may potentiate other deleterious effects due to their diverse pharmacological properties, which may alter drug and amino acid metabolism, modulate the activity of environmental genotoxicants, and alter the activity of other key metabolizing enzymes. Although there is ample evidence that a flavonoid-rich diet may promote good health and provide protection from age-related diseases, there remains uncertainty regarding the conditions and the levels of flavonoid intake necessary to pose a potential health hazard.

CHARACTERISTICS AND CONTENT OF FLAVONOIDS IN COMMON FOODS

Flavonoids are widespread in nature, occurring in all plant families, and are found in considerable quantities in fruits, vegetables, grains, cola, tea, coffee, cocoa, beer, and red wine. In the United States, the daily dietary intake of mixed flavonoids is estimated to be in the range of 500 to 1000 mg, but can be as high as several grams in those supplementing their diets with flavonoids or flavonoid-containing herbal preparations such as **ginkgo biloba**, **Pycnogenol 227**, or **grape seed extract**. Assuming that absorption from the gastrointestinal tract is effective, such intakes could provide pharmacologically active concentrations in body fluids and tissues.

Among the major groups of flavonoids in the human diet are the flavonols, proanthocyanidins (which include the catechins), isoflavonoids, flavones, and flavanones. The closely related lignans (resorcyclic acid lactones) have similar properties to flavonoids, and are therefore included here for completeness. Quercetin, a flavonol, is the most predominant flavonoid in the human diet and estimates of human consumption are in the range of 4 to 68 mg/day. Among these studies, the Japanese population had the highest levels of flavonol intake, which was mainly attributed to their green tea consumption.

Biological activities ascribed to quercetin include its antiviral, anti-inflammatory, antiproliferative, and antimicrobial properties. Quercetin is found in high concentrations in commonly consumed foods such as onions, apples, kale, red wine, and green and black teas, and other foods shown below.

Flavonoid	Common Sources
Quercetin	Onion
	Apple
	Kale
	Red wine
	Green/black teas
Genistein/diadzein	Tofu
	Soy milk
	Soy milk formula
Coumestrol	Alfalfa sprouts
	Clover sprouts
Secoisolariciresinol	Flaxseed
	Flaxmeal
Apigenin/luteolin	Millet (fonio)
Naringin	Grapefruit juice

Genistein has been considered the primary anticancer constituent in soy, based on putative *in vitro* activities that include its ability to inhibit topoisomerase I and II activity, inhibit protein tyrosine phosphorylation, induce differentiation of cancer cell lines, and act as an estrogen agonist.

Coumestans, such as **coumestrol**, are found in high concentrations in alfalfa and clover and have been recognized, along with **formononetin**, as the cause of infertility in grazing herbivores. While coumestrol acts as an estrogen mimic directly, formononetin has to be metabolized to the estrogenically active compounds **diadzein**, **equol**, or *O*-demethylangolensein.

Lignans are present in many fruits, vegetables, tea beverages, and cereal grains such as flaxseed. **Flaxseed** is one of the richest sources of **phytoestrogens**. It contains very high concentrations of the lignan 2,3-bis(3-methoxy-4-hydroxybenzyl)butane-1,4-diol (**secoisolarici- resinol**) which is a precursor to the primary mammalian lignans **enterolactone** and **enterodiol**. **Lignans** are known for their antioxidant activity, and have recently been associated with a reduction in hypercholesterolemic atherosclerosis in animal feeding studies.

Unlike the phytoestrogens mentioned above, flavones do not possess estrogenic activity. However, the flavones **apigenin** and **luteolin** act as potent inhibitors of **aromatase** and **17β-hydroxysteroid oxidoreductase**, enzymes involved in estrogen metabolism. Studies have also demonstrated these flavones and several **glycosylflavones** are potent **goitrogens**, particularly in association with **millet consumption**.

SAFE FLAVONOID INTAKE

Recent epidemiological studies indicate that the intake of flavonoids is associated with a reduced risk of coronary heart disease, stroke, diabetes, and cancer, including cancers of the breast, prostate, lung, colon, and stomach, and is associated with increased consumption of fruits, vegetables, and soy products (see Anila and Vijayalakshmi, 2000). Populations at lowest risk are Asians and vegetarians. Based on the average daily intake of flavonols (68 mg) and isoflavones (20 to 240 mg) in Asian populations, dietary exposures at these doses are not likely to cause adverse health effects. To date, no human data on the long-term effects of high-dose supplementation are available. The level of flavonoids required to induce mutations and cytotoxicity may not be physiologically achievable through dietary sources. However, the use of supplements, particularly antioxidant formulas and herbal mixtures that are commonly recommended in terms of gram rather than milligram doses, could result in exposure to potentially toxic levels. For example, typical manufacturers' recommended doses of quercetin supplements range between 500 and 1000 mg/day, which is 10 to 20 times what can be consumed in a typical vegetarian diet. This suggests that unregulated, commercially available flavonoid-containing supplements may have biologic activity that can adversely affect human health.

A significant number of studies provide evidence that the biologic activities of flavonoids may play a dual role in mutagenesis and carcinogenesis. They can act as antimutagens/promutagens and antioxidants/pro-oxidants, which is largely dependent upon the levels consumed as well as the physiological conditions in the body. Exposure to increased levels of flavonoids, whether through the diet or by supplementation, may potentially overwhelm the system, leading to the formation of ROS, and ultimately DNA damage. Furthermore, these effects may be enhanced in fetal development where there is rapid cell growth, which may increase sensitivity to phytochemical exposure. Indeed, little is actually known about the toxicology of excess flavonoid intake, while beneficial attributes are commonly overemphasized (see Skibola and Smith, 2000).

30

FOXGLOVE, CARDIAC GLYCOSIDES, AND CONGESTIVE HEART FAILURE

DIGITALIS

At the end of the 18th century a most important addition to our therapeutic resources occurred with the demonstration of the efficiency of digitalis in certain cases of **dropsy**. The drug had been used for different purposes long before, but its specific action was proved by **William Withering, "the flower of English physicians"** (Figure 76), as he was spoken of at his death bed. Withering (1741–1799) was born in Wellington, England. He studied medicine in Edinburgh and graduated there in 1766 with a paper "Malignant Putrid Sore Throat." He then worked for some years in Stafford as a physician at the infirmary for the poor. From 1775, he was in Birmingham, attached to the staff of the general hospital. He was also much appreciated as a consultant. In 1786, he moved to Edgbaston outside Birmingham.

The cardiac glycosides have been used from time immemorial principally because of their toxic effects. **Strophanthus**, for example, is in the **Ebers papyrus** (1500 B.C.) and was used by the ancient Egyptians in many illnesses. Withering, in 1785, first introduced the use of digitalis in his classic monograph, **"An Account of the Foxglove and Some of Its Medical Uses: with Practical Remarks on Dropsy, and Other Diseases."** He was led to his discovery by investigating the remarkable effects obtained by an old woman in the treatment of dropsy. Withering found that the beneficent effects were due to the presence of digitalis among the sundry herbs that the "old woman of Shropshire" used in her treatment.

Because of the lack of information regarding the mechanism of action of digitalis, and because of its toxic effects when used indiscriminately, the use of the drug gradually fell into disrepute until **MacKenzie**, in 1905, reestablished the correctness of Withering's teachings. Subsequent physiological studies elucidated the mechanism of its action in human disease. Many of the active principles derived from the digitalis bodies have been isolated and made available in pure form for clinical use, in addition to the galenical preparations of the crude drugs (see Grollman, 1962).

Figure 76 William Withering.

THE DIGITALIS SERIES

The digitalis series embraces a considerable number of substances that are characterized by their action on the heart. They are widely distributed in the vegetable kingdom in very different botanical families and have long been in use for various purposes in civilized countries. Some of them were employed as remedies by the laity long before their virtues were recognized by the medical profession, while others have been used as **arrow poisons** and **ordeal poisons** by the natives of different parts of Africa and the Eastern Archipelago.

The most important plants containing bodies belonging to this group that are used therapeutically are ***Digitalis pupurea*** (**purple foxglove**), ***D. lanata***, ***Strophanthus hispidus*** and ***S. kombè***, ***A. ouabia***, ***A. venenata***, and ***Scilla maritima*** (**squills**). Others which are now used are ***Helleborus niger*** (**Christmas rose**), ***Convallaria majalis*** (**lily of the valley**), ***Apocynum cannabinum*** (**Canadian hemp**), ***Adonis vernalis*** (**pheasant's eye**), ***Erythrophlaeum*** (**sassy** or **Casca bark**), ***Thevetia*** (**yellow oleander**), ***Cheiranthus*** (**wallflower**), ***Periploca graca*** (**silk vine**), and ***Coronilla*** (**axseed**). Other plants resemble digitalis in their effects, and moreover these bodies are not confined to the

Figure 77 Foxglove *(Digitalis purpurea)* in bloom. Note the thimble (finger)-shaped flowers from which it derives its name and the leaves from which medicinal digitalis is derived.

vegetable kingdom, but have also been isolated from various species of toads (Figure 77).

CHEMISTRY

The distribution of the cardiac glycosides in the plant varies in different species. In the case of **digitalis**, the leaves and seeds are utilized; in **strophanthus**, the seeds; in **squill**, the bulbs; in *Convallaria*, the flowers; and in **ouabaio**, the wood and bark. In addition, there are saponin bodies, which do not show the typical digitalis effect on the heart but which are extracted along with the glycosides, which possess the cardiac actions.

The digitalis bodies are present in the plant in the form of **glycosides**, which are readily altered by enzymatic processes unless special precautions from digitalis are taken in their isolation. Purpurea may be separated by physical methods into

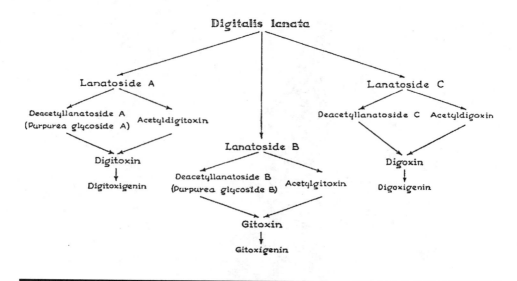

Figure 78 Interrelationship of the *Digitalis* glycosides.

two fractions that are designated as **purpurea glycosides A** and **B**, while from **Digitalis lanata** three glycosides designated as **lanatosides A, B,** and **C** are obtainable, and these are called **acetyldigitoxin, acetylgitoxin,** and **acetyl-digoxin,** respectively. The acetyl derivatives when hydrolyzed with mild alkali or their corresponding deacetyllanatosides when hydrolyzed enzymatically, split-off an acetyl group or glucose, respectively, to give the simple derived glycosides **digitoxin, gitoxin,** and **digoxin.** Digitoxin and gitoxin are also obtainable from *Digitalis purpurea*; digoxin is a specific glycoside of *Digitalis lanata* (Figure 78).

By splitting off the molecule of glucose, purpurea glycoside A is converted to digitoxin, which consists of a steroid nucleus attached to three molecules of **digitoxose**. Purpurea glycoside B in the same way may be hydrolyzed to give one molecule of glucose and gitoxin. The digitalis bodies may be hydrolyzed further to form steroidal compounds designated as "**aglycones**" or "**genins**" and sugars, which in some cases possess unusual character. Many of the aglycones are chemically very closely related and possess the same number of carbon atoms. The structures of several typical members of this group are given in Figure 79. They are steroid compounds possessing the **pentanoperhydro-phenanthrene nucleus** with a four or five carbon lactone ring attached at the 17-carbon atom. The hydroxyl group at carbon-3 is attached to one or more sugar molecules to give the glycoside. The differences between the various aglycones are due mainly to the number, arrangement, and function of the oxygen atoms present in the molecule. Thus, **digoxigenin** differs from **digitoxigenin** in having an –OH group at C_{14}. **Strophanthidin** possesses, in addition to the tertiary hydroxyl group at C_{14}, a second tertiary hydroxyl group at C_5. The interest in and importance of these aglycones stem from the fact that it is in this portion of the glycosidal molecule that the characteristic cardiac action resides.

The sugar portion of the cardiac glycosides, while itself inert, through its union with the aglycone gives the special character to the molecule that affects its

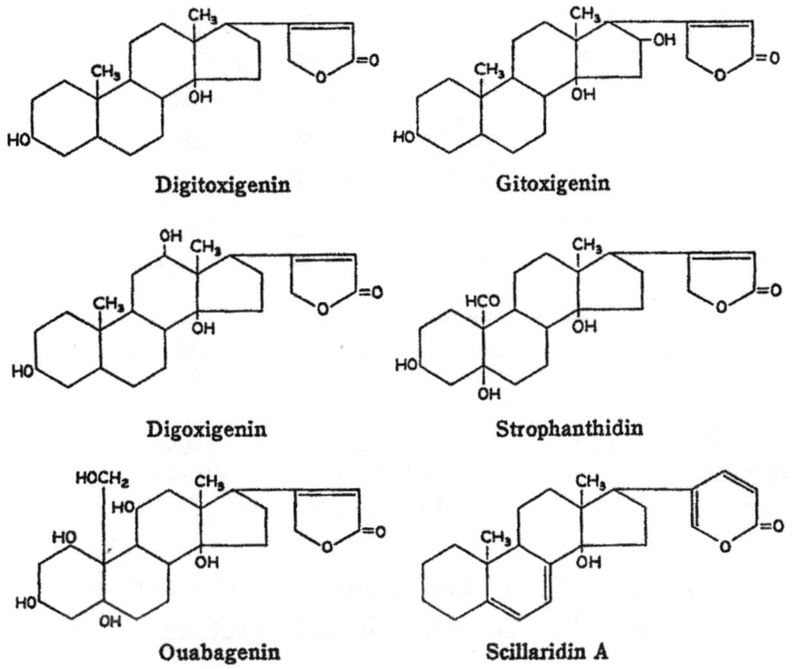

Figure 79 Structural formula of the aglycones of the cardiac glycosides.

absorption and transport in the organism and so influences its specific affinity for cardiac muscle. Several of the sugars, e.g., **digitoxose, cymarose, sarmentose,** and **digitalose,** have thus far been found to occur naturally only in the cardiac glycosides. Several synthetic glycosides have also been prepared from glucose and the naturally occurring genins. The β-glycosides thus prepared are in some cases more potent than the naturally occurring drugs, but have not been applied clinically.

The other members of this series also consist of glycosides that undergo reactions comparable to those already described. Thus, when hydrolyzed, the prinicipal glycosides of strophanthus (cymarin, $C_{30}H_{44}O_9$) are split into its aglycone, strophanthidin (and a sugar, cymarose) (Figure 80). The other glycosides of strophanthus consist of cymarin in combination with one or more molecules of glucose.

From **Strophanthus gratus** a crystalline glycoside, **ouabain,** has been obtained. It was so named because it was first isolated from the root and bark of the **Ouabaio tree** which is used as a source of an arrow poison by the Somalis of East Africa. Ouabain, $C_{29}H_{44}O_{12}$, because of its well-defined physical character-istics and crystalline structure served for a time as a standard in the United States for the assay of the digitalis group of drugs.

The close relationship of the various glycosides of the digitalis series to each other is especially well shown when the aglycones into which they are hydrolyzed are arranged in a series, as follows:

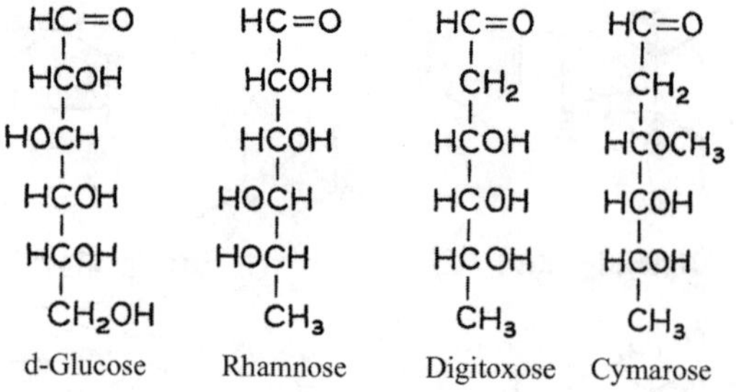

d-Glucose	Rhamnose	Digitoxose	Cymarose
HC=O	HC=O	HC=O	HC=O
HCOH	HCOH	CH$_2$	CH$_2$
HOCH	HCOH	HCOH	HCOCH$_3$
HCOH	HOCH	HCOH	HCOH
HCOH	HOCH	HCOH	HCOH
CH$_2$OH	CH$_3$	CH$_3$	CH$_3$

Figure 80 Structural formula of some simple sugars of the cardiac glycosides.

Strophanthidin $C_{23}H_{32}O_6$ **Digoxigenin** $C_{23}H_{34}O_5$
Digitoxigenin $C_{23}H_{34}O_4$ **Periplogenin** $C_{23}H_{34}O_5$
Gitoxigenin $C_{23}H_{34}O_5$ **Ouabagenin** $C_{23}H_{34}O_8$

CARDIAC GLYCOSIDES IN THE TREATMENT OF CONGESTIVE HEART FAILURE

The most important and often-used drugs in the treatment of congestive heart failure are the cardiac glycosides, which may exist and occur naturally in the body.

When the heart can no longer pump an adequate supply of blood to meet the metabolic needs of the tissues or in relation to venous return, cardiac failure may ensue. The causes of cardiac failure are complex, but stem from **mechanical abnormalities** (e.g., pericardial tamponade), **myocardial failure** (e.g., cardiomyopathy and inflammation), and **arrhythmias.** In high-output failure, the cardiac output, which may be normal or even higher than normal, is not sufficient to meet the metabolic requirement of the body. Cardiac failure may predispose to **congestive heart failure**, which is a state of circulatory congestion. **Toxic injury**, caused by agents such as **doxorubicin**, the alkaloid **emetine** in ipecac syrup, **cocaine**, or **ethyl alcohol**, is one of the ways that the functional integrity of the heart may also be compromised.

A compensatory mechanism is brought into play in the event of **congestive heart failure.** This consists of:

- **Cardiac dilatation** and hypertrophy — taking advantage of the Frank–Starling relationship to utilize more contractile elements.
- **Sympathetic stimulation** — increasing the heart rate to maintain contractility and cardiac output.
- **Increasing oxygen consumption** through the arterial venous oxygen difference — increasing extraction of oxygen from limited blood flow.
- **Production of aldosterone** — increasing sodium and fluid retention, which may not be advantageous to the organism.

Agents with **positive inotropic actions** that may be used in the management of congestive heart failure include the **cardiac glycosides** (e.g., **digoxin** and

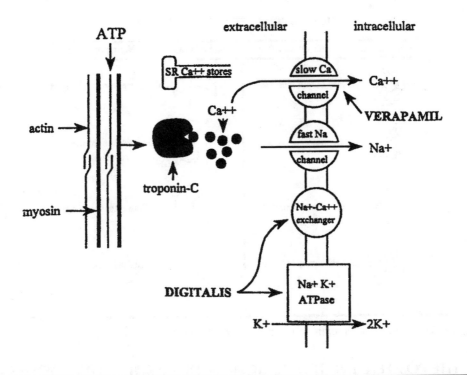

Figure 81 The action of **verapamil** on myocardial contraction. ATP = adenosine triphosphate; SR = sarcoplasmic reticulum.

digitoxin), **dopaminergic analogues** (e.g., **dobutamine**), **phosphodiesterase inhibitors** (e.g., **amrinone** and **milrinone**), **angiotensin antagonists** (e.g., **captopril**, **enalapril**, and **lisinopril**), and **vasodilators** (**nitrates** and **hydralazine**).

BIOCHEMICAL BASIS OF MYOCARDIAL CONTRACTION

The myocardium develops force by forming **cross-bridges** between **actin** and **myosin myofilaments.** By using the energy contained in adenosine triphosphate (ATP), myosin and actin are engaged and cardiac muscle shortens. The contraction of the myocardial cell is initiated by **membrane depolarization** through the rapid influx of **sodium** through the fast sodium channels. The myofilament interaction is regulated by **troponin**, which contains three subunits — **troponin-C, troponin-I,** and **troponin-T.** Troponin-C has a specific site for **calcium.** The subsequent activation of contractile protein is brought about by the influx of extracellular calcium through the slow calcium channels and the subsequent release of calcium from the sarcoplasmic reticulum. Na^+-K^+-ATPase, sarcolemmal Ca^{2+}-ATPase, and the sodium–calcium exchanger play important roles in maintaining the steady-state intracellular level of calcium, sodium, and potassium (Figure 81).

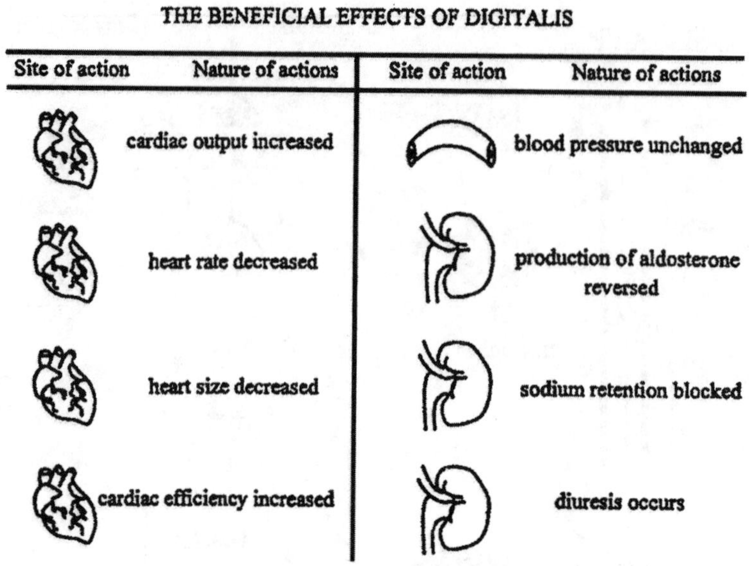

THE BENEFICIAL EFFECTS OF DIGITALIS

Site of action	Nature of actions	Site of action	Nature of actions
	cardiac output increased		blood pressure unchanged
	heart rate decreased		production of aldosterone reversed
	heart size decreased		sodium retention blocked
	cardiac efficiency increased		diuresis occurs

Figure 82 Actions of digitalis in congestive heart failure.

THE POSITIVE INOTROPIC ACTION OF CARDIAC GLYCOSIDES

Cardiac glycosides (**digitalis**) potentiate the coupling of electrical excitation with mechanical contraction, and, by augmenting the myoplasmic concentration of calcium, they provoke a more forceful contraction. It is thought that digitalis inhibits sodium–calcium exchanges by inhibiting **Na$^+$-K$^+$-ATPase.** This results in an enhanced intracellular concentration of sodium, which in turn leads to a greater sodium influx that then elicits stronger systolic contraction.

The Modes of Action of Cardiac Glycosides

The cardiac glycosides increase **cardiac output** through their positive inotropic effect. They slow **heart rate** by relieving the sympathetic tone and through their vagotonic effects. They reduce the **heart size** by relieving the Frank–Starling relationship. They increase **cardiac efficiency** by increasing cardiac output and decreasing oxygen consumption (decreased heart size and rate).

Blood pressure remains unchanged following the administration of cardiac glycosides. In **congestive heart failure**, the cardiac output is reduced but the total peripheral resistance is increased, and these effects are reversed by cardiac glycosides.

Cardiac glycosides bring about **diuresis** by increasing both cardiac output and renal blood flow; the latter in turn reverses the **renal compensatory mechanism** activated in congestive heart failure. Consequently, the production of **aldosterone** is reduced, sodium retention is reversed, and the excretion of edematous fluid is enhanced (Figure 82).

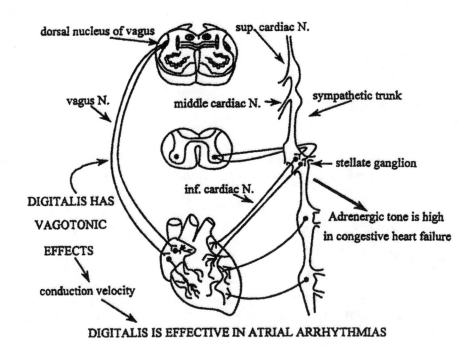

Figure 83 The vagotonic effects of digitalis are useful in treating atrial arrhythmias. N = nerve; sup. = superior; inf. = inferior.

Electrophysiological Effects of Cardiac Glycosides

Cardiac glycosides have a **vagotonic effect** and may decrease impulse formation in the sinoatrial node. Although **automaticity** is not directly influenced by digitalis, **conduction velocity** is decreased. This effect of digitalis on the atrioventricular node is more prominent in the context of congestive heart failure, where the vagal tone is low and the adrenergic tone is high. Digitalis shortens the refractory period, in part due to enhanced intracellular calcium levels, decreasing membrane resistance, and increasing membrane potassium conductance, which lead to shortening of the action potential and contribute to shortening of atrial and ventricular refractoriness. The electrophysiological properties of digitalis make it a useful compound in the treatment of **atrial arrhythmias** (for its vagotonic effect), **atrial flutter** (for its depressant effect on atrioventricular conduction), and **atrial fibrillation** (also for its vagotonic effect) (Figure 83).

DIGITALIS TOXICITY

The toxic effects of digitalis are frequent and may be fatal. Toxicity may result from **overdosage**, **decreased metabolism and excretion**, and **hypokalemia** stemming from the use of thiazide diuretics, diarrhea, and vomiting. Digitalis toxicity has several manifestations:

- Any arrhythmia occurring *de novo*
- Renal insufficiency
- Electrolyte disturbances
- Visual symptoms
- Headache
- Psychotic symptoms
- Pulmonary disease
- Anorexia

Digitalis toxicity should also be closely watched in **elderly** patients and those who have had a **recent myocardial infarction**, as these predispose to toxic reactions.

Symptoms of Digitalis Toxicity

Cardiac Effects

The most commonly reported cardiac signs of toxicity are **dysrhythmias** such as **ventricular ectopic depolarization, second-** and **third-degree heart block, junctional tachycardia, atrial tachycardia with block, ventricular tachycardia, sinoatrial block**, and **sinus arrest**.

Gastrointestinal Effects

Anorexia is seen as a gastrointestinal complication of cardiac glycode use, and is followed by nausea and vomiting.

Visual Effects

The most common visual side effects are blurring, dimness of vision, flickering or flashing lights, color vision (yellow, green, red, and white), **cycloplegia**, and **diplopia**.

Neuropsychiatric Symptoms

A few of the neuropsychiatric symptoms that have been reported in conjunction with cardiac glycoside use are agitation, apathy, aphasia, ataxia, belligerence, changes in affect or personality, confusion, delirium, delusions, depression, disorientation, dizziness, drowsiness, euphoria, excitement, fatigue, hallucinations, headache, insomnia, irritability, mania, muscle pain, nervousness, neuralgias, nightmares, paresthesias, vertigo, violence, and weakness.

Treatment of Digitalis Toxicity

General Treatment

Once digitalis toxicity is diagnosed, digitalis and diuretic use should be stopped. Furthermore, the patient should be monitored closely for any alteration in the pharmacokinetic profile of the cardiac glycoside being used.

Specific Treatment

Treatment with **potassium** and magnesium may be indicated. Potassium is recommended for patients with digitalis-induced ectopic beat or tachycardia, provided the patient is not hyperkalemic, uremic, or oliguric. It is the drug of choice if the patient is hypokalemic.

To manage digitalis-induced arrhythmia, **lidocaine** with its fast onset and short duration of action is the drug of choice. Because lidocaine is metabolized, it should be used cautiously in patients with liver disease. **Phenytoin** may be used if potassium or lidocaine prove ineffective. **Propranolol** is effective in treating ventricular tachycardia. **Atropine** is effective if digitalis-induced conduction delay is at the atrioventricular node and is mediated via the vagus. Calcium channel-blocking agents such as **verapamil** are effective if the arrhythmia is due to reentry, increased diastolic depolarization in the Purkinje fibers, or **oscillatory after-potential**. In addition to these drugs, a temporary pacemaker may be indicated.

The following interventions are contraindicated: **Quindine** should not be used because it displaces digoxin from binding sites, and **bretylium** should not be used because it releases norepinephrine. **Carotid sinus stimulation** should be discouraged, as it may precipitate ventricular fibrillation.

Antidigoxin Antibodies

The antidigitoxin or the **antidigoxin antibodies** (**Digibind**) have been used to control digitalis intoxication. The antibody mobilizes depot digoxin and is excreted by the kidney as an antibody–digoxin complex.

DRUGS ACTING ON PERIPHERAL DOPAMINE RECEPTORS

Dopamine Receptors

Peripheral dopaminergic receptor agents are useful in the treatment of congestive heart failure (Figure 84). Two distinct subtypes of dopamine receptors have been identified. The **dopamine$_2$** (DA$_2$) receptors are located at various sites within the sympathetic nervous system and their activation results in inhibition of sympathetic nervous system. In contrast, activation of the postsynaptic **dopamine$_1$** (DA$_1$) receptors, which are located on vascular smooth muscles, causes vasodilation in the renal, mesentery, cerebral, and coronary vascular beds. Thus, the pharmacological response to activation of the DA$_2$ and DA$_1$ receptors is hypotension, bradycardia, diuresis, and naturesis.

Dopamine stimulates dopamine, α-, and β-adrenergic receptors. The use of dopamine in congestive heart failure is limited because it causes nausea and vomiting, becomes inactive when given orally, increases afterload (α-adrenergic receptor-mediated peripheral vasoconstriction), and enhances oxygen demand on the left ventricle.

Dobutamine, which is available only for parenteral administration, stimulates β$_1$-adrenergic receptors, producing a strong inotropic effect.

Ibopamine, which is active orally, is capable of eliciting peripheral and renal vasodilation and a positive vasodilation and a positive inotropic action. Ibopamine is converted to **epinine**, which is the active drug.

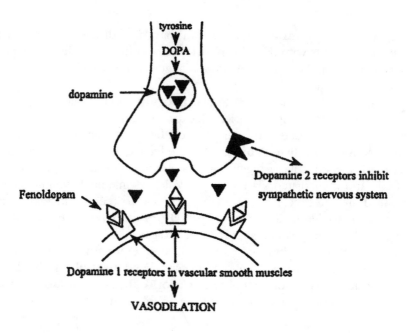

Figure 84 Fenoldopam causing vasodilation is useful in congestive heart failure.

Fenoldopam is an orally active DA_1-receptor agonist. It is more potent than dopamine in causing renal vasodilation without having adrenergic, cholinergic, or histaminergic properties.

DRUGS INHIBITING PHOSPHODIESTERASE

Amrinone, **milrinone**, and **enoximone** differ from aminophylline in that they exhibit a certain degree of selectivity for peak III phosphodiesterase, which is found predominantly in myocardial and vascular tissues. These agents exert both **positive inotropic** and **direct vasodilating actions** (Figure 85).

VASODILATORS AND ANGIOTENSIN ANTAGONISTS

Nitrates and **hydralazine** have been used in patients with congestive heart failure. An **angiotensin-converting enzyme** inhibitor such as **lisinopril** increases the left ventricular ejection fraction in patients with congestive heart failure and the effectiveness of the drug is not diminished in the presence of impaired renal function.

In addition, a **vasodilator** in combination with an angiotensin-converting enzyme inhibitor has been used in congestive heart failure. The vasodilators may be classified as **venodilators**, **arterial dilators**, or **balanced-type vasodilators.**

The rationale for vasodilation in the management of congestive heart failure is based on the increased arteriolar vasotone that occurs. This initiates a vicious circle in which cardiac function is further depressed by an increase in afterload and in resistance to ejection (Figure 86).

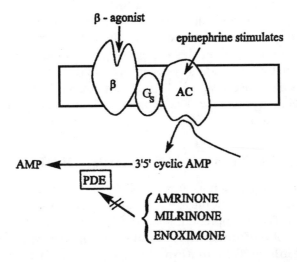

Figure 85 Amrinone exerts positive inotropic and vasodilating actions. AMP = adenosine monophosphate; AC = adenylate cyclase; PDE = phosphodiesterase; G_s = stimulatory guanine nucleotide-binding regulatory protein.

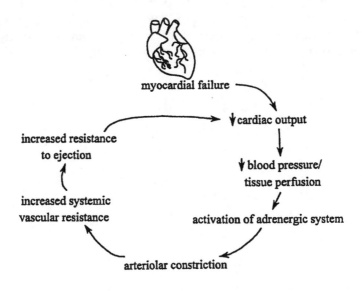

Figure 86 Neurohumoral activation during myocardial failure.

RECEPTOR-DEPENDENT VASODILATION

Vascular tone is regulated by the cytosolic **calcium** level, the interaction of calcium and **calmodulin** with myosin light-chain kinase, and subsequent myosin light-chain phosphorylation, which promotes the interaction of myosin with actin and finally leads to contraction.

Receptor-dependent vasodilation may also take place in a more indirect way through the presynaptic modulation of the release of neurotransmitters such as **norepinephrine** and **acetylcholine**. In addition to its effects on postsynaptic receptors, norepinephrine stimulates the presynaptic α_2-receptor, thereby inhibiting further transmitter release. Moreover, the activation of other presynaptic receptors such as the muscarinic cholinergic, dopaminergic, purinergic, serotoninergic, and histaminergic receptors leads to diminished norepinephrine release and subsequent vasodilation.

ENDOTHELIUM-DEPENDENT VS. INDEPENDENT VASODILATORS

A number of vasodilators, such as acetylcholine, bradykinin, adenine nucleosides, thrombin, histamine, or serotonin, require an **intact vascular endothelium** to exert their effects. For example, stimulation of endothelial cholinergic receptors causes the release of **endothelium-derived relaxing factors** (EDRF), which may involve **arachidonic acid formation** and compartmentalization via the lipoxygenase pathway. EDRF, which is identical to **nitric oxide**, activates guanylate cyclase and enhances the formation of **cyclic guanosine monophosphate** (cyclic GMP) in smooth muscle. **Tetranoic acid** (a vasoconstrictor), **thromboxane A$_2$** (a vasoconstrictor), and **prostacyclin** (a vasodilator) are formed through the lipoxygenase pathway.

The vasodilating properties of **captopril** or **hydralazine** (antihypertensive agents) are mediated by the formation of EDRF or prostaglandin, or both. On the other hand, the vasodilating properties of **nitroprusside** (an antihypertensive agent) result directly from the formation of cyclic GMP.

31

GINKGO BILOBA

- Is an antioxidant
- Causes vasoregulation
- Inhibits platelet aggregation
- Reduces brain edema
- Enhances functional recovery following ischemia

INTRODUCTION

Ginkgo biloba **extract** enhances, accelerates, and prolongs the activation of microglia and astrocytosis at the site of brain injury. Extract of *Ginkgo biloba* (EGb 761) is a complex mixture of different compounds, containing about 25% **flavonoidic substances** such as **kaempferol**, **quecetine**, **isorhamnetine** and **proanthocynidines**, **terpenes heterosides** (**ginkgolides**, **bilobalides**), and some organic acids. Systemic administration of *Ginkgo biloba* extract has been shown to enhance functional recovery after a variety of lesions in locations in the rat brain. There are many possible mechanisms proposed for its protective effects, including its **anti-ischemic–antiedemic action**, **free radical-scavenging activity**, **neurotransmitter activity**, and **neurotrophin gene expression**. In addition, *Ginkgo biloba* extract could manifest its beneficial effects through modification of the injury-induced proliferation, labeling, and staining of glia.

When tissue damage occurs, astrocytes and microglia are the first cells to be activated. Until recently, the role of astrocytes in the injured brain was believed to be limited to scar tissue formation, separating dying from healthy tissue. Microglial cells were thought to be confined to a primarily phagocytic function, i.e., removing tissue debris after trauma. Now it is widely recognized that astrocytes and microglia play a more pivotal role in the recovery process by reestablishing homeostasis, synthesizing and releasing neuroactive substances, and triggering an orchestrated immune reaction within the brain itself. These events are called "glial activation." Furthermore, activated astrocytes can enhance **axonal sprouting** and **reactive synaptogenesis**, events that are known to correlate well with behavioral recovery. Under the appropriate conditions, glial cells can intervene in neuronal signal transduction by absorbing or releasing neurotransmitter molecules such as glutamate and aspartate. Although the activation of both glial populations occurs after trauma, the reported time delays between the appearance of microglial and astroglial cells led to the assumption that microglia respond first and then they, in turn, trigger the increased metabolic and biological activity of astrocytes (see Dietrich et al., 2000).

GINKGO BILOBA EXTRACT DOES NOT INHIBIT MONOAMINE OXIDASE A AND B IN LIVING HUMAN BRAIN

Extracts of **Ginkgo biloba** contain dozens of chemical compounds including **flavonoids** and **terpene lactones**, which are considered to be the active compounds. Although the pharmacological actions of these compounds are not yet known, a number of studies suggest that *Ginkgo biloba* has the following pharmacodynamic effects:

- Vasoregulation
- Platelet-activating factor antagonism
- Neurotransmitter regulation
- Antioxidant properties

The increasing use of alternative and complementary medicines presents many challenges, including the need for careful clinical trials to demonstrate efficacy, the need to understand mechanisms of action, and an awareness of potential adverse interactions between alternative medicines and conventional drugs. The extract of *Ginkgo biloba*, EGb 761, is a popular alternative medicine. Its main use has been for treating symptoms of **cerebral insufficiency**. In a double-blind, placebo-controlled trial, EGb 761 was reported to stabilize and to improve cognitive function in patients suffering with **dementia**. There is evidence that some of its effects stem from its antioxidant properties, which may confer neuroprotective effects. This is supported by *in vitro* studies showing that EGb 761 reduces lipid peroxidation. In addition, it has been recently reported that EGb 761 is neuroprotective in a dose-dependent fashion against MPTP-induced neurotoxicity (see Chapter 12) and evidence was presented suggesting that the inhibitory effect of EGb 761 on brain monoamine oxidase (MAO) may be involved in the neuroprotective effect.

Fowler et al. (2000) used positron emission tomography (PET) to measure the effects of *Ginkgo biloba* on human brain MAO-A and MAO-B in ten subjects treated for 1 month with 120 mg/day of the *Ginkgo biloba* extract EGb 761, using [^{11}C]clorgyline and [^{11}C]L-deprenyl-D$_2$ to measure MAO A and B, respectively. A three-compartment model was used to calculate the plasma-to-brain transfer constant K_1 which is related to blood flow, and λk_3, a model term that is a function of the concentration of catalytically active MAO molecules. *Ginkgo biloba* administration did not produce significant changes in brain MAO-A or MAO-B suggesting that mechanisms other than MAO inhibition need to be considered as mediating some of its central nervous system (CNS) effects.

GINKGO BILOBA AND NITRIC OXIDE

The **Ginkgo biloba** extract (EGb 761) protects and rescues hippocampal cells against **nitric oxide-induced toxicity** (Bastianetto et al., 2000). An excess of the free radical nitric oxide (NO) is viewed as a deleterious factor involved in various CNS disorders. Numerous studies have shown that the ***Ginkgo biloba* extract EGb 761 is a NO scavenger with neuroprotective properties**. The effects of different constituents of *Ginkgo biloba* extracts, i.e., **flavonoids** and **terpenoids**, against toxicity induced by NO generators on cells of the **hippocampus**, a brain area particularly susceptible to neurodegenerative damage, have been studied.

Exposure of rat primary mixed hippocampal cell cultures to either **sodium nitroprusside** (SNP; 100 µ*M*) or **3-morpholinosydnonimine** resulted in both a decrease in cell survival and an increase in free radical accumulation. These SNP-induced events were blocked by either EGb 761 (10 to 100 µg/ml) or its flavonoid fraction CP 205 (25 µg/ml), as well as by inhibitors of protein kinase C (PKC; **chelerythrine**) and L-type calcium channels (**nitrendipine**). In contrast, the terpenoid constituents of EGb 761, known as **bilobalide** and **ginkgolide B**, as well as inhibitors of **phospholipases A [3-[4-octadecyl)benzoyl]acrylic acid (OBAA)]** and C (**U-73122**), failed to display any significant effects. Moreover, EGb 761 (50 µg/ml), CP 205 (25 µg/ml), and **chelerythrine** were also able to rescue hippocampal cells preexposed to SNP (up to 1 m*M*). Finally, EGb 761 (100 g/ml) was shown to block the activation of PKC induced by SNP (100 µ*M*). These data suggest that the protective and rescuing abilities of EGb 761 are not only attributable to the antioxidant properties of its flavonoid constituents but also via their ability to inhibit NO-stimulated PKC activity (Figure 87).

NO is a neuronal messenger molecule whose overproduction can initiate neurotoxic events under pathological conditions. NO production has clearly been linked to neurodegeneration in animal models of ischemia and *in vitro* cultured cells. The final cellular pathways that lead from the generation of NO to neuronal death include the formation of the potent oxidant **peroxynitrite**, the release of intracellular Ca^{2+}, and the activation of PKC and **phospholipase C**. NO also mediates the neurotoxic effects of glutamate that are implicated in hypoxic/ischemic brain injury and possibly in Alzheimer's disease.

The *Ginkgo biloba* extract EGb 761 is a standardized mixture of active substances, including 24% **flavonoid glycosides** and 6% **terpenoids**, obtained from green leaves of the *G. biloba* tree. EGb 761 is a polyvalent agent capable of scavenging free radicals such as NO, reducing Ca^{2+}- stimulated intracellular events, and modulating intracellular signal transduction events, including those involving phospholipases A and C and PKC. All of these signal transduction molecules are likely involved in brain ischemia and neurodegenerative diseases.

It is interesting that *in vivo* experiments have revealed that a treatment with EGb 761 inhibits ischemia-induced activation of total PKC activity in rat and decreases, partly via the inhibition of phospholipase C, the accumulation of hippocampal lipid-derived second messengers in rats subjected to electroconvulsive shock. In addition, systemic administration of **bilobalide**, one of the terpenoid constituents of EGb 761, protects hippocampal slices against hypoxia-induced phospholipid breakdown, presumably owing to an inhibition of phospholipase A. Moreover, **ginkgolide B**, another terpenoid constituent, has been reported to display an antagonistic activity against platelet-activating factor, a potent phospholipid derivative that is produced during cerebral ischemia through **calcium-activated phospholipase A$_2$**. Such intracellular changes may contribute to the

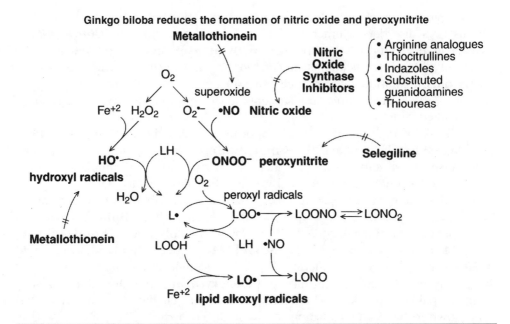

Figure 87 Nitric oxide regulates critical lipid membrane and lipoprotein oxidation events by (1) contributing to the formation of more potent secondary oxidants from $O_2^{\bullet-}$ (i.e., ONOO⁻), (2) catalyzing the redirection of $O_2^{\bullet-}$ and H_2O_2-mediated cytotoxic reactions to other oxidative pathways, and (3) termination of lipid radicals to possibly less reactive secondary nitrogen-containing products. Moreover, metallothionein isoforms I and II protect against neurotoxic effects of superoxide anions, hydroxyl radicals, and peroxynitrite radicals. *Ginkgo biloba* reduces the formation of nitric oxide and peroxynitrite.

neuroprotective effects of EGb 761 in animal models of focal and global cerebral ischemia, of retinal damage induced by ischemia, and in hypoxia. Clinical trials support the potential therapeutic usefulness of EGb 761 in the treatment of **cerebral insufficiency** and mild **cognitive impairments** in elderly patients, as well as in Alzheimer's disease and vascular dementia.

The precise mechanisms underlying the neuroprotective effects of EGb 761, particularly with respect to ischemia, have yet to be clearly established, but they have been reported to be associated with either its terpenoid and/or its flavonoid fractions. The present evidence indicates that EGb 761, acting principally via its flavonoid constituents, is able to protect and rescue hippocampal neuronal cells against NO-induced toxicity. This is consistent with earlier data reporting protective effects of EGb 761 in animal models of cerebral ischemia, in *in vitro* models of oxidative stress-induced toxicity and, more recently, in a model of β-amyloid-induced toxicity. This is of particular interest given that ischemic neurodegeneration is associated with a loss of hippocampal neurons and that this event is likely the result of an increased production of NO (see Figure 87).

The production of **superoxide anions** is one of the major factors involved in NO toxicity because superoxide anions can react with NO to form the highly toxic free radical peroxynitrite. A pivotal role for superoxide anions in NO-related insults is emphasized by results showing that transgenic mice overexpressing superoxide dismutase (SOD) are resistant to brain ischemia. Superoxide can protect against SNP-induced toxicity. Thus, the superoxide-scavenging properties of EGb 761 are likely to explain, at least in part, its ability to block cell death and the increase in reactive oxygen species accumulation induced by the two NO donors used here, SNP and SIN-1.

EGb 761 might also directly scavenge peroxynitrites and inhibit **lipid peroxidation** because it has been reported both to block the cytotoxicity induced by peroxynitrites and to inhibit **cyclosporin A-induced peroxidation**. This hypothesis is supported by the finding that the purported anti-ischemic agent **ebselen** completely protected and rescued hippocampal cells against SNP-induced toxicity. In contrast, the hydroxyl radical-scavenging properties of EGb 761 do not seem able to account for its protective effects because catalase failed to display any neuroprotective properties in this model.

Aside from decreases in reactive oxygen species accumulation, inhibition of PKC, whose levels are increased in the hippocampal CA1 region during cerebral ischemia, it has also been shown that *Ginkgo biloba* blocks the toxic effects of NO, which is linked to neuroprotection during global ischemia. In this respect, EGb 761 shares with the PKC inhibitor **chelerythrine chloride**, but not with phospholipase C (U-73122) or A_2 (OBAA) inhibitors, the ability to block the toxic effects of SNP. These results are consistent with previous studies and suggest that the protective effects of EGb 761, together with its antioxidant properties, may be related to its inhibitory actions on PKC. The fact that EGb 761 was able to block SNP-induced toxicity suggests that the blockade of L-type voltage-dependent calcium influx maybe an early event that may inactivate PKC.

The rescuing effect of EGb 761 against NO-induced toxicity suggests that it can also inhibit active, but reversible, molecular pathways that are induced by NO and that lead to cell death. Among those, PKC is certainly a strong candidate because chelerythrine chloride mediated both the neuroprotective and neurorescuing effects of EGb 761 against NO-induced toxicity. In contrast, inhibitors of phospholipase C (U-73122) and A_2 (OBAA) failed to mimic the actions of EGb 761, indicating that EGb 761 regulates PKC through phospholipase A_2 and C independent mechanisms (Calpai et al., 2000; Bastianetto et al., 2000).

32

GINSENG AND CHOLINERGIC TRANSMISSION

INTRODUCTION

Ginseng contains 20 different **ginsenoids**, which belong to a family of steroids named **steroidal saponins**. They have been named **ginsenoside saponins**, **triterpenoid saponins**, or **dammarane derivatives** under previous classifications. Ginsenosides possess the four *trans*-ring rigid steroid skeleton, with a modified side chain at C-20. The classical steroid hormones have a truncated side chain (**progesterone, cortisol,** and **aldosterone**) or no side chain (**estradiol** and **testosterone**).

Ginsenosides belong to a family of steroids and share their structural characteristics. Like steroids, they can traverse cell membranes freely. Moreover, their presence has been demonstrated within cells, particularly the nucleus. Steroid hormone action, steroids that bind nuclear receptors, are thought to affect primarily the transcription of mRNA and subsequent protein synthesis. Intracellular steroid binding proteins present possible attractive targets for ginsenosides.

Ginseng:

- Enhances memory
- Protects neurons against ischemic damage
- Exhibits antineoplastic actions
- Possesses immunomodulating actions

Ginseng is a highly valued herb in the Far Fast and has gained popularity in the West during the last decades. There is extensive literature on the beneficial effects of ginseng and its constituents. The major active components of **ginsenosides** are a diverse group of **steroidal saponins** (Attele et al., 1999).

The **ginseng root** has been used for over 2000 years, in the belief that it **is a panacea and promotes longevity**. Recently, there has been renewed interest in investigating ginseng pharmacology using biochemical and molecular biological techniques. Pharmacological effects of ginseng have been demonstrated in the central nervous system (CNS) and in cardiovascular, endocrine, and immune systems. In addition, ginseng and its constituents have been ascribed antineoplastic, antistress, and antioxidant activity. It is an herb with many active components,

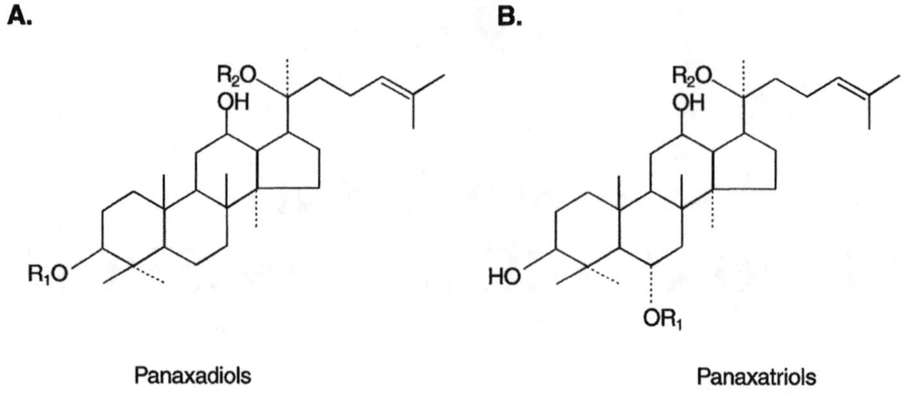

A.   **B.**

Panaxadiols   Panaxatriols

Figure 88 Structures of ginsenoids based on chemical structure; there are two major groups: panaxadiols (A) and panaxatriols (B).

and there is evidence from numerous studies that ginseng does have beneficial effects.

Ginseng has three species:

- ■ *Panax ginseng* (Asian ginseng)
- ■ *Panax quinquefolius* (American ginseng)
- ■ *Panax japonicus* (Japanese ginseng)

Active constituents found in most ginseng species include **ginsenosides**, **polysaccharides**, **peptides**, **polyacetylenic alcohols**, and **fatty acids**. There is a wide variation (2 to 20%) in the ginsenoside content of different species of ginseng. Moreover, pharmacological differences within a single species cultivated in two different locations have been reported. For example, the potency of extracts from *P. quinquefolius,* cultivated in the United States, for modulating neuronal activity is significantly higher than for the same species cultivated in China (Figure 88). Most pharmacological actions of ginseng are attributed to ginsenosides. More than 20 ginsenosides have been isolated, and novel structures continue to be reported, particularly from *P. quinquefolius* and *P. japonicus.*

Ginseng by increasing the uptake of **choline** and facilitating the release of **acetylcholine** prevents **scopolamine**-induced memory impairment (Figure 89).

CHOLINERGIC TRANSMISSION

Division and Functions of the Autonomic Nervous System

The autonomic nervous system consists of central connections, visceral afferent fibers, and visceral efferent fibers. The hypothalamus is where the principal integration of the entire autonomic nervous system takes place and it is involved in the regulation of blood pressure, body temperature, water balance, metabolism of carbohydrates and lipids, sleep, emotion, and sexual reflexes. To a lesser extent,

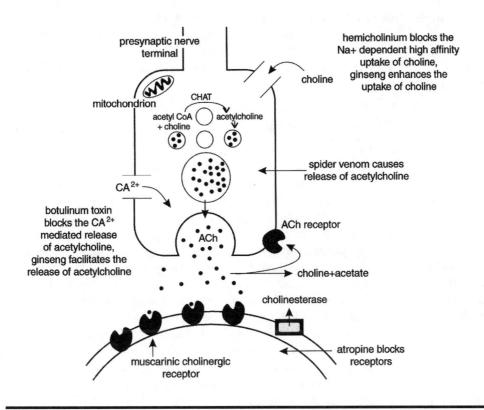

Figure 89 The metabolism and actions of acetylcholine (ACh). CoA = coenzyme A; CHAT = choline acetylase.

the medulla oblongata, limbic system, and cerebral cortex also integrate, coordinate, and adjust the functions of the autonomic nervous system.

Most blood vessels, the sweat glands, and the spleen are innervated only by one division of the autonomic nervous system. In the salivary glands, the two divisions of the autonomic nervous system supplement one another. In the bladder, bronchi, gastrointestinal tract, heart, pupils, and sex organs, the two divisions of the autonomic nervous system have opposing effects.

Neurochemical Basis of Cholinergic Transmission

Acetylcholine, an ester of choline and acetic acid, is synthesized in cholinergic neurons according to the following scheme:

$$\text{Acetyl CoA + choline} \xrightarrow{\substack{\text{Choline} \\ \text{acetyltransferase}}} \text{acetylcholine}$$

The acetylcholine, in turn, is hydrolyzed by both **acetylcholinesterase** and plasma **butyrylcholinesterase**. Choline is actively transported into nerve terminals (synaptosomes) by a high-affinity uptake mechanism. Furthermore, the availability of choline regulates the synthesis of acetylcholine (see Figure 89).

Hemicholinium blocks the transport of choline into synaptosomes, whereas botulinum toxin blocks the calcium-mediated release of acetylcholine. The released acetylcholine is hydrolyzed rapidly by acetylcholinesterase to choline and acetate. **Ginseng enhances the uptake of choline and facilitates the release of acetylcholine**.

Classification of Cholinergic Receptors

Acetylcholine receptors are classified as either muscarinic or nicotinic. The alkaloid **muscarine** mimics the effects produced by stimulation of the parasympathetic system. These effects are postganglionic and are exerted on exocrine glands, cardiac muscle, and smooth muscle. The alkaloid nicotine mimics the actions of acetylcholine, which include stimulation of all autonomic ganglia, stimulation of the adrenal medulla, and contraction of skeletal muscle.

Dimethylphenylpiperazinium stimulates the autonomic ganglia, **tetraethylammonium** and **hexamethonium** block the autonomic ganglia, **phenyltrimethylammonium** stimulates skeletal motor muscle end plates, **decamethonium** produces neuromuscular blockade, and ***d*-tubocurarine** blocks both the autonomic ganglia and the motor fiber end plates.

Among the agents cited, only *d*-tubocurarine is useful as a drug (skeletal muscle relaxant); the rest are useful only as research tools. Cholinesterase, found in liver and plasma, can hydrolyze other esters such as **succinylcholine** (a skeletal muscle relaxant). Cholinergic peripheral receptors are located on: (1) postganglionic parasympathetic fibers, (2) postganglionic sympathetic fibers, (3) all autonomic ganglia, and (4) skeletal end plates.

In addition, cholinergic receptors are distributed extensively in the central nervous system (CNS) and participate in diversified functions such as audition, vision, learning and memory, ingestive behaviors (thirst and hunger), thermoregulation, locomotor activity, diurnal rhythms, sleep, and sexual activity. Changes in cholinergic neurons have been observed in neurological syndromes such as catalepsy, stereotypy, and tremor, and in psychiatric disorders such as schizophrenia. Furthermore, cholinergic neurons have been implicated in addiction to opiates and alcohol, and in physiological dependence and withdrawal syndromes. The role of acetylcholine in schizophrenia has been extensively studied. For example, anticholinergic drugs such as **atropine**, **benztropine**, and **trihexyphenidyl** result in an acute exacerbation of schizophrenic psychosis; they may also provoke toxic psychosis in nonpsychotic patients, similar to paranoid schizophrenia. On the other hand, **cholinomimetic agents** such as **physostigmine** ameliorate schizophrenic symptoms.

Cholinergic (Cholinomimetic) Receptor Agonists

Methacholine, **carbachol**, and **bethanechol** are all agents that mimic the effects of stimulation of cholinergic nerves.

The two currently used derivatives of acetylcholine are bethanechol (Urecholine chloride) and carbachol (Miostat). Unlike acetylcholine, both agents are resistant to hydrolysis by cholinesterase. Both agents are muscarinic agonists. The nicotinic action of carbachol is greater than that of acetylcholine, whereas bethanechol is devoid of nicotinic action. The cardiovascular actions of acetylcholine are vasodilation and negative chronotropic and inotropic effects. The cardiovascular effects of methacholine are more pronounced than those of acetylcholine, which in turn are greater than those of carbachol or bethanechol. The gastrointestinal effects (increase in tone, amplitude of contractions, and peristalsis) of bethanechol and carbachol are equal but greater than those of acetylcholine. The effects of carbachol and bethanechol on the urinary tract, consisting of ureteral peristalsis, contraction of the detrusor muscle of the urinary bladder, and an increase in voluntary voiding pressure, are equivalent and exceed those produced by acetylcholine.

The miotic effects of carbachol and bethanechol are greater than those of acetylcholine. Atropine is able to antagonize all cholinergic (muscarinic) effects produced by acetylcholine, methacholine, carbachol, and bethanechol. However, this antagonism is least evident with carbachol.

Bethanechol is of value in the management of postoperative abdominal distention, gastric atony or stasis, and urinary retention. Carbachol (0.25 to 3.00%) may be used for the long-term therapy of noncongestive **wide-angle glaucoma**.

Pilocarpine is a naturally occurring (active ingredient of poisonous mushrooms, **Amanita muscaria**) cholinomimetic agent possessing both muscarinic and nicotinic properties (stimulates autonomic ganglia). This agent causes miosis, reduces intraocular pressure, and is used in the treatment of wide-angle glaucoma. In addition, it may be applied topically in the eye in the form of a drug reservoir (**Ocusert**).

Anticholinesterase Agents

Anticholinesterases are drugs that inhibit or inactivate acetylcholinesterase, causing the accumulation of acetylcholine at the cholinergic receptors.

Structure of Acetylcholinesterase. The active site of acetylcholinesterase consists of an anionic site, which interacts with the quaternary group on the choline moiety of acetylcholine, and an esteratic site, which interacts with and hydrolyzes the ester grouping of acetylcholine. A serine hydroxyl group at the esteratic site becomes acetylated and is subsequently regenerated by interaction with water. The inhibition of acetylcholinesterase by drugs such as **physostigmine**, **neostigmine**, or **edrophonium** produces effects similar to those seen with the prolonged stimulation of cholinergic fibers. Anticholinesterases are classified as reversible and irreversible inhibitors.

Classification of Cholinesterase Inhibitors

The reversible inhibitors, which have a short to moderate duration of action, fall into two categories. Type one, exemplified by edrophonium, forms an ionic bond at the anionic site and a weak hydrogen bond at the esteratic site of acetylcholinesterase. Type two, exemplified by neostigmine, forms an ionic bond at the anionic site and a hydrolyzable covalent bond at the esteratic site. The irreversible inhibitors, exemplified by organophosphorus compounds (**Diisopropyl fluorophosphate**, **parathion**, **malathion**, **diazinon**), have long durations of action and form a covalent bond with acetylcholinesterase, which is hydrolyzed very slowly and negligibly, but the inhibition may be overcome by cholinesterase activators such as pralidoxime.

Cholinesterase inhibitors may also be classified according to agents that possess tertiary nitrogens (e.g., **physostigmine** and most organophosphorus compounds) and those that contain quaternary nitrogens (e.g., **neostigmine**, **pyridostigmine**, and some organophosphorus compounds such as **echothiophate**). The following summarizes the comparative properties of these agents.

Property	Physostigmine	Neostigmine
Oral absorption	Good	Poor
Passing across blood–brain barrier	Well	No
Stimulating nicotinic receptors (skeletal muscle)	Yes	Yes
Used to combat the CNS toxicity of numerous anticholinergic drugs	Yes	No

Physostigmine (eserine sulfate) causes miosis and spasm of accommodation; it also lowers intraocular pressure and hence can be used in the treatment of wide-angle glaucoma. As it is lipid soluble, it penetrates into the brain rapidly, raises the acetylcholine concentration, and, in toxic amounts, may cause cholinergic CNS toxicity, which is characterized by restlessness, insomnia, tremors, confusion, ataxia, convulsions, respiratory depression, and circulatory collapse. These effects are reversed by **atropine**.

Neostigmine, which is unable to penetrate the blood–brain barrier, does not cause CNS toxicity. However, it may produce a dose-dependent and full range of muscarinic effects, characterized by miosis, blurring of vision, lacrimation, salivation, sweating, increased bronchial secretion, bronchoconstriction, bradycardia, hypotension, and urinary incontinence. Atropine is able to oppose these muscarinic effects. In addition, neostigmine, which has both a direct action as well as an indirect action that is mediated by acetylcholine on end-plate nicotinic receptors, may produce muscular fasciculation, muscular cramps, weakness, and even paralysis. These effects are not countered by atropine. Furthermore, neostigmine enhances gastric contraction and secretion. Neostigmine itself is metabolized by plasma acetylcholinesterase.

The therapeutic uses of neostigmine include the treatment of atony of the urinary bladder and postoperative abdominal distention. In addition, it antagonizes the action of **d-tubocurarine** and curariform drugs. Edrophonium, neostigmine, or pyridostigmine may be used to diagnose **myasthenia gravis**. Because edrophonium has the shortest duration of action, it is most often used for this purpose.

Antidote to Irreversible Cholinesterase Inhibitors

The irreversible cholinesterase inhibitors, such as DFP (**isoflurophate**), are used only for local application in the treatment of wide-angle glaucoma. Their pharmacological effects, which are similar to those produced by physostigmine, are intense and of long duration. As organophosphorus insecticides, they are of paramount importance in cases of accidental poisoning and suicidal and homicidal attempts. They produce a cholinergic crisis which must be treated by (1) decontaminating the patient, (2) supporting respiration, (3) blocking the muscarinic effects by atropine, and (4) reactivating the inhibited cholinesterase by treatment with **pralidoxime**.

Cholinergic Receptor Blocking Agents

Atropine and scopolamine, which are obtained from **belladonna alkaloids**, as well as other synthetic anticholinergic drugs, inhibit the actions of acetylcholine and cholinomimetic drugs at muscarinic receptors in smooth muscles, heart, and exocrine glands. In addition to these peripheral effects, anticholinergic drugs, by blocking the acetylcholine receptor sites in the CNS, have pronounced CNS effects such as restlessness, irritability, excitement, and hallucinations. **Scopolamine**, on the other hand, depresses the CNS and, in therapeutic doses, produces fatigue, hypnosis, and amnesia. Therefore, it is used extensively in numerous medications, often in combination with antihistamines.

Dose-Dependent Effects of Atropine

The pharmacological effects of atropine in general are dose dependent. For example, in small doses, atropine depresses sweating, elevates body temperature, decreases salivary and bronchial secretions, and relaxes bronchial smooth muscles. In somewhat larger doses (1 to 3 mg), it produces **mydriasis** (blockade of the iris sphincter muscle), **cycloplegia** (blockade of the ciliary muscle), and cardiovascular effects characterized by transient bradycardia (central vagal stimulation) and tachycardia (vagal blockade at the sinoatrial node). Lacking any significant effects on circulation, atropine is often used as a preanesthetic medication to depress bronchial secretion and prevent pronounced bradycardia during abdominal surgical procedures. In still larger doses, it depresses the tone and motility of the gastrointestinal tract, the tone of the urinary bladder, and gastric secretion. Therefore, the effective doses for use in acid-pepsin diseases are preceded by numerous side effects.

Atropine is absorbed orally and crosses the placental barrier, whereupon it causes fetal tachycardia. Atropine has been used to examine the functional integrity of the placenta. Atropine toxicity is characterized by dry mouth, burning sensation in the mouth, rapid pulse, mydriasis, blurred vision, photophobia, dry and flushed skin, restlessness, and excitement.

Physostigmine, given intravenously, counteracts both the peripheral and central side effects of atropine and other anticholinergic drugs such as **thioridazine** (neuroleptic), **imipramine** (antidepressant), and **benztropine** (antiparkinsonian medication).

Synthetic Muscarinic Receptor Antagonists

The various applications of synthetic muscarinic receptor antagonists are listed below:

Ophthalmology

- Homatropine. Produces mydriasis and cycloplegia
- Eucatropine. Produces only mydriasis

Preoperative uses

- Atropine. To prevent excess salivation and bradycardia
- Scopolamine. In obstetrics, to produce sedation and amnesia

Cardiology

- Atropine
  To reduce severe bradycardia in hyperactive carotid sinus reflex
  Diagnostically in Wolff–Parkinson–White syndrome, to restore the PRS complex to normal duration

Gastroenterology

- Propantheline, oxyphenonium, pirenzepine
  To diminish vagally mediated secretion of gastric juices and slow gastric emptying in peptic ulcer
  To reduce diarrhea associated with dysenteries and diverticulitis
  To reduce excess salivation associated with heavy metal poisoning or parkinsonism

Neurology

- Trihexyphenidyl, benztropine. In parkinsonism and drug-induced pseudoparkinsonism
- Scopolamine. In vestibular disorders such as motion sickness

Methantheline and **propantheline** are synthetic derivatives that, in addition to their antimuscarinic effects, are ganglionic blocking agents and block the skeletal neuromuscular junction. Propantheline and **oxyphenonium** reduce gastric secretion, and **pirenzepine**, in addition to reducing gastric secretion, also reduces gastric motility.

Contraindication to Anticholinergic Agents

Conditions that are contraindications to the use of atropine and related drugs are glaucoma and prostatic hypertrophy, in which they cause urinary retention.

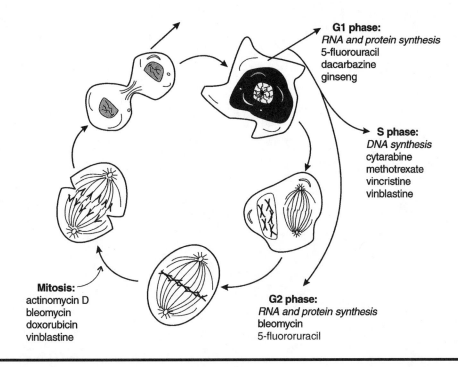

G1 phase:
RNA and protein synthesis
5-fluorouracil
dacarbazine
ginseng

S phase:
DNA synthesis
cytarabine
methotrexate
vincristine
vinblastine

G2 phase:
RNA and protein synthesis
bleomycin
5-flororuracil

Mitosis:
actinomycin D
bleomycin
doxorubicin
vinblastine

Figure 90   The actions of antineoplastic agents on different phases of the cell cycle.

OTHER CNS ACTIONS OF GINSENG

- Ginseng protects **hippocampal neurons** from **ischemic damage**.
- Ginseng inhibits the uptake of **GABA**, **glutamate**, **norepinephrine**, **dopamine**, and **serotonin**.
- *Panax quinquefolius* extracts are agonists at **GABA A** receptors.
- Ginseng prolongs the **hexobarbital** sleeping time.
- **Ginseng** inhibits **calcium channels** on primary sensory neurons.

ANTINEOPLASTIC AND IMMUNOMODULATORY EFFECTS OF GINSENG

- **Ginseng** extracts exert antineoplastic effects and have cytotoxic properties and growth inhibitory properties *in vivo*.
- **Ginseng** arrests cell cycle progression at the G1 stage and inhibits **cyclin-dependent kinase** activity (Figure 90).
- Ginseng enhances and activates both **polymerase** and **exonuclease activities of DNA polymerase δ**.
- **Ginseng** possesses some immunomodulatory properties, primarily associated with NK cell activity. Ginsenoside was shown to increase both humoral and **cell-mediated immune responses** (see Attele et al., 1999).

33

GREEN AND BLACK TEAS

- Antioxidant
- Antineoplastic agent
- Dental anticariogenic

INTRODUCTION

Approximately 78% of the 2.5 million metric tons of tea leaves produced worldwide was used for the preparation of black tea, which is mainly consumed in Western nations, and about 20% was used for the preparation of green tea, which is mainly consumed in Asian countries and in some parts of North Africa. Green tea mainly contains polyphenols such as **flavanols (catechins)**, **flavonols**, **flavandiols**, and **phenolic acids**. These polyphenols account for about 25 to 30% of the solids in water extracts of green tea leaves. The major green tea catechins are **(–)-epigallocatechin-3-gallate** (EGCG), **(–)-epigallocatechin** (EGC), **(–)-epicatechin-3-gallate** (ECG), **(–)-epicatechin** (EC), and **(+)-catechin**. In the manufacture of black tea, the polyphenols in tea leaves undergo polyphenol oxidase–catalyzed oxidative polymerization, which leads to the formation of **bisflavanols**, **theaflavins**, **thearubigins**, and other oligomers in a process commonly known as "tea fermentation." Theaflavins (about 1 to 2% of the solids in water extracts of black tea leaves) include **theaflavin**, **theaflavin-3-gallate**, **theaflavin-3′-gallate**, and **theaflavin-3,3′-digallate**, and these substances contribute to the characteristic color and taste of black tea. A substantial portion of the solids in water extracts of black tea leaves represents **thearubigens**, which are even more extensively oxidized and polymerized than the aflavins. The thearubigens have a wide range of molecular weights and are poorly characterized. Thus, it has been assumed that there are significant differences in antioxidant properties between green and black teas.

ANTIOXIDANT PROPERTIES

Wei et al. (1999) compared the effects of different extractable fractions of green and black teas on scavenging hydrogen peroxide H_2O_2, and ultraviolet (UV) irradiation-induced formation of **8-hydroxy 2′-deoxyguanosine (8-OHdG)** *in* *vitro*. Green and black teas were extracted by serial chloroform, ethyl acetate, and *n*-butanol, and divided into four subfractions designated as GT1 to GT4 for green tea and BT1 to BT4 for black tea, respectively. The total extracts from green and black teas exhibited a potent scavenging capacity of exogenous H_2O_2 in a dose-dependent manner. It appeared that the total extracts from black tea scavenging H_2O_2 were more potent than those from green tea. When tested individually, the potency of scavenging H_2O_2 by green tea subtractions was GT2 > GT3 > GT1 > GT4, whereas the order of efficacy for black tea was BT > 2 > BT3 > BT4 > BT1. In addition, they demonstrated that total fractions of green and black teas substantially inhibited induction of 8-OHdG in calf thymus by all three portions of UV spectrum (UVA, B, and C). Consistent with the capacity of scavenging H_2O_2, the subfractions from black tea showed a greater inhibition of UV-induced 8-OHdG than those from green tea. At low concentrations, the order of potency of quenching of 8-OHdG by green tea subfractions was GT2 > GT3 > GT4 > GT1 and the efficacy of all subfractions became similar at high concentrations. All subfractions of the black tea except BT1 strongly inhibited UV-induced 8-OHdG and the order of potency was BT2 > BT3 > BT4 > BT1. Addition of EGCG, an ingredient of green tea extract, to low concentrations of green and black tea extracts substantially enhanced the scavenging of H_2O_2 and quenching of 8-OHdG, suggesting the important role of EGCG in the antioxidant activities of tea extracts. The potent scavenging of oxygen species and blocking of UV-induced oxidative DNA damage may, at least in part, explain the mechanism(s) by which green teas inhibit **photocarcinogenesis**.

In recent years, evidence has been accumulated showing that tea has anticancer properties. Although the epidemiological data are lacking, the experimental studies have consistently demonstrated that drinking of tea, either black or green, substantially inhibited carcinogen- or UV irradiation-induced animal tumors including skin, lung, stomach, colon, and breast cancer. In spite of exciting results in animal studies, the mechanisms of the anticarcinogenic action of tea have remained largely unclear. Some studies have attributed the inhibitory effects of green or black tea on UV-induced carcinogenesis to the antioxidant and free radical-scavenging activities of the teas, since numerous antioxidants, such as **ascorbic acid**, **α-tocopherol**, **β-carotene**, **selenium**, **butylated hydroxytoluene**, and a mixture of dietary antioxidants, have been reported to inhibit UV-induced skin carcinogenesis. Several **polyphenolic substances** in green and black teas have been shown to have antioxidant activities.

Oral administration of black or green tea was reported to have a similar inhibitory effect on **N-nitrosomethyl-benzylamine-induced esophageal tumorigenesis**, and oral administration of decaffeinated black or green tea was reported to have a similar strong inhibitory effect on **4-(methylnitrosamino)-1-(3-pyridyl)-1-butanone-induced lung tumorigenesis**, as well as UV-induced skin carcinogenesis in mice. **(−)-Epigallocatechin-3-gallate** is considered a potent antioxidant and major anticarcinogenic component in tea. It has been hypothesized that blockage of UV-induced oxidative DNA damage may be the mechanism(s) by which tea inhibits UV-induced skin tumorigenesis. Several investigators reported that drinking of green tea significantly inhibited the formation of carcinogen-induced elevation of 8-OHdG in lung, liver, pancreas, and colon, and thereby reduced the toxicity and carcinogenicity in those organs.

In summary, aqueous extracts from green and black teas have potent antioxidant activities and exceptional quenching capacity of UV-induced oxidative DNA damage. These effects could be, at least in part, the mechanisms of action by which green and black teas counteract carcinogenic processes (Wei et al., 1999).

THE EFFICACY OF TEA IN ATTENUATING SKIN CANCER

It is estimated that 1.2 million new cases of **skin cancer** occur each year in the United States, and the majority of these are believed to result from heavy **exposure to UV** light from the sun. Although most of these skin cancers are curable **squamous cell carcinomas** and **basal cell carcinomas**, deaths do occur from these cancers and from more fatal sunlight-related **malignant melanomas**. In the past two decades, the skin cancers (both nonmelanoma and melanoma) have been increasing alarmingly. Thus, elucidation of etiologies of skin cancers and development of preventive strategy have been greatly advocated.

Tea is widely used as a beverage in Asian countries, including China, Japan, and India. Although **green tea** is also used in North America and Europe, **black tea** is a more popular beverage in the Western countries.

The topical application of cation of **EGCG**, a major **polyphenolic catechin** in green tea, inhibited **teleocindin-induced tumor promotion** in the skin of mice previously initiated with **7,12-dimethylbenz[a]anthracene**. In subsequent studies, it was found that oral administration of EGCG or tea fractions inhibited the carcinogen-induced tumors of various organs in animals, including the duodenum, stomach, lung, esophagus, and colon. Topical application of a green tea polyphenol fraction to CD-1 mice inhibited the tumor-initiating activities of **benzo[a]pyrene** and **7,12-dimethylbenz[a]anthracene** and the tumor promoting activity of **tissue plasminogen activator** (TPA). The oral administration of a green tea polyphenol fraction or an aqueous extract of green and black teas to SKH-1 mice as their sole source of drinking fluid was shown to inhibit UVB-induced skin tumorigenesis. Tumors from animals treated with UVB together with green tea were much smaller in volume as compared to UVB-treated positive controls. In addition, green and black teas were found to inhibit the growth of established skin papillomas in mice.

ANTICARIOGENIC EFFECTS OF TEA

Pathogenesis of Dental Caries

Dental caries is a very common chronic disease, arising from interplay among oral flora, the teeth, and dietary factors. The major etiological players are thought to be the two α-hemolytic "mutans group" streptococci, **Streptococcus mutans** and **S. sobrinus**, although several other types of bacteria (notably **lactobacilli** and **actinomyces**) may also be involved. The nature of the condition does not readily lend itself to conventional antibacterial therapy (i.e., the use of an antibiotic active against the causal agent), and prevention is largely directed toward the individual host — e.g., diet, mechanical factors — and the public health measure of fluoridation of drinking water.

The pathogenesis of dental caries may involve three distinct processes: (1) adherence of the bacteria to the tooth; (2) formation of glycocalyx due to synthesis of a sticky glucan by the action of the bacterial enzyme **glucosyl transferase** on sucrose; and (3) accumulation of **biobilm** (plaque), within which there is continuing acid production by constituent bacteria (including **streptococci** and **lactobacilli**) able to metabolize carbohydrates at low pH values. This acid demineralizes an enamel.

The second and third of these steps depend on a supply in the mouth of appropriate carbohydrate substrates, most favorably sucrose. The latter can become available either directly (sugar ingested in food or drink) or be derived from dietary starch by the action of bacterial or salivary amylases, or both. Of particular relevance in this context is the trapping of carbohydrates as or on food particles remaining in the mouth for considerable periods.

Potential Anticariogenic Actions of Tea

Various components in green and black tea, the beverages made by infusing appropriately processed dried leaves of ***Camellia sinensis***, notably simple cat-echins, have properties *in vitro* that suggest an anticariogenic activity. These include a direct bactericidal effect against *S. mutans* and *S. sobrinus*; prevention of bacterial adherence to teeth; inhibition of **glucosyl transferase**, thus limiting the biosynthesis of sticky glucan; inhibition of human and bacterial **amylases**. Studies in animal models show that these *in vitro* effects can translate into caries prevention. A limited number of clinical trials in humans suggest that regular tea drinking may reduce the incidence and severity of caries. If substantiated, this could offer a very economical public health intervention (see Hamilton-Miller, 2001).

34

HORSERADISH AND ANTIPLATELET ACTIONS

INTRODUCTION

Platelets provide the initial hemostatic plug at sites of vascular injury. They also participate in reactions that lead to atherosclerosis and pathological thrombosis. Antagonists of platelet function have thus been used in attempts to prevent thrombosis and to alter the natural history of atherosclerotic vascular disease.

Japanese domestic **horseradish**, *wasabi*, containing **isothiocyanates** has **antiplatelet activity**. **Glucosinolates** are found in cell vacuoles of various plants in the family **Cruciferae** such as **horseradish**, **mustard**, **broccoli**, and **wasabi** (***Wasabia japonica*, syn. *Eutrema wasabi***). When plant cells are damaged, glucosinolates are hydrolyzed by **myrosinase** (**thioglucoside glucohydrolase**, EC 3.2.3.1), which is also produced in the same family, and produce **isothiocyanates** (Figure 91). Most of these Crucifer isothiocyanates are well known for having antimicrobial, fungicidal, and pesticidal activities. A major wasabi flavor compound, **6-methylthiohexyl isothiocyanate** (MS-ITC), has been shown to possess antimicrobial and platelet aggregation inhibition activities (Morimitsu et al., 2000).

Similarly, it has been shown that consumption of cruciferous vegetables is associated with a lower incidence of cancers. Induction of phase II enzymes such as the **glutathione *S*-transferase** (GST) and **quinone reductase** (QR) have been demonstrated in **broccoli**, **cabbage**, and **brussel sprouts**. Many natural isothiocyanates derived from cruciferous vegetables and some fruits have been shown to cause induction of phase II enzymes in cultured cells and rodents.

Morimitsu et al. (2000) isolated **6-methylsulfinyhexyl isothiocyanate** (MS-ITC) from wasabi (*W. japonica*, Japanese domestic horseradish) as a potential inhibitor of human platelet aggregation *in vitro* through extensive screening of vegetables and fruits (Figure 92).

In the course of another screening for the induction of **GST** activity in RL34 cells, MS-ITC was inadvertently isolated from wasabi as a potential inducer of GST. MS-ITC administered to rats or mice also showed both activities *in vivo*. As a result of elucidation of the platelet aggregation inhibition and the GST induction

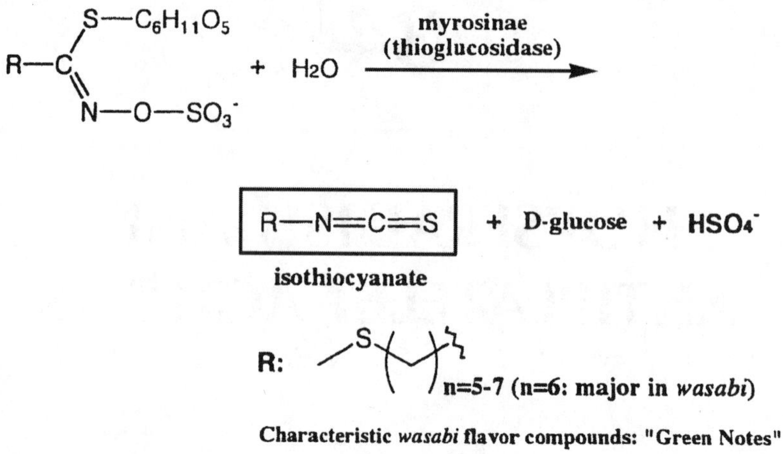

Figure 91 The enzymatic formation of isothiocyanates from *Brassica* sp.

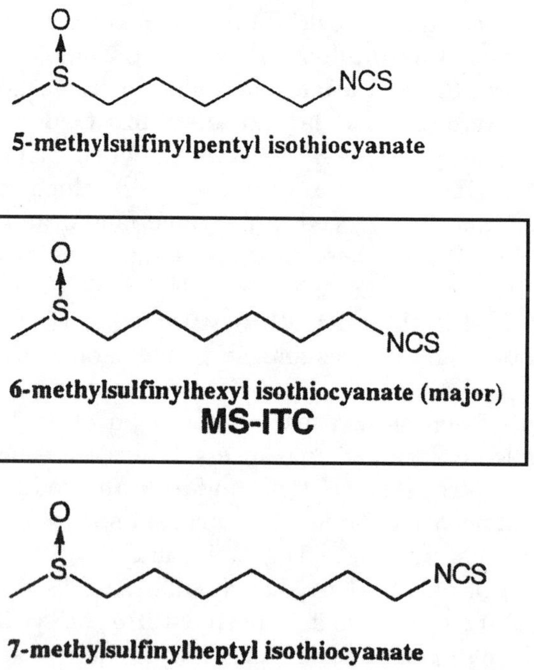

Figure 92 The chemical structures of platelet aggregation inhibitors, which were also identified as potential inducers of GST in RL34 cells.

mechanisms of MS-ITC, the isothiocyanate moiety of MS-ITC plays an important role for antiplatelet and anticancer activities because of its high reactivity with **sulfhydryl** (RSH) groups in biomolecules (GSH, cysteine residue in a certain protein, etc.) (Figure 93).

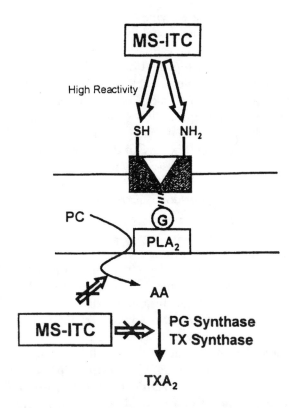

Figure 93 The isothiocyanate moiety of MS-ITC plays an important role for antiplatelet and anticancer activities because of its high reactivity with sulfhydryl (RSH) groups in biomolecules (GSH, cysteine residue in a certain protein).

Morimitsu et al. (2000) have shown that after incubation with MS-ITC (5 or 10 minutes), the treated platelets were washed two times with phosphate-buffered saline, and MS-ITC potently inhibited platelet aggregation induced by thrombin. Then, MS-ITC could not show any effects on **arachidonic acid cascade (phospholipase A$_2$, prostaglandin synthase**, and **thromboxane synthase)** in human platelets.

ANTIPLATELET DRUGS

Aspirin

Processes including thrombosis, inflammation, wound healing, and allergy are modulated by oxygenated metabolites of arachidonate and related polyunsaturated fatty acids that are collectively termed **eicosanoids**. Interference with the synthesis of eicosanoids is the basis for the effects of many therapeutic agents, including analgesics, anti-inflammatory drugs, and antithrombotic agents.

In platelets, the major cyclooxygenase product is **thromboxane A$_2$**, a labile inducer of platelet aggregation and a potent vasoconstrictor. **Aspirin** blocks production of thromboxane A$_2$ by covalently acetylating a serine residue near the active site of cyclooxygenase, the enzyme that produces the cyclic endoperoxide precursor of thromboxane A_2. Since platelets do not synthesize new proteins, the action of aspirin on platelet cyclooxygenase is permanent, lasting for the life of the platelet (7 to 10 days). Thus, repeated doses of aspirin produce a cumulative effect on platelet function. Complete inactivation of platelet cyclooxygenase is achieved when 160 mg of aspirin is taken daily. Therefore, aspirin is maximally effective as an antithrombotic agent at doses much lower than required for other actions of the drug. Numerous trials indicate that aspirin, when used as an antithrombotic drug, is maximally effective at doses of 160 to 320 mg/day. Higher doses do not improve efficacy; moreover, they potentially are less efficacious because of inhibition of prostacyclin production, which can be largely spared by using lower doses of aspirin. Higher doses also increase toxicity, especially bleeding.

Dipyridamole

Dipyridamole (Persantine) is a vasodilator that, in combination with **warfarin**, inhibits embolization from prosthetic heart valves and, in combination with aspirin, reduces thrombosis in patients with thrombotic diseases. Dipyridamole by itself has little or no benefit; in fact, in trials where a regimen of dipyridamole plus aspirin was compared with aspirin alone, dipyridamole provided no additional beneficial effect. Dipyridamole interferes with platelet function by increasing the cellular concentration of **adenosine 3′,5′-monophosphate** (cyclic AMP). This effect is mediated by inhibition of **cyclic nucleotide phosphodiesterase** and by blockade of uptake of **adenosine**, which acts at A_2 receptors for adenosine to stimulate **platelet adenylyl cyclase**. The only current recommended use of dipyridamole is for primary prophylaxis of thromboemboli in patients with prosthetic heart valves; the drug is given in combination with warfarin.

Ticlopidine

Ticlopidine (Ticlid) is a thienopyridine that inhibits platelet function by inducing a thrombasthenia-like state. It interacts with platelet glycoprotein IIb/IIIa in an unknown way to inhibit the binding of fibrinogen to activated platelets. Glycoprotein IIb/IIIa is a fibrinogen receptor that links platelets via fibrinogen to form an aggregated plug; this action allows clot retraction. Thus, ticlopidine inhibits platelet aggregation and clot retraction. Ticlopidine prolongs the template bleeding time, with a maximal effect seen only after several days of therapy; abnormal platelet function persists for several days after treatment is discontinued. It is possible that some metabolite of ticlopidine is the active antithrombotic agent, since the drug is relatively ineffective in inhibiting platelet aggregation when added to platelets *in vitro* compared with its effects on platelets taken from patients who are ingesting the drug. The drug has no effect on eicosanoid metabolism, and it should act independently of aspirin; to date, however, these two agents have not been tested in combination. Ticlopidine currently is used for prevention of thrombosis in cerebral vascular and coronary artery disease. Side effects include bleeding, nausea, and diarrhea in 10% of patients, and severe neutropenia in approximately 1% of patients. Ticlopidine is recommended for patients unable to tolerate aspirin.

THERAPEUTIC USES OF ANTICOAGULANT, THROMBOLYTIC, AND ANTIPLATELET DRUGS

Venous Thromboembolism

Treatment. The goal in the treatment of deep venous thrombosis and pulmonary embolism is the prevention of recurrent, fatal embolism.

Myocardial Infarction

All patients with acute myocardial infarction should be considered for intravenous thrombolytic therapy with **streptokinase**, **tissue plasminogen activator (TPA)**, or anistreplase because these agents are effective in both preserving cardiac function and reducing mortality.

Unstable Angina

Aspirin (160 mg/day) combined with heparin followed by warfarin (INR 2.0 to 3.0) may be the optimal treatment.

Saphenous Vein Bypass Grafts

Antiplatelet therapy reduces the risk of occlusion of saphenous grafts, and a "low" dose of aspirin (325 mg/day) is as effective as a higher dose (975 mg/day) or the combination of aspirin and dipyridamole. Rates of occlusion are similar for aspirin begun before or 6 hours after surgery, but the incidence of hemorrhage is increased with preoperative administration.

Percutaneous Transluminal Coronary Angioplasty

Full-dose heparin therapy, beginning with an intravenous bolus of 10,000 U and continued for 4 to 24 hours, is standard practice in patients undergoing angioplasty. Aspirin begun 1 day prior to surgery reduces periprocedural thrombosis and should be continued indefinitely due to its beneficial effect on coronary artery disease.

Atrial Fibrillation

More than 50% of patients with cerebral embolism have atrial fibrillation. In the majority of these patients, the underlying cardiac disease is nonvalvular. The risk of ischemic stroke with atrial fibrillation increases with age, with a cumulative risk of 35% during a patient's lifetime. Combined results from several randomized trials show that warfarin reduces the risk of stroke in patients with nonrheumatic atrial fibrillation by 68% (to 1.4% per year), with an excess incidence of major hemorrhage (including intracranial) of only 0.3% per year.

Prosthetic Heart Valves

The risk of embolism associated with mechanical heart valves is 2 to 6% per patient per year despite anticoagulation and is highest with valves in the mitral position. Warfarin therapy (INR 2.5 to 3.5) is recommended in these patients. The addition of enteric-coated aspirin (100 mg/day) to warfarin (INR 3.0 to 4.5) in high-risk patients (preoperative atrial fibrillation, coronary artery disease, history of thromboembolism) with mechanical valves decreases the incidence of systemic embolism and death from vascular causes (1.9 vs. 8.5% per year), but increases the risk of bleeding.

Valvular Heart Disease

Rheumatic mitral valve disease is associated with thromboembolic complications at reported rates of 1.5 to 4.7% per year; the incidence in patients with mitral stenosis is approximately 1.5 to 2 times that in patients with mitral regurgitation. The presence of atrial fibrillation is the single most important risk factor for thromboembolism in valvular disease, increasing the incidence of thromboembolism in both mitral stenosis and regurgitation four- to sevenfold. In current practice, patients with nonrheumatic atrial fibrillation at low risk for thromboembolism based on clinical characteristics frequently are treated with aspirin. Warfarin therapy is considered in higher-risk patients, especially those with previous thromboembolism and in whom anticoagulation is not contraindicated due to preexisting conditions.

Cerebrovascular Disease

Most ischemic strokes are due to atherosclerotic cerebrovascular disease. Patients with symptomatic carotid stenosis of >70% should be considered for **endarterectomy**. Aspirin is used for prophylaxis following transient ischemic attacks and minor stroke.

Peripheral Vascular Disease

Antithrombotic therapy for acute peripheral occlusive disease is largely empirical. Thrombolytic therapy typically is reserved for patients in whom the occlusion is not amenable to surgery and for those in whom a possible delay between the initiation of therapy and thrombolysis would not jeopardize the viability of the limb. Evidence that antithrombotic therapy changes the natural course of the peripheral disease is sparse, but these patients are at an increased risk of cardiovascular mortality and should receive long-term **aspirin** therapy. Initial trials suggest that ticlopidine may improve the symptoms of chronic arteriosclerotic arterial insufficiency and also reduce fatal and nonfatal cardiovascular events, but further studies are needed.

Primary Prevention of Arterial Thromboembolism

In view of the perceived benefit of aspirin in the secondary prevention of stroke and myocardial infarction, two large trials involving physicians as subjects were initiated to study the effect of aspirin in the primary prevention of arterial thrombosis. In the American study, 22,000 volunteers (age 40 to 84 years) were randomly assigned to take 325 mg of aspirin every other day or placebo. The trial was halted early, after a mean follow-up of 5 years, when a 45% reduction in the incidence of myocardial infarction and a 72% reduction in the incidence of fatal myocardial infarction were noted with aspirin treatment. However, total mortality was reduced only 4% in the aspirin group, a difference that was not statistically significant, and there was a trend for a greater risk of hemorrhagic stroke with aspirin. Thus, the prophylactic use of aspirin in an apparently healthy population is not recommended at this time, unless there are risk factors for cardiovascular disease.

35

PLANTS
AND DIABETES

- *Acacia arabica*
- *Artemisia herba alba*
- *Cleome droserifolia*
- *Eugenia jambolana*
- *Ficus bengalensis*
- *Glossostemon bruguieri*
- *Lagerstraemia speciosa*
- *Lythrum salicaria*

- *Momordica charantia*
- *Opuntia cactus*
- *Panax ginseng*
- *Petiveria alleaceae*
- *Phaseolus vulgaris*
- *Swertia chirayita*
- *Swertia japonica*

INTRODUCTION

Hypoglycemic natural products comprise **flavonoids**, **xanthones**, **triterpenoids**, **alkaloids**, **glycosides**, **alkyldisulfides**, **aminobutyric acid derivatives**, **guanidine**, **polysaccharides**, and **peptides** (see Wang and Ng, 1999). The mechanisms of actions of these hypoglycemic plants are uncertain. They either enhance the release of insulin or enhance the peripheral utilization of glucose.

ANTIDIABETIC AGENTS

Diabetes mellitus results from disturbances in the metabolism of carbohydrates, lipids, and proteins. Normally, the blood glucose level is maintained within a range of 80 to 130 mg/ml. When the level rises above 180 mg/ml, the glucose spills into the urine, causing glucosuria. The utilization of glucose by most tissues, including muscle and adipose tissue, is insulin dependent. The brain is an exception in that its utilization of glucose is insulin independent. In the absence of insulin, the organs other than brain are able to make use of amino acids and fatty acids as alternative sources of energy.

The release of insulin is closely coupled with the glucose level. Hypoglycemia results in a low level of **insulin** and a high level of **glucagon**, and hence favors the processes of **glycogenolysis** and **gluconeogenesis**.

Growth hormone is one of the glucose counterregulatory hormones. It is released in response to hypoglycemia and has intrinsic hyperglycemic actions as well as causes insulin resistance.

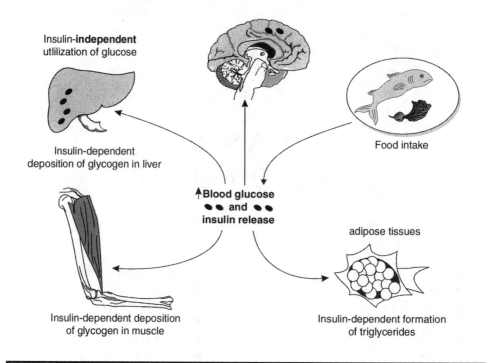

Figure 94 Glucose intake enhances glycogen storage.

THE EFFECTS OF FOOD ON INSULIN RELEASE

Following ingestion of a meal or the administration of glucose (e.g., in a glucose tolerance test), the glucose level rises, causing the release of insulin and inhibiting the release of hyperglycemic glucagon. Excess glucose is transformed into glycogen in the liver and the muscles. The high level of amino acids and free fatty acid fosters the respective formation of proteins in the muscles and triglycerides in the adipose tissues (Figure 94).

THE EFFECTS OF FASTING ON INSULIN RELEASE

In a nondiabetic fasting subject, the ensuing hypoglycemia not only discourages the release of insulin but also activates the homeostatic mechanisms that block the action of insulin and convert the storage forms of fuel into utilizable glucose. Consequently, a number of hormones including glucagon, epinephrine, and glucocorticoid are released, and these convert glycogen into glucose, triglyceride into free fatty acid, and proteins into amino acids (gluconeogenesis), respectively (Figure 95). Furthermore, the uptake and utilization of glucose in the peripheral tissue decrease. The muscles and other tissues utilize amino acids and free fatty acid, thus providing the brain with an adequate supply of glucose.

METABOLIC CONSEQUENCES OF DIABETES MELLITUS

In an individual with diabetes who suffers from a deficiency of insulin, all of the aforementioned measures that apply to a fasting individual may also occur.

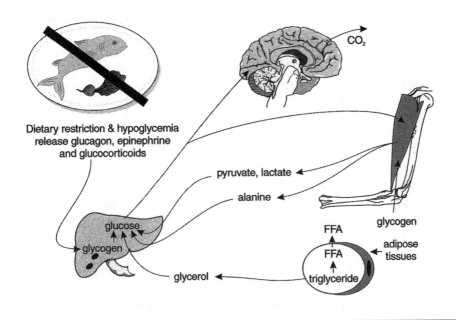

Figure 95 During fasting, glycogen is converted into glucose. FFA = free fatty acid.

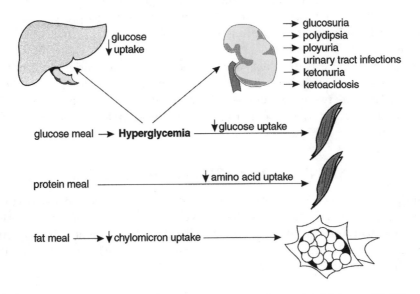

Biochemical abnormalities in an untreated patient with Diabetes Mellitus

Figure 96 Consequences of hyperglycemic states.

However, the consumption of a meal or the administration of glucose will instead cause pronounced hyperglycemia because the insulin-dependent utilization of glucose by muscles and adipose tissues is lacking (Figure 96). The elevated glucose level thus surpasses the renal threshold and glucose may appear continuously in the urine. The osmotic diuretic effects of glucose cause **polyuria** and **polydipsia** and the chronic **glucosuria** may lead to urinary tract infection. Because the conversion to triglyceride does not take place, free fatty acids are metabolized to ketone bodies, causing ketonuria and ketoacidosis (acetone or **fruity breath**). The continuous destruction of muscular proteins may ultimately lead to muscle wasting and weight loss.

DIABETES MELLITUS AND PREGNANCY

Early in pregnancy, glucose homeostasis is altered by the increasing levels of **estrogen** and **progesterone**, which lead to β-cell hyperplasia and an increased insulin response to a glucose load. During the second half of pregnancy, rising levels of human placental lactogen and other contrainsulin hormones synthesized by the placenta modify maternal utilization of glucose and amino acids. The actions of **lactogen** are responsible, in part, for the diabetogenic state associated with pregnancy.

CLASSES OF DIABETES MELLITUS

There are two types of diabetes mellitus: Type I, or **insulin-dependent diabetes mellitus**, and Type II, or **non-insulin-dependent diabetes mellitus**.

Patients with Type I diabetes may have islet cell antibodies and human leukocyte antigens (HLA). They are dependent on insulin to prevent ketosis and hence have **insulinopenia**. Affected individuals consist mostly of children and young adults.

Patients with Type II diabetes, who are non-insulin-dependent, are not prone to ketosis. This type of diabetes is not an autoimmune disorder nor is it associated with HLA. The patients are generally older (>40 years), may or may not be obese, and may or may not have been treated with insulin for the control of their hyperglycemia.

SIGNS AND SYMPTOMS OF DIABETES

The signs and symptoms of diabetes consist of thirst, anorexia, nausea, vomiting, abdominal pain, headache, drowsiness, weakness, coma, severe acidosis, **air hunger (Kussmaul's breathing)**, sweetish odor of the breath, hyperglycemia, decreased blood bicarbonate level, decreased blood pH, and plasma that is strongly positive for ketone bodies.

COMPLICATIONS OF DIABETES

There are a number of complications that arise as the result of poorly treated or unstabilized diabetes.

Vascular Complications

Vascular complications may be manifested by **microangiopathy** (thickening of the capillary basement membrane), **intracapillary glomerulosclerosis** (thickening of the glomerular capillary basement membrane, which leads to a nephrotic syndrome characterized by edema, albuminuria, or renal failure), and microangiopathy of the blood vessels supplying the retina (**diabetic retinopathy**). In fact, diabetes is still the leading cause of blindness in the world. In addition, there may be atherosclerosis of the peripheral arteries.

Neuropathy

Diabetic neuropathy may be associated with **neuropathic ulcer**, ptosis, **diplopia**, **strabismus**, loss of deep tendon reflexes, ankle drop, wrist drop, **paresthesia**, **hyperalgesia**, **hyperesthesia**, and orthostatic hypotension (because of autonomic dysfunction).

THE INSULIN RECEPTOR

Structure and Function

When insulin binds to specific membrane receptors on target cells, this initiates a wide spectrum of biologic activities:

- Enhancing the transport of sugar and amino acids
- Stimulating anabolic pathways
- Stimulating growth and development by triggering RNA and DNA synthesis

The insulin receptor is a disulfide-linked oligomer consisting of two α and two β chains, with molecular weights of 130,000 and 95,000, respectively (Figure 97). Cross-linking studies using iodine-125-labeled insulin have shown that the insulin-binding domain is situated primarily in the α subunit. In addition, it has been observed that proteolysis of the β subunit does not appreciably influence insulin binding. Insulin binding to the α subunit was found to induce rapid phosphorylation of the intracellular domain of the β subunit. The β subunit contains a putative adenosine triphosphate (ATP)-binding site and an intrinsic tyrosine-specific kinase as part of the receptor. The enzymatic activity of the receptor is activated by insulin binding, which results in increased tyrosine phosphorylation of the β subunit as well as the production of a number of other cellular proteins.

Termination of the Insulin Signal

The insulin receptor is internalized, and this action terminates the insulin signal at the surface of the cell. Once internalized, some of the receptors are degraded and others are recycled back to the membrane. In addition, phosphatases are able to dephosphorylate the phosphorylated insulin receptor. This dephosphorylation reduces kinase activity and decreases the responsiveness to insulin.

Clinical Syndromes Associated with Cellular Defects in the Insulin Receptor

A number of disorders are associated with the development of insulin resistance. Although some cases are due to autoimmune responses such as the development

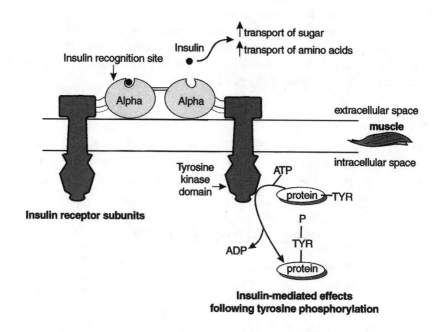

Figure 97   The insulin receptor. ATP = adenosine triphosphate; ADP = adenosine diphosphate; TYR = tyrosine.

of antiinsulin or **anti-insulin receptor antibodies**, insulin resistance often results from defects at the cellular level in the insulin receptor or in postreceptor function.

TREATMENT OF DIABETES MELLITUS

Dietary Treatment

The American Diabetes Association currently makes the following nutritional recommendations for people with diabetes:

1. Achieve and maintain ideal body weight
2. Derive 55 to 60% of total caloric intake from carbohydrates
3. Consume foods containing unrefined carbohydrate with fiber, attempting to take in 40 g of soluble fiber per day
4. Consume only modest amounts of sucrose

Drug Treatment

The drug therapy of diabetes includes eliminating obesity (which causes resistance to both endogenous and exogenous forms of insulin), exercising (to promote glucose utilization and reduce insulin requirement), dieting (to restrict intake of excess amounts of carbohydrates), and taking insulin (primarily in polyuric, polydipsic, and ketonuric patients).

Table 34   Properties of Insulin Preparations

Type	Added Protein	Action (hr)		
		Onset	Peak	Duration
Rapid				
Regular (crystalline)	None	0.3–0.7	2–4	5–8
Semilente	None	0.5–1.0	2–8	12–16
Intermediate				
NPH (Isophane)	Protamine	1–2	6–12	18–24
Lente	None	1–2	6–12	18–24
Slow				
Ultralente	None	4–6	16–18	20–36
Protamine zinc	Protamine	4–6	14–20	24–36

Insulin preparations that are commercially available differ in their relative onset of action, maximal activity, and duration of action. Conjugation of the insulin molecule with either zinc or protamine, or both, will convert the normally rapidly absorbed parenterally administered insulin to a preparation with a more prolonged duration of action. The various formulations of insulin are usually classified as short-acting (0.5–14 h), intermediate-acting (1–28 h), and long-acting (4–36 h). The duration of action can vary, however, depending on injection volume, injection site, blood flow at the site of administration.

Insulin

Insulin preparations are fast, intermediate, or long acting, as summarized in Table 34. Crystalline (regular) insulin may be used as a supplemental injection or for instituting corrective measures in the management of infection and trauma, for postoperative stabilization, and for the rehabilitation of patients recovering from ketoacidosis and coma. In addition, NPH (isophane) contains regular insulin.

Ultralente or semilente insulin is used to eliminate nocturnal and early morning hyperglycemia.

Complications of Insulin Therapy

Hypoglycemia is a primary complication of insulin therapy and may result from either an excess of insulin or a lack of glucose, or both. Severe hypoglycemia may cause headache, confusion, double vision, drowsiness, and convulsions. The treatment of this hypoglycemia may include the administration of glucose or glucagon.

Lipodystrophy can also result from insulin therapy and is characterized by atrophy of subcutaneous fat. Insulin edema is manifested by a generalized retention of fluid. Insulin resistance arises when there is an excess insulin requirement that exceeds 200 units per day.

Agents That Alter the Release of Insulin

The release of insulin is enhanced by certain physiological substances (glucose, leucine, arginine, gastrin, secretin, and pancreozymin) and by certain pharmacological agents (oral **hypoglycemic agents**). The release of insulin is also inhibited by some physiological substances (epinephrine and norepinephrine) as well as by some pharmacological substances (**thiazide diuretics**, **diazoxide**, and **chlorpromazine**).

Oral Hypoglycemic Agents

Oral hypoglycemic agents have advantages over insulin, because, by releasing insulin and by decreasing the release of glucagon, they mimic physiological processes and cause fewer allergic reactions. Furthermore, they are effective in an oral form, thus eliminating the need for daily injections. The properties of these agents are described in Table 35.

The mechanisms that underlie the hypoglycemic actions of sulfonylureas are:

- Pancreatic
    Improved insulin secretion
    Reduced glucagon secretion

- Extrapancreatic
    Improved tissue sensitivity to insulin
    Direct
      Increased receptor binding
      Improved postbinding action
    Indirect
      Reduced hyperglycemia
      Decreased plasma concentrations of free fatty acids
      Reduced hepatic insulin extraction

RECEPTORS FOR SULFONYLUREA AND THEIR MECHANISM OF ACTION

Sulfonylureas such as **glyburide** and **glipizide** bind to **sulfonylurea receptors** located on the surface of β cells and trigger insulin releases at nanomolar concentrations (Figure 98). Sulfonylureas bind to **ATP-sensitive potassium channels** and inhibit potassium efflux through these channels. The inhibition of ATP-sensitive potassium channels then leads to depolarization of the β cell; this opens voltage-dependent calcium channels and allows the entry of extracellular calcium. The rising level of cytosolic-free calcium next triggers the release of insulin. An increase in the cyclic adenosine monophosphate levels in the cells can also open the voltage-dependent calcium channels, thus increasing calcium-influx into the cells.

Table 35   Comparison of Orally Administered Sulfonylurea Hypoglycemic Agents

Characteristic	Tolbutamide	Acetohexamide	Tolazamide	Chlorpropamide	Glipizide	Glyburide
Relative potency	1	2.5	5	6	100	150
Duration of action (h)	6–10	12–18	16–24	24–72	16–24	18–24
Extent of protein binding (%)	>98	~90	>98	~95	>98	>98
Hepatic metabolism	Yes	Yes	Yes	Yes	Yes	Yes
Urinary excretion	Yes	Yes	Yes	Yes	Yes	Yes
Fecal excretion (% of dose)	Negligible	Negligible	Negligible	Negligible	12	50
Dose (mg) range	500–3000	250–1500	100–1000	100–500	2.5–40	1.25–20
Diuretic	Yes	Yes	Yes	No	No	Yes
Antidiuretic	Yes	No	No	Yes	No	No
Disulfiram-like effects	No	No	No	Yes	No	No

Although insulin has the disadvantage of having to be injected, it is without question the most uniformly effective treatment of diabetes mellitus available. Although insulin remains the drug of choice in severe cases of diabetes and in IDDM, some milder forms of diabetes mellitus that do not respond to diet management alone can be treated with oral hypoglycemic agents. The success of oral hypoglycemic drug therapy is usually based on a restoration of normal blood glucose levels and the absence of glycosuria.

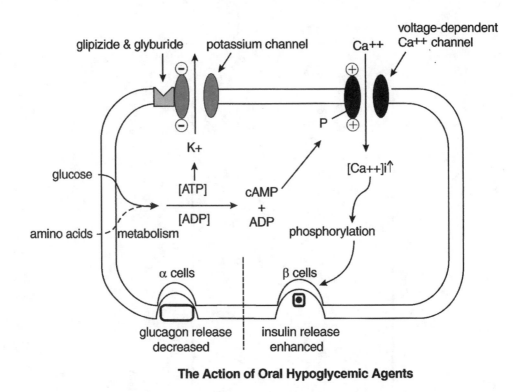

Figure 98 The actions of glipizide. ATP = adenosine triphosphate; ADP = adenosine diphosphate; cAMP = cyclic adenosine monophosphate.

TREATMENT OF DIABETIC KETOACIDOSIS

Diabetic ketoacidosis may either result from or be aggravated by infection, surgery, trauma, shock, emotional stress, or failure to take sufficient amounts of insulin. Treatment is focused on reversing the hypokalemia by administering potassium chloride and on offsetting the acidosis by providing bicarbonate. The dehydration and electrolyte imbalance are treated with appropriate measures and **crystalline zinc insulin** is administered to counter the hyperglycemia.

36

HYPERICUM (ST. JOHN'S WORT) AND DEPRESSION

- *Hypericum* extract has been used for the treatment of mild to moderate mental depression.
- *Hypericum* is thought to affect the synaptosomal uptake of serotonin, dopamine, and norepinephrine.
- From the existing literature, St. John's wort appears to be a safe and effective alternative in the treatment of depression. Tricyclic antidepressants and monoamine oxidase inhibitors can produce serious cardiac side effects, such as tachycardia and postural hypotension, and many unwanted anticholinergic side effects, including dry mouth and constipation. St. John's wort has proved to be free of any cardiac, as well as anticholinergic, side effects normally seen with antidepressant medications. Based upon limited studies, St. John's wort appears to be an acceptable alternative to traditional antidepressant therapy.

INTRODUCTION

More than 3 million people over the age of 85 currently reside in the United States and the geriatric population is continually rising. One of the most common disabilities seen in elderly people is major depression, which affects 17% of the population. Between 35 and 45% of hospitalized elderly patients experience a depressive syndrome at some point. Cognitive symptoms of depression may include a loss of interest in activities, suicidal thoughts, depressed mood, and feelings of worthlessness, as well as a number of symptoms that are indistinguishable from other disease states. Depression also presents with somatic symptoms, including weight loss, fatigue, loss of concentration, disturbed sleep, and an overconcern with health problems.

Because suicide is one of the leading causes of death in elderly people and in other populations, rapid and effective treatment of depression is warranted. Current therapies include the use of electroconvulsive (shock) therapy, psychiatric intervention, and antidepressant drugs such as **tricyclic antidepressants** (TCAs), **monoamine oxidase inhibitors** (MAOIs), and **serotonin-selective reuptake**

inhibitors (SSRIs). Recently, the use of **St. John's wort** (*Hypericum perforatum*) in the United States has become more prevalent, especially in the treatment of depression.

Hypericum perforatum is a herbaceous perennial plant widely distributed throughout Europe, Asia, Northern Africa, Canada, and British Columbia. In the United States, it grows wild in northern California and southern Oregon.

St. John's wort was believed in the Middle Ages to have magical powers that would protect one from evil. The plant was originally named after **John the Baptist** and is commonly collected on June 25, **St. John's Day**, and soaked in olive oil, eventually producing a red oil symbolizing the "**Blood of St. John.**" This oil, exposed to sunlight for several weeks, has a reputation as an anti-inflammatory and healing agent used especially in the treatment of wounds. *Hypericum* has also been used in Europe as a soothing agent and is often used in the treatment of hemorrhoids. More recently, St. John's wort has been tested for use as an antiviral agent, most specifically in the treatment of AIDS.

Today, *Hypericum* is used widely in Germany for the treatment of depression where it is prescribed approximately 20 times more often than **fluoxetine**, one of the most highly prescribed antidepressants in the United States. In the United States, *Hypericum* is increasingly used as an over-the-counter remedy by a significant portion of the lay population for the treatment of depression. In 1998, *Hypericum* was expected to garner $400 million in sales in the United States and an estimated $6 billion in Europe, despite a lack of consensus regarding its efficacy among the medical community and the absence of standardization guidelines in the United States. Furthermore, the specific modes of purported antidepressant activity are not well defined. There is some evidence that *Hypericum* may exert a significant influence on catecholamine neurotransmission via known pathways, including (1) inhibition of neurotransmitter metabolism, (2) modulation of neurotransmitter receptor density and sensitivity, and (3) synaptic reuptake inhibition. Similar to conventional antidepressant pharmacology, these mechanisms may ultimately lead to increased synaptic availability of the neurotransmitters believed to be implicated in clinical depression, namely, serotonin (5-HT), norepinephrine (NE), and dopamine (DA) (Greeson et al., 2001).

ST. JOHN'S WORT

Botanical Name

Hypericum perforatum L.

Family

Hypericacae (*alt.* Clusiaceae = Guttiferae)

Synonyms

Hyperici herba

Common Names

Amber, Amber Touch-and-Heal, Blutkraut, Chassediable, Corazoncillo, Goatweed, Hardhay, Hartheu, Herba di S. Giovanni, Herbe de Millepertuis, Herbe St. Jean, Herrgottsblut, Hexenkraut, Heirba de San Juan, Hipericon, Hypericum, Iperico, Johannesort, Johannisblut, Johanniskraut, Johnswort, Klamath weed, Millepertiuis, Pelatro, Perforata, Qian Ceng Lou, Rosin Rose, Sankt Hans Urt, St. Jan's Kraut, St. John's wort, St. Johnswort, Tipton weed, Toutsaine, Tupfelhartheu, Walpurgiskraut, Zwieroboij

Commercial Products

The commercial products of **St. John's wort** are derived from the dried flowering tops or aerial parts of ***Hypericum perforatum*** L.; they are harvested shortly before or during the flowering period. Hypericum preparations include the dried herb (chopped or powdered), tea infusion, liquid extract, dried (hydroalcoholic) extract, oil, and tincture.

Active Ingredients

The active ingredients for hypericum include **cyclopseudohypericin**, **hypericin**, **hyperforin**, **isohypericin**, **protohypericin**, and **pseudohypericin**. Other active ingredients include the following flavonoids: **hyperin**, **hyperoside**, **isoquercitrin**, **kaempferol**, **luteolin**, **quercetin**, **quercitrin**, **rutin**, and the following biflavonoids: **amentoflavone** and **I3,II8-biapigenin**. Although the flavonoids, the essential oils, and the phenolic acids are common plant constituents, the **hypericins**, **1,3,6,7-tetrahydroxy-xanthone**, and **hyperforin** are characteristic constituents of hypericum (Figure 99).

HYPERICUM (ST. JOHN'S WORT) IN FOLK MEDICINE AND TRADITIONAL SYSTEMS OF MEDICINE

In folk medicine and traditional systems of medicine, various species of *Hypericum* have been used orally to treat anxiety, bed-wetting, dyspepsia, excitability, exhaustion, fibrositis, gastritis, gout, hemorrhage, hysteria, insomnia, irritability, jaundice, migraine headaches, neuralgia, pulmonary complaints, rheumatism, sciatica, and swelling. It also has been used orally as an anthelmintic, an antidiarrheal, and a diuretic. Various dosage forms of hypericum have been used topically as an astringent and to treat injuries or conditions such as blisters, burns, cuts, hemorrhoids, inflammation, insect bites, itching, redness, sunburns, and wounds. *Hypericum* with or without light therapy may be useful for the treatment of **seasonal affective disorder** (SAD), but studies to date have used too few subjects to determine its efficacy. In addition, light therapy combined with hypericum intake may increase the risk of phototoxicity.

Hypericum is being combined with other products such as **ma huang** (ephedra) and promoted for weight loss as an alternative to prescription weight loss medications.

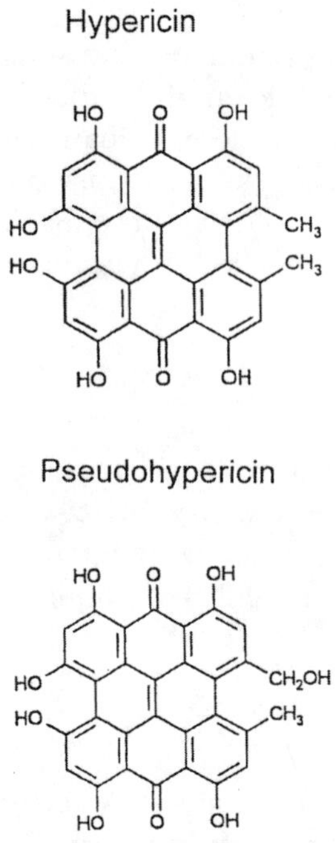

Hypericin

Pseudohypericin

Figure 99 Principal constituents of *Hypericum* (St. John's wort).

ANTIDEPRESSANTS

Depressive disorders are the most prevalent of psychiatric illnesses, occurring more often among women (16%) than among men (8%). Depression, a recurrent but self-limiting disorder, is classified as (1) exogenous or reactive depression or as (2) endogenous depression. Depression is further classified as unipolar and bipolar. Unipolar patients display one type of affective disorder, most commonly depression; bipolar patients usually experience mania and depression (Figure 100).

Symptoms of major depression and dysthymia:

- Depressed mood most of the day, nearly every day
- Markedly diminished interest or pleasure in all or almost all activities
- Marked weight loss or gain, or a decrease or increase in appetite
- Insomnia or hypersomnia nearly every day
- Psychomotor agitation or retardation nearly every day
- Fatigue or loss of energy nearly every day
- Feelings of worthlessness or excessive or inappropriate guilt
- Diminished ability to think or concentrate, or indecisiveness
- Recurrent thoughts of death, or suicidal ideation or attempt
- Feelings of hopelessness

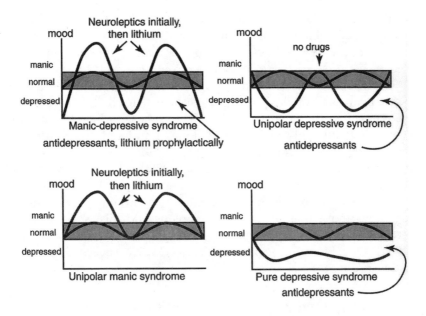

Figure 100   The use of antidepressants in bipolar depression.

Characteristics of melancholic depression:

- Loss of interest or pleasure in all or almost all activities
- Lack of reactivity to usually pleasurable stimuli
- Depression regularly worse in the morning
- Early morning awakening
- Psychomotor retardation or agitation (not merely subjective reports)
- Marked anorexia or weight loss
- No marked personality disturbance before the first major depressive episode
- One or more previous major depressive episodes followed by complete or nearly complete recovery
- Previous good response to specific and adequate somatic antidepressant therapy (e.g., tricyclic drugs, electroconvulsive therapy, monoamine oxidase inhibitors, and lithium)

NEUROTRANSMITTERS AND THEIR ROLE IN DEPRESSION

Depression has been noted in 10 to 15% of all patients who take **reserpine** for the treatment of hypertension. When used for the treatment of **tuberculosis**, **isoniazid** brings about mood elevation. Reserpine was found to diminish the levels of brain neurotransmitters such as dopamine, serotonin, and norepinephrine, whereas isoniazid, through its **monoamine oxidase inhibitory actions**, augments the levels of these substances. It was thus theorized that too much of one or more of these neurotransmitters might be associated with states of hypomania or mania, whereas too little would correlate with depression.

Other neurotransmitters besides dopamine, norepinephrine, or serotonin may be involved in depression. These include acetylcholine and γ-aminobutyric acid (GABA). Acetylcholine stimulates the **locus ceruleus** through the **muscarinic cholinergic receptors**. Clinical data have shown that cholinergic agonists can cause depression in that the response to muscarinic agonists is augmented in euthymic patients with a history of depression. In addition, **iprindole**, an antidepressant that does not alter the uptake or metabolism of monoamines, has potent anticholinergic properties. Moreover, electroshock treatment, one of the most efficacious of treatments for managing depression, also has little, if any, effect on available measures of the functioning of neurotransmitter systems.

GABA inhibits the action of the locus ceruleus and this is mediated by the $GABA_A$ receptor–chloride channel system. The plasma and cerebrospinal fluid levels of GABA are low, and **progabide**, a $GABA_A$ **receptor agonist**, has antidepressant properties.

CLASSIFICATION OF ANTIDEPRESSANTS

Drugs used in the treatment of depression are referred to as **thymoleptics**, **thymoanaleptics**, **psychoanaleptics**, **psychic energizers**, and antidepressants.
Antidepressants are divided into the following classes:

- The dibenzapine derivatives are called tricyclic antidepressants and include **imipramine** (Tofranil), **desipramine** (Norpramin), **amitriptyline** (Elavil), **nortriptyline** (Aventyl), **protriptyline** (Vivactil), and **doxepin** (Adapin).
- The monoamine oxidase inhibitors are used occasionally to treat depression. The hydrazine derivatives consist of **isocarboxazid** (Marplan) and **phenelizine sulfate** (Nardil). The nonhydrazine derivatives include **tranylcypromine** (Parnate).
- L-Tryptophan is the only member of the monoamine precursors used to treat depression.
- Newer agents include amoxapine, doxepin, fluoxetine, maprotiline, trazodone, mianserin, alprazolam, and bupropion.

DIBENZAPINE DERIVATIVES

The three-ring nucleus characteristic of the dibenzapine derivatives has given rise to their name of tricyclics or heterocyclics. They resemble the phenothiazine derivatives such as chlorpromazine in structure and function.

Like the phenothiazine derivatives (e.g., chlorpromazine), tricyclic antidepressants (e.g., imipramine) may:

- Reduce the seizure threshold and precipitate seizures in patients with epilepsy
- Cause cholestatic jaundice
- Cause movement disorders
- Cause hematological side effects

Unlike the phenothiazine derivatives, the tricyclic antidepressants:

- May increase motor activity
- Are antidepressant
- Have a very slow onset and long duration of action
- Have a relatively narrow margin of safety
- Have strong anticholinergic effect; in fact, dry mouth is the most common side effect and other anticholinergic effects such as tachycardia, loss of accommodation, constipation, urinary retention, and paralytic ileus have been reported

Conversion to Pharmacologically Active Metabolites

Imipramine is demethylated to desipramine, and amitriptyline is demethylated to nortriptyline. Both metabolites are active antidepressants.

Various Properties of the Tricyclic Antidepressants

Tricyclic antidepressants, like some of the phenothiazine derivatives, are sedative in nature. Those compounds containing a tertiary amine (**imipramine, amitriptyline**, and **doxepin**) are the most sedative. Those compounds containing a secondary amine (**nortriptyline** and **desipramine**) are less so, and protriptyline has no sedative effect.

Tricyclic antidepressants, like some of the phenothiazine derivatives (e.g., **thioridazine**), have an anticholinergic property. **Amitriptyline** is the strongest in this regard and desipramine is the weakest.

The tricyclic antidepressants also have cardiovascular actions. In particular, they cause orthostatic hypotension by obtunding the various reflex mechanisms involved in maintaining blood pressure.

Antidepressants may block the uptake of norepinephrine or serotonin (Table 36).

The Sites of Action of Antidepressants

The precise mechanism by which the first-generation tricyclic antidepressants, monoamine oxidase inhibitors, and the newer-generation antidepressants exert their effects is uncertain. However, it is clear that antidepressants exert their effects at both pre- and postsynaptic receptor sites (Figures 101 and 102).

The blockade of norepinephrine and serotonin uptake occurs immediately, but antidepressants are not effective clinically for 2 to 4 weeks. Long-term treatment with antidepressants causes:

- Decreased sensitivity of postsynaptic β_1-adrenergic receptors
- Decreased sensitivity of postsynaptic serotonin receptors
- Decreased sensitivity of α_1-adrenergic presynaptic receptors
- Increased sensitivity of α_1-adrenergic postsynaptic receptors (see Table 36)

Table 36 The Inhibition of Monoamine Uptake by Antidepressants

| Drug | Inhibition of Reuptake | |
	Norepinephrine	Serotonin
Older drugs		
Amitriptyline	+	++
Nortriptyline	++	+
Imipramine	+	+
Desipramine	+++	0
Clomipramine	+	+++
Newer drugs		
Fluoxetine	0	+++
Bupropion	±	0

0 = no effect; ± = an equivocal effect; + = slight effect; ++ = moderate effect; +++ = large effect.

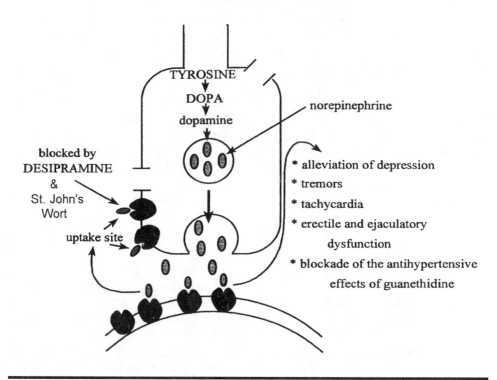

Figure 101   Desipramine blocks the uptake sites for norepinephrine.

Side Effects of Antidepressants

The first-generation tricyclic antidepressants, the monoamine oxidase inhibitors, and the newer agents can cause sedation, insomnia, orthostatic hypotension, or nausea. Because of their anticholinergic properties, they may also produce cardiac toxicities (Table 37).

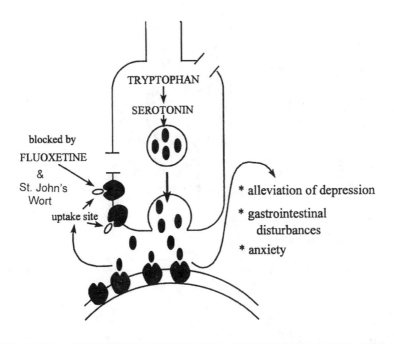

* alleviation of depression
* gastrointestinal
  disturbances
* anxiety

Figure 102   Fluoxetine blocks the uptake sites for serotonin.

Table 37   Selected Side Effects of Antidepressant Drugs

Drug	Sedation	Anticholinergic Effects	Orthostatic Hypotension
Tricyclic drugs			
Amitriptyline	+++	+++	+++
Desipramine	+	+	+
Doxepin	+++	++	+++
Imipramine	++	++	++
Nortriptyline	++	+	+
Protriptyline	+	++	+
Monoamine oxidase inhibitors			
Tranylcypromine	0	0	+++
Newer agents			
Amoxapine	++	+	++
Fluoxetine	0	0	0
Maprotiline	++	+	++
Trazodone	++++	0	++
Alprazolam	+	0	0
Bupropion	0	0	0

0 = no side effect; + = minor side effect; ++ = moderate side effect; +++ = major side effect; ++++ = extreme side effect.

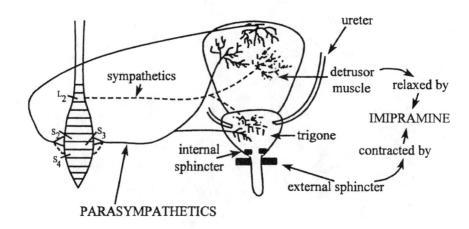

Figure 103 Imipramine is used to manage enuresis. L_2 = second lumbar nerves; S_2, S_3, S_4 = second, third, and fourth sacral nerves.

Therapeutic Indications

A primary indication for the use of tricyclic antidepressants is endogenous depression. Before treating an endogenous depression, however, it should first be differentiated from sadness. Disabling depression and its vegetative symptoms generally respond to tricyclic antidepressants but sadness does not, although it responds to changes in environmental events. The effective dose of tricyclic antidepressants, which are equivalent drugs, is chosen empirically. The less sedative agents are chosen for apathetic and withdrawn patients. Because the margin of safety for these agents is very narrow, they should not be prescribed in large quantities for a depressed patient who may use them to attempt suicide.

The anticholinergic effect of imipramine has been used successfully in managing enuresis (Figure 103).

The pain associated with diabetic peripheral neuropathy, trigeminal neuralgia, or cancer may predispose such patients to depression. Tricyclic antidepressants have been shown to be an effective adjunct in managing these and other similar conditions.

Some episodic phobias are regarded as "masked" depression and thus respond to treatment with tricyclic antidepressants.

Fluoxetine, in addition to its antidepressant property, has been used as an appetite suppressant. Imipramine and desipramine have been used as **antibulimic substances**.

Desipramine has been used as part of the **treatment of alcoholism**. Because depression has led to relapsed drinking in people suffering alcoholism striving to maintain sobriety, treatment with antidepressants may reverse or prevent these depressive symptoms. They may also correct the biochemical abnormalities hypothesized to underlie both depression and alcoholism, thus helping to ensure abstinence in recovering alcoholics.

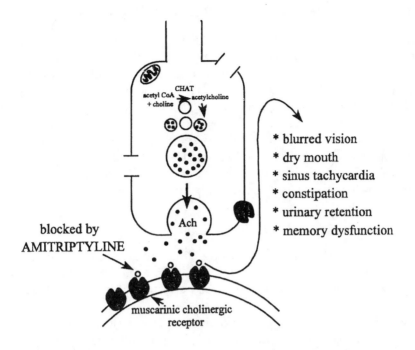

* blurred vision
* dry mouth
* sinus tachycardia
* constipation
* urinary retention
* memory dysfunction

Figure 104 Amitriptyline has potent anticholinergic properties. CHAT = choline acetylase; CoA = coenzyme A; ACh = acetylcholine.

Fatal Overdoses

Currently, overdoses of tricyclics are one of the most serious types of poisoning encountered in clinical practice, because the depressed patients who are treated with these drugs are also those who are the most prone to using them for suicidal purposes.

The diagnostic triad of coma, seizures, and cardiac arrhythmias should raise the suspicion of tricyclic overdose, if there is otherwise no verified history of drug intake. Cardiac arrhythmias and conduction abnormalities are the major distinguishing features. A trial dose of intravenously administered **physostigmine** (1 to 4 mg) may suggest the diagnosis because this will awaken the comatose patient or mitigate the arrhythmias. The effects of physostigmine are transient, however, and it is not a definitive treatment. Other problems encountered in such poisonings include neuromuscular irritability, delirium, hyperpyrexia, hypotension, and bladder or bowel paralysis (Figure 104).

The cardiac arrhythmias are life-threatening, so the patient must be closely monitored, with facilities available for possible resuscitation. Drugs such as **quinidine** and **procainamide** are contraindicated, but **lidocaine, propranolol**, or **phenytoin** has been used safely and effectively. The arterial blood gas levels, pH, and electrolyte concentrations should be monitored so that metabolic acidosis or hypokalemia can be identified that would further aggravate the arrhythmias. Electrical pacing may be required if the antiarrhythmic drugs fail. Hyperpyrexia is treated by cooling. Seizures may be managed by intravenous doses of diazepam.

MONOAMINE OXIDASE INHIBITORS

Monoamine oxidase can metabolize monoamines by oxidative deamination and convert them to inactive acidic derivatives. In general, not only do these agents inhibit the oxidase that metabolizes amines but they also inhibit the oxidase that metabolizes drugs and essential nutrients. Hence, the incidence of drug–drug and drug–food interactions is extremely high with these agents. These agents should be used with extreme caution in conjunction with sympathomimetic amines, ganglionic blocking agents, procaine, and anesthetic agents. They are contraindicated in patients with hyperthyroidism and in combination with tricyclic antidepressants. In the event of poisoning, adrenergic blocking agents such as **phentolamine** may be effective for combating the hypertensive crisis.

Indications for Monoamine Oxidase Inhibitors

The high incidence of drug–food and drug–drug interactions rules out monoamine oxidase inhibitors as antidepressants of first choice. However, there are circumstances in which these agents may be used effectively and successfully:

- When a patient has not responded to a tricyclic antidepressant for an adequate trial period and with an appropriate dosage
- When a patient has developed allergic reactions to tricyclics
- When a patient has had previous depressive episodes that responded well to monoamine oxidase inhibitors

NEWER, NOVEL, OR ATYPICAL ANTIDEPRESSANTS

The newer agents that may have actions that are novel or atypical include maprotiline, amoxapine, fluoxetine, trazodone, bupropion, mianserin, and alprazolam.

Maprotiline has relatively minor anticholinergic properties compared to amitriptyline. It has a very large elimination half-life of 36 to 48 hours and in large doses produces convulsions.

Amoxapine has neuroleptic properties that stem from its dopamine receptor-binding affinity. Similar to the neuroleptics, it may produce movement disorders. In toxic doses, amoxapine may provoke difficult-to-control convulsions.

Fluoxetine selectively blocks the uptake of serotonin. It is devoid of anticholinergic properties and hence has little or no effect on the cardiovascular system, including orthostatic hypotension or arrhythmias.

Trazodone is perhaps the most sedative antidepressant available. It is devoid of anticholinergic effects and causes postural hypotension as well as rarely serious priapism.

Bupropion is free of anticholinergic, antiadrenergic, and cardiotoxic properties. Its structure is related to that of amphetamine and it possesses stimulating effects; hence it may be useful in hyperphagic or obese individuals.

Mianserin does not alter the uptake of norepinephrine, serotonin, or dopamine. It is devoid of anticholinergic properties and thus is not cardiotoxic. It blocks both α_1- and α_2-adrenergic receptors, and mianserin is sedative in nature.

Alprazolam causes pronounced sedation but is devoid of anticholinergic properties. Alprazolam and, to a certain extent, clonazepam are effective in the treatment of panic disorder.

CONTROL OF MANIC EPISODES IN MANIC-DEPRESSIVE PSYCHOSIS USING LITHIUM

Pharmacokinetics

Lithium is given orally as a salt, and the particular salt does not affect the therapeutic action. The anionic partner of lithium — carbonate, chloride, acetate, citrate, or sulfate — serves only as an inert vehicle. Carbonate is by far the most widely used lithium salt. In addition, lithium carbonate contains more lithium, weight for weight, than do the other lithium salts.

Because lithium is not bound to any plasma or tissue proteins, it is widely distributed throughout the body. Lithium ions are eliminated mainly by the kidneys. There is a direct relationship between the amount of sodium chloride ingested and the fraction of filtered lithium resorbed, in that, the lower the sodium intake, the greater is the lithium retention. The contraindications are significant cardiovascular or renal diseases that would compromise its excretion.

Side Effects

Lithium is unique among the psychopharmacological compounds in that it rarely has any undesirable effects on emotional and intellectual functioning. A few unwanted effects are seen in the somatic sphere, and these fall into three overlapping categories:

1. Initially, when the maintenance dose of lithium is being established, the patient may experience gastrointestinal discomforts such as nausea, vomiting, diarrhea, stomach pain, muscular weakness, unusual thirst, frequent urination, a slight feeling of being dazed, tiredness, and sleepiness. These early side effects disappear once the patient is stabilized.
2. From the beginning of treatment, patients exhibits light and barely noticeable hand tremors, which do not respond to antiparkinsonian agents.
3. After several months of continuous therapy with lithium, **diabetes insipidus** and goiter may develop. The kidney tubules then become insensitive to the action of antidiuretic hormone and its administration is ineffective. Either a dose reduction or discontinuation of the lithium corrects this side effect without leaving any residual pathology. In the presence of goiter, the patient remains euthyroid. It has been reported that the administration of small amounts of thyroxine may counteract this side effect.

Pharmacodynamics of Lithium

Lithium is thought to exert its effect by interfering with the calcium-mediated release of norepinephrine, increasing the uptake of norepinephrine, and decreasing the sensitivity of postsynaptic receptor sites to norepinephrine.

Table 38   Regulatory Status of Hypericum

Country	Regulatory Status
Canada	Authorization has been granted for hypericum preparations as traditional herbal medicines intended to be used as a sedative for relief of restlessness (or nervousness) due to overwork, tiredness, and fatigue. Crude dried herbs of hypericum intended for sale as food, without medicinal representations, may be sold if they conform with the pertinent requirements of the Food Regulations.
France	The flowering tip has traditional use externally for chapped skin, insect bites, minor burns, pruritus, and rash. It is permitted for use in temporary relief of sore throat and hoarseness.
Germany	Hypericum is registered for mild to moderate depressive episodes as specified in the *International Classification of Diseases*, 10th revision. Commission E has approved the oily hypericum preparations for topical use to treat incised and contused wounds, myalgia, and first-degree burns.

St. John's wort is stated to possess sedative and astringent properties. It has been used for excitability, neuralgia, fibrositis, sciatica, wounds, and specifically for menopausal neurosis. St. John's wort is used extensively in homeopathic preparations as well as in herbal products.

In addition, there is increasing evidence that lithium exerts its therapeutic action by interfering with the polyphosphoinositide metabolism in the brain and by preventing inositol recycling through the uncompetitive inhibition of inositol monophosphatase.

Therapeutic Uses of Lithium

The uses of lithium fall into two categories: established and innovative. Among its established uses, lithium salts are used to treat acute mania and as a prophylactic measure to prevent the recurrence of bipolar manic-depressive illness. As an innovative agent, lithium salts have been used with certain success in the management of the following illnesses or conditions.

In combination with tricyclic antidepressants, lithium is used in treating recurrent endogenous depression. In combination with neuroleptics, it is used in the management of schizoaffective disorders. In combination with neuroleptics, it is used to control schizophrenia. Lithium is also used in the case of patients with alcoholism associated with depression and has been used to correct the neutropenia that occurs during cancer chemotherapy.

Lithium has been investigated for use in subduing aggressive behaviors in nonpsychotic but possibly brain-damaged patients. Its use has also been investigated in the management of inappropriate secretion of antidiuretic hormone.

REGULATORY STATUS OF HYPERICUM

Hypericum (as *Hypericum perforatum* L.) has official status (i.e., government approval in some capacity) in several countries throughout the world. In some cases, the status is based on traditional use rather than on controlled trials. Table 38 shows the status of hypericum.

37

IPECAC

■ An amebicidal and emetic agent

IPECACUANHA AND EMETINE

Ipecac, the root of a Brazilian plant, contains several alkaloids, of which two, **emetine** and **cephaëline**, produce local irritation and nausea and emesis, by central and local action, without danger of side effects.

Ipecac is the root of ***Cephaëlis ipecacuanha***, or of *C. acuminata*, a perennial shrub growing in Brazil and other South American states (Figure 105). It contains three alkaloids — ***emetin, cephaëlin***, and ***psychotrin***. The dose of the powdered drug as an expectorant is from ½ to 2 grain (0.03 to 0.13 g); as an emetic, 15 to 30 grain (1.0 to 2.0 g) (Table 39).

IPECAC

Ipecac ("**Brazil root**") was long employed by the native people of Brazil in the treatment of diarrheas. It was sold as a secret remedy to the French government in 1658 and its use in dysenteries rapidly spread throughout Europe and India. Its employment was entirely empirical until 1912 when Vedder demonstrated the *in vitro* efficacy of emetine against *E. histolytica* and suggested that ipecac be used in amebic infections. The source of ipecac is the dried root or rhizome of *C. ipecacuanha* or *C. acuminata*, plants native to Brazil and Central America, but also cultivated in India, the Straits Settlements, and the Federated Malay States (see Grollman, 1962).

Active Principles

The efficacy of ipecac in amebic infections depends upon its content of alkaloids, the principal ones being ***emetine*** and ***cephaëline***. Both are **amebicidal**, but emetine is much more active. Cephaëline is more toxic than emetine, except for the heart, and causes more nausea and vomiting. Emetine constitutes more than one half of the total alkaloidal content of ipecac.

Figure 105 Ipecac plant [*Cephaëlis (Uragoga) ipecacuanha*]: (A), flowering shoot; (B), flower in longitudinal section; (C), fruit; (D), fruit in transverse section; (E), seed; (F), annulate root.

Table 39   The Emetic Doses of Ipecac Preparations

Preparations	Dose
Fluidextractum Ipecacuanhae	As an expectorant 2–5 min (0.1–0.3 cc); as an emetic, 15–30 min (1.0–2.0 cc)
Syrupus Ipecacuanhae	As an expectorant, 10–60 min (0.6–4.0 cc); as an emetic, 2–4 fl. dr. (8.0–15.0 cc)
Vinum Ipecacuanhae	As an expectorant, 10–60 min (0.6–4.0 cc); as an emetic, 2–4 fl. dr. (8.0–15.0 cc)

The most commonly used emetics are ipecac and apomorphine. Induced emesis is the preferred means of emptying the stomach in awake patients who have ingested a toxic substance or have recently taken a drug overdosage. Emesis should not be induced if there is central nervous system depression or ingestion of certain volatile hydrocarbons and caustic substances. *Ipecac syrup* is prepared from the dried rhizome and roots of *Cephaelis ipecacuanha* or of *C. accuminata*, plants from Brazil and Central American in which the alkaloid emetine is its active principal ingredient.

Status and Use

Ipecac is official in the U.S.P. in powder form and as a fluid extract and syrup. It is rarely used in the modern therapy of **amebiasis** because its administration results in severe gastrointestinal irritation, nausea, and vomiting. The drug is not employed unless the patient has proved refractory to other methods of treatment. If prescribed, ipecac is given as the powder in salol-coated pills, each containing 0.03 g. The dose is 10 to 15 such pills taken at night before retiring. Ipecac should not be given in divided doses. The full course consists of 100 pills. The patient should be at absolute bed rest during the treatment. All the precautions outlined below for emetine therapy must be observed. **Tincture of opium** may be needed to control nausea and vomiting, which, nevertheless, maybe so severe as to necessitate cessation of ipecac administration.

History

Emetine was first described in 1817 by Pelletier, the discoverer of **quinine**. However, Pelletier was actually dealing with a mixture of the alkaloids of ipecac, that is, **emetine**, **cephaëline**, and **psychotrine** (see Grollman, 1962).

DRUGS THAT PRODUCE VOMITING (EMETICS)

There are a few important drugs that cause nausea and vomiting as undesirable side actions when used in therapeutic dosage, and perhaps a larger number that induce the state as part of a serious toxic reaction to them in overdosage, but the number that is used intentionally to cause the patient to vomit is relatively small. Here, the following four — ipecac, apomorphine, mustard, copper sulfate — are considered as a group.

Source, Nature, and Preparations

Ipecac, which is a mixture of the alcohol-soluble alkaloids from the underground parts of the South American plant, *C. ipecacuanha,* is used almost solely as the U.S.P. ipecac syrup.

Apomorphine hydrochloride (Figure 106), an alkaloid obtained in the laboratory by rearrangements within the morphine molecule, is a grayish-white, glistening, light-sensitive, and air-sensitive crystalline powder, soluble in both water and alcohol in the proportions of about 1:50; it is available in small amounts in bulk and more easily as 5 mg (1/12 grain) tablets.

Black mustard is a light olive-brown powder obtained by grinding the dried ripe seeds of several varieties of plants of the **Brassica** genus; moisture yields from this, through enzymatic hydrolysis of a contained glycoside, are about 0.6% of the volatile oil of mustard, which has as its main ingredient the irritant compound **allyl isothiocyanate**. Mustard is available in bulk and as the official plaster, the latter used only for the local irritant (**counterirritant**) properties it develops when applied to the skin after moistening with tepid water.

Copper sulfate ($CuSO_4 \cdot 5H_2O$) occurs as deep blue crystals or granules that effloresce slowly in dry air; it is freely soluble in water and in glycerin (1:3),

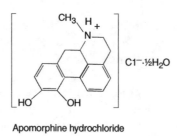

Apomorphine hydrochloride

Figure 106 The structure of apomorphine hydrochloride.

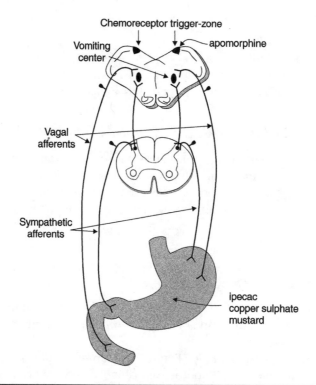

Figure 107 Sites of action of emetic drugs as employed clinically.

slightly soluble in alcohol (1:500), the solutions being acidic; and it is available in bulk.

Nature of Emetic Action of Ipecac

There are three sites at which chemical irritation with emetic drugs can initiate vomiting: the gastroduodenal mucosa, the chemoreceptor trigger zone in the caudal portion of the fourth ventricle (area postrema), and the lateral reticular formation of the medulla (vomiting center) (Figure 107). The intimate nature of the interferences in cellular metabolism effected by the drugs has not been determined, but it is known at which of these sites the individual drugs act. In Figure 107 are shown the points of attack when the drugs are given under clinical circumstances, but this picture must be amplified somewhat with words to bring in the occurrences under experimental conditions as well. When given orally in therapeutic dosage **ipecac** and **copper sulfate** fire reflex discharges in the gastroduodenal mucosa whose pathways to the vomiting center are afferent fibrils carried in the vagal and sympathetic nerves. In the laboratory, copper sulfate in high doses orally or intravenously also acts at the **chemoreceptor trigger zone**. And **emetine**, an alkaloid obtained from ipecac, likewise has central action when given parenterally. **Apomorphine** acts solely at the chemoreceptor trigger zone; mustard is not known to act elsewhere than locally on the gastroduodenal mucosa.

Clinical Effects

Emetics are used rather infrequently because the occasions for making a patient vomit do not arise very often. In the several types of food poisoning it is usually considered advisable, if emptying of the stomach is indicated, to use the tube rather than emetic drugs. This is the case also when potent drugs have been ingested in excessive dosage by accident or with suicidal intent. Even in the case of **phosphorus poisoning**, lavage is most important, although it is undeniable that copper sulfate will precipitate an impermeable coating of **copper phosphide** on the encountered phosphorus particles.

Actually, in the past perhaps the only three really important uses of emetics under ordinary circumstances were to treat:

1. The **fulminating-asthmatic attack**
2. The recalcitrant attack or **paroxysmal atrial tachycardia**
3. The midnight attack of **catarrhal croup** in an infant

In all three instances ipecac is the drug of choice and it is used not to rid the stomach of noxious contents but to profit by the act of vomiting itself. In the asthmatic attack, with the bronchi blocked with accumulations of mucus, forceful vomiting will release the plugs and often afford truly dramatic relief. It is possible that the retching and vomiting induce a peristaltoid action in the trachea and release the obstructing plugs through what has been called "**tracheal vomiting**." In the case of the arrhythmia, the action comes about through an increase of cardiac vagal influence simply because central, vagal stimulation is an accompaniment of the act of vomiting. It is also this central vagal stimulation that appears to relieve the **glottic spasm in croup**.

PREVENTION OF FURTHER ABSORPTION OF POISON

Emesis

Although emesis is indicated after poisoning by oral ingestion of most chemicals, it is contraindicated in certain situations:

- If the patient has ingested a **corrosive poison**, such as a strong acid or alkali (e.g., drain cleaners), emesis increases the likelihood of gastric perforation and further necrosis of the esophagus.
- If the patient is comatose or in a state of stupor or delirium, emesis may cause aspiration of the gastric contents.
- If the patient has ingested a central nervous system (CNS) stimulant, further stimulation associated with vomiting may precipitate convulsions.
- If the patient has ingested a **petroleum distillate** (e.g., **kerosene, gasoline**, or petroleum-based liquid furniture polish), regurgitated hydrocarbons can be aspirated readily and cause **chemical pneumonitis**.

In contrast, emesis should be considered if the solution that is ingested contains potentially dangerous compounds, such as **pesticides**. There are marked differences in the capabilities of various petroleum distillates to produce **hydrocarbon pneumonia**, which is an acute, **hemorrhagic necrotizing process**. In general, the ability of various hydrocarbons to produce **pneumonitis** is inversely proportional to the viscosity of the agent: if the viscosity is high, as with oils and greases, the risk is limited; if viscosity is low, as with mineral seal oil found in liquid furniture polishes, the risk of aspiration is high.

Vomiting can be induced mechanically by stroking the posterior pharynx. However, this technique is not as effective as the administration of ipecac or apomorphine.

Ipecac

The most useful household emetic is **syrup of ipecac** (not **ipecac fluid extract**, which is 14 times more potent and may cause fatalities). Syrup of ipecac is available in 0.5- and 1-fluid ounce containers (approximately 15 and 30 ml), which may be purchased without prescription. The drug can be given orally, but it takes 15 to 30 minutes to produce emesis; this compares favorably with the time usually required for adequate gastric lavage. The oral dose is 15 ml in children from 6 months to 12 years of age and 30 ml in older children and adults. Because emesis may not occur when the stomach is empty, the administration of ipecac should be followed by a drink of water.

Monitoring

In dealing with poisoning, serial measurement and charting of vital signs and important reflexes help to judge progress of intoxication, response to therapy, and the need for additional treatment. This monitoring usually requires hospitalization. The classification in Table 40 often is used to indicate the severity of CNS intoxication. Treatment with large doses of stimulants and sedatives often can cause more harm than the poison. Chemical antidotes should be used judiciously; heroic measures seldom are necessary.

AMEBIASIS

Amebiasis is an infection of the large intestine produced by ***Entamoeba histolytica*** that causes symptoms that range from mild diarrhea to fulminant dysentery.

Table 40   Signs and Symptoms of CNS Intoxication

Degree of Severity	Characteristics
Depressants	
0	Asleep, but can be aroused and can answer questions
I	Semicomatose, withdraws from painful stimuli, reflexes intact
II	Comatose, does not withdraw from painful stimuli, no respiratory or circulatory depression, most reflexes intact
III	Comatose, most or all reflexes absent, but without depression of respiration or of circulation
IV	Comatose, reflexes absent, respiratory depression with cyanosis or circulatory, failure and shock, or both
Stimulants	
I	Restlessness, irritability, insomnia, tremor, hyperreflexia, sweating, mydriasis, flushing
II	Confusion, hyperactivity, hypertension, tachypnea, tachycardia, extrasystoles, sweating, mydriasis, flushing, mild hyperpyrexia
III	Delirium, mania, self-injury, marked hypertension, tachycardia, arrhythmias, hyperpyrexia
IV	As in III, plus convulsions, coma, and circulatory collapse

As with therapeutic effects, the toxicity of drugs and other chemicals is a function of the concentration of the compound achieved at the site of action. For locally acting external irritants this means that the magnitude of toxicity produced is proportional to the concentration of the substance in the external medium (air or water); for systemically acting agents, the magnitude of the adverse effects produced is proportional to the dose of the toxicant.

Therapy includes treating the asymptomatic carrier of the cysts, as well as the acute amebic dysentery, and the amebic hepatitis and abscess in the symptomatic patient.

Treatment of the Asymptomatic Carrier

Effective drugs for treating the asymptomatic carrier are the **8-hydroxyquinoline derivatives**, such as **iodoquinol** (the preferred drug) and **difoxanide furoate** (an alternative drug).

Treatment of Acute Amebic Dysentery

Dehydroemetine is given for 5 days and effects rapid relief of the symptoms of acute amebic dysentery. The patient is then switched to **metronidazole**. If the response to metronidazole is not satisfactory, **dehydroemetine plus tetracycline** or **dehydroemetine plus paromomycin** are given along with the metronidazole.

Treatment of Amebic Hepatitis and Abscess

Amebic hepatitis and abscess are best treated with metronidazole. Dehydroemetine or chloroquine can serve as alternative drugs.

38

ALGAE

- Treat roundworm disease
- Are potent neurotoxins
- Cause amnesic shellfish poisoning

INTRODUCTION

The **red alga _Digenea simplex_** has been used for the treatment of **roundworm disease** for centuries. Its active principle is **kainic acid**. The related **domoic acid** is a constituent of another red alga, **_Chondria armata_**, used for the same purpose. These compounds known as **kainoids** are potent **neurotoxins** and **excitatory amino acids**. Kainoids are important tools in neurophysiological research. Domoic acids are also produced by diatoms and were responsible for the **shellfish poisonings** known as **amnesic shellfish poisonings** that occurred in Canada in 1987.

Caulerpin and **caulerpicin** have been described as toxic constituents of edible species of the **green algal genus _Caulerpa_**, but evidences in later studies indicate that they have no acute toxicity. Caulerpin, which has a structure related to **auxin**, promotes plant growth. **Caulerpenyne**, a toxic constituent of _Caulerpa taxifolia_ and other inedible species, has been evaluated for its ecotoxicological effect in the Mediterranean where **_C. taxifolia_** bloomed explosively. Three different classes of compounds have been identified in the poisonings with species in the genus **_Gracilaria_**. They are **prostaglandin E$_2$** from **_G. verrucosa_** in Japan, **aplysiatoxins** and related compounds from **_G. coronopifolia_** in Hawaii, and **polycavernosides** from **_G. tsudai_** in Guam (Higa and Kuniyoshi, 2000).

As a result, a large number of secondary metabolites have been described from various species of algae. Just the two best-studied groups, the **red algal genus _Laurencia_** and the brown algal family **Dictyotaceae**, account for more than 500 compounds. Many of these compounds are known to be **cytotoxic**, **ichthyotoxic**, or **antimicrobial**.

Some 400 species of seaweed are used as food, feed, medicines, fertilizers, and industrial raw materials in the world. The use of seaweed as food is perhaps more common in Asia and the Pacific regions than the rest of the world. In Japan, the majority of people take one or more kinds of seaweed in their daily diet. There are no records of poisonings with most frequently eaten seaweed such as "**kombu**" and "**wakame**" belonging to the brown algal family **Laminariaceae** and "nori" to the red algal genus *Porphyra*. However, there are several recorded cases of poisonings with some other seaweed, i.e., brown algae *Nemacystus decipiens* and *Cladosiphon okamuranus* and red algae of the genus *Gracilaria*.

Not many species of seaweed are widely known among folk remedies. A Chinese booklet, the title of which may be translated as "Medicinal Marine Organisms of the South Sea," published in 1978 by Ocean Research Institute of Chinese Academy of Science lists 38 species of algae. Among them the best-known medicinal seaweed is perhaps *Digenea simplex*, an anthelmintic drug widely used in Asia over many centuries. This alga and the related *Chondria armata* are among a few species whose active ingredients have been fully identified, because these active constituents, known collectively as **kainoids**, are **neurotoxins**.

KAINIC ACID

The red alga *D. simplex* has been used as a folk remedy for the **treatment of roundworm disease** (**ascariasis**) in Asia for many centuries. It is described in a Japanese pharmacopoeia published in the 9th century. The disease had been quite common in postwar Japan until around 1960 because of poor sanitary conditions caused by the war devastation and also because of the then common practice of using human feces as fertilizer. After World War II, **santonin** produced from some species of **mugworts** was introduced in tablet form, but the decoction of the alga was still available. An active constituent of the alga, α-**kainic acid**, has been developed into an **anthelmintic drug**. It is ten times more active than **santonin**. Later, kainic acid was found to have both **neurotoxic** and **neuroexcitatory properties**. Today, together with other **kainoid amino acids**, it is more important as a research reagent in neurophysiology than as a human medicine.

Glutamate and **aspartate** are found in very high concentrations in brain, and both amino acids have extremely powerful **excitatory effects** on neurons in virtually every region of the central nervous system. Their widespread distribution tended to obscure their roles as transmitters, but there is now broad acceptance of the view that glutamate and possibly aspartate function as the principal fast ("classical") excitatory transmitters throughout the central nervous system. Furthermore, over the past decade, multiple subtypes of receptors for excitatory amino acids have been characterized pharmacologically, based on the relative potencies of synthetic agonists and the discovery of potent and selective antagonists.

Glutamate receptors are classified functionally as ligand-gated ion channels (also called "**ionotropic**" receptors) or as "**metabotropic**" (G protein-coupled) receptors. The metabotropic receptors couple via GTP-binding proteins to a variety of effector mechanisms, whereas the ligand-gated ion channels contain an integral cation channel that gates Na^+ and in some cases Ca^{2+}. Metabotropic receptors are composed of a single seven-transmembrane-spanning protein, whereas ligand-gated channels are multisubunit complexes. Neither the precise number of subunits that assemble to generate a functional glutamate receptor ion channel *in vivo* nor the topography of each subunit has been established unequivocally.

The ligand-gated ion channels are further classified according to the identity of agonists that selectively activate each receptor subtype. These receptors include **α-amino-3-hydroxy-5-methyl-4-isoxazole propionic acid (AMPA)**, **kainate**, and **N-methyl-D-aspartate (NMDA) receptors**. A number of selective antagonists for these receptors are now available. In the case of NMDA receptors, noncompetitive antagonists acting at various sites on the receptor protein have been described in addition to competitive (glutamate site) antagonists. These include open-channel blockers such as **phencyclidine (PCP** or "**angel dust**"), antagonists such as **5,7-dichlorokynurenic acid** that act at an allosteric **glycine-binding site**, and the novel antagonist **ifenprodil**, which may act as a closed-channel blocker. In addition, the activity of NMDA receptors is sensitive to pH and also can be modulated by a variety of endogenous modulators, including Zn^{2+}, some **neurosteroids, arachidonic acid, redox reagents**, and **polyamines** such as **spermine**. In some cases, the effects of endogenous modulators and of competitive and noncompetitive antagonists are selective for certain subtypes of NMDA receptors composed of particular subunit combinations. Multiple cDNAs encoding metabotropic receptors and subunits of **NMDA, AMPA,** and **kainate receptors** have been cloned in recent years. The diversity of gene expression and, consequently, of the protein structure of glutamate receptors also arises by alternative splicing and in some cases by single-base editing of mRNAs encoding the receptors or receptor subunits. Alternative splicing has been described for metabotropic receptors and for subunits of NMDA, AMPA, and kainate receptors. A remarkable form of endogenous molecular engineering occurs with some subunits of AMPA and kainate receptors in which the RNA sequence differs from the genomic sequence in a single codon of the receptor subunit and determines the extent of Ca^{2+} permeability of the receptor channel. This RNA-editing process alters the identity of a single amino acid (out of about 900 amino acids) that dictates whether or not the receptor channel gates Ca^{2+}. The glutamate receptor genes seem to be unique families with only limited similarity to other ligand-gated channels such as the nicotinic acetylcholine receptor or, in the case of metabotropic receptors, to members of the G protein-coupled receptor superfamily.

AMPA and kainate receptors mediate fast depolarization at most glutamatergic synapses in the brain and spinal cord. NMDA receptors are also involved in normal synaptic transmission, but activation of NMDA receptors is more closely associated with the induction of various forms of synaptic plasticity rather than with fast point-to-point signaling in the brain. AMPA or kainate receptors and NMDA receptors may be co-localized at many glutamatergic synapses. A well-characterized phenomenon that involves NMDA receptors is the induction of **long-term potentiation** (LTP). LTP refers to a prolonged (hours to days) increase in the size of a postsynaptic response to a presynaptic stimulus of given strength. Activation of NMDA receptors is obligatory for the induction of one type of LTP that occurs in the hippocampus. NMDA receptors normally are blocked by Mg^{2+} at resting membrane potentials. Thus, activation of NMDA receptors requires not only binding of synaptically released glutamate but simultaneous depolarization of the postsynaptic membrane. This is achieved by activation of AMPA/kainate receptors at nearby synapses from inputs from different neurons. Thus, NMDA receptors may function as "coincidence detectors"; that is, they are activated only when there is simultaneous firing of two or more neurons. LTP has been proposed as a cellular model for some forms of learning and memory. Interestingly, NMDA receptors can also induce long-term depression (LTD; the flip side of LTP) at central nervous system synapses.

Exposure of neurons to high concentrations of glutamate for only a few minutes can lead to neuronal cell death. The cascade of events leads to excessive activation of NMDA receptors and influx of Ca^{2+} into the neurons. This process is thought to be similar to the neurotoxicity that occurs after ischemia or hypoglycemia in the brain, where a massive release and impaired reuptake of glutamate in the synapse leads to excess stimulation of glutamate receptors and subsequent cell death. NMDA receptor antagonists can attenuate or block neuronal cell death induced by activation of these receptors. There is considerable therapeutic potential for the use of NMDA antagonists as neuroprotectants. NMDA receptors also may be involved in the development of susceptibility to epileptic seizures and in the occurrence of seizure activity. In animal models, NMDA receptor antagonists have **anticonvulsant activity** and may find clinical use as anticonvulsants.

Because of the widespread distribution of glutamate receptors in the central nervous system, it is likely that these receptors ultimately will become the targets for diverse therapeutic interventions. It also is conceivable that abnormal expression, regulation, or function of glutamate receptors may be involved in the etiology of some neurological disorders. A role for glutamate receptors or disordered glutamatergic transmission in the etiology of chronic neurodegenerative diseases and in **schizophrenia** has been postulated. Cases of **Rasmussen's encephalitis**, a childhood disease leading to **intractable seizures** and dementia, were found to correlate with levels of serum antibodies to a glutamate receptor subunit. Thus, an autoimmune reaction against some glutamate receptors may underlie this disease.

DOMOIC ACIDS

Another anthelmintic folk medicine used in Japan is the red alga *Chondria armata* belonging to the same family (**Rhodomelaceae**) with *Digenea simplex*. However, its use in this purpose has only been known from **Tokunoshima**, a small island located in the northeast of **Okinawa**. Its **insecticidal property** against flies has been noted by the residents of **Yakushima**, another island farther north.

Domoic acid has been demonstrated to be responsible for the insecticidal activity of *C. armata*. It was 14 times more potent than **DDT** when administered subcutaneously into the abdomen of American cockroach.

Domoic acid had never been known to be a causative agent of seafood poisoning until 1987 when an outbreak of serious poisoning with **blue mussels** (*Mytilus edulis*) occurred in Canada. More than 150 people were intoxicated by eating blue mussels cultured in eastern Prince Edward Island during a period from November to December 1987. The symptoms included nausea, vomiting, diarrhea, memory loss, disorientation, and coma. Three elderly patients died. In some cases neurological symptoms such as memory loss persisted more than 5 years. About a month after the first outbreak, Canadian scientists identified domoic acid as the causative agent by analysis of the contaminated mussels.

The term **amnesic shellfish poison** (ASP) has been coined for this new shellfish toxin. The causative organism of ASP was then identified to be the pennate diatom *Nitzschia pungens* Grunow f. multiseries Hasle, a bloom of which contaminated the cultured mussels in eastern Prince Edward Island in 1987.

In early September 1991, another outbreak of domoic acid poisoning was recorded on the other side of the American continent, **Monterey Bay**, California. This time, it was not human intoxication but poisoning of **wild pelicans** and **cormorants**. The symptoms of the affected birds were unusual head movements, scratching, vomiting, the loss of righting reflex, and even death. The presence of domoic acid was found in another species of pennate diatom, *Pseudonitzschia australis* Frenguelli, which was blooming in the bay at that time. Remnants of the diatom and high levels of domoic acid were found in the stomach contents of poisoned birds and in the viscera of local anchovies, a principal food of the birds.

TOXICITY OF *CAULERPA*

Some species of the **green algal genus *Caulerpa*** are edible and largely consumed as salad in the Philippines and other tropical Pacific regions. The edible species become peppery to the taste during the rainy months in the Philippines, and other species are not eaten because of the extremely peppery taste. Chemical investigation on the toxic constituents of *Caulerpa* was initiated in the 1960s and a compound named **caulerpicin** was reported to be was responsible for the manifestation of such toxic symptoms as a mild, anesthetizing sensation, numbness of the tongue, and cold sensation in the feet and fingers.

TOXICITY OF *GRACILARIA*

The red alga *Gracilaria verrucosa* known as *"Ogonori"* is widely consumed as a garnish of salad or "**Sashimi**" in Japan. Other species of this genus are also eaten in wide regions of the Pacific. Some species of Gracilaria are important raw material for the production of **agar**. In Taiwan a species of *Gracilaria* is used as feed in the **abalone aquaculture**.

In general, *Gracilaria* species are nontoxic and seldom cause poisoning. However, there are several cases of intoxication reported from Japan, Guam, and Hawaii. Two cases in Japan were involved with the ingestion of *G. verrucosa* and one case with *G. chorda*. In each of these cases one person was killed. The symptoms of the victims included nausea, vomiting, stomach pain, and diarrhea. The onsets of the poisonings were fast, often within 30 minutes after ingestion (Higa and Kuniyoshi, 2000).

39

TOMATOES

Tomatoes contain lycopene, which is useful in medicine.

TOMATO AND MEDICINE

- Tomato-based food products such as tomato paste, tomato sauce, and tomato-based soups are rich in **carotenoid compounds** and are frequently consumed in the United States. Foods such as these, which are high in carotenoid content, are of interest because of the demonstrated association between consumption of fruits and vegetables and reduced risk of lung and other epithelial cancers in humans.
- **Lycopene**, the predominant carotenoid in tomatoes (***Solanum lycopersicum***) exhibits the highest antioxidant activity and **singlet oxygen quenching** ability of all dietary carotenoids.
- The carotenoids are substances with very special and remarkable properties that no other groups of substances possess and that form the basis of their many, varied functions and actions in all kinds of living organisms. Without carotenoids, **photosynthesis** and all life in an oxygen atmosphere would be impossible.
- Some common carotenoids found in nature are **all-*trans*-β-carotene, lycopene, α-carotene, β-cryptoxanthin, canthaxanthin, zeaxanthin, antheraxanthin, violaxanthin, lutein, 9-*cis*-β-carotene**, and **15,15′-*cis*-phytoene.**
- Lycopene, in addition to possessing antioxidant activities induces **gap conjunction communication**, inhibits the growth of chemically transformed cells and retards cell proliferation. Unlike β-carotene, lycopene does not have provitamin A activity. Carotenoids have recently been implicated in the prevention of or protection against serious human health disorders such as cancer and heart disease. At present, available data point to a possible role of lycopene in human health, which may be related to its antioxidant activities and the influence on cell growth and intercellular communication. However, more information is required to firmly establish its role in health protection and to identify the underlying biochemical mechanisms. Lycopene might interact with other food components providing additive or synergistic effects. One important aspect of lycopene research is the bioavailability of this compound and its metabolism to

derivatives which might also exhibit biological activities contributing to its potential cancer-preventing property.

HISTORY OF CAROTENOIDS

Lycopene was introduced into the Western world after the Spanish conquistador **Hernando Cortes** was offered some tomatoes by the **Aztec emperor Montezuma**. Cortes disregarded this courteous gesture and went on to conquer Mexico from 1529 to 1531. In the states bordering Mexico — Texas, New Mexico, Arizona, and California — tomatoes and other Mexican-Indian foods soon began to be adopted but also made their way to Europe. In Italy, tomatoes were mentioned as early as in 1554 and were then slowly assimilated. Their initial color was yellow (**pomo d'oro**); they obtained their red appearance only after years of cultivation. Today, tomatoes, the major source of lycopene, are an important part of the Mediterranean diet.

Awareness in the nutritional and medical sciences of lycopene as a potentially beneficial carotenoid is less than 10 years old. Therefore, comparatively few scientific data are available from animal and human studies. But in this short time, a number of features of lycopene have emerged that are unique among the common carotenoids, quite apart from its exceptionally high singlet oxygen quenching capacity *in vitro*. The most obvious distinction is that unlike the other hydrocarbon carotenoids, α- and β-carotene, lycopene has no provitamin A activity.

PHYSICOCHEMICAL NATURE OF CAROTENOIDS

Like other carotenoids, lycopene is a natural pigment synthesized by plants and microorganisms, but not by animals. More than 600 carotenoids have been identified in nature. Inasmuch as they all possess many conjugated double bonds — usually 9 to 13, each can form many geometric isomers. β-Carotene, for example, with nine double bonds in its polyene chain that are free to assume *cis/trans* configurations, can theoretically form 272 isomers, whereas its asymmetric isomer, α-carotene, can form 512. Thus, the total possible number of compounds in the class, including all possible isomers, easily exceeds 200,000. As they are nonpolar, they are largely associated with lipophilic parts of the cell, including membranes.

LYCOPENE, AN IMPORTANT CAROTENOID IN TUMOR SUPPRESSION

Carotenoids are important plant pigments found in the photosynthetic pigment–protein complex of plants, photosynthetic bacteria, fungi, and algae and are responsible for the bright colors of various fruits and vegetables. Among the more than 600 different carotenoids, there are well-known compounds such as lycopene and α- and β-carotene (Figures 108 and 109), which are widely used as sources for vitamin A or as food colorants.

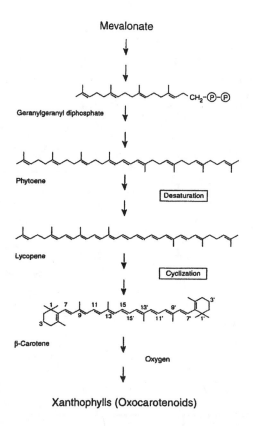

Mevalonate

Geranylgeranyl diphosphate

Phytoene

Desaturation

Lycopene

Cyclization

β-Carotene

Oxygen

Xanthophylls (Oxocarotenoids)

Figure 108 Starting with mevalonate, carotenoids are biosynthesized by a special branch of the terpenoid pathway. The first C-40 hydrocarbon unit formed is phytoene, a carotenoid with three conjugated double bonds, which then is enzymatically desaturated to yield successively β-carotene, neurosporene, and lycopene. Other carotenoids such as β-carotene and oxocarotenoids are produced from lycopene following cyclization and hydroxylation reactions. Thus, lycopene is a central molecule in the biosynthesis pathway of carotenoids.

Carotenoids are tetraterpenes formed by tail-to-tail linkage of two C-20 units, and in many carotenoids the end groups are modified into five- or six-membered rings giving monocyclic or dicyclic compounds. Lycopene is an acyclic carotenoid that contains 11 conjugated double bonds arranged linearly in the all-*trans* form. It belongs to the subgroup of carotenes consisting only of hydrogen and carbon atoms.

CAROTENOID CONTENT OF FRUITS AND VEGETABLES

Carrots, tomatoes, and dark green vegetables are rich in specific carotenoids. Numerous epidemiological studies have shown that the increased intake of fruits and vegetables is associated with a reduced risk of lung and other epithelial cancers. Accurate food composition data have been prepared to test the association between dietary intake of specific carotenoids and disease incidence (Table 41).

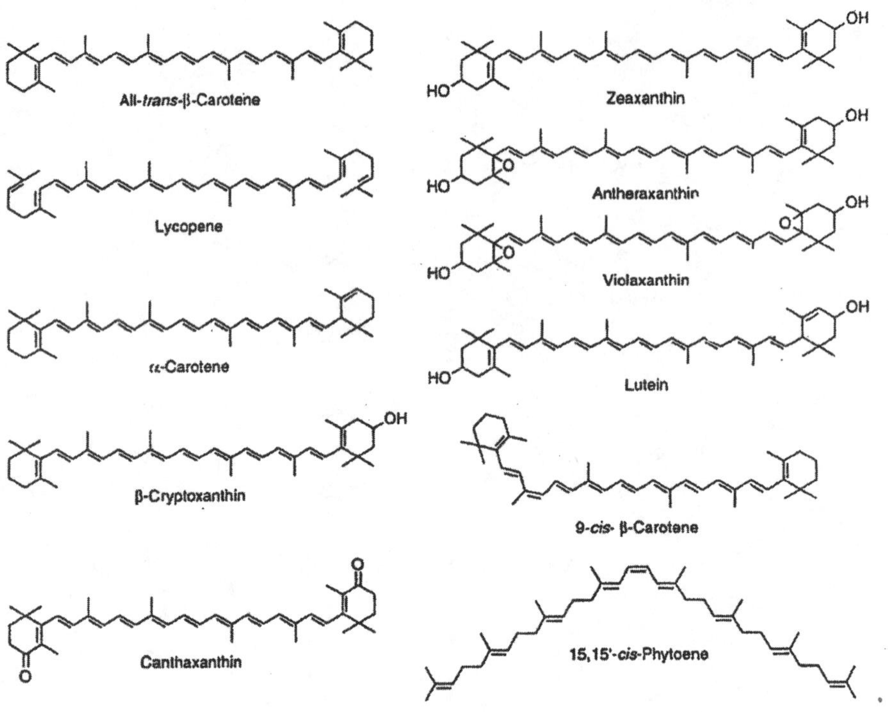

Figure 109 Essentially all carotenoids, which are widespread in nature, possess certain common chemical features: a polyisoprenoid structure, a long conjugated chain of double bonds in the central portion of the molecule, and near symmetry around the central double bond. This basic structure can be modified in a variety of ways, most prominently by cyclization of the end groups and by the introduction of oxygen functions, to yield a large family of >600 compounds, exclusive of *cis/trans* isomers.

DISTRIBUTION OF LYCOPENE IN HUMANS

It is conceivable that the different carotenoids have specialized functions in different tissues as has been shown for the xanthophylls lutein and zeaxanthin, which are virtually the only carotenoids occurring in the macular area of the retina. The distribution of lycopene in human tissues is shown in Tables 42 and 43.

CONCENTRATIONS OF CAROTENOIDS IN HUMAN ORGANS

The uneven but wide tissue distribution of most dietary carotenoids may indicate an active biological role for these compounds (see Table 43). The organs with the greatest number of low-density lipoprotein (LDL) receptors and the highest rates of lipoprotein uptake (adrenals, testes, and liver) generally contain the greatest amounts of carotenoids. It is generally believed that the carotenoids are primarily transported in blood by the LDL and that the LDL particles themselves take up the carotenoids by a passive mechanism.

Table 41   Lycopene Content of Foods

Foods	States	Content (mg/100 g)[a]
Tomatoes	Fresh, raw	2937
	Fresh, cooked	3703
	Sauce, canned	6205
	Paste, canned	6500 (range 5400–15,000)
	Fresh, raw	3100 (range 879–4200)
	Juice, canned	8580 (range 5000–11,600)
	Catsup	9900
Apricot	Dried	864
	Canned, drained	65
	Raw	5
Grapefruit	Pink, raw	3362
Guava	Juice	3340
	Raw	5400 (range 5340–5500)
Rosehip	Puree, canned	780
Watermelon	Fresh, raw	4100 (range 2300–7200)
Papaya	Fresh	2000–5300

[a] Edible portion as wet weight.

Most of the organs (thyroid, spleen, kidney, liver, pancreas, and heart) investigated appear similar in that lycopene and β-carotene are the predominant carotenoids found in these organs, with an approximately equal percentage distribution between them. Selectively, the cellular uptake of individual carotenoids may be based upon the selective transport of the individual carotenoids.

TOMATOES, LYCOPENE, AND CANCER RISK

Dietary recommendations to increase intake of citrus fruits, cruciferous vegetables, green and yellow vegetables, and fruits and vegetables high in vitamin A and C to lower cancer risk have been made by several organizations, including the National Research Council of the National Academy of Sciences, the National Cancer Institute, the American Cancer Society, the World Cancer Research Fund, and the American Institute for Cancer Research.

Lung and Pleural Cancers

One of the cancer sites for which a benefit of fruits and vegetables has been most apparent is for cancers of the lung (the leading cause of cancer death worldwide). Although the focus has been on β-carotene, the literature shows that several groups, including leafy green and yellow/orange vegetables, are associated with a lower risk of lung cancer.

- ■ **Stomach Cancer**. Although it has become relatively uncommon in most economically developed countries, stomach or gastric cancer remains one of the major causes of cancer death in the world. Inverse associations between tomato consumption and risk of gastric cancer were observed in the United

Table 42 Concentration of Lycopene in Human Tissues (nmol/g wet tissue)

	Testis	Adrenal	Liver	Ovary	Fat	Kidney	Brainstem
Lycopene (range)	4.34	1.90	1.28	0.25	0.20	0.15	Not detectable
	(0.41–9.38)	(0.19–5.60)	(0.10–4.08)	(0.14–0.38)	(0.00–0.51)	(0.03–0.34)	
	11.43	11.56	1.31	0.15	0.699	0.21	
	(3.26–63.9)	(2.04–58.5)	(0.16–10.3)	(0.09–5.56)	(0.02–3.70)	(0.05–1.20)	
Total carotenoids	7.6 ± 6.9^a	9.7 ± 7.8	5.1 ± 3.6^a	0.9 ± 0.5^a	0.8 ± 0.8^a	0.9 ± 0.7^a	<0.04
	14.3	18.3	2.7	1.4	1.8	0.53	

[a] SEM.

Table 43 Median Levels of Tissue Carotenoids (ptg/g wet tissue range)

Organs	n	Total Carotenoids	Zeaxanthin	Crypto-xanthin	Lycopene	α-Carotene	β-Carotene
Liver	13	2.7	0.292	0.035	1.31	0.112	0.977
Pancreas	13	2.0	0.443	0.040	0.687	0.106	0.670
Fat	12	1.8	0.788	0.025	0.699	0.050	0.202
Kidney	12	0.53	0.090	0.019	0.210	0.030	0.167
Adrenal	11	18.3	0.908	0.160	11.56	0.617	5.04
Spleen	11	0.52	0.097	0.020	0.100	0.050	0.235
Heart	7	0.44	0.035	0.025	0.188	0.050	0.123
Testes	5	14.3	0.211	0.068	11.430	0.257	2.34
Thyroid	5	0.43	0.047	0.020	0.155	0.050	0.135
Ovary	3	1.4	0.425	0.080	0.150	0.164	0.530

Total carotenoids (mg/g wet tissue), as the sum of the major carotenoid fractions listed in the table.

States, Israel, Italy, Spain, Poland, Belgium, and Sweden. The consistent inverse association observed in diverse populations strongly suggests a protective effect of tomato and lycopene consumption on gastric cancer.

- **Colorectal Cancer**. Cancers of the colorectum are common in economically developed areas. Five studies have reported on tomato intake in relation to colorectal cancer risk. One study in the United States reported statistically significant inverse associations between tomato consumption and colon cancer risk for men and women.
- **Oral/Laryngeal/Pharyngeal Cancer**. Only three case-control studies have reported on tomato intake in relation to oral cancers. One study in China reported that high consumption of tomatoes was related to approximately half the risk of oral cancer. A similar finding was observed between tomato consumption and cancers of the oral cavity and pharynx in Italy.
- **Esophageal Cancer**. Esophageal cancers have received little study regarding tomatoes and lycopene. One study in Iran, which has extremely high rates of esophageal cancer particularly in men, found a 39% statistically significant reduction in risk for men who consumed tomatoes frequently, but no relationship was apparent for women.
- **Pancreatic Cancer**. Four studies have examined tomato or lycopene status in relation to risk of pancreatic cancer; all of these studies support an inverse association.
- **Prostate Cancer**. Three studies report data on the relationship between tomato or lycopene consumption and reduced prostate cancer.
- **Bladder Cancer**. Tomato or lycopene does not reduce the risk of bladder cancer.
- **Breast Cancer**. Dietary-based studies do not support an association between tomato intake and risk of breast cancer.
- **Ovarian Cancer**. No association between tomato consumption and ovarian cancer exists.

QUENCHING OF REACTIVE OXYGEN BY LYCOPENE

Oxidative stress may be defined as that state in which "exposure to free radicals or other oxidants represent a challenge to normal function, or even to survival." Such situations may arise from an increase in exposure to radicals/oxidants or may be the result of a decreased antioxidant capacity.

A "**radical**" is defined as an atom or group of atoms with an unpaired electron. Because free radicals are missing an electron in their outer shell, they are highly reactive, and in excess they can damage or kill cells in culture or cause lesions in an organism. The formation of oxygen radicals and their metabolites is summarized below:

$$O_2 \longrightarrow O_2^{\bullet -} \text{ Superoxide Anion}$$

$$O_2^{\bullet -} \xrightarrow{e^-} H_2\text{-}O_2 \text{ Hydrogen Peroxide}$$

$$H_2O_2 \xrightarrow{e^-} OH \text{ Hydroxyl Radical}$$

$$^{\bullet}OH \xrightarrow{e^-} H_2O \text{ Water}$$

Therefore, the sequential addition of a single electron results in the formation of superoxide anion, hydrogen peroxide, hydroxyl radical, and water.

Hydroxyl radicals ($^\bullet$OH) are the most damaging free radicals. Although they exist only for a fraction of a second, they are able to destroy vital enzymes, cause linking of DNA, unleash proteolytic enzymes, tear apart polysaccharides, and cause lipid peroxidation altering membrane permeability and associated functions.

There are at least nine separate steps where free radicals can be generated, which include the following items:

Haber–Weiss reaction

$$O_2 + H_2O_2 \longrightarrow O_2 + {}^1O_2 + HO^- + HO^\bullet$$

Fenton reaction

$$O_{2-} + Fe^{3+} \longrightarrow Fe^{2+} + O_2$$

$$Fe^{2+} + H_2O_2 \longrightarrow Fe^{3+} + HO^- + HO^\bullet$$

$$O_2 + H_2O \longrightarrow HO^- + HO^\bullet$$

A number of enzymes (antioxidants) prevent oxidative damage from the reactive products of oxygen that are formed during normal metabolic events. These are:

$$2O_2^{\bullet-} + 2H \xrightarrow{\text{Superoxide dismutase}} H_2O_2 + O_2$$

$$2H_2O_2 \xrightarrow{\text{Catalase}} 2H_2O + O_2$$

$$H_2O_2 + \text{Glutathione} \xrightarrow{\text{Glutathione peroxidase}} 2H_2O + GSSG$$

In addition, a number of compounds, such as vitamin E, β-carotene, uric acid, vitamin C, retinoic acid, iron chelators, vitamin E, glutathione, and zinc metallothionein (see Figure 31) have been shown to have antioxidant activities. Many of the putative biologic effects and health benefits of lycopene and other carotenoids are hypothesized to occur via protection against oxidative damage. The ability of lycopene to quench $^1O_2$ or oxygen radicals is only beginning to be investigated in experimental systems. The quenching of $^1O_2$ by lycopene and other carotenoids can occur through physical or chemical means. Physical quenching predominates and leaves the carotenoid intact, whereas chemical quenching is responsible for the decomposition of the carotenoid ("bleaching"). Physical quenching involves the transfer of excitation energy from $^1O_2$ to the carotenoid, thereby producing ground-state oxygen and a carotenoid in the excited triplet state. The excess energy is dissipated as heat through rotational and vibrational interactions with the surrounding solvent and structures. The regeneration of the ground-state carotenoid allows it to function as a catalyst and undergo additional cycles of $^1O_2$ quenching. The quenching capacity of carotenoid depends primarily on the number of conjugated double bonds, which accounts for the exceptionally high capacity exhibited by lycopene compared with many other carotenoids.

Lycopene may also interact with reactive oxygen species such as hydrogen peroxide and nitrogen dioxide. A recent study showed that lycopene was exceptionally potent and twice as effective as β-carotene in protecting lymphocytes from NO_2 radical cell death and membrane damage. β-Carotene is an efficient scavenger of peroxyl radicals, especially at low oxygen tension. Although less research has focused on lycopene, current studies suggest that lycopene is a more potent scavenger of oxygen radicals than other major dietary carotenoids.

40

MARINE
THERAPEUTICS

More than 10,000 new compounds have been isolated from bacteria, fungi, microalgae, seaweed, sponges, soft corals, opisthobranch mollusks, bryozoans, echinoderms, and ascidians.

There are two serious obstacles that prevent development of drugs from marine natural products. Obviously, the most important is the "supply issue." The majority of promising compounds have complex structures, which limits the ability to supply large amounts of samples by chemical synthesis. Usually, yields from organisms are quite low as well. Many researchers believe that these compounds are actually produced by **"symbiotic" microorganisms**. Yet bacteria or fungi that produce highly promising metabolites have not been isolated from parent organisms.

Most of the highly bioactive marine metabolites are also highly toxic, which is very serious. If enough materials are available, parent compounds could be modified to reduce their toxicity or to improve their therapeutic efficacy. However, such attempts have not been accomplished because of limited amounts of materials and their highly complex structures, except for readily synthesized substances such as didemnins and dolastatins.

Of course, the discovery of new structures is very important. Marine microbes are relatively new targets for such research. Bacteria and fungi isolated from marine environments, i.e., seawater, sediments, and marine organisms, often produce metabolites identical or similar to those from terrestrial species.

Phytoplanktons produce a wide variety of unusual metabolites with strong bioactivities and frequently unprecedented structural features. However, there are some problems; low growth rates, low productivity of metabolites, and complexity of structures are negative factors (Fusetani, 2000). Tables 44 and 45 list a few of the more than 10,000 potential drugs that are being isolated from the sea.

Table 44 Potential Drugs from Marine Bacteria

Producing Strains	Sources	Compounds
Anticancer		
Actinomycete	Sediment	Lagunapyrones
Alteromonas haloplanktis	Sediment	Bisucaberin
Alteromonas sp.	Sponge	Alteramide
Bacillus cereus	Mollusk	Homocereulide
Bacillus sp.	Sediment	Halobacillin
Bacillus sp.	Sediment	Isocoumarin
Chaina purpurogena	Sediment	SS-228Y
LL-141352	Tunicate	LL-141352β
Micromonospora sp.	Soft coral	Thiocoraline
Pelagiobacter sp.	Alga	Pelagiomicins
Streptomyces hygroscopicus	Fish gut	Halichomycin
Streptomyces sioyaensis	Sediment	Allemicidin
Streptomyces sp.	Soft coral	Octalacins
Streptomyces sp.	Sediment	γ-Indomycinone
Streptomyces sp.	Sediment	δ-Indomycinone
Streptomycete	Mollusk	Aburatubolactam C
Vibrio sp.	Driftwood	Acyldepsipeptide
Antibacterial		
Actinomycete	Sediment	Marinone
Alcaligenes faecalis	Mollusk	B-1015
Alteromonas rava	Seawater	Thiomarinol
Bacillus sp.	Marine worm	Loloatins
Chromobacterium sp.	Seawater	Bromopyrroles
Maduromycete	Sediment	Maduralide
Pseudoalteromonas	Alga	Korormicin
Pseudomonad	Seawater	Quinolinol
Pseudomonas aeruginosa	Sponge	Diketopiperazine
Pseudomonas bromoutilis	Seagrass	Pentabromopseudilin
Pseudomonas fluorescens	Ascidian	Andrimid, noiramides
Pseudomonas sp.	Alga	Massetolides
Streptomyces griseus	Sediment	Aplasmomycins
Streptomyces sp.	Sediment	Phenazines
Streptomyces sp.	Sponge	Urauchimycins
Streptomyces sp.	Sediment	Bioxalomycins
Streptomyces tenjimariensis	Sediment	Istamycins
Streptomycete	Sediment	Wailupemycins
Vibrio gazogenes	Sediment	Magnesidins
Vibrio sp.	Sponge	Phenolic
Vibrio sp.	Sponge	Trisindoline
Antiviral		
Unidentified Gram-positive	Sediment	Macrolactins
Unidentified Gram-positive	Sediment	Caprolactins
Anti-inflammatory		
Actinomycete	Jellyfish	Salinamides

Table 45 Potential Drugs from Marine Fungi

Producing Strains	Sources	Compounds
Anticancer		
Aspergillus fumigatus	Fish	Fumiquinazolines
Aspergillus fumigatus	Sediment	Tryprostatins
Aspergillus niger	Sponge	Asperazine
Aspergillus sp.	Sediment	Aspergillamides
Aspergillus versicolor	Alga	Sesquiterpene esters
Fusarium sp.	Wood	Neomangicols
Gymnasella dankaliensis	Sponge	Gymnastatins
Gymnasella dankaliensis	Sponge	Gymnasterones
Leptosphaeria sp.	Alga	Leptosins
Penicillium fellutanum	Fish	Fellutamides
Penicillium sp.	Alga	Cummunesins
Penicillium sp.	Alga	Penochalasins
Penicillium sp.	Alga	Penostatins
Penicillium waksmanii	Alga	Pyrenocines
Periconia sp.	Sea hare	Pericosines
Phomopsis sp.	Coral reef	Phomopsidin
Trichoderma harzianum	Sponge	Trichodenones, harzialactones spiroxins
Antibacterial		
Coniothyrium sp.	Sponge	Hydroxyphenyl
Corollospora pulchelia	Sand	Melinacidins, gancidin
Exophiala pisciphila	Sponge	Exophilin A
Microsphaeropsis sp.	Sponge	Microsphaeropsisin
Preussia aurantiaca	Mangrove	Auranticins
Unidentified fungus	Sponge	Secocurvularin
Unidentified fungus	Sponge	Hirsutanols
Antiviral		
Scytalidium sp.	Seagrass	Halovirs
Antifungal		
Aspergillus sp.	Alga	Mactanamide
Hypoxlon oceanicum	Wood	15G256γ

41

MORPHINE

Morphine is a narcotic analgesic.

MORPHINE SULFATE

(Astramorph PF, Duramorph, MS Contin, Infumorph, Kadian, MSIR, OMS Concentrate, RMS, Roxanol, Roxanol Rescudose Roxanol SR, Roxanol 100, Roxanol UD)

Morphine relieves pain by stimulating opiate receptors in the central nervous system (CNS). It is used in moderate to severe acute and chronic pain and is used as a preoperative sedative and adjunct to anesthesia. Moreover, morphine is used in the management of dyspnea associated with left ventricular failure and pulmonary edema. Morphine is contraindicated in upper airway obstruction and acute asthma.

OPIOID ANALGESICS IN HISTORY

Although the psychological effects of opium may have been known to the ancient **Sumerians**, the first undisputed reference to poppy juice is found in the writings of **Theophrastus** in the third century B.C. The word *opium* itself is derived from the Greek name for juice, the drug being obtained from the juice of the poppy, *Papaver somniferum*. Arabian physicians were well versed in the uses of opium; Arabian traders introduced the drug to the Orient, where it was employed mainly for the control of dysenteries. **Paracelsus** (1493–1541) is credited with repopularizing the use of opium in Europe after it had fallen into disfavor because of its toxicity. By the middle of the 16th century, many of the uses of opium were appreciated. In 1680, **Sydenham** wrote, "Among the remedies which it has pleased Almighty God to give to man to relieve his sufferings, none is so universal and so efficacious as opium" (see Grollman, 1962).

Opium contains more than 20 distinct alkaloids. In 1806, **Sertürner** reported the isolation of a pure substance in opium that he named morphine, after **Morpheus**, the **Greek god of dreams**. The discovery of other alkaloids in opium quickly followed that of morphine (**codeine** by Robiquet in 1832, **papaverine** by Merck in 1848). By the middle of the 19th century, the use of pure alkaloids rather than crude opium preparations began to spread throughout the medical world.

In the United States, opioid abuse was accentuated by the unrestricted avail-ability of opium that prevailed until the early years of the 20th century and by the influx of opium-smoking immigrants from the Orient. In addition, the invention of the hypodermic needle led to the parenteral use of morphine and to a more severe variety of compulsive drug abuse.

The problem of addiction to opioids stimulated a search for potent analgesics free of addictive potential. Just prior to and following World War II, synthetic compounds such as **meperidine** and **methadone** were introduced into clinical medicine, but proved to have typical morphinelike actions. **Nalorphine**, a derivative of morphine, was an exception. Nalorphine antagonized the effects of morphine and was used to reverse morphine poisoning in the early 1950s. Higher doses of nalorphine are analgesic in postoperative patients, but the drug is not used clinically as an analgesic because of side effects such as anxiety and dysphoria. However, its unusual pharmacological profile ushered in the development of new drugs, such as the relatively pure antagonist **naloxone** and compounds with mixed actions (e.g., **pentazocine**, **butorphanol**, and **buprenorphine**). Such agents enlarged the range of available therapeutic entities and provided tools needed to explore the mechanisms of opioid actions (Goodman and Gilman, 1955).

The complex interactions of morphine and drugs with mixed agonist/antagonist properties, such as nalorphine, led to the proposal of the existence of multiple classes of opioid receptors. This proposal has now been confirmed, first by receptor-binding studies and more recently with the cloning of four distinct but closely related opioid receptors. Soon after the demonstration of the existence of opioid-binding sites, three classes of endogenous opioid peptides were isolated. They are encoded by different genes, are expressed in distinct neuronal pathways or cell types, and have differing selectivities for the various classes of opioid receptors.

FRIEDRICH WILHELM ADAM SERTÜRNER AND OPIUM

Friedrich Wilhelm Adam Sertürner (1783–1841) was born in Neuhaus close to the German town Paderborn in Westphalia. At the age of 16 he became an apprentice in the pharmacy of Paderborn, moving from there to Einbeck (1806) and later to Hameln (1820). From 1809, he had his own pharmacy. Already during his apprenticeship he published his first paper on opium (1805), in which he described the discovery of **meconic acid** in the drug. The next year, a further publication followed, announcing another new substance derived from opium. It was char-acterized by its solubility in acid water, from which it could be precipitated by ammonia. It thus had the character of a weak base. Since it provoked sleep in a dog, Sertürner supposed it to be "**der eigentliche betäubende Grundstoff**" of opium (the specific narcotic element of opium). In his own laboratory he followed up the study and now called the new substance **morphine** after the god of sleep, **Morpheus**. Its stupendous effects were clearly shown in experiments on Sertürner himself and three young men. Sertürner had in fact discovered a new group of highly active substances, to which the name *alkaloids* was attached in 1818 by W. Meissner. The priority of the discovery was disputed, because Derosne in 1804 had isolated from opium juice crystals, which he considered to be a salt. The crystals were probably a mixture of morphine and narcotine, but Derosne had completely missed that the substances had the nature of a base. Sertürner's work at first was neglected in Germany, but thanks to the French chemist J. L. Gay-Lussac its importance was soon recognized, and in 1831 he received from the Institut de France the Monthyon prize (2000 francs) "pour avoir reconnu la nature alcaline de la morphine et avoir ainsi ouvert une voie, qui a produit de grandes découvertes médicales" (for having found the alkaline nature of morphine and thereby opened a way which has led to great medical discoveries). By this time the methods of Sertürner had been successfully applied to different drugs, and especially through the efforts of P. J. Pelletier (1784–1841) and J. B. Caventou (1788–1842) a series of alkaloids like **quinine**, **strychnine**, **brucine**, **veratrine**, and **emetine** had been isolated, and many more were to follow. Other work by Sertürner, such as that on the composition of corrosive alkalis and on the nature of cholera (which he considered to be caused by a living organism that might be combated by disinfectants or by boiling the water), proved his unusual gifts (see Potter, 1910; Gathercoal and Wirth, 1947).

OPIUM

Opium or **gum opium** is the air-dried, milky exudation obtained by incising the unripe capsules of ***Papaver somniferum*** Linné or its variety *P. album* DeCandolle. The term *opium* is from the Greek *opion*, meaning poppy juice; *papaver* is the Latin name for the poppy and *somniferum* is Latin meaning to produce sleep. The opium poppy is an annual herb with large, showy, solitary flowers varying in color from white (var. *album*) to pink or purple (Figure 110).

Opium, which is native to Asia Minor, was introduced into India by the Mohammedans in the 15th century and cultivation was begun in Macedonia and Persia about the middle of the 19th century. Opium is commercially produced now in Turkey, the Balkan States, Persia, India, and China. The discovery of the medicinal qualities of opium is lost in antiquity. **Theophastus** (third century B.C.) mentions it, and **Dioscorides** (A.D. 77) distinguishes between the juice of the poppy and an extract of the entire plant. In 1806, Sertüner first isolated the alkaloid morphine from opium.

CULTIVATION, COLLECTION, AND COMMERCE

The seeds of the opium poppy are sown in October in well-cultivated soil. The seeds germinate in the fall and the seedlings may be an inch high when snow falls; this protects them from freezing. In the spring when the plants have attained the height of 6 inches, the fields are cultivated and the plants thinned to stand about 2 feet apart. The poppy blossoms in April or May and the capsules mature in June or July. Each plant bears from five to eight capsules.

The ripening capsules, about 4 cm in diameter, change from bluish green to yellowish in color. This is a critical time for collecting the latex. The capsules are incised with a knife, usually three-bladed, with the incision made around the circumference of the capsule (see Figure 110). The latex tubes open into one another so that it is not necessary to incise them all. Great skill, however, is required so as not to cut through the endocarp, in which case the latex would flow into the interior of the capsule and be lost. The latex, which is at first white, rapidly coagulates and turns brown. This is removed early the following morning,

Figure 110 The Oriental poppy (*Papaver somniferum*) from which opium is derived. The unripe pod from which opium juice is obtained is shown together with the flowering plant.

being scraped off with a knife and transferred to a poppy leaf. When sufficient latex is collected it is kneaded into balls that are wrapped in poppy leaves and shade-dried. The opium is then inspected and usually packed with the brown winged fruits of a *Rumex,* which prevents cohering. In some districts the latex is molded or pressed into cakes or other forms that are then wrapped in paper (Gathercoal and Wirth, 1947).

Opium contains 20 alkaloids in combination with meconic, lactic, and sulfuric acids; the neutral principles **Meconin** and **Meconoiasin**; also glucose, mucilage, resin, pectin, caoutchouc, fats, essential oil, odorous substances, salts of ammonium, magnesium, and calcium, and water. Its principal alkaloids are the following six:

- **Morphine**, $C_{17}H_{19}NO_3$, 2½ to 2% — The principal alkaloid, occurring in the drug in the form of the tribasic meconate. Its properties are anodyne, hypnotic, and narcotic. From it by a process of dehydration by heat and hydrochloric acid is prepared the artificial alkaloid ***apomorphine***, a powerful emetic and expectorant.
- **Codeine**, $C_{18}H_{21}NO_3$, 0.3 to 0.5% — A calmative and, when pure, not very active alkaloid, but is frequently contaminated with other alkaloids. *Apomorphine* may be prepared from it.

- **Narceine**, $C_{22}H_{29}NO_9$, 0.2 to 0.7% — Now believed to have little or no action.
- **Narcotine**, $C_{22}H_{23}NO_7$, 2 to 10% — An antiperiodic and a tetanizer, but wholly devoid of narcotic properties.
- **Thebaine**, or ***Paramorphine***, $C_{19}H_{21}NO_3$, 0.2 to 1% — A powerful spinal exaltant and tetanizer, resembling **strychnine** in its action.
- **Papaverine**, $C_{20}H_{21}NO_4$, 1% — Stands midway between morphine and codeine in its action on the CNS, but is a comparatively weak poison.

Other alkaloids are: **Codamine, Cryptopine, Gnoscopine, Hydrocotarnine, Lanthopine, Laudanine, Laudanosine, Meconidine, Oxynarcotine, Papaveramine, Protopine, Pseudomorphine, Rhoeadine**, and **Tritopine**. Many of them occur only in traces, and some are regarded as probable derivatives of morphine. ***Porphyroxin*** is said to be a complex combination of several of the alkaloids, and not a proximate principle (Potter, 1910).

MORPHINE AND OTHER NARCOTIC ANALGESICS

Analgesia means lack of pain, and analgesics are substances that obtund the perception of pain without causing loss of consciousness. However, analgesics should not be given until the cause of pain has been determined. If a proper diagnosis is made, the use of an analgesic may not be necessary. A few examples are cited.

If individuals become dehydrated in hot weather without fluid replenishment, they will suffer painful muscular cramp, known as **stoker's cramp**. This condition may be prevented or corrected by administering sodium chloride and by replenishing the lost fluid. The severe muscular spasm occurring in tetanus is managed by muscle relaxants. The painful epigastric pain associated with acid-pepsin disease is managed with antacids and other medications. The pain associated with gouty arthritis is managed with anti-inflammatory drugs and agents that reduce the synthesis and the pool of uric acid in the body. The headache associated with malignant hypertension is managed with antihypertensive agents.

Although analgesics may ameliorate the pain in all of these conditions, their use should not be implemented until the pathology has been discerned and appropriate rehabilitative measures instituted.

MULTIPLE MEDICATIONS TO TREAT PAINFUL CONDITIONS

In a pathological condition that causes pain, drugs may be used to either care for the acute attack of pain or as prophylaxis to prevent the occurrence of pain. For example, **colchicine** is used during an acute attack of gout, and after the pain has subsided initially, the patient is switched to uricosuric agents such as **probenecid** or an inhibitor of uric acid synthesis such as **allopurinol**.

Drugs used in migraine therapy can be divided into two groups: those that abort an established migraine attack and those used prophylactically to reduce the number of migraine attacks. Each group has drugs that are specific for migrainous headaches and those that are nonspecific but treat the accompanying

Table 46 Endogenous and Synthetic Opioid Peptides

Selected Endogenous Opioid Peptides

[Leu^5]enkephalin	Tyr-Gly-Gly-Phe-Leu
[Met^5]enkephalin	Tyr-Gly-Gly-Phe-Met
Dynorphin A	Tyr-Gly-Gly-Phe-Leu-Arg-Arg-Ile-Arg-Pro-Lys-Leu- Lys-Trp-Asp-Asn-Gln
Dynorphin B	Tyr-Gly-Gly-Phe-Leu-Arg-Arg-Gln-Phe-Lys-Val-Val-Thr
α-Neoendorphin	Tyr-Gly-Gly-Phe-Leu-Arg-Lys-Tyr-Pro-Lys
β-Neoendorphin	Tyr-Gly-Gly-Phe-Leu-Arg-Lys-Tyr-Pro
β$_h$-Endorphin	Tyr-Gly-Gly-Phe-Met-Thr-Ser-Glu-Lys-Ser-Gln-Thr- Pro-Leu-Val-Thr-Leu-Phe-Lys-Asn-Ala-Ile-Ile- Lys-Asn-Ala-Tyr-Lys-Lys-Gly-Glu

Selected Synthetic Opioid Peptides

DAMGO	[D-Ala^2, MePhe^4,Gly(ol)^5]enkephalin
DPDPE	[D-Pen^2, D-Pen^5]enkephalin
DSLET	[D-Ser^2, Leu^5]enkephalin-Thr^6
DADL	[D-Ala^2, D-Leu^5]enkephalin
CTOP	D-Phe-Cys-Tyr-D-Trp-Orn-Thr-Pen-Thr-NH$_2$
FK-33824	[D-Ala^2, N-MePhe^4, Met(O)^5-ol]enkephalin
[D-Ala^2]Deltorphin I	Tyr-D-Ala-Phe-Asp-Val-Val-Gly-NH$_2$
[D-Ala^2,Glu^4]Deltorphin (Deltorphin II)	Tyr-D-Ala-Phe-Glu-Val-Val-Gly-NH$_2$
Morphiceptin	Tyr-Pro-Phe-Pro-NH$_2$
PL-017	Tyr-Pro-MePhe-D-Pro-NH$_2$
DALCE	[D-Ala^2, Leu^5, Cys^6]enkephalin

Three distinct families of peptides have been identified: the *enkephalins*, the *endorphins*, and the *dynorphins*. Each family is derived from a distinct precursor polypeptide and has a characteristic anatomical distribution. These precursors are now designated as proenkephalin (also proenkephalin A), proopiomelanocortin (POMC), and prodynorphin (also proenkephalin B).

headache (analgesics), vomiting (antiemetics), anxiety (sedatives and anxiolytics), or depression (antidepressants). The main drugs with specific actions on migraine include ergot alkaloids (**ergotamine**, **dihydroergotamine**), specific serotonin receptor agonists (**sumatriptan**), β-adrenergic receptor antagonists (**propranolol** and **metoprolol**), calcium antagonists (**flunarizine**), and anti-inflammatory agents (**indomethacin**).

ENDOGENOUS OPIOID PEPTIDES

For many years, pharmacologists considered the possibility that opioids mimic a naturally ongoing process. Investigations isolated opiate-like peptides from the brain that consisted of two similar pentapeptides with the following sequences (Tables 46 through 48).

Table 47   Proposed Functions of the Opioid Receptor Types

Function	Receptor Types	Anatomy
Appetite modulation, eating behavior	μ, δ, and κ	Ventral tegmental area
Cardiovascular regulation	μ, δ, and κ	Nucleus tractus solitarii
Fluid balance	κ: diuresis	Hypothalamus and/or pituitary; also possibly kidney (κ)
Endocrine responses	μ: antidiuresis	Hypothalamus, possible pituitary
Stimulatory effects:		
Growth hormone	δ	
Corticotropin	μ and κ	
Prolactin	μ and κ	
Inhibitory effects:		
Luteinizing hormone	μ and δ	
Vasopressin	κ	(Also nucleus tractus solitarii)
Oxytocin	μ and κ	
Pain inhibition	μ and δ	Supraspinal
	δ	Spinal medullary reticular formation
	κ	Spinal
Respiration	μ and δ may mediate respiratory depression	Brainstem
Locomotor behavior	μ: increased activity	A9, A10 DA systems
	κ: sedation	A10 DA systems
Thermoregulation	μ: may mediate hypothermia	Hypothalamus
	δ: may mediate hyperthermia	

DA = dopaminergic.

OPIOID RECEPTORS

The Evidence for Multiple Opioid Receptors

The opioids produce a large variety of pharmacological responses by interacting with multiple opioid receptors. The indirect evidence for the presence of these multiple receptors is as follows:

- **Nalorphine** produces many of the pharmacological effects of morphine, yet it blocks or reverses the action of morphine.
- **Naloxone**, which does not produce any of the pharmacological effects of morphine, blocks or reverses the action of morphine.
- Subjects who are tolerant to the psychomimetic effects of morphine are not cross-tolerant to the psychomimetic effects of **cyclazocine**.
- The abstinence syndrome associated with morphine is qualitatively different from the abstinence syndrome caused by cyclazocine.
- Agonists–antagonists nalorphine and cyclazocine can induce tolerance to their agonistic effects but not to their antagonistic effects.

Table 48   Actions and Selectivities of Opioids at the Various Opioid Receptor Classes

	Receptor Types			
	μ	δ	κ_1	κ_3
Drugs				
Morphine	+++		+	+
Methadone	+++			
Etorphine	+++	+++	+++	+++
Levorphanol	+++		NA	+++
Fentanyl	+++			
Sufentanil	+++	+	+	
DAMGO	+++			+
Butorphanol	P	NA	+++	NA
Buprenorphine	P	NA	– –	NA
Naloxone	– – –	–	– –	– –
Naltrexone	– – –	–	– – –	– –
CTOP	– – –			–
Diprenorphine	– – –	– –	– – –	– – –
β-Funaltrexamine	– – –	–	++	NA
Naloxonazine	– – –	–	–	–
Nalorphine	– – –		+	+++
Pentazocine	P		++	+
Nalbuphine	– –		++	++
Naloxone benzoylhydrazone	– – –	–	–	+++
Bremazocine	+++	++	+++	++
Ethylketocyclazocine	P	+	+++	+++
U50,488			+++	
U69,593			+++	
Spiradoline	+		+++	
nor-Binaltorphimine	–	–	– – –	–
Naltrindole	–	– – –	–	–
DPDPE		++		
[D-Ala^2,Glu^4]deltorphin		++		
DSLET	+	++		
Endogenous Peptides				
Met-enkephalin	++	+++		
Leu-enkephalin	++	+++		
β-Endorphin	+++	+++		
Dynorphin A	++		+++	NA
Dynorphin B	+	+	+++	NA
α-Neoendorphin	+	+	+++	NA

Activities of drugs are given at the receptors for which the agent has reasonable affinity. +, agonist; –, antagonist; P, partial agonist; NA, data not available or inadequate; DAMGO, CTOP, DPDPE, DSLET. The number of symbols is an indication of potency; the ratio for a given drug denotes selectivity. These values were obtained primarily from animal studies and should be extrapolated to humans with caution. Both β-funaltrexamine and naloxonazine are irreversible μ antagonists, but β-funaltrexamine also has reversible κ agonist activity.

Distribution of Opioid Receptors

The opioid receptors are distributed widely throughout the neuraxis, but the highest density is found in the limbic structures, thalamic nuclei, and neural areas important for visceral functioning.

Proposed Functions of the Opioid Receptor Subtypes

The precise functions of the opioid receptor subtypes remain to be delineated. In many studies, it has been established that they modulate the functions of the dopaminergic system. For example, administration of morphine into the area 10 region results in a dopamine-dependent increase in locomotor activity. Electrophysiological studies of area 9 neurons suggest that μ and κ opioids have opposite effects on motor behaviors — activation or sedation, respectively. κ and μ agonists have also been found to exert opposite actions with respect to fluid regulation: diuretic and antidiuretic effects, respectively. The μ, δ, and κ opioid receptor types have all been implicated in the mediation of analgesia (see Table 47).

MORPHINE AND RELATED COMPOUNDS

The opium alkaloids, which are obtained from **Papaver somniferum**, contain two groups of compounds: compounds with phenanthrene derivatives, consisting of **morphine**, **codeine**, and **thebaine** and compounds with isoquinoline derivatives, consisting of **papaverine** and **noscapine**.

Narcotics are divided into naturally occurring, semisynthetic, and synthetic derivatives. The naturally occurring analgesics consist of morphine and codeine (**methylmorphine**). The semisynthetic analgesics include **hydromorphone** (Dilaudid) and **hydrocodone** (Dicodid). The synthetic analgesics consist of **meperidine** (Demerol), **alphaprodine** (Nisentil), **methadone** (Dolophine), **propoxyphene** (Darvon), and pentazocine (Talwin).

The narcotic antagonists are naltrexone and naloxone (Narcan). **Dextromethorphan** (Romilar) is used as an antitussive preparation. Apomorphine is used as an emetic agent.

Narcotic analgesics may have either a high potency (morphine, hydromorphone, oxymorphone, methadone, meperidine, fentanyl, and levorphanol) or low potency (codeine, oxycodone, hydrocodone, propoxyphene, and diphenoxylate). These agents may be a pure agonist (morphine), pure antagonist (naloxone), or mixed agonist–antagonist (pentazocine).

In discussing the pharmacology of narcotic analgesics, morphine is considered in greater detail as a prototype drug, and all the other compounds are compared with it.

Morphine, A Naturally Occurring Analgesic

Effects on the Central Nervous System

The majority of the analgesic agents exert their main action within the CNS particularly at the cerebral cortical levels of integration of the pain impulse (Figure 111). Many act primarily on the hypothalamic centers, which are also the

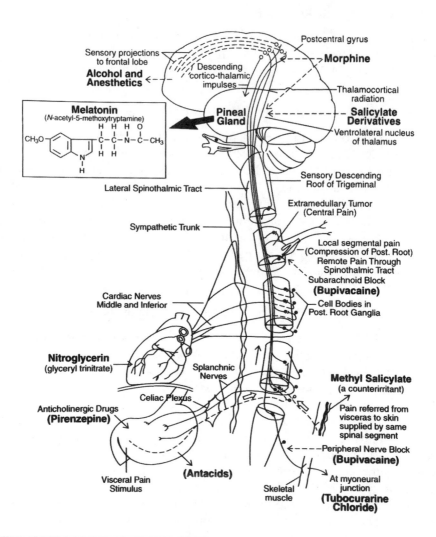

Figure 111 Schematic representation of the pain tracts and their potential interruption at various levels. The figure is organized in 5 sections in relation to the spinal cord, and depicts from bottom up, pain originating in skeletal muscle, stomach (with referral of pain to the skin), heart, spinal cord, and head, respectively. The action of drugs is indicated as of three types: (1) abolition of the pain stimulus; (2) the local interruption of the conducting pathways; and (3) elevation of the pain threshold. The spinothalamic tract is shown as a combination of both the right and left sides above the thoracic segment.

site of temperature control, and hence exert an antipyretic as well as an analgesic action. The **salicylates**, for example, are effective antipyretics as well as analgesics. The antirheumatics, which are used primarily as analgesics in gout and rheumatism, are also included here. In addition to the analgesic property, many drugs are used because of their capacity to relieve pain but their analgesic action is not dependent on any action on the CNS but rather on the end organ. For example, **amyl nitrite** relieves the pain of **angina pectoris** by its action on the coronary blood flow and alleviates the pain of biliary colic by reducing the intrabiliary pressure. It, therefore, is not included among the analgesics, which act directly on the higher centers involved in the perception of pain. **Morphine**, on the other hand, although not affecting coronary blood flow and actually increasing intra-biliary pressure, nevertheless provides relief from pain in these conditions by its action on the higher cerebral centers and hence is classified as an analgesic.

Pain may be relieved by the following:

- Removing its point of origin by drugs (e.g., by regional anesthesia, relief of muscular spasm, etc.) or by surgery (e.g., by excision or denervation of the painful focus)
- By the application of physical measures or anesthetics to the point to which the pain is referred or to the hyperirritable segments of the cord
- By interrupting the impulses that convey the pain by anesthetizing or destroying the nerve tracts
- By modifying the central reception of pain by blocking the facilitating reflexes and thereby raising the pain threshold with analgesic drugs
- By modifying the central perception, interpretation, and reaction to pain by interrupting the pathways for their transmission in the brain by such drugs as morphine or by such surgical procedures as lobotomy, leucotomy, etc.
- By modifying the central perception of pain of peripheral origin and mitigating psychogenic pain by depressing reflex activity with such drugs as alcohol, the barbiturates, etc.

The ideal analgesic should obliterate pain and diminish the anxiety associated with it; exert a minimal degree of narcosis and stupefaction; be free of such undesirable side effects as constipation, nausea, respiratory depression, etc.; have no tendency to addiction or the development of tolerance; have a rapid onset and long duration of action; be effective and well tolerated when administered orally; and be relatively inexpensive (Grollman, 1962).

Pain is probably the symptom that most often and most promptly brings the patient to the attention of the physician. The nature of the pain is important in arriving at a correct diagnosis but, once this has been made, relief is in order. The most suitable analgesic to be used in a given patient is dependent upon the nature of the pain, its origin and severity, as well as upon the temperament and personality of the patient. In many cases, narcotics, despite their effectiveness as analgesics, must be excluded because of their potential to cause addiction. The use of a sedative in conjunction with the analgesic is often indicated particularly in nervous individuals.

Experimental evaluation of the analgesic activity of a given drug in humans is difficult, for the reactivity to a painful stimulus varies in different individuals and even in the same individual from time to time. The analgesic action depends on the alteration of the transmission of the pain impulse from the periphery to the cerebral cortex but is modified by psychic factors.

It appears from the available evidence that analgesics generally act by altering the reaction of the patient to the pain rather than by interfering with the pain impulse. This accounts for the great variability in the response elicited in different patients from the same analgesic and the effectiveness of placebos and suggestion in susceptible individuals.

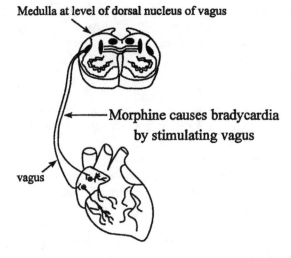

Medulla at level of dorsal nucleus of vagus

Morphine causes bradycardia
by stimulating vagus

vagus

Figure 112   Morphine causes bradycardia.

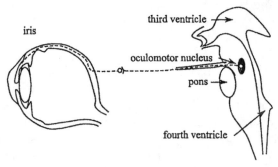

MORPHINE causes miosis by stimulating OCULOMOTOR NERVE

iris

third ventricle

oculomotor nucleus

pons

fourth ventricle

Figure 113   Morphine causes miosis by stimulating oculomotor nerve.

Morphine Stimulates Parts of the Central Nervous System

In horses, morphine stimulates the spinal cord in a predictable fashion. This effect is short-lived in humans, and is seldom seen when given in therapeutic doses. Initially, morphine stimulates the vomiting center, and emesis occurs early in cases of intoxication. Depression of the vomiting center then ensues late in intoxication. Morphine stimulates the vagus nerve, causing bradycardia, and stimulates the nucleus of the third cranial nerve (oculomotor), causing miosis (Figures 112 and 113).

Analgesic Effects

The relief of pain brought about by morphine is selective, and other sensory modalities such as touch, vibration, vision, hearing, and the like are not obtruded.

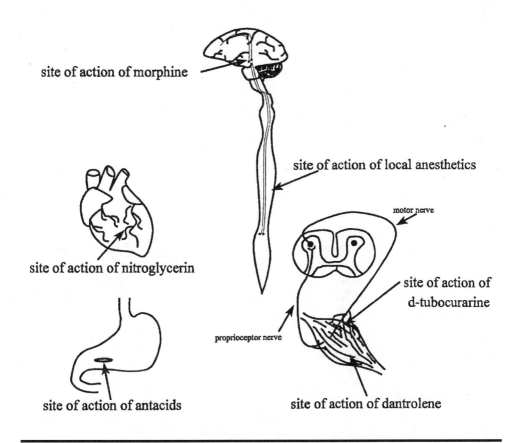

site of action of morphine

site of action of local anesthetics

motor nerve

site of action of nitroglycerin

site of action of
d-tubocurarine

proprioceptor nerve

site of action of antacids

site of action of dantrolene

Figure 114   Morphine alters the patient's reaction to pain.

Morphine does not reduce the responsiveness of nerve endings to noxious stimuli, nor does it impair the conduction of nerve impulses along the peripheral nerves, as seen following the administration of local anesthetics.

Morphine exerts its analgesic effects by elevating the pain threshold and especially by altering the patient's reactions to pain (Figure 114). Morphine induces analgesia by activating the opioid, adrenergic, and serotoninergic systems. Analgesia results from the activation of those systems within the dorsal horn of the spinal cord that depresses the transmission of pain sensation to the brain. This is accomplished by decreasing the release of pain transmitters such as substance P or by hyperpolarizing the interneurons within the dorsal horn, or both. Morphine activates these mechanisms by interacting with the μ receptors located on neurons in the dorsal cord. There is considerable evidence suggesting that this action depends on the activation of the adrenergic system within the dorsal horn. This, in turn, suggests that morphine analgesia could be potentiated by the addition of drugs such as **clonidine** that activate the adrenergic system.

In addition, morphine appears to activate the endogenous supraspinal system that is normally activated by pain to protect the body from excessive nonessential pain stimulation. Patients report that the sensation of pain often exists, but, under the influence of morphine, they feel more at ease and comfortable. This euphoria is present in 90 to 95% of patients. Morphine may cause dysphoria in the remaining 5 to 10%.

Therapeutic Uses of Morphine

Local Uses of Opiates

With the exception of the local application of **ethylmorphine** to the conjunctiva or the tympanum to produce hyperemia, there are no rational local uses of opium alkaloids. These drugs are not local anesthetics and relieve pain only by central actions. Indeed, irritation and vesication may result from opium plasters, ointment, or washes applied to the skin or mucosa.

Sudorific Action of Morphine

Although morphine produces some degree of sweating, it is a very poor **diaphoretic** when compared with **pilocarpine** or **methacholine**. The sweating caused by **Dover's powder** is mainly due to the ipecac content. The use of this powder as a diaphoretic for aborting the onset of "colds" due to exposure may conceivably be based on the same rationale as suggested with alcohol.

Analgesic Action of Morphine

The opium alkaloids have no rival for the relief of pain, and when suffering cannot be allayed by non-narcotic analgesics or other therapeutic measures, morphine may be considered indispensable. In a relatively small dose of 5 to 10 mg, morphine relieves the constant but dull pain originating from the viscera, such as that of coronary, pulmonary, and biliary origin. In somewhat larger doses (10 to 20 mg), morphine relieves the sharp, lancinating, and intermittent pain resulting from bone fractures and other physical injuries. Inoperable and terminal causes of neoplastic diseases usually require the administration of morphine or other narcotics in increasing doses that eventually lead to both tolerance and addiction.

Pain from all parts of the body is dulled by the central action of the morphine alkaloids. The more severe the pain, the larger is the dose of opiate required and the greater the amount of drug that can be tolerated. Thus, for the severe pain of coronary thrombosis, two to four times the usual therapeutic dose of morphine may be given over a suitable period without danger of seriously embarrassing the respiration. This is in keeping with the general principle that the degree of stupefaction caused by a given amount of a depressant drug is directly proportional to the level of reflex excitability of the nervous system. It is not unusual to administer within an hour from 45 to 60 mg of morphine in 15-mg portions for patients with excruciating pain. However, caution is required in the administration of these large total doses, especially when they are used for severe pain of a type that may suddenly lessen in severity or even disappear entirely. For example, the pain of coronary occlusion or of renal or biliary colic may rapidly decline spontaneously. If patients have already been given a large amount of morphine, they may quickly show signs of morphine poisoning. Close observation of the patient and proper spacing of the doses of morphine will usually prevent such an occurrence.

Pain is a chief diagnostic sign for the physician, and its total abolition may seriously handicap the discovery of the nature of the patient's illness. It has been said of morphine that, "It not only puts the patient to sleep but also puts the physician to sleep." Although it is generally unwarranted to make the diagnosis at the cost of unbearable pain, nevertheless there are certain conditions in which opiates are not usually prescribed until the diagnosis has been made and the permission obtained for operation, should it be necessary. These conditions especially comprise what is known as the acute surgical abdomen. For example, no single cause, other than the use of a cathartic, is responsible for more complications or deaths from acute appendicitis than the misuse of opiates.

Nevertheless, it has been claimed that the skillful use of morphine may sometimes aid considerably in the diagnosis of acute abdominal conditions. The patient who is restless, apprehensive, and in severe pain often cannot give a satisfactory or correct history. Rigidity of the abdominal muscles may prevent an adequate physical examination. This rigidity may be due not only to the reflex spinal component as a result of an inflamed viscus but also to a cortical component secondary to intense pain. Morphine in analgesic doses abolishes the cortical component and leaves the spinal reflexes unaffected. In expert hands, the drug may thus be employed to aid in abdominal diagnosis, for which purpose it has been injected intravenously (10 mg) to obtain analgesia and cooperation quickly. Naturally, a baseline history and physical examination are necessary and the patient must be followed closely.

In patients with pain due to **neoplastic growths** or other chronic diseases, the use of morphine alkaloids should be reserved until non-narcotic drugs no longer give relief. Then codeine should be employed in combination with non-narcotic sedatives and analgesics and the doses increased until pain can no longer be controlled without morphine, metopon, meperidine, or methadone. The doses should be kept as low as possible so that maximal relief can be given for the longest possible period. Since tolerance develops, one wishes to reserve a margin of susceptibility to cope with the terminal stages when pain may be excruciating. In other words, "The physician should use morphine as a miser spends his gold." If absolutely necessary, of course, morphine is given without stint. When opiates and other analgesics are no longer satisfactory, nerve blocking by injections of alcohol, chordotomy, lobotomy, or other type of neurosurgical intervention may be required if the nature of the lesion allows. In most cases the euphoria, tranquility, and relief of pain afforded by the wise use of narcotics are a blessing to the patient and family and death is made easy and comfortable. Narcotic addiction, even in terminal cases, should be avoided if possible; nevertheless, it often occurs despite precautionary measures (Goodman and Gilman, 1955).

For the relief of pain arising from spasm of smooth muscle, as in **renal** or **biliary colic**, morphine is frequently employed. Other measures including anti-spasmodics such as atropine, atropine substitutes, theophylline, nitrites, and heat may be employed first; however, if they are ineffective, meperidine, methadone, or opiates must be used. Morphine relieves pain only by a central action and may aggravate the condition producing the pain by exaggerating the smooth muscle spasm. Morphine may also be indispensable for the relief of pain due to acute vascular occlusion, whether this be peripheral, pulmonary, or coronary in origin. In painful acute pericarditis, pleurisy, and spontaneous pneumothorax, morphine is likewise indicated. Carefully chosen and properly spaced doses of codeine or morphine may occasionally be necessary in pneumonia to control pain, dyspnea, and restlessness. Traumatic pain arising from fractures, burns, etc. frequently requires morphine. In shock, whether due to trauma, poisons, or other causes, morphine may be required to relieve severe pain.

The use of opiates in **obstetrical analgesia** is a highly specialized field requiring considerable experience and sound judgment to ensure safety. Morphine has been combined with **scopolamine** in "**twilight sleep**," but this mixture is not used nearly as much as formerly. Various combinations of meperidine, barbiturate, scopolamine, paraldehyde, and the inhalation anesthetics have made morphine less popular in obstetrical work. The opiates are powerful respiratory depressants. The fetus is especially susceptible to morphine, which greatly increases the incidence of asphyxia in the newborn. Morphine and its derivatives are particularly contraindicated in premature labor due to the untoward effect of such medication on the premature infant.

Postoperatively, narcotics may be employed to control pain and discomfort for the first 1 to 3 days, but if prescribed indiscriminately, they may obscure the outcome of surgery and the course of recovery, as well as prevent the early recognition of complications. For headache, codeine may be effectively combined with analgesics such as salicylate or acetanilid. Morphine is not used in migraine unless all other measures, including ergotamine, have failed. Great care should be exercised to prevent addiction.

Although morphine is the most potent and effective analgesic known, it possesses many properties and actions that are undesirable, including spasm of smooth muscle, nausea and vomiting, interference with gastrointestinal function, urinary retention, respiratory depression, stupefaction, pruritus and urticaria, tolerance to the analgesic effect, physical dependence, and addiction liability. If one lists the obvious criteria desirable in an analgesic, morphine does not receive a very high score when measured by such criteria. Nevertheless, its proper use often permits maximal benefit with minimal untoward effect, but this requires considerable knowledge of the drug, the patient, and the disease or syndrome being treated.

Sedation and Sleep

Whenever possible, drugs other than narcotics are to be prescribed for sedation, tranquility, or sleep. In many cases, however, the opiates must be resorted to in order to provide the necessary relief from insomnia, restlessness, or excitement and thus conserve the patient's strength. The ordinary non-narcotic somnifacients such as bromides, chloral hydrate, and barbiturates are not analgesic and, when sleeplessness is due to pain or cough, opiates are often required. In certain forms of delirium, mania, and psychoses, and in the hyperexcitable states caused by thyrotoxicosis and certain encephalopathies, opium alkaloids may be needed. In these conditions smaller doses of morphine produce sedation than are necessary when pain is also present. When sleep and relief from restlessness and apprehension are absolutely essential, as in threatened abortion, internal hemorrhage, congestive heart failure, and massive pulmonary hemorrhage, morphine provides a valuable therapeutic agent.

Cough and Dyspnea

Cough, especially when unproductive and irritating, interferes with rest and sleep and may be painful. It can also result in cardiac strain and cause pulmonary or cardiovascular accidents. The opiates comprise the so-called **anodyne expectorants**, and codeine in particular deserves preference for quieting an overactive cough reflex. The antitussive dose is kept small (5 to 10 mg) because repeated medication may be necessary. The uniformly effective oral dose of codeine for cough is about 10 mg. Large doses of codeine (over 60 mg) may actually increase the irritability of the respiratory and cough mechanism. Codeine can be effectively incorporated in an expectorant cough mixture for a dry, irritating cough with the object of stimulating secretion of mucus and yet reserving the cough reflex for productive expectoration. Coughing should not be unnecessarily depressed, especially if bronchial secretion is profuse, because drainage will then be insufficient and complications may ensue. Small amounts of codeine, or of morphine (2 to 4 mg), will decrease the frequency of coughing without altering significantly the respiratory minute volume.

Certain forms of dyspnea yield only to opiates. Especially in this category is the dyspnea of acute left ventricular failure and pulmonary edema. Most authorities agree that morphine is contraindicated in patients with pulmonary edema caused by chemical respiratory irritants. If needed in such cases for severe pain, its use should be combined with oxygen inhalation and positive-pressure therapy. In bronchial asthma, morphine is usually contraindicated because there is danger of addiction, the drug tends to depress respiration and to constrict bronchioles, and patients with asthma may be allergic to the drug. Deaths have occurred from the use of morphine in asthma.

There is no good evidence that the use of codeine or papaverine shortens the duration of the "common cold." Opiates tend to dry the mucosal lining of the throat and nose, but other measures such as atropine can accomplish this end. If headache and pain require relief, non-narcotic analgesics should first be employed. If the symptomatic improvement afforded by codeine encourages patients to be ambulatory when they should be in bed, the medication has done harm by increasing the chances for complications and the infection of other individuals (Goodman and Gilman, 1955).

Side Effects of Morphine

Morphine Depresses Respiration

Morphine depresses all phases of respiration (respiratory rate, tidal volume, and minute volume) when given in subhypnotic and subanalgesic doses (Figure 115). In humans, a morphine overdose causes respiratory arrest and death. Therefore, morphine and other narcotic analgesics should be used with extreme caution in patients with asthma, emphysema, and cor pulmonale, and in disorders that may involve hypoxia, such as chest wound, pneumothorax, or bulbar poliomyelitis.

Morphine Causes Hypotension

Morphine releases histamine and may cause peripheral vasodilation and orthostatic hypotension (Figure 116). The cutaneous blood vessels dilate around the "blush

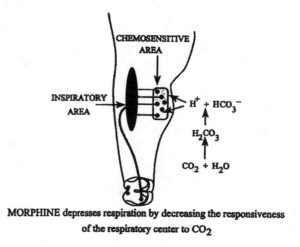

MORPHINE depresses respiration by decreasing the responsiveness
of the respiratory center to CO_2

Figure 115 Morphine depresses respiration by decreasing the responsiveness of the respiratory center to CO_2.

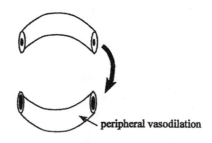

peripheral vasodilation

MORPHINE by releasing HISTAMINE causes
ORTHOSTATIC HYPOTENSION

Figure 116 Morphine causes vasodilation.

areas" such as the face, neck, and upper thorax. Morphine causes cerebral vasodilation (due to increased carbon dioxide retention secondary to respiratory depression), and hence it increases the cerebrospinal fluid pressure. Therefore, morphine should be used cautiously in patients with either meningitis or recent head injury. When given subcutaneously, morphine is absorbed poorly whenever there is either traumatic or hemorrhagic shock.

Morphine Causes Constipation

Morphine reduces the activity of the entire gastrointestinal tract in that it reduces the secretion of hydrochloric acid, diminishes the motility of the stomach, and increases the tone of the upper part of the duodenum. These actions may delay

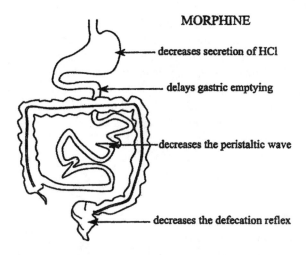

MORPHINE

- decreases secretion of HCl
- delays gastric emptying
- decreases the peristaltic wave
- decreases the defecation reflex

Figure 117   Morphine causes constipation.

passage of the stomach contents into the duodenum. Both pancreatic and biliary secretions are diminished, and this may also hinder digestion. In the large intestine, the propulsive peristaltic wave in the colon is reduced, the muscle tone including that of the anal sphincter is increased, and the **gastrocolic reflex** (defecation reflex) is reduced. These actions, in combination, cause constipation, which seems to be a chronic problem among the addicts (Figure 117).

Morphine-Induced Addiction

The narcotic alkaloids of opium head the list of addictive drugs, and this fact is a major consideration in their therapeutic use. Opium addiction has existed for centuries in the East, where the drug is habitually smoked by millions of people and where the cultural attitude toward its use is quite different than from that in the West. In the United States and Europe, the pure alkaloids are used, and the addiction takes a more serious form. The following discussion applies mainly to morphine and its congeners. Significant variations from the morphine pattern are exhibited by meperidine and methadone. The etiology of narcotic addiction varies, but addicts can seldom be relied upon for accurate information regarding the origin of their addiction. It is likely that only a small percentage becomes addicted as a result of the therapeutic use of opiates; however, many female addicts do acquire their addiction in this way. The vast majority of addicts are persons classified as neurotic or constitutional psychopathic inferiors, and addiction is only one manifestation of their fundamental personality defect. In this sense, they do not differ from chronic alcohol or barbiturate addicts. Since motivations related to hunger, sexual urges, and pain are reduced by morphine, use of the drug provides an escape mechanism from reality, a way of release from the failures and disappointments of everyday life, a means of bridging the gap between ambition and accomplishment. Some addicts start their abuse of narcotics out of curiosity or as a means of dissipation with evil associates. When a "thrill"or "kick" is desired, according to the language of the addict, the dose must constantly be increased as tolerance develops, in order to duplicate the original thrill, dreamy state, or euphoria. Many heroin and cocaine addicts are in this category. On the other hand, the few persons who develop addiction accidentally from the therapeutic use of a narcotic usually increase their dose as little as possible and continue to take the drug mainly to prevent unpleasant withdrawal symptoms. Aside from personality factors, addicts do not differ from the rest of the population with respect to intelligence or physical fitness or the incidence of psychoses.

There are on record a few authenticated cases of **congenital morphinism**, that is, morphinism in the newborn of addicted mothers; convulsions are a prominent feature. Unless the syndrome is promptly recognized and properly treated, the infant may succumb.

Types of Drug Used

The opium narcotics used in the United States are, particularly, morphine and heroin, especially the former; but juveniles definitely prefer heroin. Heroin has greater addiction liability than other narcotics, produces more euphoria and "stimulation," requires smaller doses, and is easier to traffic in illicitly. Codeine addiction is rare because the drug produces relatively little euphoria, and is thus less desirable to the potential addict, and because it is quite expensive and bulky in effective amounts. Codeine addicts are usually persons who originally received the drug for clinical purposes. Many narcotic addicts also use cocaine, usually in combination with heroin, and resort to whisky or barbiturate when their supply of narcotic is low. Every conceivable method and route are used to get the narcotic into the body. Morphine and heroin, however, are ordinarily taken hypodermically or intravenously; cocaine and also heroin are employed by snuffing. The use of galenical preparations of opium is rare in the United States. Occasionally, however, a **paregoric addict** is discovered.

Time Required for Addiction

Addiction occurs after a variable period of time. Although it usually requires more than 2 weeks of repeated use of a narcotic, addiction may develop in a few days and in some unstable personalities after only a few doses. By the use of nalorphine, physical dependence has been shown to develop within 2 days. Great care should therefore be exercised not to prescribe a narcotic for a longer period of time than is absolutely necessary, and to avoid clocklike regularity in its administration. The continued employment of opiate to relieve pain of terminal illnesses or neoplastic diseases does not come under the scope of the present discussion, and addiction under such circumstances is often unavoidable.

Amount of Drug Taken

The amount of narcotic taken by the addict may be remarkably large. De Quincey in his *Autobiography and Confessions of an Opium Eater* remarked that at one time he took 133 dr (over 500 ml) of tincture of opium daily. This is the equivalent of 4.7 g of morphine, or 300 times the single therapeutic dose. A case has been verified of a patient who employed 4.5 g of heroin daily by subcutaneous injection.

More often, the total daily dose of heroin, cocaine, or morphine used by addicts ranges between 0.5 and 2 g. The daily dose of paregoric may reach a quart. Addicts cannot be trusted to give an honest report on the amount used because they feel that by exaggerating their needs they will more easily succeed in obtaining what they desire. In estimating the daily requirements of an addict it is necessary to establish the amount which keeps the person comfortable and prevents withdrawal symptoms.

Symptoms and Effects of Morphine Addiction

The popular conception of the morphine addict, as a cunning, cringing, malicious, and degenerate criminal who is shabbily dressed, physically ill, and devoid of the social amenities could not be farther from the truth. When addicts are properly dressed and fed and receive their daily drug requirement through legitimate medical channels, they cannot easily be distinguished from other persons. They may remain in good health and suffer little inconvenience or physical deficiency unless they cease to take the drug. However, certain serious consequences do occur even under favorable circumstances. To cite but two examples, decrease in libido may lead to marital difficulties, and masking of pain may result in neglect of a pathological process such as malignant neoplasm. The euphoria experienced at first soon disappears unless the dose is continuously elevated. Some addicts take morphine for the euphoria, thrill, or gratification it affords. Such an effect is particularly observed when the drug is taken intravenously; the resulting thrill has the attributes of a sexual orgastic experience, localized in the abdomen rather than the genitalia. Yet an appreciable number continue their addiction mainly to escape from the distress that quickly follows withdrawal of the narcotic. Brilliant and famous persons have been narcotic addicts and their affliction often has remained unknown even to their closest associates. The only adequate test of morphine addiction is the occurrence of characteristic withdrawal symptoms upon cessation of the narcotic or upon administration of nalorphine.

Good health and productive work are thus not incompatible with addiction to morphine. As an example may be cited the case of a physician who was a morphine addict for 62 years and who exhibited no evidence of mental or physical deterioration due to the drug, when carefully studied at the age of 84.

Ill-health, crime, degeneracy, and a low standard of living are the result not of the pharmacological effects of morphine but of the sacrifice of money, social position, food, and self-respect in order to obtain the drug. Inasmuch as narcotics are ordinarily obtained through illicit channels, the cost is high. It is difficult for a normal person to appreciate how completely the necessity and compulsion to maintain an adequate supply of narcotic dominate the entire thought, action, and daily life of the addict. The major purpose of existence is to obtain sufficient narcotic for the daily needs.

Withdrawal Symptoms

The character and the severity of the untreated abstinence syndrome depend on many factors, including the particular narcotic and total daily dose used, the interval between doses, the duration of addiction, and the health and personality of the addict. The major factor, however, is the intensity of physical dependence on the drug. If it is absolutely clear to addicts that regardless of their complaints, pleas, and demands they will not receive an opiate, the psychogenic aspect of the withdrawal syndrome is usually less pronounced. For example, those forced against their will to undergo therapy are difficult to manage. On the other hand, many confirmed addicts "take the cure" to be able to start over again with a dose that is small and, therefore, cheap, and which again produces the euphoria that tolerance prevented them from experiencing. Such patients will sometimes plead for a small "pick-up" dose of morphine about the third or fourth day after withdrawal is started. If euphoria is once again experienced from this small dose, they know that their tolerance is broken and that their original susceptibility to morphine is restored. They then continue quietly and cooperatively with the remainder of the treatment, their objective having been attained.

Although psychic factors are undoubtedly important in determining certain features of the withdrawal syndrome, most of the signs and symptoms have a physiological basis, and represent a fundamental imbalance in the homeostatic adaptive mechanisms of the body that developed in response to the continued use of morphine. At the height of the syndrome, tolerance is still present and injection of the dose to which the patient was accustomed will quickly relieve all subjective distress and physical signs and completely restore the equanimity of the patient. By the time withdrawal symptoms have ceased, tolerance has disappeared.

The first changes are noted shortly before the time of the next regularly scheduled dose. Patients volubly express thier craving and need for the drug, state that they feel sick, and are apprehensive and irritable. If conditions permit, they will do various things to obtain the drug. This type of activity has been termed purposive in that it is directed toward the goal of obtaining the narcotics. Purposive behavior reaches its peak intensity in 36 to 72 hours after the last dose of morphine, and then subsides at a variable rate; but craving may persist for months or years. About 8 hours after withdrawal, addicts fall into a deep sleep, known as a "yen," which may last for several hours, but from which they awaken more restless and miserable than before. Nonpurposive abstinence changes are not goal-directed, appear about 15 to 20 hours after the last dose, reach their peak intensity in 48 to 72 hours after abrupt withdrawal, and are usually entirely gone in 10 to 14 days. This category of behavioral alteration has an organic basis. Prominent features are lacrimation, yawning, sneezing, sweating, anorexia, tremor, restlessness, irritability, muscular weakness, mydriasis, fever, and increased respiratory rate. There then follow increasing irritability, insomnia, marked anorexia, violent yawning, severe sneezing, excessive lacrimation and coryza, and hoarseness. Weakness and depression are pronounced. Vomiting may occur, and colic and diarrhea are usually prominent. Heart rate and blood pressure are elevated. Marked chilliness alternating with flushing and excessive sweating is characteristic. The chilliness is associated with pilomotor activity and the skin resembles that of a plucked turkey. This feature is the basis of the expression "cold turkey" to signify abrupt morphine withdrawal without substitution therapy. Patients lie on their side in a flexed position and covers themselves with blankets even on hot days.

The failure to take fluid and food, coupled with vomiting, sweating, and diarrhea, results in weight loss, dehydration, ketosis, and disturbance in acid–base balance, usually acidosis. Abdominal cramps and pains in the back and extremities are also quite characteristic; pain is particularly severe in the popliteal region. There may be rigidity in the right lower quadrant; this, combined with abdominal pain and a commonly occurring leucocytosis, may simulate acute appendicitis. Severe muscle tremors and twitchings, headache, diplopia, delirium, occasionally mania and seminal emissions in men and orgasms in women complete the dramatic picture. Sometimes there is cardiovascular collapse; this complication requires the administration of morphine to prevent death.

As symptoms subside, patients begin to eat, drink, and sleep and to regain their composure and weight. Physical recovery from addiction is an irregular process requiring approximately 6 months; of the many physical and laboratory features studied, weight gain proved to be the most important single index of recovery.

The former opinion that the abstinence syndrome is largely psychogenic in origin and merely the manifestation of anxiety and terror is no longer tenable. The typical nonpurposive signs and symptoms of withdrawal occur in addicts who are kept under heavy sedation as well as in addicts who, for other purposes, have undergone frontal lobotomy prior to narcotic withdrawal (see Goodman and Gilman, 1955).

Antidiarrheal Effects

Opiate preparations, usually given as **paregoric**, are effective and fast acting antidiarrheal agents. These agents are also useful postoperatively to produce solid stool following an **ileostomy** or **colostomy**. A meperidine derivative, **diphenoxylate**, is usually dispensed with atropine and sold as **Lomotil**. The atropine is added to discourage the abuse of diphenoxylate by narcotic addicts who are tolerant to massive doses of narcotic but not to the CNS stimulant effects of atropine.

Oliguric Effect

Morphine causes oliguria, which results from (1) pronounced diaphoresis, (2) the relative hypotension and decreased glomerular filtration rate, and (3) the release of antidiuretic hormone from the neurohypophysis. In an elderly patient with prostatic hypertrophy, morphine may cause acute urinary retention. Morphine may reduce the effectiveness of a diuretic when both drugs are used in combination in the treatment of congestive heart failure.

Morphine-Induced Tolerance

Tolerance develops to the narcotic and analgesic actions of morphine, so that increasingly larger doses are needed to render patients pain free. Tolerance develops to many effects of morphine such as analgesia, euphoria, narcosis, respiratory depression, hypotension, and antidiuresis. Morphine-induced bradycardia may be experienced. However, no tolerance develops to morphine-induced miosis or constipation. If the administration of morphine is discontinued, the tolerance is lost and the preaddiction analgesic doses of morphine become effective once more.

NATURAL OPIUM ALKALOIDS

Codeine

Codeine given in moderate quantities resembles morphine in its action in humans but is much weaker. Thus, 60 mg of codeine induces sleep and relieves pain in about the same degree as 10 mg of morphine, but the sleep is not so deep and restful and the subject awakens unrefreshed. Larger quantities may increase the restlessness and exaggerate the reflex excitability. The respiration is slowed but to a lesser degree than by morphine. The pupil is slightly contracted, but dilates if the excitement stage follows its administration. Codeine thus depresses the CNS in humans, although there are indications of stimulation also when larger quantities are used. In animals, these symptoms of excitation are more obvious, however, especially in the spinal cord, in which the reflexes are rendered more acute and may finally give rise to spasms.

Codeine acts less on the stomach and bowel than morphine but when given in large doses also causes constipation. It is excreted unchanged in the urine. Tolerance for codeine is difficult to develop but cases of addiction to it have been reported. They are usually secondary to the use of morphine, that is, the patient, addicted to morphine and unable to secure this drug, resorts to codeine.

Codeine, when incubated with the livers of several mammalian species, is converted to morphine and formaldehyde. The biotransformation products identified in the urine following the administration of codeine include morphine, norcodeine, and free and bound codeine.

Dihydrocodeine is somewhat less analgesic, whereas dihydroisocodeine is approximately as analgesic as morphine in comparable doses. Dihydrocodeine is essentially free of such side effects as respiratory depression, nausea, and mood changes induced by codeine and **dihydroisocodeine**.

Papaverine

Papaverine, as already indicated, is unrelated chemically to morphine and exerts no analgesic or narcotic action. It is essentially free of any effect on the CNS, its principal actions being to depress conduction and prolong the refractory period of the heart and to relax smooth muscle. It also exerts a mild local anesthetic action counteracting, for example, the pruritus induced by the application of **cowhage** to the skin. Itching apparently is not transmitted by the same nerve fibers as are sensations of pain, since analgesics fail to alleviate the itching induced by cowhage.

Papaverine has a greater tendency to slow the heart rate than morphine, acting directly on the heart muscle to produce this effect. In addition, it depresses the smooth muscles of arterioles so that when it is perfused through them, the blood pressure is somewhat lowered. However, ordinary amounts administered systemically cause little if any change. In the dog, papaverine significantly raises the fibrillation threshold of the ventricle when administered in therapeutic doses. Large doses lower the threshold for fibrillation especially if enough is given to cause a sharp drop in blood pressure. However, there is no conclusive proof that papaverine will bring about spontaneous recovery in a fibrillating heart. Its action upon the intestinal tract of the intact animal is without therapeutic importance. Upon excised tissues (ureter, gall bladder, etc.) papaverine and the other **benzyl isoquinoline alkaloids** relax the tone, slow the contractions, and thus antagonize the action of morphine. Papaverine undergoes complete destruction in the tissues. Tolerance and addiction to papaverine are unknown.

Papaverine, because of its general depressant effect on smooth muscle, has been used in doses of 30 to 60 mg, subcutaneously and intravenously, in peripheral thrombosis and embolism, acute myocardial infarction, angina pectoris, bronchial asthma, renal and biliary colic, and other conditions in which relaxation of smooth muscle is desired. However, the therapeutic effectiveness of papaverine is questionable and there is not established indication for its use.

The **tetraethyl homologue of papaverine (Perparin®, Perperine®, Ethaverine®, Diquinol®)** is more active than papaverine and is used for its spasmolytic effects on the uterus, ileum, and circulatory system.

Noscapine

Noscapine (Nectadon®), 2-methyl-8-methoxy-6,7-methylenedioxy-*l*-(6,7-dimethoxy-3-phthalidyl)-1,2,3,4-tetrahydroisoquinoline, is also one of the isoquinoline derivatives present in opium. Next to morphine, it is the most abundant of the opium alkaloids, being present to the extent of 6% in the seed capsules. This drug was formerly known as **narcotine** but, since it is neither chemically nor pharmacologically related to the narcotics, is more appropriately designated as **noscapine**.

Noscapine resembles papaverine in its action on smooth muscle and in large doses induces bronchodilation in animals. It is readily absorbed from the gastrointestinal tract but its metabolic fate is unknown. In experimentally induced cough in animals and humans, noscapine exerts an antitussive action approximately equal to that of codeine without any of the unpleasant side effects of codeine. It is administered orally in doses of 15 to 30 mg, three or four times daily as an antitussive.

SYNTHETIC DERIVATIVES OF THE OPIUM ALKALOIDS

Many attempts have been made to modify the morphine molecule or break it down to some constituent that might retain its analgesic action without its tendency to develop tolerance and addiction. However, only very closely related derivatives of morphine have been found to maintain its analgesic action, and still retain to a considerable degree the undesirable potentiality to develop tolerance and addiction.

Heroin (Diamorphine)

Heroin, a synthetic alkaloid formed from morphine by substituting acetyl for its two hydroxyl groups, resembles morphine in its general effects, but acts more strongly on both cerebrum and medulla than does morphine, and is therefore more poisonous, with the usual dose about one fourth that of morphine. The action on the respiration is the same kind as that of morphine, and the advantages claimed for heroin by its advocates have not been confirmed by impartial investigation. It is effective, but not more so than is morphine, in the treatment of cough, but is less effective in the treatment of dyspnea. Heroin is excreted mainly in the urine unchanged, but some is found in the stools. In animals, large doses cause excitement and convulsions, and in humans these have been observed in cases of poisoning.

Many cases of heroin habit have occurred and this aspect of the action of the drug has brought it into serious disrepute. The fact that it can be easily administered and that its dose is small makes trafficking with it easier than with some of the other narcotics. The social danger from it, moreover, seems to be greater than with morphine in that it produces a change in the personality as shown by an utter disregard for the conventions and morals of civilization. Degenerative changes in the individual progress faster than with any of the other narcotic drugs, and all the higher faculties of the mind, such as judgment, self-control, and attention, are weakened and the addict rapidly becomes a mental and moral degenerate. The heroin habit is most difficult to cure, not only in the active withdrawal period, but also in the convalescent stage, and relapse is frequent. Habitués take it either by snuffing or by hypodermic injection. Because of its marked tendency to addiction and its minor therapeutic value, the manufacture and importation of heroin has been outlawed in the United States.

Dihydrocodeinone

In this synthetic derivative of codeine, a ketone group replaces the –OH of codeine at position 6 and two H atoms are added at positions 7 and 8. It thus bears the same relation to codeine as **dihydromorphinone (Dilaudid®)** does to morphine. It is marketed as the tartrate under the trade names **Dicodid®** and **Hycodan®** and is used chiefly for the relief of cough.

The risk of addiction is greater in the case of dihydrocodeinone than with codeine, but small doses (5 to 10 mg) are often of value in tuberculosis, measles, atypical pneumonia, etc. where cough is a troublesome symptom.

Dihydromorphinone

Dihydromorphinone hydrochloride (Dilaudid) is five to ten times as potent and more toxic than morphine, but its duration of action is less, as are also its effects on the gastrointestinal tract. Chemically, it differs from morphine in that the alcoholic hydroxyl group of that alkaloid is replaced by ketonic oxygen and the adjacent double bond is removed by hydrogenation. Dihydromorphinone produces analgesia and narcosis and acts upon the respiration in a manner similar to morphine, with the difference that it is effective in about one quarter the dosage. In the dog, restlessness and vomiting are followed by depression, somnolence, analgesia, and slowing in the respiration. In the rabbit, the respiration is slowed very markedly with some deepening of the respiration so that the effect of the slowing is partially compensated. In humans, the drug is powerfully analgesic and also markedly depressing to the respiration while nausea, vomiting, and constipation are not so marked as with morphine. Tolerance and addiction occur readily, and the same care should be exercised in prescribing it as is used in the case of the natural opium alkaloids.

Dihydromorphinone is used in the same manner as morphine for the relief of pain but in much smaller doses — usually 1 to 2 mg. For cough, a dose about half that size is used. Administered orally, dihydromorphinone is more effective than morphine and it may also be administered in a rectal suppository. Its principal indication is for acute pain of short duration.

Oxymorphone (Numorphan®), the *1*-14-hydroxyderivative of dihydro-morphinone hydrochloride, is also about ten times as active as morphine and has a rapid onset of action that lasts for about 6 hours. It is administered subcutaneously or intramuscularly in doses of 1 ml (1.5 mg.) or as a rectal suppository (2 and 5 mg).

Metopon (methyldihydromorphinone) manifests the properties of toler-ance and liability to addiction, but it is about twice as effective as an analgesic and its duration of action is equal to that of morphine. Tolerance and physical dependence to metopon develop more slowly than to morphine. In therapeutic doses, it produces little or no respiratory depression and less mental dullness. Metopon thus possesses many advantages over morphine, but is no longer available commercially.

Dextromethorphan Hydrobromide

The dextrorotatory isomer of **3-methoxy-N-methylmorphinan**, a synthetic mor-phine derivative (**Romilar hydrobromide®, Tusilan®**) is employed exclusively as an antitussive agent. Its levorotary isomer (in the form of the 3-hydroxy tartrate) is described next. These compounds differ from morphine chemically in lacking the –O– bridge between carbons 4 and 5, the OH group at C_6, and the double bond between carbons 7 and 8. Dextromethorphan hydrobromide exhibits little or no central depressant activity and does not produce analgesia, nor does it manifest any addicting effects.

Dextromethorphan hydrobromide is administered orally in doses of 10 to 20 mg one to four times daily to adults. This dose is reduced to one half for children over 4 years of age and to one quarter for children under 4 years of age.

Levorphanol Tartrate

The levorotatory isomer of **3-hydroxy-N-methylmorphinan tartrate dihydrate (Levo-dromoran tartrate,®)** is a more potent analgesic than morphine with an approximately equal margin of safety, but is longer acting and has a lesser constipating action. Contraindications to its use, including the danger of addiction, are similar to those of morphine.

Levorphanol tartrate is used for the relief of severe pain in cancer, trauma, colic, myocardial infarction, and pre- and postoperatively. It is administered orally or subcutaneously in doses of 2 to 3 mg according to the age and weight of the patient and the severity of the pain.

Ethylmorphine

Ethylmorphine hydrochloride (Dionine®) resembles codeine more closely in its action than it does morphine. It is used primarily as a chemotic to produce vasodilatation and edema of the conjunctiva in corneal ulcer and other inflammatory conditions of the eye. For this purpose it is instilled topically as a 1 to 5% solution in the eye. It has also been used as an antitussive in doses of 15 mg.

Apomorphine

When morphine is acted on by acids and other dehydrating agents, it loses a molecule of water, and a new alkaloid is formed, apomorphine ($C_{17}H_{17}NO_2$), the structural formula of which is shown in Figure 106. Through this change, the action of the original alkaloid is considerably modified; apomorphine preserves the stimulant but loses to a great degree the depressant action of morphine on the CNS. This stimulant action extends over the whole CNS but is most developed in the "vomiting center" of the medullar oblangata.

Symptoms. In humans, apomorphine in doses of 5 to 10 mg induces within 10 to 15 minutes nausea and vomiting. Very often the nausea passes off immediately after the evacuation of the stomach, but when larger quantities have been administered, repeated vomiting and retching may occur. Occasionally, depression and sleep follow the emesis after even small doses.

The attendant symptoms are profuse salivation, increased secretion of the mucous glands of the nose, throat, and bronchial passages, tears, and a cold perspiration. A feeling of depression and muscular weakness and acceleration of the pulse are also well-known symptoms accompanying nausea and vomiting and are present after apomorphine. These are all to be regarded as sequelae of the emetic action, however, and not as due to the direct action of the drug on the glands and other organs. In a few instances, the depression and weakness have passed into alarming collapse, but no actual fatality is recorded from the use of apomorphine.

Very small doses of apomorphine may induce the secondary symptoms without actual vomiting. Thus the saliva, perspiration, tears, and other secretions may be augmented by quantities which are too small to act as emetics.

Apomorphine induces vomiting through changes in the medulla oblongata and not by irritation of the stomach. This is shown by the fact that it acts much more quickly and in smaller doses when it is infected hypodermically or intramuscularly than when it is swallowed and also by the fact that if the medulla is brushed with apomorphine solution, vomiting follows immediately. The movements of vomiting may also be induced in animals after the removal of the stomach and intestines. Large doses may cause such depression of the CNS as to prevent vomiting.

Apomorphine is used chiefly as an emetic (dose, 5 mg) and presents several advantages over other drugs employed for this purpose, inasmuch as it acts more promptly and can be administered by injection while most of the other emetics (ipecac and emetine excepted) have to be given by mouth, which is a serious drawback in cases of poisoning. However, vomiting is not now as important a method of treatment as it was formerly, and the emetics are less frequently employed to evacuate the stomach than other measures, such as repeated washing of the stomach by means of the stomach tube. In narcotic poisoning, apomorphine not infrequently fails to act, owing to the depression of the vomiting center, and in corrosive poisoning, a certain amount of danger attends its use as the pressure on the walls of the stomach exerted by the contraction of the diaphragm and abdominal muscles may lead to the rupture of the weakened walls of the organ.

Emetics, when administered in small doses, act as expectorants and are used in inflammatory conditions of the respiratory tract to increase the bronchial secretion and render it less tenacious. The most commonly used expectorants are ipecac, ammonium chloride, and apomorphine. The last-named is administered in doses of 1 mg in the form of an elixir or syrup. Apomorphine injected in subemetic doses of 1 to 2 mg is also used as a sedative in the delirium following anesthesia, in acute alcoholic psychosis, and prior to anesthesia in patients manifesting severe agitation.

NONOPIATE ADDICTING ANALGESICS

The first success in obtaining synthetic substances approaching morphine or its derivatives was **meperidine** (Demerol®) or **isonipecaine** as a substitute for morphine. Following World War II, examination of the files of the German dye trust revealed several related products, of which **methadone** has proved to be a valuable analgesic. These compounds are more potent per unit of weight in their analgesic action than morphine and have encouraged the synthesis of other related compounds in the search for an ideal analgesic.

The synthetic substitutes for the opium alkaloids, or opioids as they have been designated, include the **morphinans, meperidines, methadones, hexamethyleneimines**, and **dithienylbutenyl-amines**. The two last-named groups bear only a slight chemical resemblance to the morphine derivatives but possess the branched methyl and tertiary amino group characteristic of methadone. Some of these are as potent as morphine as analgesics and all of them possess addictive properties. In addition to the drugs to be described, related synthetic agents have been introduced into medicine as analgesics and antitussives. These include **phenadoxone hydrochloride**, 6-(*N*-morpholino)-4,4-diphenyl-3-heptanone hydrochloride (**Heptalgin®**, **Heptazone®**, **Morphodone®**), and **dipipanone**, the hydrochloride of *dl*-6-piperidine-4,-diphenyl-heptan-3-one.

Meperidine (Pethidine)

Meperidine (isonipecaine, Demerol®, Dolantin®) is the methyl ester of **methyl-4-phenylpiperidine-4 carboxylic acid**. It is used in the form of its soluble hydrochloride. The piperidine nucleus is also present in many alkaloids, such as **atropine** or **lobeline**, which are not analgesics, as well as in morphine.

Action

Meperidine shows certain relationships to both morphine and atropine in its action. Like morphine, it exerts a generalized depression of the CNS but in ordinary doses does not affect the cough reflex. Like morphine, it induces euphoria and its continued use leads to tolerance and addiction. However, its euphoric and sedative effects are less than that of morphine and addiction to meperidine is more serious than that which follows the use of morphine. Meperidine is an effective analgesic, although less so than morphine. Following its parenteral administration, relief of pain is experienced in about 15 minutes and lasts for several hours. Like atropine, meperidine exerts a spasmolytic action on the gastrointestinal tract except on the duodenum and jejunum. Like morphine, it increases the intrabiliary pressure.

In ordinary doses, meperidine exerts inappreciable effects on the respiration and circulation. In large doses, it interferes with the facilitatory function of the pneumotaxic center and the vagal afferent impulses, thereby reducing the rate and depth of respiration. With still higher dose, meperidine produces an irregular respiratory rhythm and ultimately induces apnea. It causes relaxation of the ureter, gall bladder, and bronchi. This action is partly a direct one on the smooth muscle of these structures and in part a result of its anticholinergic activity. It does not affect the size of the pupil or alter the tonus of the uterus.

Undesirable Side Effects

Dizziness, sweating, nausea and vomiting, headache, and occasionally syncope are sometimes noted following the administration of therapeutic doses of meperidine. Excessive or repeated doses of the drug may induce more serious toxic effects marked by cerebral stimulation similar to that seen in atropine poisoning, with dilatation of the pupils, tachycardia, disorientation, muscle twitchings, convulsions, and depression of the respiration. Maximum tolerance to the analgesic properties of the drug develops in about 8 weeks. Because of its tendency to addiction, the sale of the drug is subject to the Federal Narcotic Regulations.

Metabolism

Meperidine is in part demethylated in the body appearing as **normeperidine** in the urine. It is, however, for the most part hydrolyzed and excreted as free and conjugated **meperidinic acid**. The normeperidine is also partially hydrolyzed to normeperidinic acid, which is partly conjugated. Only a small part of the drug is excreted in an unchanged form. In humans, meperidine is metabolized at the rate of about 17% per hour, so that a single dose exerts its analgesic effect for 3 to 4 hours. In the dog, on the other hand, its rapid metabolism (70 to 90% per hour) makes it difficult to induce the tolerance or addiction.

Therapeutic Use

Meperidine is available in the form of tablets (50 mg) for oral use or in solution (50 mg/ml) for intramuscular administration. It is administered in doses of 50 to 100 mg at 3- to 4-hour intervals. Because of its antispasmodic effects, it is particularly useful in pain due to colic. It is also used prior to anesthesia, particularly cyclopropane. Meperidine is used in asthma, but its tendency to addiction limits its use in this condition to acute attacks.

Meperidine has replaced morphine to a large extent in medical practice because of the physician's reluctance to use an opiate and the belief that meperidine manifests less undesirable side effects than does morphine. However, both of these assumptions are ill-founded. Addiction to meperidine is much less amenable to treatment than is addiction to morphine. Meperidine, like morphine and codeine, causes spasm of the upper gastrointestinal tract and typical attacks of biliary colic in biliary tract disease. Meperidine, in doses giving an equal analgesic effect, induces as much respiratory depression as does morphine. Like the latter, it also crosses the placental barrier and must therefore be used cautiously in the latter stages of labor.

Methadone

Of the synthetic compounds related to meperidine, **methadone** (**Adanon**®, **Amidone**®, Dolophine®) is most active and has proved to be an effective substitute for morphine. Chemically, it is **6-dimethylamino-4,4-diphenyl-3-heptanone** and is used as the hydrochloride. Methadone is related to meperidine, morphine, and atropine, and contains an asymmetric carbon atom. It was the first synthetic opium substitute not containing a piperidine ring. The levorotatory optical isomer has twice the analgesic power of *d,l*-methadone. **Isomethadone** (**6-dimethyl-amino-5-methyl-4,4-diphenyl hexanone-3**) resembles methadone in action, but is less active.

Action

Methadone, in general, exerts a pharmacological action similar to that of morphine. Its action on smooth muscle resembles that of meperidine. *d,l*-Methadone, per unit of weight, is equivalent to morphine in analgesic potency, but *l*-methodone is twice as effective. When used in comparable analgesic doses, methadone is as toxic as morphine and exerts the same effects on the pulse rate and respiration as does morphine. Methadone exerts no sedative effect and hence is of no value as a preanesthetic agent.

Like morphine and meperidine, methadone has a definite addiction liability but withdrawal symptoms after prolonged administration come on more slowly and are of less intensity than after morphine. Dogs receiving gradually increasing doses of methadone develop tolerance to its analgesic and narcotic actions but no tolerance to the cardiac slowing and effects on intestinal motility. The para-sympathomimetic effects of methadone on the heart and intestine are of central origin and resemble those of morphine.

Side reactions to methadone include nausea, diaphoresis, and dizziness, but these are uncommon when minimal doses (2.5 to 5 mg) are used. Such doses have no effect on the temperature, pulse, blood pressure, or respiration. Miosis is produced with doses of 10 mg or more but is less than with morphine.

Mechanism of Action

Although resembling morphine in its analgesic properties, the mechanism of its action as inferred from *in vitro* studies on brain metabolism appears to be different. Small concentrations of methadone cause an initial period of stimulation of cellular metabolism, which is not observed with morphine. Larger doses (0.005 M) reduce the oxygen uptake, a phenomenon not observed with comparable doses of morphine.

Therapeutic Uses

Methadone is used as an analgesic for the relief of moderate to severe pain. It is effective in conditions in which the analgesic effect of morphine is desirable and it exerts a satisfactory antitussive action.

In adults, a dose of 2.5 to 10 mg is satisfactory for the relief of most forms of pain. This is administered orally every 3 to 4 hours. For suppression of nonproductive cough, 1.25 to 2.5 mg orally every 3 hours is usually satisfactory. Where oral administration is undesirable, the drug may be administered subcutaneously, but the oral route is almost as effective as its hypodermic injection. It may also be administered intramuscularly, but should not be injected intravenously.

Dextromoramide tartrate (**Palfium**®), which is related chemically to methadone, is as effective when administered orally in doses of 5 to 10 mg as when injected.

Alphaprodine Hydrochloride

Alphaprodine, **1,3-dimethyl-4-phenyl-4-piperidyl propionate hydrochloride** (**Nisentil**®, **Prisilidene**®), resembles meperidine chemically. Its analgesic and depressant actions are less intense, but more prompt and of shorter duration, than those of morphine. It is a potent, short-acting narcotic used in intense pain of brief duration as in obstetrics, urologic examinations, and minor surgical procedures. Although producing little or no cumulative effect, tolerance and the liability of addiction can develop.

Alphaprodine hydrochloride is administered by subcutaneous injection in doses of 40 to 60 mg depending upon the weight of the patient. Analgesia occurs within 5 minutes and lasts about 2 hours. Nalorphine is an effective antidote to its overdosage.

Anileridine

Anileridine, ethyl 1-(4-amino phenethyl)-4 phenylisonipecotate is available in the form of its dihydrochloride for oral administration and as the phosphate for injection (**Leritine**®). Its analgesic potency is intermediate between that of meperidine and morphine. Like meperidine it exerts mild antihistaminic and spasmolytic effects but it is devoid of the constipating effect of the opiates. Sedative and direct hypnotic effects are similar to those of meperidine but less than with morphine.

Anileridine hydrochloride is administered orally in the form of 25 mg tablets repeated every 6 hours if necessary, but dosages up to 50 mg or more frequent doses may be desirable for more severe pain. Anileridine phosphate is administered by subcutaneous, intramuscular, or intravenous injection in doses of 25 to 50 mg (1 to 2 ml) at intervals of 4 to 6 hours.

Phenazocine

Phenazocine (**Prinadol**®), **1,2,3,4,5,6-hexahydro-8-hydroxy-6,11-dimethyl-3-phenethyl-2,6-methano-3-benzazocine hydrobromide**, was synthesized at the National Institutes of Health in a search for new analgesics with low addictive liability. Addiction to phenazocine develops more slowly and is less intense when dosages equivalent to morphine are administered chronically for 30 days. The incidence of side effects (respiratory depression, hypotension, nausea and vomiting) are less with phenazocine than with morphine.

Phenazocine is administered intramuscularly in doses of 1 to 3 mg every 4 to 6 hours. Its depressant effects are potentiated by CNS depressants as the barbiturates, phenothiazines, and anesthetics so that smaller doses are indicated when these are used concomitantly with the phenazocine.

Piminodine Ethanesulfonate

Piminodine (**Alvodine**®), **ethyl-4-phenyl-*l*-(3-phenylaminopropyl)piperidine-4-carboxylate ethanesulfonate**, is reputed to be equal to morphine in analgesic potency, but relatively free of the latter's narcotizing action. It is administered in doses of 25 to 50 mg orally every 4 to 6 hours, or in doses of 10 to 20 mg subcutaneously or intramuscularly every 4 hours as needed. Piminodine is habit forming, requires a narcotic prescription, and should be used with the same precautions as other members of this group. **Levallorphan tartrate**, **nalorphine**, **naloxone**, or **naltrexone** are antidotes to it.

Dextropropoxyphene

Propoxyphene is structurally very similar to methadone and possesses four stereoisomers. **Dextropropoxyphene** is an analgesic with a potency two thirds that of codeine. **Levopropoxyphene** is an antitussive, but lacks analgesic properties.

Adverse reactions to dextropropoxyphene include nausea, vomiting, sedation, dizziness, constipation, and skin rash, with a frequency of incidence somewhat less than that seen with codeine use. Although respiratory depression is a cardinal sign of acute dextropropoxyphene poisoning, the drug apparently does not affect respiration in the usual therapeutic doses of 32 to 65 mg.

OPIOIDS WITH MIXED ACTIONS: AGONISTS–ANTAGONISTS AND PARTIAL AGONISTS

The analgesia produced by 30 mg of **pentazocine** (**Talwin**), a mixed-narcotic agonist and a weak antagonist, is comparable to that elicited by 10 mg of morphine. Its onset and duration of action are shorter than those of morphine. Pentazocine will antagonize some of the respiratory depression and analgesia produced by morphine and meperidine. However, the analgesic action and respiratory depression produced by pentazocine can be reversed by a narcotic antagonist. Pentazocine causes tolerance and addiction, but their emergences are very slow compared with those induced by morphine.

Nalbuphine is structurally related to both naloxone and oxymorphone. It is an agonist–antagonist opioid possessing a spectrum of effects that resemble those of pentazocine; however, nalbuphine is a more potent antagonist at µ receptors and is less likely to produce dysphoria than is pentazocine.

Butorphanol is a **morphinan congener** with a profile of actions similar to that of pentazocine.

Buprenorphine is a semisynthetic, highly lipophilic opioid derived from thebaine. It is 25 to 50 times more potent than morphine.

OPIOID RECEPTOR ANTAGONISTS

Naloxone (Narcan) and **naltrexone hydrochloride (Trexan)** reverse the respiratory depressant action of the narcotics related to morphine, meperidine, and methadone. They differ from the other narcotic analgesics in several respects. Naloxone does not cause respiratory depression, pupillary constriction, sedation, or analgesia. However, it does antagonize the actions of pentazocine. Naloxone neither antagonizes the respiratory depressant effects of barbiturates and other hypnotics nor aggravates their depressant effects on respiration. Similar to nalorphine, naloxone precipitates an abstinence syndrome when administered to patients addicted to opiate-like drugs.

ANTITUSSIVE PREPARATIONS

Dextromethorphan (Romilar) is the dextroisomer of the methyl ether of **levorphanol**. Unlike its levorotatory congener, it possesses no significant analgesic property, exerts no depressant effects on respiration, and lacks addiction liability. It is an antitussive agent with a potency approximately one half that of codeine.

Therapeutic doses of dextromethorphan (15 to 30 mg) produce little or no side effects, whereas excessively high doses (300 to 1500 mg) have been reported to produce a state resembling intoxication accompanied by euphoria.

42

OATS (*AVENA SATIVA* L.) AND THEIR ANTIOXIDANT ACTIVITY

- Oat lowers cholesterol.
- Oat combats heart attack.
- Oat controls diabetes mellitus.

Several classes of compounds with antioxidant activity have been identified in oat (*Avena sativa* L.) including **vitamin E tocols, flavonoids,** and **nonflavonoid phenolic acids** (see Handelman et al., 1999).

Grains are the seeds of food grasses and other plants and include barley, corn, millet, oats, rice, rye, the hybrid triticale, and wheat. Buckwheat, usually grouped with grains, is actually a botanical relative of rhubarb. Grains were the first cultivated food. Once humans learned to plant crops they could give up the nomadic way of life and settle down to farming. Usually, the locally grown grain became the staple food of the area, rice in the Far East, oats in Scotland, wheat in the United States. Grains can be eaten whole, such as brown rice, or processed into cereals or flour for a multitude of other food products.

Grains are:

- An excellent source of starchy carbohydrate and dietary fiber
- A good source of **niacin, riboflavin,** other **B vitamins, iron,** and **calcium**

Oats are used in breakfast cereals and baked goods. Oat bran is high in soluble fiber, which can help lower blood cholesterol levels, thus possibly reducing the risk of **heart attacks**. It also helps the body utilize insulin more efficiently, an important asset in controlling **diabetes**. Products made with oats contain little or no **gluten** and can usually be tolerated by people with **celiac disease**.

Whole oats, called **groats**, consist of the bran, germ, and endosperm of the whole grain with the outer hull removed. Whole oats are softer than wheat grains and can be crushed at home. They are available in specialty food stores.

Rolled oats. The most common form of oats, rolled oats are so-called because the whole groats are steamed and then flattened between rollers before being made into flakes. Rolled oats are popularly called **oatmeal**. They are available in three thicknesses, and the thicker the flake, the longer the oats will take to cook. **Scotch** or **steel-cut oats** are the thickest, and are sometimes called **Scotch** or **Irish oatmeal**. **Table-cut** or **regular old-fashioned oats**, sold in specialty stores, are nuttier and chewier than quick-cooking oats but are otherwise the same. **Quick-cooking or instant oats** are the thinnest flakes. Some think they have a raw taste compared with other oats.

Handelman et al. (1999) treated milled oat groat pearlings, trichomes, flour, and bran with methanol and tested the fractions for antioxidant capacity against low-density lipoprotein (LDL) oxidation and **R-phycoerythrin protein oxidation** in the oxygen radical absorbance capacity (ORAC) assay. The oxidative reactions were generated by **2,2′-azobis(2-amidinopropane) HCl** (AAPH) or Cu^{2+} in the LDL assay and by AAPH or Cu^{2+} + H_2O_2 in the ORAC assay and calibrated against a Trolox standard to calculate **Trolox equivalents** (1 Trolox equivalent = 1 TE = activity of 1 mol of Trolox). The antioxidant capacity of the oat fractions was generally consistent with a potency rank of pearlings (2.89 to 8.58 TE/g) > flour (1.00 to 3.54 TE/g) > trichome (1.74 TE/g) = bran (1.02 to 1.62 TE/g) in both LDL and ORAC assays regardless of the free radical generator employed. A portion of the oat antioxidant constituents may be heat labile as the greatest activity was found among non-steam-treated pearlings. The contribution of oat toculs from the fractions accounted for <5% of the measured antioxidant capacity. AAPH-initiated oxidation of LDL was inhibited by the oat fractions in a dose-dependent manner, although complete suppression was not achieved with the highest doses tested. In contrast, Cu^{2+}-initiated oxidation of LDL stimulated peroxide formation with low oat concentrations but completely inhibited oxidation with higher doses. Thus, oats possess antioxidant capacity most of which is likely derived from polar **phenolic compounds** in the **aleurone**.

In conclusion, phenolic-rich fractions of oats possess an antioxidant capacity that can be assessed quantitatively through their ability to inhibit LDL oxidation and protein oxidation. The greatest degree of antioxidant capacity was associated with compounds extracted with methanol from the aleurone. The identification of the oat constituents from these fractions should be investigated, although candidate compounds would include **caffeic acid**, **ferulic acid**, and **avenanthramides**. These compounds may be bioavailable and contribute to the health effects associated with dietary antioxidants and oats.

43

OLIVES
AND OLIVE OIL

Olives possess potent antioxidants causing a lower incidence of coronary heart disease and cancer in the Mediterranean area.

The impact of the environment on human physiology and pathology has been recognized for decades. In the Mediterranean basin several components of the diet are derived from plants and fruits and, among them, **olives** and **grapes** have been, at least originally, rather unique to this area.

Because of the particular climate, plants in the Mediterranean basin are subjected to prolonged exposure to sunlight. Thus, many fruits, including olives and grapes, have developed an array of protective compounds that limit the effects of ultraviolet light and the subsequent oxidative damage. Accordingly, the levels, dependent upon the rates of synthesis, of **anthocyanins**, **flavonoids**, and **phenols** in dark-colored fruits are enhanced by sunlight. Wine and olive oil, which originate from the Mediterranean area, are obtained by pressure techniques from ripe fruits, the compounds, responsible for the **organoleptic** and other properties of the fruits, are thus transferred to the fluids (Galli and Visioli, 1999).

The observation of a lower incidence of coronary heart disease (CHD) and of certain types of cancers in the Mediterranean area led to the hypothesis that a diet rich in grain, legumes, fresh fruits and vegetables, wine in moderate amounts, and olive oil had beneficial effects on human health. To date, this effect has been mainly attributed to the low saturated fat intake of the Mediterranean diet and to its high monounsaturates proportion, which indeed may favorably affect the plasma **lipid/lipoprotein profiles**. Nevertheless, other components of the diet, such as fiber, vitamins, flavonoids, and phenols, may play an important role in disease prevention, acting on different cardiovascular variables.

Olive oil, in particular, is the principal source of fat of the Mediterranean diet. The production of olive oil for 1997–1998 in the European Community is estimated to be 1,941,000 tons, which represents 85% of the world production. Because of to the increasing popularity of the Mediterranean diet, its consumption is expanding to nonproducer countries such as the United States, Canada, and Japan. Abundance of **oleic acid**, a monounsaturated fatty acid, is the feature that sets olive oil apart from other vegetable oils. In particular, oleic acid (18:1n-9) ranges from 56 to 84% of total fatty acids, while **linoleic acid** (18:2n-6), the major essential fatty acid and the most abundant polyunsaturate in the U.S. diet, is present in concentrations between 3 and 21%. Depending on its chemical and **organoleptic properties**, olive oil is classified into different grades that also serve as guidelines for the consumer's preference. In many producing countries, **extra virgin olive oil**, i.e., the one with the highest quality, accounts for just 10% of the whole production. In addition to triglycerols and free fatty acids, olive oil contains a variety of nonsaponifiable compounds that add to 1 to 2% of the oil and are important for its stability and unique flavor and taste. In contrast, other edible seed oils lose most of their minor compounds during the refining stages.

44

ONION AND GARLIC

Onion (*Allium cepa*) and garlic (*Allium sativum*) exhibit:

- Antimicrobial activities
- Antifungal activities
- Antiviral activities
- Antithrombotic actions
- Antineoplastic effects
- Hypolipidemic properties
- Antiarthritic actions
- Hypoglycemic functions

INTRODUCTION

Garlic (*Allium sativum*) and **onion (*A. cepa*)** are among the oldest of all cultivated plants with their origin in central Asia. Garlic has been used as a spice, food, and folklore medicine for over 4000 years, and is the most widely researched medicinal plant (see Ali et al., 2000).

Codex Ebers, an Egyptian **medical papyrus**, dating to about 1550 B.C., gives more than 800 therapeutic formulas of which 22 mention garlic as an effective remedy for a variety of ailments including heart problems, headache, bites, worms, and tumors.

Aristotle, **Hippocrates**, and **Aristophanes** recommended garlic for its medicinal effects. The Roman naturalist **Pliny the Elder** cited numerous therapeutic uses for both garlic and onions. **Discorides**, a chief physician to the Roman army in the first century A.D., prescribed garlic as a vermifuge or expeller of intestinal worms. During the first Olympic games in Greece, **athletes** ingested garlic as a stimulant. **Louis Pasteur** reported the antibacterial properties of garlic as early as 1858. In India, garlic has been used for centuries as an **antiseptic lotion** for washing wounds and ulcers. During the two world wars, garlic was used as an antiseptic for the **prevention of gangrene**. Many workers have researched its insecticidal, antimicrobial, antiprotozoal, and antitumor activities.

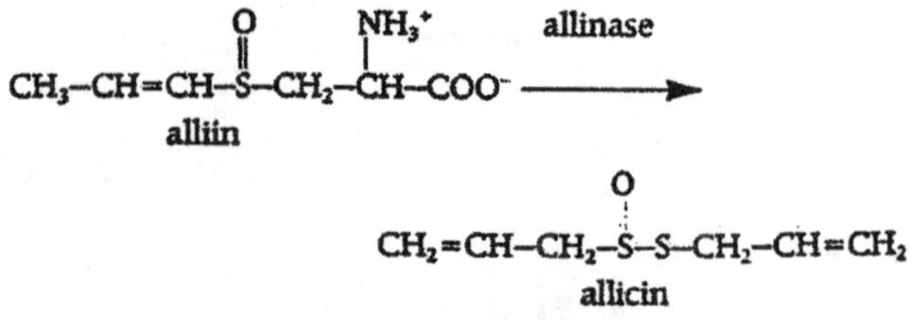

$$CH_3-CH=CH-\overset{\overset{\text{O}}{\|}}{S}-CH_2-\overset{\overset{NH_3^+}{|}}{CH}-COO^- \xrightarrow{\ \ \text{allinase}\ \ }$$

alliin

$$CH_2=CH-CH_2-\overset{\overset{\text{O}}{\cdot}}{S}-S-CH_2-CH=CH_2$$

allicin

Figure 118   The structure of allicin.

More recently, garlic has been reported to be effective in various ailments such as cardiovascular diseases because of its ability to lower serum **cholesterol**. A component of both garlic and onion, **S-methylcysteine sulfoxide** (SMCS), has been shown to reduce both blood cholesterol and the severity of atherosclerosis. Garlic has protective effects against **stroke, coronary thrombosis, atherosclerosis, platelet aggregation**, as well as **infections** and **vascular disorders**. Furthermore, fibrinolytic activity increases following administration of garlic thus preventing **atherosclerosis** and ischemic heart disease. Many claims of an **antibiotic action, a hypoglycemic effect, antitumor, antioxidant**, and **antithrombotic properties** have also been attributed to garlic (and onion) extracts.

ACTIVE INGREDIENTS IN ONION AND GARLIC

When garlic is cut and the parenchyma is destroyed, **alliin** is the major cysteine sulfoxide liberated. Alliin is acted upon by the enzyme **allinase (alliin lyase)** to produce **allicin**. **Allicin [S-(2-propenyl)2-propene-1-sulfinothioate** or **diallylthiosulfinate]** is an odoriferous compound and the main component of freshly crushed garlic homogenates (Figure 118).

Garlic also contains **S-propylcysteinesulfoxide** (PCSO) and **S-methylcysteine-sulfoxide** (MCSO). PCSO can generate over 50 compounds depending on temperature as well as water content. The action of allinase on the mixture of alliin, PCSO, and MCSO can produce a number of other molecules including: **allyl methane thiosulfinate**, **methyl methanethiosulfinate**, and other mixed or symmetrical **thiosulfinates** (R-S-S-R'), where R and R' are methyl, propyl, and allyl groups.

ONION AND GARLIC INHIBIT PLATELET AGGREGATION BY ALTERING THE METABOLISM OF ARACHIDONIC ACID

Anticoagulants and Thrombolytic Agents

The clotting of blood, which protects against hemorrhage, involves the sequential initiation, interaction, and completion of several stages in hemostasis. The adhesion and **aggregation of platelets** are mediated via the release of **adenosine diphosphate** (ADP). An extensive number of pharmacological agents, such as **acetylsalicylic acid, indomethacin, phenylbutazone, sulfinpyrazone**, and

dipyridamole, and foods, such as garlic and onion, inhibit both platelet aggregation and thrombus formation, and thus may be of value in the treatment of thrombotic disorders. The formation of fibrin itself takes place by means of a cascading group of reactions involving numerous blood-clotting factors and is accomplished in several stages (see Table 6).

Hemorrhage may result from several causes: an abnormality or deficiency of platelets (**thrombocytopenic purpuras**), a **deficiency of clotting factors** (factors II, VII, IX, or X), **vitamin K deficiency** (vitamin K is necessary for synthesizing clotting factors), or liver diseases that involve the synthesis of clotting factors. Increased clotting may occur in the **presence of thrombosis** (enhanced formation of fibrin), **stasis and phlebitis** (diminished circulation), or **embolism** (dislocation and lodging of blood clots). Although **heparin** is an extremely effective anticoagulant, it has certain limitations that are not shared by the newer thrombin inhibitors. As a result, these novel inhibitors may have advantages over heparin for use in certain clinical settings.

Thrombin activates platelets, converts fibrinogen to fibrin, activates factor XIII, which stabilizes fibrin, and activates factors V and VIII, which accelerate the **generation of prothrombinase**. Therefore, the inhibition of thrombin is essential in preventing and treating **thromboembolic disorders**.

Aggregin and Platelet Aggregation

Platelets circulate in blood without adhering to other platelets or to the endothelium. However, when the endothelial cells are perturbed, the platelets adhere and undergo a change in shape, and aggregate. ADP is known to induce the platelet shape change, aggregation, and exposure of fibrinogen-binding sites.

The platelet surface contains aggregin, a membrane protein with a molecular weight of 100 kDa, which has physical and immunochemical properties that differ from those of platelet glycoprotein IIIa.

Binding to aggregin is required in order for **epinephrine-induced platelet aggregation** to take place. In turn, epinephrine increases the affinity of ADP for its receptor. Thrombin stimulates platelet aggregation independent of ADP, but, by raising the level of calcium in the cytoplasm, it activates **platelet calpain**, which in turn cleaves aggregin.

Vitamin E and Platelet Aggregation

α-Tocopherol, a natural antioxidant, inhibits platelet aggregation and release. The effect of vitamin E is due to a reduction in platelet **cyclooxygenase** activity and inhibition of **lipid peroxide** formation. It is believed that supplementing the diet with **vitamin E** could play a role in the treatment of thromboembolic disease, especially if it is given in conjunction with an inhibitor of platelet aggregation.

Platelet-Activating Factor

A platelet-activating factor, **1-0-alkyl-2-(R)acetyl-*sn*-glyceryl-3-phosphocholine**, is released in the presence of shock and ischemia. The platelet-activating factor antagonist can protect the heart and brain against ischemic injury.

Platelet-inhibiting Drugs

A combination of **acetylsalicylic acid** and **dipyridimole** has been found to be effective in preventing myocardial reinfarction and occlusion of aortocoronary grafts.

EICOSANOIDS

All naturally occurring prostaglandins are derived through the cyclization of 20-carbon unsaturated fatty acids such as arachidonic acid, which in turn is synthesized from the essential fatty acid linoleic acid. The prostaglandins are named according to their ring substitutions and the number of additional side-chain double bands, as seen in prostaglandins E_1, $F_{1\alpha}$, and $F_{2\alpha}$.

In addition to serving as a precursor for the synthesis of prostaglandins, arachidonic acid is also a precursor for the synthesis of **prostacyclin, thromboxanes**, and **leukotrienes**.

Pharmacokinetic Properties of the Eicosanoids

The prostaglandins are inactivated rapidly (half-life < 1 minute) by the pulmonary, hepatic, and renal vascular beds through the actions of **prostaglandin dehydrogenase** and **prostaglandin reductase**. In addition, the aliphatic side chain undergoes relatively slow β and ω oxidation. The rapid metabolism and removal logically protect the cardiovascular system from the smooth muscle-stimulating effects of prostaglandins. Prostacyclins are also rapidly metabolized by prostaglandin dehydrogenase. Thromboxane A_2 is hydrated in the blood to thromboxane B_2. The thromboxanes are metabolized extensively and their metabolites appear in the urine.

Physiology of the Eicosanoids

The most important known effects of the prostaglandins and other eicosanoids are the contraction or relaxation of smooth muscles.

- **Smooth Muscles and Vascular Smooth Muscles.** Prostaglandins E_2 and I_2 cause arteriolar dilation in the systemic and pulmonary vascular beds. Prostaglandins $F_{2\alpha}$ and E_2 as well as thromboxane B_2 constrict human umbilical cord. Leukotrienes C_4 and D_4, which release prostaglandin E, decrease peripheral vascular resistance.
- **Respiratory Smooth Muscles.** Prostaglandins E_1, E_2, and I_2 cause bronchodilation and oppose the actions of acetylcholine, histamine, and bradykinin. On the other hand, prostaglandin $F_{2\alpha}$, leukotrienes C_4 and D_4, thromboxane A_2, and 5-HPETE are all bronchoconstricting substances possessing different potencies.
- **Gastrointestinal Smooth Muscles.** Prostaglandin E_2 contracts longitudinal but relaxes circular smooth muscles, whereas prostaglandin $F_{2\alpha}$ contracts both.

- **Reproductive Smooth Muscles.** Prostaglandins E_1, E_2, and $F_{2\alpha}$ cause contraction of the pregnant and nonpregnant human uterus, and produce laborlike contractions. In contrast to **oxytocin**, this effect is possible in all stages of pregnancy.

In general, the constricting and relaxing effects of the eicosanoids on the smooth muscles are not mediated by means of the classic neurotransmitters such as acetylcholine, catecholamine, or histamine, as proved by the fact that they are not altered by antihistaminic substances, anticholinergic agents, or either α- or β-adrenergic receptor-blocking agents.

Prostaglandins exhibit the following effects:

- **Cardiovascular System.** Prostaglandins E_2 and $F_{2\alpha}$ have positive isotropic effects and, in general, elicit increases in the cardiac output.
- **Blood.** Prostaglandin E_1 inhibits platelet aggregation, whereas thromboxane A_2 induces platelet aggregation. Prostaglandins A_2, E_1, and E_2 enhance **erythropoiesis** by augmenting the renal cortical release of **erythropoietin**.
- **Kidneys.** Prostaglandins E_1, E_2, and I_2 increase renal blood flow and produce diuresis, natriuresis, and kaliuresis. Prostaglandins E_1 and E_2 antagonize the action of antidiuretic hormone.
- **Endocrine System.** Various prostaglandins have the following effects in the endocrine system:

Prostaglandins E_1 and $F_{2\alpha}$ stimulate the release of **corticotropin**.
Prostaglandins E_1 and E_2 enhance the release of **growth hormone**.
Prostaglandin $F_{2\alpha}$ increases the release of **prolactin**.
Prostaglandin E_2 increases the release of **luteinizing hormone**.
Prostaglandin $F_{2\alpha}$ reduces progesterone output and causes regression of the **corpus luteum**.

- **Metabolic Effects.** Prostaglandin E_1 inhibits basal and **catecholamine-stimulated lipolysis**.
- **Nociception.** Prostaglandins E_1 and E_2 bring about the sensation of pain by sensitizing the afferent nerve endings to noxious chemical and physical stimuli.
- **Inflammatory and Immune Responses.** The prostaglandins are involved in both the genesis and manifestation of inflammation. Furthermore, prostaglandins are thought to regulate the functions of B and T lymphocytes.
- **Central Nervous System.** Prostaglandins E_1 and E_2 cause sedation.
- **Pyrexia.** Prostaglandin E_1 precipitates fever when it is injected intracerebroventricularly.

The Pharmacodynamics of Eicosanoids

Some, but not all, of the pharmacological effects of eicosanoids are mediated through alterations in the concentration of cyclic adenosine monophosphate (cyclic AMP). For example, prostaglandins E_1 and E_2 inhibit platelet aggregation by increasing the cyclic AMP concentration. Conversely, thromboxane A_2, which induces platelet aggregation, inhibits the synthesis of cyclic AMP. Prostaglandins also regulate the functions of B and T lymphocytes.

The Therapeutic Uses of Eicosanoids

Prostaglandins are mostly used as **abortifacients**. They may be administered by vaginal suppository (**Dinoprostone**), which contains prostaglandin E_2, by intramuscular injection (**carboprost** and **tromethamine**), which contains **15-methyl prostaglandin $F_{2\alpha}$**, or by intraamnionic administration (**dinoprost** and **tromethamine**), which contains prostaglandin $F_{2\alpha}$. Other possible uses of prostaglandins may include the treatment of ductus arteriosus (prostaglandin E_1) to maintain patency and as a vasodilator (prostaglandin E_1) in the management of peripheral vascular diseases. High levels of prostaglandin $F_{2\alpha}$ may cause dysmenorrhea, as substances such as indomethacin and ibuprofen are effective in relieving these symptoms.

Agents That Alter Metabolism of Eicosanoids

Additional deleterious or beneficial actions of agents that alter the metabolism of eicosanoids are described in the following sections.

- **Interleukin-1, Pancreatic β Cells, and Insulin-Dependent Diabetes Mellitus.** Insulin-dependent diabetes mellitus is an autoimmune disease that causes the gradual destruction of insulin-producing pancreatic β cells. It has been postulated that the infiltration of macrophages into the pancreatic islets plays a key role in the destruction of β cells and that cytokines, especially interleukin-1, which is released locally from macrophages, may be the toxic molecule causing this destruction.
- **Prostaglandin, Renin, and the Control of Arterial Pressure.** Arachidonic acid, prostacyclin (prostaglandin I_2), and prostaglandin E_2 stimulate the secretion of renin, and this effect is blocked by **indomethacin** or **meclofenamate**. Because stimulation of the sympathetic nervous system increases the release of renin and prostaglandin, it is postulated that a portion of the neurogenically activated renin secretion is governed by prostaglandin-dependent pathways.
- **Aspirin, Prostaglandin, and Platelet Aggregation.** Thromboxane A_2 is a potent vasoconstrictor and a stimulus to platelet aggregation; prostacyclin, on the other hand, is the principal cyclooxygenase product of the vascular endothelium and has the opposite effect on platelet function and vascular tone. Aspirin inhibits the arachidonic acid-induced platelet aggregation. It and other nonsteroidal anti-inflammatory drugs block the biosynthesis of prostaglandins and thromboxane A_2 by inhibiting cyclooxygenase.
- **Indomethacin, Prostaglandin, and Ductus Arteriosus.** The patency of the **ductus arteriosus** is maintained in part by a prostaglandin. Indomethacin induces constriction of the ductus during the neonatal period, whereas infusion of prostaglandin E_1 maintains its patency.

- **Aspirin, Cyclooxygenase, and the Provocation of Asthma.** A few people with asthma suffer from a unique syndrome in which the ingestion of 40 to 300 mg of aspirin produces rhinorrhea and acute bronchoconstriction. All cyclooxygenase inhibitors will precipitate bronchospasm in patients with aspirin-evoked asthma.
- **Aspirin, Prostaglandin, and Hypercalcemia.** Certain solid tumors (**renal cell carcinoma**) cause hypercalcemia, which may be mediated by prostaglandin E_2. Taking a high dose of aspirin or other nonsteroidal anti-inflammatory agent will lower the serum calcium level.
- **Indomethacin, Cyclooxygenase, and Hypertension.** Arthritis and hypertension frequently coexist, and certain nonsteroidal anti-inflammatory agents antagonize the action of antihypertensive medications. For example, indomethacin will elevate blood pressure in hypertensive patients being treated with diuretics, β-adrenergic receptor antagonists, or converting-enzyme inhibitors. Similar effects are seen with **ibuprofen** and **piroxicam**, but not with aspirin or **sulindac**.
- **Aspirin, Prostaglandins, and Mastocytosis.** Mastocytosis is a disease of unknown cause that is characterized by the increased proliferation of tissue mast cells, which leads to an episodic release of mediators and the manifestation of prominent symptoms such as vasodilation, flushing, hypotension, and tachycardia. Increased intestinal motility, which is also seen in the disorder, may result in abdominal cramping and diarrhea. Additional manifestations of mastocytosis are chest pain, dyspnea, pruritus, and headache. The production of prostaglandin D_2 is elevated in mastocytosis. Histamine is a potent vasodilator and causes contraction of gastrointestinal smooth muscle. However, the symptoms of mastocytosis do not respond to an antihistaminic alone, although the episode can be prevented by an antihistaminic agent or with high doses of aspirin or some other nonsteroidal anti-inflammatory agent that inhibits the synthesis of prostaglandins.

EFFECT OF GARLIC AND ONIONS ON EICOSANOID METABOLISM

The effects of onion and garlic on eicosanoid metabolism have mainly been studied by observing the effects of garlic and onion preparations on platelet aggregation and subsequent eicosanoid synthesis.

ADP, thrombin, collagen, and the **calcium ionophore A23187** can induce aggregation of platelets followed by secretion of the contents of storage granules. With the exception of collagen, these agonists can also shift the metabolic balance in platelets toward rapid mobilization. Arachidonic acid (AA) from phospholipid stores and the conversion of AA into eicosanoids occurs. Therefore, platelet aggregation and activation are not always linked.

After liberation, AA is converted into **prostaglandin endoperoxides** by the action of the enzyme cyclooxygenase. Endoperoxides are rapidly converted into either stable prostaglandins, PGE_2, $PGF_{2\alpha}$, and PGD_2, or thromboxane A_2. AA can also be converted into lipoxygenase-dependent products such as **12-hydroperoxyeicosatetraenoic acid** (12-HPETE) and **12-hydroxyeicosatetraenoic acid** (12-HETE).

Onion oil has been shown to inhibit both the cyclooxygenase and 12-lipoxy-genase in platelets as well as inhibiting platelet aggregation induced by epinephrine, ADP, or AA. Inhibitors of lipoxygenase and cyclooxygenase have been isolated and characterized from onion extracts. More specifically, "**cepaenes**" found in onions have been found to be potent inhibitors of sheep seminal **microsomal cyclooxygenase** and porcine **leukocyte 5-lipoxygenase** (see Ali et al., 2000).

Both aqueous and organic garlic extracts have been found to inhibit several steps of the AA cascade in platelets. Garlic extracts inhibit incorporation of labeled AA into platelet phospholipids, and also inhibit deacylation of platelet phospholipids upon stimulation with the calcium ionophore A23187, resulting in less eicosanoid synthesis. In addition, a direct inhibitory effect of garlic extracts on AA-metabolizing enzymes has been observed. More recently, Ali et al. (2000) observed noncompetitive and irreversible inhibition of cyclooxygenase by aqueous extracts of raw garlic. This inhibitory activity was destroyed by boiling the garlic.

CARDIOVASCULAR AND LIPID-LOWERING EFFECTS OF GARLIC

Diseases related to atherosclerosis such as ischemic heart disease, stroke, and peripheral arterial diseases are associated with elevated serum lipids. Hyperlipidemia is one of the major risk factors for atherosclerosis. The lipid-lowering properties of garlic and onion have been studied in great detail. Intervention studies have reported the effect of garlic on various risk factors with inconsistent results, reflecting variations in protocol design, as well as dosage types and garlic preparations used.

It was demonstrated previously that **allicin** extracted from garlic had a lipid-lowering effect on long-term feeding to healthy rats. These studies reported a significant decrease in total serum lipids, phospholipids, and cholesterol in the animals fed allicin compared with control animals.

The **cholesterol-lowering effect of garlic** was also seen in rabbits that were fed 2 g cholesterol/day for 16 weeks. A dose-dependent reduction of serum cholesterol in garlic-fed rats was observed in these experiments.

A study carried out in the United States showed a 56% decrease in serum total lipid radioactivity in rats fed the equivalent of 5 g of fresh garlic bulbs/day for 7 days. Rats fed an experimental diet containing cholesterol were treated either with intraperitoneal injections of C-14 labeled acetate or a diet containing C-14 sucrose.

In summary, the consumption of garlic appears to reduce serum cholesterol in experimental animals in a dose-dependent fashion. This may be due to decreased synthesis and/or increased excretion of cholesterol through the intestinal tract. It has been reported that garlic consumption increases high-density lipoprotein (HDL) levels, which may help to remove excess cholesterol from arterial tissue.

GARLIC AND CANCER PREVENTION

The anticancer properties of garlic have been recognized since ancient times. The ancient Egyptians used garlic externally for the treatment of tumors, and Hippocrates and physicians in ancient India are reported to have used garlic externally for cancer treatment.

Although the anticancer properties of garlic have been recognized for centuries, the most recent studies involving garlic have focused on several aspects including **chemoprevention**. A compound that can inhibit the carcinogenic process is called an anticarcinogen or chemopreventive. The field describing the effects of these agents in inhibiting cancer in animals and humans is called chemoprevention.

Studies on the chemopreventive activity of garlic have utilized a number of different garlic preparations including fresh garlic extract, aged garlic, garlic oil, and several organosulfur compounds derived from garlic. In addition to studying the anticarcinogenic activity of garlic components, a number of researchers have recently focused on the antimutagenic activity of garlic.

It is clear that the chemopreventive activity of garlic is related to the **organo-sulfur compounds** (OSCS) derived from garlic. Although how garlic achieves chemoprevention is not fully understood, several modes of action have been proposed on the basis of recent studies.

ANTICLASTOGENIC EFFECTS OF GARLIC AND ONION

Aqueous garlic extract in combination with **mustard oil** reduces the chromosomal aberrations caused by **sodium arsenate**. Furthermore, both onion and garlic oil decrease the **phorbol-myristate-acetate**-induced tumor promotion.

ANTIMICROBIAL EFFECTS OF GARLIC AND ONIONS

Garlic extracts have been shown to inhibit the growth of a variety of bacteria and fungi. In addition, garlic extracts have been shown to have antiviral properties. Onion extracts have also been shown to have antibacterial properties.

A number of studies have focused on the identification of the active antimicrobial principles in fresh garlic extract. **Allicin** has been identified as the active agent in garlic extracts. The antifungal activity of allicin has been shown to depend upon the sulfhydryl moiety since activity is destroyed by thiols such as L-**cysteine**, **glutathione**, and **mercaptoethanol**. **Ajoene**, a compound found in oil-macerated garlic, has also been shown to have antimicrobial activity (Ali et al., 2000).

THE ANTIOXIDANT ACTIONS
AND OTHER THERAPEUTIC EFFECTS OF GARLIC

Garlic has been recognized since ancient times not only as a flavoring agent for food but also for its medicinal properties. The intrinsic antioxidant activities of **garlic**, **garlic extracts**, and some **garlic constituents** have been widely documented *in vivo* and *in vitro*, as well as the ability of aged garlic extract to decrease H_2O_2 levels *in vitro*.

Studies have shown that 2% garlic feedings for 2 weeks:

- Decrease H_2O_2 generation
- Increase the activity of **superoxide dismutase**

- Increase the activity of **glutathione peroxidase**
- Increase catalase activity in vascular endothelial cells

Garlic oil and ***S*-allyl-cysteine sulfoxide (Alliin)**, a garlic compound, exhibited similar antioxidant properties. Aged garlic extract prevents ischemia and reperfusion injuries to the heart and brain.

CARDIOPROTECTIVE ACTIONS OF WILD GARLIC (*ALLIUM URSINUM*) IN ISCHEMIA AND REPERFUSION

Ventricular arrhythmias arising predominantly from ischemic heart disease are a major case of death in the Western industrialized countries, and the prevention and the pharmacological management of life-threatening arrhythmias remain a great challenge. It is, therefore, important to search for alternative dietary and phytotherapeutic strategies. In the framework of these studies it was found that feeding of garlic powder (***A. sativum***) as well as of pulverized dried wild garlic (***A. ursinum***) leaves for several weeks has significant protective effects with respect to the size of the ischemic zone and the incidence of arrhythmias.

Rietz et al. (1993) investigated the susceptibility to ventricular arrhythmias under the conditions of cardiac ischemia and reperfusion in the Langendorff heart preparation of rats fed for eight weeks a standard chow enriched with 2% of pulverized wild garlic leaves. The isolated hearts were perfused with a modified Krebs–Henseleit solution. The incidence of ventricular fibrillation (VF) during 20-minute occlusion of the descending branch of the left coronary artery (LAD) was significantly reduced in the wild garlic group as compared with untreated controls (20 vs. 88%). The same holds for the size of the ischemic zone (33.6 vs. 40.9% of heart weight). In the reperfusion experiments (5 minutes after 10-minute ischemia), ventricular tachycardia (VT) occurred in 70% of the wild garlic group vs. 100% in untreated controls and VF in 50 vs. 90%. The time until occurrence of extrasytoles, VT, or VR was prolonged. No significant alterations in cardiac fatty acid composition could be observed. Although the **prostacyclin** production was slightly increased in hearts of the wild garlic group, inhibition of **cyclooxygenase** by **acetylsalicylic acid** (ASA; **aspirin**) could not completely prevent the cardioprotective effects suggesting that the prostaglandin system does not play a decisive role in the cardioprotective action of wild garlic. Furthermore, a moderate **angiotensin-converting enzyme** (ACE) inhibiting action of wild garlic was found *in vitro* as well as *in vivo* that could contribute to the cardioprotective and blood pressure-lowering action of wild garlic (Figure 119).

ACE INHIBITORS

In Myocardial Infarction

The use of ACE inhibitors in myocardial infarction is rapidly evolving. Several large, prospective, randomized clinical studies involving thousands of patients have been published, and these studies provide convincing evidence that ACE

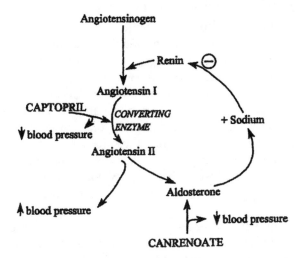

Figure 119 Captopril and garlic inhibit angiotensin-converting enzyme.

inhibitors reduce overall mortality when treatment is begun during the peri-infarction period. In this regard, clinical trials with ACE inhibitors in myocardial infarction can be divided into two groups: (1) treatment for several years post-myocardial infarction of patients with left ventricular systolic dysfunction (with or without overt heart failure); and (2) treatment for several weeks postmyocardial infarction of patients regardless of ventricular function. These studies suggest that, in selected high-risk patients (i.e., those with systolic dysfunction), ACE inhibitors save 40 to 70 lives per 1000 patients. Short-term treatment of unselected patients postmyocardial infarction saves about 5 lives per 1000.

An important issue is when to initiate therapy and how long to maintain treatment. In the CONSENSUS II study, **enalaprilat** was administered intrave-nously within 24 hours of myocardial infarction, and no benefit was demonstrated. This study raised serious doubts about giving ACE inhibitors too soon after myocardial infarction. However, in the SMILE study, treatment with oral **zofenopril** was initiated within the first 24 hours, and a reduction in mortality was observed within a few days. Most likely, the symptomatic hypotension induced by intrave-nous enalaprilat explains the negative finding in the CONSENSUS II study. Given the present state of knowledge, an oral ACE inhibitor should be initiated as soon as possible, but no later than 16 days after infarction. In all patients, therapy should be maintained for at least 6 weeks, and in patients with left ventricular systolic dysfunction, ACE inhibitor therapy should be maintained long term. Whether long-term ACE inhibitor therapy would benefit patients without ventric-ular dysfunction is an open question that is being addressed by two ongoing clinical trials. Doubtless, the dosage of ACE inhibitor and duration of treatment for various categories of patients with ischemic heart disease will be continually refined in coming years.

In Progressive Renal Impairment

The combination of diabetes mellitus and hypertension inexorably leads to diabetic nephropathy and is the major cause of **end-stage renal failure**. ACE inhibitors have been shown in numerous animal studies and in several small clinical trials to retard significantly the loss of kidney function associated with diabetic nephropathy. A large, prospective, placebo-controlled study has clearly established that **captopril** slows the progression of diabetic nephropathy in patients with **insulin-dependent diabetes mellitus** and established renal disease. In this regard, captopril significantly reduced proteinuria, the rate of decline in **creatinine clearance**, and the combined end points of dialysis, transplantation, and death. These results support the conclusion that, unless contraindicated, patients with diabetic nephropathy (whether normotensive or hypertensive) should be treated with an ACE inhibitor.

A broader issue is whether or not ACE inhibitors should be used routinely in nondiabetic patients with chronic renal failure. No large, long-term, prospective, and placebo-controlled study has yet addressed this issue; however, when the animal studies and small, less-than-adequate, clinical trials are viewed in aggregate, the data suggest that ACE inhibitors should be tried in hypertensive patients with progressive renal impairment who are at high risk of developing end-stage renal failure. Since renal impairment predisposes toward ACE inhibitor-induced hyperkalemia, serum K^+ levels should be monitored carefully in such patients. The development of ACE inhibitors that do not inhibit the **intraadrenal renin–angiotensin system** may alleviate this concern. It is likely that future studies will broaden the indications for ACE inhibitors in renal disease.

Several mechanisms participate in the renal protection afforded by ACE inhibitors. Increased glomerular capillary pressure induces glomerular injury, and ACE inhibitors reduce this parameter both by decreasing arterial blood pressure and by dilating renal efferent arterioles. ACE inhibitors increase the permeability selectivity of the filtering membrane, thereby diminishing exposure of the mesangium to proteinaceous factors that may stimulate mesangial cell proliferation and matrix production, two processes that contribute to expansion of the mesangium in diabetic nephropathy. Since angiotensin II is a growth factor, reductions in the intrarenal levels of angiotensin II may further attenuate mesangial cell growth and matrix production.

In Scleroderma Renal Crisis

Before the use of ACE inhibitors, patients with scleroderma renal crisis generally died within several weeks. A few small, observational studies have suggested that captopril markedly improved this otherwise grim prognosis. As stated previously, **garlic behaves like captopril** and inhibits the activity of ACE. Considering the well-known difficulties in the prevention and pharmacological treatment of life-threatening cardiac arrhythmias, it seems appropriate also to investigate systematically dietary and phytotherapeutic interventions. A number of studies demonstrated cardioprotective actions of diets enriched with polyunsaturated fatty acids. In addition, the antiarrhythmic effects of various oil diets, particularly fish oil and linseed oil, are well known. Indeed, a dietary supplement of wild garlic can significantly reduce the size of the ischemic zone as well as ischemia and reperfusion-induced arrhythmias.

Oxygen free radical generation has been proposed to be a major mechanism particularly in the pathogenesis of reperfusion injury. In this context, studies on antioxidant and free radical scavenging activity of garlic and alliin are of great interest. As the compounds of garlic and wild garlic are very lipophilic, they might be incorporated into the myocardial membranes acting as radical scavengers.

ANTIOXIDANT ACTIVITY OF ALLICIN, AN ACTIVE PRINCIPLE IN GARLIC

Use of garlic is rising in the health-conscious population. Reports suggest that it has beneficial effects in diseases such as ischemic-reperfusion arrhythmias and infarction, ischemic heart disease, hypertension, hyperlipidemia, peripheral arterial occlusive disease, and hyper-cholesterolemic atherosclerosis. Reactive oxygen metabolites (ROMs), which include superoxide anion ($O_2^{\bullet-}$), hydrogen peroxide (H_2O_2), and hydroxyl radicals ($\bullet OH$), have been implicated in the pathophysiology of hypercholesterolemic atherosclerosis and ischemia-reperfusion cardiac injury. Antioxidants are effective against ischemia-reperfusion cardiac injury, and hyper-cholesterolemic atherosclerosis. The beneficial effects of garlic in the above diseases and other undocumented health problems may be due to its antioxidant property. However, it is not known if garlic scavenges $O_2^{\bullet-}$, H_2O_2, or $\bullet OH$. $\bullet OH$ is the most toxic of all the ROMs. The **allicin** (active ingredient of garlic) content of commercial preparations of garlic varies tremendously. **Garlicin** preparations contain 1800 g of allicin per tablet.

Prasad et al. (1995) investigated the ability of **allicin** contained in the commercial preparation **Garlicin** to scavenge **hydroxyl radicals** ($\bullet OH$) using high-pressure liquid chromatography (HPLC). $\bullet OH$ was generated by photolysis of H_2O_2 (1.25 to 10 μmol/ml) with ultraviolet light and was trapped with salicylic acid, which is hydroxylated to produce $\bullet OH$ adduct products **2,3-** and **2,5-dihydroxybenzoic acid** (DHBA). H_2O_2 produced a concentration-dependent $\bullet OH$ as estimated by $\bullet OH$ adduct products 2,3-DHBA and 2,5-DHBA. Allicin equivalents in Garlicin (1.8, 3.6, 7.2, 14.4, 21.6, 28.8, and 36 μg) produced concentration-dependent decreases in the formation of 2,3-DHBA and 2,5-DHBA. The inhibition of formation of 2,3-DHBA and 2,5-DHBA with 1.8 μg/ml was 32.36 and 43.2%, respectively, and with 36.0 μg/ml the inhibition was approximately 94.0 and 90.0%, respectively. The decrease in $\bullet OH$ adduct products was due to scavenging of $\bullet OH$ and not by scavenging of formed $\bullet OH$ adduct products. Allicin prevented the lipid peroxidation of liver homogenate in a concentration-dependent manner. These results suggest that **allicin scavenges $\bullet OH$** and **Garlicin has antioxidant activity**.

There are numerous commercial garlic preparations available on the market in the form of **garlic powder tablet**, **steam distilled garlic oils**, and **oil macerated garlic**. The best garlic powder tablets have been reported to be as active as clove homogenate, whereas steam-distilled oils were 35% as active and oil macerates only 12% as active in inhibition of whole blood platelet aggregation.

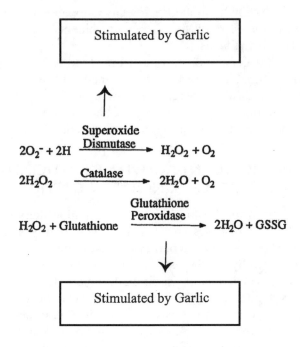

Figure 120   Garlic increases antioxidant mechanisms.

Allicin, an active principle in garlic, is present only in garlic powder and fresh garlic but not in garlic oil or oil-macerated garlic. Allicin content of commercial garlic powder tablets varies from 160 to 3600 µg/g. The results suggest that the claimed beneficial effects of garlic in various diseases could be due to •OH scavenging property of allicin content of garlic (Figure 120).

ATTENUATION OF RAT ISCHEMIC BRAIN DAMAGE BY AGED GARLIC EXTRACTS

Through numerous studies on the mechanism of pathogenesis of cerebral ischemia, various compounds were proposed that could effectively protect the brain against ischemia. Since free radicals are one of the focuses of study in the mechanism of cerebral ischemic injury, a variety of agents that have antioxidant capacity have been developed:

- **Lazaroid** compounds, which do not have glucocorticoid activity, but do have antioxidant properties
- **Transgenic mouse models**, which have enhanced superoxide dismutase (SOD) activity
- **Prostaglandin oligomeric** compounds
- **Aged extra of garlic (*Allium sativum*)** containing biochemically active substances including thioallyl compounds

Numagami et al. (1996) examined the effects of an aged garlic extract and its thioallyl components on rat brain ischemia using a middle cerebral artery occlusion model and a transient global ischemia model. In focal ischemia, an aged garlic extract, **S-allyl cysteine** (SAC), **allyl sulfide** (AS), or **allyl disulfide** (ADS), was administered 30 minutes prior to ischemic insult. At 3 days after ischemic insult, water contents of both ischemic and contralateral hemispheres were measured to assess the degree of ischemic damage. The water content of the ischemic control (no drug treatment) group was 81.50 ± 0.07% (mean ± SEM). It was significantly reduced with the administration of 300 mg/kg of SAC; the water content was 80.66 ± 0.11% ($p < 0.001$). The histological observation using **2,3,5-triphenyltetrazolium chloride** staining demonstrated that the administration of SAC reduced infarct volume. Neither AS nor ADS was effective. In global ischemia, the production of reactive oxygen species (ROS) was measured *ex vivo* using a spin-trapping agent, **α-phenyl-N-tert-butylnitrone**, and electron paramagnetic resonance spectroscopy. The production of ROS had two peaks; first at 5 minutes and second at 20 minutes after reperfusion. Both SAC and **7-nitro indazole**, a **nitric oxide synthase inhibitor**, did not attenuate the amount of ROS produced at the first peak, but did the amount of the second peak. A possible involvement of **peroxynitrite**, which may be formed from superoxide and nitric oxide and is known to be highly toxic in ischemia/reperfusion injury of the brain, was suggested.

GARLIC INHIBITS FREE RADICAL GENERATION AND AUGMENTS ANTIOXIDANT ENZYME ACTIVITY IN VASCULAR ENDOTHELIAL CELLS

Oxygen-free radicals such as **hydrogen peroxide** (H_2O_2), **superoxide anion** ($O_2^{\bullet-}$), and **hydroxyl radical** (OH$^{\bullet}$) have been implicated in mediating various pathological processes such as ischemia, inflammatory diseases, diabetes, and atherosclerosis. Endothelial cell injury is often the first stage of these disorders. Free radicals produced by **activated polymorphonuclear neutrophils** (PMN) or endothelial cells are critical factors in causing endothelial cell injury. The antioxidant enzymes (see Figure 119) **superoxide dismutase** (SOD), **catalase** (CAT), and **glutathione peroxidase** (GPX) either present intracellularly or released into the extracellular milieu can directly scavenge these oxidants or prevent their conversion to toxic species. They also inhibit PMN adhesion to endothelial cells and therefore play an important role in preventing endothelial cell injury.

Garlic has been demonstrated to prevent atherosclerosis and cancer and to retard the aging process. Alliin from garlic can scavenge OH$^{\bullet}$, and garlic powder has been shown to scavenge both OH$^{\bullet}$ and **1,1-diphenyl-2-picrylhydrazyl radicals**.

Aged garlic extract (AGE) has been shown to prevent oxidant-induced injury of endothelial cells. Wei and Lau (1998) determined the effects of AGE on the generation of hydrogen peroxide and superoxide anion ($O_2^{\bullet-}$) and the activity of three antioxidant enzymes in bovine pulmonary artery endothelial cell (PAEC). Confluent monolayers of PAEC were incubated with AGE, and oxidative stress was triggered by **hypoxanthine** and **xanthine oxidase** or H_2O_2. AGE exhibited both concentration- and time-dependent suppression of H_2O_2 and $O_2^{\bullet-}$ generation, and it also significantly increased the activities of SOD, CAT, and GPX. The results suggest that AGE may be an effective antioxidant in preventing or treating disorders related to endothelial cell injury associated with free radicals.

Oxygen is required for many life-sustaining metabolic reactions. Normal oxygen utilization in mammalian tissues requires an enzymatically controlled four-electron reduction to form water. If oxygen accepts fewer than four electrons, highly reactive free radicals (**oxyradicals**) are generated. One-electron reduction of molecular oxygen yields superoxide anion ($O_2^{\bullet-}$); two-electron reduction results in hydrogen peroxide (H_2O_2); and three-electron reduction generates hydroxyl radical ($OH^{\bullet}$). Both activated PMN and reoxygenated endothelial cells can accelerate oxyradical generation. The primary reactant that serves as the ultimate source for these oxyradicals is $O_2^{\bullet-}$ derived from the one-electron reduction of oxygen catalyzed by the enzyme **NADPH-oxidase** in PMN. Another source of $O_2^{\bullet-}$ is xanthine oxidase (XO) localized within the endothelial cells that utilizes HPX as a substrate. Free radicals contribute to endothelial cell injury, increase microvascular permeability and tissue damage, and play a critical role in mediating various pathological processes such as atherosclerosis, ischemia and reperfusion, and inflammatory diseases.

The antioxidant enzymes SOD, GPX, and CAT play an important role in maintaining physiological levels of $O_2^{\bullet-}$ and H_2O_2. The SODs are a group of enzymes that rapidly catalyze the conversion of $O_2^{\bullet-}$ to H_2O_2 and oxygen. GPX is a major enzyme that inactivates a variety of organic peroxides and thus controls the cellular peroxide levels. It also reduces H_2O_2 to water with generation of **oxidized glutathione**. Several studies have indicated that treatment of endothelial cells with SOD and/or CAT decreased damage caused by free radicals. Other data show that AGE increases the activities of SOD, GPX, and CAT. By inducing activities of antioxidant enzymes, AGE may hasten dismutation of $O_2^{\bullet-}$ and decomposition of H_2O_2. The effects of AGE on H_2O_2 and $O_2^{\bullet-}$ generation observed in this study may, therefore, be due to enhanced activities of these antioxidant enzymes. Recently, Wei and Lau (1998) reported that AGE modulates glutathione redox cycle with the particular result of increasing the intracellular glutathione levels. The increase of intracellular glutathione may also account for the decrease of H_2O_2 observed in this study.

In summary, Wei and Lau (1998) have shown that AGE inhibited H_2O_2 and $O_2^{\bullet-}$ generation and augmented SOD, CAT, and GPX activities. The results suggest that AGE may be an effective antioxidant in preventing endothelial cell damage and may therefore play a significant role in the defense against free radical-mediated disorders.

ATTENUATION OF RENAL OXIDATIVE STRESS WITH GARLIC OIL

Allium sativum, commonly known as garlic, is a common constituent of the diet. It is used with spices to give foods a special flavor and fragrance. It is also used as a medicinal herb in many countries. A number of pharmacological effects of garlic constituents have been reported. These include bactericidal, antibiotic, antitumor, hypolipidemic, hypoglycemic, and antiatherosclerotic activities. Garlic also inhibits **7,12-dimethyl benz[*a*]anthracene** and **benzo[*a*]pyrene-induced skin carcinogenesis** in mice. In other organs, too, garlic oil reduces the tumor induction response.

Nitrilotriacetic acid (NTA) is a constituent of various domestic and hospital detergents and is a common water contaminant. NTA forms water-soluble chelate complexes with various metal ions including iron at neutral pH. Its iron complex, Fe-NTA, is a known potent nephrotoxic agent. The renal toxicity is assumed to be caused by the elevation of serum free-iron concentration following the reduction of Fe-NTA at the luminal side of proximal tubule, which generates reactive oxygen species and leads to enhancement of lipid peroxidation.

Iqbal and Athar (1998) showed that Fe-NTA-mediated nephrotoxicity is diminished by 1 week of oral daily pretreatment of male albino Wistar rats with garlic oil given by gavage at 50 or 100 mg/kg body weight/ml corn oil. Intraperitoneal Fe-NTA treatment at a dose level of 9 mg Fe/kg body weight/10 ml enhanced renal microsomal lipid peroxidation and hydrogen peroxide generation, which are accompanied by a decrease in the activities of renal antioxidant enzymes (e.g., catalase, glutathione peroxidase, glutathione reductase, and glutathione *S*-transferase), and a depletion in the level of renal glutathione. Parallel to these changes, a sharp increase in blood urea nitrogen and serum creatinine has been observed. In addition, Fe-NTA treatment also enhanced renal **ornithine decarboxylase** (ODC) activity and increases **[³H]thymidine incorporation** into renal DNA. Prophylactic treatment of animals with garlic oil before the administration of Fe-NTA resulted in the diminution of Fe-NTA-mediated injury. The enhancement of renal lipid peroxidation and hydrogen peroxide generation was decreased. In addition, there was recovery of glutathione depletion and inhibition of the activities of antioxidant enzymes. Similarly, in animals given the higher dose of garlic oil (100 mg/kg body weight) the enhanced blood urea nitrogen and serum creatinine levels, which are indicative of renal injury, showed a reduction of about 30 and 40%, respectively, in comparison with the group treated with Fe-NTA alone. Pretreatment with garlic oil also ameliorated the Fe-NTA-mediated induction of ODC activity and enhancement of [³H]thymidine incorporation into DNA in a dose-dependent manner. These data suggest that garlic oil is a potent chemopreventive agent and may suppress Fe-NTA-induced nephrotoxicity.

A large number of chemopreventive agents, particularly those that are a part of our diet, afford protection against the onset of various diseases including cancer. Many of these actions have been related to the abilities of the agent to enhance the activities of carcinogen-metabolizing enzymes or to bind with toxicants thus reducing their effective critical concentrations. Alternatively, chemopreventive agents act as antioxidants and counteract the increased amount of oxidants generated by the toxicants. The medicinal uses and chemical composition of garlic have been widely studied. Both garlic and its constituents have been shown to inhibit skin, forestomach, esophageal, and colon carcinogenesis in animal model systems. Garlic oil, which contains a large number of sulfur-containing compounds, has been shown to act as a potent modifier of tissue oxidant response, and therefore the protection afforded by garlic oil against Fe-NTA-mediated renal damage may be due to the counteracting effect of its antioxidant constituents on Fe-NTA-mediated generation of oxidative stress.

In the study carried out by Iqbal and Athar (1998) garlic oil ameliorated Fe-NTA-induced lipid peroxidation, enhanced generation of H_2O_2, and depleted the level of glutathione. In addition, garlic oil reversed Fe-NTA-mediated inhibition of the activities of antioxidant enzymes such as catalase, glutathione peroxidase, glutathione reductase, glucose-6-phosphate dehydrogenase, and GST.

The exact mechanism by which garlic oil may suppress Fe-NTA-mediated renal damage is not known. However, onion and garlic extracts have been shown to inhibit **platelet aggregation** and **prostaglandin synthesis**.

GARLIC AMELIORATES HYPERLIPIDEMIA IN CHRONIC AMINONUCLEOSIDE NEPHROSIS

Garlic (**A. sativum L.**) has long been widely used not only as a flavoring but also as a folk medication. Some of the beneficial effects of garlic have been attributed to its hypolipidemic and antioxidant properties. The hypolipidemic effect of garlic in experimental animals is generally consistent. Long-term exposure to garlic, water-soluble garlic extracts, or garlic oil resulted in decreased serum levels of cholesterol and triglycerides in rats and rabbits fed with high lipid diet. Perhaps the hypolipidemic action is primarily due to the decrease in hepatic **cholesterolgenesis** in garlic-treated rats probably by the inhibition of **3-OH-3-methyl glutaryl CoA reductase** (HMG-CoA), the rate-limiting enzyme, early in cholesterol synthesis; whereas the triacylglycerol-lowering effect of garlic appears to be due to inhibition of fatty acid synthesis.

Pedraza-Chaverri et al. (2000a) studied **nephrotic syndrome** (NS), which is characterized by proteinuria, oxidative stress, and endogenous hyperlipidemia. Hyperlipidermia and oxidative stress may be involved in coronary heart disease and the progression of renal damage in these patients. Garlic has been suggested to be beneficial in various disease states. Some of the beneficial effects of garlic may be secondary to its hypolipidemic and antioxidant properties. Therefore, the effect of a 2% garlic diet on acute and chronic experimental NS induced by **puromycin aminonucleoside** (PAN) was studied in this work. Acute NS was induced by a single injection of PAN to rats, which were sacrificed 10 days later. Chronic NS was induced by repeated injections of PAN to rats which were sacrificed 84 days after the first injection. Garlic treatment was unable to modify proteinuria in either acute or chronic NS and was unable to modify hypercholesterolemia and hypertriglyceridemia in acute NS. However, garlic treatment diminished significantly total cholesterol, LDL cholesterol, and triglycerides, but not HDL cholesterol in chronic NS. Garlic induced no change in the percentage of sclerotic glomeruli in chronic NS and a significative decrease on the percentage of sclerotic area of these glomeruli ($33 \pm 3\%$ in NS + Garlic group vs. $47 \pm 4\%$ in NS group, $p = 0.0126$). The enhanced *in vivo* renal H_2O_2 production and the diminished renal Cu,Zn-SOD and catalase activities in acute NS, and the decreased renal catalase activity in chronic NS were not prevented by garlic treatment. These data indicate that garlic treatment ameliorates hyperlipidemia and renal damage in chronic NS, which is unrelated to proteinuria or antioxidant enzymes.

GARLIC AMELIORATES GENTAMICIN NEPHROTOXICITY: RELATION TO ANTIOXIDANT ENZYMES

Gentamicin (GM) is an **aminoglycoside antibiotic** commonly used in treating life-threatening Gram-negative bacterial infections. However, 30% of the patients treated with GM for more than 7 days show some signs of nephrotoxicity, and serious complications resulting from GM-induced nephrotoxicity are a limiting factor for its clinical usage. GM is not metabolized in the body but is essentially eliminated by glomerular filtration and partially reabsorbed by proximal tubular cells. Lysosomes have been shown to be the unique site of GM accumulation into proximal cells. GM induces a **lysosomal phospholipidosis** by inhibiting the activity of **phospholipases A_1 and C** and of **sphingomyelinase**.

A potential therapeutic approach to protect or reverse renal GM damage would have very important clinical consequences. GM administration into rats provides an excellent model of acute renal failure for studying therapeutic potential in preventing the side effects of acute renal failure pathogenesis in humans.

The pathophysiology of GM nephropathy is incompletely understood. However, **reactive oxygen species** (ROS) have been involved in GM-induced nephropathy. It has been found that renal cortical lipoperoxidation, *in vivo* renal H_2O_2 generation, and *in vitro* mitochondrial H_2O_2 generation are increased in GM-treated rats.

Pedraza-Chaverri et al. (2000b) studied the efficacy of garlic in preventing GM-induced nephrotoxicity. Four groups of rats were studied: (1) fed normal diet (CT); (2) treated with GM (GM); (3) fed 2% garlic diet (GA); and (4) treated with GM and 2% garlic diet (GM + GA). Rats were placed in metabolic cages and GM nephrotoxicity was induced by injections of GM (75 mg/kg every 12 hours) for 6 days. Lipoperoxidation and enzyme determinations were made in renal cortex on day 7. GM nephrotoxicity was made evident on day 7 by (1) tubular histological damage; (2) enhanced BUN and urinary excretion of **N-acetyl-β-D-glucosaminidase**; and (3) decreased creatinine clearance. These alterations were prevented or ameliorated in the GM + GA group. The rise in lipoperoxidation and the decrease in Mn-SOD and GPX activities observed in the GM group were prevented in the GM + GA group. Cu,Zn-SOD activity and Mn-SOD and Cu,Zn-SOD content did not change. CAT activity and content decreased in the GM, GA, and GM + GA groups. CAT mRNA levels decreased in the GM group. The protective effect of garlic is associated with the prevention of the decrease of Mn-SOD and GPX activities and with the rise of lipoperoxidation in renal cortex (see Figure 120).

45

PILOCARPINE
AND GLAUCOMA

Pilocarpine, an antiglaucoma agent, is the chief alkaloid derived from the leaves of *Pilocarpus jaborandi, P. microphyllus,* and other tropical American shrubs of the family Rutaceae.

PILOCARPUS JABORANDI

The habitat of *Pilocarpus jaborandi* is Brazil. The plant was introduced into Europe in 1847, and now is cultivated. The names *Jaborandi, Jamborandi,* and *Iaborandi* are applied natively, in both generic and specific sense, to several dissimilar pungent plants having sialagogic, diaphoretic, and sudorific properties, as **Serro'nea Jaborandi, Piper Jaborandi** (possibly the true Jaborandi), **P. unguicula'tum, P. citrifo'lium, P. reticula'tum, P. Mollico'mum, Erte'la (Auble'tia) trifo'lia,** and **Xanthroxylum el'egans**. Leaves should be collected when grown, after the rainy season and, because they incline to mustiness, should be dried thoroughly before packing. The once official species are high-priced, scarce, and subject to much substitution, while the Rio Jaborandi (*P. selloanus*), also once official and popular, continues to have a limited demand in spite of great irregularity in characteristics and constituents.

PILOCARPINE

Pilocarpine HCl

Adsorbocarpine, Akarpine, Isoptocarpine, Pilocar, Piloptic, Pilopine HS, Pilostat

Pilocarpine Nitrate

Pilagan

Pilocarpine Ocular Therapeutic System

Ocusert Pilo-20, Ocusert Pilo-40

Pilocarpine decreases intraocular pressure by constricting pupil and stimulating ciliary muscles to open trabecular meshwork spaces and facilitate outflow of aqueous humor (see Figure 41).

INDICATIONS

Pilocarpine is used ophthalmically in the treatment of **chronic simple glaucoma**, **chronic angle-closure glaucoma**, **acute angle-closure glaucoma**, **pre-** and **postoperative management of intraocular tension**, and **treatment of mydriasis**. Pilocarpine also is used in the treatment of glaucoma, where it is instilled into the eye as a 0.5 to 4.0% solution; up to a 10% solution is available. It is usually better tolerated than are the anticholesterases, and pilocarpine is the standard cholinergic agent for initial treatment of open-angle glaucoma. Reduction of intraocular pressure occurs within a few minutes and lasts 4 to 8 hours. The miotic action of pilocarpine is useful in overcoming the mydriasis produced by atropine; alternated with mydriatics, pilocarpine is employed to break adhesions between the iris and the lens.

Pilocarpine is used orally in the treatment of **xerostomia** in patients with malfunctioning salivary gland due to radiotherapy for cancer of head and neck. Pilocarpine is administered orally in 5- to 10-mg doses for the treatment of xerostomia that follows head and neck radiation treatments or that is associated with **Sjögren's syndrome**. The latter is an autoimmune disorder occurring primarily in women. Secretions, particularly those from the salivary glands, are compromised. Provided salivary parenchyma maintain a residual function, enhanced salivary secretion, ease of swallowing, and subjective improvement in hydration of the oral cavity are achieved. Side effects are those typical of cholinergic stimulation, with sweating the most common complaint. **Bethanechol** offers an alternative oral agent, which some feel produces less diaphoresis.

The three natural cholinomimetic alkaloids are **pilocarpine**, **muscarine**, and **arecoline**. Pilocarpus is the leaflets of *P. microphyllus* and of *P. jaborandi*, shrubs growing in South America. It contains three alkaloids — *pilocarpin*, *pilocarpidin*, and *isopilocarpin* — which produce similar effects, but pilocarpin is the most active, and is present in larger quantity (¼ to ½%) than the others (Figure 121).

Arecoline and pilocarpine are tertiary amines. Muscarine, a quaternary ammonium compound, shows more limited absorption. **Oxotremorine** is a synthetic compound that is used as an investigative tool. In the periphery, it acts as a muscarinic agonist or partial agonist at low concentrations. Its parkinsonian-like central effects include tremor, ataxia, and spasticity, which result apparently from activation of muscarinic receptors in the basal ganglia and elsewhere.

PHARMACOLOGY OF PILOCARPINE

Pilocarpine Causes Miosis

Pilocarpine, when applied locally to the eye, causes pupillary constriction, spasm of accommodation, and a transitory rise in intraocular pressure, followed by a more persistent fall (see Figure 41). Miosis lasts from several hours to a day, but

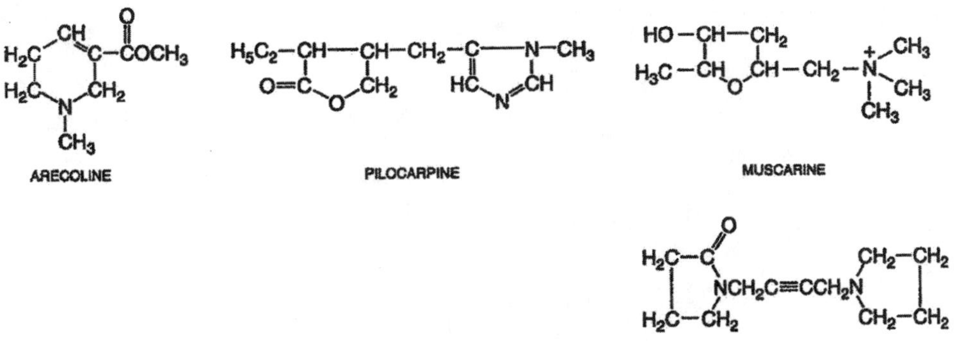

Figure 121   Structural formulas of cholinomimetic natural alkaloids and synthetic analogues.

the effect on accommodation disappears in about 2 hours. Pilocarpine causes contraction of the pupil, and of the ciliary muscle, with a lowering of intraocular pressure. The pupil of the rat is exceptional in being dilated by pilocarpine. Pilocarpine has no effect on the pupil in birds, in which the muscle of the iris is striated.

Pilocarpine May Precipitate Asthmatic Attack

The muscarinic alkaloids stimulate the smooth muscles of the intestinal tract, thereby increasing tone and motility; large doses cause marked spasm and tenesmus. The bronchial musculature also is stimulated; patients with asthma uniformly respond to pilocarpine with a reduction in vital capacity, and a typical asthmatic attack may be precipitated. Pilocarpine and muscarine also enhance the tone and motility of the ureters, urinary bladder, gallbladder, and biliary ducts.

Pilocarpine Causes Central Nervous System Stimulation

The intravenous injection of relatively small doses of pilocarpine, muscarine, and arecoline evokes a characteristic cortical arousal or activation response in cats, similar to that produced by injection of anticholinesterase agents or by electrical stimulation of the brain stem reticular formation. The arousal response to all of these drugs is reduced or blocked by atropine and related agents.

Pilocarpine Causes Diaphoresis and Salivation

Pilocarpine (10 to 15 mg, subcutaneously) causes marked diaphoresis in humans; 2 to 3 l of sweat may be secreted. Salivation also is increased markedly. Oral pilocarpine appears to cause a more continuous production of saliva. Muscarine and arecoline also are potent diaphoretic agents. Accompanying side effects may include hiccough, salivation, nausea, vomiting, weakness, and occasionally collapse. These alkaloids also stimulate the lacrimal, gastric, pancreatic, and intestinal glands, and the mucous cells of the respiratory tract.

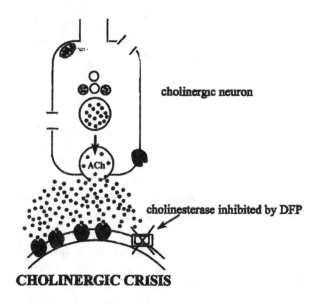

cholinergic neuron

cholinesterase inhibited by DFP

CHOLINERGIC CRISIS

Figure 122 Mushrooms and organophosphorus-induced cholinergic crisis. ACh = acetylcholine; DFP = diisopropyl fluorophosphate.

Pilocarpine Causes Hypotension

The most prominent cardiovascular effects following the intravenous injection of extremely small doses (0.01 to 0.03 µg/kg) of muscarine in various species are a marked fall in the blood pressure and a slowing or temporary cessation of the heart beat. An intravenous injection of 0.1 mg/kg of pilocarpine produces a brief fall in blood pressure. However, if this is preceded by an appropriate dose of a nicotinic blocking agent, pilocarpine produces a marked rise in pressure. Both the vasodepressor and pressor responses are prevented by atropine; the latter effect also is abolished by α-adrenergic blocking agents. These actions of pilocarpine have not been fully explained, but may arise from ganglionic and adrenomedullary stimulation.

Pilocarpine Poisoning Causes Cholinergic Crisis Resembling Mushroom Poisoning

Poisoning from pilocarpine, muscarine, or arecoline is characterized chiefly by exaggeration of their various parasympathomimetic effects (**cholinergic crisis**) and resembles those produced by consumption of mushrooms of the genus *Inocybe*, or administration of the irreversible cholinesterase inhibitors (Figure 122).

Treatment of pilocarpine poisoning consists of the parenteral administration of atropine in doses sufficient to cross and the blood–brain barrier and adequate measures to support the respiration and the circulation and to counteract pulmonary edema.

SUMMARY

The irreversible cholinesterase inhibitors, such as DFP (**isofluorophate**), are used only for local application in the treatment of wide-angle glaucoma. Their pharmacological effects, which are similar to those produced by **physostigmine**, are intense and of long duration. As organophosphorus insecticides, they are of paramount importance in cases of accidental poisoning and suicidal and homicidal attempts. They produce a cholinergic crisis, which must be treated by (1) decontaminating the patient, (2) supporting respiration, (3) blocking the muscarinic effects by atropine, and (4) reactivating the inhibited cholinesterase by treatment with **pralidoxime**.

MUSHROOM POISONING (MYCETISM)

Various species of mushrooms contain many toxins, and species within the same genus may contain distinct toxins. Although *Amanita muscaria* is the source from which muscarine was isolated, its content of the alkaloid is so low (approximately 0.003%) that muscarine cannot be responsible for the major toxic effects. Much higher concentrations of muscarine are present in various species of *Inocybe* and *Clitocybe*. The symptoms of intoxication attributable to muscarine develop rapidly, within 30 to 60 minutes of ingestion; they include salivation, lacrimation, nausea, vomiting, headache, visual disturbances, abdominal colic, diarrhea, bronchospasm, bradycardia, hypotension, and shock. Treatment with atropine (1 to 2 mg intramuscularly every 30 minutes) effectively blocks these effects.

Intoxication produced by *A. muscaria* and related *Amanita* species arises from the neurological and hallucinogenic properties of **muscimol**, **ibotenic acid**, and other isoxazole derivatives. These agents stimulate excitatory and inhibitory amino acid receptors. Symptoms range from irritability, restlessness, ataxia, hallucinations, and delirium to drowsiness and sedation. Treatment is mainly supportive; **benzodiazepines** are indicated when excitation predominates, whereas atropine often exacerbates the delirium.

Mushrooms from *Psilocybe* and *Panaeolus* species contain **psilocybin** and related derivatives of **tryptamine**. They also cause short-lasting hallucinations. *Gyromitra* sp. (false morrels) produce gastrointestinal disorders and a delayed hepatotoxicity. The toxic substance is **acetaldehyde methylformylhydrazone**, which is converted in the body to reactive hydrazines. Although fatalities from liver and kidney failure have been reported, they are far less frequent than with amatoxin-containing mushrooms discussed below.

The most serious form of mycetism is produced by *Amanita phalloides*, and other *Amanita* species, *Lepiota*, and *Galerina* species. These species account for over 90% of all fatal cases. Ingestion of as little as 50 g of *A. phalloides* (**deadly nightcap**) can be fatal. The principal toxins are the **amatoxins** (α- and β-amanitin), a group of cyclic octapeptides that inhibit RNA polymerase II and hence block the synthesis of mRNA. This causes cell death, manifested particularly in the gastrointestinal mucosa, liver, and kidneys. Initial symptoms, which often are unnoticed or, when present, are due to other toxins, include diarrhea and abdominal cramps. A symptom-free period lasting up to 24 hours is followed by hepatic and renal malfunction. Death occurs in 4 to 7 days from renal and hepatic failure. Treatment is largely supportive; **penicillin**, **thiotic acid**, and **silibinin** may be effective antidotes, but the evidence is based largely on anecdotal studies.

In cases of atropine poisoning, the use of pilocarpine is quite unjustified as the danger arises from the central nervous system in which the action of atropine is not antagonized by pilocarpine. On the other hand, in poisoning from pilocarpine or muscarine small quantities of atropine are the antidote recommended alike by pharmacological experiment and by clinical experience, since in this case, the toxic effects are the result of stimulation of parasympathetic effector organs and are antagonized by the peripheral action of atropine.

46

PLANTAIN HAS IMMUNOENHANCING ACTIONS

INTRODUCTION

Humans have a long history of the use of higher plant extracts for the therapy of diverse maladies. Medicinal plants play a major role in the lives of many people worldwide, and their usage has increased significantly. Epidemiological studies have associated a reduced risk of infectious diseases and cancer with a diet high in fruits and vegetables, and have determined that molecules such as **β-carotene**, **tocopherols**, **vitamin C**, and **flavonoids** confer some of this protective benefit. Finding additional agents for human or agricultural use based upon higher plant extracts may contribute to increasing the number of plant compounds of potentially beneficial application.

Plantago major, also known as **plantain**, **waybread**, or **dooryard plantain**, is found on roadsides, fields, lawns, and waste places in temperate zones worldwide. *Plantago major* has been used in traditional medicine as an astringent, anesthetic, anthelmintic, analgesic, analeptic, antiviral, antihistaminic, anti-inflammatory, antirheumatic, antitumor, antiulcer, diuretic, hypotensive, and expectorant. The fresh leaves of *P. major* are crushed and applied to wounds to prevent or cure infection and hasten healing. Moreover, plantain is a rapid pain reliever of stings, bites, and poison ivy. The intracellular fluid from *P. major* has been also shown to possess prophylactic activity against the development of mammary tumors in mice.

Gomez-Flores et al. (2000) designed experiments to evaluate the *in vitro* effects of methanol extracts of *P. major* leaves on the peritoneal macrophage and **thymic lymphocyte functions**. Macrophages play a central role in modulating humoral and cellular immunity against infectious diseases and cancer. Immunomodulatory agents such as **interferon-γ** (IFNγ) and **lipopolysaccharide** (LPS) are capable of activating them. Activated macrophages produce mediators of cytotoxicity such as **nitric oxide** (NO) and **tumor necrosis factor-α** (TNFα) which protect the host against the development of infections and tumors. On the other hand, T lymphocytes respond to antigen challenge by proliferating and expanding the antigen-specific lymphocyte clones thus amplifying immune responses. Functional T-cell proliferating activity can be studied by the use of polyclonal mitogens such as **concanavalin A** (Con A) and **phytohemagglutinin** (PHA), which bind to certain sugar residues on T-cell surface glycoproteins, including the T-cell receptor and CD_3 protein, and stimulate T-cell proliferative response.

Gomez-Flores et al. (2000) found that endotoxin-free methanol extracts from *P. major* leaves, at doses of 50, 100, 250, and 500 µg/ml, were associated with 4.4 ± 1, 6 ± 1, 12 ± 0.4, and 18 ± 0.4-fold increases of NO production, and increased TNFα production (621 ± 31, 721 ± 36, 727 ± 36, and 1056 ± 52 U/ml, respectively) by peritoneal macrophages, in the absence of IFNγ or LPS. NO and TNFα production by untreated macrophages was negligible. In addition, *p. major* extracts potentiated ConA-induced lymphoproliferation (3- to 12-fold increases) in a dose-dependent fashion, compared with the effect of ConA alone.

IMMUNOPHARMACOLOGY

The immune system protects the body from bacteria, viruses, and other harmful microorganisms. It is also able to attack a healthy body and cause life-threatening diseases such as multiple sclerosis. In the past, immunosuppressive agents were used solely in patients undergoing allotransplantation. However, a new understanding of the role of interleukins in the pathophysiology of diseases has spawned new applications for these agents. Moreover, because tissue transplantation, including even bone marrow transplantation, is associated with complications, and, because the pharmacokinetics of agents are modified in organ transplant patients, an even greater understanding of the therapeutic refinements of immunomodulating agents is required.

The principal elements in the cellular and molecular cascades that participate in activating immune response and the sites of actions of various immunosuppressive agents are depicted in Figure 123. In this process, the immune system is first provoked by either autoantigens or alloantigens. This activates the phagocytic cells, which, in turn, communicate their message via helper T cells by elaborating several mediators collectively called the **cell differentiation complex$_3$** (CD_3). This culminates in **interleukin 2-mediated cell proliferation** (see Figures 9 and 123).

THERAPIES THAT ENHANCE OR SUPPRESS THE IMMUNE SYSTEM

Radioimmunotherapy

Radioimmunotherapy or monoclonal antibody-mediated radioimmunotherapy is a process that couples an antibody to a radioactive isotope to enhance its tumoricidal activity. The long-range β-emitters rhenium-90Y and -188 have now replaced iodine-131 as isotopes.

Cell Transfer Therapy and Cancer

Cell transfer therapy is a new approach to strengthening the innate ability of the immune system to fight against cancer (Figure 124). In this therapy, lymphocytes are isolated and cultured with interleukin-2 for 3 days to yield lymphokine-activated killer cells, which are then administered to patients along with interleukin-2.

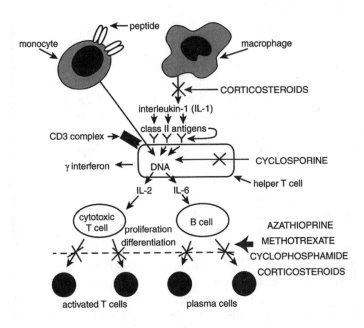

Figure 123 The sites of actions of immunosuppressive agents.

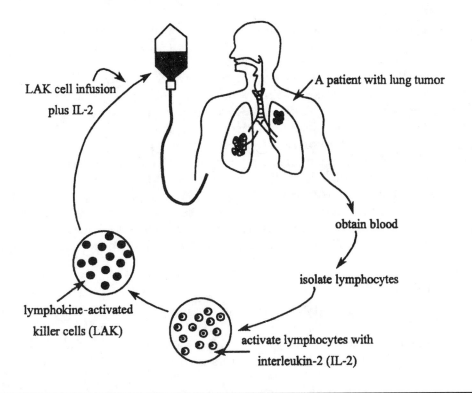

Figure 124 Cell transfer therapy with lymphokine-activated killer cells.

Immunosuppressive Agents

The following agents and measures are used for immunosuppressive therapy: corticosteroids, cytotoxic agents (alkylating agents and antimetabolites), cyclosporin A and dihydrocyclosporin C, antilymphocyte globulin and Rho (D) immune globulin (RhoGAM), lymphoid irradiation and thoracic duct drainage, and immunomodulating agents (interferons and their inducers).

Immunosuppressive agents are used in patients undergoing organ transplantation, such as of the liver, heart, and kidney. One or a combination of agents may be given, including **glucocorticosteroids**, **azathioprine**, and **cyclophosphamide**. In addition, a combination of drug therapy and other ameliorative techniques that bring about lymphocyte depletion, such as thoracic duct drainage or total lymph node irradiation, may be indicated.

Immunosuppressive agents are used in the treatment of autoimmune diseases. For example, the treatment approach to chronic active hepatitis not attributable to drugs, Wilson's disease, or α_1-antitrypsin deficiency consists of prednisone combined with azathioprine. Azathioprine by itself is not effective in this disorder. **Idiopathic thrombocytopenic purpura** is treated with corticosteroids and splenectomy, and immunosuppressive agents are used in refractory cases. Hemolysis due to warm-reacting autoimmune antibodies (**autoimmune hemolytic anemia**) involving **immunoglobulin G** (predominantly IgG_1 and IgG_3) is initially treated with prednisone. If this proves unsuccessful, splenectomy (in younger patients) or immunosuppression using azathioprine or cyclophosphamide (in older patients) is carried out next.

Acute rheumatic fever is treated with anti-inflammatory agents such as aspirin. Corticosteroids are reserved for those patients with severe carditis who do not respond to or are unable to take salicylates.

Myasthenia is treated with cholinergic drugs, which actually have no influence on the course of the disease itself, or with therapeutic regimens such as thymectomy, corticosteroids, or immunosuppressive agents, which are intended to induce remission of the disease. Immunosuppressive agents are used in isoimmune disorders such as **Rh hemolytic disease of the newborn**.

Immunomodulating agents are being investigated for their potential application in immunodeficiency disorders, chronic infections, and neoplasms.

The immunosuppressive agents are not without their problems, however. With the exception of **RhoGAM**, the currently available immunosuppressive agents are nonspecific in their actions, and generalized and prolonged immunosuppression increases the susceptibility to infections and the risk of **lymphoreticular cancer**.

The Corticosteroids

Corticosteroids such as prednisone or dexamethasone have lympholytic properties. They reduce the lymphoid contents of the lymph nodes and the spleen without influencing the myeloid or erythroid stem cells. Corticosteroids have the following pharmacological effects on immunosuppression:

- They inhibit prostaglandin E_2 and leukotriene synthesis.
- They reduce the macrophage-mediated lysosomal contents.

- They reduce the activity of the lymphocyte-mediated chemotactic factor and lymphotoxin.
- They increase the catabolism of immunoglobulins such as IgG.
- They are able to lyse T-helper cells.
- They interfere with the ability of reticuloendothelial macrophages to attack and destroy antibody-coated cells.

Prednisone is used in the treatment of autoimmune diseases, both organic-specific autoimmune diseases such as **myasthenia gravis** and idiopathic thrombocytopenic purpura and non-organ-specific autoimmune diseases such as **lupus erythematosus**, **rheumatoid arthritis**, and **periarteritis nodosa**. In addition, prednisone is used in patients receiving organ transplants and is of paramount value in countering the problems associated with organ rejection.

The Cytotoxic Compounds Alkylating Agents

As an alkylating agent, **cyclophosphamide** (Cytoxan) and its active metabolites destroy the rapidly proliferating lymphocytes, and hence are potent immunosuppressive agents. Those metabolites that are devoid of alkylating properties also lack an immunosuppressive capability. Because the biotransformation to active metabolites takes place in the liver, inducers of microsomal enzymes influence both the therapeutic spectrum and the toxicity of cyclophosphamide. When used in combination with prednisone and antilymphocyte globulin, cyclophosphamide prolongs the survival of organ and skin allografts. It is also effective in the treatment of **T-cell-dependent autoimmune diseases** such as **chronic lymphocytic thyroiditis**.

Antimetabolites

When injected, **azathioprine** (Imuran) is rapidly converted to **6-mercaptopurine**. The half-life of azathioprine after intravenous injection is 10 to 20 minutes, and that of 6-mercaptopurine is somewhat longer. The cytotoxic activity of these thiopurines is due to the conversion of mercaptopurine to **6-thiouric acid**, a noncarcinostatic metabolite. This action is thought to block the excess synthesis of inosinic acid from its precursors, glutamine and **phosphoribosylpyrophosphate.** In addition, unlike cyclophosphamide, azathioprine is a potent anti-inflammatory substance that can cause a reduction in the number of monocytes and neutrophils at inflammatory sites. Antibody responses are also inhibited by azathioprine. Studies in humans have shown that azathioprine decreases the γ-globulin and antibody levels, thus influencing IgG rather than IgM production. This makes azathioprine an effective immunosuppressant in the early phases of immune responses. It is less effective or completely ineffective in altering either the effector phase or already established reactivities.

Azathioprine is an effective agent in suppressing the immune system in patients undergoing renal transplantation and in patients suffering from **acute glomerulonephritis**, the renal component of **systemic lupus erythematosus**, prednisone-resistant **idiopathic thrombocytopenic purpura**, and functioning **autoimmune hemolytic anemia**. Azathioprine depresses bone marrow functioning, which is its chief side effect.

Cyclosporin A and Dihydrocyclosporin C

Cyclosporin A, a fungal metabolite, is a cyclic polypeptide that consists of 11 amino acids. It has a biologic half-life of 4 to 6 hours and displays a preferential T-cell cytotoxic property, in that it inhibits the factors that stimulate T-lymphocyte proliferation. Cyclosporin A has been used as the sole immunosuppressant (without prednisone or other drugs) for cadaveric transplants of the kidney, pancreas, and liver. Cyclosporin A has been observed to cause reversible hepatic toxicity and nephrotoxicity.

Dihydrocyclosporin C, another fungal metabolite, is even more selective than cyclosporin A, in that it suppresses T-lymphocyte production with only marginal effects on the antibody response.

The Antilymphocyte Globulin

The antilymphocyte globulin is obtained through the immunization of horses with human lymphoid cells or with fetal thymus cells. The antilymphocyte antibody destroys the T cells and impairs delayed hypersensitivity and cellular immunity without altering humoral antibody formation. The pattern of immunosuppression obtained with antilymphocyte globulin is identical to that brought about following thoracic duct drainage that depletes the numbers of small lymphocytes.

Rho (D) Immune Globulin

RhoGAM is a solution of human IgG-containing antibodies against the Rho (D) antigen of the red blood cells. It is administered to a mother after the birth of her **Rh-negative baby** as a prophylactic measure to prevent **erythroblastosis fetalis** (hemolytic disease of the newborn) in infants born in succeeding pregnancies. Treatment with RhoGAM may also be indicated in Rh-negative mothers who have had ectopic pregnancies, miscarriages, or abortions in which the blood type of the fetus is unknown.

Immunomodulating Agents

In addition to its antiviral actions, interferon has an antiproliferative effect and modifies the functions of macrophages and natural killer cells. **Thymosin**, a protein synthesized by the epithelioid component of the thymus, may be potentially valuable in patients with **DiGeorge's syndrome** or other **T-cell deficiency states**. **Levamisole** augments T-cell-mediated immunity and may be of value in the immunodeficiency associated with **Hodgkin's disease**.

SUMMARY

The regulation of immune parameters by *P. major* leaf extract (**Plantain**) may be clinically relevant in numerous diseases including chronic viral infections, tuberculosis, AIDS, and cancer.

47

PYCNOGENOL

Obtained from the bark of the French maritime pine ***Pinus maritima***, has been used by Hippocrates and

- Is an antioxidant
- Is a vasorelaxant
- Inhibits angiotensin-converting enzyme
- Improves capillary circulation

INTRODUCTION

The term ***pycnogenols*** describes an entire class of **flavonoids** composed of **flavan-3-ol derivatives**, which is now utilized throughout the world as a nutritional supplement and as a phytochemical remedy for various diseases ranging from chronic inflammation to circulatory dysfunction. ***Thesaurus medicaminum*** (pine bark) was considered helpful for **wound healing**. In old Europe pine bark was utilized against **inflammation** and to overcome the symptoms of **scurvy**. Other uses of pycnogenol were suggested by the naturalist **Hieronymus Boch** and included topical application on **skin ulcers** and general use against **skin disorders**. In the New World, **Native Americans** utilized the bark of the pine as a food, a beverage, and a remedy for various conditions, such as inflamed wounds or ulcers, now recognized to have free radical involvement. Pine bark was used for many conditions that are now known to involve **vitamin C deficiency**, such as **scurvy**, skin disorders, and wound healing. This indicates an interaction between **vitamin C** and flavonoids, which was also suggested by **Albert Szent-Gyorgyi**, who discovered vitamin C and won a Nobel prize for it.

Pycnogenol is exclusively prepared from the bark of the ***P. maritima***, from **Bay of Biscay** in the **Landes de Gascogne** in France, where climatic conditions exist that strongly influence the characteristics of this subspecies of *Pinus*. Various efforts to cultivate the same subspecies in other geographic areas, such as in the **Iberian Peninsula** or in **Korea**, have failed, indicating the uniqueness of the environmental conditions necessary for the growth of the French maritime pine (Packer et al., 1999).

Grape seed extract, which is another widely used plant extract similar to pycnogenol, contains **procyanidins**, which have a strong antioxidant activity.

ANTIOXIDANT ACTIVITY OF PYCNOGENOL

Phenolic acids, **polyphenols**, and in particular **flavonoids** are composed of one or more aromatic rings bearing one or more hydroxyl groups and are therefore potentially able to quench free radicals by forming resonance-stabilized phenoxyl radicals. **Superoxide radical anion** ($O_2^{\bullet-}$), **hydroxyl radical** (HO), **lipid peroxyl radical**, and the **reactive nitrogen species**, such as **nitric oxide radical** (NO), and **peroxynitrite** ($ONOO^-$), which are among the most important free radicals in the biologic environment and in human health and disease, have been investigated either *in vitro* or *in vivo*. Moreover, as suggested by the early reports from the laboratory of Szent-Gyorgyi (Benthsath et al., 1936), flavonoids may have a fundamental role in the antioxidant network interplaying together with other antioxidants and in particular with ascorbic acid, thus significantly contributing to cellular and extracellular defenses against oxidative stress.

One of the earliest studies addressing the antioxidant activity of various **procyanidins** *in vitro* reported strong scavenging activity against free radicals such as the stable radical, **1,1-diphenyl-2-picryl-hydrazyl** (**DPPH**), and the oxygen free radicals $O_2^{\bullet-}$ and HO, as assessed by ESR spectroscopy.

The capacity of pycnogenol to protect cellular systems has also been investigated, in particular in cultured endothelial cells and macrophages. In a bovine line of cultured normal endothelial cells (pulmonary artery endothelial, PAEC), preincubation with 20 to 80 µg/ml pycnogenol is associated with significant protection from both lipid peroxidation and cell damage induced by ***tert*-butyl-hydroperoxide** (t-BHP). At 60 µg/ml pycnogenol completely inhibited the release of **lactate dehydrogenase** after T-BHP treatment, suggesting significant protection from oxidative stress-induced cytotoxicity. The generation of **thiobarbituric acid reactive substances** (TBARS) was also significantly decreased, indicating that the protective effect was due, at least in part, to the antioxidant activity of pycnogenol. In the same cellular system, pycnogenol induced a dose-dependent decrease in the steady-state production of both $O_2^{\bullet-}$ and hydrogen peroxide (H_2O_2), and decreased the rate of H_2O_2 accumulation following treatment with xanthine/xanthine oxidase (X/XO) as a superoxide-generating system. The effect on both the steady state and the clearance of reactive oxygen species ($O_2^{\bullet-}$ and H_2O_2) after X/XO treatment was attributed to the effects of pycnogenol on **glutathione** (GSH) levels and on the enhancement of the activity of the enzymatic machinery that regulates GSH redox status. A significant increase in GSH levels, an increased activity of the GSH redox enzymes (glutathione reductase and glutathione peroxidases), and an increase in the enzymatic activity of both superoxide dismutase (SOD) and catalase were observed.

PYCNOGENOL HAS VASORELAXANT ACTIVITY

Pycnogenol is commonly suggested to lower the risk of cardiovascular diseases and also as a therapeutic alternative to established pharmacology for mild pathological conditions of blood vessels. The rationale behind this is the link between cardiovascular disease and free radical-induced stress. However, besides its strong antioxidant capacity, pycnogenol has been reported to have other activities related to cardiovascular functionality, such as a **vasorelaxant activity**, **inhibition of**

angiotensin-converting enzyme (see Figure 119), and the ability to **enhance microcirculation** by **increasing capillary resistance**.

Pycnogenol effects on circulation can be seen as the outcome of two different activities: relaxation of arterial walls and increase of capillary resistance. The combination of these effects results in an increase of peripheral blood flow and a facilitation of the microcirculation. Early studies suggested a strong vasoactivity of **procyanidins** extracted from sources such as **wine** and **grape seed**, which have recently been confirmed for pycnogenol.

PYCNOGENOL INHIBITS PLATELET AGGREGATION

Pycnogenol inhibits platelet aggregation induced by cigarette smoking without the adverse effect on bleeding time that characterizes aspirin use.

PYCNOGENOL ENHANCES MICROCIRCULATION

Pycnogenol enhances microcirculation and reduces capillary resistance in spontaneous hypertensive animals.

PYCNOGENOL INHIBITS THE ACTIVITY OF ELASTASE

The fragmentation of elastic fibers is a classical histological hallmark of the inflammatory response associated with several pathological conditions and the aging process. The dramatic proteolytic activity results from an imbalance between natural inhibitors, such as **α-1 antitrypsin** and **α-2 macroglobulin**, and leads to **elastolysis**. **Elastin**, together with **collagen**, is an important component of blood vessels, contributing to their integrity, elasticity, and permeability. Studies have reported an inhibitory effect of **procyanidins** on elastase activity *in vitro* and *in vivo*.

PYCNOGENOL REGULATES INFLAMMATORY RESPONSE

Inflammation is a multifactorial cellular response that is of fundamental importance in the maintenance of homeostasis when the organism is challenged by noxious agents (e.g., bacteria, viruses, irritating agents) or by tissue mechanical injury. Inflammation is associated with a dramatic rise in the number of polymorphonuclear leukocytes and monocytes in the affected tissue and with the release of inflammatory mediators such as **prostaglandins** and **cytokines**. Under ideal conditions, inflammation results in the complete recovery of the integrity of the affected tissue, but if the response to the triggering stimulus is not subjected to tight regulation, cellular and extracellular components of the organism adjacent to the inflammation site can be injured, inducing a condition known as chronic inflammatory disease. Indeed, a chronic inflammatory-like environment characterizes the pathogenesis of various diseases such as **atheriosclerosis**, **arthritis**, and **Crohn's disease**, and it is thought to be among the causative factors of more than 30% of human **cancers**. Thus, it is tremendously important to localize the inflammatory response, which on one hand increases the ability of the immune machinery to destroy the noxious agent, and on the other hand provides an efficient defense from self-inducing damage.

Pycnogenol reduces the production of **interleukin-6**, and restores the activity of **natural killer cells** in retrovirus-infected animals. Pycnogenol delays the development of immune dysfunctions secondary to retrovirus infection by restoring the imbalanced cytokine secretion by **T-helper 1** and **T-helper 2 cells**.

PYCNOGENOL MODULATES MACROPHAGE ACTIVITY

Pycnogenol affects the activity of **nitric oxide synthase** and the production of nitric oxide in murine macrophages (RAW 264.7 cell line) activated by **endotoxin** (the lipopolysaccharide) and **tumor necrosis factor** (TNF).

SUMMARY

In conclusion, pycnogenol is a complex mixture mainly composed of **oligomeric procyanidins** and other **flavonoids** and **polyphenols**. A wide spectrum of biologic activities has been described for the mixture. Pycnogenol has been reported to significantly affect circulation, inflammation, and the immune response. The molecular bases of pycnogenol activity are manyfold, but appear mainly to depend on its capacity to efficiently scavenge reactive oxygen and reactive nitrogen species. Other major biologic effects are probably due to its ability to specifically bind to proteins, therefore affecting both structural and functional characteristics. Finally, pycnogenol has been demonstrated to participate in the cellular antioxidant network and to be able to affect the expression of those genes that are regulated by cell redox status (see Packer et al., 1999).

48

PYGEUM AFRICANUM AND PREMIXON FOR THE TREATMENT OF PATIENTS WITH BENIGN PROSTATIC HYPERPLASIA

INTRODUCTION

Symptomatic **benign prostatic hyperplasia** (BPH) is a common medical problem in older men. As many as 40% of men aged 60 years or older have lower urinary tract symptoms consistent with **bladder outlet obstruction**. Treatment goals in the vast majority of men are to relieve bothersome symptoms that reduce quality of life. In the United States, treatment for benign prostatic hyperplasia costs more than $2 billion per year and accounts for 1.7 million physician office visits annually (Ishani et al. 2000).

The management of lower urinary tract symptoms suggestive of clinical symptomatic BPH remains somewhat controversial. Treatment options range from watchful waiting, for those patients wishing to delay any active therapy, to minimally invasive treatment, such as **transurethral needle ablation of the prostate** and **transurethral microwave therapy**, to surgical interventions in the form of **transurethral prostatectomy** or **open enucleation of the prostate.** Most patients present with difficulties in urination for which a variety of medical therapies are available, including synthetic **5-α-reductase inhibitors**, α-blockers, and plant extracts (Boyle et al. 2000).

Medicinal herbs, or phytotherapy, have been used extensively for benign prostatic hyperplasia in Europe and are being used more commonly in the United States. Sales of herbal medicine reached $4 billion in the United States in 1998, and sales of **saw palmetto extract** for treatment of symptoms attributable to benign prostatic hyperplasia exceeded $20 million, making it the seventh most commonly purchased medicinal herbal preparation.

A recent survey demonstrated that one third of men choosing nonsurgical therapy for benign prostatic hyperplasia were using herbal preparations alone or in combination with prescription medications. There is emerging evidence that several plant extracts are well tolerated and provide at least short-term improvement in urologic symptoms and flow.

Pygeum africanum, an extract from the bark of the **African prune tree**, has been used in Europe since 1969 to treat men with mild-to-moderate symptoms of benign prostatic hyperplasia. The mechanism of action of *P. africanum* is not known. In animal models, *P. africanum* modulates bladder contractility, has anti-inflammatory activity, decreases production of leukotrienes and other 5-lipoxygenase metabolites, inhibits fibroblast production, affects adrenal androgens, and restores the secretory activity of prostate epithelium.

In a study involving 1562 men, Ishani et al. (2000), by using *P. africanum* (100 mg/day) for at least 30 days showed that *P. africanum* modestly, but significantly, improved urologic symptoms and flow measures.

URINARY SYMPTOMS AND FLOW MEASURES

Pygeum africanum improved specific urinary symptoms and flow measures. In six double-blind trials involving 430 participants, men receiving *P. africanum* were more than twice as likely to be rated by their physician as having overall improvement in symptoms compared with men taking placebo. *Pygeum africanum* reduced **nocturia** compared with placebo. *Pygeum africanum* also increased **peak urine flow** compared with placebo. Additionally, *P. africanum* reduced **residual urine volume.**

PERMIXON IN THE TREATMENT OF SYMPTOMATIC BENIGN PROSTATIC HYPERPLASIA

Permixon is a standardized lipid-sterolic extract of ***Serenoa repens.*** Permixon is a registered trademark of Pierre Fabre Médicament, Castres, France. Its major mechanism of action is still uncertain; however, several activities have been demonstrated. It repeatedly exerts antiandrogenic activity in the form of a non-competitive inhibition of **5-α-reductase types I and II**, resulting in decreased prostatic **dihydrotestosterone** content in patients with BPH treated with this compound. Permixon also appears to inhibit ***in vitro* basic fibroblast growth factor** and **epidermal growth factor–induced prostate epithelial cell proliferation** and decrease epidermal growth factor concentrations in human prostate tissue. It also exerts an anti-inflammatory effect through the inhibition of the enzymes responsible for **prostaglandin** and **leukotriene synthesis**.

To assess the effects of permixon in men with lower urinary tract symptoms and BPH, a meta analysis of all available clinical trial data was undertaken. Boyle et al. (2000), in randomized clinical trials involving 2859 patients, found that permixon in the treatment of men with BPH revealed a significant improvement in peak flow rate and reduction in nocturia.

49

RESERPINE

Reserpine is a neuroleptic and an antihypertensive medication obtained from *Rauwolfia serpentina*.

Rauwolfia (Raudixin, Rauserpa, etc.), the powdered whole root of the *Rauwolfia* plant, is a tan, bitter, fine amorphous powder with slight odor, sparingly soluble in alcohol and only very slightly soluble in water.

Alseroxylon (Rauwiloid), an alkaloidal fraction from *Rauwolfia* root, is a reddish-brown amorphous powder with a characteristic odor.

Deserpidine (Harmonyl) is an alkaloid from *Rauwolfia* root.

Rescinnamine (Moderil) is an alkaloid of the **alseroxylon fraction** of *Rauwolfia* root.

Syrosingopine (Singoserp) is a preparation made from reserpine by hydrolysis and reesterification.

THE *RAUWOLFIA* ALKALOIDS

Rauwolfia serpentina **Benth**, which derives its name from **Leonhart Rauwolf**, a botanist of the 16th century, and its serpentine root (Figure 125) has long been used in India for a variety of ailments. The discovery of its tranquilizing action, particularly in lowering the blood pressure, led to its introduction into Western medicine. The *Rauwolfia* alkaloids are derived from a family of tropical and semitropical plants related to **oleander** and **periwinkle**. They vary from small shrubs to tall trees. The important species from which the alkaloids are derived include *Rauwolfia serpentina* (*Ophioxylon serpentinum* or **Indian snake root**), *R. micrantha*, *R. vomitoria*, and *R. birsuta* (*Canescens heterophylla*).

CHEMISTRY

The first alkaloid of this group to be isolated and synthesized was **reserpine** (Serpasil®), which has the structure shown in Figure 126. Reserpine is an indole derivative related to **yohimbine** which is also present in preparations of *Rauwolfia*. A number of related alkaloids have been isolated from *Rauwolfia*, some of which are used clinically, e.g., **rescinnamine**, the 3:4:5-trimethoxycinnamic acid ester of methyl reserpate and **deserpidine** (11-desmethoxyreserpine), which

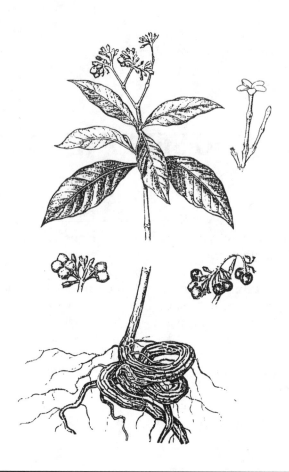

Figure 125 A flowering *Rauwolfia serpentina* plant. Note the serpentine root from which the plant obtains its name (snake root).

lacks the methoxy group of ring A of reserpine. **Deserpidine** has also been designated as **canescine**, **recanescine**, and **raunormine**. Other alkaloids present in *Rauwolfia* include **serpine**, **serpagine** (**serpagene** or **raupine**), **rauhimbine**, **reserpinine**, **ajmaline** (or rauwolfine), **serpentine**, **ajmalicine**, etc. **Syrosingopine**, a synthetic derivative prepared from reserpine by hydrolysis and reesterification, is used in hypertension (Grollman, 1962).

PHARMACOLOGY

Reserpine and its related alkaloids exert their principal action on the posterior hypothalamus and the reticular formation of the brain stem, inducing a facilitative effect on the transmission of impulses in these areas. This effect on the central nervous system is readily demonstrated experimentally when the drugs are administered to the monkey. Following its injection, a state of repose develops gradually during which it is possible to handle and stroke a previously untamed animal safely with the bare hands. The animal shows less natural curiosity and interest than it normally displays toward its surroundings. Although it appears to be asleep,

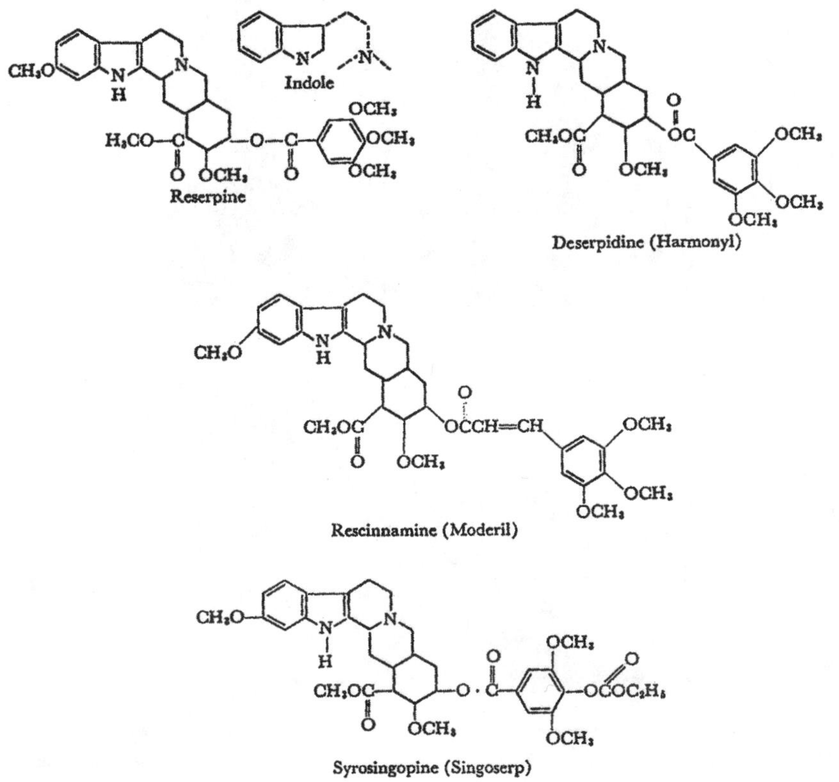

Figure 126 The structures of reserpine and related compounds.

it may be aroused readily; it eats normally and does not appear to sleep more than usual. The electroencephalogram remains normal in contrast to the sleep pattern seen following the administration of barbiturates. The drug thus acts primarily as a tranquilizer and inactivating agent rather than as a soporific or hypnotic agent. In patients also the drug induces quiet and ease with an inner calm and a less active response to external stimuli. Spontaneous activity is reduced and although the patients often complain of being sleepy they are not actually put to sleep. However, drowsiness, mild dizziness, and fatigue often accompany reserpine therapy (Figure 127).

The peripheral effects of reserpine are similar to those resulting from stimulation of the parasympathetic nervous system. Thus, there is bradycardia, constriction of the pupil, increased motor activity of the intestinal tract, and an increase in the volume and hydrochloric acid content of the gastric juice following its intravenous administration. This increased secretion may be effectively prevented for several hours by the use of an anticholinergic drug. However, the tranquilizing effect of well tolerated doses of reserpine may bring about a secondary decrease in gastric acidity. Hence, no consistent change in volume, free hydrochloric acid concentration, or its

Figure 127 The sedative and tranquilizing effects of reserpine. The photographs show a normal cat before (upper) and two hours after (lower) the injection intramuscularly of reserpine (1 mg/kg of body weight).

output is observed after prolonged oral administration. The increased motility of the bowel may cause diarrhea, which is noted in the unanesthetized dog following administration of the drug. When administered parenterally, reserpine induces a gradual increase in bladder tone that is not blocked by ganglionic blocking agents. The effect of reserpine on the cardiovascular activity, intestinal motility, and pupillary size suggest that the posterior hypothalamus is its primary site of action since it does not antagonize the effects of sympaticomimetic substances such as epinephrine or norepinephrine nor does it potentiate the effects of acetylcholine.

In the unanesthetized dog, reserpine selectively depresses the sympathetic and induces facilitation of the parasympathetic centers in the diencephalon. The latter effect accounts for the bradycardia, miosis, aggravation of bronchial asthma, renal and biliary colic, and ulcerative colitis observed in some patients receiving the drug.

MECHANISM OF ACTION

Large single doses of reserpine (1 to 5 mg/kg) in animals cause a progressive decline in the **serotonin** content in the intestine. After 16 hours about 90% of it has disappeared, the level remaining low for about 48 hours after which it gradually rises to its normal level again in the course of about 7 days. Similar changes occur in the platelets. The brain is more sensitive to the effects of reserpine so that 80% of the serotonin originally present disappears within an hour and the liberation of serotonin may be observed in doses as low as 0.05 mg/kg.

It appears therefore that the main action of reserpine is to produce a biochemical change so that the cells no longer retain serotonin at a high concentration. In other words, the binding of serotonin is prevented. Thus, after the administration of **5-hydroxytryptophan** (a precursor of serotonin) to rabbits pretreated with reserpine, serotonin is rapidly formed but remains in a free form. Presumably, free serotonin, before it is metabolized to **5-hydroxyindole acetic acid** by amine oxidase, acts to produce the central action of reserpine. Reserpine affects the serotonin binding sites by an unknown mechanism releasing serotonin but it also causes a depletion from the body stores of **epinephrine** and **norepinephrine**.

Reserpine enters the brain rapidly after its intravenous administration reaching a maximum level almost immediately. The drug is metabolized rapidly, presumably in the liver, and disappears from the brain in 8 hours. However, the inability to take up serotonin persists with its pharmacological effect since apparently new binding sites for serotonin must be produced. This explains the persistence of the tranquilizing effect after reserpine medication is discontinued and the cumulative effect of the customary small daily doses of this drug. Although the drug is present in the body for only a few hours and a single dose is too small to affect serotonin or exert a tranquilizing action, each dose inactivates a sufficient number of the binding sites of serotonin so that the typical effects of free serotonin are produced. This phenomenon also explains the appearance of signs of overdosage in patients who have been maintained on the *Rauwolfia* drugs for a long time.

Like serotonin, the *Rauwolfia* alkaloids potentiate the action of barbiturates and alcohol. This potentiating action is mediated through serotonin. The alkaloids also exert an antiveratrinic and antiarrhythmic action on the isolated heart.

RESERPINE PREVENTS THE STORAGE OF NOREPINEPHRINE

Neurochemical Basis of Adrenergic Transmission

Dopamine, norepinephrine, and epinephrine are classified as catecholamines and are synthesized according to the scheme depicted in Figure 70.

Tyrosine is converted to dopa by the rate-limiting enzyme **tyrosine hydroxylase**, which requires **tetrahydrobiopterin**, and is inhibited by **α-methyltyrosine**. Dopa is decarboxylated to dopamine by **L-aromatic amino acid decarboxylase**, which requires **pyridoxal phosphate (vitamin B$_6$)** as a coenzyme. **Carbidopa**, which is used with levodopa in the treatment of parkinsonism, inhibits this enzyme. Dopamine is converted to norepinephrine by **dopamine β-hydroxylase**, which requires **ascorbic acid** (vitamin C), and is inhibited by **diethyldithiocarbamate**. Norepinephrine is converted to epinephrine by **phenylethanolamine N-methyltransferase** (PNMT), requiring **S-adenosylmethionine**. The activity of PNMT is stimulated by corticosteroids.

The catecholamine-synthesizing enzymes are not only able to synthesize dopamine and norepinephrine from a physiologically occurring substrate such as levodopa, but also from exogenous substrates such as **α-methyldopa**, which is converted to **α-methyldopamine** and in turn to **α-methylnorepinephrine**. α-Methyldopamine and α-methylnorepinephrine are called **false transmitters** and, in general (except for α-methylnorepinephrine), are weaker agonists. α-Methyldopa is used in the management of hypertension.

In addition to being synthesized in the peripheral nervous system, dopamine is synthesized in the corpus striatum and in the mesocortical, mesolimbic, and tuberoinfundibular systems. Norepinephrine is synthesized and stored primarily in sympathetic noradrenergic nerve terminals, as well as in the brain and the adrenal medulla. Epinephrine is synthesized and stored primarily in the adrenal medulla and, to a certain extent, in the hypothalamic nuclei.

In sympathetic nerve terminals, as well as the brain, the adrenal medulla, and sympathetic postganglionic terminals, there are osmophilic granules (synaptic vesicles) that are capable of storing high concentrations of catecholamine (a complex with adenosine triphosphate, or ATP, and protein). The stored amines are not metabolized by the intersynaptosomal mitochondrial enzyme (monoamine oxidase).

In addition to releasing norepinephrine (through exocytosis), the stimulation of sympathetic neurons also releases ATP, storage protein, and dopamine β-hydroxylase. The released norepinephrine interacts with receptor sites located postsynaptically (α_1) to produce the desired effects.

The action of norepinephrine is terminated by reuptake mechanisms, two of which have been identified: **Biogenic amine Uptake 1** is located in the presynaptic membrane, requires energy for the transport, is sodium and temperature dependent, and is inhibited by ouabain (a cardiac glycoside), cocaine (a local anesthetic), and imipramine (an antidepressant). **Biogenic amine Uptake 2** is located extraneuronally in various smooth muscles and glands, requires energy, and is temperature dependent. Approximately 20% of the amine is either taken up by the Uptake 2 mechanism or is metabolized.

Catecholamine Metabolism

There are two enzymes capable of metabolizing catecholamines. The first is **monoamine oxidase** (MAO), a mitochondrial enzyme that oxidatively deaminates catecholamines, tyramine, serotonin, and histamine. MAO is further subclassified as either monoamine oxidase A, which metabolizes norepinephrine and is inhibited by tranylcypromine, and monoamine oxidase B, which metabolizes dopamine and is inhibited by L-deprenyl. **Catechol-O-methyltransferase (COMT)**, a soluble enzyme present mainly in the liver and kidney, is also found in postsynaptic neuronal elements. About 15% of norepinephrine is metabolized postsynaptically by COMT.

ADRENERGIC NEURONAL BLOCKING DRUGS

Reserpine (Serpasil) depletes the store of catecholamine peripherally and centrally and attenuates but does not abolish sympathetic reflexes. Reserpine is useful in the management of mild to moderate hypertension. Its onset of action is very slow (2 to 3 weeks) when given orally. The side effects of reserpine are manifested by cholinergic hyperactivity such as diarrhea, bradycardia, and nasal stuffiness.

Reserpine can activate a peptic ulcer (cholinergic dominance) and cause depression (depletes norepinephrine stores). Reserpine and propranolol have potential cardiac depressant activity and should not be used together.

Although the *Rauwolfia* compounds were the pioneers among **tranquilizing agents (neuroleptics)**, they are very little used for that purpose any more for two main reasons: first, they tend to cause serious depressions; second, the **phenothiazine derivatives** and even newer neuroleptics such as **clozapine** are fully as effective, act much more rapidly, and are infrequently followed by depression.

SIDE EFFECTS

Although the *Rauwolfia* alkaloids show a low degree of acute toxicity, their continued use may be attenuated by serious side effects. One of the most troublesome symptoms observed in the therapeutic use of the *Rauwolfia* alkaloids is nasal congestion and stuffiness, which may be so severe as to necessitate discontinuance of therapy. Increased motility of the bowel, diarrhea, and increased gastric secretion resulting from its action on the autonomic nervous system are observed frequently. Skin eruptions, epistaxis, and peptic ulceration are rare complications.

The more serious side effects observed in patients receiving reserpine for long periods involve the nervous system and include withdrawal from the environment, unhappiness, depression, lack of ambition, crying spells, introspection, and lethargy. Bizarre dreams and nightmares may necessitate cessation of therapy. **Loss of libido** is a frequent complaint in the male. A parkinsonian-like syndrome may result from prolonged therapy or excessive dosage.

50

RHUBARB

Rhubarb (the "wondrous drug") is a purgative agent, and is an effective remedy in "Oketsu Syndrome," a blood stagnation syndrome.

INTRODUCTION

Rhubarb, the rhizomes of **Rheum palmatum L., R. tanguticum Maxim., R. officinale Baill., Baill., R. coreanum Nakai,** and **R. undulatum L.,** is used in remedies for blood stagnation syndrome (called **"Oketsu syndrome"** in Japanese traditional medicine) as well as a **purgative agent** in Japanese, Korean, and Chinese traditional medicines. Among them, the rhizome of *R. undulatum*, a Korean rhubarb, is considered to have less purgative effect but more potent effect on Oketsu syndrome than other kinds of rhubarbs.

RHUBARB AND ITS LAXATIVE PROPERTIES

The medicinal parts are the dried underground parts and the underground parts freed from the stem remnants, the smaller roots, and most of the root bark in the dried form. **Garden rhubarb** is *Rheum ponticum*.

The inflorescence is an erect panicle foliated to the tip. The flowers consist of narrow, red, pink, or whitish yellow tepals, which are curved and located far back in the mature flowers to facilitate wind pollination. The fruit is red-brown to brown, and oval. The fruit is angular, about 10.2 to 7.8 mm wide and usually has scarious wings. The nutlet is 6 to 10 mm long and 7 mm in diameter.

The plant is a large, sturdy herbaceous perennial. The stem grows to over 1.5 m high. The leaves are orbicular-cordate, palmate lobed, somewhat rough on the upper surface or smooth and three to five ribbed. The lobes are oblong-ovate to lanceolate, dentate or pinnatisect. The root system consists of a tuber, which after a number of years measures 10 to 15 cm in diameter and has arm-thick lateral roots.

The plant is indigenous to the western and northwestern provinces of China and is cultivated in many regions around the world. The main producers are China and Russia.

Rhubarb consists of the dried underground parts of *R. paimatum, R. officinale,* or of both species. Stem parts, roots, and most of the bark are removed from the rhizomes.

Rhubarb contains:

- **Anthracene derivatives** (3 to 12%): chief components 1- or 8β-glucosides of the **aglycones rheumemodin, aloe-emodin, rhein, chrysophanol, physcion** (together 60 to 80%), 8,8′-diglucosides of dianthrones (10 to 25%), including, among others, **sennosides A and B**
- **Tannins:** gallo tannins, including, among others, **galloylglucose, galloyl-saccharose, lindleyine, isolindleyine**
- **Flavonoids** (2 to 3%)
- **Naphthohydroquinone glycosides**

Rhubarb is a laxative, primarily due to the influence of the herb on the motility of the colon, inhibiting stationary and stimulating propulsive contractions. This results in an accelerated intestinal passage and, because of the active chloride secretion, increases the water and electrolyte content of stool.

Rhubarb is indicated in constipation, bowel movement relief with anal fissures, hemorrhoids, and after recto-anal surgery and in preparation for diagnostic interventions of the gastrointestinal tract.

Rhubarb is contraindicated in cases of intestinal obturation, acute inflammatory intestinal disease, appendicitis, and abdominal pain of unknown origin. Spasmodic gastrointestinal complaints can occur as a side effect to the purgative effect of the drug. Long-term use leads to losses of electrolytes, in particular K^+ ions, and as a result of this to hyperaldosteronism, inhibition of intestinal motility, and enhancement of the effect of cardioactive steroids; in rare cases it also leads to heart arrhythmias, nephropathies, edemas, and accelerated bone deterioration.

The question of the increase in probability of the appearance of carcinomas in the colon following long-term administration of anthracene drugs has not yet been fully clarified. Recent studies show no connections between the administration of anthracene drugs and the frequency of carcinomas in the colon. Potassium deficiency can cause an increase in the effect of cardiac glycosides.

CONSTIPATION

Constipation may be defined as the passage of excessively dry stools, infrequent stools, or stools of insufficient size. Constipation is a symptom and not a disease. It may be of brief duration (e.g., when one's living habits or diet change abruptly) or it may be a life-long problem, as occurs in congenital aganglionosis of the colon (**Hirschsprung's disease**). The causes of constipation are multiple and include the following:

- Functional causes
    Fiber-deficient diets
    Variants of irritable bowel syndrome
    Debilitation and extreme old age
- Colonic diseases
    Chronic obstructive lesions (e.g., tumors or strictures)
    Ulcerative colitis
    Collagen vascular diseases

- Rectal diseases
    Stricture (e.g., ulcerative colitis)
    Painful conditions (fissure or abscess)
- Neurologic diseases
    Hirschsprung's disease
    Spinal cord injuries and disease
    Parkinson's disease
    Cerebral tumors and cerebrovascular disease
- Metabolic diseases
    Porphyria
    Hypothyroidism
    Hypercalcemia
    Pheochromocytoma
    Uremia

Use of the following drugs may also lead to constipation:

- Anticholinergic drugs contained in many of the over-the-counter medications
- Antiparkinsonian drugs possessing anticholinergic properties (e.g., **trihexyphenidyl** and **ethopropazine**)
- Antihistaminic drugs with anticholinergic properties (e.g., **diphenhydramine**)
- Neuroleptics with anticholinergic properties (e.g., **thioridazine**)
- Antidepressants with anticholinergic properties (e.g., **amitriptyline**)
- Anticonvulsants with anticholinergic properties (e.g., **carbamazepine**)
- Analgesics (e.g., morphine, codeine, and **diphenoxylate**)
- Ganglionic blocking agents (e.g., **mecamylamine hydrochloride** and **pempidine**)
- Antacids (calcium- or aluminum-containing compounds)

LAXATIVES AND CATHARTICS

Although often used interchangeably, the terms *laxative* and *cathartic* have slightly different meanings. A laxative effect refers to the excretion of a soft, formed stool; catharsis implies a more fluid and complete evacuation.

Irritants

Irritant agents used in the treatment of constipation include **cascara sagrada**, **castor oil**, **senna**, **rhubarb**, **phenolphthalein**, and **acetphenolisatin**. Phenolphthalein is a constituent of many over-the-counter preparations, including Correctol, Ex-Lax, and Feen-A-Mint. Most of these agents, with the exception of castor oil, are slow in their onset of action (24 hours).

Phenolphthalein is thought to exert its effect by inhibiting the movement of water and sodium from the colon into the blood and by stimulating mucus secretion. If misused on a prolonged basis, a consequential loss of mucus may lower the plasma protein level. Caster oil is hydrolyzed to **ricinoleic acid**, the active cathartic. It has an onset of action of 2 to 6 hours.

The misuse of any of these agents has been shown to cause hypokalemia, dehydration, and a cathartic colon (resembling ulcerative colitis). Phenolphthalein-containing products may color alkaline urine red.

Bulk Saline Laxatives

Bulk saline laxatives fall into two categories: inorganic salts — magnesium sulfate, magnesium citrate, milk of magnesia, sodium sulfate, and sodium phosphate — and organic hydrophilic colloids — **methylcellulose**, **carboxymethylcellulose** (Metamucil), plantago seed, agar, psyllium, bran, and fruits. They exert their effects by absorbing and retaining water, increasing bulk, stimulating colonic peristaltic movements, and lubricating and hydrating the desiccated fecal materials.

These agents are more effective when administered with water. The onset of action of organic salts is relatively fast (2 to 6 hours) and that of colloids is relatively slow (1 to 3 days). These agents, which are very effective and safe, should not be used when the intestinal lumen has been narrowed. The prolonged use of saline cathartics may create problems for certain individuals. For example, magnesium salts have been known to cause hypermagnesemia, coma, and death in patients with renal insufficiency. Sodium salts may also be responsible for causing congestive heart failure.

Lubricants

The lubricants consist of mineral oil and **dioctyl sodium sulfosuccinate** (Colace). Colace is used in the pharmaceutical industry as an emulsifying and dispersing substance. Both agents are taken orally. These agents, which do not influence peristalsis, soften desiccated stools or delay the desiccation of fecal materials. They are especially useful in patients with painful bowel movements resulting from inspissated stools or inflammation of the anal sphincter such as occurs with hemorrhoids or anal fissures. Colace is also useful for patients in whom the consequences of "straining at stool" may be harmful.

When used for a long time, mineral oil may come to interfere with the absorption of fat-soluble vitamins and other essential nutrients. Lipid pneumonitis may evolve if mineral oil is used as a vehicle for drugs that are taken nasally.

OTHER USES OF LAXATIVES

- **Poisoning**. Laxatives are used to hasten the elimination and reduce the absorption of a poison that has been taken.
- **Anthelmintics.** Laxatives are used before and after treatment with anthelmintic drugs.
- **Radiology.** Laxatives are used to clean the gastrointestinal tract before radiographic techniques are performed.

ANTIOXIDANT PROPERTIES OF RHUBARB

The Japanese scientists believe that the **anti-oketsu effects** of rhubarb are brought about by rhubarb inhibiting the production of nitric oxide, inhibiting platelet aggregation, having antiallergic effects, and anti-inflammatory properties. Matsuda et al. (2001) studied and reported that the methanolic extracts from five kinds of rhubarb were found to show scavenging activity for **1,1-diphenyl-2-picrylhydrazyl** (DPPH) **radical** and **superoxide anion radical** ($O_2^{\bullet-}$) generated by the **xanthine/xanthine oxidase** system and/or on lipid peroxidation by ***tert*-butyl hydroperoxide** (*t*-BuOOH) in the erythrocyte membrane ghost system.

Two new **anthraquinone glucosides** were isolated from the rhizome of *R. undulatum* L. together with two anthraquinone glucosides, a naphthalene glucoside, and 10 **stilbenes**. In the screening test for radical-scavenging activity of rhubarb constituents, stilbenes and a **naphthalene glucoside** showed activity, but **anthraquinones** and **sennosides** did not. In addition, most stilbenes inhibited lipid peroxidation of erythrocyte membrane by *tert*-butyl hydroperoxide. Detailed examination of the scavenging effect on various related compounds suggested the following structural requirements; (1) phenolic hydroxyl groups are essential to show the activity; (2) galloyl moiety enhances the activity; (3) glucoside moiety reduces the activity; (4) dihydrostilbene derivatives maintain the scavenging activity for the DPPH radical, but they show weak activity for superoxide anion. In addition, several stilbenes with both the 3-hydroxyl and 4'-methoxyl groups inhibited xanthine oxidase.

51

SAFFRON

- Has antitumor effects
- Scavenges free radicals
- Exhibits hypolipidemic properties
- Attenuates alcohol-induced memory impairment
- Is an aromatic and tasty spice in preparing many foods

INTRODUCTION

Crocus sativus L., commonly known as **saffron**, is used in folk medicine as an **antispasmodic, eupeptic, gingival sedative, anticatarrhal, nerve sedative, carminative, diaphoretic, expectorant, stimulant, stomachic, aphrodisiac,** and **emmenagogue**. Furthermore, modern pharmacological studies have demonstrated that saffron extract or its active constituents have **antitumor effects, radical scavenger properties**, and **hypolipemic effects** (see Rios et al., 1996).

ANTIOXIDANT EFFECT OF SAFFRON

Saffron extract contains many **carotenoids** such as **crocetin, crocetin di-glucose ester, crocetin gentiobiose glucose ester**, and **crocin** (crocetin di-gentiobiose ester), whose chemical structures are shown in Figure 128. These carotenoids scavenge free radicals, especially superoxide anions, and thereby may protect cells from oxidative stress. Indeed, it has been demonstrated that these carotenoids are useful in **sperm cryoconservation** and in protecting heptocytes from toxins.

Accumulating evidence suggests that cellular stress induced by free radicals is responsible for a variety of central nervous system (CNS) neurodegenerative disorders, including **Alzheimer's disease, Parkinson's disease**, and **amyotrophic lateral sclerosis**, and in pathological conditions such as ischemia and excitotoxicity. The importance of preventing oxidative stress-induced neuronal damage has recently been emphasized. Because of its powerful antioxidant activity, saffron extract, crocetin, or crocin could be useful in the therapy of brain neurodegenerative disorders, although their effects have not yet been evaluated.

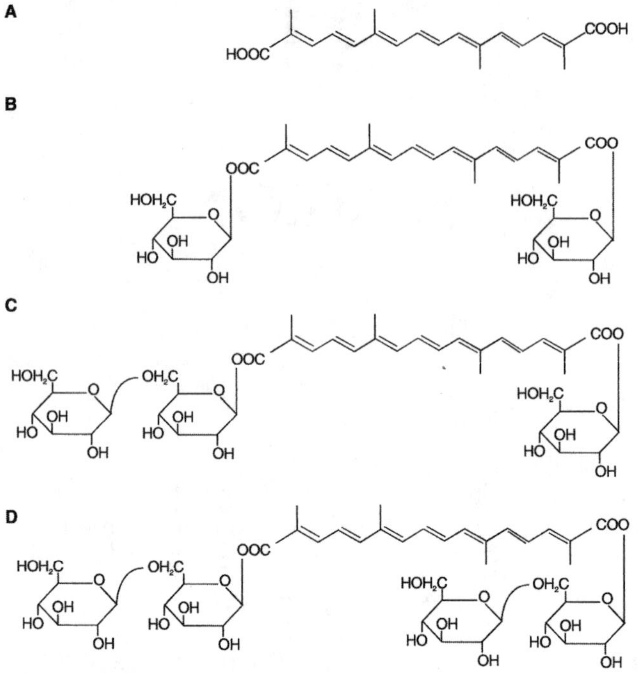

Figure 128 Chemical structures of crocetin (A), crocetin diglucose ester (B), crocetin gentiobiose glucose ester (C), and crocin (D).

EFFECT OF SAFFRON ON LEARNING AND MEMORY

CNS neurodegenerative disorders often accompany the impairment of memory and other cognitive functions. This impairment is probably due to selective neuronal death in the cerebral cortex and hippocampus, brain regions that are crucially involved in learning and memory. It has recently been found that the alcohol extract of pistils of *C. sativus* **L.** (CSE) affects learning and memory in mice. Oral administration of CSE (125 to 500 mg/kg) alone had no effect on learning behavior of mice in passive avoidance tests, but significantly improved ethanol-induced impairment of memory acquisition.

The effect of CSE was reproduced by **crocin**, a major constituent in CSE. Oral administration of crocin (50 to 200 mg/kg) alone had no effect, but significantly improved ethanol-induced impairment of memory acquisition in mice. Picrocrocin, another constituent in CSE, was ineffective in the same experiment. These results indicate that **crocin** has a specific, preventive effect on ethanol-induced impairment of learning and memory.

POSSIBLE MOLECULAR TARGET OF CROCIN

The induction of **long-term potentiation** (LTP) in CA1 and dentate gyrus regions of the hippocampus essentially requires the activation of the **N-methyl-D-aspartate (NMDA)** type of glutamate receptors. Ca^{2+} influx into postsynaptic cells through NMDA receptors channels the activation of enzymes linked to the induction of LTP. Furthermore, ethanol has been considered to inhibit the induction of LTP by suppressing the activity of the NMDA receptor channel complex. In rat hippocampal slices and single hippocampal neurons, ethanol inhibited NMDA receptor-mediated synaptic excitation, and the inhibitory effect of ethanol was significantly attenuated by crocin, but not by crocetin diglucose ester. Crocin is likely to prevent ethanol-induced inhibition of hippocampal LTP by antagonizing the inhibitory effect of ethanol on NMDA receptors, although it is not clear whether crocin acts directly on the NMDA receptor channel complex or indirectly modulates NMDA receptor functions (see Figure 31). In addition, since the antagonizing effect of crocin against ethanol on NMDA receptor functions was smaller than that observed in behavioral and LTP experiments, this action of crocin cannot account solely for the antagonism against **ethanol-induced memory impairment**. Further investigations are necessary to explore other molecular targets of crocin. Saffron extract contains many carotenoids with powerful antioxidant effects, and may protect CNS neurons from oxidative damage. Saffron extract or its constituents, especially crocin, improves the impairment of certain types of learning and memory. Therefore, saffron and its active constituents should be useful for the therapy of CNS neurodegenerative disorders accompanying the impairment of memory and cognitive functions.

In addition, the finding that crocin specifically prevents ethanol-induced memory impairment and inhibition of hippocampal LTP is of great significance in the following respects. First, the consistency of effects of crocin on learning behavior and LTP further supports the hypothesis that hippocampal LTP is involved in memory formation. Second, crocin should be useful as a new pharmacological tool for studying the mechanism of ethanol-induced memory impairment. Further investigation on molecular target(s) of crocin may give new insight into the mechanism underlying memory formation and LTP (Abe and Saito, 2000).

52

SALICYLATE

Salicylate (aspirin) is found in willow, wintergreen, and sweet birch, and

- Is an antirheumatic agent
- Causes analgesia
- Reduces fever
- Has anti-inflammatory activity
- Inhibits platelet aggregation
- Has neuroprotective effects in an animal model of Parkinson's disease
- May bring about chemoprevention of colorectal cancer
- May prevent Alzheimer's disease
- Could be used in the treatment of reflux esophagitis

SOURCE

Salicinum, Salicin, $C_{13}H_{18}O_7$, is a glucoside obtained from several species of *Salix*, the **willow**, and *populus*, the **poplar**, trees of the nat. ord. **Salicaceae**. It is found also in *Gaultheria procumbens*, the **wintergreen**, nat. ord. **Ericaceae**; and in *Betula lenta*, the **sweet birch**, nat. ord. **Betulaceae**; the volatile oils of which, distilled from the leaves of the former and from the bark of the latter, consist almost entirely of **methyl salicylate**.

Salicin occurs in colorless or white and silky, shining crystalline needles, or a crystalline powder, odorless, of very bitter taste, permanent in the air, of neutral reaction, soluble in 28 parts water and in 30 parts alcohol, in 0.7 parts of boiling water and in 2 parts of boiling alcohol; almost insoluble in ether or chloroform.

Salix nigra, the *pussy willow*, grows along streams in the Southern states. It has been used in the late 19th century as a "**sexual sedative**."

Acidum salicylicum, *salicylic acid*, $HC_7H_5O_3$, is a monobasic organic acid, existing naturally in combination in various plants, but generally prepared synthetically from phenol. It occurs in light, fine, white prismatic needles, or a crystalline powder, odorless, of sweetish, afterward acrid taste and acid reaction, permanent in the air; soluble in about 450 parts cold water, but readily soluble in water containing 8% of borax or 10% of sodium phosphate. It is soluble in 2½ parts alcohol, in 14 parts boiling water, in 2 parts ether, in 80 parts chloroform, and is very soluble in boiling alcohol.

OFFICIAL SALICYLATES USED IN THE EARLY 20TH CENTURY

- **Lithii Salicylas,** *Lithium Salicylate*, $LiC_7H_5O_3$, is a white, or grayish-white powder, odorless, sweetish, very soluble in water and in alcohol.
- **Sodii Salicylas,** *Sodium Salicylate*, $NaC_7H_5O_3$, is a white, amorphous powder, soluble in ¾ part water and in 6 parts alcohol, also in glycerin.
- **Strontii Salicylas,** *Strontium Salicylate*, is a white, crystalline powder, soluble in 18 parts water and in 66 parts alcohol.
- **Methylis Salicylas,** *Methyl Salicylate*, is an ester, produced synthetically, and is the principal constituent of Oil of Gaultheria and Oil of Betula. It is soluble in all proportions in alcohol or glacial acetic acid.
- **Phenylis Salicylas,** *Phenyl Salicylate, Salol*, $C_{13}H_{10}O_3$, is the salicylic ester of phenyl, and occurs as a white, crystalline powder, odorless and almost tasteless, nearly insoluble in water, soluble in 10 parts alcohol, and very soluble in ether, chloroform, and oils. On warming with an alkali, it splits up into salicylic acid 60, and phenol 40, frequently repeated, in compressed tablets or in cachets, or suspended by mucilage of acacia or of tragacanth.
- **Oleum Betulae,** *Oil of Betula (Oil of Sweet Birch)*, is a volatile oil distilled from the bark of *Betula lenta*, the sweet birch. It is identical with methyl salicylate and nearly identical with Oil of Gaultheria.
- **Oleum Gaultheriae,** *Oil of Gaultheria, Oil of Wintergreen*, consists almost entirely of methyl salicylate, and is nearly identical with the preceding.

ANALGESICS–ANTIPYRETICS AND ANTI-INFLAMMATORY AGENTS

The development of aspirin was a significant landmark in the history of medicine because it stimulated the development of a family of medicines, collectively known as the **nonsteroidal anti-inflammatory drugs** (NSAIDs). NSAIDs such as **ibuprofen, naproxen**, and **sulindac** are valuable drugs for the alleviation of pain, inflammation, and fever and they are commonly prescribed for the treatment of rheumatoid disorders such as arthritis. The world market now exceeds $6 billion for NSAIDs (Vainio and Morgan, 1997).

Despite the introduction of many new drugs, **aspirin (acetylsalicylic acid)** is still the most widely prescribed analgesic–antipyretic and anti-inflammatory agent and is the standard for the comparison and evaluation of the others. Prodigious amounts of the drug are consumed in the United States; some estimates place the quantity as high as 10,000 to 20,000 tons annually. Aspirin is the common household analgesic; yet, because the drug is so generally available, its usefulness often is underrated. Despite the efficacy and safety of aspirin as an analgesic and antirheumatic agent, it is necessary to be aware of its role in **Reye's syndrome** and as a common cause of lethal drug poisoning in young children, as well as its potential for serious toxicity if used improperly.

Salicylates and allied compounds have analgesic, antipyretic, uricosuric, and anti-inflammatory properties. Their mechanisms of action differ from those of the anti-inflammatory steroids and the opioid analgesics. They are classified into the following categories:

Salicylate derivatives
   Acetylsalicylic acid (aspirin)
Pyrazolone derivatives
   Phenylbutazone (Butazolidin)
   Oxyphenbutazone (Oxalid, Tandearil)
   Sulfinpyrazone (Anturane)
Paraaminophenol derivatives
   Acetaminophen (Tylenol, Datril)
   Phenacetin (Acetophenetidin)
Propionic acid derivatives
   Ibuprofen (Motrin)
   Naproxen (Naprosyn)
   Fenoprofen (Nalfon)
   Flurbiprofen (Ansaid)
   Ketoprofen (Orudis)
Other related compounds
   Indomethacin (Indocin)
   Sulindac (Clinoril)
   Mefenamic acid (Ponstel)
   Tolmetin (Tolectin)
   Piroxicam (Feldene)
   Diclofenac sodium (Voltaren)
Drugs used in the treatment of gout
   Colchicine
   Allopurinol
   Uricosuric agents
Drugs used in the treatment of arthritis
   Aurothioglucose (Solganal)
   Gold sodium thiomalate (Myochrysine)
   Auranofin (Ridaura)

The pharmacology of acetylsalicylic acid (aspirin) is discussed in detail as a prototype drug, and all the other drugs are compared to it.

MECHANISM OF ACTION OF ASPIRIN

The mechanism by which these drugs act to reduce mild to moderate pain is based on the relationship between drugs such as aspirin and prostaglandin synthesis. Studies in humans have shown that intravenous administration of certain **prostaglandins** elicits headache and pain and produces hyperalgesia, sensitizing the individual to stimuli that normally would not produce pain. Aspirin and related compounds inhibit the enzyme **cyclooxygenase** and prevent the formation of **prostaglandin endoperoxides**, PGG_2 and PGH_2, which are normally formed from **arachidonic acid**.

ASPIRIN AND ITS USES

Aspirin-Induced Analgesia

Unlike the narcotic analgesics such as **morphine**, aspirin does not depress respiration, is relatively nontoxic, and lacks addiction liability. Aspirin is a weak or mild analgesic that is effective for ameliorating short, intermittent types of pain such as neuralgia, myalgia, and toothache. It does not have the efficacy of morphine and cannot relieve the severe, prolonged, and lancinating types of pain associated with trauma such as burns or fractures. Like morphine, it produces analgesia by raising the pain threshold in the thalamus, but, unlike morphine, it does not alter the patient's reactions to pain. Because aspirin does not cause hypnosis or euphoria, its site of action has been postulated to be subcortical (see also Figure 111). In addition to **raising the pain threshold**, the anti-inflammatory effects of aspirin may contribute to its analgesic actions. However, no direct association between the anti-inflammatory and analgesic effects of these compounds should be expected. For example, aspirin has both analgesic and anti-inflammatory properties, whereas **acetaminophen** has analgesic but not anti-inflammatory properties. Furthermore, potent anti-inflammatory agents such as **phenylbutazone** have only weak analgesic effects.

Aspirin-Induced Antipyresis

Aspirin does not alter the normal body temperature, which is maintained by a balance between heat production and dissipation. In a fever associated with infection, increased oxidative processes enhance heat production. Aspirin acts by causing cutaneous vasodilation, which prompts perspiration and enhances heat dissipation. This effect is mediated via the hypothalamic nuclei, as proved by the fact that a lesion in the preoptic area suppresses the mechanism through which aspirin exerts its antipyretic effects. The antipyretic effects of aspirin may be due to its inhibition of hypothalamic prostaglandin synthesis. Although aspirin-induced diaphoresis contributes to its antipyretic effects, it is not an absolutely necessary process, since antipyresis takes place in the presence of atropine.

Uricosuric Effects of Aspirin

Small doses (600 mg) of aspirin cause hyperuricemia, but large doses (>5 g) have a uricosuric effect. Aspirin inhibits uric acid resorption by the tubules in the kidneys.

Anti-inflammatory Effects of Aspirin

Aspirin has an anti-inflammatory action as well as antirheumatic and antiarthritic effects, and may therefore be used in the treatment of rheumatic fever. However, it cannot alter the cardiac lesion and other visceral effects of the disease. Aspirin is extremely effective in managing rheumatoid arthritis and allied diseases involving the joints, such as ankylosing spondylitis and osteoarthritis. It is thought that aspirin and indomethacin exert their anti-inflammatory effects by inhibiting prostaglandin synthesis through the inhibition of cyclooxygenase. The presynthesized prostaglandins are released during a tissue injury that fosters inflammation and pain.

Furthermore, aspirin reduces the formation of prostaglandin in the platelets and leukocytes, which is responsible for the reported hematological effects associated with aspirin.

Aspirin and Cardiovascular Disease

The current thinking concerning the role of aspirin in the prevention of cardio-vascular disease is that it is beneficial in the event of myocardial infarction and stroke. It is effective because, in platelets, small amounts of aspirin acetylate irreversibly and bind to the active site of **thromboxane A$_2$**, a potent promoter of platelet aggregation.

Aspirin and the Premenstrual Syndrome

The menstrual cycle is associated with two potentially incapacitating events: dysmenorrhea and the premenstrual syndrome. Substantial evidence indicates that the excessive production of prostaglandin F$_{2\alpha}$ is the major source of painful menstruation. The NSAIDs approved for the treatment of dysmenorrhea are aspirin, **ibuprofen**, **mefenamic acid**, and **naproxen**.

SIDE EFFECTS OF ASPIRIN

Effects of Salicylate on Respiration

Aspirin both directly and indirectly stimulates respiration. In analgesic doses, aspirin increases oxygen consumption and carbon dioxide production. However, increased alveolar ventilation balances the increased carbon dioxide production, thus the partial pressure of CO$_2$ (PCO$_2$) in plasma does not change. In the event of salicylate intoxication (e.g., 10 to 12 g of aspirin given in 6 to 8 hours in adults, and an even smaller dosage in children, whose brains are far more sensitive to salicylate intoxication), salicylate stimulates the medullary centers directly and this causes hyperventilation characterized by an increase in the depth and rate of respiration. The PCO$_2$ level declines, causing hypocapnia, and the blood pH increases, causing respiratory alkalosis. The low PCO$_2$ then decreases the renal tubular resorption of bicarbonate and compensates for the alkalosis.

If the salicylate level continues to rise, the respiratory centers become depressed, the PCO$_2$ level becomes elevated, and the blood pH becomes more acidic, causing respiratory acidosis. Dehydration, reduced bicarbonate levels, and the accumulation of **salicylic acid**, **salicyluric acid** resulting from metabolism of aspirin, and lactic and pyruvic acid resulting from deranged carbohydrate metabolism may cause metabolic acidosis.

Treatment of Aspirin Poisoning

The supportive treatment of aspirin poisoning may include gastric lavage (to prevent the further absorption of salicylate), fluid replenishment (to offset the dehydration and oliguria), alcohol and water sponging (to combat the hyperther-mia), the administration of vitamin K (to prevent possible hemorrhage), sodium bicarbonate administration (to combat acidosis), and, in extreme cases, peritoneal dialysis and exchange transfusion.

The Gastrointestinal Effect of Aspirin

Although innocuous in most subjects, therapeutic analgesic doses of aspirin may cause epigastric distress, nausea, vomiting, and bleeding. Aspirin can also exacerbate the symptoms of peptic ulcer, characterized by heartburn, dyspepsia, and erosive gastritis. Furthermore, compounds possessing anti-inflammatory properties (aspirin, phenylbutazone, and oxyphenbutazone) are associated with a higher incidence of gastrointestinal toxicity than compounds devoid of anti-inflammatory properties (phenacetin and acetaminophen).

The Hematopoietic Effect of Aspirin

Aspirin reduces the leukocytosis associated with acute rheumatic fever. When given on a long-term basis, it also reduces the hemoglobin level and the hematocrit. Aspirin use can cause reversible hypoprothrombinemia by interfering with the function of vitamin K in prothrombin synthesis. Therefore, aspirin should be used with caution in patients with vitamin K deficiency, preexisting hypoprothrombinemia, or hepatic damage; in patients taking anticoagulants; and in patients scheduled for surgery. Aspirin leads to hemolytic anemia in individuals with **glucose 6-phosphate dehydrogenase deficiency**. An aspirin tolerance test is used diagnostically in **von Willebrand's disease**, because aspirin will further prolong the bleeding time if the disease exists. Aspirin prevents platelet aggregation and may be helpful in the treatment of thromboembolic disease. In addition to aspirin, indomethacin, phenylbutazone, sulfinpyrazone, and dipyridamole prevent platelet aggregation, whereas epinephrine, serotonin, and prostaglandins promote platelet aggregation and hence are procoagulants. The erythrocyte sedimentation rate is often elevated in infections and inflammations, but aspirin therapy will yield a false negative.

NEW USES OF ASPIRIN

Although the uses of NSAIDs have been identical to the uses of the plant extracts that contained **salicylic acid**, namely, inflammation, pain, and fever, these drugs are increasingly used for other indications.

■ In the case of aspirin, there are numerous reports of beneficial effects of low-dose aspirin in the secondary **prophylaxis of cardiovascular disease** (myocardial infarction and stroke). Aspirin clearly reduces the risk of myocardial infarction and stroke among patients who already have manifestations of cardiovascular disease.

■ More recently, growing evidence has suggested that aspirin and other NSAIDs might have another public health indication: **colorectal cancer** might be prevented by regular aspirin use. Administration of **sulindac** to affected individuals induced polyp regression that was reversed by cessation of sulindac administration. A large body of epidemiological evidence point to an inverse association between aspirin use and colorectal cancer risk.

■ In the case of **reflux esophagitis**, the combination of low-dose NSAIDs coupled to an acid inhibitor (to simultaneously reduce gastrotoxicity and the reflux of gastric juice) may have potent healing properties. The presence of the NSAID may also help to reduce the risk of reflux-related complications such as esophageal carcinoma.

■ Other potential indications for NSAID use is prevention and treatment of **Alzheimer's disease**. In a recent review of 15 published studies, 14 suggest that NSAID treatment prevents or ameliorates symptoms of Alzheimer's disease (see Vainio and Morgan, 1997).

SALICYLATE PROTECTS AGAINST MPTP-INDUCED PARKINSONISM

Since the chance discovery of the neurotoxic effects of 1-methyl-4-phenyl-1,2,3,6-tetra-hydropyridine (**MPTP**) in humans, it has been thought that a similar type of environmental toxin might be responsible for idiopathic Parkinson's disease and that an understanding of the neurotoxic effects of MPTP might lead to prevention of this debilitating disorder. MPTP is metabolized to the 1-methyl-4-phenylpyridinium ion (**MPP$^+$**) by **monoamine oxidase (MAO) B**, and this highly toxic metabolite is selectively taken up into dopaminergic neurons via the dopamine transporter (see Figure 30). MPP$^+$ is a **mitochondrial toxin** that selectively inhibits **complex I** of the respiratory chain, leading evidently to energy compromise and to the production of potentially cytotoxic **free radicals**. Accumulating evidence suggests that oxidative stress is also a feature of Parkinson's disease neuropathology, as is mitochondrial complex I deficiency. MPTP produces a selective lesion of the nigrostriatal dopamine system that closely mimics the neuropathological and symptomatic sequelae of Parkinson's disease. Blockade either of MAO-B or of **dopamine uptake** is known to protect against the neurotoxic effects of MPTP in laboratory animals.

MPTP decreases glutathione levels and increases the levels of reactive oxygen species and the degree of lipid peroxidation in mouse brain slices *in vitro* and increases the levels of reactive oxygen species in mouse brain *in vivo*. MPTP neurotoxicity *in vitro* is reduced by glutathione. *In vitro* studies have shown that MPP$^+$ neurotoxicity can be reduced by vitamin E, vitamin C, **coenzyme Q**, and mannitol (but not by superoxide dismutase, catalase, allopurinol, or dimethyl sulfoxide). β-Carotene, vitamin C, and *N*-acetylcysteine partially protect against the neurotoxic effects of MPTP in mice, as do nicotinamide, coenzyme Q, and the free radical spin trap ***N-tert*-butyl-α-(sulfophenyl) nitrone**.

Aubin et al. (1998) reported that the neurotoxic effects of the dopamine-selective neurotoxin MPTP (15 mg/kg, s.c.), in mice, were totally prevented by systemic administration of **salicylate** (ED$_{50}$ = 40 mg/kg, i.p.), **aspirin** (ED$_{50}$ = 60 mg/kg, i.p.), or the soluble lysine salt of aspirin, **Aspegic** (ED$_{50}$ = 80 mg/kg, i.p.). The protective effects of aspirin are unlikely to be related to cyclooxygenase inhibition because **paracetamol** (100 mg/kg, i.p.), **diclofenac** (100 mg/kg, i.p.), **ibuprofen** (20 mg/kg, i.p.), and **indomethacin** (100 mg/kg, i.p.) were ineffective.

Dexamethasone (3 to 30 mg/kg, i.p.), which, like aspirin and salicylate, has been reported to inhibit the **transcription factor NF-κβ**, was also ineffective. Aspirin or salicylate (100 μM) had no effect on dopamine uptake into striatal synaptosomes or on monoamine oxidase B activity. The neuroprotective effects of salicylate derivatives could perhaps be related to hydroxyl radical scavenging. This was suggested by the fact that hydroxylated metabolites of salicylate (2,3- and 2,5-dihydrobenzoic acid) were recovered in brain tissue following the combined administration of MPTP and aspirin to a greater extent than following aspirin alone. The surprising neuroprotective effects of aspirin in an animal model of Parkinson's disease warrant further clinical investigation.

HEALTH EFFECTS OF SALICYLATES IN FOODS

Salicylates are found in fruits and vegetables. Salicylate-elimination diets have been used successfully in the treatment of asthma and chronic urticaria in **aspirin-sensitive individuals** (Swain et al., 1985; Perry et al., 1996).

ASPIRIN-INDUCED ASTHMA

The drugs most commonly associated with the induction of acute episodes of asthma are aspirin, coloring agents such as tartrazine, β-adrenergic antagonists, and sulfiting agents. It is important to recognize drug-induced bronchial narrowing because its presence is often associated with great morbidity. Furthermore, death sometimes has followed the ingestion of aspirin (or other nonsteroidal anti-inflammatory agents) or β-adrenergic antagonists. The typical **aspirin-sensitive respiratory syndrome** primarily affects adults, although the condition may be seen in childhood. This problem usually begins with **perennial vasomotor rhinitis** that is followed by a **hyperplastic rhinosinusitis** with **nasal polyps**. Progressive asthma then appears. On exposure to even very small quantities of aspirin, affected individuals typically develop ocular and nasal congestion and acute, often severe episodes of airways obstruction. The prevalence of aspirin sensitivity in subjects with asthma varies from study to study, but many authorities feel that 10% is a reasonable figure. There is a great deal of cross-reactivity between aspirin and other nonsteroidal anti-inflammatory compounds. **Indomethacin, fenoprofen, naproxen, zomepirac sodium, ibuprofen, mefenamic acid**, and **phenylbutazone** are particularly important in this regard. On the other hand, **acetaminophen, sodium salicylate, choline salicylate, salicylamide**, and **propoxyphene** are well-tolerated. The exact frequency of cross-reactivity with tartrazine and other dyes in aspirin-sensitive subjects with asthma is also controversial; again, 10% is the commonly accepted figure (Table 49). This peculiar complication of aspirin-sensitive asthma is particularly insidious, however, in that tartrazine and other potentially troublesome dyes are widely present in the environment and may be unknowingly ingested by sensitive patients.

Patients with aspirin sensitivity can be desensitized by daily administration of the drug. Following this form of therapy, cross-tolerance also develops to other NSAIDs. The mechanism by which aspirin and other such drugs produce bronchospasm has not yet been elucidated but may be related to aspirin-induced preferential generation of **leukotrienes**. Immediate hypersensitivity does not seem

Table 49 Naturally Occurring Salicylates in Food

	Type	mg/100 g
Fruits		
Apples	Granny Smith	0.59
	Jonathan	0.38
	Red Delicious	0.19
	Golden Delicious	0.08
Apricot, fresh		2.58
Avocado, fresh		0.60
Cantaloupe, fresh	Edible portion	1.50
Cherry, fresh	Sweet	0.85
Dates, dried		4.50
Figs, dried	Calamata	0.64
Grapes, fresh	Sultana	1.88
Grapefruit, fresh		0.68
Guava, canned		2.02
Kiwi fruit, fresh		0.32
Lemon, fresh		0.18
Loquat, fresh		0.26
Mandarin, fresh		0.56
Mango, fresh		0.11
Mulberry, fresh		0.76
Nectarine, fresh		0.49
Orange, fresh		2.39
Passionfruit, fresh		0.14
Peach, fresh		0.58
Pear, fresh	With skin	0.31
Persimmon, fresh		0.18
Pineapple, fresh		2.10
Plum, fresh	Red	0.11
Pomegranate, fresh		0.07
Raisins, dried		6.62
Raspberries, fresh		5.14
Rhubarb, fresh		0.13
Strawberry, fresh		1.36
Tangelo, fresh		0.72
Watermelon, fresh		0.48
Vegetables		
Alfalfa, fresh		0.70
Asparagus, fresh		0.14
Beans, fresh	Broad	0.73
	Green (French)	0.11
Bean sprouts, fresh		0.06
Beets, fresh		0.18
Broccoli, fresh		0.65

Table 49   Naturally Occurring Salicylates in Food (Continued)

	Type	mg/100 g
Brussels sprouts, fresh		0.07
Carrot, fresh		0.23
Cauliflower, fresh		0.16
Chicory, fresh		1.02
Chickpeas, dried		0.00
Chives, fresh		0.03
Cucumber	No peel	0.78
Eggplant	With peel	0.88
Endive, fresh		1.90
Leek, fresh		0.08
Mushroom, fresh		0.24
Olive, canned	Black	0.34
	Green	1.29
Onion, fresh		0.16
Parsnip, fresh		0.45
Peas, fresh	Green	0.04
Peppers, fresh	Green chili	0.64
	Red chili	1.20
	Yellow chili	0.62
	Green sweet	1.20
Potato, fresh	White, with peal	0.12
Pumpkin, fresh		0.12
Radish, fresh	Red, small	1.24
Shallots, fresh		0.03
Spinach		
Fresh		0.58
Frozen		0.16
Squash	Baby	0.63
Sweet corn, fresh		0.13
Sweet potato, fresh	White	0.50
	Yellow	0.48
Tomato		
Fresh		0.13
Juice	Heinz	0.12
Paste	Campbell	0.57
Sauce	Heinz	2.48
Turnip, fresh		0.16
Watercress, fresh		0.84
Zucchini, fresh		1.04
Spices		
Allspice	Powder, dry	5.2
Aniseed	Powder, dry	22.8

Table 49 Naturally Occurring Salicylates in Food (Continued)

	Type	mg/100 g
Bay leaf	Leaves, dry	2.52
Basil	Powder, dry	3.4
Canella	Powder, dry	42.6
Cardamom	Powder, dry	7.7
Caraway	Powder, dry	2.82
Cayenne	Powder, dry	17.6
Celery	Powder, dry	10.1
Chili	Flakes, dry	1.38
	Powder, dry	1.3
Cinnamon	Powder, dry	15.2
Cloves	Whole, dry	5.74
Coriander	Leaves, fresh	0.2
Cumin	Powder, dry	45.0
Curry	Powder, dry	218.0
Dill	Fresh	6.9
	Powder, dry	94.4
Fennel	Powder, dry	0.8
Fenugreek	Powder, dry	12.2
Five spice	Powder, dry	30.8
Garam masala	Powder, dry	66.8
Garlic	Bulbs, fresh	0.1
Ginger	Root, fresh	4.5
Mace	Powder, dry	32.2
Mint	Garden, fresh	9.4
Mixed herbs	Leaves, dry	55.6
Mustard	Powder, dry	26.0
Nutmeg	Powder, dry	2.4
Oregano	Powder, dry	66.0
Paprika	Hot powder, dry	203.0
	Sweet powder, dry	5.7
Parsley	Leaves, fresh	0.08
Pepper	Black powder, dry	6.2
	White powder, dry	1.1
Pimiento	Powder, dry	4.9
Rosemary	Powder, dry	68.0
Sage	Leaves, dry	21.7
Soy sauce	Liquid	0.00
Tabasco sauce	Liquid	0.45
Tarragon	Powder, dry	34.8
Turmeric	Powder, dry	76.4
Thyme	Leaves, dry	183.0
Vanilla	Essence, liquid	1.44
Worcestershire	Sauce, liquid	64.3

Table 49   Naturally Occurring Salicylates in Food (Continued)

	Type	mg/100 g
Miscellaneous		
Alcohol		
Benedictine		0.94
Brandy	Hennessy	0.40
Cointreau		0.66
Drambuie		1.68
Gin	Gilbey's	0.00
Rum	Captain Morgan	1.28
Tia Maria		0.83
Wine	Cabernet Sauvignon	0.86
	Champagne	1.02
	Claret	0.90
	Claret (reserve)	0.35
	Dry white	0.10
	Riesling	0.81
	Rose	0.37
	Vermouth	0.46
Beverages		
Coca-Cola	Liquid	0.25
Coffee	International Roast	0.96
	Maxwell House Instant (powder)	0.84
	Nescafe Instant (granules)	0.59
Herbal tea	Chamomile, bag	0.06
	Fruit, bag	0.36
	Peppermint, bag	1.10
	Rose hip, bag	0.40
Tea	Burmese green, leaves	2.97
	Bushells, bag	4.78
	Golden Days decaf, bag	0.37
	Harris, bag	4.0
	Indian green, leaves	2.97
	Peony Jasmine, leaves	1.9
	Old Chinese, leaves	1.9
	Tetley, bag	5.57
	Twinings:	
	Earl Grey, bag	3.0
	English Breakfast, bag	3.0
Negligible		
Cereals		
Maize	Meal, dry	0.43
Dairy		
Cheese	Blue Vein	0.05
	Camembert	0.01
	Mozzarella	0.02

Table 49 Naturally Occurring Salicylates in Food (Continued)

	Type	mg/100 g
Meats		
Liver	Fresh	0.05
Prawn	Fresh	0.04
Scallop	Fresh	0.02

Pain serves the useful function of alerting an individual that some component of a physiological system has gone awry. Ideally, the pain can be alleviated by removal of the underlying cause. In many situations, however, the stimulus of the pain is either not easily defined or is not readily susceptible to removal. Therefore, the physician is often faced with the necessity of treating pain as a symptom. Aspirin, a member of the salicylate family, is the most expensively employed analgesic agent. Of the remaining non-narcotic analgesic drugs, only two aniline derivatives, acetaminophen and phenacetin, and one arylalkanoic acid derivative, ibuprofen, are sufficiently safe to be used in general, nonprescription preparations. Although other agents have important therapeutic applications, toxicity limits their usefulness.

Data modified from Swain, AR, et al., 1985.

to be involved. The availability of effective inhibitors of leukotriene synthesis or receptor activity should aid materially in the treatment of this problem. β-Adrenergic antagonists regularly obstruct the airways in people with asthma as well as in others with heightened airway reactivity and should be avoided by such individuals. Even the selective β_1 agents have this propensity, particularly at higher doses. In fact, the local use of β_1 blockers in the eye for the treatment of glaucoma has been associated with worsening asthma.

Sulfiting agents, such as **potassium metabisulfite**, **potassium** and **sodium bisulfite**, **sodium sulfite**, and **sulfur dioxide**, which are widely used in the food and pharmaceutical industries as sanitizing and preserving agents, also can produce acute airways obstruction in sensitive individuals. Exposure usually follows ingestion of food or beverages containing these compounds, e.g., salads, fresh fruit, potatoes, shellfish, and wine. Exacerbation of asthma has been reported following the use of sulfite-containing topical ophthalmic solutions, intravenous glucocorticoids, and some inhalational bronchodilator solutions. The incidence and mechanism of action of this phenomenon are unknown. When suspected, the diagnosis can be confirmed by either oral or inhalational provocations (see Habtemariam, 2000).

ASPIRIN AND OTHER NATURAL INHIBITORS OF TUMOR NECROSIS FACTOR-α PRODUCTION, SECRETION, AND FUNCTION

Tumor necrosis factor (**TNF**; also known as **cachectin**) was originally discovered by its antitumor activity, but it is now recognized to be one of the most pleotropic cytokines acting as a host defense factor in immunological and inflammatory responses. Among its many different activities, TNF has effects on the vascular endothelium, which lead to upregulation of various cell adhesion molecules such as **intercellular adhesion molecule-1 (ICAM-1), vascular cell**

adhesion molecule-1 (VCAM-1), and **endothelial-leucocyte adhesion molecule-1 (ELAM-1)**. The regulated expression of these adhesion molecules and their counterreceptors on leucocytes mediates the adhesion and extravasation of white blood cells during inflammatory reaction. While mediation of inflammation and antitumor activity by TNF could be beneficial to the host, overproduction of TNF is the basis for the development of various diseases. There is now overwhelming evidence to suggest that TNF mediates the **wasting disease**, cachexia, associated with chronic diseases such as cancer and AIDS. TNF also plays pivotal roles in the development of pathologies such as **disseminated intravascular coagulation** and death in septic shock and **cerebral malaria**, and a range of inflammatory diseases including **asthma, dermatitis, multiple sclerosis, inflammatory bowel disease, cystic fibrosis, rheumatoid arthritis, multiple sclerosis**, and **immunological diseases**. It is thus clear that suppression of TNF production/release or inhibition of its function could benefit in the treatment of these TNF-mediated diseases (Janeway and Travers, 1994).

Potential target sites for inhibitory small-molecular-weight natural products could be broadly divided into three areas:

1. Inhibition of TNF production and secretion
2. TNF receptor antagonism
3. Inhibition of TNF function through modulation of its signal transduction pathway(s)

Inhibition of TNF Production and Secretion

Monocytes and macrophages are the principal TNF factories in the body, but it has now emerged that other cell types including T lymphocytes, mast cells, neutrophils, keratinocytes, astrocytes, microglial cells, smooth muscle cells, and many others are known to produce TNF. The induced (e.g., by **bacterial endotoxins** and **proinflammatory cytokines**) release of TNF from most of these cells requires *de novo* protein synthesis, which can be targeted by drugs at transcriptional, translational, and post-translational levels. It is now well established that one crucial step in the transcriptional activation of TNF gene is the transcription factor, NF-κB mobilization. NF-κB is sequestered in the cytoplasm by **inhibitory proteins IκB**, which are phosphorylated by a cellular kinase leading to degradation and translocation of NF-κB to the nucleus. The potential target of NF-κB in TNF-mediated diseases is evident, because inappropriate regulation of NF-κB has been shown to be associated with diseases such as septic shock, graft vs. host reaction, acute inflammatory condition, acute-phase response, radiation damage, atherosclerosis, and cancer. Of the known natural inhibitors of TNF production (Figure 129), the two classical anti-inflammatory agents **aspirin** and **salicylic acid** have been shown to act *in vitro* through inhibition of IκB degradation and the subsequent inhibition of NF-κB activation. Other anti-inflammatory agents reported for inhibition (at μ*M* concentration range) of NF-κB activation and the subsequent TNF release include **resveratrol** and **quercetin, nordihydroguaiaretic acid, butylated hydroxyanisol** and **tetrahydropapaveralione, curcumin, myricetin, epigallocatechin-gallate, 4-hydroxy-2-nonenal**, and **capsaicin**.

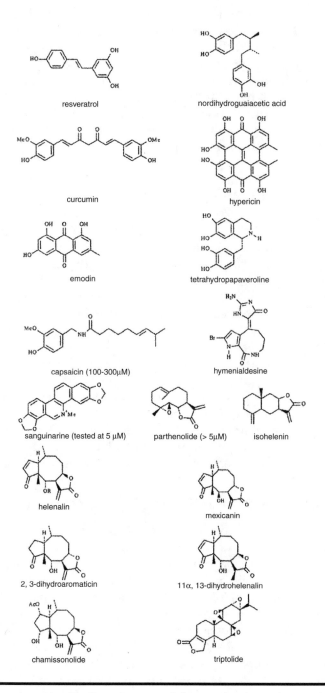

resveratrol

nordihydroguaiacetic acid

curcumin

hypericin

emodin

tetrahydropapaveroline

capsaicin (100-300μM)

hymenialdesine

sanguinarine (tested at 5 μM)

parthenolide (> 5μM)

isohelenin

helenalin

mexicanin

2, 3-dihydroaromaticin

11α, 13-dihydrohelenalin

chamissonolide

triptolide

Figure 129   Structure of naturally occurring inhibitors of TNF production/function.

Although the underlying mechanism has not yet been clearly established, many other natural products have been shown to inhibit TNF production. These include the antipsoriatic phenolic compound **anthraline**, **lignans-woorenosides II-V**, **lariciresinol glycoside**, **pinoresinol**, **eudesmin**, and to some extent **magnolin** and **lirioresinol-B dimethyl ether**; flavonoids — **naringin, epicatechin, some isoprenoid-substituted flavones** of *Artocarpus* spp., namely, **heterophyllin, artonin E, artobiloxanthone, cycloartobiloxanthone, cycloheterophyllin,** and **morusin,** and **silymarin,** which is known to be a mixture of **silydianin** and **silibinin; alkaloids — anisodamine,** quinine (at concentrations that block potassium channels), **fangchinoline, isotetrandrine,** and **colchicine;** terpenoids — **cynaropicrin, reynosin, santamarin,** and **dehydrocostus lactone** and the anti-inflammatory **saponin, esculentoside A,** both of which have also been shown to suppress the serum level of TNF *in vivo*; long-chain fatty acids - ω-fatty acids of fish oil origin, **docoshexaenoic acid** (DHA) and **eicosahexaenoic acid** (EPA), but also to some extent saturated fatty acids, **stearic acid, palmitic acid,** and **anandamide.** The hepatoprotective compound, **tetrahydroswertianolin,** which inhibits the D-galactoseamine (DGaIN)/LPS liver injury has been shown to suppress the serum level of TNF. Of the fungal metabolites, the immunosuppressive agent **cyclosporin A** has been demonstrated to suppress TNF production both *in vitro* and *in vivo*. Another fungal metabolite of interest is **trichodimerol,** as well as **tetracycline,** both of which have been shown to inhibit TNF secretion. Inhibition of TNF release at higher concentrations (m*M* range) of the antibiotic tetracycline has been shown to be associated with retention of membrane-associated TNF, suggesting possible metalloproteinase blocking activity. Processing of the TNF precursor (pro-TNF) to mature biologically active TNF by metalloproteinases has been one of the anti-inflammatory target sites for synthetic agents. The other possible natural metalloproteinase inhibitor is the immunosuppressive and psychoactive agent, Δ^9-**tetrahydrocannabinol,** which has been reported to inhibit TNF production through action at a post-transcriptional level. Other natural products of fungal origin with TNF production inhibitory effects are the antibiotic **erythromycin, cytochalasin D,** the toxic metabolite **aflatoxin B$_1$,** and known kinase inhibitors **herbimycin A** and **staurosporin.**

INHIBITORS OF TNF-MEDIATED CYTOTOXICITY

Although TNF toxicity to tumor cells and augmentation of this effect by drugs could be beneficial, the unwanted systemic cytotoxicity of TNF associated with various diseases needs to be modulated. As with TNF release, the cytotoxic response of TNF could be targeted at the various levels of TNF signaling. Experimentally, potential inhibitors of TNF cytotoxicity could easily be studied using TNF-sensitive cell lines, including by far the most commonly used, L929 cells. In the presence of protein synthesis inhibitors such as **actinomycin D** (usually 2 μg/ml), TNF induces DNA fragmentation within 6 hours followed by cell death within 24 hours. Based on this L929 cell-based bioassay, a number of phenolic compounds (**3,4-dihydroxybenzoic acid** and **caffeic acid** and also neurotransmitters **dopamine** and **noradrenaline**), which possess the catecholic functional moiety, have been shown to suppress TNF cytotoxicity. These effects of catechols, including that of **nordihydroguaiaretic acid,** are not related to their antioxidant action, because related antioxidants failed to protect cells from TNF cytotoxicity. Further studies on the mechanism of action of catechols revealed that inhibition of TNF cytotoxicity by catechols is related to their iron-chelating activity and the subsequent inhibition of **lipoxygenase enzymes.** On the basis of these findings, large numbers of flavonoids were tested for their inhibitory effect on TNF cytotoxicity *in vitro*. It appears that all **flavonols** tested (**galangin, kaempferol, kaempferide, quercetin, myricetin, morin,** and **rutin**) inhibited TNF cytotoxicity and the C-3 free hydroxy group of these structural groups appears to play a pivotal role in the observed protective effect. In contrast to this protective effect of flavonols, no protective activity was observed for the **flavones** (**chrysin, apigenin,** and **luteolin**) tested. In fact, **apigenin** and **chrysin**, which possess one (C-4′) hydroxy group or none on the B-ring, respectively, enhanced the TNF cytotoxicity. **Epicatechin** and **flavanones** that bear the catecholid functional moiety (**eriodictyol** and **taxifolin**) were protective, whereas other flavanones without this *o*-dihydroxy functionality either failed to protect (**pinocembrin, isosakuranetin, hesperetin**) or enhanced (**naringenin**) the TNF cytotoxicity. It is worth noting that all protective flavonoids and other catecholic compounds were effective when they were added as post-TNF treatment. Because some of the flavonoids showed enhancement of the TNF cytotoxicity, the clinical use of such compounds for inhibition of TNF cytotoxicity needs detailed analysis of their structural moiety *ex vivo* and *in vivo*.

Other agents that have been reported to inhibit TNF cytotoxicity *in vitro*, but with as yet unknown mechanism of action, include butylated **hydroxyanisol, pyridoxin, retinoic acid** (**vitamin A**, both the *cis* and *trans* form), and to a lesser extent **carotene**. Two **lignanamides, tribulusamides A** and **B**, have been shown to inhibit the TNF cytotoxicity in DGalN-sensitized cultured mouse hepatocytes, and the **sesquiterpene lactones germacrone, neocurdione, curdione,** and **curcumenol** have been reported to inhibit *in vivo* liver injury induced by coadministration of TNF and DGalN. These latter compounds, as well as **gentiopicroside** and **sweroside**, have also been shown to inhibit the endotoxin/DGalN-induced liver injury *in vivo*, which is now known to be TNF dependent. Whether the compounds act through direct inhibition of the TNF-induced cytotoxicity in liver cells, however, remains to be proved.

In conclusion, while the complete biochemical pathways for TNF production and action have yet to be fully elucidated, several target sites have been identified. Of these targets, the transcription factor NF-κB is most interesting because NF-κB regulates TNF production and, in turn, TNF regulates NF-κB reciprocally to produce its biological effects. A number of natural products widely known for their anti-inflammatory activities has been shown to modulate TNF release and function through suppression of NF-κB activation. Other targets for natural products that have been demonstrated to inhibit TNF production are kinase enzymes and the cAMP system. There appears to be very few compounds inhibiting TNF production/function at the post-translational level and the much-in-demand TNF receptor antagonist is still not available (see Habtemariam, 2000).

53

SHENG JING AND MALE INFERTILITY

Sheng Jing is a Chinese herbal formula used in male infertility and erectile dysfunction (impotence).

INTRODUCTION

Erectile dysfunction (ED) is a condition defined by the inability to attain or maintain penile erection sufficient for satisfactory sexual intercourse. In 1995, it was estimated that approximately 152 million men worldwide suffered from ED, with projections for 2025 growing to a prevalence of 322 million affected men. In the past, ED was believed to be caused by nonspecific psychological causes; however, in the past two decades, the majority of cases have been attributed to an organic etiology. Although ED patients can have a number of medical conditions, organic ED is usually associated with vascular risk factors such as arteriosclerosis, hypertension, diabetes mellitus, Peyronie's disease, and renal disease. In addition, pelvic trauma and pelvic surgery (radical prostatectomy or radical cystectomy) can cause ED by either vascular or nerve damage.

Since the early 1980s, understanding about the pharmacology of the erectile mechanism has advanced significantly. Basic research in corporal cavernosal smooth muscle (CCSM) physiology and identification of the central mediators involved in the erectile process has contributed to the development of pharmacological agents that can effectively treat ED patients. At present, the diagnosis and treatment of ED has evolved to the point where virtually every patient suffering from ED can be successfully treated. **Vacuum erection devices**, **intracavernosal injection therapy**, **intraurethral suppositories**, oral medications, **penile vascular procedures**, and **surgical implantation of prosthetic devices** offer most men a viable option to correct their ED. However, despite the overall success and efficacy of the aforementioned therapies, there are implicit side effects, complications, and contraindications. Therefore, the development of future therapeutic options for the treatment of ED should focus on those strategies with fewer adverse effects and an absence of contraindications. Gene therapy for the treatment of ED may become a viable and relatively noninvasive therapeutic option (Bivalacqua and Hellstrom, 2001).

IMPOTENCE

A variety of endocrine, vascular, neurological, and psychiatric diseases disrupt normal sexual and reproductive function in men. Furthermore, sexual dysfunction may be the presenting symptom of systemic disease.

NORMAL SEXUAL FUNCTION

Penile erection is initiated by neuropsychological stimuli that ultimately produce vasodilation of the sinusoidal spaces and arteries within the paired corpora cavernosa. Erection is normally preceded by **sexual desire** (or **libido**), which is regulated in part by androgen-dependent psychological factors. Although nocturnal and diurnal spontaneous erections are suppressed in men with androgen deficiency, erections may continue for long periods in response to erotic stimuli. Thus, continuing action of testicular androgens appears to be required for normal libido but not for the erectile mechanism itself.

The penis is innervated by sympathetic, parasympathetic, and somatic fibers. Somatic fibers in the dorsal nerve of the penis form the afferent limb of the erectile reflex by transmitting sensory impulses from the penile skin and glans to the S2–S4 dorsal root ganglia via the pudendal nerve. Unlike the corpuscular-type endings in the penile shaft skin, most afferents in the glans terminate in free nerve endings. The efferent limb begins with parasympathetic preganglionic fibers from S2–S4 that pass in the pelvic nerves to the pelvic plexus. Sympathetic fibers emerging from the intermediolateral gray areas of T11–L2 travel through the paravertebral sympathetic chain ganglia, superior hypogastric plexus, and hypogastric nerves to enter the pelvic plexus along with parasympathetic fibers. Somatic efferent fibers from S3–S4 that travel in the **pudendal nerve** to the ischiocavernosus and bulbocavernosus muscles and postganglionic sympathetic fibers that innervate the smooth muscle of the epididymis, vas deferens, seminal vesicle, and internal sphincter of the bladder mediate rhythmic contraction of these structures at the **time of ejaculation**.

Autonomic nerve impulses, integrated in the pelvic plexus, project to the penis through the cavernous nerves that course along the posterolateral aspect of the prostate before penetrating the pelvic floor muscles immediately lateral to the urethra. Distal to the membranous urethra, some fibers enter the corpus spongiosum, while the remainder enter the corpora cavernosa along with the terminal branches of the pudendal artery and exiting cavernous veins. If disruption of the cavernous nerves occurs following pelvic trauma or surgery, erectile impotence may ensue.

The brain exerts an important modulatory influence over spinal reflex pathways that control **penile function**. A variety of visual, auditory, olfactory, and imaginative stimuli elicit erectile responses that involve cortical, thalamic, rhinencephalic, and limbic input to the medial preoptic-anterior hypothalamic area, which acts as an integrating center. Other areas of the brain, such as the **amygdaloid complex**, may inhibit sexual function.

Although the parasympathetic nervous system is the primary effector of erection, the transformation of the penis to an erect organ is a vascular phenomenon.

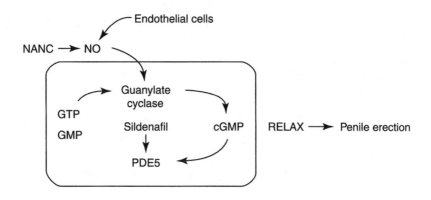

Figure 130 Mechanism of action of sildenafil in corpus cavernosal smooth muscle cells. NANC: Nonadrenergic, noncholinergic; NO: nitric oxide; PDE5: Phosphodiesterase Type 5.

In the flaccid state the arteries, arterioles, and sinusoidal spaces within the corpora cavernosa are constricted due to sympathetic-mediated contraction of smooth muscle in the walls of these structures. The venules between the sinusoids and the **densetanica albuginea** surrounding the cavernosa open freely to the emissary veins. Erection begins when relaxation of the sinusoidal smooth muscles leads to dilation of the sinusoids and a decrease in peripheral resistance, causing a rapid increase in arterial blood flow through internal **pudendal** and **cavernosa arteries**. Expansion of the sinusoidal system compresses the venules against the interior surface of the tunica albuginea, resulting in venous occlusion. The increase in intracorporeal pressure leads to rigidity; less than complete expansion of the sinusoidal spaces leads to less than complete rigidity.

Erection occurs when adrenergic-induced sinusoid tone is antagonized by sacral parasympathetic stimulation that produces sinusoidal relaxation primarily by synthesis and release of the nonadrenergic-noncholinergic (NANC) **neurotransmitter nitric oxide** (NO). The contribution of acetylcholine-dependent release of NO from the vascular endothelium is uncertain. *In vitro* electrical stimulation of isolated corpus cavernosum strips (with or without endothelium) produces sinusoidal relaxation by release of neurotransmitters within nerve terminals that is resistant to adrenergic and cholinergic blockers. Inhibitors of the synthesis of NO or of **guanosine monophosphate** (GMP), as well as **nitric oxide scavengers**, block sinusoidal relaxation (Figure 130). A variety of neuropeptides found in corporal tissues, including **vasoactive intestinal peptide** (VIP) and **calcitonin gene-related peptide** (CGRP), produce tumescence when injected into the penis but have uncertain physiological roles. Norepinephrine plays an important role in the adrenergic mechanism of detumescence (Gingell and Lockyer, 1999).

Seminal emission and **ejaculation** are under control of the sympathetic nervous system. Emission results from α-adrenergic-mediated contraction of the **epididymis**, **vas deferens**, **seminal vesicles**, and **prostate**, which causes seminal fluid to enter the prostatic urethra. Concomitant closure of the bladder neck prevents retrograde flow of semen into the bladder, and antegrade ejaculation results from contraction of the muscles of the pelvic floor including the bulbocavernosus and *ischiocavernosus muscles.*

Orgasm is a psychosensory phenomenon in which the rhythmic contraction of the pelvic muscles is perceived as pleasurable. Orgasm can occur without either erection or ejaculation or in the presence of retrograde ejaculation (Gingell and Lockyer, 1999).

FAILURE OF ERECTION

The organic causes of erectile impotence can be grouped into endocrine, drug, local, neurological, and vascular causes (Table 50).

THERAPY

Sheng Jing (Table 51) has been tested in 202 patients, given twice a day for 60 days. The results showed "a significant improvement in sperm density, motility and grade, levels of **follicle-stimulating hormone** (FSH), **luteinizing hormone** (LH), and **testosterone** (T) and reduction in serum anti-sperm antibody titers. Sperm density increased from 16.2×10^6/ml to 56.1×10^6/ml. Sperm density in men with severe oligospermia (2.1×10^6/ml) also experienced significant improvements (17.9×10^6/ml). Sperm motility increased from 34% to 46%. They reported a 78% pregnancy rate in the 148 couples available for follow-up" (Crimmel et al., 2001).

In addition, the efficacy of **acupuncture** on penile erection has been tested. Organic disease was excluded by a combination of nocturnal penile tumescence monitoring and pharmacologically enhanced duplex ultrasonography. Twenty-nine patients with a mean age of 40 years received a series of ten treatments over a 4-week period. If no improvement was observed, they received a second course of ten treatments. Mean follow-up was 8 months. Of the patients, 69% demonstrated successful results, defined as having two or more erections per week that were satisfactory for intercourse (Crimmel et al., 2001).

DRUGS THAT STIMULATE SEXUAL BEHAVIOR

Various clinically used or experimental drugs enhance sexual interest or potency as a side effect in humans.

Levodopa

Levodopa (L-dopa) is a natural intermediate in the biosynthesis of catecholamines in the brain and peripheral adrenergic nerve terminals. In the biologic sequence of events it is converted to dopamine, which in turn serves as a substrate of the neurotransmitter norepinephrine. Levodopa is used successfully in the treatment of **Parkinson's syndrome**, a disease characterized by dopamine deficiency. When levodopa is administered to an individual with this syndrome, the symptoms

Table 50   Some Organic Causes of Erectile Impotence in Men

Endocrine Causes
Testicular failure
Hyperprolactinemia

Drugs
Antiandrogens
   Histamine (H_2) blockers (e.g., cimetidine)
   Spironolactone
   Ketoconazole
   Finasteride
Antihypertensives
   Central-acting sympatholytics (e.g., clonidine and methyldopa)
   Peripheral-acting sympatholytics (e.g., guanadrel)
   β-blockers
   Thiazides
Anticholinergics
Antidepressants
   Monoamine oxidase inhibitors
   Tricyclic antidepressants
Antipsychotics
Central nervous system depressants
   Sedatives (e.g., barbiturates)
   Antianxiety drugs (e.g., diazepam)
Drugs of habituation or addiction
   Alcohol
   Methadone
   Heroin
   Tobacco

Penile Diseases
Peyronie's disease
Previous priapism
Penile trauma

Neurological Diseases
Anterior temporal lobe lesions
Diseases of the spinal cord
Loss of sensory input
   Tabes dorsalis
   Disease of dorsal root ganglia
Disease of nervi erigentes
   Radical prostatectomy and cystectomy
   Rectosigmoid operations
Diabetic autonomic neuropathy and various polyneuropathies

Table 50   Some Organic Causes of Erectile Impotence in Men

Vascular Diseases
Aortic occlusion (Leriche syndrome)
Atherosclerotic occlusion or stenosis of the pudendal or cavemosa arteries
Arterial damage from pelvic radiation
Venous leak
Disease of the sinusoidal spaces

Sexuality is an integral part of one's identity; it is a reflection of how one feels about oneself and how one interacts with others. Sexual function refers to the psychological ability to perform in a sexually satisfying manner, with or without a partner. Drugs can influence both sexuality and sexual function.

Table 51   Ingredients of the Chinese Herbal Formula Sheng Jing

Western Name	Chinese Name
Antler gelatin	Lu jiao jiao
Epimedium	Yin yang huo
Curculigo	Xian mao
Cherokee rosa	Jin ying zi
Schizandra	Wu wei zi
Loranthus	Sang Ji Sheng
Rehmannia	Shu di huang
Lycium	Gou qi zi
Cuscuta	Tu si zi
Dendrobium	Shi hu
Morinda	Ba Ji Tian
Rubus	Fu Pen Zi
Achyranthes	Huai niu xi
Scripus	San leng
Zedoaria	E zhu

of Parkinson's disease are ameliorated, presumably because the drug is converted to dopamine and thereby counteracts the deficiency. Individuals treated with levodopa, especially older men, have been observed to experience a **sexual rejuvenation**. This effect has led to the belief that levodopa stimulates sexual powers. Consequently, studies with younger men complaining of decreased erectile ability have shown that levodopa increases libido and the incidence of penile erections. Overall, however, these effects are short-lived and do not reflect continued satisfactory sexual function and potency. Thus, levodopa is not a true aphrodisiac. The increased sexual activity experienced by parkinsonian patients treated with levodopa may reflect improved well-being and partial recovery of normal sexual functions that were impaired by Parkinson's disease.

Amyl Nitrite

Amyl nitrite, a drug used in the past to treat **angina pectoris**, is alleged to enhance sexual activity in humans. As a vasodilator and smooth muscle stimulant, amyl nitrite has been reported to intensify the orgasmic experience for men if inhaled at the moment of orgasm. This effect is probably the result of relaxation of smooth muscles and consequent vasodilation of the genitourinary tract. No effects of amyl nitrite on libido have been reported, but a loss of erection or delayed ejaculation may result. Women generally experience negative effects on orgasm when taking this drug.

Vitamin E

Much has been said about the positive effects of vitamin E (**α-tocopherol**) on sexual performance and ability in humans. Unfortunately, there is little scientific rationale to substantiate such claims. The primary reasons for attributing a positive role in sexual performance to vitamin E come from experiments on vitamin E deficiency in laboratory animals. In such experiments the principal manifestation of this deficiency is infertility, although the reasons for this condition differ in males and females. In female rats there is no loss in ability to produce apparently healthy ova, nor is there any defect in the placenta or uterus. However, fetal death occurs shortly after the first week of embryonic life, and fetuses are reabsorbed. This situation can be prevented if vitamin E is administered any time up to day 5 or 6 of embryonic life. In the male rat the earliest observable effect of vitamin E deficiency is immobility of spermatozoa, with subsequent degeneration of the germinal epithelium. Secondary sex organs are not altered and sexual vigor is not diminished, but vigor may decrease if the deficiency continues.

Because of experimental results such as these, vitamin E has been conjectured to restore potency or to preserve fertility, sexual interest, and endurance in humans. No evidence supports these contentions, but because sexual performance is often influenced by mental attitude, a person who believes vitamin E may improve sexual prowess may actually find improvement. The only established therapeutic use for vitamin E is for the prevention or treatment of vitamin E deficiency, a condition that is rare in humans.

Sildenafil (Viagra)

Sildenafil increases the release of **nitric oxide** and increases the levels of **cyclic guanosine monophosphate (cGMP)**, a smooth muscle relaxant. Sildenafil enhances the effects of nitric oxide by inhibiting **phosphodiesterase 5**, an enzyme found primarily in the penis that degrades cGMP. As a result, increased levels of cGMP in the corpus cavernosum enhance smooth muscle relaxation, the inflow of blood, and erection. Sildenafil has no effect in the absence of sexual stimulation (see Figure 130).

This is an exciting new era in the treatment of ED. Since the development of such a highly effective oral preparation as sildenafil, much more research activity will be generated to produce competing therapies. Locally applied, rapidly absorbed preparations will almost certainly have a role to play. Combination therapy has to date not been extensively studied. The potential of utilizing centrally acting drugs or sildenafil in combination with locally applied preparations in the more clinically challenging patient is an exciting therapeutic opportunity. Much as been achieved in the past 15 years in the field of andrological research and much will be achieved in the very near future for the benefit of patients (Gingell and Lockyer, 1999).

EMERGING PHARMACOLOGICAL THERAPIES FOR ERECTILE DYSFUNCTION

In addition to **Sheng Jing** and acupuncture, which are old (Crimmel et al., 2001), and gene therapy, which is novel and new (Bivalacqua and Hellstrom, 2001), other emerging therapeutics for erectile dysfunction are the following:

- Phosphodiesterase inhibitors (sildenafil)
- Cyclic AMP activators (**vasoactive intestinal polypeptide**)
- α-Adrenergic receptor antagonists (**phentolamine**)
- Dopaminergic receptor agonists (**apomorphine**)
- α-Melonocyte stimulating hormone
- Potassium channel modulators
- Endothelin antagonists
- Nitric oxide donors
- Drugs topically applied to penis (transdermal or transglandular nitroglycerine, capsaicine, papaverine, or prostaglandin E_1)

54

SPONGES
AND STEROIDS

- All marine organisms have been proved to be a veritable cornucopia of unusual steroid metabolites, but some believe that marine sponges may provide the most diverse and biogenetically unprecedented array of unconventional steroids in the entire animal kingdom.
- These steroids are potent inhibitors of histamine release, are anti-inflammatory agents, are immunosuppressants, are antineoplastic agents, and are antiviral agents. However, their full therapeutic potential remains to be delineated.

INTRODUCTION

Unconventional steroids often co-occur with conventional ones and are sometimes present in small amounts; however, many exceptions are reported for some sponges that are found with unusual structures as the predominant steroids rather than cholesterol or the conventional 3,β-hydroxy sterols. It is, therefore, particularly interesting when a sponge contains unusual steroids in large quantities, as these very likely play a functional (rather than metabolic) role in maintaining the integrity of membranous structures. It has been hypothesized and, to some extent, documented that the uniqueness of sterols in cell membranes of sponges is related to the other membrane components, particularly the phospholipids. These latter compounds seem to have head groups and fatty acids very different from those of higher animals; therefore, the structural modifications exhibited by the sponge sterols may be a sort of structural adjustments for a better fit with other membrane components.

The highly functionalized steroids have recently attracted considerable attention because of their biological and pharmacological activities. Remarkable examples are **herbasterol** (which is **ichthyotoxic**) **xestobergsterols**, and **contignasterol**, potent inhibitors of histamine release from rat mast cells induced by anti-IgE, and **halistanol disulfate B**, an endothelin-converting enzyme inhibitor. Most of the oxygenated cholesterol derivatives have been shown to exert a cytotoxic activity on human cancer cells lines *in vitro*. Also, enzymatic transformations leading to **secosterols**, i.e., ring cleavage products of cholesterol, may result in the formation of products with cell division-inhibitory properties. Finally, the recent discovery of the antiviral properties of **sulfated polyhydroxysterols** has increased interest in these compounds (see Aiello et al., 1999).

EXAMPLES OF STEROIDS FOUND IN SPONGES

Polyoxygenated Steroids

Six new **3β,5,6β-trihydroxysterols** with a saturated nucleus have been isolated from two different populations of the sponge *Cliona copiosa*, collected from two different sites of the **bay of Naples**. Successively, several new **Δ⁵-3β-hydroxy-7-ketosteroids** and the related **Δ⁵-3β, 7β**, and **Δ⁵-3β,7α-dihydroxysteroids** have been described as constituent of the same species.

Steroid Sulfates

Several **sulfated polyhydroxysterols** have been recently isolated from sponges. Most of these natural products are characterized by the **2β,3α,6α-tri-O-sulfate** functions together with additional alkylation in the side chain. These steroids are of interest not only because of their structures but also because of their physiological activities, which include an **anti-human immunodeficiency virus** (HIV) effect and a high inhibitory action on some enzymes.

Steroids with Unconventional Side Chains

Steroids isolated from sponges frequently contain substantially modified side chains, such as those with high degrees of alkylation or other unusual functionalization. Biosynthetic studies have been performed on mechanism and scope of sterol side-chain dealkylation in sponges.

- **25S-26-Methyl-24-methylenecholest-4-en-3-one** has been isolated from the marine fossil sponge *Neosiphonia supertes*.
- **Baikalosterol** is a novel steroid with a **24-ethyl-26-nor-22,25-diene** group in the side chain isolated from the sponge *Baicalospongia bacilifera*, a common species of the **Baikal lake**.
- **Sutinasterol** contains a C_{12} side chain that is presumably the product of quadruple bioalkylation. It represents the 94% of the steroid fraction of a *Xestospongia* sponge.
- A new cytotoxic sterol, **bienmasterol**, possessing the rare 22,25-diene side chain, has been isolated from the Okinawan sponge *Bienma* sp.
- **Topsentinols** A through J, ten new 7-hydroxysterols with unusual polyalkylated side chains, have been isolated from the Okinawan marine sponge *Topsentia* sp.
- Two new sterols, **epipolasterol** and **22,23-dihydroepipolasterol**, have been found in *Epipolasis* sp.
- There are 22 steroid components present in the deep-water sponge *Stelodoryx chlorophylla*, from New Caledonia.

ADRENAL STEROIDS

The adrenal gland, which is located at the cap of the kidney, is divided histologically into three zones: the outer zone or zona glomerulosa, the middle zone or zona fasciculata, and the inner zone or zona reticularis. The adrenal cortex synthesizes cholesterol and pregnenolone through the interaction of a group of enzymatic reactions (Figure 131).

The Synthesis of Cholesterol

Cholesterol, which is essential for the synthesis of adrenal, ovarian, and testicular steroid hormones, originates from two sources. The body synthesizes approximately 2 g of cholesterol per day, according to the following pathway:

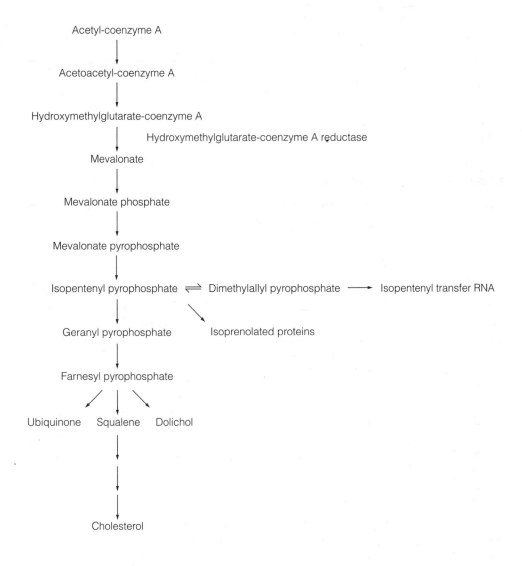

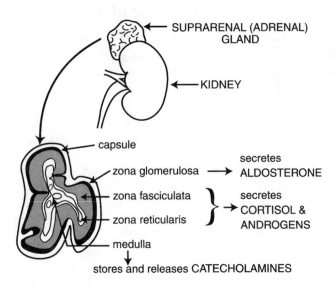

Figure 131   The sites of synthesis of aldosterone, cortisol, and androgens.

In addition, between 300 and 800 mg of cholesterol is ingested per day, depending on a person's diet. On average, between 300 and 1500 mg of cholesterol is excreted per day.

Classification of Adrenal Steroids

The adrenal steroids are divided into three major categories: **glucocorticoids**, **mineralocorticoids**, and **sex hormones**.

The Glucocorticoids

The glucocorticoids mainly influence carbohydrate metabolism and, to a certain extent, protein and lipid metabolism. The main glucocorticoid is cortisol, with a daily secretion of 15 mg. Cortisol is synthesized through the 11-β-hydroxylation of **11-deoxycortisol**. Besides cortisol, the adrenal gland also synthesizes and releases a small amount of **corticosterone**, whose synthesis from **11-deoxycorticosterone** is catalyzed by 11-β-hydroxylase. A deficiency of 11-β-hydroxylase causes:

- Diminished secretion of cortisol
- Diminished secretion of corticosterone
- Enhanced compensatory secretion of adrenocorticotropic hormone (ACTH)
- Enhanced secretion of 11-deoxycortisol and 11-deoxycorticosterone
- Enhanced secretion of androgens

The clinical manifestations of 11-β-hydroxylase deficiency are virilization, resulting from the overproduction of androgen, and hypertension, stemming from the overproduction of deoxycorticosteroids.

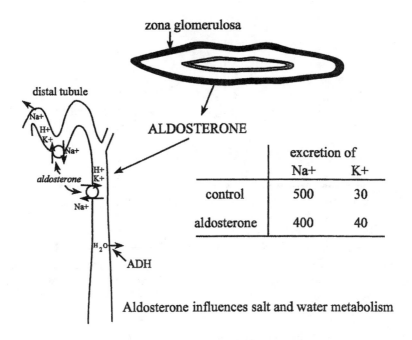

| | excretion of | |
	Na+	K+
control	500	30
aldosterone	400	40

Aldosterone influences salt and water metabolism

Figure 132 The action of aldosterone. ADH = antidiuretic hormone.

The glucocorticoids work by binding to specific intracellular receptors in target tissues. The receptor hormone complex is then transported into the nucleus where the complex interacts with the DNA, thus augmenting the synthesis of specific RNAs.

Cortisol, corticosterone, aldosterone, and the synthetic steroids used in steroid therapy (e.g., prednisolone, dexamethasone, and triamcinolone) are glucocorticoid agonists and therefore elicit glucocorticoid responses. A number of other steroids bind to the glucocorticoid receptor and thus suppress glucocorticoid responses.

The Mineralocorticoids

The mineralocorticoids influence salt and water metabolism and in general conserve sodium levels. They promote the resorption of sodium and the secretion of potassium in the cortical collecting tubules and possibly the connecting segment. They also elicit hydrogen secretion in the medullary collecting tubules (Figure 132).

The main mineralocorticoid is aldosterone, with a daily secretion of 100 μg. Aldosterone is synthesized from 18-hydroxycorticosterone by a dehydrogenase. The consequence of 18-hydroxycortisterone dehydrogenease deficiency is diminished secretion of aldosterone, and the clinical manifestations consist of sodium depletion, dehydration, hypotension, potassium retention, and enhanced plasma renin levels (Figure 133).

The Sex Hormones

Small quantities of progesterone, testosterone, and estradiol are also produced by the adrenal gland. However, they play a minor role compared to the testicular

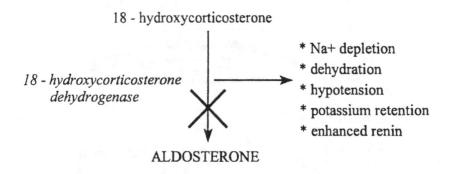

Figure 133 The consequences of hydroxycorticosterone dehydrogenase deficiency.

and ovarian hormones. Progesterone, which is the precursor of cortisol, aldosterone, testosterone, and estradiol, is synthesized from **5-pregnenolone** by **3-β ol-dehydrogenase**. Deficiency of this enzyme results in cortisol and aldosterone deficiencies. Such patients require replacement therapy with both glucocorticoids and mineralocorticoids.

Regulation of Adrenal Steroid Biosynthesis

Adrenal glucocorticoid and androgen production is controlled predominantly by the hypothalamic–pituitary axis, whereas the production of aldosterone by the zona glomerulosa is predominantly regulated by the **renin-angiotensin system** and potassium concentration. The hypothalamus, pituitary, and adrenal form a neuroendocrine axis whose primary function is to regulate the production of both cortisol and some of the adrenal steroids (Figure 134).

Adrenocorticotropic Hormones

Corticotropin-releasing factor and arginine vasopressin, which are released predominantly by the paraventricular nucleus of the hypothalamus, are important regulators of corticotropin (ACTH) release, which in turn triggers the release of cortisol and other steroids by the adrenal gland. Both the administration of certain psychoactive agents and emotional arousal originating from the limbic system are able to modify the functions of the pituitary–adrenal axis and to stimulate the synthesis of cortisol.

ACTH elicits the following effects:

- ACTH enhances the synthesis of pregnenolone.
- ACTH activates adenylate cyclase and elevates the cyclic adenosine monophosphate level.
- ACTH enhances the level of adrenal steroids, especially cortisol.
- ACTH reduces the level of ascorbic acid.

The level of cortisol is thought to control directly the secretion of ACTH through a negative feedback mechanism that may be directed at both the hypothalamus and the anterior pituitary gland. Conversely, a reduced concentration of cortisol

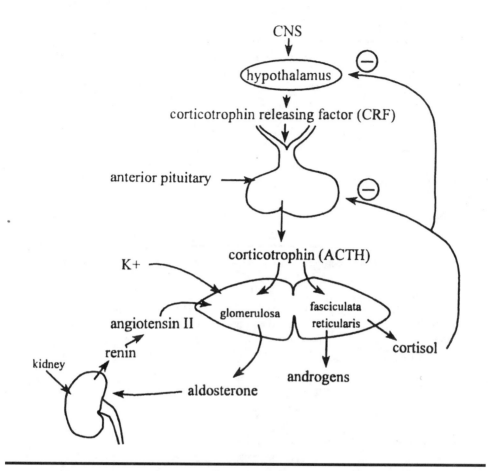

Figure 134 The action of aldosterone on renin–angiotensin system.

or cortisol-like substances eliminates the negative effect and enhances the release of ACTH (see Figure 134).

The **metyrapone test** may be used diagnostically to evaluate the proper functioning of the anterior pituitary gland. When administered orally, metyrapone:

- Inhibits the activity of 11-β-hydroxylase, which is necessary for the synthesis of cortisol, corticosterone, and aldosterone
- Promotes the release of corticotropin, which in turn increases production of the precursors (**11-deoxycortisol** and **11-deoxycorticosterone**)
- Enhances the appearance of **17-hydroxycorticosteroids** and **17-ketogenic steroids**

In the event that the pituitary gland is nonfunctional, and therefore cannot stimulate ACTH secretion, the levels of these urinary metabolites will not increase.

Actions and Pharmacological Applications of Glucocorticoids

The glucocorticoids possess a plethora of physiological actions, including a role in differentiation and development. They are vital in the treatment of adrenal

Table 52 Actions of Glucocorticoids

Steroid	Anti-inflammatory Potency (% compared to cortisol)	Sodium Retention (% compared to cortisol)
Betamethasone	20	0
Dexamethasone	20	0
Fludrocortisone	12	100
Paramethasone	6	0
Triamcinolone	5	0
Methylprednisolone	4	0
Prednisolone	3	0.8
Prednisone	2.5	0.8
Cortisol	1	1

The glucocorticoids mainly influence carbohydrate metabolism and, to a certain extent, protein and lipid metabolism.

insufficiency and are used extensively in large pharmacological doses as anti-inflammatory and immunosuppressive agents. Some of the nonendocrine conditions for which they may be used include arthritis, tenosynovitis, systemic lupus erythematosus, acute rheumatic carditis, bronchial asthma, organ transplantation, ulcerative colitis, cerebral edema, and myasthenia gravis.

The relative anti-inflammatory potency and sodium-retaining properties of several steroids are listed in Table 52.

- **Glucose Metabolism.** Cortisol has an anti-insulin effect and aggravates the pathological consequences of diabetes mellitus. It increases gluconeogenesis, inhibits the peripheral utilization of glucose, and causes hyperglycemia and glucosuria.
- **Protein Metabolism.** Cortisol promotes the breakdown of proteins and inhibits protein synthesis. This leads to muscle wasting in the quadriceps–femoris groups, and muscular activities may become difficult as a result. The effect of cortisol is opposite to that of insulin.
- **Glycogen Metabolism.** The effects of glucocorticoids on glycogen accumulation appear to be predominantly, although not exclusively, insulin dependent, because glycogen accumulation is markedly reduced in pancreatectomized animals. Glucocorticoid-stimulated increases in insulin secretion promote further glycogen accumulation.
- **Lipid Metabolism.** Cortisol causes the abnormal deposition of a fat pad called "**buffalo hump**."
- **Electrolytes and Water Metabolism.** Cortisol use can bring about hypernatremia, hypokalemia, and hypercalciuria.
- **Uric Acid.** Cortisol use causes hyperuricemia by suppressing the renal tubular resorption of uric acid.
- **Gastric Hydrochloric Acid.** Cortisol promotes the production of gastric hydrochloric acid.
- **Blood Coagulation.** Cortisol, like epinephrine, augments the coagulability of blood.

Table 53 Immunological Mediators Whose Production or Actions Can Be Blocked by Glucocortisoids

Bradykinin
Cachectin (tumor necrosis factor)
Colony-stimulating factor
Histamine
Interleukin-1 (lymphocyte-activating factor)
Interleukin-2 (T-cell growth factor)

The glucocorticoids possess a plethora of physiological actions, including a role in differentiation and development. They are vital in the treatment of adrenal insufficiency and are used extensively in large pharmacological doses as anti-inflammatory and immunosuppressive agents.

- **Anti-inflammatory Action.** Cortisol exerts its anti-inflammatory effect in part by blocking the release and action of histamine. In addition, it decreases the migration of polymorphonuclear leukocytes.
- **Hematological Effects.** Cortisol produces eosinophilia and causes the involution of lymphoid tissues.
- **Leukocytes.** The administration of glucocorticoids to human subjects brings about lymphocytopenia, monocytopenia, and eosinopenia. In addition, glucocorticoids block a number of lymphocytic functions (Table 53).
- **Immunosuppression.** Although considered to be immunosuppressive, therapeutic doses of glucocorticoids do not significantly decrease the concentration of antibodies in the circulation. Furthermore, during glucocorticoid therapy, patients exhibit a nearly normal antibody response to antigenic challenge.

Central Nervous System Effects

Glucocorticoids, which do penetrate the blood–brain barrier, affect behavior, mood, and neural activity, and are able to regulate the permeability of the blood–brain barrier to other substances. Hence, they are used to treat brain edema. Both glucocorticoid deficiency and excess may cause mood swings and rarely psychosis. Patients receiving glucocorticoids have a feeling of well-being. On the other hand, patients with spontaneously evolving **Cushing's syndrome**, which involves the overproduction of glucocorticoids, are commonly depressed. Patients with **Addison's disease**, which is caused by a deficiency of cortisol and aldosterone, tend to be depressed, negativistic, irritable, seclusive, and apathetic. Patients with Addison's disease also suffer from anorexia, whereas a glucocorticoid excess stimulates the appetite. In addition, high doses of glucocorticoids can affect sleep, with a trend toward increased wakefulness and a reduction in rapid-eye-movement (REM) sleep, an increase in stage II sleep, and an increase in the time to the first REM sleep.

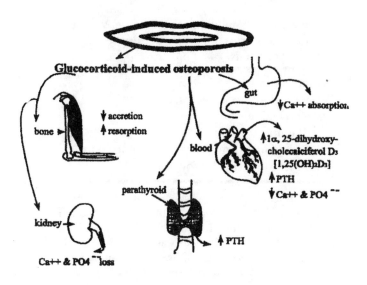

Figure 135 Actions of vitamin D on calcium metabolism. PTH = parathyroid hormone.

Calcium Metabolism

Glucocorticoids affect bone metabolism in a variety of ways (Figure 135). Mild hyperkalemia can occur in Addison's disease. Conversely, glucocorticoids are used to treat certain hypercalcemias, largely granulomatous conditions such as sarcoidosis in which the steroid blocks the formation of 1-α,25-dihydroxycholecalciferol [1-α,25-$(OH)_2D_3$] by the granulomatous tissues. However, most hypercalcemias are not responsive to glucocorticoid and, in general, glucocorticoid excess does not result in lower serum calcium levels. Serum phosphate levels are lowered and urinary calcium and phosphorus concentrations are elevated in glucocorticoid excess states. Glucocorticoid excess ultimately leads to osteoporosis, the major limitation to their long-term use (see Figure 135).

Contraindications to Glucocorticoid Therapy

Besides the adverse effects just described, glucocorticoid therapy is contraindicated under the following circumstances: diabetes mellitus, digitalis therapy, glaucoma, hypertension, infection, osteoporosis, peptic ulcer, tuberculosis, and viral infection.

ADRENAL DISORDERS — CUSHING'S SYNDROME

Cushing's syndrome is a rare condition, occurring in one of 1000 individuals but seen primarily in the female population, that appears in the third or fourth decade of life. A basophilic adenoma of the anterior pituitary gland, hyperplasia, adenoma, and carcinoma of the adrenal cortex may all cause Cushing's syndrome. Patients typically have red cheeks, a moon face, a buffalo hump, thin skin, high blood pressure, a pendulous abdomen with red striae, poor wound healing, osteoporosis, and some signs of virilism (enlarged clitoris). The treatment of Cushing's syndrome may include:

- Removal of the microadenoma using the transsphenoidal route to approach the pituitary gland
- Removal of the entire adenohypophysis and life-long treatment with cortisol, thyroxine, and sex hormones
- Bilateral adrenalectomy

REPRODUCTIVE PHARMACOLOGY

In the absence of pituitary gonadotropins, the gonads fail to develop properly and removal of the pituitary gland causes reproductive failure.

Steroid Hormones

Receptors

All classes of steroid hormones bind to specific cytoplasmic receptors in their respective target tissues, and are then translocated to the nucleus. For example, testosterone, a lipid-soluble substance, enters the cell and is enzymatically reduced to **dihydrotestosterone by 5-α reductase**. **Dihydrotestosterone** then becomes bound to a specific androgen receptor site located in the cytoplasm. This complex becomes activated and is then translocated to the nucleus, where it binds to the chromatin acceptor site consisting of DNA and nonhistone chromosomal proteins. This interaction results in the transcription of a specific messenger RNA that is then relocated to the cytoplasm and translated on the cytoplasmic ribosomes, resulting in the synthesis of a new protein that sponsors the androgenic functions (Figure 136).

Receptor Antagonists

Similar to neurotransmitters, steroid hormones are classified as either agonists or antagonists, as the following shows:

Hormone	Agonist	Antagonist(s)
Estrogen	Estrogen	Nafoxidine, Tamoxifen
Androgen	Testosterone	Flutamide, Finazteride
Progesterone	Progesterone	Mifepristone

Tamoxifen decreases the rate of proliferation of breast cancer cells *in vitro* by inhibiting the estrogen-dependent production of specific proteins and growth factors that exert autocrine effects on cell division. Antiandrogenic compounds antagonize the effects of testosterone and are of value in the treatment of hirsutism and other **masculinizing syndromes**.

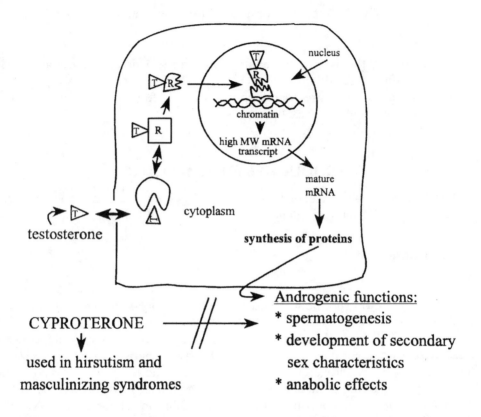

Figure 136 The steroid hormones and their receptors. The vitamin D, thyroid hormone, and estrogen receptors are all concentrated in the cell nucleus. However, the importance of steroid (S) and receptor (R) complex formation in the cytoplasm can be judged by the fact that antiestrogenic substances such as clomiphene and nafoxidine reduce the specific uptake of estrogen and hence reduce estrogen-induced uterine growth. Progesterone receptors are present in female reproductive tissues. Androgen receptors exist in male genital and accessory sex tissues. Glucocorticoid receptors are present in most mammalian tissues. Aldosterone receptors are found in mineralocorticoid target tissues such as the kidney, bladder, parotid gland, and gut.

Estrogens

Estrogens are synthesized mainly in the ovaries, the placenta, and the adrenal glands. A minute amount of estradiol is synthesized in the testes. Estrogens are synthesized according to the following scheme.

Cholesterol → pregnenolone → progesterone → androstenedione → estradiol 17 β → estrone → estriol

Physiological Actions

Estrogen dramatically influences the growth and development of the female reproductive organs, as summarized in Table 54.

Table 54 Summary of the Physiological Actions of Estrogen

Organs	Action
Uterus	Estrogens participate in the growth and development of muscular and mucosal elements of the uterus, the oviducts, and the fallopian tubes
Myometrium	Estrogens stimulate the growth of the myometrium at puberty and during pregnancy
Endometrium	Estrogens increase the endothelium and "rebuild" the endometrial lining in the proliferative phase
Vagina	Estrogens increase the epithelial layer in the vaginal tract and enhance glandular secretion
Breast	Estrogens influence the breast tissue at puberty, during each ovulatory cycle, and during pregnancy; they participate in duct formation, and, in conjugation with progesterone, oxytocin, and other hormones, prepare the breasts for lactation

Estrogens cause the growth of the uterus, fallopian tubes, and vagina. Estrogens also are responsible for the expression of female secondary sex characteristics during puberty. These include breast enlargement, the female distribution of body hair, female body contours as determined by subcutaneous fat deposition, and skin texture. Estrogens can stimulate the release of growth hormone and exert a positive effect on nitrogen balance. These effects contribute to the growth spurt during puberty. Closure of the bone epiphyses signaling the end of long bone growth is also estrogen mediated. Estrogens maintain bone mass by inhibiting bone resorption. This action of estrogen may be mediated by stimulating calcitonin production. Additionally, progestins antagonize loss of bone. In males, estrogens stimulate the growth of the stromal cells in the accessory sex organs.

Biochemical Actions

The uterus and vagina are sensitive to the biochemical actions of estrogens, which are as follows:

- *Early events* — release of histamine, synthesis of cyclic adenosine monophosphate (cyclic AMP), stimulation of RNA polymerase, and increased excitability of the myometrium.
- *Intermediate events* — synthesis of RNA and DNA, imbibition of water, and stimulation of certain enzymes.
- *Late events* — increased secretary activity, morphological changes, increased protein synthesis, stimulation of lipid and carbohydrate metabolism, and increased gravimetric responses.

Estrogen Preparations

The estrogen preparations are basically divided in tree groups: natural steroids, semisynthetic steroids, and nonsteroidal chemical compounds possessing estrogenic activities (Table 55). Transdermal estrogen (**Estraderm**) equals the efficacy of the oral preparation.

Table 55 Progesterone and Selected Progestins

Agent	Actions and Indications
Progesterone (Progekan)	Must be given intramuscularly, brief duration of action
Ethisterone (17α-ethinyltestosterone)	Effective orally
Ethynodiol acetate	Potent progestin
Hydroxyprogesterone caproate (Delalutin)	Long-acting, given intramuscularly
Medroxyprogesterone acetate (Provera)	Effective orally
Megestrol acetate (Megace)	Orally effective antifertility agent
Northindrone (Norethisterone)	Potent oral progestin, mild androgenic effect
Norethynodrel	Used as an antifertility agent
Norgestrel (18-homonorethisterone)	Potent progestin

Therapeutic Uses

Estrogens are used extensively in the treatment of endocrine and non-endocrine diseases, a few of which are cited below:

- Menopause — as a replacement therapy.
- Atrophic vaginitis — to thicken epithelial cells and to cause mucosal cells to proliferate.
- Primary hypogonadism — to correct ovarian failure.
- Osteoporosis — estrogen by itself or with hypercalcemic steroid is used in the treatment of osteoporosis.
- Primary amenorrhea — to cause endometrium to proliferate.
- Uterine bleeding — to reverse estrogen deficiency (in this case, oral contraceptives containing 80 to 100 μg of estrogen are recommended).
- Postpartum lactation — to relieve postpartum painful breast engorgement and prevent postpartum lactation; bromocriptine is also effective.
- Control of height — to cause closure of the epiphyses in unusually tall young girls.
- Dermatological problems — used with some success in the treatment of acne (Figure 137).

Adverse Effects

Low-dose estrogens are safe only when taken for a limited period. The most often reported side effects are breakthrough bleeding, breast tenderness, and very infrequent gastrointestinal upsets. When estrogens are used in large doses or injudiciously, they may cause thromboembolic disorders, hypertension in susceptible individuals, and cholestasis.

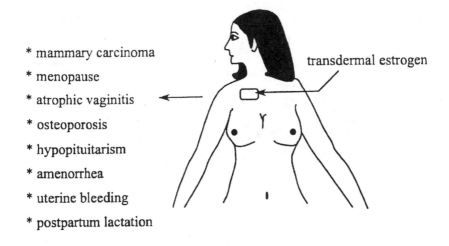

* mammary carcinoma
* menopause
* atrophic vaginitis
* osteoporosis
* hypopituitarism
* amenorrhea
* uterine bleeding
* postpartum lactation

transdermal estrogen

Figure 137 The uses of estrogen.

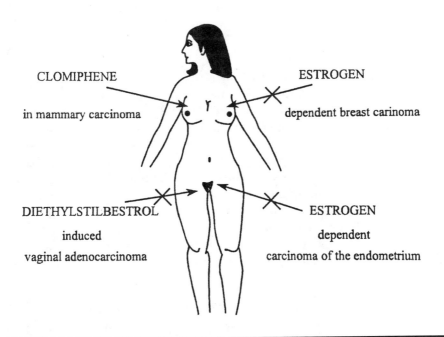

CLOMIPHENE

in mammary carcinoma

ESTROGEN

dependent breast carinoma

DIETHYLSTILBESTROL

induced

vaginal adenocarcinoma

ESTROGEN

dependent

carcinoma of the endometrium

Figure 138 The contraindications for estrogen.

Estrogens are contraindicated in patients with estrogen-dependent neoplasms such as carcinoma breast or endometrium (Figure 138). Vaginal adenocarcinoma has been reported in young women whose mothers were treated when pregnant with them with diethylstilbestrol in an effort to prevent miscarriage.

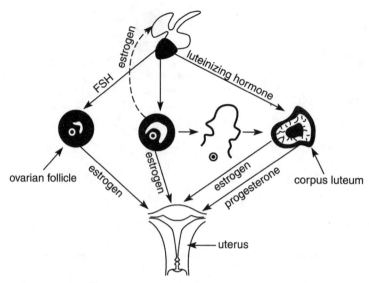

Progesterone prepares the uterus for implantation of fertilized ovum

Figure 139 The actions of progesterone. FSH = follicle-stimulating hormone.

Antiestrogens

Clomiphene (Clomid) and **tamoxifen** (Nolvadex) modify or inhibit the actions of estrogens. They accomplish this by binding to the cytoplasmic estrogen receptors that are then translocated to the nucleus. By diminishing the number of estrogen-binding sites, they interfere with the physiological actions of estrogens. Furthermore, by interfering with the normal hypothalamic and hypophyseal feedback inhibition of estrogen synthesis, these agents cause an increased secretion of **luteinizing hormone-releasing hormone, follicle-stimulating hormone-releasing hormone**, and **gonadotropins**. This leads to ovarian stimulation and ovulation. Clomiphene has been used successfully in some cases of infertility but causes multiple births. Antiestrogens are able to arrest the growth of estrogen-dependent malignant mammary cells. Clomiphene has been used in certain cases of disseminated breast cancer (see Figure 138).

Progesterone

Progesterone is synthesized by the ovaries, the adrenal glands, and the placenta. In a nonpregnant woman, it is produced by the corpus luteum during the latter part of the menstrual cycle under the influence of luteinizing and luteotropic hormones. In a pregnant woman, it is produced initially by the corpus luteum under the influence of chorionic gonadotropins and is synthesized by the placenta after failure of the corpus luteum (Figure 139).

Progesterone is not only an important progestin, but is also an important precursor for androgen. It is synthesized according to the following scheme:

Acetate → cholesterol → pregnenolone → progesterone → testosterone → estradiol

Progesterone is absorbed rapidly when given orally and has a plasma half-life of 5 minutes. It is completely metabolized in the liver and is cleared completely during first passage through the liver.

Physiological Actions. Progesterone initially prepares the uterus for implantation of the fertilized egg and prevents uterine contraction that would expel the fetus (see Figure 139). Progesterone has been used in the past to prevent **threatened abortion**. In addition, progesterone exerts effects on the secretory cells of the mammary glands. Progesterone competes with aldosterone and causes a decrease in sodium resorption; therefore, it antagonizes **aldosterone-induced sodium retention**. Progesterone increases the body temperature and decreases the plasma level of many amino acids.

Progesterone and Progestins

In the treatment of progesterone-related disorders, progesterone, which must be injected and has a short duration of action, has been replaced by the **progestins**. These newer synthetic derivatives of progesterone are effective orally and have a longer duration of action. Unlike progesterone, some of these agents have androgenic, estrogenic, and even glucocorticoid-like effects (see Table 55).

Therapeutic Uses

Progestins are used as **antifertility agents** and in the treatment of dysfunctional uterine bleeding, which may occur as a result of insufficient estrogen or because of continued estrogen secretion in the absence of progesterone.

- **Amenorrhea.** Progestins such as medroxyprogesterone are useful in the diagnosis and treatment of amenorrhea.
- **Dysmenorrhea.** Because prostaglandin $F_{2\alpha}$ is capable of inducing contraction in the uterus, agents that are able to block the synthesis of prostaglandin, such as aspirin or aspirin-like substances, have been shown to be effective in easing dysmenorrhea. For sexually active women, oral contraceptives have been found to be effective in relieving dysmenorrhea.
- **Endometriosis.** Endometriosis, which was formerly treated by surgical removal of the ovaries and uterus, is now treated with the continuous administration of progestin, or with progestin combined with estrogen. In addition, progestin may be useful in the management of **endometrial carcinoma**.
- **Suppression of Postpartum Lactation.** Estrogen, progesterone, and bromocriptine (a dopamine receptor agonist) are all effective in suppressing postpartum lactation.

Antifertility Agents

Oral contraceptive medications are among the most prescribed and effective drugs. The failure rates associated with various methods of contraception are cited below:

Method	Failure Rate (%)
Sterilization	0.0
Oral contraceptive	2.5
IUD	4.8
Condom	9.6
Diaphragm	14.4
Spermicide	17.7
Rhythm method	18.8

The most common type of oral contraceptive is the combination preparation and these are listed in Table 56.

Mechanism of Action

The antifertility agents suppress ovulation by:

■ Inhibiting the release of **hypophyseal ovulation-regulating gonadotropin**
■ Producing thick mucus from the cervical glands and hence impeding the penetration of sperm cells into the uterus
■ Impeding the transfer of the ovum from the oviduct to the uterus
■ Preventing implantation of the fertilized ovum should fertilization take place (Figure 140)

Adverse Metabolic Effects

The various adverse metabolic effects of oral contraceptives are listed below:

■ The incidence of venous and arterial thrombosis is related to the dose of estrogen.
■ Blood pressure is increased.
■ The incidence of myocardial infarction is higher among older women.
■ Adverse alterations in the levels of high-density lipoproteins, cholesterol, and low-density lipoproteins are produced by progestins.
■ The estrogen component of the preparation accelerates the development of gallbladder disease.

Potential Beneficial Effects

The potential benefits of oral contraceptive use consist of:

■ Reduced menstrual blood loss and thus less iron deficiency anemia
■ Reduced incidence of menorrhagia and irregular bleeding
■ Lower incidence of breast and endometrial cancers
■ Less incidence of salpingitis and increased bone density

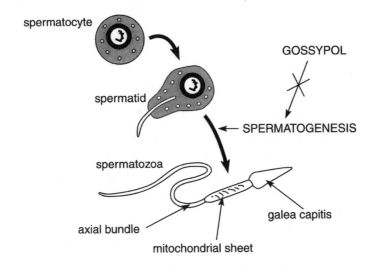

Figure 140   Actions of oral contraceptive medications.

Androgens

Testosterone, the male sex hormone, is responsible for the development and maintenance of the male sex organs (the penis, prostate gland, seminal vesicle, and vas deferens) and secondary sex characteristics. In addition, testosterone has anabolic effects. Similar to progesterone, testosterone is metabolized very rapidly by the liver by the first-pass mechanism, and hence requires structural modifications in order to be effective. For example, the 17-OH group of testosterone may be modified by the addition of propionic acid, which yields **testosterone propionate**, cyclopentyl propionic acid, which yields testosterone cypionate, or enanthate, which yields testosterone enanthate. In addition, the 17 position may be methylated to yield methyltestosterone or a fluorine and a methyl group may be inserted to yield fluoxymesterone. In general, these agents are more effective when given orally and have a longer duration of action than testosterone itself (Table 57).

Therapeutic Uses

Testosterone and its derivatives are used in the treatment of **hypogonadism (eunuchoidism)**, hypopituitarism, accelerated growth, aging in men, osteoporosis, anemia, endometriosis, promotion of anabolism, suppression of lactation, and breast carcinoma.

Hormonal therapy with testosterone should be reserved primarily for patients with hypogonadal disorders. There are two important warnings about the indiscriminate use of intramuscular testosterone in patients with serum testosterone levels in the normal range. First, many impotent patients are older and may have **adenocarcinoma of the prostate**, thus exogenous testosterone may accelerate the growth of the neoplasm. Second, although testosterone may induce a marked increase in libido, patients may still be unable to achieve adequate erection.

Table 56 Composition and Doses of Some Oral Contraceptives

Estrogen (mg)	Progestin^a (mg)	Representative Trade Names
Combinations^b		
0.02 Ethinyl estradiol	1.0 Norethindrone acetate	Loestrin 1/20
0.03 Ethinyl estradiol	0.3 Norgestrel	Lo/Orval
0.03 Ethinyl estradiol	1.5 Norethindrone acetate	Loestrin 1.5/30
0.03 Ethinyl estradiol	0.15 Levonorgestrel	Nordette
0.035 Ethinyl estradiol	0.4 Norethindrone	Ovcon 35
0.035 Ethinyl estradiol	0.5 Norethindrone	Brevicon
0.035 Ethinyl estradiol	1.0 Ethynodiol diacetate	Demulen 1/35
0.035 Ethinyl estradiol	1.0 Norethindrone	Ortho-Novum 1/35
0.05 Mestranol	1.0 Norethindrone	Ortho-Novum 1/50
0.05 Ethinyl estradiol	0.5 Norgestrel	Ovral
0.05 Ethinyl estradiol	1.0 Ethynodiol diacetate	Demulen 1/50
0.05 Ethinyl estradiol	1.0 Norethindrone	Ovcon 50
0.05 Ethinyl estradiol	1.0 Norethindrone acetate	Norlestrin 1/50
0.05 Ethinyl estradiol	2.5 Norethindrone acetate	Norlestrin 2.5/50
Sequentials^c		
0.03, 0.04, 0.03 Ethinyl estradiol	0.05, 0.75, 0.125 Levonorgestrel	Tri-Levlen
0.035 Ethinyl estradiol	0.5, 1.0, 0.5 Norethindrone	Tri-Norinyl
0.035 Ethinyl estradiol	0.5, 0.75, 1.0 Norethindrone	Ortho-Novum 7/7/7
0.035 Ethinyl estradiol	0.5, 1.0 Norethindrone	Ortho-Novum 10/11

Minipills[d]

a   0.35   Norethindrone   Micronor

Postcoital Medications

Ethinyl estradiol and norgestrel (combination available as Ovral)	100 µg ethinyl estradiol and 1 mg norgestrel (two Ovral tablets) taken twice 12 hours apart
Ethinyl estradiol	2.5 mg twice daily for 5 days
Conjugated estrogens	10 mg three times daily for 5 days
Estrone	5 mg three times daily for 5 days
Diethylstilbestrol	25 mg twice daily for 5 days

[a] Of the progestins used, norgestrel is strongly androgenic; the others have moderate androgenic activity.

[b] Combination tablets are taken for 21 days and are omitted for 7 days. These preparations are listed in order of the increasing content of estrogen.

[c] These preparations include fixed-dose tablets with the same or different amounts of estrogen and variable amounts of progestin. With biphasic preparations, the first set of tablets is taken for 10 days and the second for 11 days, followed by 7 days of no medication. With triphasic preparations, each set of tablets is taken for 5 to 10 days in three sequential phases, followed by 7 days of no medication.

[d] Minipills are taken daily continually.

Table 57   Examples of Anabolic and Androgenic Steroids

Steroids with anabolic activities
  Dromostanolone propionate (Drolban)
  Ethylestrenol (Maxibolin)
  Methandrostenolone (Dianabol)
  Nandrolone decanoate (Deca-Durabolin)
  Nandrolone phenpropionate (Durabolin)
  Oxandrolone (Anavar)
  Oxymetholone (Adroyd)
  Stanozolol (Winstrol)
  Testolactone (Teslac)
Steroids with androgenic properties
  Fluoxymesterone (Halotestin)
  Methyltestosterone (Metandren)
  Testosterone (Android-T)
  Testosterone propionate cypionate (Depotestosterone)
  Testosterone enanthate (Delatestryl)

Androgens are steroid hormones that are secreted primarily by the testis but also by the mammalian adrenal gland and ovaries. Testosterone, the principal androgen secreted by the testis, and other androgenic compounds possess virilizing activities that serve to regulate both differentiation and secretory function of the male sex accessory organs. Androgens also exhibit protein anabolic activity in skeletal muscle, bone, and kidneys; this effect results in a positive nitrogen balance and increased protein synthesis. In recent years, compounds have been synthesized that possess a preponderance of either virilizing or protein anabolic activity. As a class, the androgens are reasonably safe drugs, having limited and relatively predictable side effects.

Side Effects

One of the side effects of testosterone compounds is masculinization in women (such as hirsutism, acne, depression of menses, and clitoral enlargement) and of their female offspring. Therefore, androgens are contraindicated in pregnant women. Prostatic hypertrophy may occur in males, which leads to urinary retention. Therefore, androgens are contraindicated in men with prostatic carcinoma.

Antiandrogens

Cyproterone inhibits the action of androgens and **gossypol** prevents spermatogenesis without altering the other endocrine functions of the testis (Figure 141).

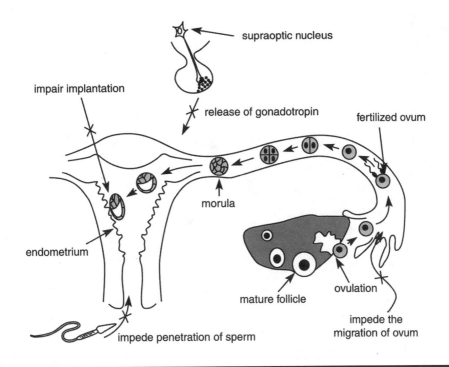

Figure 141 The action of gossypol on spermatogenesis.

55

TAXOL® (PACLITAXEL) AND CANCER CHEMOTHERAPY

Taxol is an antineoplastic agent.

INTRODUCTION

This compound, first isolated from the bark of the **Western yew tree** in 1971, exhibits unique pharmacological actions as an inhibitor of mitosis, differing from the **vinca alkaloids** and **colchicine derivatives** in that it promotes rather than inhibits microtubule formation. Following its introduction into clinical trial, the drug was approved for treatment of **cisplatin-refractory ovarian cancer** in 1992 and has promising activity against **cancers of the breast**, **lung**, **esophagus**, **and head and neck**.

Malignant neoplastic diseases may be treated by various approaches: surgery, radiation therapy, immunotherapy, or chemotherapy, or a combination of these. The extent of a malignant disease (staging) should be ascertained in order to plan an effective therapeutic intervention.

CANCER CHEMOTHERAPY

Before discussing the specific pharmacokinetics and pharmacodynamics of each class of antineoplastic agent, several fundamental concepts and therapeutic objectives will be considered first.

Because a single cancerous cell is capable of multiplying rapidly and eventually causing the host's death, one of the therapeutic objectives is to eradicate the last neoplastic cell. Unlike normal cells, cancerous cells multiply ceaselessly, and, unless arrested, they will kill the host. In the early phase, cancerous cells grow exponentially. However, as the tumor grows in mass, the time needed for the number of cells to double also increases. The kinetics of cell multiplication are said to follow a **Gompertzian growth curve**. Tumor growth may be divided into three phases: (1) the subclinical phase, in which 10×10^4 cells are present, (2) the clinical phase, in which 10×10^8 cells (1-cm^3 nodule) are present, and (3) the fatal phase, in which the number of cancerous cells equals or exceeds 10×10^{12}.

Most human cancerous cells evolve from the single clone of a malignant cell. As a tumor grows, significant mutation takes place, producing cells that exhibit diversified morphological and biochemical characteristics.

During the subclinical phase, the rapidly growing cell population is uniform in character and thus highly sensitive to drug treatment; this is the reason for the importance of early diagnosis and treatment. During the clinical phase, the nondividing and slowly growing cells are nonuniform in character and less sensitive to drug treatment, thus necessitating the need for multiple-drug treatment.

Cell destruction with antineoplastic agents follows a first-order kinetic, indicating that the drugs kill a constant fraction of cells and not a constant number of cells. This concept is depicted mathematically in the following table.

No. of Treatments	% Killed	No. of Tumor Cells Killed each Treatment	No. of Surviving Tumor Cells
Start	—	—	1,000,000
1	90	900,000	100,000
2	90	90,000	10,000
3	90	9,000	1,000
4	90	900	100
5	90	90	10
6	90	9	1

Cytotoxic drugs are not specific in their actions. They arrest not only cancerous cells, but also normal cells, especially those of the rapidly proliferating tissues such as the bone marrow, lymphoid system, oral and gastrointestinal epithelium, skin and hair follicles, and germinal epithelium of the gonads. Consequently, the therapeutic regimen must be carried out using high-dose intermittent schedules and not a low-dose continuous approach. Succeeding doses are given as soon as the patient has recovered from the previous treatment.

Antineoplastic agents may be teratogenic, carcinogenic, or immunosuppressant, and they exert their lethal effects on different phases of cell cycle by being either cell-cycle specific or nonspecific (see Figure 4).

The management of cancer includes treatment with **alkylating agents** (nitrogen mustards and alkyl sulfonates), **antimetabolites** (methotrexate and purine analogues), **natural products** (vinca alkaloids and taxol), miscellaneous compounds (hydroxyurea, procarbazine, and *cis*-platinum), **hormones** (estrogens and corticosteroids), and **radioactive isotopes**.

ALKYLATING AGENTS

The alkylating agents exert their antineoplastic actions by generating highly reactive **carbonium ion intermediates** that form a covalent linkage with various nucleophilic components on both proteins and DNA. The 7 position of the purine base guanine is particularly susceptible to alkylation, resulting in miscoding, depurination, or ring cleavage. **Bifunctional alkylating agents** are able to cross-link either two nucleic acid molecules or one protein and one nucleic acid molecule. Although these agents are very active from a therapeutic perspective, they are also notorious for their tendency to cause carcinogenesis and mutagenesis. Alkylating agents that have a nonspecific effect on the cell-cycle phase are the most cytotoxic to rapidly proliferating tissues.

Nitrogen Mustards

The activity of nitrogen mustards depends on the presence of a ***bis*-(2-chloro-ethyl) grouping**:

$$
\begin{array}{c}
\diagup CH_2\text{-}CH_2Cl \\
N \\
\diagdown CH_2\text{-}CH_2Cl
\end{array}
$$

This is present in **mechlorethamine** (Mustargen), which is used in patients with Hodgkin's disease and other lymphomas, usually in combination with other drugs, such as in **MOPP therapy** (mechlorethamine, Oncovin [vincristine], procarbazine, and prednisone). It may cause bone marrow depression.

Chlorambucil

Chlorambucil (Leukeran) is the least toxic nitrogen mustard, and is used as the drug of choice in the treatment of chronic lymphocytic leukemia. It is absorbed orally, is slow in its onset of action, and may cause bone marrow depression.

Cyclophosphamide

Cyclophosphamide (Cytoxan and Endoxan) is used in the treatment of Hodgkin's disease, lymphosarcoma, and other lymphomas. It is employed as a secondary drug in patients with acute leukemia and in combination with doxorubicin in women and breast cancer. A drug combination effective in the treatment of breast cancer is **cyclophosphamide**, **methotrexate**, **fluorouracil**, and **prednisone** (CMFP). Cyclophosphamide is also an immunosuppressive agent. The toxicity of cyclophosphamide causes alopecia, bone marrow depression, nausea and vomiting, and hemorrhagic cystitis.

Alkyl Sulfonate

The **alkyl sulfonate busulfan** (Myleran) is metabolized to an alkylating agent. Because it produces selective myelosuppression, it is used in cases of chronic myelocytic leukemia. It causes pronounced hyperuricemia stemming from the catabolism of purine.

Nitrosoureas

Carmustine (BCNU), lomustine (CCNU), and semustine (methyl-CCNU) generate alkyl carbonium ions and isocyanate molecules and hence are able to interact with DNA and other macromolecules. These agents, which are lipid soluble, cross the blood–brain barrier and are therefore effective in treating brain tumors. They are bone marrow depressants.

Triazenes

Dacarbazine (DTIC-Dome) is metabolized to an active alkylating substance. It is used in the treatment of malignant melanoma and causes myelosuppression.

ANTIMETABOLITES

Antimetabolites are structural analogues of naturally occurring compounds and function as fraudulent substances for vital biochemical reactions.

Folic Acid Analogues

Methotrexate (Amethopterin) is a folic acid antagonist that binds to **dihydrofolate reductase**, thus interfering with the synthesis of the active cofactor **tetrahydrofolic acid**, which is necessary for the synthesis of thymidylate, purine nucleotides, and the amino acids serine and methionine. Methotrexate is used for the following types of cancer:

- **Acute lymphoid leukemia.** During the initial phase, vincristine and prednisone are used. Methotrexate and mercaptopurine are used for maintenance therapy. In addition, methotrexate is given intrathecally, with or without radiotherapy, to prevent meningeal leukemia.
- **Diffuse histiocytic lymphoma.** Cyclophosphamide, vincristine, methotrexate, and cytarabine (COMA).
- **Mycosis fungoides**. Methotrexate.
- **Squamous cell, large-cell anaplastic, and adenocarcinoma**. Doxorubicin and cyclophosphamide, or methotrexate.
- **Head and neck squamous cell.** *Cis*-platinum and bleomycin, or methotrexate.
- **Choriocarcinoma.** Methotrexate.

Tumor cells acquire resistance to methotrexate as the result of several factors:

- The deletion of a high-affinity, carrier-mediated transport system for reduced folates.
- An increase in the concentration of dihydrofolate reductase.
- The formation of a biochemically altered reductase with reduced affinity for methotrexate.

To overcome this resistance, higher doses of methotrexate need to be administered.

The effects of methotrexate may be reversed by the administration of **leucovorin**, the reduced folate. This leucovorin "rescue" prevents or reduces the toxicity of methotrexate, which is expressed as mouth lesions (stomatitis), injury to the gastrointestinal epithelium (diarrhea), leukopenia, and thrombocytopenia.

Pyrimidine Analogues

Fluorouracil and **fluorodeoxyuridine** (floxuridine) inhibit pyrimidine nucleotide biosynthesis and interfere with the synthesis and actions of nucleic acids. To exert its effect, fluorouracil (5-FU) must first be converted to nucleotide derivatives such as 5-fluorodeoxyuridylate (5-FdUMP). Similarly, floxuridine (FUdR) is also converted to FdUMP by the following reactions:

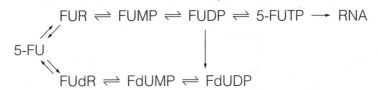

FdUMP inhibits **thymidylate synthetase**, and this in turn inhibits the essential formation of DTTP, one of the four precursors of DNA. In addition, 5-FU is sequentially converted to 5-FUTP, which becomes incorporated into RNA, thus inhibiting its processing and functioning.

Fluorouracil is used for the following types of cancer:

- **Breast carcinoma.** Cyclophosphamide, methotrexate, fluorouracil, and prednisone (CMP + P). The alternate drugs are doxorubicin and cyclophosphamide.
- **Colon carcinoma.** Fluorouracil.
- **Gastric adenocarcinoma.** Fluorouracil, doxorubicin (Adriamycin), and mitomycin (FAM), or fluorouracil and semustine.
- **Hepatocellular carcinoma.** Fluorouracil alone or in combination with lomustine.
- **Pancreatic adenocarcinoma**. Fluorouracil.

Resistance of 5-FU occurs as the result of one or a combination of the following factors:

- Deletion of uridine kinase
- Deletion of nucleoside phosphorylase
- Deletion of orotic acid phosphoribosyltransferase
- Increased thymidylate kinase

Because 5-FU is metabolized rapidly in the liver, it is administered intravenously and not orally. 5-FU causes myelosuppression and mucositis.

Deoxycytidine Analogues

Cytosine arabinoside (**Cytarabine**, **Cytosar**, and **Ara-C**) is an analogue of deoxycytidine, differing only in its substitution of sugar arabinose for deoxyribose. It is converted to Ara-CTP, and thereby inhibits DNA polymerase according to the following reactions:

$$\text{Ara-C} \xrightarrow{\text{Deoxycytidine kinase}} \text{Ara-CMP} \xrightarrow{\text{dCMP kinase}} \text{Ara-CDP}$$

$$\text{Ara-CDP} \xrightarrow{\text{NDP kinase}} \text{Ara-CTP}$$

$$\text{Deoxynucleotides} \xrightarrow[\text{DNA-polymerase}]{} \text{DNA}$$

Cytosine arabinoside is used in the treatment of acute granulocytic leukemia. Doxorubicin; daunorubicin and cytarabine; cytarabine and thioguanine; or cytarabine, vincristine, and prednisone are the combinations of agents employed.

Resistance to cytosine arabinoside may stem from the following factors:

- The deletion of deoxycytidine kinase
- An increased intracellular pool of dCTP, a nucleotide that competes with Ara-CTP
- Increased cytidine deaminase activity, converting Ara-C to inactive Ara-U

The toxic effects of cytosine arabinoside are myelosuppression and injury to the gastrointestinal epithelium, which causes nausea, vomiting, and diarrhea.

Purine Antimetabolites

6-Mercaptopurine (6MP) and 6-thioguanine (6TG) are analogues of the purines hypoxanthine and guanine, which must be activated by nucleotide formation, according to the following scheme:

$$\text{6–MP} + \text{phosphoribosylpyrophosphate (PRPP)} \xrightarrow{\substack{\text{Hypoxanthine-guanine} \\ \text{phosphoribosyl} \\ \text{transferase}}} \text{6 ThioIMP}$$

$$\text{6TG} + \text{PRPP} \xrightarrow{\hspace{2cm}} \text{6 ThioGMP}$$

ThioIMP and ThioGMP are feedback inhibitors of **phosphoribosylpyrophosphate amidotransferase**, which is the first and rate-limiting step in the synthesis of purine. In addition, these analogues inhibit the *de novo* biosynthesis of purine and block the conversion of inosinic acid to adenylic acid or guanylic acid. The triphosphate nucleotides are incorporated into DNA, and this results in delayed toxicity after several cell divisions. **6-Mercaptopurine** is used in the treatment of acute lymphoid leukemia. Maintenance therapy makes use of both methotrexate and 6-mercaptopurine. Mercaptopurine is absorbed well from the gastrointestinal tract. It is metabolized through: (1) methylation of the sulfhydryl group and subsequent oxidation; and (2) conversion to 6-thiouric acid by the aid of xanthine oxidase, which is inhibited by allopurinol. Mercaptopurine may cause hyperuricemia. Its chief toxicities are hepatic damage and bone marrow depression.

Thioguanine is used in patients with acute granulocytic leukemia, usually in combination with cytosine arabinoside and daunorubicin.

NATURAL PRODUCTS

Vinca Alkaloids

The vinca alkaloids (**vinblastine**, **vincristine**, and **vindesine**), which bind to tubulin, block mitosis with metaphase arrest. Vinca alkaloids are used for the following types of cancer:

- **Acute lymphoid leukemia.** In the induction phase, vincristine is used with prednisone.
- **Acute myelomonocytic or monocytic leukemia.** Cytarabine, vineristine, and prednisone.
- **Hodgkin's disease.** Mechlorethamine, Oncovin (vincristine), procarbazine, and prednisone (MOPP).
- **Nodular lymphoma.** Cyclophosphamide, Oncovin (vincristine), and prednisone (CVP).
- **Diffuse histiocytic lymphoma.** Cyclophosphamide, Adriamycin (doxorubicin), vincristine, and prednisone (CHOP); bleomycin, Adriamycin (doxorubicin), cyclophosphamide, Oncovin (vincristine), and prednisone (BACOP); or cyclophosphamide, Oncovin (vincristine), methotrexate, and cytarabine (COMA).
- **Wilms' tumor.** Dactinomycin and vincristine.
- **Ewing's sarcoma.** Cyclophosphamide, dactinomycin, or vincristine.
- **Embryonal rhabdomyosarcoma.** Cyclophosphamide, dactinomycin, or vincristine.
- **Bronchogenic carcinoma.** Doxorubicin, cyclophosphamide, and vincristine.

The chief toxicity associated with vinblastine use is bone marrow depression. The toxicity of vincristine consists of paresthesia, neuritic pain, muscle weakness, and visual disturbances. In addition, both vinblastine and vincristine may cause alopecia.

Dactinomycin

The antibiotics that bind to DNA are nonspecific to the cell-cycle phase. Dactinomycin (actinomycin D and Cosmegen) binds to double-stranded DNA and prevents RNA synthesis by inhibiting DNA-dependent RNA polymerase. It is administered intravenously in the treatment for pediatric solid tumors such as **Wilms' tumor** and **rhabdomyosarcoma** and for gestational **choriocarcinoma**. Dactinomycin causes skin reactions, gastrointestinal injury, and delayed bone marrow depression.

Mithramycin

The mechanism of action of mithramycin (Mithracin) is similar to that of dactinomycin. It is used in patients with advanced disseminated tumors of the testis and for the treatment of hypercalcemia associated with cancer. Mithramycin may cause gastrointestinal injury, bone marrow depression, hepatic and renal damage, and hemorrhagic tendency.

Daunorubicin and Doxorubicin

Daunorubicin (**Daunomycin** and **Cerubidine**) and doxorubicin (**Adriamycin**) bind to and cause the intercalation of the DNA molecule, thereby inhibiting DNA template function. They also provoke DNA chain scission and chromosomal damage.

Daunorubicin is useful in treating patients with acute lymphocytic or acute granulocytic leukemia. Adriamycin is useful in cases of solid tumors such as sarcoma, metastatic breast cancer, and thyroid cancer. These agents cause stomatitis, alopecia, myelosuppression, and cardiac abnormalities ranging from arrhythmias to cardiomyopathy.

Bleomycin

Bleomycin (**Blenoxane**) causes chain scission and fragmentation of DNA. With the exception of the skin and lungs, most tissues can enzymatically inactivate bleomycin. Bleomycin is used in the management of squamous cell carcinoma of the head, neck, and esophagus in combination with other drugs in patients with testicular carcinoma, and in the treatment of Hodgkin's disease and other lymphomas. Bleomycin causes stomatitis, ulceration, hyperpigmentation, erythema, and pulmonary fibrosis.

Paclitaxel and Docetaxel

Paclitaxel (**Taxol**) is a diterpenoid compound that contains a complex taxane ring as its nucleus (Figure 142). The side chain linked to the taxane ring at carbon 13 is essential for its antitumor activity. Modification of the side chain has led to identification of a more potent analogue, docetaxel (Taxotere), which has clinical activity against breast and ovarian cancers. Originally purified as the parent molecule from yew bark, paclitaxel can now be obtained for commercial purposes by semisynthesis from **10-desacetylbaccatin**, a precursor found in yew leaves. It also has been successfully synthesized from simple off-the-shelf reagents in a complex series of reactions.

Taxol binds specifically to the β-tubulin subunit of microtubules and appears to antagonize the disassembly of this key cytoskeletal protein, with the result that bundles of microtubules and aberrant structures derived from microtubules appear in paclitaxel-treated cells. Arrest in mitosis follows. Cell killing is dependent on both drug concentrations and duration of cell exposure. Paclitaxel has undergone initial phases of testing in patients with metastatic ovarian and breast cancer; it has significant activity in both diseases, including diseases in patients that have

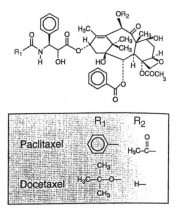

Figure 142 Chemical structures of paclitaxel (Taxol) and its more potent analogue, docetaxel (Taxotere).

progressed on standard primary combination regimens. Response rates in relapsed patients range from 20 to 50%, depending on the treatment history and the regimen employed. Early trials indicate significant response rates in lung, head and neck, esophageal, and bladder carcinomas as well.

Pactitaxel exerts its primary toxic effects on the bone marrow. Neutropenia usually occurs 8 to 11 days after a dose and reverses rapidly by days 15 to 21. Many patients experience myalgias for several days after receiving paclitaxel. In high-dose schedules, a stocking-glove sensory neuropathy can be disabling, particularly in patients with underlying diabetic or alcoholic neuropathy (see Suffness, 1995).

MISCELLANEOUS ANTINEOPLASTIC AGENTS

Asparaginase

Normal cells are able to synthesize asparagine but neoplastic tissues must obtain it from external sources. By metabolizing asparagine, asparaginase (Elspar) deprives the neoplastic tissues of asparagine, and in turn inhibits protein and nucleic acid synthesis. The resistant tumors are thought to possess higher than ordinary amounts of asparagine synthetase. Asparaginase, which is prepared from *Escherichia coli*, is used to induce remission of acute lymphocytic leukemia. Asparaginase causes malaise, anorexia, chills, fever, and hypersensitivity reactions. In general, it does not damage the bone marrow or other rapidly growing tissues as much as other antineoplastic agents do.

Hydroxyurea

Hydroxyurea suppresses DNA synthesis by inhibiting ribonucleoside diphosphate reductase, which catalyzes the reduction of ribonucleotides to deoxyribonucleotides. Hydroxyurea is used in chronic cases of granulocytic leukemia that are unresponsive to busulfan. In addition, it is used for acute lymphoblastic leukemia. Hydroxyurea may cause bone marrow depression.

Cis-Platinum

Cis-platinum (cisplatin) binds to intracellular DNA, causing both interstrand and intrastrand cross-linking. It is a cell-cycle phase nonspecific agent. *Cis*-platinum, which is ineffective orally, it used for testicular, bladder, and head and neck cancers. It precipitates nephrotoxicity, ototoxicity, and gastrointestinal injury.

Carboplatin

Carboplatin is a promising second-generation platinum agent. Because it is less reactive, it causes less nephrotoxicity, myelosuppression, and thrombocytopenia.

Procarbazine

Procarbazine (Matulane) inhibits DNA, RNA, and protein synthesis through the operation of an unknown mechanism. It is effective in patients with Hodgkin's disease when given in combination with mechlorethamine, vincristine, and prednisone (MOPP). Procarbazine causes neurotoxicity, bone marrow depression, and gastrointestinal injury.

Etoposide

Etoposide is used to combat several types of tumors, including testicular and small-cell lung cancers, lymphoma, leukemia, and Kaposi's sarcoma.

CANCER CHEMOTHERAPY-INDUCED EMESIS

Nausea and vomiting are frequent side effects of radiotherapy and cancer chemotherapy. The incidence of this is relatively low for bleomycin, vincristine, and chlorambucil but is high for the remaining agents. Besides **prochlorperazine** and **metoclopramide** (dopamine receptor-blocking agents), **nabilone** (a cannabinoid), **batanopride**, **granisetron**, and **ondansetron** (all serotonin receptor-blocking agents) have been shown to be effective in ameliorating these symptoms.

56

HERBAL PLANTS AND TUBERCULOSIS

The plants with the highest antimycobacterial activity include:

- *Allium sativum*
- *Borrichia frutescens*
- *Ferula communis*
- *Heracleum maximum*
- *Karwinskia humboldtiana*
- *Leucas vokensii*
- *Moneses uniflora*
- *Oplopanax horridus*
- *Salvia multicaulis*
- *Strobilanthus cusia*

ANTITUBERCULOSIS EFFICACY OF NATURAL PRODUCTS

Plant species from a wide range of families have been shown to have significant *in vitro* antimycobacterial activities, and a number of active plant-derived compounds belonging to various chemical classes have been isolated.

The infectious killer disease, **tuberculosis** (TB), is the leading cause of death worldwide from a single human pathogen, claiming more adult lives than diseases such as **acquired immunodeficiency syndrome** (AIDS), **malaria**, **diarrhoea**, **leprosy**, and all other tropical diseases combined. The organism usually responsible, the **tubercle bacillus**, *Mycobacterium tuberculosis* (MT), was discovered by Robert Koch in 1882. However, *M. bovis*, which infects cattle, may also infect humans and **M. africanum** is a cause of TB in West Africa. Furthermore, a number of normally nonpathogenic mycobacteria, especially **M. avium**, **M. intracellulare**, and **M. scrofulaceum**, cause opportunistic infectious disease in patients with AIDS. Pulmonary TB, the most common type of the disease, is usually acquired by inhalation of the bacillus from an infectious patient and causes irreversible lung destruction (Newton et al., 2000).

About one third of the world population is currently infected with *M. tuberculosis*; 10% of those infected will develop clinical disease, particularly those who also have the human immunodeficiency virus (HIV) infection. With the discovery of effective antimycobacterial agents (including **ethambutol, isoniazid, pyrazinamide, rifampicin**, and **streptomycin**) and a reduction in poverty, there was a drastic decline in the number of TB cases, especially in developed nations. However, since the late 1980s, the number of cases of TB throughout the world has been increasing rapidly due partly to the emergence of multidrug-resistant *M. tuberculosis*. Thus, the TB problem requires urgent attention. Short-course anti-TB regimens initially using at least three first-line drugs (including **isoniazid, rifampicin**, and **pyrazinamide**) are effective. The major problems faced in tuberculosis control are poor infrastructures for diagnosis and drug supply and failure of patients to complete their course of drugs. This is usually due to poor supervision and medical care and, as a result, drug resistance develops. Second-line drugs (e.g., **capreomycin, kanamycin, cycloserine, ethionamide**), which are often more toxic, have to be used in this case. Furthermore, drugs with broader ranges of activity are also required to target emerging pathogens, such as those of the **M. avium complex**. Hence, there is a great need to search for and develop new, affordable, anti-TB agents.

A number of plant extracts and compounds have potent antimycobacterial properties. Examples of the species that appear to be among the most active include **Allium sativum, Borrichia frutescens, Ferula communis, Heracleum maximum, Karwinskia humboldtiana, Leucas volkensii, Moneses uniflora, Oplopanax horridus, Salvia multicaulis**, and **Strobilanthus cusia**. (Newton et al., 2000).

MYCOBACTERIAL DISEASES

Tuberculosis

Tuberculosis is a necrotizing infection caused by *Mycobacterium tuberculosis* that primarily affects the lungs, but can also involve the kidneys, bones, lymph nodes, and meninges. Treatment takes two forms: chemoprophylaxis and active treatment. The recommended drug for chemoprophylaxis is isoniazid (300 mg/day for 1 year). In cases of active tuberculosis, the primary drugs are isoniazid, rifempin, streptomycin, ethambutol, and pyrazinamide. Secondary drugs consist of *p*-aminosalicylic acid, ethionamide, viomycin, kanamycin, capreomycin, and cycloserine. The drug combination is determined by the stage of the disease.

During the initial phase, isoniazid is always used in combination with one other drug: rifampin, streptomycin, or ethambutol. In advanced or cavitary pulmonary tuberculosis, often three drugs are used: isoniazid, rifampin, and streptomycin or ethambutol. The pharmacological properties of the most often used drugs are summarized in Table 58.

Leprosy

Leprosy (**Hansen's disease**) is a chronic granulomatous disease that attacks superficial tissues such as the skin, nasal mucosa, and peripheral nerves. There are two types of leprosy: lepromatous and tuberculoid. The **sulfones**, which are derivatives of **4,4′-diaminodiphenylsulfone**, are bacteriostatic. **Dapsone** (DDS)

Table 58   Select Pharmacological Properties of Drugs Most Often Used in Tuberculosis

Drug	Properties
Isoniazid	Isoniazid is bactericidal for growing tubercle bacilli, is absorbed orally, and is metabolized by acetylation. It is a structural analogue of pyridoxine and may cause pyridoxine deficiency, peripheral neuritis, and, in toxic doses, pyridoxine-responsive convulsions. Its mechanism of action is not known.
Streptomycin	Streptomycin is given intramuscularly. It exerts its effects only on extracellular tubercle bacilli. When combined with other drugs, it delays the emergence of streptomycin-resistant mutants. It is ototoxic and may cause deafness.
Rifampin	Rifampin is absorbed from the gastrointestinal tract and excreted mainly into bile. It binds to DNA-dependent RNA polymerase and inhibits RNA synthesis. In higher than therapeutic doses, rifampin may cause a flulike syndrome and thrombocytopenia.
Ethambutol	Ethambutol suppresses the growth of isoniazid- and streptomycin-resistant tubercle bacilli. The most important but not common side effects are optic neuritis, decreased visual acuity, and inability to perceive the color green.

The treatment of mycobacterial infections has become an even more important and challenging problem because of the emergence of multiple-drug-resistant organisms and because of the acquired immunodeficiency syndrome (AIDS) pandemic, which has been associated with a marked increase in tuberculosis and infection caused by the *M. avium* complex. Because the microorganisms grow slowly and the diseases often are chronic, patient compliance, drug toxicity, and the development of microbial resistance present special therapeutic problems.

and **sulfoxone sodium** are the most useful and effective agents currently available. They should be given in low doses initially, and then the dosage gradually increased until a full dose of 300 to 400 mg/week is reached. During this period, the patient must be monitored carefully. With adequate precautions and appropriate doses, sulfones may be used safely for years. Nevertheless, side effects such as anorexia, nervousness, insomnia, blurred vision, paresthesia, and peripheral neuropathy do occur. Hemolysis is common, especially in patients with **glucose 6-phosphate dehydrogenase deficiency**. A fatal exacerbation of **lepromatous leprosy** and an infectious mononucleosis-like syndrome rarely occur. **Clofazimine** (Lamprene) may be effective in patients who show resistance to the sulfones and may also dramatically reduce an exacerbation of leprosy. Red discoloration of the skin and eosinophilic enteritis have occurred following clofazimine therapy.

Other Diseases

Not all mycobacterial infections are caused by *M. tuberculosis* or *M. leprae*. These atypical mycobacteria require treatment with secondary medications as well as other chemotherapeutic agents. For example, *M. marinum* causes skin granulomas, and effective drugs in the treatment of infection are **rifampin** or **minocycline**. *Mycobacterium fortuitum* causes skin ulcers and the medications recommended for treatment are **ethambutol**, **cycloserine**, and **rifampin** in combination with **amikacin**.

ANTIMICROBIAL ACTIVITY OF PLANTS

Screening plant extracts for antimycobacterial activity is usually carried out using mycobacteria cultured in various types of broth- and agar-based media. *Mycobacterium tuberculosis* has the disadvantages of being slow growing so that tests take several weeks and containment facilities are needed because it is a dangerous pathogen. Many investigators have therefore used nonpathogenic species of mycobacteria such as *M. avium*, *M. intracellulare*, and *M. kansaii*, which like *M. tuberculosis* are slow growing, and other species including *M. chelonei*, *M. fortuitum*, and *M. smegmatis*, which are faster growing, allowing tests to be completed in a few days (see Table 59 and Newton et al., 2000, for a few examples).

Table 59   Antimicrobial Activities of Natural Plants

Plants	Origin	Part Used	Active Constituents/Comments
(Ranunculaceae) *Adhatoda vasica* (Acanthaceae)	India	Leaf	**Bromhexine** **Ambroxol** (semisynthetic derivatives of alkaloid vasicine)
Allium sativum (Liliaceae)	—	Bulb	**Garlic** has been used in traditional Chinese and Egyptian medicine for many centuries; positive controls included **isoniazid, streptomycin, ethambutol, rifampicin**
Alnus rubra Bong. (Betulaceae)	Br C	Bark	**Isoniazid** was used as positive control
Amyris elemifera L. (Rutaceae)	—	—	**Taxalin** (Oxazole)
Antirrhinum majus (Scrophulariaceae)	—	—	Active
Balsoamorhiza sagittata	Br C	Root	Active
Borrichia frutescens L. (Asteraceae)	USA	Flower	Active
Chrysoma pauciflosculosa Michx. (Asteraceae)	USA	Root	Active
Chrysanthum sinense Sab. (Asteraceae)	China	—	Active

Table 59 Antimicrobial Activities of Natural Plants

Plants	Origin	Part Used	Active Constituents/Comments
Ferula communis (Umbelliferae)	Saudi Arabia	Rhizome	Active
Lavandula angustifolia Mill. (lavender) (Labiatae)	France	Flower	Active
Leucas volkensii Gurke (Labiatae)	Kenya	Aerial	Active
Mammea americana (Guttiferaceae)	Puerto Rico	Leaf	Active
Melia volkensii Gurke (Meliaceae)	Kenya	Seeds	Active
Piper cubeba (Piperaceae)	Indonesia	—	Active
Propolis	Chile	Resin	Active
Solanum sodomaeum L. (Solanaceae)	Libya	Berry	Active

57

ULCER THERAPY
WITH HERBAL MEDICINE

INTRODUCTION

Phytogenic agents have traditionally been used by herbalists and indigenous healers for the prevention and treatment of **peptic ulcer**. Botanical compounds with antiulcer activity include **flavonoids** (i.e. quercetin, naringin, silymarin, anthocyanosides, sophoradin derivatives), **saponins** (i.e. from *Panax japonicus* and *Kochia scoparia*), **tannins** (i.e. from *Linderae umbellatae*), and **gums and mucilages** (i.e. gum guar and myrrh). Among herbal drugs, **licorice**, **aloe gel**, and **capsicum** (chili) have been used extensively and their clinical efficacy documented. Also, ethnomedical systems employ several plant extracts for the treatment of peptic ulcer. Despite progress in conventional chemistry and pharmacology in producing effective drugs, the plant kingdom might provide a useful source of new antiulcer compounds for development as pharmaceutical entities or, alternatively, as simple dietary adjuncts to existing therapies (Borrelli and Izzo, 2000).

ACID-PEPSIN DISEASE

The medical treatment of esophageal, gastric, and duodenal ulcer includes relieving the symptoms, accelerating healing, preventing complications, and preventing recurrence. Drug treatment includes the use of antacids, anticholinergic drugs, histamine H_2-receptor antagonists, and inhibitors of H^+K^+-ATPase.

PATHOGENESIS OF PEPTIC ULCER

The normal gastroduodenal mucosa has three defense mechanisms for resisting injury arising from the acid and peptic activity in gastric juice:

1. The surface epithelial cells, which secrete mucus and bicarbonate
2. The gastric mucosal cells, which have a specialized apical surface membrane that resists the diffusion of acid back into the cells
3. The mucosal cells, which possess intrinsic mechanisms that resist injury, for example, by extruding back-diffused hydrogens ions using a basolateral carrier (e.g., sodium-hydrogen or sodium-bicarbonate exchange) (Figure 143)

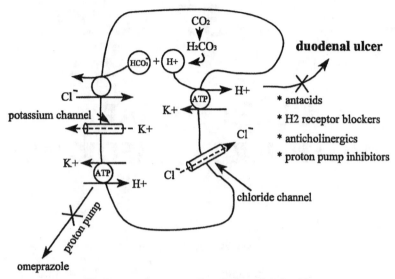

Hydrogen ion secretion by parietal cells

Figure 143   Omeprazole inhibits proton pump. H2 = histamine$_2$.

About a third of the patients with duodenal ulcer suffer from an excess secretion of gastric acid. This may be due to an increase in the mass of acid-secreting gastric mucosa or to vagal hyperactivity, or both.

PHYSIOLOGY OF GASTRIC ACID SECRETION

The parietal cells release hydrogen ions through the diversified mechanisms outlined in Figure 75. These mechanisms include:

- **Histamine** stimulation of parietal cell function by increasing the formation of cyclic adenosine monophosphate (cyclic AMP)
- **Acetylcholine** and **gastrin** stimulation of parietal cell function by increasing the level of cystosolic calcium
- **Prostaglandin** inhibition of gastric acid secretion by blocking the formation of cyclic adenosine monophosphate
- **Omeprazole** inhibition of H^+K^+-ATPase

DRUG THERAPY

Antacids

Because acid-pepsin disease rarely occurs in the absence of gastric acid and pepsin, antacids are highly effective in its overall management. Antacids consist of a mixture of magnesium, aluminum, and calcium compounds. Their efficacy is based on their inherent ability to react with and neutralize gastric acid. Sodium bicarbonate, which may leave the stomach rapidly, can cause alkalosis and sodium retention. Calcium salts may produce hypercalcemia, which can be detrimental in patients with impaired renal function. **Aluminum salts** may decrease the absorption of **tetracyclines** and anticholinergic drugs.

Anticholinergic Drugs

Vagal impulses elicit the release of acetylcholine in the parietal cells and in the gastric mucosal cells containing gastrin, a peptide hormone. Both the directly released acetylcholine and the indirectly released gastrin then stimulate the parietal cells to secrete hydrogen ions into the gastric lumen.

The most useful anticholinergic drugs are **propantheline** (Pro-Banthine), **pirenzepine**, and **telenzepine**, which antagonize muscarinic cholinergic receptors (M_1 receptors). All three agents depress gastric motility and secretion. The production of pepsin is also reduced. Propantheline may be used as adjunctive therapy with antacids but not as a sole agent. The side effects and contraindications of propantheline use are identical to those of atropine (prostatic hypertrophy, urinary retention, glaucoma, and cardiac arrhythmias).

The timing of medication is critical in ulcer therapy. Anticholinergic drugs should be given about 30 minutes before meals, and antacids about 1 hour after meals. A double-dose of an antacid is often taken just before bedtime.

Histamine-Receptor Antagonists

There are two types of histamine receptors: H_1 receptors, which are blocked by agents such as **diphenhydramine** and other antiallergic compounds, and H_2 receptors, which are blocked by **cimetidine** (Tagamet), **ranitidine** (Zantac), **famotidine** (Pepcid), and **nizatidine** (Axid). Cimetidine has no effect on most H_1 receptor-mediated effects, such as bronchoconstriction.

The clinical use of H_2-receptor antagonists stems from their capacity to inhibit gastric acid secretion, especially in patients with peptic ulceration. Cimetidine, which is far more efficacious than anticholinergic drugs, is used in the treatment of duodenal ulcers and gastrinoma, and in patients suffering from gastroesophageal reflux. It is absorbed orally, has a plasma half-life of 2 hours, and is excreted mainly unchanged by the kidney. The doses of cimetidine must be reduced in the presence of impaired renal function. The few and infrequent adverse effects of cimetidine use include:

- Gynecomastia (may bind to androgen receptor sites)
- Galactorrhea (especially in patients with gastrinoma)
- Granulocytopenia, agranulocytosis (very rare)
- Mental confusion (especially in elderly people)
- Restlessness, seizures, and reduced sperm count

Ranitidine is more effective than cimetidine and allegedly has fewer side effects.

Inhibitors of H⁺K⁺-ATPase

The proton pump (H^+K^+ATPase) of the apical membrane of the parietal cells is the ultimate mechanism that governs acid secretion. Among a family of benzimidazole derivatives, **omeprazole** (Losec) promotes the healing of ulcers in the stomach, duodenum, and esophagus, and is of special value in patients who do not respond to H_2-receptor antagonists.

ACTIVE HERBAL PRODUCTS WITH ANTIULCER ACTIVITY

Flavonoids

Flavonoids are a group of about 4000 naturally occurring compounds with a wide range of biological effects, including antiulcer activity. They are important constituents of the human diet (a daily diet contains approximately 1 g of flavonoids per day) and are also found in several medicinal plants used in folk medicine around the world. Several mechanisms have been proposed to explain the gastroprotective effect of flavonoids:

- Increase of mucosal **prostaglandin** content
- Decrease of **histamine** secretion from mast cells by inhibition of histidine decarboxylase
- Inhibition of *Helicobacter pylori* growth

In addition, flavonoids have been found to be free radical scavengers, and free radicals play an important role in ulcerative and erosive lesions of the gastrointestinal tract. In relation to their low toxicity and to the properties reported, flavonoids could have a therapeutic potential ideal for treatment of gastrointestinal diseases associated with *H. pylori* infection, i.e., type B gastritis and duodenal ulcer (Table 60).

Quercetin

Quercetin is the most abundant of the flavonoid molecules and it is found in many medicinal botanicals, including ***Thea sinensis, Glycyrrhiza glabra, Hypericum perforatum, Ginkgo biloba***, and many others. It has been reported to prevent gastric mucosal lesions produced by cold-restraint stress, pylorus ligation, reserpine, and ethanol or acidified ethanol. Quercetin increases the amount of neutral glycoproteins in the gastric mucosa. As these proteins are the most abundant and possibly the most important in the gastric mucosa, it can be assumed that their quantitative replacement is a return to normality of the mucosa, and thus a recovery of the defensive capacity against aggression from absolute ethanol.

Stimulation of local **prostaglandin** production is possibly one protective mechanism as quercetin stimulates the enzyme cyclooxygenase and, in addition, the **cyclooxygenase** inhibitor, **indomethacin**, reverses the protective effect of quercetin on ethanol-induced ulceration. The increase in prostaglandin synthesis

Table 60 Flavonoids with Anticancer Activity

Flavonoids	Ulcer Model
Anthocyanosides	Pylorus-ligated, reserpine, phenylbutazone
Catechin	Stress
Genistin	Phenylbutazone, serotonin pylorus-ligated, stress, reserpine
Hypolaetin-8-glucoside	Stress, ethanol, acetylsalicyclic acid
Kaempferol	Ethanol
Leucocyanidin	Aspirin
Luteolin-7-glycoside	Pylorus-ligated, stress, reserpine, phenylbutazone, serotonin
5-Methoxyflavone	Indomethacin
Myricetin 3-O-D-galactoside	Stress, pylorus-ligated, ethanol
Naringin	Ethanol, stress, pylorus-ligated
Quercetin	Stress, ethanol, reserpine
Rutin	Ethanol
Silymarin	Stress
Ternatin	Ethanol, indomethacin, stress
Vexibinol	HCl, ethanol

may explain the increase in the amount of mucus observed and its participation in ulcer prevention. Other possible mechanisms include:

- Inhibition of the **gastric proton pump**
- Inhibition of the **lipoxygenase** pathway
- Inhibition of **platelet activating** synthesis
- Inhibition of **lipid peroxidation**
- Scavenging of free radicals associated with a significant enhancement in **glutathione peroxidase** activity

An interesting aspect of the antiulcer effect of quercetin is that it has been shown to inhibit the growth of ***Helicobacter pylori*** in a dose-dependent manner *in vitro*.

Naringin

Naringin has been shown to prevent gastric mucosal ulceration in several animal models, including restraint stress, pyloric occlusion, and ethanol-induced chronic ulcer. In ethanol-treated rats, naringin significantly reduced the ulcer index and increased the **hexosamine** content of the gastric mucus without affecting prostaglandin E_2 and the total protein content. Thus, the gastroprotective action of naringin could be explained, at least in part, through a complex non-prostaglandin-dependent mechanism that involves an increase in the glycoprotein content and viscosity of the gastric mucosa. Naringin also possesses antioxidant and superoxide anion scavenger properties that could contribute to its gastroprotective effect.

Anthocyanosides

Anthocyanosides, extracts from **Vaccinium myrtillus**, exert a significant preventive and curative antiulcer activity. Protective effects have been observed not only on ulcer induced by ligature of the pylorus, by reserpine, and by phenylbutazone, in which oral treatment was given before or simultaneously with ulcer formation, but also in ulcer induced by restraint or by acetic acid, in which the animal received treatment when the lesions were already established. The antiulcer activity is not exerted through a blockade of gastric secretion, but it is, in part, attributed to the increase in the mucus of the stomach wall. Anthocyanosides are thought to act by influencing the biosynthesis of the **mucopolysaccharides**, thus improving the efficiency of the mucus barrier at the gastric level.

Sophoradin Derivatives

Sophoradin has been isolated from the root of the ancient Chinese medical plant **Sophora subprostrata**, a plant used in China for the treatment of digestive diseases. Flavonoids derived from sophoradin are known to exhibit gastroprotective and ulcer-healing properties. **Solon**, a synthetic flavonoid derived from sophoradin, has been shown to have antiulcer and gastroprotective effects by influencing the formation and metabolism of prostaglandins in the gastric mucosa.

Saponins

Saponins are widely distributed in plants and are a particular form of **glycosides**. They are so-called because of their soaplike effect, which is due to their **surfactant properties**. They also have hemolytic properties and, when injected into the bloodstream, are highly toxic. When taken by mouth, saponins are comparatively harmless. According to the structure of the **aglycone** or **sapogenin** two kinds of saponin are recognized, the steroidal and triterpenoid type.

Plant materials often contain triterpenoid saponin in considerable amounts. With regard to plants with antiulcer activity, **licorice root** contains about 2 to 12% of **glycyrrhizic acid** and the seeds of the **horse-chestnut** up to 13% of aescin. Several plants containing high amounts have been shown to possess antiulcer activity in several experimental ulcer models (Table 61).

Among these, saponins isolated from the rhizome of *Panax japonicus* and the fruit of *Kochia scoparia* (which contain approximately 20% of saponins) have been demonstrated to possess gastroprotective properties. Some **oleanolic acid oligoglycosides**, extracted from *P. japonicus* and *K. scoparia*, showed protective effects on ethanol- and indomethacin-induced gastric damage. Moreover, a methanol extract of *P. japonicus* rhizome was demonstrated to possess protective activity also on stress- or HCl-induced ulcers. The protective activities of all these active saponins are not due to inhibition of gastric acid secretion but probably due to activation of mucous membrane protective factors.

Aescin is a mixture of saponins obtained from the seeds of **Aesculus hippocastanum** (horse-chestnut). Aescin has been shown to possess antiulcer activity in various ulcer models (cold restraint and pylorus-ligated), an effect which is, in

Table 61   Plants Containing Saponins with Antiulcer Activity

Botanical Name	Part Plant	Ulcer Model
Calendula officinalis	Rhizone	Caffeine-arsenic, butadione, pylorus-ligated
Calliandra portoticensis	Leaves	Stress, pylorus-ligated, Escherichia coli
Glycyrrhiza glabra	Root	Acetic acid, pylorus-ligated
Kochia scoparia	Fruit	Ethanol, indomethacin
Panax binnatifidus	Rhizome	Psychological stress
P. japonicus	Rhizome	Stress, HCl
Pyrenacantha staudtii	Leaves	Indomethacin, serotonin, stress
Rhigiocarya racemifera	Leaves	Indomethacin, reserpine, serotonin
Spartium junceum	Flowers	Ethanol
Veronica officinalis	Aerial parts	Indomethacin, reserpine

part, due to inhibition of gastric acid and pepsinogen secretion. However, aescin also prevents gastric lesions due to absolute ethanol, a model of gastric ulceration in which acid and pepsin do not play a significant role; thus, the protective effect of aescin involves other mechanisms, such as improvement of blood flow. It is unlikely that prostaglandins or the maintenance of surface mucus gel could play a role in aescin-induced gastroprotection as this saponin does not enhance prostaglandin E_2 levels, nor does it increase the amount (or the composition) of mucus in ethanol-induced gastric ulceration.

Tannins

Tannins are used in medicine primarily because of their astringent properties, which are due to the fact that they react with the proteins of the layers of tissue with which they come into contact. Tannins are known to "tan" the outermost layer of the mucosa and to render it less permeable and more resistant to chemical and mechanical injury or irritation; however, the correlation between the molecular structures of tannins and the astringent/antiulcer activity is not known.

When a low concentration of tannins is applied to the mucosa, only the outermost layer is tanned, becoming less permeable and affording an increased protection to the subjacent layers against the action of bacteria, chemical irritation, and, to a certain extent, against mechanical irritation. High concentrations of tannins cause coagulation of the proteins of the deeper layer of the mucosa, resulting in inflammation, diarrhea, and vomiting.

Several plants with antiulcer active which contain tannins have been reported (Table 62).

A crude extract of *Linderae umbellatae* exhibited antipeptic and antiulcerogenic activity, and these effects were considered ascribable to the presence of tannins or related compounds. Nine condensed tannins (monomers, dimers, trimers, and tetramers) have been isolated and their antipeptic and antiulcer activity confirmed experimentally (pylorus-ligated in rats and stress-induced gastric lesions in mice).

Marked differences were observed among monomers, dimers, trimers, and tetramers. Monomers and dimers did not inhibit peptic activity *in vitro*, whereas

Table 62   Plants Containing Tannins with Anti-ulcer Activity

Botanical Name	Part Plant	Ulcer Model
Calliandra portoticensis	Leaves	Stress, pylorus-ligated, *E. coli*
Entandrophragma utile	Bark	Ethanol
Linderae umbellatae	Stem	Stress
Mallotus japonicus	Bark	(Clinical study)
Rhigiocarya racemifera	Leaves	Indomethacin, reserpine, serotonin
Veronica officinalis	Aerial parts	Indomethacin, reserpine

trimers displayed higher inhibition of peptic activity than tetramers (although tetramers showed higher astringency than trimers). In pylorus-ligated mice, trimers and tetramers clearly suppressed the peptic activity of gastric juice and also monomers and dimers slightly suppressed the peptic activity of mouse gastric juice *in vivo.* As monomers and dimers are inactive *in vitro,* it is possible that their activity *in vivo* is not related to the direct inhibition of pepsin, but related to influence on the secretion mechanism of pepsin.

Gums and Mucilages

Gums and mucilages are usually brittle, amorphous, transparent, or translucent substances, which readily absorb water to form gelatinous masses or viscous colloidal solutions. The colloidal character of gums, mucilages, and other mucoids accounts largely for their use as therapeutic agents. Mucilaginous drugs have the property of covering and protecting the mucosa of the stomach and are used in the treatment of gastric ulcer. Plants containing mucilages traditionally used in several countries in the treatment of gastric ulcer include **Althaea officinalis**, **Cetraria islandica, Malva sylvestris, Matricaria chamomilla**, and **Aloe** species.

Guar gum is obtained from the endosperm of the seed of **Cyamopsis tetragonolobus**, a plant long cultivated in India and Pakistan and nowadays also grown in the United States. The main constituent is a **galactomannan**, a β-1,4-linked D-mannose linear polysaccharide with an α-1,6-linked D-galactose residue attached to every other D-mannose unit. Proposed mechanisms are reduced acidity, increased local mucosal supply of energy, and mechanical protection. Moreover, guar gum, in duodenal ulcer patients, decreases gastric acidity and the rate of emptying of gastric contents, probably because of its effects on viscosity and neutralization of gastric acidity.

Myrrh, an oleo-gum-resin obtained from *Commiphora molmol*, contains up to 60% gum and up to 40% resin. Myrrh pretreatment produced a dose-dependent protection against the ulcerogenic effects of different necrotizing agents such as ethanol, indomethacin, sodium hydroxide, or hypertonic sodium chloride. The protective effect of myrrh is attributed to its effect on mucus production or increase in nucleic acid and nonprotein sulfydryl concentration, which appears to be mediated through its free radical-scavenging, thyroid-stimulating, and prostaglandin-inducing properties.

HERBAL DRUGS WITH ANTI-ULCER ACTIVITY

Licorice

Licorice, the root and rhizome of different varieties of **Glycyrrhiza glabra** (Leguminosae), has been extensively used in medicine for its antiulcer activity. The principal constituent of licorice is **glycyrrhizinic acid**, a triterpenoid saponin. This chemical class is known to offer protection against ulcers.

Carbenoxolone derives from the hydrolysis of glycyrrhizinic acid after its extraction from licorice root. Carbenoxolone was the first drug found to accelerate peptic ulcer healing by a mechanism not involving the inhibition of acid secretion. Several mechanisms of action have been proposed to explain the pharmacological activity of carbenoxolone:

- Stimulates gastric mucus production
- Enhances the rate of incorporation of various sugars into gastric mucosal glycoproteins
- Promotes mucosal cell proliferation
- Inhibits mucosal cell exfoliation
- Inhibits prostaglandin degradation
- Increases the release of PGE_2 and reduces the formation of thromboxane B_2
- Regulates DNA and protein synthesis in gastric mucosa

More recently, nitric oxide has been claimed to contribute to the antiulcer effect of carbenoxolone.

In excess, carbenoxolone produces effects similar to those of aldosterone excess (sodium retention and hypokalaemia leading to hypertension, edema, and cardiac failure).

Aloe Gel

Two distinct preparations of *Aloe* plants are most used medicinally. The resinous leaf exudate (named aloe) is used as a laxative, and the mucilaginous gel (named aloe vera or aloe gel) from the leaf parenchyma is used as a remedy against a variety of skin disorders. Aloe gel had a prophylactic effect and was also curative if given as a treatment for stress-induced gastric ulceration in rats. A **lectin fraction** (glycoprotein) from **Aloe arborescens**, Aloctin A, had antiulcer effect, and another high-molecular-weight fraction, not containing glycoprotein, was very effective in heating ulcer induced by mechanical or chemicals stimuli but not those induced by stress. This fraction contained substances with molecular weights between 5000 and 50,000 Da, which were considered both to suppress peptic ulcers and to heal chronic gastric ulcers. In addition, a component from Cape Aloe exudate, named **aloe ulcin**, suppressed ulcer growth and L-histidine decarboxylase.

Capsicum (Chili)

Capsicum, also known as chili or **paprika**, is the fruit of various **Capsicum** species. It is widely used as a spice and, traditionally, it has been used internally for **colic**, **flatulent dyspepsia**, **chronic laryngitis**, **insufficiency of peripheral circulation**, and externally for neuralgia. **Capsaicin** (the active pungent ingredient) has been used extensively as a probe to elucidate the function of sensory neurons in various organs and systems (including the stomach), because of its ability to excite and later defunctionalize a subset of primary afferent neurons.

Chili, as well as capsaicin, has a protective effect on ethanol- or aspirin-induced lesion formation in the gastric mucosa; in addition, capsaicin and long-term chili intake (360 mg daily for 4 weeks) protects against hemorrhagic shock-induced gastric mucosal injury, an effect that may be mediated by capsaicin-sensitive afferent neurons. The pathophysiological basis of gastric protection by chili (capsaicin) remains unclear. An increase in gastric mucosal blood flow has been described and this may be related to the release of **calcitonin gene-related peptide** and **nitric oxide** rather than to the production of **prostaglandins**. The protective effect of capsicum could also involve **vanilloid receptors** because **resiniferatoxin**, an ultrapotent analogue of capsaicin present in the latex of *Euphorbia resinifera* (Euphorbium), also displays antiulcer activity and both capsaicin and resiniferatoxin act on vanilloid (capsaicin) receptors.

Chili causes dyspepsia in patients with or without ulcer, and patients with ulcer are often advised to avoid its use. Therefore, patients with peptic ulcer consume less **chili** than controls. Nevertheless, epidemiological and clinical data suggest that chili ingestion may have a beneficial effect on human peptic ulcer disease. For examples, chili has a gastroprotective effect on aspirin-induced gastric mucosal injury in humans. In addition, chili ingestion has no detrimental effect on the healing rates of patients with duodenal ulcer and it does not cause macroscopic in humans. By contrast, chili increases DNA loss and microbleeding, suggesting the occurrence of cellular damage (see Borrelli and Izzo, 2000).

58

VALERIAN FOR SLEEP DISORDERS

Valerian inhibits the uptake of GABA and enhances the release of GABA.

VALERIAN IN THE MANAGEMENT OF INSOMNIA

Insomnia is one of the most common complaints in general medical practice and its treatment is predicated upon proper diagnosis. A variety of pharmacological agents are available for the treatment of insomnia. The "perfect" hypnotic would allow sleep to occur, with normal sleep architecture, rather than produce a pharmacologically altered sleep pattern. It would not cause next-day effects, either of **rebound anxiety** or continued sedation. It would not interact with other medications. It could be used chronically without causing dependence or rebound insomnia on discontinuation. Regular moderate exercise meets the criteria, but often is not effective by itself, and patients with significant cardiorespiratory disease may not be able to exercise. However, even small amounts of exercise often are effective in promoting sleep. Although the precise function of sleep is not known, adequate sleep improves the quality of daytime wakefulness, and hypnotics should be used judiciously to avoid its impairment.

The National Institute of Mental Health Consensus Development Conference divided insomnia into three categories:

1. ***Transient insomnia*** lasts less than 3 days and usually is caused by a brief environmental or situational stressor. It may respond to attention to sleep hygiene rules. If hypnotics are prescribed, they should be used at the lowest dose and for only 2 to 3 nights.
2. ***Short-term insomnia*** lasts from 3 days to 3 weeks and usually is caused by an ongoing personal stressor such as illness, grief, or job problems. Again, sleep hygiene education is the first step. Hypnotics may be used adjunctively for 7 to 10 nights. Hypnotics are best used intermittently during this time, with the patient skipping a dose after 1 to 2 nights of good sleep.
3. ***Long-term insomnia*** is insomnia that lasts for more than 3 weeks; no specific stressor may be identifiable. A more complete medical evaluation is necessary in these patients, but most do not need an all-night sleep study.

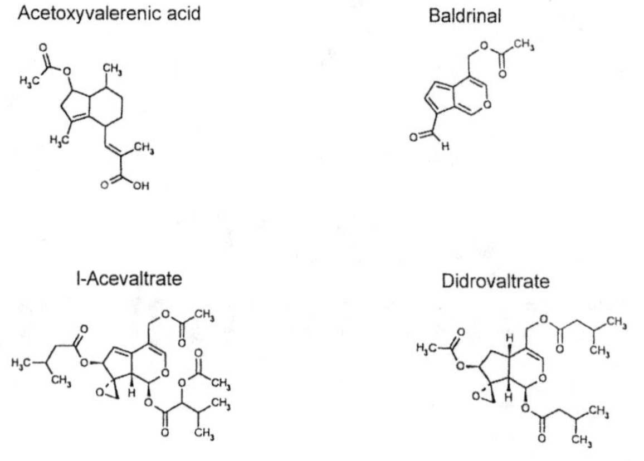

Figure 144 Principal constituents of valerian.

THE USE OF VALERIAN IN FOLK MEDICINE AND TRADITIONAL SYSTEMS OF MEDICINE

In folk medicine and traditional systems of medicine, various species of valerian have been used as an oral infusion to treat **migraine headaches**, **hysteria**, **nervous unrest**, **neurasthenia**, **fatigue**, **seizures**, **stomach cramps** that cause vomiting, and other nervous conditions. Valerian also has been used internally as a spasmolytic, a carminative, and a stomachic and externally on cuts, sores, and acne. Some herbalists recommend it as a possible vermifuge. In Chinese medicine, valerian is used to treat the conditions described above, as well as chronic headache, numbness due to rheumatic conditions, colds, menstrual difficulties, and bruises. Because there has been no extensive research conducted on uses other that the sedative properties of valerian and its constituents, these other uses cannot be scientifically evaluated.

VALERIAN

The commercial valerian products consist of, or are derived from, the rhizome, roots, and/or stolons of ***Valeriana officinalis L.*** The crude herb is dried and may be used as is or to prepare an extract that can be used to make an oral solution, tablet, capsule, or tea. Active constituents of valerian include **acetoxy-valerenic acid**, **1-acevaltrate**, **baldrinal**, **didrovaltrate**, **hydroxyvalerenic acid**, **kessane derivatives**, **valeranone**, **valerenal**, **valerenic acid**, and **valtrate** (Figure 144).

Valerian is used in the short-term treatment of insomnia characterized by difficulty in falling asleep and poor sleep quality.

Valeriana officinalis L. inhibits reuptake of and stimulates the release of γ-aminobutyric acid (**GABA**), contributing to its sedative properties.

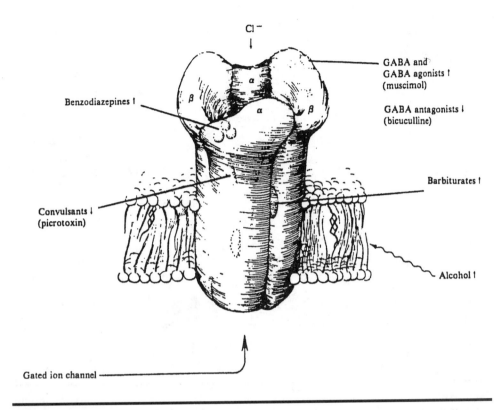

Figure 145 Schematic model of the GABA$_A$ receptor. The receptor spans the cell membrane. GABA binds to the outside of the receptor, causing an influx of Cl$^-$ ions through the channel. Benzodiazepines and barbiturates interact with different recognition sites on the receptor and increase the effectiveness of GABA.

γ-AMINOBUTYRIC ACID

GABA is the major inhibitory neurotransmitter in the mammalian central nervous system (CNS). GABA is primarily synthesized from glutamate by the enzyme **L-glutamic acid-1-decarboxylase (GAD)**; it is subsequently transaminated with **α-oxoglutarate** by **GABA-α-oxoglutarate transaminase (GABA-T)** to yield glutamate and succinic semialdehyde.

Two types of GABA receptors have been identified in mammals, a GABA$_A$ and a GABA$_B$ receptor. The GABA$_A$ receptor (or recognition site), when coupled with GABA, induces a shift in membrane permeability (primarily to chloride ions), causing a hyperpolarization of the neuron. This GABA receptor appears to be part of a macromolecule that contains, in addition to the GABA$_A$ receptor, a **benzodiazepine receptor** and the **chloride ionophore** (chloride channel) (Figure 145).

BASIC MECHANISMS OF EPILEPSY

The basic mechanisms underlying convulsive disorders are so diverse that to seek a single cause would be a gross oversimplification. Rather, a number of different

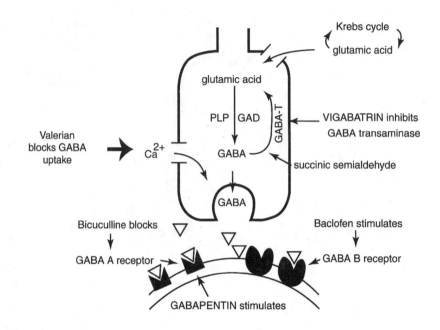

Figure 146 The actions of drugs on GABAergic transmission. GABA = γ-aminobutyric acid; GAD = glutamic acid decarboxylase; PLP = pyridoxal phosphate.

disorders, singly or in combination, may result in seizure discharge. However, one of the better-characterized defects concerns GABA and its role as a neurotransmitter. Biochemical lesions that interrupt the synthesis, storage, release, or postsynaptic actions of inhibitory neurotransmitters such as GABA lead to disinhibition of neurons.

The synthesis of GABA is achieved by decarboxylation of glutamate, which is mediated by GAD, a rate-limiting enzyme. The importance of this **pyridoxal phosphate**-dependent enzyme in the etiology of epilepsy is illustrated by a patient who suffered from seizures and who was found at autopsy to have reduced levels of pyridoxal phosphate as well as reduced GAD activity and GABA levels in the occipital cortex, which exhibited gliosis. Several compounds that have been found to lower the threshold for seizures or frankly to elicit seizures have also been found to inhibit GAD. Hence, reducing available GABA levels appears to represent a possible epileptogenic mechanism.

Several agents related to GABA metabolism and its receptors have been found to possess some degree of antiepileptic activity. Specific benzodiazepine and barbiturate receptors have been identified in the postsynaptic membrane, which, when activated by a benzodiazepine or barbiturate, enhance GABA binding to postsynaptic GABA receptors, resulting in prolonged chloride conductance and increased inhibition.

The anticonvulsant agents have been developed based on the knowledge of such epileptogenic mechanisms as illustrated by GABA deficiency (Figure 146).

- Substances that stimulate GABA$_A$ receptors (**GABApentin**) are anticonvulsants.
- Substances blocking GABA$_A$ receptors (**Bicuculline**) cause convulsive seizures.
- Substances inhibiting GABA transaminase (**Valproate**) are anticonvulsants.

THE ANXIOLYTIC AGENTS

The anxiolytic agents consist of benzodiazepine derivatives and azaspirodecanedione derivatives.

The benzodiazepine derivatives class of antianxiety agents shares the property of binding to a benzodiazepine receptor, part of the GABA receptor–chloride channel complex whose function it modulates allosterically. Not only the anxiolytic effects of the benzodiazepines, but also the anticonvulsant, sedative, or muscle relaxant effects seem to be mediated by the GABA-related mechanism. Besides the direct involvement of the GABA system, in parallel or more downstream to this, several other neurotransmitters such as serotonin have been suggested to participate in different aspects of benzodiazepine action.

The azaspirodecanedione derivatives include **buspirone, gepirone**, and **ipsapirone**.

PHARMACODYNAMICS OF BENZODIAZEPINES

The neurochemical basis for anxiety is only partially understood. Various manifestations of anxiety, such as palpitations and tremulousness, may be viewed as hyperactivity of the adrenergic system, and β-adrenergic receptor-blocking agents are effective for the treatment of acute stress reactions, adjustment disorders, generalized anxiety, panic disorder, and agoraphobia. The discovery of the benzodiazepine–GABA receptor–chloride ionophore complex furnished additional evidence that this complex participates in the etiology and manifestation of anxiety (Figure 147).

GABA RECEPTORS

GABA is the major neurotransmitter in the brain that mediates inhibition. Virtually every neuron in the brain is inhibited by GABA. GABA receptors have been divided into two types: A and B. The GABA$_A$ receptors are stimulated by GABA, **muscimol**, and **isoguvacine**, and are inhibited by the convulsants **bicuculline** and **picrotoxin**. GABA$_A$ receptors are associated with a chloride channel. GABA$_B$ receptors are stimulated by GABA and (−)**baclofen**, and are inhibited by **phaclofen**. GABA$_B$ receptors seem to be coupled to calcium and potassium channels, and possibly to second-messenger systems.

BENZODIAZEPINE-BINDING SITES

The binding sites for benzodiazepines are divided into central and peripheral types. The central or neuronal benzodiazepine receptors are influenced by GABA,

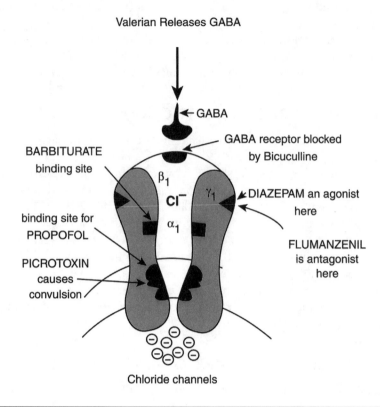

Valerian Releases GABA

← GABA

GABA receptor blocked
by Bicuculline

BARBITURATE
binding site

β₁

Cl⁻

γ₁

DIAZEPAM an agonist
here

binding site for
PROPOFOL

α₁

FLUMANZENIL
is antagonist
here

PICROTOXIN
causes
convulsion

Chloride channels

Figure 147   GABA–benzodiazepine–chloride channel complex. GABA = γ-aminobutyric acid.

linked to chloride channel, and responsible for mediating the anxiolytic, sedative, and anticonvulsant properties of the benzodiazepines. The peripheral or non-neuronal benzodiazepine receptors are found in a variety of tissues, including the kidney, heart, lung, liver, adrenal gland, testis, intestinal smooth muscles, mast cells, platelets, several cell lines, and non-neuronal elements in the central nervous system (CNS).

The released GABA may modulate by inhibiting the functions of other neurons such as the noradrenergic, cholinergic, dopaminergic, and serotoninergic neurons (Figure 148). Therefore, because benzodiazepine facilitates GABAergic functions, the anticonvulsant, muscle relaxant, sedative, hypnotic, and antianxiety effects of benzodiazepine may be elicited by GABAergic neurons and active metabolites, with longer half-lives. Consequently, it may be necessary to reduce the doses of those benzodiazepines with active metabolites.

Benzodiazepines may depict two- or three-compartment disposition kinetics. For example, a benzodiazepine derivative given in a suitable dose at night may impose dose-dependent effects, in that the concentration attained at night may be in the hypnotic range, whereas the concentration of the drug or its active metabolite or metabolites during the following day may be sufficient to produce antianxiety effects.

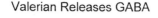

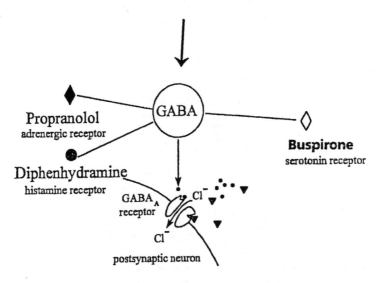

Valerian Releases GABA

Propranolol
adrenergic receptor

Diphenhydramine
histamine receptor

GABA

Buspirone
serotonin receptor

GABA$_A$
receptor

Cl$^-$

Cl$^-$

postsynaptic neuron

Antianxiety effects brought about by DIAZEPAM

DIAZEPAM, BUSPIRONE, PROPRANOLOL AND DIPHENHYDRAMINE
ALL REDUCE ANXIETY

Figure 148 The anxiolytic effect of buspirone. GABA = γ-aminobutyric acid.

It is generally accepted that the pharmacology of benzodiazepine derivatives is identical qualitatively but varies quantitatively. In other words, the sedative, hypnotic, anticonvulsant, muscle relaxant, and anxiolytic properties reside to various degrees in all of them. Nevertheless, they do exhibit pharmacological specificity, making it necessary to select a particular drug for its desired therapeutic effect.

MODE OF ACTION OF BENZODIAZEPINE DERIVATIVES

Antianxiety Effects

Benzodiazepines, by facilitating GABAergic actions, exert their anxiety-reducing effects by depressing the hyperactivity of neuronal circuits in the limbic system. In addition, they may inhibit the hyperexcitability of hippocampal cholinergic and serotoninergic neurons (see Figure 148).

Muscle Relaxant Activity

Benzodiazepines enhance presynaptic inhibition by releasing GABA from interneurons and by facilitating the action of glycine in the spinal cord and brain stem.

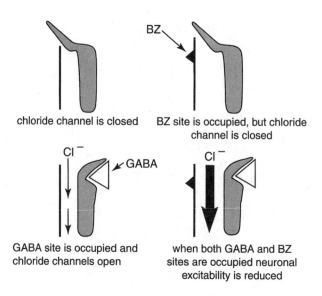

chloride channel is closed

BZ site is occupied, but chloride channel is closed

GABA site is occupied and chloride channels open

when both GABA and BZ sites are occupied neuronal excitability is reduced

Figure 149 Diazepam opens the chloride channel. BZ = benzodiazepine; GABA = γ-aminobutyric acid.

Glutamate and aspartate have excitatory effects on neurons, whereas GABA, glycine, and taurine have inhibitory actions in the CNS. Besides existing in spinal interneurons, excluding the cuneate nucleus, glycine is an inhibitory transmitter in the reticular formation.

Sedative Effects

The sleep-promoting effects of benzodiazepines may relate to their ability to reduce hyperarousability and emotional tension. The anterograde amnesia produced by high doses may be due to their interference with hippocampal functions.

Anticonvulsant Activity

Benzodiazepine derivatives are thought to exert their anticonvulsant activity by facilitating GABAergic transmission. Substances that inhibit the synthesis of GABA (e.g., **isoniazid**) or block the GABA recognition site (e.g., **bicuculline**) cause convulsions (Figure 149).

MODE OF ACTIONS OF THE GENERAL ANESTHETICS

General anesthetics alter the excitation of the neuronal membrane and modify impulse conduction. Specifically, the general anesthetics have the following common properties. (1) They decrease the activity of neurons by increasing their threshold to fire; (2) by interfering with sodium influx, they prevent the action potential from rising to a normal rate (Figure 150).

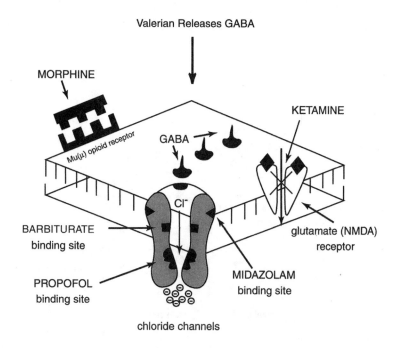

Figure 150 The sites of action of general anesthetics. GABA = γ-aminobutyric acid.

Barbiturates and benzodiazepines act at two distinct recognition sites on the GABA$_A$ receptor/chloride channel molecular complex to potentiate GABA-mediated Cl⁻ conductance and neuronal inhibition. The depressant effects of morphine-like opioids on neuronal activity are mediated by the μ-opioid receptors. **Ketamine** appears to act not only by enhancing neuronal inhibition but also by blocking neuronal excitation. Ketamine binds to the **phencyclidine receptor**, a site located within the cation channel that is gated by the **N-methyl-D-aspartate type of glutamate receptor**. Binding to this receptor blocks cation conductance through the channel, thereby blocking the actions of glutamic acid, the principal excitatory neurotransmitter in the brain.

GLYCINE

Glycine is another inhibitory CNS neurotransmitter. Whereas GABA is located primarily in the brain, glycine is found predominantly in the ventral horn of the spinal cord. Relatively few drugs are known to interact with glycine; the best-known example is the convulsant **strychnine**, which appears to be a relatively **specific antagonist of glycine** (Figure 151).

SEDATIVES AND HYPNOTICS

Sedatives and hypnotics may be divided into two categories: barbiturates and nonbarbiturates.

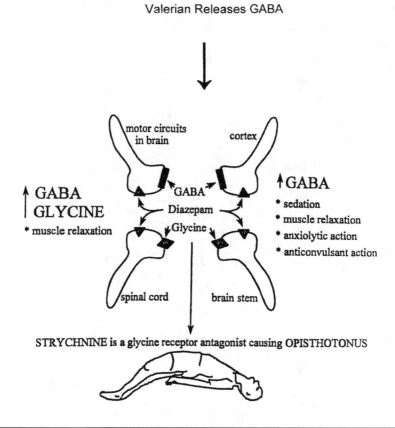

Valerian Releases GABA

Figure 151 The actions of GABA and glycine. GABA = γ-aminobutyric acid.

The Barbiturates

The barbiturates were used extensively in the past as hypnotic sedatives, but have been replaced by the much safer benzodiazepine derivatives. They do continue to be used as anesthetics and as anticonvulsants. The primary mechanism of action of barbiturates is to increase inhibition through the **GABA system.** Anesthetic barbiturates also decrease excitation via a decrease in calcium conductance.

Classification

The most commonly used barbiturates are:

- **Thiopental** (Pentothal)
- **Methohexital** (Brevital)
- **Secobarbital** (Seconal)
- **Pentobarbital** (Nembutal)
- **Amobarbital** (Amytal)
- **Phenobarbital** (Luminal)

Barbiturates are classified according to their duration of action; these are ultrashort acting (thiopental and methohexital), short to intermediate acting (pentobarbital, secobarbital, and amobarbital), and long acting (phenobarbital).

In general, the more **lipid soluble** a barbiturate derivative is, the greater is its plasma and tissue-binding capacity, the extent of its metabolism, and its storage in adipose tissues. In addition, very lipid-soluble substances have a faster onset of action and a shorter duration of action.

Barbiturates do not raise the **pain threshold** and have no **analgesic** property. In anesthetic doses, they depress all areas of the CNS, including the hypothalamic thermoregulatory system, respiratory center, and vasomotor centers, as well as the **polysynaptic pathways in the spinal column**. In addition, some, such as phenobarbital, but not all, are anticonvulsants. In toxic doses, barbiturates cause oliguria.

Pharmacokinetics

Barbiturates are absorbed orally and distributed widely throughout the body. They are metabolized in the liver by aliphatic oxygenation, aromatic oxygenation, and N-dealkylation.

The inactive metabolites are excreted in the urine. The administration of **bicarbonate** enhances the urinary excretion of barbiturates that have a pK_α of 7.4 (**phenobarbital** and **thiopental**). This generalization is not true of other barbiturates. The long-term administration of barbiturates activates the cytochrome P-450 drug-metabolizing system.

Acute Toxicity

Acute barbiturate toxicity is characterized by **automatism**, or a state of drug-induced confusion, in which patients lose track of how much medication they have taken and take more. Death results from respiratory failure. The treatment of poisoning consists of supporting the respiration, prevention of hypotension, diuresis, hemodialysis, and in the event of phenobarbital poisoning, the administration of **sodium bicarbonate**. Tolerance does not develop to lethal doses.

Addiction

The abrupt withdrawal from barbiturates may cause tremors, restlessness, anxiety, weakness, nausea and vomiting, **seizures**, delirium, and cardiac arrest.

Selection of Drugs

The selection of a barbiturate is in part determined by the duration of action desired and by the clinical problems at hand. An ultrashort-acting drug is used for inducing anesthesia. For treating epilepsy, a long-acting drug is used, whereas, in a sleep disorder, a short-acting or an intermediate-type drug is used, depending on whether patients have difficulty falling asleep or if they have difficulty staying asleep.

Table 63   Characteristics of Benzodiazepines

Benzodiazepines	Active Metabolites	Half-Life	Marketed As
Long-acting			
Chlordiazepoxide	Yes	2–4 days	—
Diazepam	Yes	2–4 days	—
Flurazepam	Yes	2–3 days	Hypnotic
Chlorazepate	Yes	2–4 days	—
Clonazepam	Yes	2–3 days	Anticonvulsant
Prazepam	Yes	2–4 days	—
Halazepam	Yes	2–4 days	—
Intermediate-acting			
Oxazepam	No	8–12 h	Anxiolytic
Lorazepam	No	10–20 h	—
Temazepam	No	10–20 h	Hypnotic
Alprazolam	No	14 h	—
Short-acting			
Triazolam	No	2–5 h	Hypnotic
Midazolam	No	2–3 h	—

Benzodiazepines have only a limited capacity to produce profound and potentially fatal CNS depression. Although coma may be produced at very high doses, benzodiazepines cannot induce a state of surgical anesthesia by themselves and are virtually incapable of causing fatal respiratory depression or cardiovascular collapse unless other CNS depressants also are present. Because of this measure of safety, benzodiazepines have substantially replaced older agents for the treatment of insomnia or anxiety.

Nonbarbiturate Sedatives and Hypnotics — Benzodiazepine Derivatives

Flurazepam (Dalmane), **temazepam** (Restoril), and **triazolam** (Halcion) are all marketed as hypnotic agents, but other benzodiazepine derivatives are also effective as hypnotic agents (Table 63).

REGULATORY STATUS

Valerian (as *Valeriana officinalis* L.) has official status (i.e., government approval in some capacity) in many countries throughout the world. In some cases, the status is based on traditional use rather than controlled trials. Table 64 shows the regulatory status of valerian.

Table 64   Regulatory Status of Valerian

Country	Regulatory Status
Australia	*Valeriana officinalis* L. is acceptable as an active ingredient in the "listed products" category of the Therapeutic Goods Administration
Belgium	Subterranean parts, powder, extract, and tincture are permitted as traditional tranquilizers
Canada	The Health Protection Branch of Health Canada allows products containing valerian as a single medicinal ingredient in the form of crude dried root in tablets, capsules, powders, extracts, tinctures, drops, or tea bags intended to be used as sleeping aids/sedatives
France	Valerian root (from *Valeriana officinalis* L.) has traditional use in symptomatic treatment of neurotonic conditions of adults and children, particularly in cases of minor sleep disturbances
Germany	Commission E has approved valerian for restlessness and nervous disturbances of sleep
United Kingdom	Valerian is included in the General Sale List of the Medicines Control Agency and permitted in "traditional herbal remedies" as a sedative and to promote natural sleep

Valerian consists of the dried rhizome and roots of *Valeriana officinalis* Linné (Fam. Valerianaceae). It has been employed as an antianxiety agent and sleep aid for more than 1000 years. The drug contains from 0.3 to 0.7% of an unpleasant-smelling volatile oil containing bornyl acetate and the sesquiterpenoids, valerenic acid, and acetoxyvalerenolic acid. Also present is a mixture of lipophilic iridoid principles known as valepotriates. These bicyclic monoterpenoids are quite unstable and occur only in the fresh plant or in material dried at temperatures under 40°C. Although the specific active principle(s) of valerian have not been determined, it is possible that a combination of the sesquiterpenoids and the valepotriates may be involved. The drug may be administered as a tea prepared from 2 to 3 g of the dried herb or equivalent amounts of a tincture or extract may be employed.

59

WINE

Epidemiologists have suggested that the countries in which atherosclerosis incidence is low are those countries in which red wine consumption is high.

INTRODUCTION

The term **alcohol** usually implies ethyl alcohol, although actually there are a large number of alcohols that vary in physical structure from liquids to waxy solids. Of the host of known alcohols, only two are frequently consumed by humans. They are **glycerol**, from the digestion of fats, and **ethyl alcohol**, from alcoholic beverages. The quantity of glycerol in our food is usually insufficient to cause any pharmacological effect. On the other hand, the amounts of ethyl alcohol commonly ingested do produce pharmacological effects of varying degree. The major portion of this chapter is devoted to ethyl alcohol.

The ethyl alcohol in all alcoholic beverages results from the **fermentation of a sugar by yeast**. If a cereal is the raw material for the beverage, it must first be malted to convert the starch to **maltose**, since yeast will not ferment starch. Malt is produced by moistening barley and allowing it to sprout. At this stage, the grain contains **maltose** and a large amount of the enzyme **diastase**. The sprouted barley is next dried at a temperature sufficient to kill the sprout without injuring the enzyme, and is then ground. A **mash** is prepared by incubating a mixture of ground cereal, water, and malt. Mash for brewing beer is filtered, and the liquid, or **wort**, is then treated with yeast. For making **whisky**, yeast is added directly to the malted mash. After fermentation, the liquid portion, or distiller's beer, is run through a simple still. The distillate, or raw whisky, is mostly water and alcohol, plus traces of other volatile products of the fermentation. If a more efficient fractionating still is used, only water and alcohol are recovered. This product, **neutral spirits**, contains 90 to 95% of alcohol by volume. Information regarding some common alcoholic beverages is given in Table 65.

Wines containing more than 16% of alcohol are fortified by the addition of neutral spirits. In making beer, a small amount of **hops** is added to give it a bitter flavor. Whisky must be aged by storage in a **charred oak barrel**. This causes esterification of some of the organic acids with the higher alcohols and adds extractives plus color from the charred wood. Blended whisky is a mixture of aged whisky with neutral spirits and water. Some blended whisky contains less

Table 65 Alcoholic Beverages

Beverage	Raw Material	Malting	Distillation	Alcohol Content, (percent by volume)	Calories[a] (per 100 cc)
Wine	Grape juice	No	No	10–22	60–175
Hard cider	Apple juice	No	No	8–12	35–60
Beer	Cereals	Yes	No	3.5–6	40–60
Ale	Cereals	Yes	No	6–8	55–70
Brandy	Wine	No	Yes	40–55	225–300
Whisky	Cereals	Yes	Yes	40–55	225–300
Rum	Molasses	No	Yes	40–55	225–300
Gin	Neutral spirits plus orange peel and juniper berries	No	Usually	40–55	225–300
Vodka	Neutral spirits	No	No	40–55	225–300
Milk (for comparison)					60–80

[a] Almost half the calories in beer and about one third of the calories in ale are furnished by unfermented carbohydrate. Sweet wine may contain 4 to 10% sugar.

than half aged whisky. Whisky, brandy, and wine contain a small percent of **fusel oil**, which is composed of the higher homologues of ethyl alcohol, chiefly **isoamyl alcohols**, and acids such as **capric acid**.

With wines and distilled liquors, the bottle label must state the percentage of alcohol expressed by volume or as a number followed by the word proof. In the United States the proof number is twice the percentage of alcohol by volume. The term *proof* originated in an old English custom of testing the alcohol content of whisky. The whisky was poured over gunpowder and a flame applied to it. If the gunpowder exploded, the whisky was said to be of "**proof strength**." For the whisky to ignite it must contain at least 50% of alcohol by volume. In Great Britain and Canada proof whisky must contain a little over 57% alcohol by volume.

THE PHARMACOLOGY OF ALCOHOL

Effects of Ethyl Alcohol

Central Nervous System

As a CNS depressant, ethyl alcohol (ethanol) obeys the law of descending depression, in that it first inhibits the cerebral cortex, then the cerebellum, spinal cord, and medullary center.

Acute Intake of Alcohol

Taken in small quantities, alcohol brings about a feeling of well-being. When consumed in large quantities, alcohol produces more boisterous behavior. Self-control

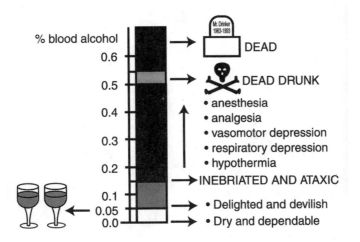

Figure 152 The dose-dependent actions of ethanol.

is lost and judgment is impaired. Alcohol works by depressing the inhibitory control mechanism and the reticular activating system. If a large amount of alcohol is consumed within a short time, unconsciousness and general anesthesia ensue. Death is due to respiratory and cardiac failure (Figure 152).

Cardiovascular System

Alcohol produces dilation of the skin vessels, flushing, and a sensation of warmth. Alcohol also interferes with the normal cutaneous vasoconstriction in response to cold. The body heat is therefore lost very rapidly and the internal temperature consequently falls. At toxic alcohol levels, the hypothalamic temperature-regulating mechanism becomes depressed and the fall in body temperature becomes pronounced. For these reasons, consuming alcoholic beverages for the purpose of keeping warm in cold weather is obviously irrational.

Gastrointestinal Effects

As a gastric secretagogue, alcohol stimulates the secretion of gastric juice, which is rich in acid and pepsin. Therefore, the consumption of alcohol is contraindicated in subjects with untreated acid-pepsin disease (Figure 153). In addition, alcohol releases **histamine**, which in turn releases gastric juice. This effect is not blocked by **atropine**.

In toxic doses (20%), gastric secretion is inhibited and peptic activity is depressed. From this, it is easy to deduce that small amounts of alcohol stimulate appetite and aid digestion, but large amounts may produce indigestion. Alcohol is also a carminative substance, in that it facilitates the expulsion of gas from the stomach.

Liver

Alcohol enhances the accumulation of fat in the liver. In alcoholics, this fat accumulation continues and **cirrhosis of the liver** may ensue; however, the two phenomena are not related (Figure 154).

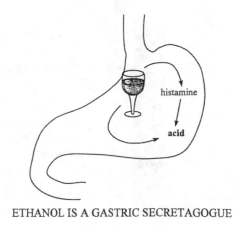

ETHANOL IS A GASTRIC SECRETAGOGUE

Figure 153 Ethanol aggravates acid-pepsin diseases.

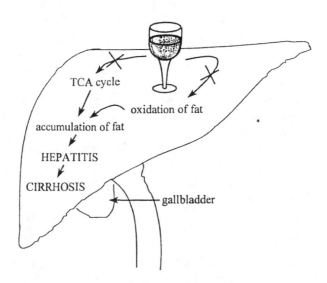

ETHANOL CAUSES CIRRHOSIS

Figure 154 Ethanol causes hepatic failure. TCA = tricarboxylic acid.

Endocrine Glands

Alcohol may release epinephrine, which leads to transient hyperglycemia and hyperlipemia. Therefore, alcohol consumption is contraindicated in those with diabetes. Alcohol causes diuresis by increasing fluid intake and by inhibiting the secretion of **antidiuretic hormone** elaborated by the posterior pituitary gland (Figure 155).

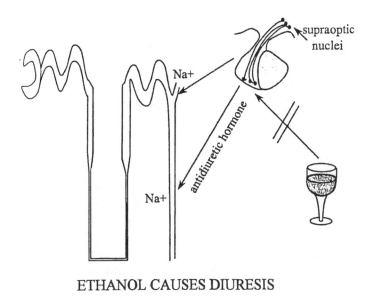

ETHANOL CAUSES DIURESIS

Figure 155   Ethanol inhibits the secretion of antidiuretic hormone.

Fetal Alcohol Syndrome

Alcohol has teratogenic effects that are manifested by CNS dysfunction.

Pharmacokinetics of Ethanol

Alcohol is absorbed from the small intestine. Patients who have undergone gastrectomy may therefore become intoxicated relatively fast. The absorption that takes place through unbroken skin is negligible. As a water-soluble substance with a small molecular weight, alcohol is distributed uniformly throughout all tissues and tissue fluids. It passes across the placental barrier, is found in spinal fluid, and accumulates in the brain. Consequently, any physiological fluids (urine, blood, spinal fluid, breast milk, or saliva) are suitable for determining the concentration of alcohol. The metabolism of ethanol, which shows genetic polymorphism, is catalyzed primarily by **alcohol dehydrogenase** with zero-order kinetics, according to the following scheme.

$$\text{Ethanol} \xrightarrow[\text{dehydrogenase}]{\text{Alcohol}} \text{acetaldehyde} \xrightarrow[\text{dehydrogenase}]{\text{Acetaldehyde}} \text{acetic acid}$$

$$\text{NAD}^+ \qquad\qquad \text{NADH} \qquad\qquad\qquad \text{NAD}^+ \qquad\qquad \text{NADH}$$

The rate-limiting factor in the metabolism of ethanol is the availability of NAD^+.

Ethanol is not metabolized by **cytochrome P-450 enzymes** (**microsomal drug-metabolizing systems**, or MEDS). However, it is metabolized to a certain extent by the **microsomal ethanol-oxidizing system** (MEOS).

Although ethanol is not metabolized by the microsomal drug-metabolizing system, it inhibits it and increases the rate of its synthesis. This effect may create a significant alcohol–drug interaction in both nonalcoholics and alcoholics who are taking medications.

POISONING AND TOXICITY

Acute Poisoning

Poisoning may be characterized by inebriation, muscular incoordination, blurred vision, impaired reaction time, excitement due to loss of inhibitions, impairment of consciousness, coma, tachycardia, and slow respiration. A blood alcohol level of 80 mg/dl can produce recognizable features of drunkenness; a level above 300 mg/dl is life-threatening. In children, severe hypoglycemia and convulsions may also occur.

Acute poisoning is treated with gastric aspiration and lavage combined with intensive supportive therapy, including thorough assessment of the patient plus measures to prevent respiratory failure. In cases of very severe poisoning, peritoneal dialysis or hemodialysis may be necessary.

Delirium Tremens

Delirium tremens usually arises in chronic alcohol abusers. The clinical features may include hallucinations, intense fear, sleeplessness, restlessness, agitation, delirium, and sometimes grand mal convulsions. In addition, tachycardia, hypotension, and clover-shaped ST changes in the electrocardiogram are evident.

The treatment of patients during a delirium tremens episode includes the intravenous administration of another CNS depressant (usually diazepam) during the acute phase, followed by the oral administration of chlordiazepoxide or oxazepam. In addition, other medications plus dietary management may become essential.

Chronic Toxicity

Prolonged alcohol consumption in malnourished individuals causes far more deleterious effects. These include:

- **Neuropathy**, which is characterized by dysesthesia, paralysis, and trophic skin changes.
- **Myopathy**, which is manifested by rhabdomyolysis, muscle pain and cramps, elevated serum creatine and phosphokinase levels, and myoglobinuria.
- **Wernicke–Korsakoff syndrome**, stemming from thiamine deficiency and characterized by confusion, ataxia, and ocular abnormalities (nystagmus and lateral rectus muscle palsy).
- **Cerebellar degeneration**, degeneration of the cerebellar cortex, especially the Purkinje cells, causing ataxia and truncal instability.
- **Amblyopia**, decreased visual acuity and color perception.

- **Marchiafava–Bignami disease**, resulting from central demyelination of the corpus callosum (also seen in chronic cyanide intoxication, which causes progressive dementia).
- **Central pontine myelinolysis**, characterized by progressive bulbar weakness and quadriparesis.

EFFECTS OF RED WINE

Moderate consumption of red wine may decrease the risk of heart disease and certain cancers. The pigments found in grapes, such as **anthocyanins** and **tannins**, may protect against viruses and inhibit the formation of dental plaque.

Red wine is made from **purple grapes**, but white wine is not necessarily made from white grapes. Many white wines are made from purple grapes, but the skins are removed before they color the fermenting juice, called **must**. The grape skins contain most of the **bioflavonoids**, **phenols**, **tannins**, and other compounds that give wine its flavor and possibly healthful properties. The protectors against atherosclerosis in red wine are its **polyphenolic flavonoids**, **quercitin**, and **catechin**. It was interesting to learn from epidemiologists that a high consumption of **green tea** was also geographically associated with low rates of cancer, and to observe that the induction by **demethyl benzanthracene** of skin squamous carcinomas was reduced in mice by green tea or its polyphenolic fraction or agents, especially **mezerein**. These green tea polyphenols also protect against *N*-**nitrosorbis (oxoprophyl)** amine-induced pancreatic duct epithelial hyperplasia and carcinoma (Mathé, 1999).

60

YOGURT AND
THE IMMUNE SYSTEM

- A probiotic agent
- An immune enhancing agent

INTRODUCTION

Yogurt is defined as a coagulated milk product that results from fermentation of lactic acid in milk by ***Lactobacillus bulgaricus*** and ***Streptococcus thermophilus***. Other **lactic acid bacteria** (LAB) species can be combined with *L. bulgaricus* and *S. thermophilus*. In the finished product, the LAB must be alive and in substantial amounts. LAB have been used for thousands of years to produce fermented food and milk products. Fermented products contain a variety of fermenting microorganisms belonging to various genera and species, all of which produce lactic acid (Meydani and Ha, 2000).

With few exceptions, milk and yogurt have similar vitamin and mineral compositions. During fermentation, **vitamin B-12** and vitamin C are consumed and **folic acid** is produced. The differences in other vitamins between milk and yogurt are small and depend on the strain of bacteria used for fermentation. Although milk and yogurt have similar mineral compositions, some minerals, e.g., calcium, are more bioavailable from yogurt than from milk. In general, yogurt also has less lactose and more lactic acid, galactose, peptides, free amino acids, and face fatty acid than does milk.

IMMUNE SYSTEM FUNCTIONS

The main functions of the immune system are to eliminate invading viruses and foreign microorganisms, to rid the body of damaged tissue, and to destroy neoplasms in the body. Healthy humans have two immune mechanisms: **acquired (specific) immunity**, which responds to specific stimuli (antigens) and is enhanced by repeated exposure; and **innate (nonspecific) immunity**, which does not require stimulation and is not enhanced by repeated exposure. Innate immune mechanisms consist of physical barriers, such as mucous membranes, and the phagocytic and cytotoxic function of neutrophils, monocytes, macrophages, and lymphatic cells (NK cells). **Acquired immunity** can be classified into two types on the basis of the components of the immune system that mediate the response. i.e., **humoral immunity** and **cell-mediated immunity**. Humoral immunity is mediated by immunoglobulins produced by **bone marrow-derived lymphocytes** (B lymphocytes) and is responsible for specific recognition and elimination of extracellular antigens. Cell-mediated immunity is mediated by cells of the immune system, particularly **thymus-derived lymphocytes** (T lymphocytes). Cell-mediated immunity is responsible for delayed-type hypersensitivity (DTH) reactions, foreign graft rejection, resistance to many pathogenic microorganisms, and tumor immunosurveillance. In addition to their involvement in nonspecific immunity, macrophages are important in cell-mediated immunity as antigen-presenting cells and through the production of regulatory mediators such as cytokines and eicosanoids. Several *in vitro* and *in vivo* tests were developed to assess the function of immune cells. Although the study of immune response in animals and humans is based on similar principles, the methods used to separate cells, the types of stimuli used *in vitro*, and the antigen used for *in vivo* challenge vary. In addition, the type of antibody used to measure different mediators or to determine cell-surface proteins is species specific.

IN VITRO INDEXES OF IMMUNE FUNCTION

To study immune function *in vitro*, immune cells are first separated from whole blood, lymphoid tissues, and gut-associated immune cells. The cells are then maintained and cultured with and without various immune cell stimuli. To measure the activity of isolated phagocytes, the cells are incubated with bacteria or other engulfable materials with or without opsonin for a limited time and then stained for uptake of foreign bodies. Lymphocytes are usually stimulated for varying lengths of time by a variety of stimuli (mitogens, antigens, and other stimulator or target cells) for measurement of their proliferative or cytotoxic activity or release of immunologically active molecules such as antibodies, cytokines, and eicosanoids.

PHAGOCYTIC ACTIVITY

The ability to perform phagocytosis and kill microbes, including bacterial pathogens, is a major effector function of **macrophages**. These properties of macrophages are particularly important for host defense against facultative intracellular organisms, which can replicate within macrophages. The pathogenesis of facultative intracellular bacteria is determined by their ability to survive within macrophages. Several organisms were used previously as targets to determine macrophage killing. These include *Staphylococcus aureus*, *Escherichia coli*, *Pseudomonas aeruginosa*, *Salmonella typhimurium*, *Listeria monocytogenes*, and *Candida albicans*.

Bacteria bind to complement components and the bacterium–complement complexes bind complement receptors on the surface of macrophages. Phagocytosis may also be mediated by specific antibodies that function as **opsonins**, which bind to particles, rendering them susceptible to phagocytosis. The bacterium–antibody complex then binds the macrophages via the **Fc receptor** and phagocytosis begins.

Measurement of phagocytic activity of macrophages was among the earliest techniques for evaluating the immunological effects of LAB. This assay measures the ability of macrophages to bind, internalize, and phagocytose bacteria. Monocytes or macrophages isolated from human peripheral blood mononuclear cells (PBMCs) or from the peritoneal cavity of animals are mixed with bacteria in suspension and incubated at 37°C. Extracellular bacteria are then removed through washing and centrifugation or through washing only over sucrose. The degree of phagocytosis is determined by examining stained cells under oil-immersion microscopy and quantifying the number of internalized bacteria in each cell. This method takes into account not only the percentage of phagocytic cells but also the strength of the phagocytic ability of these cells, i.e., how many bacteria are internalized by each cell.

LYMPHOCYTE PROLIFERATION ASSAY

Measurement of the proliferative response of lymphocytes is the most commonly used technique for evaluating cell-mediated immune response. Quantitative analysis of proliferative response involves measuring the number of cells in culture in the presence and absence of a stimulatory agent such as an antigen or a mitogen. The most common polyclonal mitogens used to test the proliferation of lymphocytes are **concanavalin A** (ConA), **phytohemagglutinin**, **lipopolysaccharide** (LPS); and **pokeweed mitogen**. T and B lymphocytes are stimulated by different polyclonal mitogens. ConA and phytohemagglutinin stimulate T cells, LPS stimulates both T and B cells, and pokeweed mitogen stimulates both T and B cells. When mitogens are used, prior exposure of the host to the mitogens is not necessary. However, to measure antigen-specific proliferation, the host should be exposed to the antigen before the cells are stimulated with that antigen *in vitro*. Lymphocytes normally exist as resting cells in the G_0 phase of the cell cycle.

When stimulated with polyclonal mitogens, lymphocytes rapidly enter the G_1 phase and progress through the cell cycle. Measuring incorporation of [^{3}H]thymidine into DNA is the most commonly used method for estimating changes in the number of cells. The proliferative assay is used to assess the overall immunological competence of lymphocytes, as manifested by the ability of lymphocytes to respond to proliferation signals. Decreased proliferation, observed in chronic diseases such as cancer and HIV infection and in the aging process, may indicate impaired cell-mediated immune function.

CYTOKINE PRODUCTION

Cytokines, which are protein mediators produced by immune cells, are involved in the regulation of cell activation, growth and differentiation, inflammation, and immunity. Measurement of cytokine production, as determined by techniques such as bioassay, radioimmunoassay, and enzyme-linked immunosorbent assay, has been used to examine various immune functions.

Interleukin-2 (IL-2) is a T-cell growth factor produced by T-helper (T_H1) and NK cells. As an autocrine and paracrine growth factor, IL-2 induces proliferation and differentiation of T and B cells. IL-2 is responsible for the progress of T lymphocytes from the G_1 to the S phase in the cell cycle and also for stimulation of B cells for antibody synthesis. IL-2 stimulates the growth of NK cells and enhances the cytolytic function of these cells, producing **lymphokine-activated killer** (LAK) cells. IL-2 can also induce interferon (IFN)γ secretion by NK cells. IFNγ is an important macrophage-activating lymphokine. IL-2 secreted in culture media or biological fluids can be measured by immunoassay or bioassay, the most common of which uses the IL-2-dependent cytotoxic T lymphocyte line. Proliferation of cytotoxic T lymphocyte line cells reflects IL-2 activity. IL-2 activity in samples can be calculated according to a standard curve generated by adding varying concentrations of recombinant IL-2. Enzyme-linked immunosorbent assay is also used to measure IL-2. Although this assay is more specific than is the cytotoxic T lymphocyte line in measuring IL-2 protein concentrations, it does not differentiate between biologically active and nonactive proteins. Under most conditions, chances in IL-2 production are associated with the chance in lymphocyte proliferation, although sometimes these changes do not correlate with one another.

IFNγ is involved in the induction of other cytokines, particularly T_H2 cytokines, such as IL-4, IL-5, and IL-10. Because of its role in mediating macrophages and NK cell activation, IFNγ is important in host defense against intracellular pathogens (such as *Mycobacterium tuberculosis* and *L. monocytogenes*) and viruses and against tumors. It was suggested that mice and humans continuously produce small amounts of IFN, which may produce a state of alertness against tumor cells, pathogenic bacteria, and viruses. Because of the short half-life of IFNγ, plasma concentrations of IFNγ are low and difficult to measure, potentially making treatment-induced changes in plasma IFNγ concentrations difficult to detect. IFNγ produced by PBMCs or purified T cells has been shown to be more sensitive to yogurt-induced changes than is plasma IFNγ. In addition to *ex vivo* production of IFNγ, 2'-5' A synthetase activity was used as an index of the biological responsiveness of cells to IFNγ. 2'-5' A synthetase is an IFNγ, 2'-5' inducible enzyme. The amount of 2'-5' A synthetase in different organs of mice increases several fold after treatment of mice with IFN or IFN inducers, such as viral and synthetic double-stranded RNAs.

YOGURT COMPONENTS WITH POTENTIAL IMMUNOSTIMULATORY EFFECTS

Although yogurt has long been known to bolster host-defense mechanisms against invading pathogens, the components responsible for these effects have not been fully defined The immunostimulatory effects of yogurt are believed to be due to the bacterial components of yogurt. However, the mechanism or mechanisms responsible for these effects have not been fully determined. After entering the intestine, live or biologically active LAB particles may activate specific and nonspecific immune responses of gut-associated lymphoid tissue and the systemic immune response. The immunogenicity of intestinal bacteria depends on the degree of contact with lymphoid tissue in the intestinal lumen. Therefore, dead bacteria are generally less efficient as antigens than are live bacteria because dead bacteria are rapidly dislodged from the mucosa. Some studies, however, showed no difference in immunogenicity between viable and nonviable bacteria.

LAB are Gram-positive bacteria with cell wall components such as **peptidoglycan**, **polysaccharide**, and **teichoic acid**, all of which have been shown to have immunostimulatory properties. In addition to cell wall components, immunostimulatory effects were observed with antigens originated from the cytoplasms of some strains of LAB.

Nonbacterial milk components and components produced from milk fermentation also may contribute to the immunostimulatory activity of yogurt. Peptides and free fatty acids generated by fermentation have been shown to enhance the immune response. Milk components such as whey protein, calcium, and certain vitamin and trace elements also can influence the immune system (Nikbin et al., 2000).

Many investigators have studied the therapeutic effects of yogurt and LAB commonly used in yogurt production on diseases such as cancer, infection, gastrointestinal disorders, and asthma. Because the immune system is an important contributor to all of these diseases, the immunostimulatory effects of yogurt were studied by several investigators. Most of these studies used animal models; few human studies on the immunostimulatory effects of yogurt have been conducted.

Although the results of these studies mostly support the notion that yogurt has immunostimulatory effects, poor study design, lack of appropriate controls, and short duration of most of the studies limit the value of the conclusions that can be drawn from them. Most early animal and human studies included too few animals or subjects in each group and most did not include statistical analysis. Although more recent studies addressed these points, none provided the statistical basis for the selected number of subjects; that is, it seems that no power calculations were performed.

Most studies used short-term feeding protocols, which might induce a transient adjuvant effect rather than a long-term stimulation of the immune response. This was shown by several studies in which a maximum effect was seen with 2 to 5 days of yogurt consumption, after which the stimulatory effect of yogurt or yogurt bacteria diminished significantly. Furthermore, most studies investigated the effect of intravenous or intraperitoneal administration or *in vitro* application of yogurt bacteria on different variables of the immune response. Because yogurt is usually consumed orally and because bacterial and nonbacterial components of yogurt may be altered in the gastrointestinal tract, the results of these studies may not reflect those that would be found if the yogurt had been consumed orally. Also, many studies lacked a placebo group or did not use a randomized, blinded design.

Most animal and human studies investigated the effects of yogurt on *in vitro* indexes of the immune response, whereas very few examined variables of the immune system *in vivo*. Because a quantitative correlation between *in vitro* tests of the immune system and resistance to diseases is not yet available, care should be taken in using the *in vitro* results as supporting evidence for health benefits of yogurt. Further, although most past studies focused on the peripheral immune response, the gut-associated immune system is increasingly being recognized as playing an important role in host defense. This aspect of the immune response is particularly relevant to determining the beneficial effects of yogurt because the systemic effects of yogurt may depend on the interaction of the yogurt bacterial components with the immune cells of the gut.

Despite the design problems of previous studies, these studies provide a strong rationale for the hypothesis that increased yogurt consumption, particularly in immunocompromised populations such as elderly people, may enhance immunity. This hypothesis, however, needs to be substantiated by well-designed, randomized, double-blind, placebo-controlled human studies of adequate duration in which several *in vitro* and *in vivo* indexes of the immune response are tested. In particular, clinically relevant indexes such as response to vaccine and DTH should be included, as should a systematic evaluation of the gut-associated immune response. Future studies should use recent technical advances in fluorescent tagging of yogurt bacteria to enable an understanding of the immunostimulatory effects of yogurt. Information on the mechanisms by which yogurt protects is essential before the scientific community accepts claims regarding the health benefits of yogurt.

Although yogurt has long been believed to be beneficial for host-defense mechanisms, the components responsible for these effects or the way in which these components exert their immunological modifications are not completely understood. The presence of LAB is thought to be essential for yogurt to exert immunostimulatory effects but components of nonbacterial yogurt, such as whey protein, short peptides, and CLA, are believed to contribute to the beneficial effects of yogurt as well. It is proposed that the LAB that survive through the gastrointestinal tract, whether intact or modified, can bind to the luminal surface of M cells. LAB-bound M cells reaching to the dome region of PP cells stimulate local immune response, resulting in production of IFNγ by γδT cells. This may increase the M-cell population with subsequent rapid amplification of bacterial translocation, which can further activate the local immune system, resulting in stimulation of the local and the systemic immune response. As mentioned previously, further studies are needed to substantiate this.

Finally, once the efficacy of yogurt in improving the immune response has been shown in humans, the benefits of these effects will need to be shown in large clinical trials in which the main outcomes are the incidences and severity of infectious disease. Infectious disease rather than other immune-related diseases are suggested because such studies can be conducted in a relatively short (e.g., 1 hour) time compared with studies of diseases such as cancer (Nikbin et al., 2000).

REFERENCES

Abe, K. and Saito, H. (2000) Effects of saffron extract and its constituent crocin on learning behavior and long-term potentiation, *Phytother. Res.,* 14, 149–152.

Aiello, A., Fattorusso, E., and Menna, M. (1999) Steroids from sponges: recent reports, *Steroids,* 64, 687–714.

Ali, M., Thomson, M., and Afzal, M. (2000) Garlic and onions: their effect on eicosanoid metabolism and its clinical relevance, *Prostaglandins Leukotrienes Essential Fatty Acids,* 62, 55–73.

Anila, L. and Vijayalakshmi, N.R. (2000) Beneficial effects of flavonoids from *Sesamum indicum, Emblica officinalis,* and *Momordica charantia, Phytother. Res.,* 14, 592–595.

Asai, A., Nakagawa, K., and Miyazawa, T. (1999) Antioxidative effects of turmeric, rosemary and capsicum extracts on membrane phospholipid peroxidation and liver lipid metabolism in mice, *Biosci. Biotechnol. Biochem.,* 63, 2118–2122.

Attele, A.S., Wu, J.A., and Yuan, C.-S. (1999) Ginseng pharmacology: multiple constituents and multiple actions, *Biochem. Pharmacol.,* 58, 1685–1693.

Aubin, N. et al. (1998) Aspirin and salicylate protect against MPTP-induced dopamine depletion in mice, *J. Neurochem.,* 71, 1635–1642.

Axelrod, J. and Felder, C.C. (1998) Cannabinoid receptors and their endogenous agonist, anandamide, *Neurochem. Res.,* 23, 575–581.

Baik, H.W. and Russell, R.M. (1999) Vitamin B12 deficiency in the elderly, *Annu. Rev. Nutr.,* 19, 357–377.

Bastianetto, S., Zheng, W.-H., and Quirion, R. (2000) The *Ginkgo biloba* extract (EGb 761) protects and rescues hippocampal cells against nitric oxide-induced toxicity: involvement of its flavonoid constituents and protein kinase C, *J. Neurochem.,* 74, 2268–2277.

Benthsath, A., Rusznyak, S., and Szent-Gyorgyi, A. (1936) Vitamin nature of flavones, *Nature,* 138, p. 798.

Bivalacqua, T.J. and Hellstrom, W.J.G. (2001) Potential application of gene therapy for the treatment of erectile dysfunction, *J. Androl.,* 22, 183–190.

Borrelli, F. and Izzo, A.A. (2000) The plant kingdom as a source of anti-ulcer remedies, *Phytother. Res.,* 14, 581–591.

Boyle, P. et al. (2000) Meta-analysis of clinical trials of permixon in the treatment of symptomatic benign prostatic hyperplasia, *Urology,* 55, 533–539.

Butterworth, C.E., Jr. (1996) Folic acid and the prevention of birth defects, *Annu. Rev. Nutr.,* 16, 73–97.

Calapai, G. et al. (2000) Neuroprotective effects of *Ginkgo biloba* extract in brain ischemia are mediated by inhibition of nitric oxide synthesis, *Life Sci.,* 67, 2673–2683.

Capasso, F. et al. (1998) Aloe and its therapeutic use, *Phytother. Res.,* 12, S124–S127.

Cheng, J.-T. (2000) Review: drug therapy in Chinese traditional medicine, *J. Clin. Pharmacol.,* 40, 445–450.

Clark, A.M. (1996) Natural products as a resource for new drugs, *Pharm. Res.,* 13, 1133–1141.

Conklin, K.A. (2000) Dietary antioxidants during cancer chemotherapy: impact on chemotherapeutic effectiveness and development of side effects, *Nutr. Cancer,* 37(1), 1–18.

Cooper, D.A., Eldridge, A.L., and Peters, J.C. (1999) Dietary carotenoids and certain cancers, heart disease, and age-related macular degeneration: a review of recent research, *Nutr. Rev.,* 57, 201–214.

Crimmel, A.S., Conner, C.S., and Monga, M. (2001) Withered Yang: a review of traditional chinese medical treatment of male infertility and erectile dysfunction, *J. Androl.,* 22, 173–182.

Culbreth, D.M.R. (1927) *A Manual of Materia Medica and Pharmacology,* 7th ed., Lea & Febiger, Philadelphia.

Cupp, M.J. (2000) *Toxicology and Clinical Pharmacology of Herbal Products,* Humana Press, Totowa, NJ.

Dawson, M.I. (2000) The importance of vitamin A in nutrition, *Curr. Pharm. Design,* 6, 311–325.

De Clercq, E. (2000) Current lead natural products for the chemotherapy of human immunodeficiency virus (HIV) infection, *Med. Res. Rev.,* 20(5), 323–349.

De Flora, S., Bagnasco, M., and Vainio, H. (1999) Modulation of genotoxic and related effects by carotenoids and vitamin A in experimental models: mechanistic issues, *Mutagen.,* 14, 153–172.

Deng, X.-S. et al. (1998) Prevention of oxidative DNA damage in rats by brussels sprouts, *Free Rad. Res.,* 28, 323–333.

De Smet, P.A.G.M. (1997) The role of plant-derived drugs and herbal medicines in healthcare, *Drugs,* 54, 801–840.

Dietrich, A. et al. (2000) The effect of *Ginkgo biloba* extract (Egb 761) on gliotic reactions in the hippocampal formation after unilateral entorhinal cortex lesions, *Restorative Neurol. Neurosci.,* 16, 87–96.

Dougans, I. (1996) *Reflexology,* Time-Life Books, Alexandria, VA.

Drill, V.A. (1958) *Pharmacology in Medicine,* 2nd ed., McGraw-Hill, New York.

Eichholzer, M. et al. (2001) The role of folate, antioxidant vitamins and other constituents in fruit and vegetables in the prevention of cardiovascular disease: the epidemiological evidence, *Int. J. Vitam. Nutr. Res.,* 71, 5–17.

Fauci, A.S. et al. (1998) *Harrison's Principles of Internal Medicine,* 14th ed., McGraw-Hill, New York, 241.

Fauci, A.S. and Lane, H.C. (1998) Human immunodeficiency virus (HIV) disease: aids and related disorders, in *Harrison's Principles of Internal Medicine,* 14th ed., Braunwald, E. et al., Eds., McGraw-Hill, New York, 1791.

Fontanarosa, P.B. (2000) *Alternative Medicine,* JAMA and Archives Journals. American Medical Association.

Foust, C.M. (1992) *Rhubarb, the Wondrous Drug,* Princeton University Press, Princeton, NJ.

Fowler, J.S. et al. (2000) Evidence that *Ginkgo biloba* extract does not inhibit MAO A and B in living human brain, *Life Sci.,* 66, 141–146.

Fugh-Berman, A. (2000) Herb–drug interactions, *Lancet,* 355, 134–138.

Fusetani, N. (2000) *Drugs from the Sea,* S. Karger, Basel, Switzerland.

Galli, C. and Visioli, F. (1999) Antioxidant and other activities of phenolics in olives/olive oil, typical components of the Mediterranean diet, *Lipids,* 34, S23–S26.

Gardner, J.L. (1982) *Reader's Digest: Eat Better and Live Better,* The Reader's Digest Association, Pleasantville, NY.

Gathercoal, E.N. and Wirth, E.H. (1947) *Pharmacognosy,* 2nd ed., Lea & Febiger, Philadephia, 181.

Gingell, J.C. and Lockyer, R. (1999) Emerging pharmacological therapies for erectile dysfunction, *Exp. Opin. Ther. Patents,* 9, 1689–1696.

Gomez-Flores, R. et al. (2000) Immunoenhancing properties of *Plantago major* leaf extract, *Phytother. Res.,* 14, 617–622.

Goodman, L.S. and Gilman, A. (1955) *The Pharmacological Basis of Therapeutics,* 2nd ed., Macmillan, New York.

Greeson, J.M., Sanford, B., and Monti, D.A. (2001) St. John's wort (*Hypericum perforatum*): a review of the current pharmacological, toxicological, and clinical literature, *Psychopharmacology*, 153, 402–414.

Grollman, A. (1962) *Pharmacology and Therapeutics,* Lea & Febriger, Philadelphia.

Guyton, A.C. (1981) *Textbook of Medical Physiology,* 6th ed., W.B. Saunders, Philadelphia.

Habtemariam, S. (2000) Natural inhibitors of tumour necrosis factor-α production, secretion and function, *Planta Med.,* 66, 303–313.

Hamilton-Miller, J.M.T. (2001) Anti-cariogenic properties of tea (*Camellia sinensis*), *J. Med. Microbiol.,* 50, 299–302.

Hammond, C. (1995) *Homeopathy,* Time-Life Books, Alexandria, VA.

Handelman, G.J. et al. (1999) Antioxidant capacity of oat (*Avena sativa* L.) extracts 1. Inhibition of low-density lipoprotein oxidation and oxygen radical absorbance capacity, *J. Agric. Food Chem.,* 47, 4888–4893.

Hardman, J.G. et al. (1996) *The Pharmacological Basis of Therapeutics,* 9th ed., McGraw-Hill, New York.

Heck, A.M., DeWitt, B.A., and Lukes, A.L. (2000) Potential interactions between alternative therapies and warfarin, *Am. J. Health Syst. Pharm.,* 57, 1221–1227.

Heinrich, M. et al. (1998) Ethnopharmacology of Mexican Asteraceae (Compositae), *Annu. Rev. Pharmacol. Toxicol.,* 38, 539–565.

Higa, T. and Kuniyoshi, M. (2000) Toxins associated with medicinal and edible seaweeds, *J. Toxicol. Toxin Rev.,* 19, 119–137.

Hoffman, D. (1996) *Holistic Herbal,* Time-Life Books, Alexandria, VA.

Holmstedt, B. (1979) Historical survey, in *Ethnopharmacologic Search for Psychoactive Drugs,* Efron, D.H., Holmstedt, B., and Kline, N.S., Eds., Raven Press, New York, p. 3–32.

Howlett, A.C. (1995) Pharmacology of cannabinoid receptors, *Annu. Rev. Pharmacol. Toxicol.,* 35, 607–634.

Hsieh, C.-L. and Yen, G.-C. (2000) Antioxidant actions of du-zhong (*Eucommia ulmoides* Oliv.) toward oxidative damage in biomolecules, *Life Sci.,* 66, 1387–1400.

Iqbal, M. and Athar, M. (1998) Attenuation of iron-nitrilotriacetate (Fe-nta)-mediated renal oxidative stress, toxicity and hyperproliferative response by the prophylactic treatment of rats with garlic oil, *Food Chem. Toxicol.,* 36, 485–495.

Ishani, A. et al. (2000) *Pygeum africanum* for the treatment of patients with benign prostatic hyperplasia: a systematic review and quantitative meta-analysis, *Am. J Med.,* 109: 654–664.

Ishige, K., Schubert, D., and Sagara, Y. (2001) Flavonoids protect neuronal cells from oxidative stress by three distinct mechanisms, *Free Radical Biol. Med.,* 30, 433–446.

Janeway, C.A., Jr. and Travers, P. (1994) *Immunobiology: The Immune System in Health and Disease,* Current Biology Ltd, London.

Josey, E.S. and Tackett, R.L. (1999) St. John's wort: a new alternative for depression? *Int. J. Clin. Pharmacol. Ther.,* 37, 111–119.

Kane, G.C. and Lipsky, J.J. (2000) Drug–grapefruit juice interactions, *Mayo Clin. Proc.,* 75, 933–942.

Kovach, S. (2001) *Super Herbs.* American Media Mini Mags, Boca Raton, FL.

Lawless, J. (1995) *Aromatherapy,* Time-Life Books, Alexandria, VA.

Lewin, L. and Schuster, P. (1929) Ergebnisse von Banisterin-veruschen an Kranken, *Dtsch. Med. Wochenschr.,* 55, 419.

Liechti, E. (1998) *Shiatsu,* Time-Life Books, Alexandria, VA.

Lis-Balchin, M., Hart, S.L., and Deans, S.G. (2000) Pharmacological and antimicrobial studies on different tea-tree oils (*Melaleuca alternifolia, Leptospermum scoparium* or Manuka and *Kunzea ericoides* or Kanuka), originating in Australia and New Zealand, *Phytother. Res.,* 14, 623–629.

Liu, F. and Ng, T.B. (2000) Antioxidative and free radical scavenging activities of selected medicinal herbs, *Life Sci.,* 66, 725–735.

Lockemann, G. and Sertürner, F.W. (1924) Ein Beitrag zu seiner wissenschaftlichen Würdigung, *Z. Angew. Chem*, 37, 525–532.

MacDonald, G. (1998) *Alexander Technique,* Time-Life Books, Alexandria, VA.

Marcus, R. and Coulson, A.M. (1996) In *The Pharmacological Basis of Therapeutics,* 9th ed., McGraw-Hill, New York, 1573.

Mathé, G. (1999) Red wine, green tea and vitamins: do their antioxidants play a role in immunologic protection against cancer or even AIDS? *Biomed. Pharmacother.,* 53, 165–167.

Matsuda, H. et al. (2001) Antioxidant constituents from rhubarb: structural requirements of stilbenes for the activity and structures of two new anthroquinone glucosides, *Bioorg. Med. Chem.,* 9, 41–50.

Mechoulam, R., Vogel, Z., and Barg, J. (1994) CNS cannabinoid receptors role and therapeutic implications for CNS disorders, *CNS Drugs,* 2, 255–260.

Meydani, S. and Ha, W.-K. (2000) Immunologic effects of yogurt, *Am. J. Clin. Nutr.,* 71, 861–872.

Middleton, E., Kandaswami, C., and Theoharides, T.C. (2000) The effects of plant flavonoids on mammalian cells: implications for inflammation, heart disease, and cancer, *Pharmacol. Rev.,* 52, 673–751.

Mitchell, S. (1997) *Massage,* Time-Life Books, Alexandria, VA.

Moreira, A.S. et al. (2000) Antiinflammatory activity of extracts and fractions from the leaves of *Gochnatia polymorphia, Phytother. Res.,* 14, 638–640.

Morimitsu, Y. et al. (2000) Antiplatelet and anticancer isothiocyanates in Japanese domestic horseradish, *Wasabi, Mech. Aging Dev.,* 116, 125–134.

Münchau, A. and Bhatia, K.P. (2000) Uses of botulinum toxin injection in medicine today, *Br. Med. J.,* 320, 161–165.

Newton, S.M., Lau, C., and Wright, C.W. (2000) A review of antimycobacterial natural products, *Phytother. Res.,* 14, 303–322.

Numagami, Y., Sato, S., and Ohnishi, S.T. (1996) Attenuation of rat ischemic brain damage by aged garlic extracts: a possible protecting mechanism as antioxidants, *Neurochem. Int.,* 29, 135–143.

Packer, L., Rimbach, G., and Virgili, F. (1999) Antioxidant activity and biologic properties of a procyanidin-rich extract from pine (*Pinus martima*) bark, pycnogenol, *Free Radical Biol. Med.,* 27, 704–724.

Pedraza-Chaverri, J. et al. (2000a) Garlic ameliorates gentamicin nephrotoxicity: relation to antioxidant enzymes, *Free Radical Biol. Med.,* 29, 602–611.

Pedraza-Chaverri, J. et al. (2000b) Garlic ameliorates hyperlipidemia in chronic aminonucleoside nephrosis, *Mol. Cell. Biochem.,* 211, 69–77.

Perry, C.A. et al. (1996) Health effects of salicylates in foods and drugs, *Nutr. Rev.,* 54, 225–240.

Peroutka, S.J. (1996) Drugs effective in the therapy of migraine, in *Goodman and Gilman's The Pharmacological Basis of Therapeutics,* 9th ed., Hardman, J.G. et al., Eds., McGraw-Hill, New York.

Potter, S.O.L. (1910) *Therapeutics, Materia Medica, and Pharmacy,* 11th ed., P. Bakliston's Son & Co., Philadelphia.

Prasad, K. et al. (1995) Antioxidant activity of allicin, an active principle in garlic, *Mol. Cell. Biochem.,* 148, 183–189.

Rietz, B. et al. (1993) Cardioprotective actions of wild garlic (*Allium ursinum*) in ischemia and reperfusion, *Mol. Cell. Biochem.,* 119, 143–150.

Rios, J.L. et al. (1996) An update review of saffron and its active constituents, *Phytother. Res.,* 10, 189–193.

Russo, E. (1998) *Cannabis* for migraine treatment: the once and future prescription? An historical and scientific review, *Pain,* 76, 3–8.

Sanchez-Ramos, J.R. (1991) Banisterine and Parkinson's disease, *Clin. Neuropharmacol.,* 14, 391–402.

Skibola, C.F. and Smith, M.T. (2000) Potential health impacts of excessive flavonoid intake, *Free Radical Biol. Med.,* 29, 375–383.

Snorrason, E. and Stefansson, J.G. (1991) Galanthamine hydrobromide in mania, *Lancet,* 337, 557.

Sollmann, T. (1944) *A Manual of Pharmacology and Its Applications to Therapeutics and Toxicology,* 6th ed., W.B. Saunders, Philadelphia.

Somerville, R. (1997) *Healing Herbs,* Time Life Books, Alexandria, VA.

Soprano, D.R. and Soprano, K.J. (1995) Retinoids as teratogens, *Annu. Rev. Nutr.,* 15, 111–132.

Suffness, M. (1995) *Taxol: Science and Application,* CRC Press, Boca Raton, FL.

Swain, A.R., Dutton, A.P. and Truswell, A.S. (1985) Salicylates in foods, *J. Am. Diet Assoc.,* 85, 950–960.

Thomas, R. (1997) *Natural Ways to Health, Alternative Medicine: An Illustrated Encyclopedia of Natural Healing,* Time-Life Books, Alexandria, VA.

Vainio, H. and Morgan, G. (1997) Aspirin for the second hundred years: new uses for an old drug, *Pharmacol. Toxicol.,* 81, 151–152.

Verhagen, H. et al. (1997) Effects of brussels sprouts on oxidative DNA-damage in man, *Cancer Lett.,* 114, 127–130.

Verhoeven, D.T.H. et al. (1997a) A review of mechanisms underlying anticarcinogenicity by *Brassica* vegetables, *Chem. Biol. Interact.,* 103, 70–129.

Vinson, J.A., Proch, A., and Zubik, L. (1999) Phenol antioxidant quantity and quality in foods: cocoa, dark chocolate, and milk chocolate, *J. Agric. Food Chem.,* 47, 4821–4824.

Wang, H.X. and Ng, T.B. (1999) Natural products with hypoglycemic, hypotensive, hypocholesterolemic, antiatherosclerotic and antithrombotic activities, *Life Sci.,* 65, 2663–2677.

Warrier, G. and Gunawant, D. (1997) *Ayurveda,* Time-Life Books, Alexandria, VA.

Wei, H. et al. (1999) Scavenging of hydrogen peroxide and inhibition of ultraviolet light-induced oxidative DNA damage by aqueous extracts from green and black teas, *Free Radical Biol. Med.,* 26, 1427–1435.

Wei, Z. and Lau, B.H.S. (1998) Garlic inhibits free radical generation and augments antioxidant enzyme activity in vascular endothelial cells, *Nutr. Res.,* 18, 61–70.

Weiss, S.E. (1997) *Reader's Digest: Foods That Harm and Foods That Heal,* Reader's Digest Association, Pleasantville, NY.

Wood, G.B. and Bache, F. (1894) *The Dispensatory of the United States of America,* 17th ed., J.B. Lippincott, Philadelphia.

Zieliński, H. and Kozlowska, H. (2000) Antioxidant activity and total phenolics in selected cereal grains and their different morphological fractions, *J. Agric. Food Chem.,* 48, 2008–2016.

INDEX

A

Abortifacients, 542
Abrin, as cathartic, 177
Absinthe, 76
ACE inhibitors, *see* Angiotensin converting enzyme, inhibition of
Acetophenolisatin, 585
Acetydigoxin, 408
Acetylcholine, 141, 195-196, 242
  in gastric secretion, 660
  in vasodilation, 418
Acetylcholine precursors, 357-358
Acetylcholine receptors, 242-243
Acetylcholine-releasing agents, 358
Acetylcholinesterase, 241-242
  action of, 427
Acetylcholinesterase inhibitors, 355-357
N-Acetylcystein, as antioxidant, 102
Acetyldigitoxin, 408
Acetylsalicylic acid, 123
Achillea, therapeutic effects of, 334
Actin, 411
Actinomycin D, alpha TNF inhibition by, 608
Acupressure, 1-2
Acupuncture, 1, xvii-xviii, 13, 614
  for impotence, 614
  molecular effect of, xviii
Adaptogen, 37
Addison's disease, xxi, 40, 626
Adenocarcinoma(s), 646, 647
  gastric, 93, 647
  pancreatic, 647
Adonis vernalis, 406
Adrenal disorders, 628-641
Adrenal steroids, 621-628
  regulation of synthesis of, 624
Adrenergic neurotransmission, 579-580
Adrenocorticotropic hormones, 359, 624-625
Adrenomimetic drugs, duration of pressor activity of, 371
Aesculus hippocastanum, *see* Horse chestnut
Aflatoxins, 75
Aggregin, 121, 539
Aging, and drug metabolism, 111-112
Aglycones, 408-409
Agnus castus, 25
Agrimony, 25
Agrippina, xix
Alantolactone, 34
Alcohol, 683-689, *see also* Ethyl alcohol; Red wine

Alcohol dehydrogenase, 687
Alcohol dehydrogenases
  flavonoid inhibition of, 399
Alcoholic beverages, 684, *see also* Ethyl alcohol
Alcoholism, chronic, 688-689
Aldehyde dehydrogenases, flavonoid inhibition of, 399
Aldose reductase, flavonoid inhibition of, 398
Aldosterone, 623
  action of, on renin–angiotensin system, 625
Alexander of Tralles, 325
Alfalfa, 25
Algae, 479-480
  uses of, 479-480
Alkaloids, xx, 179-180
  classification of, 180-186
  as hypoglycemic agents, 447
  structure of, 180-186
  unknown structures of, 185
Alkyl sulfonate, 92, 645
Alkylating agents, 90-94, 567, 644-646
Alkyldisulfides, as hypoglycemic agents, 447
Allicin
  as antimicrobial agent, 545
  as antioxidant, 549-550, 551-552
Allium cepa, *see* Onion
Allium sativum, *see* Garlic
Allium ursinum, *see* Garlic, wild
Allopurinol, for gout, 326-327
Aloe barbadensis, *see* Aloe spp.
Aloe ferox, *see* Aloe spp.
Aloe spp., preparations from, 163-164, 666
  for asthma, 166-167
  for cancer therapy, 165
  as cathartic, 176-177
  chemicals in, 164-165
  for constipation, 168
  for diabetes, 167
  for HIV infection, 166
  for peptic ulcers, 659, 667
  for skin conditions, 167-168
Aloe vera, xxiii, 26, 134
  Bible reference, xvii
  clinical trials, 145
Alpha tocopherol, 60, 121, 123, *see also* Vitamin E
Alphaprodine, 507, 530
Alseroxylon, 575
Alternative therapies, 1-23
Althaea officinalis, 666
Alzheimer's disease, 232, 353-354
  acetylcholine precursors for, 357-358

Antimalarial agents, 138-139, 283
  pharmacologic properties of, 285
Antimetabolites, 92, 567, 646-649
Antineoplastic agents, 89-99, 142-144, *see also*
        Chemotherapeutic agents
  oxidative stress and, 98-104
Antioxidants, 78, 83, 85-86
  in *Brassica* spp., 257-259
  in cereal grains, 86-87
  curcuminoids as, 289-290
  du-zhong as, 87-88
  flavonoids as, 400-401
  studies of, in Alzheimer's disease, 359
  toxicity of, 88, 89
  in vegetables and fruits, 86
  vitamins as, 390
Antiplatelet drug(s)
  aspirin as, 441-442
  dipyridamole as, 442
  herbs containing, 124-128
  horseradish as source of, 439-441
  therapeutic uses of, 443-445
  ticlopidine as, 442-443
Antitussive effect, xxii
Anxiolytic agents, 673
Aplysiatoxins, 479
Apocynum cannabinum, 406
Apomorphine, 474, 526-527, 618
  as emetic, 473
Apoptosis
  dopamine induced, 221-222
  nitric oxide induced, 224-225
  selegiline's neuroprotective effect in, 226
Apricot, 26
Arachidonic acid metabolism, 441, *see also*
        Eicosanoids
  Allium compounds that interfere with, 538-539
Arctostaphylos uva ursi, see Uva ursi
Areca catechu, see Betel nut
Arecoline, 141, 182
Armoring, 4
Arnica, 26, 134
Aromatase, flavonoid inhibition of, 397
Aromatherapy, 2
Arrhythmias, cardiac glycosides for, 413
Artabotrys uncinatus, 139
Arteflene, 139
Artemesia annua, 139, 142
Artemesia vulgaris, see Mugwort
Artemisia spp., therapeutic effects of, 334-335
Arthritis, rheumatic, 85
Artichoke, 26
Asafoetida, 27
Ascariasis, 480
Ascorbic acid, 51
Ashwaganda, xxiii
Asparaginase, 96, 651
Aspergillus terreus, 144
Aspirin, xxii, *see also* Salicylate(s)
  as antiplatelet drug, 441-442
  asthma induced by, 600, 605

and eicosanoid metabolism, 542-543
in inhibition of tumor necrosis factor alpha,
        605-608
neuroprotective effects of, 599-600
pharmacologic action of, 595
poisoning by, treatment of, 597-598
Reye's syndrome and, 594
side effects of, 597-598
therapeutic effects of, 443-445
uses of, 596-597
  prophylactic, 598-599
Asteraceae, 331
Asthma
  Aloe treatment of, 166-167
  aspirin-induced, 600, 605
  pilocarpine in, 559
Astragalus leaf, xxiii
Astragalus membranaceus, see Astragalus leaf
Atherosclerosis, 85
Atherosclerotic lesions, 389-390
ATP-sensitive potassium ion channels, 454-456
ATPase inhibitors
  flavonoids as, 395
  for peptic ulcers, 662
Atracurium, in neuromuscular blockade, 346
Atrial fibrillation, 443-444
Atropa belladonna, see Belladonna
Atropine, 235
  and derivatives, 239-240
  dose-dependent effects of, 244, 431
  pharmacologic action of, 244
  structural formula of, 237
Auricular therapy, 2-3
Autogenic training, 12
Autonomic nervous system, 240-241
  cholinergic receptors in, 242-243
    blocking, 243-244
  functions of, 242, 426-427
  neuronal transmission in, 241-242
Autumn crocus, *see* Colchicine
Avarol, for HIV infection, 158
Avarone, for HIV infection, 158
Avena sativa, see Oat(s)
Avens, 27
Avicenna, 16
Axonal sprouting, 419
Ayurveda, 3
Ayurvedic medicine, 10, 16, 130
Azaspirodecanedione derivatives, 673
Azathioprine, 567
  in neuromuscular blockade, 346

B

Baby massage, 14
Baccharis, therapeutic effects of, 335
Bach, Edward, 7-8
Badger oil, 31

H

Tetrahydropapaveralione, alpha TNF inhibition by, 606
Tetranoic acid, 418
Thea sinensis, 662
Theaflavins, 435
Thearubigins, 435
Thebaine, 507
    chemistry of, 503
Theobroma cacao, *see* Chocolate
Theobromine, 185
    in chocolate, 321, 322
Theophylline, 129, 138, 139, 185
Therapeutic touch, 16
Thevetia, 406
Thiamine, 51
    nutritional value of, 61
Thioguanine, 95, 648-649
Thrombin, 121, 123, 539
Thromboembolism, 443-445
Thrombolytic agents, 120-121, *see also* Antiplatelet drug(s)
Thromboxane A2, 418, 441, 597
Thrush, 20
Thujone, 76
Thyme, 47
Thymic lymphocyte functions, 563
Thymoanaleptics, 462
Thymoleptics, 462
Tibetan medicine, 21
Ticlopidine, as antiplatelet drug, 442-443
Tithonia, therapeutic effects of, 337
Tolnaftate, 193
Tomatoes, chemicals in, 485-486
Topoisomerase, flavonoid inhibition of, 397
Trace elements, 6
Transcendental meditation, 21-22
Transforming growth factor beta, 217-218
Transpersonal therapies, 22
Travell, Janet, 14
Tree remedies, 7-8
Tretinoin, 58
Triazenes, 92, 646
Tribulusamides, alpha TNF inhibition by, 609
Trichoderma polysporum, 144
Trichosantin, for HIV infection, 160
Tricyclic antidepressants, 457-458, 463
    overdosage of, 467
Trifolium pratense, *see* Red clover
Trigeminal neuralgia, treatment of, 384-385
Trigger Point Injection Therapy, 14
Triterpenes, 334
    as hypoglycemic agents, 447
Triticum spp., *see* Wheat
Tropane, 182
Troponin, 411
Tuberculosis, xxi, 653-654
    herbs with antimycobacterial effects in, 653-654
Tubocurarine, 129, 140, *see also* Neuromuscular blocking agents
    in neuromuscular blockade, 342-343
    physiologic action of, 430

Tumor growth, 643-644
Tumor necrosis factor(s), 227
Tumor necrosis factor alpha, 605-606
    cell mediated cytotoxicity of, 608-609
    cells producing, 605-606
    inhibitors of, 606-609
    physiologic action of, 606-608
Tumor suppression
    carotenoids in, 486-487, 489-491
    paclitaxel in, 643
    teas in, 436-437
Turmeric, 289-290, *see also* Curcumin
    drug interactions with, 124
Tyramine, 81, 180

U

Ubiquinol, 224
Ulcer(s)
    duodenal, xxii
    herb extracts effective for, 659
    peptic, pathogenesis of, 659-660
    quinine for, 286
    therapeutic drugs for peptic, 660-668
Unani-Tibb, 15-16
Uncaria tomentosa, *see* Cat's claw
Undecylenic acid, 193
Unstable angina, 443
Uricosurics, for gout, 327
Urinary symptoms, treatment of, 574
Urokinase, 123
Urushiol, 75
Usui, Mikao, 18-19
Uva ursi, xxviii, 47, 135

V

Vaccinium macrocarpon, *see* Cranberry
Vaccinium myrtillus, *see* Bilberry
Valerian, vii, viii, xxviii, 48, 135, 670
    chemicals in, 670
    drug interactions with, 119
    regulatory status of, 680-681
    uses of, 670
Valeriana officinalis, *see* Valerian
Valvular heart disease, 444
Vanadium, 64
Vascular cell adhesion molecules, 605-606
Vascular disease, 444
Vasoactive intestinal polypeptide, 618
Vasodilation, receptor-dependent, 417-418
Vasodilators, 416-417
    for congestive heart failure, 411, 416-417
    endothelium-dependent vs. independent, 418
Vasopressin, 359
Vata, 3
Vedanta philosophy, 21
Vedas, 3
Vegan diet, 20

Z